ASIAN AND
PACIFIC COASTS 2017

Proceedings of the 9th International Conference on APAC 2017

ASIAN AND PACIFIC COASTS 2017

Proceedings of the 9th International Conference on APAC 2017

SMX Convention Centre, Pasay City, Philippines   19 – 21 October 2017

Edited by

Kyung-Duck Suh
Seoul National University, South Korea

Eric C Cruz
University of the Philippines Diliman, Philippines

Yoshimitsu Tajima
The University of Tokyo, Japan

World Scientific

NEW JERSEY · LONDON · SINGAPORE · BEIJING · SHANGHAI · HONG KONG · TAIPEI · CHENNAI · TOKYO

Published by

World Scientific Publishing Co. Pte. Ltd.

5 Toh Tuck Link, Singapore 596224

USA office: 27 Warren Street, Suite 401-402, Hackensack, NJ 07601

UK office: 57 Shelton Street, Covent Garden, London WC2H 9HE

British Library Cataloguing-in-Publication Data
A catalogue record for this book is available from the British Library.

ASIAN AND PACIFIC COASTS 2017
Proceedings of the 9th International Conference on APAC 2017

ISBN 978-981-3233-80-5
ISBN 978-981-3233-81-2 (ebook)

Preface

The 2017 International Conference on Asian and Pacific Coasts was held in Pasay City, Philippines. It was organized and managed by the Local Organizing Committee and PICE in coordination with APAC International Organizing Committee. This is the 9th of the conference series that aims to promote academic and technological progress and activities, international technical transfer and cooperation, and opportunities for engineers and researchers in the field of coastal engineering and allied disciplines in the Asian and Pacific regions. The conference topics include coastal engineering, coastal environment, marine ecology, coastal oceanography, marine energy engineering, and fishery science and engineering.

The first APAC conference was held in Dalian, China with a theme "Asian and Pacific Coastal Engineering". To reflect a broader scope and synergize with researchers of allied disciplines, the conference series was renamed "Asian and Pacific Coasts". After the second conference in Japan in 2004, it has been held every other year in various Asian and Pacific countries.

2001	Dalian, China
2004	Chiba, Japan
2005	Seogwipo, Korea
2007	Nanjing, China
2009	Singapore
2011	Hong Kong
2013	Bali, Indonesia
2015	Chennai, India
2017	Pasay, Philippines

APAC 2017 Paper Review Committee Members

Council

SUH, Kyung-Duck	Seoul National University, Korea (Chair)
AOKI, Shin-ichi	Osaka University, Japan
DONG, Gouhai	Dalian University of Technology, China
DOU, Xiping	Nanjing Hydraulic Research Institute, China
PYUN, Chong Kun	Myongji University, Korea
SATO, Shinji	The University of Tokyo, Japan
ZUO, Qihua	Nanjing Hydraulic Research Institute, China

International Scientific Committee

HSU, Tai-Wen	National Taiwan Ocean University, Taiwan, China
HUANG, Zhenhua	University of Hawaii at Manoa, USA
HWANG, Jin Hwan	Seoul National University, Korea
LEE, Joseph Hun-Wei	The University of Hong Kong, Hong Kong
LEE, Chang Hoon	Sejong University, Korea
LIU, Haijiang	Zhejiang University, China
LIU, Hua	Shanghai Jiaotong University, China
SHIBAYAMA, Tomoya	Waseda University, Japan
TAJIMA, Yoshimitsu	The University of Tokyo, Japan
TAKEWAKA, Satoshi	University of Tsukuba, Japan
YU, Xiping	Tsinghua University, China

Local Scientific Committee

CRUZ, Eric	University of the Philippine Diliman, Philippines
SIRINGAN, Fernando	University of the Philippines, Philippines
TANCHULING, Antonia	University of the Philippines, Philippines
DE LEON, Mario	De La Salle University, Philippines
DANAO, Louis	University of the Philippines Diliman, Philippines
BLANCO, Ariel	University of the Philippines Diliman, Philippines
CAMELO, Jeane	University of the Philippines Diliman, Philippines
MAANO, Desiree	Department of Environment and Natural Resources, Philippines

Contents

Numerical Simulation of Solitary Waves Shoaling on Two Different Sloping Beaches

Rui You, Guanghua He*, Shijun Zhang

School of Naval Architecture and Ocean Engineering,
Harbin Institute of Technology, Weihai,
Weihai, Shandong, P. R. China
* E-mail: ghhe@hitwh.edu.cn

A two-dimensional numerical wave tank based on the CIP (Constrained Interpolation Profile) method was developed to investigate the shoaling of solitary waves at different slopes of beaches, which were simulated by virtual particles. The generation of solitary waves was first verified by comparing with the analytical solution, and the numerical cases were validated by experimental results from published paper. Then, three amplitudes of incoming solitary waves were simulated on two kinds of slopes: one was a straight beach; another was a composite one which means its inclination varies. Furthermore, elevation of the water surface was measured by wave gauges to examine the distinguished features of solitary waves on diverse beaches.

Keywords: CIP; Composite beach; Shoaling; Solitary wave.

1. Introduction

Tsunami is natural disaster which can damage offshore engineering. The research on solitary wave propagating over a slope can provide theoretical guidance for tsunami forecasting. Solitary waves are always used as the physical model for tsunami evolution. It is of significant importance to implement laboratory experiment into practical application, and thus the simulation of solitary waves on composite slope which considered topography can provide more practical data for engineering.

Solitary waves on beaches have been studied by researchers utilizing different numerical model. Fully nonlinear equation is utilized by Zelt and Raichlen[1], Wei et al[2] and Madsen et al[3]. Laplace equation is used by Grilli et al[4,5]. Hsiao and Lin[6] based on RANS (Reynolds-Averaged Navier-Stokes) equations along with model to acquire the wave load on the seawall. Most researchers simplify the beach as straight and rigid slope. Researches on special slope such as permeable slope[7], parabolic slope[8] whose surface is simulated by parabolic curve, and composite slope[9] which is first proposed by Synolakis et al[10].

In the present study, the violent interaction between solitary wave and slope is solved by a numerical model based on CIP (constrained interpolation profile) scheme. CIP is first developed by Yabe et al[11,12] to solve hyperbolic equation, and then is employed to marine engineering problems by Hu and Kashiwagi[13,14]. It can solve violent flow with high accuracy. THINC (Tangent of Hyperbola Interpolation for Interface Capturing)[15] is applied for free surface capturing. THINC employ a hyperbolic tangent function which makes it a suitable interpolation for the flux computation of a VOF function, and the competitive accuracy compared with other method has been verified by Xiao et al[15]. In this paper, the accuracy of numerical model is first verified, and then the free surface elevation caused by solitary wave propagating over a straight and composite slope is studied. Totally six cases were investigated. Water depth equals to 0.2 m, three non-dimensional incident wave heights is simulated on both straight and composite slope.

2. Numerical method

We use flow solver based on CIP method and multiphase model to treat wave body interaction problems. Continuity equation and Navier-Stokes equation is employed as governing equations:

$$\frac{\partial u_i}{\partial x_i} = 0 \tag{1}$$

$$\frac{\partial u_i}{\partial t} + u_j \frac{\partial u_i}{\partial x_j} = -\frac{1}{\rho}\frac{\partial p}{\partial x_i} + \frac{1}{\rho}\frac{\partial}{\partial x_j}(\mu S_{ij}) + f_i \,, \tag{2}$$

The flow solver divides Navier-Stokes equation into three fractional steps:

(1) Advection Step

$$\frac{u_i^* - u_i^n}{\Delta t} + u_j^n \frac{\partial u_i^n}{\partial x_j} = 0 \,, \tag{3}$$

(2) Nonadvection Step (i)

$$\frac{u_i^{**} - u_i^*}{\Delta t} = \frac{1}{\rho}\frac{\partial}{\partial x_j}(\mu S_{ij}^*) + f_i \,, \tag{4}$$

(3) Nonadvection Step (ii)

$$\frac{u_i^{n+1} - u_i^{**}}{\Delta t} = -\frac{1}{\rho}\frac{\partial p^{n+1}}{\partial x_i} \,. \tag{5}$$

Where u, ρ, p, t and f denote velocity, density, pressure, time and body force, respectively; and $S_{ij} = (\partial u_i/\partial x_j + \partial u_j/\partial x_i)/2$. Advection step is solved by CIP method, diffusion term is treated by central difference method. The pressure coupled with velocity is solved by SOR (Successive Over Relaxation).

3. Results and discussion

3.1. *Numerical experiments setup*

The non-dimensional incident wave height α is 0.2, 0.3 and 0.4. Two types of slope were simulated, one is straight and another is composite. Straight slope has an inclination of 10°. Composite slope has a 10° inclination in the beginning, then convert to 4°. The first section of the slope is with a height of 0.28 m and has 0.08 m above the still water. The water depth in the wave tank is 0.20 m. We define the leading edge of wave maker as $x = 0$ m and the starting point of the slope is set at $x = 8.00$ m. Wave gauges are set at four locations, WG1 is 0.97 m before the starting point of the slope, WG2, WG3 and WG4 are 1.28 m, 1.42 m, 1.70 m after the starting point, respectively. Fig. 1 is the sketch of the straight slope, and Fig. 2 is the schematic image of the composite model.

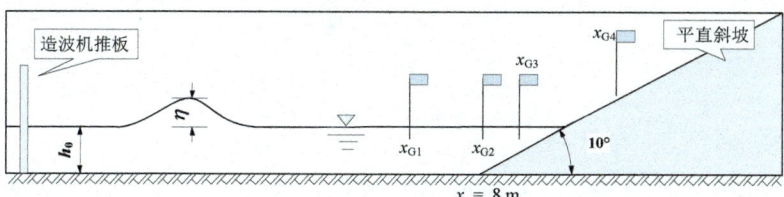

Fig. 1. Sketch of the straight slope

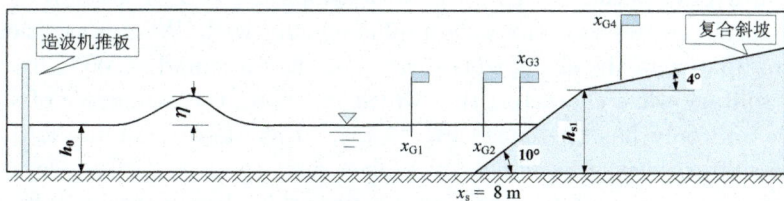

Fig. 2. Sketch of the composite slope

3.2. *Verification and validation of the numerical model*

The numerical model is first verified by the Boussinesq solitary wave solution[16]. Fig. 3 shows the comparison between Boussinesq solution and numerical results of solitary waves. X coordinate is the time series and y coordinate denotes the water surface elevation. It is show that the numerical results predicts the wave profile satisfactorily, and that when $\alpha = 0.2$, the computational results presents more accurate results since the dashed line is nearly overlapped with solid line.

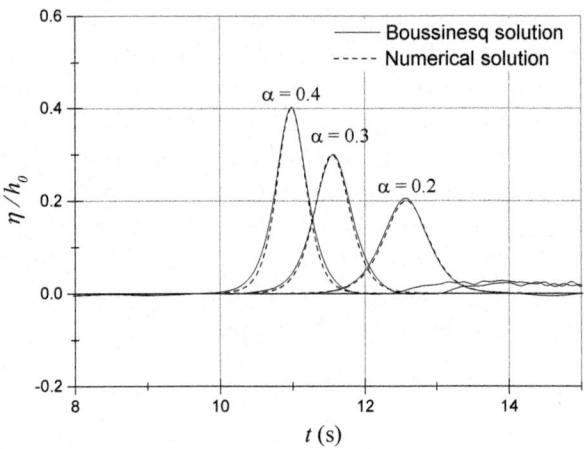

Fig. 3.   Solitary wave solution

Free surface elevation is validated by the experimental results from the published paper[9]. Fig. 4 (a), (b) and (c) are the comparison between computational results in present study and experimental data at WG1, WG2 and WG4 respectively. Solid line is the profile measured at straight slope and dashed line is the results of composite slope. Both the results from two different slopes describe the wave profile equally well. WG1 is positioned 0.97 m upwave of the slope, so the water profile measured at WG1 has exactly solitary wave characteristics, which including the symmetry of wave profile and only has one crest (Fig. 4 (a)). Fig. 4 (b) is at the location where solitary waves have already broken, and the free surface elevation on the composite slope has some retardation; this phenomenon will be discussed later in next part. The calculative free surface elevation in Fig. 4 (c) has some oscillation, compared with experimental results. Since the experimental results have not provide the wave profile at WG3, so only the CIP results at WG3 was presented in fig. 4 (d).

3.3. *Free surface elevation*

Fig. 5 demonstrates the distinction of free surface elevation on different slopes. Fig. 5 (a) presents the free surface elevation at WG2; Fig. 5 (b) is the extended image from the descending part of Fig. 5 (a), mainly focus on the tail of the solitary wave. It can be found that the free surface on the composite slope drop more quickly than on the straight slope, and the water surface will remain at a height for a while. It may due to the smaller inclination after the vertex.

Fig. 6 is the free surface elevation before and after the vertex on the composite slope. Solid line is the free surface measured at WG3 and dashed line is surface elevation measured at WG4. In the process of runup, though WG4 is set further from the beginning of the slope than WG3, the free surface elevation measured at WG4 still large than WG3 during $t = 8.0$ s to $t = 9.3$ s.

4. Conclusions

A two-dimensional numerical wave tank based on CIP method is constructed to investigate the evolution of solitary waves on straight and composite slope. The accuracy of numerical model is verified by comparing with analytical solution and experimental data. Compared with straight slope, free surface elevation due to solitary waves on composite slope will drop more quickly, and will remain at a level for a while. This may because the inclination after the vertex is smaller; the descent of free surface will have retardation due to gravity.

5. Acknowledgements

This work was supported by National Natural Science Foundation of China (51579058), Shandong Provincial Natural Science Foundation (ZR2014EEQ016), Open Research Fund Program of State Key Laboratory of Coastal and Offshore Engineering, Dalian University of Technology (LP1513), Open Research Fund Program of Key Laboratory of Water and Sediment Science and Water Hazard prevention (Changsha University of Science and Technology), Hunan Province (2015SS02).

6

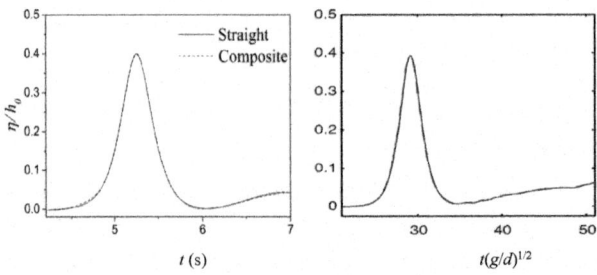

(a) Free surface elevation at WG1; left: CIP; right: experiment.

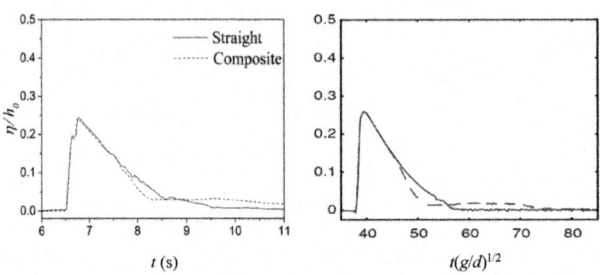

(b) Free surface elevation at WG2; left: CIP; right: experiment.

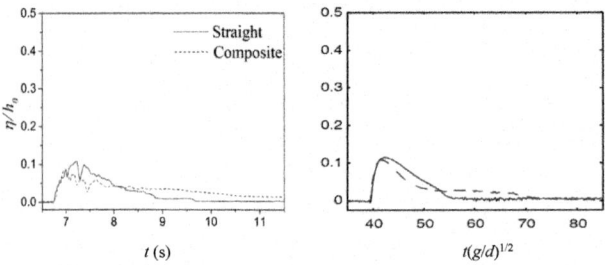

(c) Free surface elevation at WG4; left: CIP; right: experiment.

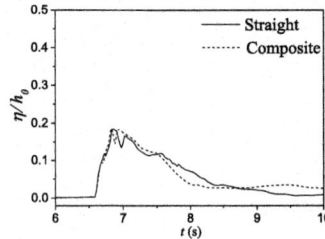

(d) Free surface elevation at WG3, CIP

Fig. 4.   Validation of free surface elevation, $\alpha = 0.4$

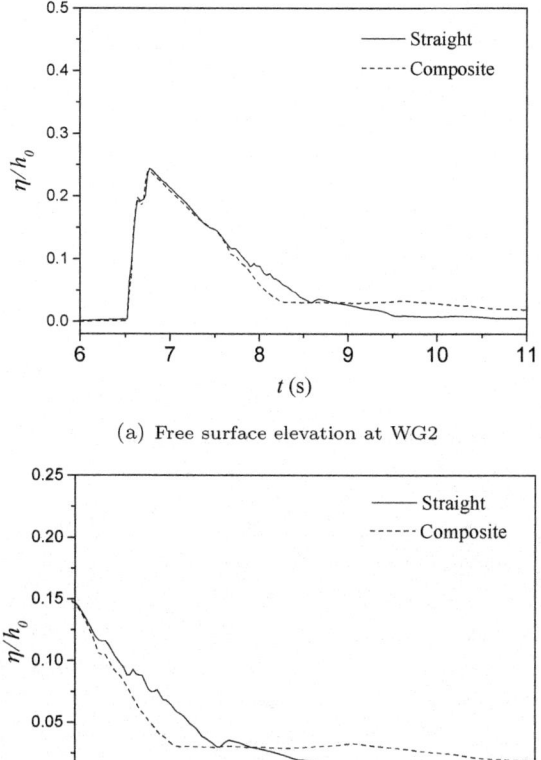

(a) Free surface elevation at WG2

(b) Extended image of the tail

Fig. 5. Comparison of free surface elevation on Straight and composite slope

References

1. J. A. Zelt and F. Raichlen, Overland flow from solitary waves, *Journal of Waterway, Port, Coastal, and Ocean Engineering* **117**, 247 (1991).
2. G. Wei, J. T. Kirby, S. T. Grilli and R. Subramanya, A fully nonlinear boussinesq model for surface waves. part 1. highly nonlinear unsteady waves, *Journal of Fluid Mechanics* **294**, 71 (1995).
3. P. A. Madsen, H. B. Bingham and H. Liu, A new boussinesq method for fully nonlinear waves from shallow to deep water, *Journal of Fluid Mechanics* **462**, 1 (2002).

8

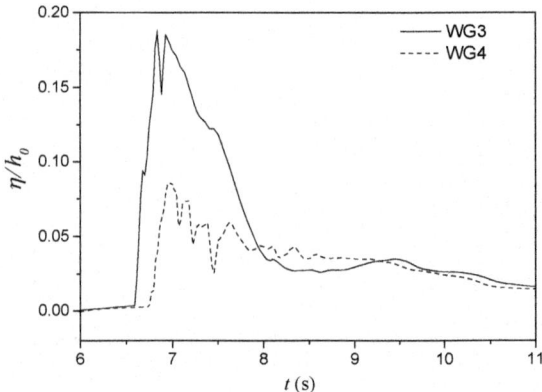

Fig. 6.   Comparison of the free surface elevation before and after the vertex

4. S. T. Grilli, I. A. Svendsen, J. Veeramony and R. Subramanya, Shoaling of solitary waves on plane beaches, *Journal of Waterway, Port, Coastal, and Ocean Engineering* **120**, 609 (1994).

5. S. Grilli, I. Svendsen and R. Subramanya, Breaking criterion and characteristics for solitary waves on slopes, *Journal of Waterway, Port, Coastal, and Ocean Engineering* **4**, 102 (1997).

6. S. C. Hsiao and T. C. Lin, Tsunami-like solitary waves impinging and overtopping an impermeable seawall: Experiment and rans modeling, *Coastal Engineering* **57**, 1 (2010).

7. C.-J. Huang, M.-L. Shen and H.-H. Chang, Propagation of a solitary wave over rigid porous beds, *Ocean Engineering* **35**, 1194 (2008).

8. B. H. Choi, E. Pelinovsky, D. C. Kim, I. Didenkulova and S. B. Woo, Two- and three-dimensional computation of solitary wave runup on non-plane beach, *Nonlinear Processes in Geophysics* **15**, 489 (2008).

9. G. Salevik, A. Jensen and G. Pedersen, Runup of solitary waves on a straight and a composite beach, *Coastal Engineering* **77**, 40 (2013).

10. C. E. Synolakis, E. N. Bernard, V. V. Titov, U. Knolu and F. I. Gonzlez, Validation and verification of tsunami numerical models, *Pure and Applied Geophysics* **165**, 2197 (2008).

11. T. Yabe and T. Aoki, A universal solver for hyperbolic-equations by cubic-polynomial interpolation. 1. one-dimensional solver, *Computer Physics Communications* **66**, 219 (1991).

12. T. Yabe, T. Ishikawa, P. Y. Wang, T. Aoki, Y. Kadota and F. Ikeda, A universal solver for hyperbolic-equations by cubic-polynomial interpo-

lation. 2. 2-dimensional and 3-dimensional solvers, *Computer Physics Communications* **66**, 233 (1991).

13. C. Hu and M. Kashiwagi, A cip-based method for numerical simulations of violent free-surface flows, *Journal of Marine Science and Technology* **9**, 143 (2004).

14. C. Hu and M. Kashiwagi, Two-dimensional numerical simulation and experiment on strongly nonlinear wavebody interactions, *Journal of Marine Science and Technology* **14**, 200 (2009).

15. F. Xiao, Y. Honma and T. Kono, A simple algebraic interface capturing scheme using hyperbolic tangent function, *International Journal for Numerical Methods in Fluids* **48**, 1023 (2005).

16. J. V. Boussinesq, Thorie des ondes et remous qui se propagent le long d'un canal rectangulaire horizontal, en communiquant au liquide contenu dans ce canal des vitesses sensiblement pareilles de la surface au fond, *Visin Electrnica* **17**, 55 (2013).

A Modified Nonlinear Schrödinger Equation for Interactions Between Waves and Shear Currents

Bo Liao, Guohai Dong, Yuxiang Ma[†], Xiaozhou Ma

State Key Laboratory of Coastal and Offshore Engineering, Dalian University of Technology, Dalian, China
[†]E-mail: yuxma@126.com (Y. Ma)

A nonlinear Schrödinger equation for the propagation of two-dimensional surface gravity waves on linear shear currents in finite water depth is derived. In the derivation, linear shear currents are assumed to be a linear combination of depth-uniform currents and constant vorticity. Therefore, the equation includes the combined effects of depth-uniform currents and constant vorticity. Furthermore, the influence of linear shear currents on the Peregrine breather is also studied. It is demonstrated that depth-uniform opposing currents can reduce the breather extension in finite water depth, but following currents has the adverse impact, indicating that a wave packets with freak waves formed on following currents contains more hazardous waves in finite water depth. However, the corresponding and coexisting vorticity can counteract the influence of currents. If the water depth is deep enough, both depth-uniform currents and vorticity have negligible effect on the characteristics of Peregrine breather.

Keywords: Nonlinear Schrödinger equation; Linear shear currents; Modulational instability; Extreme waves; Peregrine Breather solution.

1. Introduction

As waves nearly always coexist with currents in the ocean, nonlinear wave-current interactions attract the attention of researchers in ocean engineering and hydrodynamics. It is well known that currents can significantly change the properties of gravity waves [1-4]. Anomalously large waves can be created in regions with strong currents, especially for waves that propagate against currents. In particular, rogue waves have been generated frequently in these conditions [5-10].

Previous studies have proved that interactions between waves and currents are strongly dependent on the propagation direction and the vertical distribution of currents [3, 11]. Until now, however, researches on wave-current interactions are often assumed that currents are uniform with depth [7, 10-13]. However, there are many situations where currents are not uniform with depth, namely, they are vertically sheared. For example, wind driven currents and jet-like ebb flow at a river mouth [14, 15]. Therefore, the influence of vertical vorticity should be considered in the wave-current interaction. Under the condition with a linear shear current (constant vorticity), it was theoretically found that the wave motion remains irrotational and that uniform vorticity can only effect the dispersion relation at the first order. However, at higher orders of approximation, the distribution of vorticity has a significant influence on water surface elevations [16-19]. These findings were verified by several experiments [20-23]. There have been some numerical models proposed to address the problem. Dalrymple [24-26] developed a model to predict waves propagating over shear currents using a stream function. Choi [27] used a pseudo-spectral method to study the interaction of nonlinear surface waves with linear shear currents, and found that the maximum wave amplitude for a positive shear current (the current is in the same direction as the wave propagation direction) is much smaller than that in the absence of any shear, while the opposite is observed for a negative shear current (the current opposes the wave propagation direction). Nwogu [28] proposed a fully nonlinear boundary-integral approach to describe the interaction between gravity waves and shear currents with an arbitrary distribution of vorticity and found that the presence of vorticity can significantly influence the development of modulated wave trains. Recently, Moreira and Chacaltana [29] investigated the influence of uniform vorticity on a train of free-surface gravity waves interacting with underlying flows via a fully nonlinear boundary-integral method. It was shown that wave blocking and breaking could be more prominent depending on the magnitude and direction of the shear flow.

In the studies of nonlinear evolution of water waves, nonlinear Schrödinger equations (NLS) are often used as they can properly reflect the modulational instability, i.e. the Benjiamin-Feir instability, which is deemed a possible generation mechanism, for example of freak waves [30-32]. Assuming currents are vertically uniform, Gerber [33] derived a current-modified NLS equation and demonstrated that opposing currents can not only stimulate the growth of the modulational instability, but also expand the onset criterion, while following currents have the opposite effect. Stocker and Peregrine [34] extended the Dysthe equation [35] to include a prescribed potential current induced by, for example, an internal wave. Hjelmervik and Trulsen [7] derived a NLS equation suitable for waves propagating on inhomogeneous currents and studied the generation of freak waves on opposing jet currents. Later, Onorato [8] applied a transformation to the current-modified NLS equation of Hjelmervik and Trulsen [7] to obtain a NLS equation including a depth-uniform current. The numerical results of Onorato [8] showed that an initially stable wave train in quiescent water will become unstable after entering an opposing depth-uniform current region. In the real ocean, currents are always non-uniform over water depth; hence, description of waves propagating on shear currents is more realistic. Until now, there are several NLS equations considering the influence of vorticity. Johnson [36] derived a NLS equation with coefficients depending on vorticity and studied the slow modulation of the waves over a current with arbitrary vorticity. Oikawa [37] considered the instability properties of weakly nonlinear wave packets to three-dimensional disturbances in the presence of shear. In both studies, unfortunately, the resulting third-order envelope equations possess coefficients that depend in a complicated way on the shear, and thus are not simple to implement. Later, Li [38] developed a NLS equation for the motion of waves on water of infinite depth in the presence of a uniform velocity shear, and studied the resulting modulational instability. Baumstein [39] proposed a NLS equation to investigate the effect of piecewise-linear velocity profiles on the modulational instability of gravity waves in infinite depth. As the expression, which is a function of vorticity and depth of the shear layer, was not given explicitly, the coefficients of the NLS equation have to be calculated for specific values of the vorticity and depth of the shear layer. Recently, Thomas [40, 41] derived a NLS equation with constant vorticity in finite depth; this equation includes explicit coefficients as a function of vorticity. However, previous studies on vorticity-modified NLS equations only considered the influence of vorticity. Actually, vortices occur in conjunction with underlying currents. Therefore, an equation with the influence of both vorticity and depth-uniform currents is needed. This is the main objective of the present study.

Following the introduction, in Section 2, a linear shear current modified NLS equation is derived using the multiple scale method. Then, in Section 3, the influence of the shear current on the Peregrine breather is presented. Lastly, the conclusions are presented in Section 4.

2. Derivation of the Evolution Equation

Assuming that the fluid motion is inviscid and incompressible, a two-dimensional Cartesian coordinate (x, z) is adopted to derive the equation. As shown in Figure 1, the x-axis is aligned with the propagation direction of the waves and the z-axis is taken vertically upwards. Here the waves are assumed to be propagating steadily on a vertical linear-sheared current, which can be separated into a depth-uniform current U and a constant vorticity Ω. Therefore, the total velocity of the flow can be written as following:

$$\mathbf{V} = (U+\Omega z)\mathbf{i} + \nabla \varphi(x,z,t), \tag{1}$$

where U is the speed at the free surface of the linear shear current, taken positive in the wave-propagating direction; $\mathbf{i}$ is the unit vector along Ox, $\nabla = (\partial/\partial x, \partial/\partial z)$ is the two-dimensional gradient operator, and φ and t are the velocity potential and time, respectively.

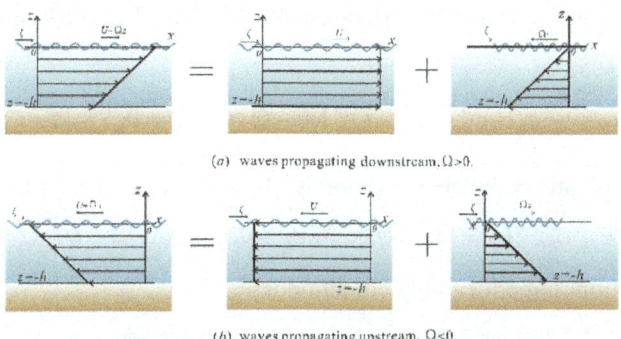

(a) waves propagating downstream, $\Omega > 0$.

(b) waves propagating upstream, $\Omega < 0$.

Fig. 1. Sketch of waves propagating on linear shear currents.

For incompressible and inviscid flow of a homogeneous fluid, the governing equations can be written as:

$$\nabla \cdot \mathbf{V} = 0, \tag{2}$$

$$\mathbf{V}_t + \mathbf{V} \cdot \nabla \mathbf{V} + \frac{\nabla p}{\rho} + g\mathbf{k} = 0, \tag{3}$$

where g is the gravitational acceleration, p is the total pressure, ρ is the density of water, and $\mathbf{k}$ is the unit vector along Oz. Substitution of Eq. (1) into Eq. (2) yields the Laplace's equation:

$$\Delta \varphi = 0, \qquad\qquad -h < z < \zeta, \tag{4}$$

where ζ is the free surface elevation. Substitution of Eq. (1) into Eq. (3) becomes

$$\nabla[\varphi_t + \frac{1}{2}(\varphi_x^2 + \varphi_z^2) + (U + \Omega z)\varphi_x - \Omega\psi + \frac{p}{\rho} + gz] = 0, \tag{5}$$

where ψ is the stream function, which is related to φ by the Cauchy-Riemann relations:

$$\psi_z = \varphi_x, \qquad\qquad \psi_x = -\varphi_z. \tag{6}$$

Eq. (5) may be spatially integrated to give a form of the Bernoulli equation

$$\varphi_t + \frac{1}{2}(\varphi_x^2 + \varphi_z^2) + (U + \Omega z)\varphi_x - \Omega\psi + \frac{p}{\rho} + gz = R. \tag{7}$$

The dynamic boundary condition states that the pressure at the free surface is continuous and equal to constant atmospheric pressure, R is the Bernoulli constant, and the dynamic boundary condition at the free surface can then be written as,

$$\varphi_t + \frac{1}{2}(\varphi_x^2 + \varphi_z^2) + (U + \Omega\zeta)\varphi_x - \Omega\psi + g\zeta = 0, \qquad z = \zeta. \tag{8}$$

The kinematic boundary condition is

$$\zeta_t + (U + \Omega\zeta + \varphi_x)\zeta_x - \varphi_z = 0, \qquad z = \zeta. \tag{9}$$

The bottom impermeability condition is

$$\varphi_z = 0, \qquad z = -h. \tag{10}$$

By taking the derivative of Eq. (8) and using the Cauchy-Riemann relations to eliminate the stream function, Eq. (8) may be converted into a free-surface boundary condition entirely in terms of φ and ζ:

$$\varphi_{tx} + \varphi_{tz}\zeta_x + \varphi_x(\varphi_{xx} + \varphi_{xz}\zeta_x) + \varphi_z(\varphi_{xz} + \varphi_{zz}\zeta_x) + g\zeta_x$$
$$+ (U + \Omega\zeta)(\varphi_{xx} + \varphi_{xz}\zeta_x) + \Omega\varphi_z = 0, \qquad z = \zeta. \tag{11}$$

Lastly, the multiple scale method [42] is adopted to derive the evolution equation. From the leading-order problem for the first harmonic, we get the linear dispersion relation

$$\omega^2(1 - \bar{u})(1 - \bar{u} + X) = gk\sigma, \tag{12}$$

where

$$X = \sigma\bar{\Omega}, \qquad \bar{\Omega} = \Omega/\omega, \qquad \sigma = \tanh q, \qquad q = kh, \qquad \bar{u} = U/c, \qquad c = \frac{\omega}{k}. \tag{13}$$

Here c denotes the wave phase velocity in the presence of current U, ω is the carrier wave angular frequency, k is the wave number in the presence of current U. Eq. (12) is actually equivalent to the results obtained by Peregrine [43]

$$(\omega - Uk)^2 = [gk - \Omega(\omega - Uk)]\sigma. \tag{14}$$

In the $O(\varepsilon^2)$ approximation, the following equation is obtained

$$A_{t_1} + c_g A_{x_1} = 0, \tag{15}$$

where A is an unknown complex amplitude, c_g denotes the group velocity in the presence of a linear shear current and can be written as:

$$c_g = \frac{\omega}{(2+X-2\bar{u})k}\left[(1+\frac{2q}{\sinh 2q}+X)-\bar{u}\frac{4q}{\sinh 2q}-\bar{u}^2(1-\frac{2q}{\sinh 2q})\right]. \tag{16}$$

At the third order $O(\varepsilon^3)$, we get a linear-shear-current-modified nonlinear Schrödinger equation (SCNLS):

$$i(A_t + c_g A_x) + \alpha A_{xx} + \beta|A|^2 A = 0. \tag{17}$$

The expressions for α, β are given in the Appendix A. We transform to moving coordinates and introduce the following variables

$$\xi = (x - c_g t), \qquad \tau = t, \tag{18}$$

$$A = B(1 + X - \bar{u})^{-1}, \tag{19}$$

the SCNLS becomes

$$iB_\tau + \alpha B_{\xi\xi} + \gamma|B|^2 B = 0, \tag{20}$$

where

$$\gamma = \beta \frac{1}{(1 + X - \bar{u})^2}. \tag{21}$$

Ignoring the influence of the linear shear currents, Eq. (20) degenerate to the classic NLS equation in finite water depth [45-47]. If only considering depth-uniform currents, the SCNLS reduces to the current-modified cubic Schrodinger equation obtained by Gerber [33]. On the other hand, if only considering vorticities, i.e. neglecting terms with depth-uniform currents, Eq. (20) becomes the vorticity modified NLS equation derived by Thomas [40].

3. Influence of Shear Currents on the Peregrine Breather

Freak waves, which appear nowhere and disappear without trace, can cause ships and offshore structures in danger due to their extraordinary large wave heights [48]. A freak wave is often identified if its height reaches to twice the significant wave height [31]. Several physical mechanisms of the freak waves are described [49]: the modulational instability, wave current interaction, geometrical and dispersive focusing. The modulational instability of gravity

16

waves can well be modeled by the NLS equations. Therefore, NLS type equations are often used to study the characteristics of freak waves [50]. Peregrine breather [51] is the simplest theoretical solution of the NLS equation; it can present a double spatial-temporal localization and the amplitude can reach three times the initial value, so it can be considered as a prototype of freak waves. Figs. 2 and 3 show the PB solution at $k_0h=2$ on different depth –uniform currents. On opposing currents, the lifetime and span of a freak wave are decreased. It is interesting to notice that the dimension of the breather is extended significantly on following currents. Fig. 3 indicates that a freak waves formed on a following currents can be more hazardous in finite water depth. The corresponding vorticity of currents has the opposite influence on the breather span, as can be seen in Figs. 4 and 5. The effect of linear shear currents on spatial and temporal profile of PB is demonstrated in Figs. 6 and 7. Due to the impact of depth-uniform currents and the corresponding vorticity is opposite, the actual influence of linear shear current on PB can be negligible even in immediate water depth, as shown in Figs. 6 and 7; for example, when $\bar{\Omega}=0.25, \bar{u}_0=0.2$, the impact of depth-uniform current counteracts absolutely the impact of the vorticity.

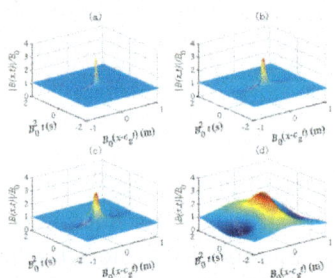

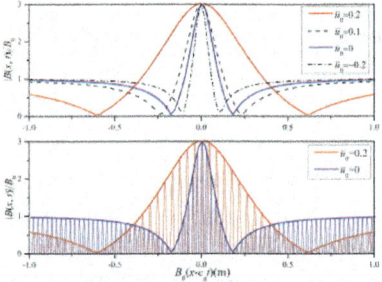

Fig. 2. The Peregrine breather with the frequency $f_0=0.8$Hz and initial amplitude $B_0=0.02$m, $k_0h=2$ $\bar{\Omega}=0$ for (a) $\bar{u}_0=-0.2$; (b) $\bar{u}_0=0$; (c) $\bar{u}_0=0.1$; (d) $\bar{u}_0=0.2$.

Fig. 3. The envelope as a function of space with different of the depth uniform currents, the wave parameters as in Fig. 2.

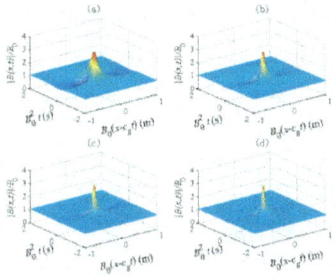

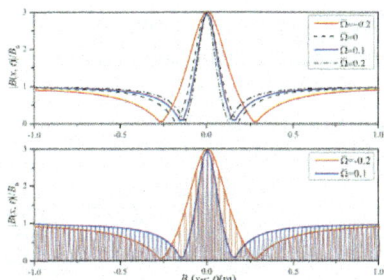

Fig. 4. The Peregrine breather with the frequency $f_0=0.8$Hz and initial amplitude $B_0=0.02$m, $k_0h=2$ $\overline{u}_0=0$ for (a) $\overline{\Omega}=-0.2$; (b) $\overline{\Omega}=0$; (c) $\overline{\Omega}=0.1$; (d) $\overline{\Omega}=0.2$.

Fig. 5. The envelope as a function of space with different of the depth uniform currents, the wave parameters as in Fig. 4.

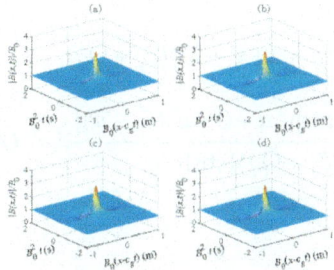

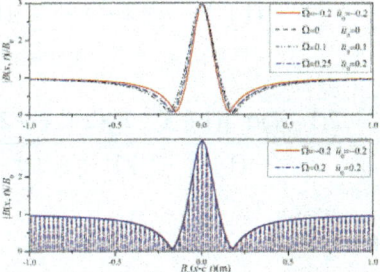

Fig. 6. The Peregrine breather with the frequency $f_0=0.8$Hz and initial amplitude $B_0=0.02$m, $k_0h=2$ for (a) $\overline{\Omega}=-0.2, \overline{u}_0=-0.2$; (b) $\overline{\Omega}=0, \overline{u}_0=0;$ (c) $\overline{\Omega}=0.1, \overline{u}_0=0.1;$ (d) $\overline{\Omega}=0.25, \overline{u}_0=0.2.$

Fig. 7. The envelope as a function of space with different of the depth uniform currents, the wave parameters as in Fig. 6.

4. Conclusion

In this paper, a cubic nonlinear Schrödinger equation for gravity waves in the presence of linear shear currents in finite water depth is derived using the multiple scale method. The present linear shear current modified nonlinear Schrödinger equation (SCNLS) contains the effects of both depth uniform currents and constant vorticity. Using this equation, the influence of linear shear currents on the Peregrine breather solution is examined. Depth-uniform opposing currents can reduce the breather extension in finite water depth, but following currents has the adverse impact, suggesting that a wave packets with freak waves formed on following currents contains more hazardous waves in finite water depth. However, the corresponding and coexisting vorticity can counteract the influence of currents.

Appendix A.

$$\alpha = -\frac{\omega}{k^2}\{\frac{1}{(2+X-2\bar{u})}\bar{c}_g^2 + \frac{q[2\sigma^2(1-\bar{u})+X]-2\bar{u}\sigma}{(2+X-2\bar{u})\sigma}\bar{c}_g - \frac{q[\bar{u}^2(1-2\sigma^2)+2\bar{u}(\sigma^2-1)+1+X]-\bar{u}^2\sigma}{(2+X-2\bar{u})\sigma}\}, \tag{A.1}$$

$$\beta = -\frac{\omega k^2(1+X-\bar{u})(M+VW)}{8(1-\bar{u})^2(2+X-2\bar{u})\sigma^4}, . \tag{A.2}$$

$$\bar{c}_g = c_g k / \omega, . \tag{A.3}$$

$$V = (1+X-\bar{u})(2+X-2\bar{u})(1-\bar{u}) \\ +\{[1+\bar{u}(\bar{u}-2-2X\cosh^2 q)]\cosh^{-2} q + X(X+2)\}(\bar{c}_g+\bar{\Omega}q), \tag{A.4}$$

$$W = \frac{2(1-\bar{u})\sigma^2[(1+X-\bar{u})(2+X-2\bar{u})(1-\bar{u})\cosh^2 q + (1-\bar{u})(\bar{c}_g-\bar{u})]}{-[(1+X-\bar{u})(1-\bar{u})q/\sigma-(\bar{c}_g-\bar{u})(\bar{c}_g+\bar{\Omega}q)]\cosh^2 q}. \tag{A.5}$$

Appendix B.

$$\frac{c}{c_0} = \frac{2+X}{2(1+X)}\bar{u}_0 + \frac{1}{2(1+X)}\left(\frac{\sigma}{\sigma_0}+\sqrt{(\bar{u}_0 X)^2+2\bar{u}_0\frac{\sigma}{\sigma_0}X+4\bar{u}_0\frac{\sigma}{\sigma_0}+(\frac{\sigma}{\sigma_0})^2}\right), \tag{B.1}$$

$$\frac{k}{k_0} = \left(\frac{c}{c_0}\right)^{-2}\frac{\sigma}{\sigma_0(1-\bar{u}_0\frac{c_0}{c})(1-\bar{u}_0\frac{c_0}{c}+X)}, \tag{B.2}$$

$$\bar{u}_0 = \frac{U}{c_0}, \quad \bar{u} = \frac{U}{c}, \quad \frac{\bar{u}}{\bar{u}_0} = \frac{c_0}{c}, \quad \sigma_0 = \tanh k_0 h, \quad c_0 = \frac{\omega}{k_0}, \tag{B.3}$$

here c_0 and k_0 denotes the wave phase velocity and wave number in the absence of current, respectively.

References

1. M. S. Longuet and R. W. Stewart, The changes in amplitude of short gravity waves on steady non-uniform currents, *J. Fluid Mech.* **10**(4), 529 (1961).

2. F. P. Bretherton and C.J.R. Garrett, Wavetrains in inhomogeneous moving media, *Proc. R. Soc. Lond.* A. **302** (1471) , 529(1968).

3. D. H. Peregrine, Interaction of water waves and currents, *Adv. Appl. Mech.* **16**, 9 (1976).

4. I. Kantardgi, Effect of depth current profile on wave parameters, *Coastal Eng.* **26** , 195(1995).

5. J. K. Mallory, Abnormal waves in the south-east coast of South Africa, *Int. Hydro. Rev.* **51**, 99 (1974).

6. R. Smith, Giant waves, *J. Fluid Mech.* **77**, 417 (1976).

7. K. Hjelmervik and K. Trulsen, Freak wave statistics on collinear currents, *J. Fluid Mech.* **637**, 267 (2009).

8. M. Onorato, D. Proment and A. Toffoli, Triggering rogue waves in opposing currents, *Phys. Rev. Lett.* **107** , 184502 (2011).

9. V. P. Ruban, On the nonlinear Schrödinger equation for waves on a non-uniform current, *JETP Letters.* **95** (9) , 486 (2012).

10. A. Toffoli, T. Waseda, H. Houtani, T. Kinoshita, K. Collins, D. Proment and M. Onorato, Excitation of rogue waves in a variable medium: an experimental study on the interaction of water waves and currents, *Phys. Rev. E.* **87**, 051201 (2013).

11. P. L.-F. Liu, M. W. Dingemans and J. K. Kostense, Long wave generation due to the refraction of short-wave groups over a shear current, *J. Phys. Oceanogr.* **20**, 53 (1990).

12. J. R. Stocker and D. H. Peregrine, The current-modified nonlinear Schrödinger equation, *J. Fluid Mech.* **399**, 335 (1999).

13. A. Toffoli, T. Waseda, H. Houtani, L. Cavaleri, D. Greaves and M. Onorato, Rogue waves in opposing currents: an experimental study on deterministic and stochastic wave trains, *J. Fluid Mech.* **769**, 277 (2015).

14. C. C. Mei and E. Lo, The effects of a jet-like current on gravity waves in shallow water, *J. Phys. Oceanogr.* **14** , 471(1984).

15. R. D. Maciver, R.R. Simons and G. P. Thomas, Gravity waves interacting with narrow jet-like current, *J. Geophys. Res.* **111**, C03009 (2006).

16. S. Tsao, Behaviour of surface waves on a linearly varying flow. Tr. Mosk. Fiz.-Tekh. Inst. Issled. Mekh. Prikl. Mat. **3**, 66(1959).

17. J. A. Simmen and P. G. Saffman, Steady deep-water waves on a linear shear current, *Stud. Appl. Maths.* **75**, 35(1985).

18. N. Kishida and R. J. Sobey, Stokes theory for waves on linear shear current, *J. Engng. Mech.* **114** , 1317(1988).

19. A. F. Teles Da Silva and D. H. Peregrine, Steep, steady surface waves on water of finite depth with constant vorticity, *J. Fluid Mech.* **195**, 281 (1988).

20. D. J. Skyner and W. J. Easson, The effect of sheared currents on wave kinematics and surface parameters, *Proc. 23rd Int. Conf. On Coast Engrg. ASCE*, New York, N.Y(1992).

21. C. Swan, An experimental study of waves on a strongly sheared current profile, In *Proc. 22nd Intl Conf. on Coastal Engng. Delft* (ed. B. L. Edge). ASCE. Vol. **1**, 489 (1990).

22. C. Swan, I. P. Cummins and R. L. James, An experimental study of two-dimensional surface water waves propagating on depth-varying currents, *J. Fluid Mech.* **428**, 273 (2001).

23. A. F. Yao and C. H. Wu, Incipient breaking of unsteady waves on sheared currents, *Phys. Fluids.* **17**, 082104 (2005).

24. R. A. Dalrymple, Water wave models and wave forces with shear currents, *Coastal & Ocean. Engng Lab. Univ. of Florida. Tech. Rep.* **20** (1973).

25. R. A. Dalrymple, A finite amplitude wave on a linear shear current, *J. Geophys. Res.* **79**, 4498 (1974).

26. R. A. Dalrymple, A numerical model for periodic finite amplitude waves on a rotational fluid. *J. Comp. Phys.* **24**, 29 (1977).

27. W. Choi, Nonlinear surface waves interacting with a linear shear current, *Math. Comput. Simul.* **80**, 29 (2009).

28. O. G. Nwogu, Interaction of finite-amplitude waves with vertically sheared current fields, *J. Fluid Mech.* **627**, 179 (2009).

29. R. M. Moreira and J. T. A. Chacaltana, Vorticity effects on nonlinear wave-current interactions in deep water, *J. Fluid Mech.* **778**, 314 (2015).

30. P. A. E. M. Janssen, Nonlinear four wave interaction and freak waves, *J. Phys. Oceanogr.* **33**, 863 (2003).

31. C. Kharif and E. Pelinovsky, Physical mechanisms of the rogue wave phenomenon, *Eur. J. Mech. (B/Fluids).* **22**(6), 603 (2003).

32. V. E. Zakharov, A. I. Dyachenko and A. O. Prokofiev, Freak waves as nonlinear stage of Stokes wave modulation instability, *Eur. J. Mech. B/Fluids.* **25**(5), 677 (2006).

33. M. Gerber, The Benjamin-Feir instability of a deep-water Stokes wave packet in the presence of a non-uniform medium, *J. Fluid Mech.* **176**, 311(1987).

34. J. R. Stocker and D. H. Peregrine, The current-modified nonlinear Schrödinger equation, *J. Fluid Mech.* **399**, (1999) 335-353.

35. K. B. Dysthe, Note on the modification to the nonlinear Schrödinger equation for application to deep water waves, *Proc. R. Soc. Lond.* A. **369**, 105 (1979).
36. R. S. Johnson, On the modulation of water waves on shear flows, *Proc. R. Soc. Lond.* A. **347**, 537 (1976).
37. M. Oikawa, K. Chow and D. J. Benney, The propagation of nonlinear wave packets in a shear flow with a free surface, *Stud. Appl. Math.* **76**, 69 (1987).
38. J .C. Li, W. H. Hui and M. A. Donelan, Effects of velocity shear on the stability of surface deep water wave trains, *In Nonlinear Water Waves* (ed. K. Horikawa & H. Maruo), 213-220. Springer (1987) .
39. A. I. Baumstein, Modulation of gravity waves with shear in water, *Stud. Appl. Math.* **100**, 365 (1998).
40. R. Thomas, L' instabilité modulationnelle en présence de vent et d' un courant cisaille uniforme. Thesis Aix-Marseille Université. (2012).
41. R. Thomas, C. Kharif, M. Manna, A nonlinear Schrödinger equation for water waves on finite depth with constant vorticity, *Phys. Fluids.* **24**, 127102 (2012).
42. C. C. Mei, The Applied Dynamics of Ocean Surface Waves. World Scientific (1989).
43. D. H. Peregrine, Interaction of water waves and currents, *Adv. Appl. Mech.* **16**, 9 (1976).
44. C. C. Mei and M. J. Hancock, Weakly nonlinear surface waves over a random seabed, *J. Fluid Mech.* **475**, 247 (2003).
45. D. J. Benney and G. J. Roskes, Wave instabilities, *Stud. Appl. Math.* **48**, 377 (1969).
46. H. Hasimoto and H. Ono, Nonlinear modulation of gravity waves, *J. Phys. Soc. Japan.* **33**, 805 (1972).
47. A. Davey and K. Stewartson, On three-dimensional packets of surface waves, *Proc. R. Soc. Lond.* A. **338** (1613), 101 (1974).
48. I. V. Lavrenov, Wind waves in ocean: dynamics and numerical simulations, Springer-Verlag, Heidelberg. (2003).
49. C. Garrett, J. Gemmrich, Rogue waves, *Phys. Today.* **62**, 62 (2009).
50. M. Onorato, A. Osborne, M. Serio and S. Bertone, Freak wave in random oceanic sea states, *Phys. Review Letters.* **86** (25), 5831(2001).
51. D. H. Peregrine, Water waves, nonlinear Schrodinger equations and their solutions, *J. Austral. Math. Soc. Ser.* B. **25**, 16 (1983).

Numerical Study on the Interactions Between the Bidirectional Wave Trains

Wei Liu, Yuxiang Ma*, Congfang Ai, Guohai Dong, Dianyong Liu

*State Key Laboratory of Coastal and Offshore Engineering,
Dalian University of Technology,
Dalian, 116024, P.R. China
E-mail: yuxma@dlut.edu.cn

Marc Perlin

*Department of Ocean Enginerring, Texas A&M university,
TX, USA*

To investigate the interactions between the bidirectional regular wave trains, the numerical 'X' configurations with different approaching angles were designed based on a fully non-hydrostatic 3D free surface flow model. The interactions of two identical regular wave trains with a relatively approaching angle of 16° were firstly simulated to verify the present model. Then, to have a deeper understanding of the influence of propagation direction on the interactions between the bidirectional wave trains, more numerical simulations with larger approaching angles were conducted. The results show that the approaching angle of wave trains is highly correlated with the interaction between bidirectional regular waves, large approaching angle can lead to intense interactions and strong three-dimensional characteristics of the wave trains after interaction. Furthermore, the maximum wave height produced by interaction was obviously observed to be reached before propagating to the center of the interaction region for large approaching angle.

Keywords: Wave-wave interaction; Non-hydrostatic model; Nonlinearity.

1. Introduction

Wave-wave interaction, as a complex process, is ubiquitous on the ocean surface and plays a key role in the process of ocean dynamics. It is generally accepted that wave-wave interaction is closely associated with wave breaking [1] and the formation of ocean freak waves [2, 3]. Therefore, wave-wave interaction is an important topic in the research fields of ocean engineering and hydrodynamics. Much endeavor has been spent to obtain a better understanding of wave-wave interactions by theoretical [4, 5], numerical [6] and experimental [7] approaches.

As mentioned above, most of the related research mainly target unidirectional progressive waves. In fact, real waves in the ocean are propagating multi-directionally and usually show obvious three-dimensional characteristics. There are also many efforts on three-dimensional waves [8, 9]. Besides multi-directional and three-dimensional characteristics, sometimes the waves are cross-sea states, which are characterized by the coexistence of two wave systems with different directions of propagation. Petrova and Soares [10] conducted an experiment to investigate the maximum wave crest and height statistics of two different long-crest irregular wave systems propagating at different angles, and then verified the experimental results by numerical simulation. Onorato et al. [11] used a higher order spectral method to simulate two long crested wave systems characterized each by JONASWAP spectra and traveling along two different directions of propagation. As the waves used in the above mentioned researches were generated by well-prescribed spectra and mainly focused on the statistical properties of cross-sea states, the interaction process of two wave systems cannot be well observed and analyzed.

Recently, Liu et al. [12] designed an ingenious experiment to study wave-wave interactions. In the experiment, two identical wave groups were generated and propagated isolated with an angle of 8° and then encountered each other and interacted in a specified interaction region. Liu et al. [12] named the interaction process is weakly three-dimensional. It is perhaps the simplest three-dimensional wave configuration that can be produced. The interactions between two identical regular wave trains were also conducted in the same configuration [13]. Although the relevant experiments based on this configuration obtained some important conclusions, due to the space limitations, only a few experiments with relative small approaching angles were conducted. To have a deeper understanding of the influence of approaching angles on wave-wave interaction, more experiments for different approaching angles should be conducted. This is the main motivation of this research. Compared with physical experiments, numerical simulations by a reliable and accurate wave model are more easily implemented. In this study, a fully non-hydrostatic three-dimensional (3D) free surface flow model proposed by Ai [14] is adopted herein, and the accuracy and efficiency of the model has been verified by some typical experiments [14].

2. Numerical Model

2.1. *Governing equations*

For non-hydrostatic free surface water flow, the governing equations are the 3D incompressible Euler equations. To reduce rounding errors, the pressure is split into hydrostatic pressure $g(\eta - z)$ and non-hydrostatic pressure q, as follows:

$$p = g(\eta - z) + q. \tag{1}$$

Then, the governing equations can be expressed as:

$$\frac{\partial u}{\partial x} + \frac{\partial v}{\partial y} + \frac{\partial w}{\partial z} = 0. \tag{2}$$

$$\frac{\partial u}{\partial t} + \frac{\partial u^2}{\partial x} + \frac{\partial uv}{\partial y} + \frac{\partial uw}{\partial z} = -g\frac{\partial \eta}{\partial x} - \frac{\partial q}{\partial x}. \tag{3}$$

$$\frac{\partial v}{\partial t} + \frac{\partial uv}{\partial x} + \frac{\partial v^2}{\partial y} + \frac{\partial vw}{\partial z} = -g\frac{\partial \eta}{\partial y} - \frac{\partial q}{\partial y}. \tag{4}$$

$$\frac{\partial w}{\partial t} + \frac{\partial uw}{\partial x} + \frac{\partial vw}{\partial y} + \frac{\partial w^2}{\partial z} = -\frac{\partial q}{\partial z}. \tag{5}$$

where t is the time; u, v, and w are the velocity components in the horizontal x, y and vertical z directions, respectively; q is the non-hydrostatic pressure component; η is the free surface elevation; g is the gravitational acceleration.

2.2. Boundary conditions

To get a unique solution, boundary conditions are required at all the boundaries of a 3D domain. At the moving free surface $\eta(x,y,t)$, the kinematic boundary is given by

$$\frac{\partial \eta}{\partial t} + u\frac{\partial \eta}{\partial x} + v\frac{\partial \eta}{\partial y} = w\big|_{z=\eta}. \tag{6}$$

It should be note that atmospheric pressure is assumed at free surface elevation, giving $q\big|_{z=\eta} = 0$.

At the impermeable bottom $z = -h(x,y)$, the kinematic boundary condition is given by

$$-u\frac{\partial h}{\partial x} - v\frac{\partial h}{\partial y} = w\big|_{z=-h}. \tag{7}$$

Using kinematic boundary conditions (6) and (7) in the integrated form of the continuity equation (2) over the water column, the free surface equation is obtained:

$$\frac{\partial \eta}{\partial t} + \frac{\partial}{\partial x}\int_{-h}^{\eta} u\,dz + \frac{\partial}{\partial y}\int_{-h}^{\eta} v\,dz = 0. \tag{8}$$

At the inflow, the normal velocity component is specified by either linear wave theory or analytical solutions. In addition, tangential velocity component are set to zero. At the outflow boundary, a combination of the Sommerfeld radiation condition and a sponger layer technique is implemented to minimize wave reflection.

2.3. *Numerical Method*

A horizontal Cartesian grid framework and vertical boundary-fitted coordinate system is presented. The 3D physical domain is bounded by the moving free surface, $z = \eta$, and the bottom, $z = -d$, and is discretized as a 2D horizontal grid with several horizontal layers. In the vertical direction, the number of vertical layers across the whole domain is constant and the layer thickness is allowed to vary respect to both the horizontal locations and time. In a layer, the combination of 2D structured grid and unstructured grid [15] is adopted, thus rendering the model can deal with geometric complexity.

A finite volume method is chosen to discretize the governing equations. An explicit projection method consists of two major steps is employed to solve the equation in two major steps. In the first step, the intermediate velocity field is achieved by means of solving the momentum equations that contain the non-hydrostatic pressure at the previous time level. In the second step, the pressure correction is computed by the discretized Poisson equation, which is obtained by the combination of the discretized continuity and the discretized momentum equation. The detailed derivation can be found in the paper of Ai [14].

3. Validation of the Model

Even though the capability of the numerical model has been verified by some classical experiments, before usage it to study the interaction of bidirectional propagating waves, the scenario of the interactions of two identical regular wave trains with a relatively approaching angle 16° are simulated to further verify the present model. The detailed experimental setup can be seen in Liu et al. [12] and Liu et al. [13]. The experiment was carried out in a wave basin located at State Key Laboratory of Coastal and Offshore Engineering, Dalian University of Technology. The wave basin is 32 m long, 24 m width, 1m depth and a water depth of 0.6m was used. The sketch layout of the experiment can be seen in Figure 1. In the experiments, an 'X' configuration was installed, formed by two 0.8m wide flumes with a relative angle of 16° (Fig. 1). Two monochromatic wave trains are simultaneously generated at one side of the flumes and propagate separately along in the two flumes, and then encounter in the interaction region of the configuration. After interactions, the two wave trains can propagate into the corresponding flumes. The detailed information of the wave parameters are shown in Table 1. All cases in the table are also contained in the experiment of Liu [13], and his experimental results demonstrate there was no wave breaking occurred in the process of the interaction for all cases.

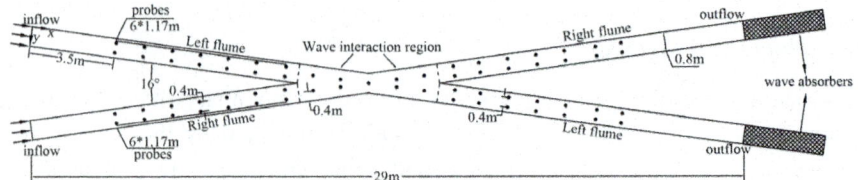

Fig. 1. Experimental setup.

Table 1. Wave parameters.

Case	f (Hz) [a]	k(m^{-1}) [b]	H_0(cm) [c]	IMS [d]
A1	0.8	2.77	8.07	0.11
A2	0.8	2.77	14.84	0.21
A3	0.8	2.77	17.11	0.24
B1	1.2	5.81	3.78	0.11
B2	1.2	5.81	4.79	0.14
B3	1.2	5.81	7.07	0.21
B4	1.2	5.81	8.43	0.24

Note: [a] Frequency. [b] Wave number. [c] Initial wave height. [d] Initial wave steepness $= H_0 k / 2$.

The 3D computational domain is discretized as a 2D unstructured horizontal grid with 10 horizontal layers. In a layer, the 2D grid is built from triangles in the interaction region and quadrilaterals in the remaining fields. The grid spacing of triangles and quadrilaterals are both set to 0.05m. The time step is taken as 0.005s and the total simulation time is 50s.

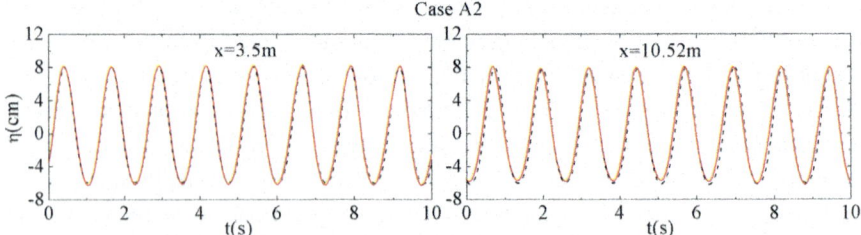

Fig. 2. Comparisons among the surface profiles measured before the interaction region between numerical results and experimental results for Case A2. The red solid lines represent the surface profiles by the numerical simulations. The black dash lines represent the surface profile measured in the experiments.

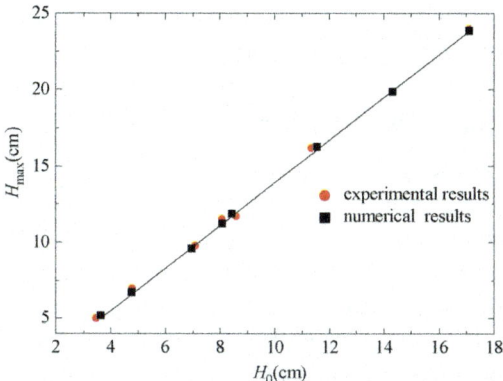

Fig. 3. Comparison of the maximum wave height against the initial wave height between numerical results and experimental results for different cases. The solid line represents a linear least-squares fit of the numerical results: $H_{max} = 1.387H_0 + 0.08$, $R^2 = 0.9998$.

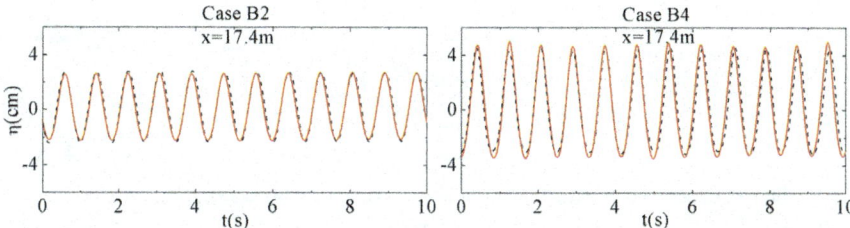

Fig. 4. Comparisons among the surface profiles measured after the interaction region between numerical results and experimental results for Case B2 and B4. The red solid lines represent the surface profiles by the numerical simulations. The black dash lines represent the surface profile measured in the experiments.

To ensure that the numerical simulation has the same initial wave heights with physical experiment, the surface elevations measured at the location $x=3.5$m were compared with the results measured at the same location in the physical experiment. Furthermore, the surface elevations measured at the locations $x=10.52$m which is close to the interaction region were also compared with the results of physical experiments. The comparison results of the wave height located at $x=3.5$m and $x=10.52$m for case A2 are shown in Figure 2.

As shown by Liu et al. [13], the maximum amplification of wave height has a linear relation with the initial wave amplitude. Figure 3 illustrates the relationship between the initial wave height and the maximum height in the middle of the interaction region. It is observed that the numerical results agree well with the experimental data, with a height amplification about 1.39 times of the initial wave height. Figure 4 shows the comparison of surface profiles after the interaction by numerical simulations and experiments for Case B2 and B4, and the comparison results are satisfactory.

4. Influence of Approaching Angle on Wave Interactions

Although in the previous experiment, Liu et al. [13] considered the influence of approaching angle between the two wave trains, because of the limitation of experimental conditions, only a small approaching angle (16°) was conducted, and the impact of approaching angles on the interaction of wave trains has been not yet known. To have a deeper understanding on the interaction process in the 'X' configuration and further study the influence of approaching angles, more numerical tests with different approaching angles are simulated. The experimental facility is similar with the previous experiment, formed by two 30m long wave channels. The included angle of two flumes is variable, from 16° to 90°, other parameters and probes arrangement are denoted in Figure 5.

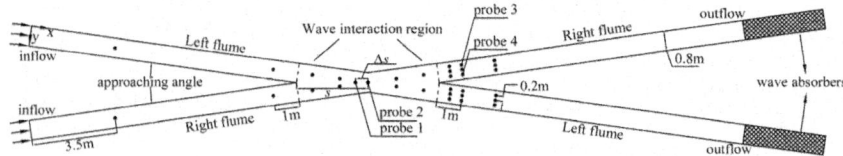

Fig. 5. Experimental setup for the surface elevation measurements. (Not to scale.)

Table 2. Numerical wave parameters.

Angle	f (Hz)	k (m^{-1})	H_0 (cm)	IMS
16°	0.8	2.77	8.15	0.11
24°	0.8	2.77	8.15	0.11
30°	0.8	2.77	8.15	0.11
36°	0.8	2.77	8.15	0.11
45°	0.8	2.77	8.15	0.11
60°	0.8	2.77	8.15	0.11
75°	0.8	2.77	8.15	0.11
90°	0.8	2.77	8.15	0.11

In the previous simulation, we observed that the interaction processes of bidirectional wave trains with 0.8Hz are more stable than that of 1.2Hz, hence the wave trains with 0.8Hz is the preferred and is detailed analyzed. To avoid the influence of wave breaking on further study, only the initial wave steepness IMS=0.11 in the physical experiments of Liu et al. [13] is considered in the next simulations, and the later analysis suggests that the maximum wave steepness reached in the interaction region is far less than the extreme wave steepness. The corresponding wave parameters are shown in Table 2, and other calculating parameters are the same as the previous simulations.

4.1. *Wave evolution*

To analyze the influence of approaching angle on the process of the interaction, the evolution of wave propagation in the 'X' configurations at different stages for different approaching angles are shown in Figures 6-8.

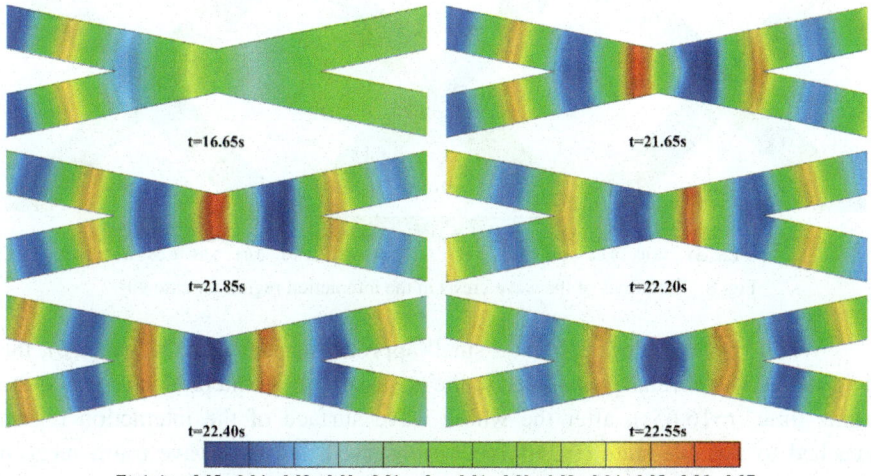

Eta(m): -0.05 -0.04 -0.03 -0.02 -0.01 0 0.01 0.02 0.03 0.04 0.05 0.06 0.07

Fig. 6. Wave interaction process with the approaching angle of 24°

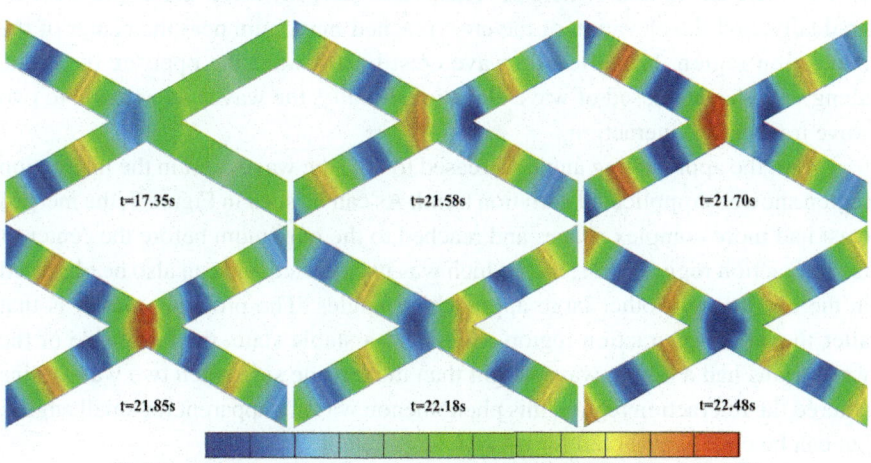

Eta(m): -0.05 -0.04 -0.03 -0.02 -0.01 0 0.01 0.02 0.03 0.04 0.05 0.06 0.07

Fig. 7. Variations of the wave crests in the interaction region for case 60°.

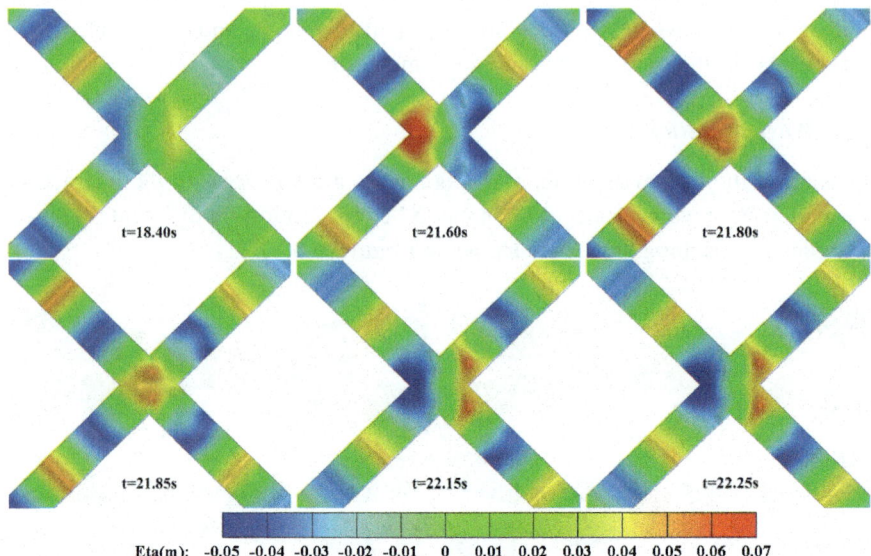

Fig. 8. Variations of the wave crests in the interaction region for case 90°.

When two wave trains have a small approaching angle of 24° (Fig. 6), the two steady wave trains propagated to the upstream of the interaction region at the same time (t=16.65s), after the whole wave surface of the interaction region reached to a stable state (t=21.65s), we observed that two wave trains merged from two straight wave crests into a smooth, crescent-shaped crest, contributing to the increase in wave height. Then, the merged wave crest straightened gradually, and the elevation of the crest reached maximum near the center of the interaction region. Hereafter the wave crest bent toward the opposite direction, along with the decreased of wave elevation. Finally, the waves separated into two wave trains after interaction.

With the approaching angle increased to 60°, the wave crest in the interaction region showed complicated variation trend. As can be seen in Figure 7, the merged crest had more complex shapes and reached to the maximum before the center of the interaction region (t=21.70s), which was unexpected and can also be observed in the condition of other large approaching angles. The probably reason is that after the whole interaction region reached to a stable state, the inner side of the wave trains had a higher wave height than the outside side when two wave trains entered the interaction region, this phenomenon was not apparent for small angles, but can be clearly observed for relative larger angles.

As the approaching angle increased to 90°, the evolution of waveform are more complicated, as shown in Figure 8. The wave packets always existed even though the wave propagated into the downstream of interaction region, suggesting there were intense interactions of wave-wave interaction.

4.2. *Maximum wave height and its position in the interaction region*

As mentioned before, when two wave trains have a large approaching angle, the position where the maximum wave height appears were observed an obvious approach toward the upstream of the interaction region. To measure the degree of the position deflection, two probes were used to measure the maximum wave height and the wave height at the center of the interaction region separately, Δs represents the distance between the two probes and s represents a half of the horizontal length of the interaction region, which are shown in Figure 5. A non-dimensional position deflection can be expressed as $\Delta s/s$. The maximum amplification of wave height has also been non-dimensionalized by the initial wave height. Figure 9 shows the maximum amplification of wave height and its position in the interaction region for different approaching angles. The non-dimensional position deflection has a positive correlation with approaching angle as shown in Figure 9(a), implying that the position shift is more significant for the condition with a larger approaching angle. Figure 9(b) indicates that the maximum wave height affected by approaching angle a lot. As the approaching angle increases, the maximum wave height decreases at a certain angle range, and then increases monotonically, reaching to 1.65 times for the condition of 90°.

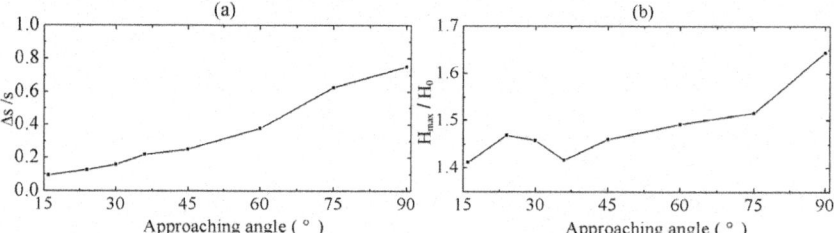

Fig. 9. The non-dimensional relative position and the maximum wave height.

4.3. *Surface Elevations*

The surface profiles measured at lateral positions in one flume, located 1m after interaction region which is shown in Figure 5. Different from the results of case 16° by the previous simulation, the surface profiles of lateral positions have some disparity when the approaching angle increase to 36° (Fig. 10(a)), and this discrepancy of surface profiles becomes more obvious for 60° (Fig. 10(b)), implying that the two wave trains may have a more violent interaction and thus lead to a stronger nonlinear characteristics of the wave trains downstream of the interaction region. If the approaching angle continue to increase (75°, Fig. 10(c)), we even find the surface profiles of left side occurred some deformation. When the approaching angle reaches to 90° (Fig. 10(d)), what is interesting is that the wave asymmetry respect to the center lines of flume seems to weaken. This phenomenon demonstrates the three-dimensional characteristics of wave trains downstream of the interaction region are not always similar for all approaching angles. For 90°, the wave trains after interaction show a horizontal asymmetry that is a larger wave crest in the center lines of two flumes , which can be clearly observed in the Figure 8 (t=22.25s). However, for other approaching angles, the wave trains after interaction present a vertical asymmetry respect to the center lines of two flumes, causing an uplift of the wave surface at one side in a flume.

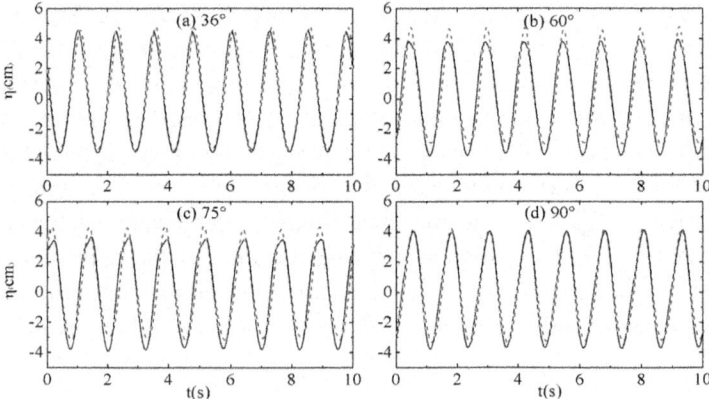

Fig. 10. Surface profiles measured in the downstream of the interaction region in the right flume for different approaching angles. The black solid lines and red dash lines represent the surface profile measured on the left and right side of the right flume respectively.

5. Conclusions

A fully non-hydrostatic three-dimensional free surface flow model is employed to investigate the interactions between two monochromatic wave trains with different approaching directions. Firstly, the physical experiment of Liu [13] is simulated to verify the accuracy of the model. The comparison results show that the model can simulate the interaction of two wave trains accurately. Then, more cases with different approaching angles from 16° to 90° were conducted to have a deeper study of the influence of approaching angles on interaction process. The further research indicates that the approaching angle of two wave trains also have an important influence on the process of interaction. Lager approaching angle can lead to more intense interactions and strong three-dimensional characteristics of the wave trains after interaction. The maximum wave height produced by interaction was observed to be reached before propagating to the center of the interaction region, and is more obvious for a larger approaching angle.

References

1. Perlin M. W., Choi W. and Tian Z., Breaking Waves in deep and intermediate waters, *Annu. Rev. Fluid Mech.* **45**, 115 (2013).
2. Kharif C. and Pelinovsky E., Physical mechanisms of the rogue wave phenomenon, *Euro. J. Mech. B/Fluids* **22(6)**, 603 (2003).
3. Dysthe K., Krogstad H. E. and Muller P., Ocean rogue waves, *Annu. Rev. Fluid Mech.* **40(1)**, 287 (2008).
4. Phillips O. M., On the dynamics of unsteady gravity waves of finite amplitude Part1, The elementary interactions, *J. Fluid Mech.* **9**, 193(1960).
5. Longuet-Higgins M. S., Resonant interactions between two trains of gravity waves, *J. Fluid Mech.* **12(3)**, 321 (1962).
6. Rapp R. J. and Melville W. K., Laboratory measurements of deep water breaking waves, *Phil. Trans. R. Soc.* **A331**, (735) 1990.
7. Chen G., Christian K., Stephane Z. and Li J., Two-dimensional Navier-Stokes simulation of breaking waves, *Phys. Fluids* **11**, 121 (1999).
8. K. She, C. A. Greated and W. J. Easson, Experimental study of three-dimensional wave breaking, *Ocean Eng.* **120(1)**, 20 (1994).
9. T. B. Johannessen and C. Swan, A laboratory study of the focusing of transient and directionally spread surface water waves, *The Royal Society*, (2000).
10. Petrova P. and Guedes S. C., Maximum wave crest and height statistics of irregular and abnormal waves in an offshore basin, *Applied Ocean Research* **30(2)**, 144 (2008).
11. Onorato M., D. Proment and A. Toffoli, Freaking waves in crossing seas, *Eur. phys. J. Spec. Top.* **185**, 45 (2010).
12. Liu D., Ma Y., Perlin M. and Dong G., An experimental Study of weakly three-dimensional non-breaking and breaking waves, *Euro. J. Mech. B/Fluids* **52**, 206 (2015).
13. Liu D., Ma Y., Perlin M. and Dong G., An experimental investigation of theinteraction between the bidirectional wave trains, *7th Chinese-German Joint Symposium on hydraulic and Ocean Engineering*, (2014).
14. Ai C., Jin S. and Lv B., A new fully non-hydrostatic 3D free surface flow model for water wave motions, *Int. J. Numer. Meth. Fluids* **66**, 1354 (2011).
15. Ai C. and Jin S., Non-hydrostatic finite volume model for non-linear waves interacting with structures, *Computers & Fluids* **39**, 2090 (2010).

A Semi-Empirical Formula for Wave Attenuation Over Muddy Bed Under Current

Xie Wenang and Tomoya Shibayama

Waseda University,
Tokyo, 169-8555, Japan

Theoretical and laboratory studies are performed to analyze the wave attenuation phenomenon over fluid mud under current. In the theoretical part, a semi-empirical formula is derived to model the attenuation process under current. Based on the linear wave theory, the change of wave amplitude is related with the averaged damping effect of the muddy bed. By solving a differential equation, the exponential decay of the wave amplitude is confirmed. A function called damping function is used in the formula, and it is expanded into a series of wave parameters and some undetermined coefficients called auxiliary damping factors. Based on the assumption that the mud properties do not change under current, the formula is combined with the dispersion relation for linear wave under uniform current to describe the attenuation process under current. Laboratory experiments are conducted first without current. The data of wave attenuation are collected and inserted to the formula obtained in the theoretical analysis to figure out the values of the auxiliary damping factors. The second set of experiments is conducted under current and the collected data are compared with the calculated values obtained by using the newly proposed formula with the dispersion relation. The result of comparison shows good applicability of the new formula to the wave attenuation under current.

Keywords: Wave attenuation; Fluid mud; Semi-empirical formula.

1. Introduction

The amplitude of water surface wave is known to be attenuated due to the energy dissipation when the wave is propagating over a muddy bed. The first rigorous study on wave attenuation phenomenon was conducted by Gade (1958) [1]. It was reported by Gade that in the central Louisiana coasts, fishermen sometimes use the areas with muddy bed as their emergency harbors when high wave and storm attacked. The wave height was found to be obviously small in these areas. Muddy bed can also be found in many river mouths and coasts all over the world. Therefore, it is necessary for coastal engineers to know how to predict the attenuation of wave amplitude over muddy bed.

After the work of Gade, the attenuation phenomenon of wave over muddy bed the has drawn attention from numbers of researchers. The rheology of mud, which is important when evaluating the energy dissipation, was studied in various ways. Viscos fluid model was applied by Dalrymple and Liu (1978) [2], and they also derived the dispersion relation of attenuated wave. Amount of more complex rheological models were also proposed to evaluate the energy dissipation (MacPherson (1980) [3], Shibayama et.al (1993) [5], Liu et.al (2007) [6], Xia (2013) [4]). These studies showed high accuracy when comparing the calculated results with the laboratory data. However, it is also notable that the equations and formulas used in these researches were complicated and, to a certain extent, inconvenient for engineers. Therefore, in this study, we tried to propose a simpler formula which is easy to be applied to engineering works.

In real coasts and river mouths, it is usual that wave motion is coupled with current. Some studies have been already conducted to consider the influences of current on wave attenuation. On wave-current-mud interaction, An and Shibayama (1994) [9] carried out laboratory and numerical study. Also, spectral analysis on the attenuation of irregular wave under current was conducted by Soltanpour et.al (2014) [7] and Samsami et.al (2015) [8]. Numerical simulations applied in these works were quite sophisticated due the complexity of the governing equations. In the present study, the target will be proposing an explicit formula for wave attenuation under current.

Semi-empirical method is usually used when the pure theoretical analysis is overwhelming but the accuracy of the calculation is required at the same time. Wave attenuation over muddy bed can also be studied by using this method. Here, we first derive an explicit formula for the wave amplitude in a relatively simple form based on linear wave theory. Then laboratory experiments where the current does not exists will be carried out, and the collected data will be used to figure out the undetermined coefficients in the derived formula. However, it will be shown by another set of laboratory experiments that this formula is also applicable when current exists as well.

2. Derivation of the formula

The first purpose of this part is to prove that, it is possible to write the attenuated wave amplitude with a constant water depth in the form of

$$a = a_0 \exp\left(\mu\left(kh\right)\cdot x\right) \tag{1}$$

where a is wave amplitude, a_0 is the initial wave amplitude, $\mu\left(kh\right)$ is an unknown function, k is wave number, h is water depth, and x is the location in the horizontal direction. In this study, we specifically call $\mu(kh)$ damping function.

Consider the horizontal velocity component, u, and vertical velocity components, v, of water particle at the water surface based on linear wave theory,

$$u = a\omega \coth(kh)\sin(kx - \omega t) \tag{2}$$

$$v = a\omega \cos(kx - \omega t) \tag{3}$$

where ω is angular velocity of wave. Therefore, the total velocity of water particle at the surface can be calculated as

$$U = a\omega\sqrt{\coth^2(kh)\sin^2(kx - \omega t) + \cos^2(kx - \omega t)} \tag{4}$$

The dispersion relation gives the relation between ω, k and h, that is, $\omega = \omega(k, h)$. By applying the dispersion relation, we can confine the discussion to fewer variables for a specific location x_0 as follows,

$$U = a \cdot f(k, h, t) \tag{5}$$

$$f(k, h, t) = \omega\sqrt{\coth^2(kh)\sin^2(kx_0 - \omega t) + \cos^2(kx_0 - \omega t)} \tag{6}$$

It is notable that, for the dispersion relation mentioned here, it is not necessary to assume non-current condition, the following dispersion relation for linear waves under uniform current can also be applied as well.

$$V + \sqrt{\frac{g}{k}\tanh(kh)} = \frac{\omega}{k} \tag{7}$$

where V is the velocity of the uniform current.

Following equation (5), the average velocity of the water particle in one wave period is therefore calculated as

$$\overline{U} = \frac{1}{T}\int_0^T a \cdot f(k, h, t)\, dt = a \cdot F(k, h) \tag{8}$$

where F is a function of k and h, since according to the dispersion relation, the wave period T is also a function of them. Up to here, we have proved that the averaged velocity of a water particle at the surface $\overline{U}$ is a function of wave amplitude a, wave number k and water depth h (even if the wave is attenuated).

While considering the wave attenuation, we assume the damping rate of wave amplitude is proportional to two factors: (i) the average velocity of water particle on the surface $\overline{U}$ (This assumption is commonly used when describing damping phenomena). (ii) a function G representing the influence from the fluid mud at the bottom. In other words,

$$\frac{da}{dx} = \overline{U} \cdot G = a \cdot F\left(k, h\right) G \qquad (9)$$

G can be a function of many parameters, for examples, wave number k, water depth h, water content ratio of mud, etc.. In this study, we assume that the properties of mud are unchanged under wave motion. For example, the water content ratio of the mud is the same in still water as under wave motion. Under this assumption, it is possible to write G as a function only depends on k and h, that is, $G = G\left(k, h\right)$, since other parameters related with mud are constants. Although the explicit form of G is unknown, it is still possible to write

$$\frac{da}{dx} = a \cdot F\left(k, h\right) \cdot G\left(k, h\right) = a\mu\left(k, h\right) \qquad (10)$$

where $\mu\left(k, h\right) = F\left(k, h\right) \cdot G\left(k, h\right)$ is called damping function. It is obvious that the amplitude can be solved as

$$a = a_0 \exp\left(\mu\left(k, h\right) \cdot x\right) \qquad (11)$$

from equation (10).

Comparing equation (11) and equation (1), it is obvious that the next step is to convert $\mu\left(k, h\right)$ into $\mu\left(kh\right)$. Since, when the current velocity is zero, equation (7) gives a relation between k, T and kh, therefore we can replace $\mu\left(k, h\right)$ by $\mu\left(kh, T, h\right)$. It can also be understood that if kh and T are specified, k will be determined. By using T and k, we can also calculate h from equation (7). Therefore, $\mu\left(kh, T, h\right)$ can be rewritten into $\mu\left(kh, T\right)$ as h is determined by kh and T.

Here, without deriving, we assumed that μ is not a function of wave period T. This statement will be proven by the laboratory data instead of mathematical derivation in the later part of this paper. Now, $\mu = \mu\left(kh\right)$ is obtained and by plugging this result into equation (11), we can have equation (1).

The next step is to figure out the explicit formula of damping function $\mu\left(kh\right)$. Due to the complex properties of the bottom fluid mud, it is difficult to apply rigorous mathematical analysis to obtain exact solution. However, semi-empirical method can be carried out to overcome this difficulty. Here, we expand the damping function into a power series of kh up to a certain order and find out the coefficients by using laboratory data.

$$\mu = \mu_1\left(kh\right) + \mu_2\left(kh\right)^2 + \mu_3\left(kh\right)^3 + O\left(\left(kh\right)^4\right) \qquad (12)$$

Here, μ_1, μ_2, μ_3 are undetermined constant coefficients specifically called auxiliary damping factors in this study, and they will be figured out by using laboratory data. We have to point out that series (12) is based on the assumption that kh is not too large. If series (12) is applied to deep water condition, higher order terms must be considered. However, in coastal areas, waves usually propagate in shallow water where kh is small value. The wave amplitude now can be approximately written as

$$a = a_0 \exp\left[\left(\mu_1(kh) + \mu_2(kh)^2 + \mu_3(kh)^3\right)x\right] \tag{13}$$

for long wave condition. By combining (13) and the dispersion relation (7), the influence of current on wave attenuation can also be studied as well.

3. Description of laboratory experiment

Wave flume used for experiment is described as following. The width of the glass-sided flume is 40.5 cm and the depth is 60 cm. The total effective length of the flume is 13.8 m. Regular waves were generated by a flap-type wave generator at one of the ends of the flume. Current can be generated by a pump system through an inlet or outlet equipped under the flume. Two capacitance wave gauges were employed to measure wave heights while the current velocity was recorded by an electromagnetic current meter. The data of gauge 1 presents the wave height before the wave start to be attenuated, while gauge 2 shows the wave height after the attenuation. Three false beds were placed in the flume to confine the 1.2m long mud section with a thickness of 10 cm. The sketch is shown in Fig.1.

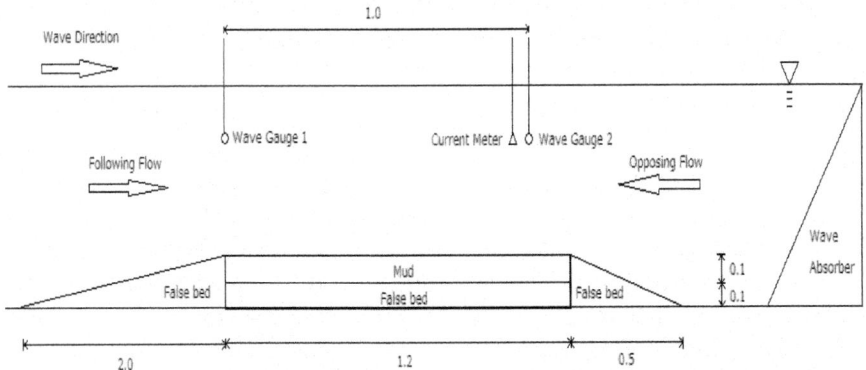

Fig. 1 Sketch of the experimental set-up (length in m, horizontally exaggerated)

A well mixture of commercial kaolinite with tap water was used as muddy bed. The chemical characteristics of the used kaolinite were $SiO2=45.9\%$, $Al2O3=37.8\%$ and $Fe2O3=0.6\%$; and the particle size distribution follows D40 and D86 of 2μm and 10μm, respectively. The water content of the mud is around 110%.

As for the procedure of the laboratory experiments, the water depth will be chosen (0.225m, 0.250m, 0.275m) firstly, and for each water depth, waves with different periods (0.8s, 0.9s,…1.5s) will be generated. The wave amplitudes will be recorded by the wave gauges. Wave attenuation is represented by the difference between the wave amplitudes recorded by gauge 1 and gauge 2. With the same wave period, initial wave heights will be chosen randomly. This experiment can be carried out with or without current by the same procedure.

4. Determining the auxiliary damping coefficients ($V = 0$)

By using the laboratory data of the condition without current ($V = 0$) obtained in the experiments, it is possible to evaluate the auxiliary damping factors in (13) by drawing fitting curve. The plots in Figure 2 are obtained from the experiments where $V = 0$, showing the values of $\ln\left(a/a_0\right)$ which is also the values of $\mu(k)$. The fitting curve suggests possible values for the auxiliary damping factors.

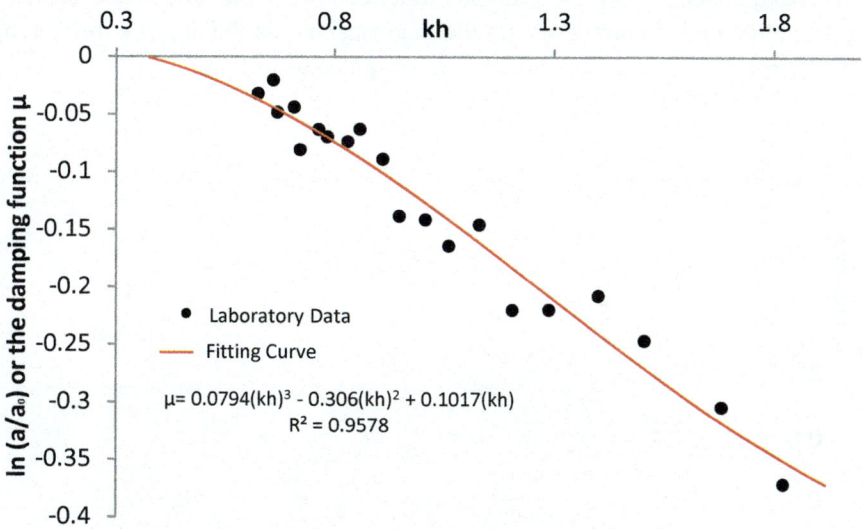

Fig. 2 Fitting curve determining the values of auxiliary damping factors

From Fig. 2, it can be seen that, although the wave period ranges from 0.8s to 1.5s, plots still distribute close to the fitting curve and give high R^2. Therefore, it can be said that, there is high possibility that damping function μ is not a function of wave period T and $\mu = \mu(kh)$ is an appropriate form for it. From the formula of the fitting curve, (13) can be approximately written as

$$a = a_0 \exp\left\{ \left[0.0794\left(kh\right)^3 - 0.306\left(kh\right)^2 + 0.1017\left(kh\right) \right] x \right\} \qquad (14)$$

for this specific case. It has to be pointed out that, the auxiliary damping factors obtained in this study is only for the certain mud used in the present laboratory work. If the characteristics of the mud change, the values of auxiliary damping factors will also be different from the present values. It means that before calculating the wave attenuation over different mud, it is necessary to conduct the same laboratory experiment again. However, since the experiment set-up and the procedure is simple, the laboratory work is actually not time consuming.

5. Application to the condition where $V \neq 0$

Formula (14) can also be applied to the condition where $V \neq 0$, although it was obtained in the condition where $V = 0$. By combining with the dispersion relation (7), formula (14) showed agreement with the laboratory data in various scenarios. Here, water depth and current velocity were changed in the present study. It is important to point out that, the properties of mud are assumed to be unchanged under current. For example, water content of the fluid mud keeps the same value under current as in the non-current condition. The results of comparison in different scenarios are demonstrated as follows.

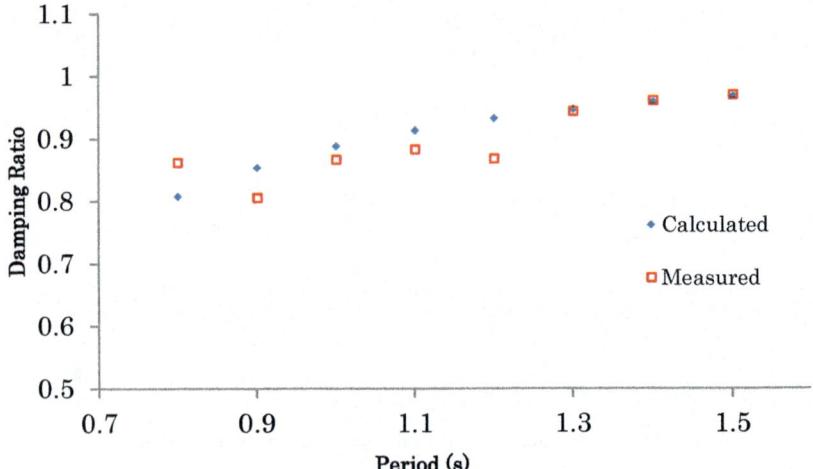

Fig. 3 Comparison of wave attenuation between calculation and laboratory data.
(Case I : Water depth H=0.225m, Current velocity V around 0.14m/s)

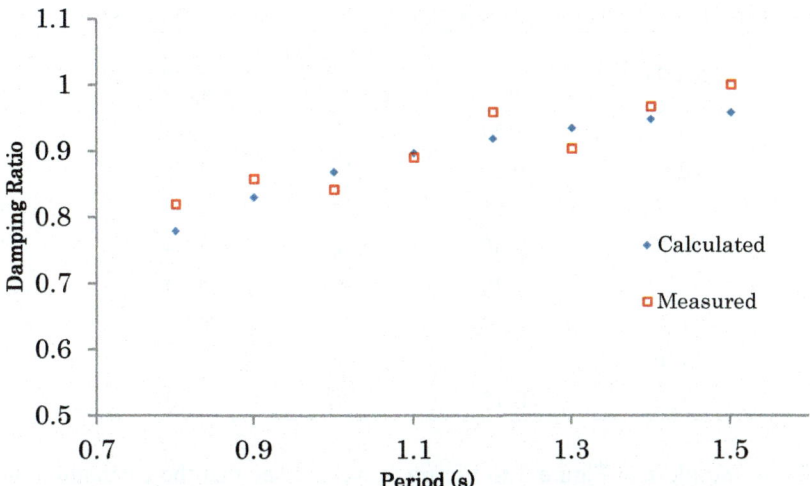

Fig. 4 Comparison of wave attenuation between calculation and laboratory data.
(Case II : Water depth H=0.250m, Current velocity V around 0.13m/s)

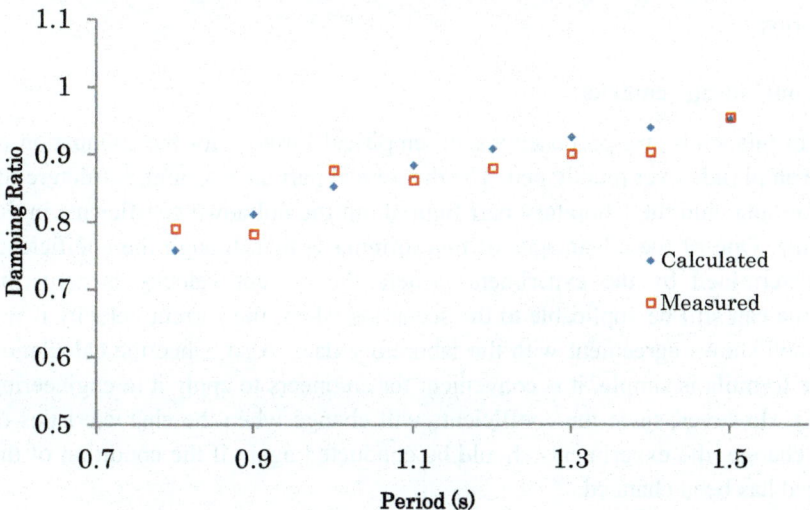

Fig. 5 Comparison of wave attenuation between calculation and laboratory data.
(Case III : Water depth H=0.275m, Current velocity V around 0.14m/s)

To demonstrate how the formula fairs the observed data in each cases above, the error between calculation and measurement are presented in the following table.

Table 1 Error between calculation and measurement in Case I, Case II and Case III

Period (s)	Error (%)		
	Case I	Case II	Case III
0.8	6.22	4.87	3.96
0.9	6.12	3.17	3.81
1.0	2.55	3.24	2.69
1.1	3.50	0.78	2.64
1.2	7.49	4.16	3.29
1.3	0.43	3.49	2.89
1.4	0.14	1.91	4.19
1.5	0.15	4.16	0.20

From the plots in Figure 3 to Figure 5, we can see that the calculation and the observed data have the same tendency when wave period increases. In Table 1, it can be seen that, for most cases, the error is under 5%. Therefore, it can be said that formula (14) successfully predicted the damping ratio in different scenarios.

6. Concluding remarks

In this study, we proposed a semi-empirical formula for the attenuation of wave amplitude over muddy bed. The theoretical part gave a simple structure for the formula, and the laboratory part figured out the unknown coefficients in the formula. One of the advantages of this formula is that, though the coefficients are determined by the experiments where the current velocity is zero, the formula can still be applicable to the scenarios where the current velocity is not zero and shows agreement with the laboratory data. Also, since the calculation of the formula is simple, it is convenient for engineers to apply it to engineering works. However, since the coefficients will change when the characteristics of mud change, the experiments should be conducted again if the condition of the sea bed has been changed.

References

1. Gade H.G., Effects of non-rigid, impermeable bottom on plane surface wave in shallow water, J. Marine Research, 16, 61-82. (1958).
2. Dalrymple R. A. Philip L.-F. Liu., Wave over Soft Mud: A Two-layer Fluid Model, Journal of Physical Oceanography, vol. 8, 1121-1131, (1978)

3. Macpherson, The attenuation of water waves over a non-rigid bed, J. Fluid Mech., vol. 97, part. 4, 721-742. (1980)

4. Xia Y-Z.. "Wave Attenuation over Seabed Mud Modelled by a Two-layered Viscoelastic Model, APAC, Indonesia, (2013)

5. Shibayama T. & An N. N., A visco-elastic-plastic model for wave-mud interaction, Coastal Engineering in Japan, 36(1), 67-89. (1993)

6. Philip L.-F. Liu, i-chi Chan, On long-wave propagation over a fluid-mud seabed, J. Fluid Mech., vol. 579, 467–480. (2007)

7. Soltanpour M., Samsami F., Shibayama T., Yamao S., Study of Irregular Wave-current-mud Interaction, Proceedings of the Coastal Engineering Conference (2014).

8. Samsami F., Soltanpour M., & Shibayama T., Spectral analysis of irregular waves in wave–mud and wave–current–mud interactions, Ocean Dynamics, 65(9-10), 1305-1320. (2015)

9. An, N.N., and T. Shibayama., Wave-current interaction with mud bed, Proc. of 24th Coastal Engineering Conference, ASCE, 2913-2927. (1994)

Trapped Waves Over the Hyperbolic-Cosine Ocean Ridge[*]

G. Wang[†] and J. Hu

Key Laboratory of Coastal Disaster and Defence (Hohai University),
Ministry of Education,
Nanjing, 210098, China
†E-mail: gangwang@hhu.edu.cn

J. H. Zheng

College of Harbour, Coastal and Offshore Engineering,
Hohai University,
Nanjing, 210098, China
E-mail: jhzheng@hhu.edu.cn

Q. H. Liang

School of Engineering, Newcastle University,
Newcastle upon Tyne, NE1 7RU, UK
E-mail: qiuhua.liang@newcastle.ac.uk

Due to refraction, the oceanic ridge acts as a waveguide forcing long-period waves to propagate along the topography. Based on the linear shallow water equations, analytical solutions of trapped waves over a hyperbolic-cosine squared ocean ridge are obtained, which are described by combining the associated Legendre functions of the first and second kinds. The spatial distribution pattern for each mode is discussed, and the wave amplitude gets the maximum at the ridge top and decays gradually towards both sides. The higher the mode number, the slower the rate of amplitude decreases, so that more energy is distributed over more of the ridge. The trapped wave number is not only related to the frequency, but also to the varying water depth parameters.

Keywords: Trapped wave; Ocean ridge; Analytical solutions; Water waves; Shallow water equations.

---

[*] This work is supported by supported by the National Key R&D Program of China (No.: 2017YFC1404205), the National Natural Science Foundation of China (NO.: 51579090) and Innovation Project of Colleges and Universities in Jiang Su Province (NO.: 2015B15714).
[†] Work partially supported by grant 2-4570.5 of the Swiss National Science Foundation.

1. Introduction

Tsunamis can be caused by earthquakes, volcanic eruptions, submarine collapses and submarine crustal movements, however, submarine earthquakes are the main cause for it. Tsunamis can be divided into transoceanic tsunamis and local tsunamis for their affected areas. The transoceanic tsunami refers to the tsunami that travels across the oceans or from far away, and the local tsunami is a tsunami that occurs when a tsunami occurs near the affected area. Due to the effect by the submarine topography, large tsunamis are usually not only a local tsunami that attacks the coastal areas, but also far-field tsunamis that can cause significant damage to tens of thousands of miles away. The results of the multi-field tsunami show that the oceanic ridge has the ability to guide the tsunami and transmit it along the ridge tens of thousands of kilometers to bring disaster to the far-field area[1-6].

Topographically trapped waves are waves that are restricted to coastal areas such as coastal, continental shelf, and other coastal areas[7-9], or oceanic ridge, seamount, submarine plateau, and other underwater raised topography. Reid[10] is divided into two types according to the ratio of the angular frequency ω and the Coriolis force parameter f. The tsunami wave ω is much larger than f belongs to the first kind of long waves or inertia-gravitational waves. Therefore, the research of this paper is without considering the Coriolis force f.

The topographically trapped waves can be divided into edge waves (or shelf waves), island trapped waves and the oceanic ridge trapped waves. The problem of oceanic ridge trapped waves is first proposed by Longuet-Higgins[11], who gives analytical solutions for oceanic ridge trapped waves with a rectangular infinite long straight ridge. Buchwald[12] discusses the rectangular ridge conditions with different water depths on both sides, and obtains the solution of the equation and gives the dispersion relation, especially the low frequency cutoff rate, phase velocity and wave group velocity. Shaw and Neu[13] gives the wave profile analytical solution, dispersion relation, cutoff frequency, phase velocity, and group velocity of the infinite straight ridge of the stepped profile. However, almost all of these studies are carried out for the ideal rectangular ridge, which is produced by the reflection of waves. In fact, most of the ridge topography is continuous, and the oceanic ridge trapped waves are generated by the wave refraction, so these studies cannot be applied to investigate the movement of the real tsunami on such ridges. Since the solution of the trapped waves is closely related to the topography, choosing an appropriate topography function is very important in the theoretical study of the trapped waves. There is a class of undisturbed ridge in the ocean, such as the Indian Ocean east longitude 90° ridge, the western Pacific Mariana ridge, etc., and these ridges are steep but the top of the hill flat, which can be approximated using the hyperbolic-cosine squared function. This paper will discuss the problem of the trapped wave on the mild slope topography of the hyperbolic-cosine squared ridge.

The present study considers a special case of a symmetrical hyperbolic-cosine squared submerged ridge and obtains an analytic trapped solution based on the linear shallow water equations. The formulation description is given in Section 2 and the dispersion relation is derived and discussed in Section 3. Then the spatial wave pattern is presented and discussed in Section 4. Results are summarized in Section 5.

2. Formulation and solution method

Figure 1 illustrates a section of an infinitely long straight ridge with a cross section of hyperbolic-cosine squared. The x axis is normal to the ridge and the y axis is parallel to the ridge with the z axis vertically upwards from the still water level. Assuming horizontal seafloor in the open ocean, the water depth of the ridge is given by

$$h(x) = \begin{cases} h_0 \cosh^2(\lambda x) & |x| \le L; \\ h_1 & |x| > L. \end{cases} \tag{1}$$

where

$$h_1 = h_0 \cosh^2(\lambda L) \tag{2}$$

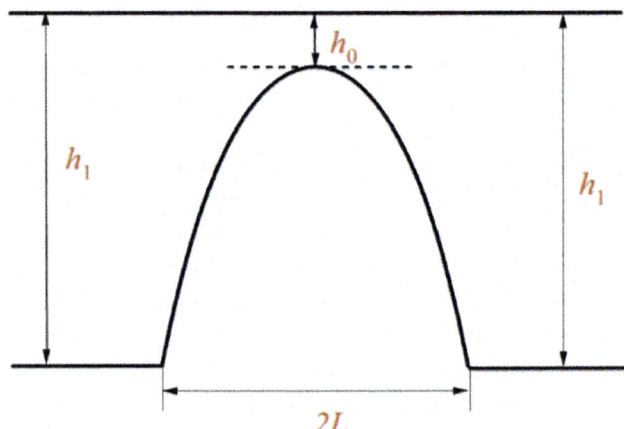

Fig. 1. Hyperbolic cosine square spine profile.

λ (m^{-1}) is a parameter determining the shape of the hyperbolic-cosine squared, $2L$ is the width of the ridge, h_0 is the distance from the top of the ridge to the still water level, h_1 is the depth of the horizontal plane beyond the ridge.

As the water depth is very small compared to the tsunami wavelength, the linear shallow water equations can be used to describe the wave motion, which is.

$$\eta_{tt} - g\nabla\left(h\nabla\eta\right) = 0 \tag{3}$$

where η is the surface elevation, $\nabla = (\partial/\partial x, \partial/\partial y)$, g is the acceleration due to gravity and t is the time.

The water surface displacement in the ridge may be written as

$$\eta(x,y,t) = \zeta(x)\exp\left[i\left(k_y y - \omega t\right)\right] \tag{4}$$

in which ω is the angular frequency of the incident wave, and k_y is the wave vector component in the y direction and i = $(-1)^{1/2}$. Inserting Eq. (4) into Eq. (3) by considering the depth profile (1) yields

$$\frac{d^2\zeta}{dx^2} + 2\lambda\tanh\left(\lambda x\right)\frac{d\zeta}{dx} + \frac{\omega^2}{gh_0}\sec h^2\left(\lambda x\right)\zeta - k_y^2\zeta = 0 \tag{5}$$

Introducing a new independent variable

$$\chi = \tanh\left(\lambda x\right) \tag{6}$$

and a new dependent variable

$$\zeta = \sqrt{1-\chi^2}\cdot f\left(\chi\right) \tag{7}$$

Eq. (5) then reduces to

$$\left(1-\chi^2\right)f_{\chi\chi} - 2\chi f_\chi + \left[\frac{\omega^2}{gh_0\lambda^2} - \frac{1+\left(k_y/\lambda\right)^2}{1-\chi^2}\right]f = 0 \tag{8}$$

The above equation is essentially the associated Legendre differential equation of degree v and order μ, where v is determined by

$$v = -\frac{1}{2} \pm \sqrt{\frac{1}{4} + \frac{\omega^2}{gh_0\lambda^2}} \tag{9}$$

and μ is determined by

$$u = \sqrt{1+\left(k_y/\lambda\right)^2} \tag{10}$$

The general solution for Eq. (8) is

$$f = A\cdot P(v,u,\chi) + B\cdot Q(v,u,\chi) \tag{11}$$

where a and b are the constants, and p and q are the two linearly independent associated Legendre functions given by

$$P(v,u,\chi) = \frac{1}{\Gamma(1-u)}\left(\frac{1+\chi}{1-\chi}\right)^{\frac{u}{2}} F(-v,v+1,1-u,\frac{1-\chi}{2}) \tag{12}$$

and

$$Q(v,u,\chi) = \frac{\pi}{2\sin u\pi}\left\{\cos(u\pi)P(v,u,\chi) - \frac{\Gamma(v+u+1)}{\Gamma(v-u+1)}P(v,-u,\chi)\right\} \tag{13}$$

in which f is a hypergeometric function defined as

$$F(a,b,c,\chi) = \sum_{s=0}^{\infty}\frac{(a)_s(b)_s}{(c)_s s!}\chi^s \tag{14}$$

where

$$(a)_s = a(a+1)(a+2)...(a+s-1) \tag{15}$$

and γ is the gamma function defined as

$$\Gamma(\chi) = \int_0^{\infty}\exp(-t)t^{\chi-1}dt \tag{16}$$

So trapped waves over the hyperbolic-cosine squared oceanic ridge can be rewritten as

$$\zeta = \sec h^2(\lambda x)\left[A \cdot P(v,u,\chi) + B \cdot Q(v,u,\chi)\right] \tag{17}$$

The associated Legendre functions behave at the singularity $x \to \infty$, where $\chi = \tanh(\lambda x) \to 1^-$, as

$$P(v,u,\chi) \sim \frac{1}{\Gamma(1-u)}\left(\frac{2}{1-\chi}\right)^{u/2} \tag{18}$$

and

$$Q(v,u,\chi) \sim \frac{1}{2}\cos(u\pi)\Gamma(u)\left(\frac{2}{1-\chi}\right)^{u/2} \tag{19}$$

To the free surface over the ridge becomes

$$\zeta(\chi \to 1^-) = 2^{\frac{u}{2}}(1+\chi)^{\frac{1}{2}}(1-\chi)^{\frac{1-u}{2}}\left[A\frac{\sin(u\pi)}{\pi} + \frac{B}{2}\cos(u\pi)\right]\Gamma(u) \tag{20}$$

As $\mu > 1$, the third term on the right-hand side of Eq. (20), $(1-\chi)^{(1-\mu)/2}$, tends to infinity unless

$$A\frac{\sin(u\pi)}{\pi} + \frac{B}{2}\cos(u\pi) = 0 \tag{21}$$

i.e

$$B = -\frac{2}{\pi}\tan(u\pi)A \tag{22}$$

So trapped waves over a hyperbolic-cosine squared ridge can be described as

$$\zeta = A \sec h^2 (\lambda x) \left[P(v,u,\chi) - \frac{2}{\pi} \tan(u\pi) Q(v,u,\chi) \right] \tag{23}$$

where a is a coefficient related to the amplitude at the top of ridge.

The relationship between μ and v is determined by the boundary conditions the top of ridge $x = 0$. Because the ridge is symmetrical topography, the wave energy distribution on both sides of the ridge should be symmetrical, which leads to no normal flux at the top, i.e

$$d\zeta / dx \big|_{x=0} = 0 \tag{24}$$

This yields

$$P(v+1,u,0) - 2/\pi \tan(u\pi) Q(v+1,u,0) = 0 \tag{25}$$

The associated Legendre functions of degree v and order μ for special value $x = 0$ are

$$P(v,u,0) = \frac{2^u \pi^{\frac{1}{2}}}{\Gamma(\frac{1}{2}v - \frac{1}{2}u + 1)\Gamma(\frac{1}{2} - \frac{1}{2}v - \frac{1}{2}u)} \tag{26}$$

and

$$Q(v,u,0) = -\frac{2^{u-1} \pi^{\frac{1}{2}} \sin\left[\frac{1}{2}(u+v)\pi\right] \Gamma\left(\frac{1}{2}v + \frac{1}{2}u + \frac{1}{2}\right)}{\Gamma\left(\frac{1}{2}v - \frac{1}{2}u + 1\right)} \tag{27}$$

Thus, Eq. (25) can be rewritten as

$$\tan\left[\frac{1}{2}(v+u)\pi\right] = \tan(u\pi) \tag{28}$$

It is equivalent to

$$v = u + 2m \qquad m = 0,1,2... \tag{29}$$

The above equation determines the relationship between the frequency and the wavenumber, which may be called the dispersion relationship for ridge trapped waves. Substituting Eqs. (9) and (10) into Eq. (29) results in

$$\omega^2 = gh_0 \lambda^2 \left(\sqrt{1+k_y^2/\lambda^2} + 2m\right)\left(\sqrt{1+k_y^2/\lambda^2} + 2m+1\right) \qquad m = 0,1,2,3\cdots \tag{30}$$

Obviously, Eq. (22) is undefined when $\mu = m + 1/2$. In this case, trapped waves can be expressed as

$$\zeta = B\sqrt{1-\chi^2} Q(v,u,\chi) \tag{31}$$

Based on the no normal flux condition at the top of ridge, it can result in the relationship between v and μ

$$v = m + \frac{1}{2}$$

From (9), it can be written as

$$\omega^2 = gh_0\lambda^2\left[(m+1)^2 - \frac{1}{4}\right]$$

(32)

3. Dispersion Relation

The relationship between the frequency and wavenumber of trapped waves over the hyperbolic-cosine squared ocean ridge is affected by h_0, λ and m. It is therefore necessary to investigate the sensitivity of the dispersion to these parameters. As it is shown in Eq. (30), ω^2 increases linearly with h_0 when k_y, λ and m are fixed. For the given h_0 and λ, the frequency ω increases with the wavenumber k_y. It also increases with m for a prescribed wavenumber (see Figure 2). Similarly, the angular frequency ω increases approximately linearly with λ for fixed k_y and h_0 (see Figure 3), and the higher the mode number m, the more quickly the rate of frequency increases. So, for the fixed trapped waves mode, the corresponding trapped waves frequency of the ridge with steep topography is larger than that of mild topography.

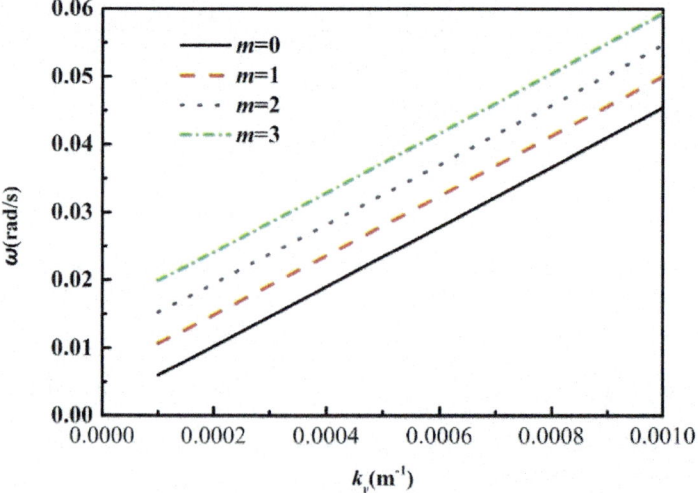

Fig. 2. The relationship between the angular frequency ω and k for trapped wave over the hyperbolic-cosine squared ridge, where L=3500 m, h_1=2000 m and h_0=200 m.

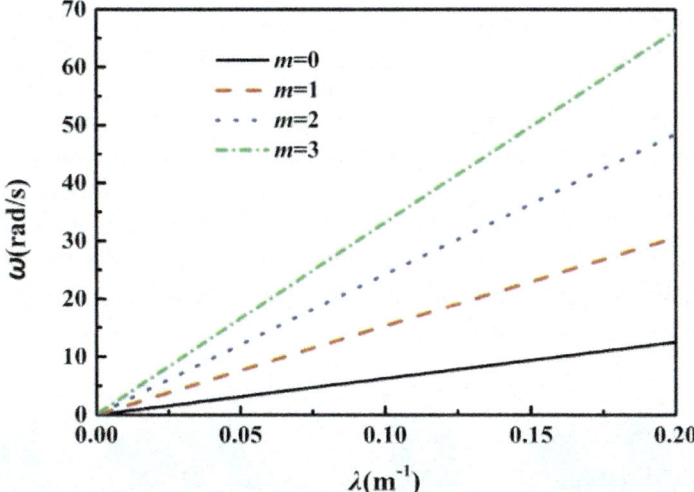

Fig. 3. The angular frequency ω versus λ for trapped wave over the hyperbolic-cosine squared ridge, where L=3500 m, and k_y=3.49×10-4m⁻¹.

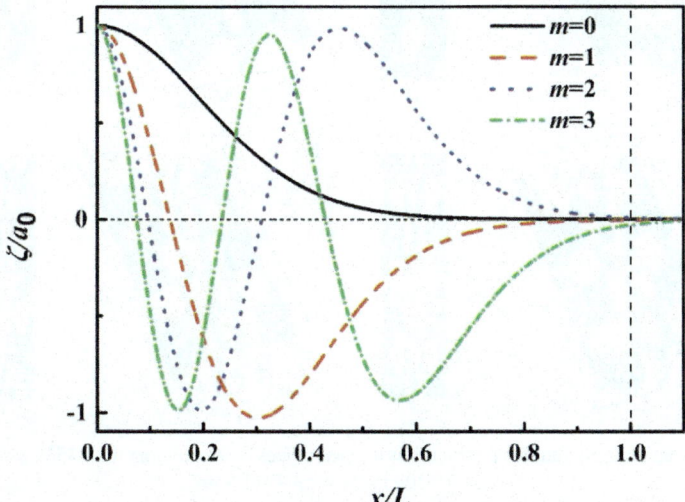

Fig. 4. Wave profiles of the first four modes over the right-half ridge, corresponding to L=3500 m, h_1=2000 m, h_0=200 m and k_y=3.49×10-4m-1.

4. Spatial structure

For a fixed k_y and m, the unique angular frequency ω can be determined by solving Eq. (30). There obviously are m different values of wavenumber k_y for the fixed frequency ω, and the corresponding solution can be called mode m. To illustrate the spatial structure, wave amplitude profiles for the first four modes are presented in Figure 4. The mode shapes of the water surface are very similar to the edge waves on a sloping beach (Ursell[14]). For the fundamental mode ($m = 0$), the maximum amplitude occurs at the top of ridge and decreases as the depth increases. For the higher modes ($m \geq 1$), the maximum amplitudes also occur at the top of ridge with amplitudes decreasing in the away from the top of the ridge on both sides. However, the higher the mode number, the slower the rate of amplitude decreases. This indicates that more energy is distributed over the larger domain of the ridge.

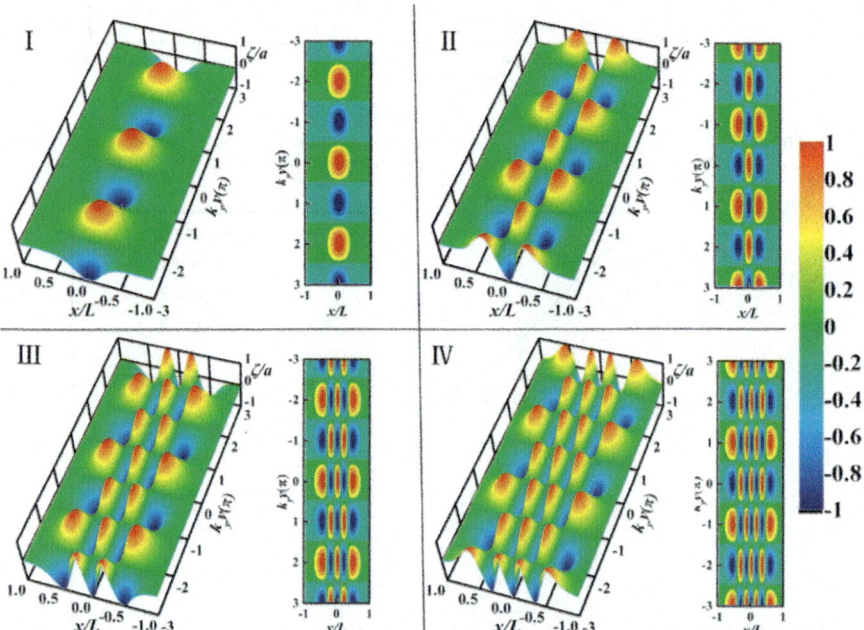

Fig. 5. Spatial structure for the first four trapped modes corresponding to L=3500 m, h_1=2000 m, h_0=200 m and k_y=3.49×10-4m^{-1}.

5. Conclusions

Based on the linear shallow water equation, analytic solutions for the trapped waves over a hyperbolic-cosine squared ridge are derived, and the spatial energy distribution characteristics of each mode are discussed. The free surface is described by the combination of the associated Legendre functions of the first and second kinds of degree v and order u. The wave amplitude gets its maximum at the top and decreases away from it. The higher the mode of the trapped waves, the slower the rate of amplitude decreases, and the more wave energy is distributed on both sides of the symmetrical axial ridge.

References

1. V. Titov, A. B. Rabinovich, H. O Mofjeld, R. E. Thomson and F. I. González, The Global Reach of the 26 December 2004 Sumatra Tsunami, *Science* **309** (5743) pp. 2045-2048 (2005).
2. A. B. Rabinovich, R. N. Candella and R. E. Thomson, Energy Decay of the 2004 Sumatra Tsunami in the World Ocean, *Pure and Applied Geophysics* **168** (11) pp. 1919-1950 (2011).
3. A. B. Rabinovich, P. L. Woodworth and V. V. Titov, Deep-sea observations and modeling of the 2004 Sumatra tsunami in Drake Passage, *Geophysical Research Letters* **38** (16) pp. 239-255 (2011).
4. R. N. Candella, A. B. Rabinovich and R. E. Thomson, Near-source observations and modeling of the Kuril Islands tsunamis of 15 November 2006 and 13 January 2007, *Advances in Geosciences* (2008).
5. Z. Kowalik, J. Horrillo, W. Knight and T. Logan, Kuril Islands tsunami of November 2006: 1. Impact at Crescent city by distant scattering, *Journal of Geophysical Research-Oceans* **113** (C1) (2008).
6. J. Horrillo, W. Knight and Z. Kowalik, Kuril Islands tsunami of November 2006: 2. Impact at Crescent city by local enhancement, *Journal of Geophysical Research-Oceans* **113** (C1) (2008).
7. T. C. Lippmann, R. A. Holman and A. J. Bowen, Generation of edge waves in shallow water, *Journal of Geophysical Research Atmospheres* **102** (C4) pp. 8663–8679 (1997).
8. S. N. Seo and P. L. F. Liu, Edge waves generated by the landslide on a sloping beach, *Coastal Engineering* **73** pp. 133-150 (2013).
9. S. N. Seo and P. L. F. Liu, Edge waves generated by atmospheric pressure disturbances moving along a shoreline on a sloping beach, *Coastal Engineering* **85** pp. 43-59 (2014).
10. R. O. Reid, Effect of Coriolis force on edge waves, I. Investigation of the normal modes. Developments in Chemical Engineering & Mineral Processing **13** (5-6) pp. 709–718 (1958).
11. M. S. Longuet-Higgins, On the trapping of waves along a discontinuity of depth in a rotating ocean, *Journal of Fluid Mechanics* **31** (03) pp. 417-434 (1968).

12. V. T. Buchwald, Long Waves on Oceanic Ridges, *Proceedings of the Royal Society of London Series a-Mathematical and Physical Sciences* **308**, pp. 343-354 (1969).

13. R. P. Shaw and W. Neu, Long-Wave Trapping by Oceanic Ridges, *Journal of Physical Oceanography* **11** (10) pp. 1334-1344 (1981).

14. F. Ursell, Edge Waves on a Sloping Beach, *Proceedings of the Royal Society of London. Series A, Mathematical and Physical Sciences*, **214** (1116) pp. 79-97 (1952).

Computer Simulation of Sadong Harbor for Investigation of Harbor Resonance[*]

Moonsu Kwak[†]

Civil Engineering, Myongji College,
Seoul, 03656, Korea
E-mail: moonsu@mjc.ac.kr

This paper presents results for studied on harbor resonance change according to harbor layout change in Sadong harbor where is located in Ullengdo island on the East of Korea. The dominant resonance modes simulated in 32 sec, 48 sec and 4.8 minute inside harbor. And the amplification factors obtained by 11.0, 11.1 and 10.4 when harbor layout is before construction, and 6.0, 6.1 and 7.3 in case after construction. The amplification factor after construction decreased in 30 ~ 45% than results of before construction. This is due to incident wave energy trapped by expanded harbor basin after construction. On the other hand, the dominant resonance modes at expanded harbor basin after construction simulated in 34 sec, 48 sec and 70 sec, and amplification factor obtained by 11.6, 22.8 and 8.9 respectively. It is need to plan harbor resonance reduction measure about 48 second wave period in expanded harbor basin after construction.

Keywords: Harbor resonance; Computer simulation; Harbor expansion; Airport construction.

1. Introduction

Sadong harbor is located in the Ullengdo island in East of Korea (Fig. 1). This harbor handles passenger ship and fishing boat which is consists of berth 628 m and breakwater 750 m in length at present state. However, Sadong harbor is presently developing to expansion of berth facility 1355 m and breakwater 640 m in length. Then this harbor is able to handles fishing boat, passenger ship and warship in the future. Meanwhile, the airport that is consists of runway 1200 m in length and 30 m in width, is plans to construct close to breakwater (Fig. 2).

Due to its location and the harbor layout, resonant motion caused by long period waves have occurred frequently which produced undesirable wave and ship oscillation in the harbor, especially during the season with waves coming

[*] This research was supported by grant 2017032990 of the National Research Foundation of Korea.
[†] This research partially supported by grant C004434431of the Korea Small and Medium Business Administration.

from the East direction. This paper presents results of computer model study for investigation of harbor resonance change inside harbor before and after harbor expansion and airport construction.

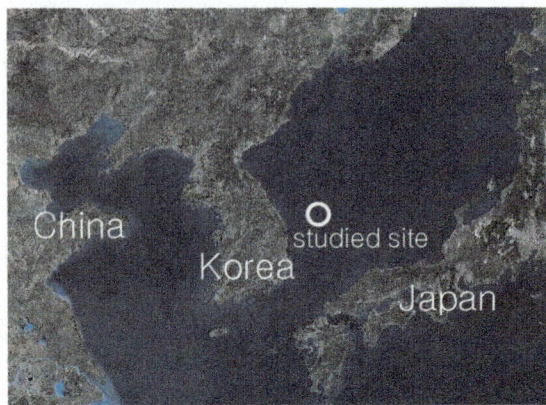

Fig. 1. Satellite map of studied site (http://map.naver.com).

Fig. 2. Close up satellite map of studied site (http://map.naver.com).

2. Numerical model for simulation of wave field

The numerical model used CGWAVE model for simulation of harbor resonance by incident long waves. The two-dimensional elliptic mild-slope wave equation is an accepted method for modeling surface gravity waves in coastal areas. This equation may be written as follows (Demirbilek and Panchang, 1998):

$$\nabla \cdot \left(CC_g \nabla \hat{\eta} \right) + \frac{C_g}{C} \sigma^2 \hat{\eta} = 0 \tag{1}$$

where

$\hat{\eta}(x, y)$ = complex surface elevation function, from which the wave height can be estimated

σ = wave angular frequency

$C(x, y)$ = phase velocity = σ/k

$C_g(x, y)$ = group velocity = $\partial\sigma/\partial k$ =nC with n = $\frac{1}{2}\left(1 + \frac{2kd}{\sinh 2kd}\right)$

$k(x, y)$ = wave number ($= 2\pi/L$), related to the local depth $d(x,y)$ through the linear dispersion relation: $\sigma^2 = gk \tanh (kd)$

Equation 1 simulates wave refraction, diffraction, and reflection (i.e., the general wave scattering problem) in coastal domains of arbitrary shape. However, various other mechanisms also influence the behavior of waves in a coastal area. The mild-slope wave equation can be modified as follows to include the effects of frictional dissipation (Dalrymple et al., 1984) and wave breaking (Dally et al., 1985):

$$\nabla \cdot \left(CC_g \nabla\hat{\eta}\right) + \left(\frac{C_g}{C}\sigma^2 + i\sigma w + iC_g\sigma\gamma\right)\hat{\eta} = 0 \qquad (2)$$

where w is a friction factor and γ is a wave breaking parameter. Following Dalrymple et al. (1984), we used the following form of the damping factor in CGWAVE:

$$\omega = \left(\frac{2n\sigma}{k}\right)\left[\frac{2f_r}{3\pi}\frac{ak^2}{(2kd+\sinh 2kd)\sinh kd}\right] \qquad (3)$$

where a ($= H/2$) is the wave amplitude and f_r is a friction coefficient to be provided by the user. The coefficient depends on the Reynolds number and the bottom roughness and may be obtained from Madsen(1976) and Dalrymple et al.(1984). Typically, values for are in the same range as for Manning's dissipation coefficient. Specifying as a function of allows the modeler to assign larger values for elements near harbor entrances to simulate entrance loss. For the wave breaking parameter, we use the following formulation (Dally et al., 1985; Demirbilek, 1994; Demirbilek et al., 1996):

$$\gamma = \frac{\chi}{d}\left(1 - \frac{\Gamma^2 d^2}{4a^2}\right) \qquad (4)$$

where χ is a constant (a value of 0.15 is used in CGWAVE following Dally et al., (1985)) and is an empirical constant (a value of 0.4 is used in CGWAVE).

58

In addition to the above mechanisms, nonlinear waves may be simulated with the mild-slope wave equation. This is accomplished by incorporating amplitude-dependent wave dispersion, which has been shown to be important in certain situations (Kirby and Dalrymple, 1986). The nonlinear dispersion relation used in place of Eq. (3) is

$$\sigma^2 = gk[1 + (ka)^2 F_1 \tanh^5 kd] \tanh(kd + kaF_2) \tag{5}$$

where

$$\left. \begin{array}{l} F_1 = \dfrac{\cosh(4kd) - 2\tanh^2(kd)}{8\sinh^4(kd)} \\[2mm] F_2 = \left(\dfrac{kd}{\sinh(kd)}\right)^4 \end{array} \right\}$$

3. Computer simulation conditions

Computer simulation carried out according to conditions of harbor layout before and after harbor expansion and airport construction (Fig. 3). In case before construction consist of East breakwater 1200 m and wharf 620 m in length. And the after construction consist of expended berth 875m and constructed airport runway 1000 m in length and 30 m in width along with breakwater. The waves period considered from 30 sec to 30 minute which is infra gravity waves and long waves.

Fig. 4 is layout of simulation domain and bottom topography. The water depth distribute from 400 m to 5m. And the triangular element have set from 60 m to 10 m in size, and 480,000 nodes. The boundary conditions applied reflectivity 1.0 at harbor structures and coast line which is a rocky coast.

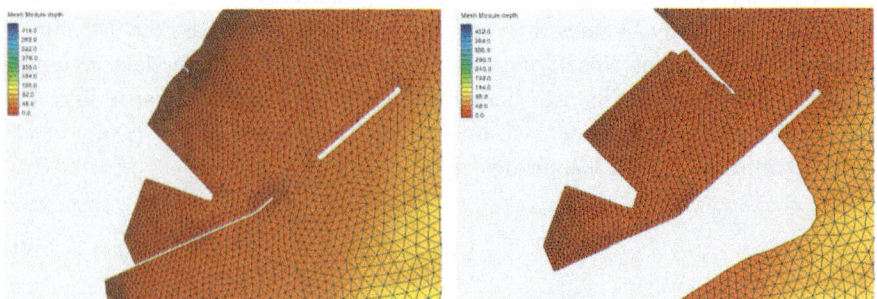

Fig. 3. Harbor layout conditions before (left) and after construction(right).

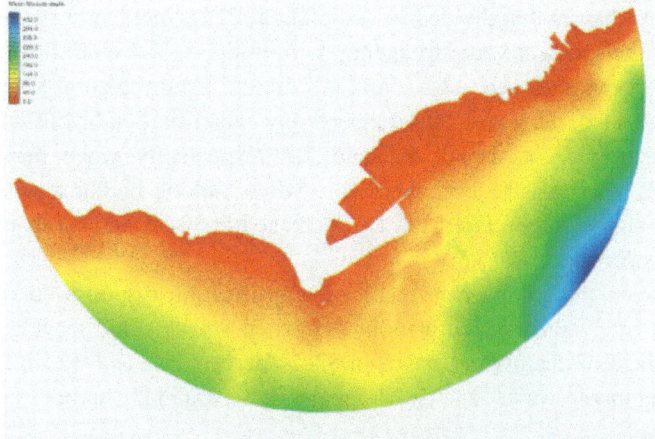

Fig. 4. Layout of simulation domain and bottom topography.

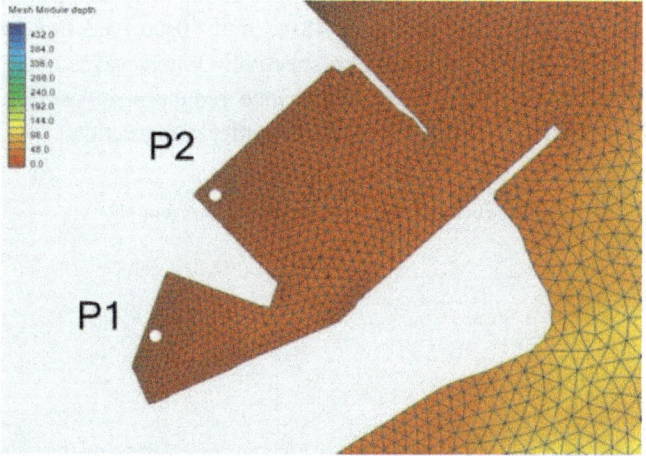

Fig. 5. Close up simulation domain with location of points P1 and P2.

4. Computer simulation results

Computer simulation results were expressed in amplification response curve with incident wave frequency at P1 and P2. Fig. 5 is the close up simulation domain with location of point P1 and P2. The point P1 located in present harbor, and the point P2 located in expended harbor basin.

Fig. 6 shows the comparison result of amplification response curve with incident waves frequency between before and after construction at point P1 when waves come from ESE direction. It covers the incident wave periods from 30 sec to 30 minute. The ordinate is the amplification factor R defined as the relative wave height that divided wave height at P1 by incident wave height. The abscissa is the incident wave frequency with unit Hz. There are three dominant resonance mode at 32 sec, 48 sec and 4.8 minute. And the amplification factor obtained by 11.0, 11.1 and 10.4 respectively when harbor layout condition is before construction, and 6.0, 6.1 and 7.3 respectively when harbor layout condition is after construction. Therefore when Sadong harbor changed harbor layout, the amplification factors at P1 decreased in 30-45 percent than results of before construction.

Fig. 7 shows the comparison result of amplification response curve between before and after construction at point P1 when waves come from E direction. The amplification factors decreased in 32-54 percent like as ESE direction. It is thought incident wave energy trapped by expanded harbor basin after construction.

Fig. 8 shows the comparison results of amplification response curve between incident wave direction E and ESE after construction at point P2. The dominant resonance modes simulated in 34sec, 48sec and 70sec, and the amplification factor obtained by 11.6, 22.8 and 8.9 respectively when waves come from ESE direction. It is need to plan harbor resonance reduction measure about wave period of 48 second in expanded harbor basin after construction.

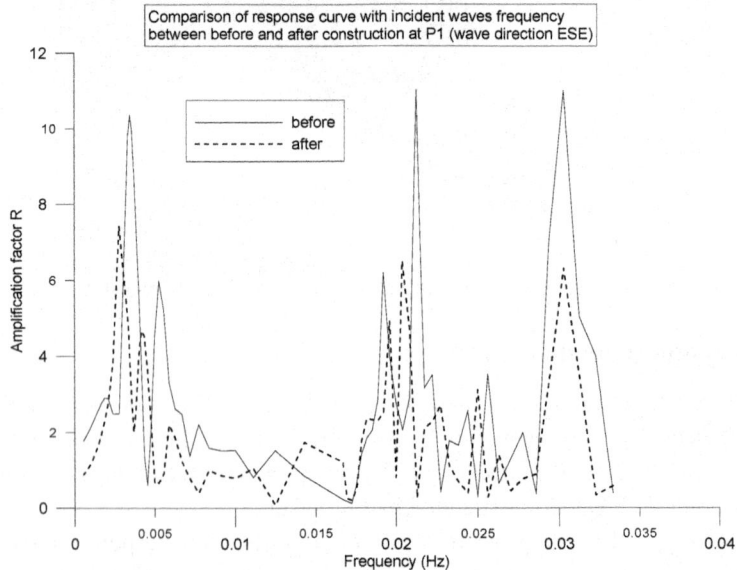

Fig. 6. Comparison result of amplification response curve between before and after construction at point P1 when waves come from ESE direction.

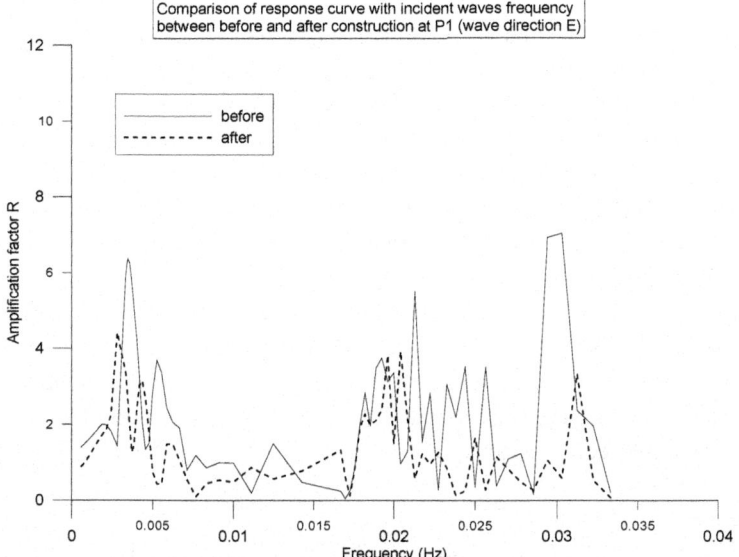

Fig. 7. Comparison result of amplification response curve between before and after construction at point P1 when waves come from E direction.

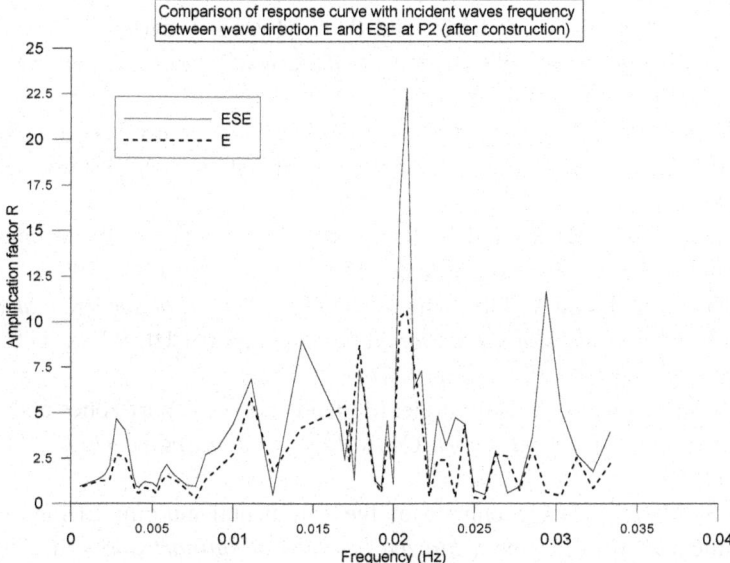

Fig. 8. Comparison result of amplification response curve between incident wave direction E and ESE at point P2 when harbor layout is after construction.

5. Conclusions

This paper presents results of computer model study for investigating harbor resonance change between present state and after harbor expansion and airport construction in Sadong harbor. The dominant resonance modes inside harbor at present state layout simulated in 32 sec, 48 sec and 4.8 minute. When Sadong harbor is completed harbor expansion and airport construction, the amplification factors inside harbor decreased in 30~45% than results of present state harbor layout. This is due to incident wave energy trapped by expanded harbor basin where will be constructed in front of present harbor.

On the other hand, the dominant resonant modes at expanded harbor basin after construction simulated in 34 sec, 48 sec and 70 sec, and amplification factor obtained by 11.6, 22.8 and 8.9 respectively. It is need to plan harbor resonance reduction measure about 48 second in wave period at expanded harbor basin after construction.

References

1. W. R. Dally, R. G. Dean and R. A. Dalrymple, Wave height variation across beaches of arbitrary profile, *J. of Geophysical Research*, **90**, C6, 11917-11927 (1985).
2. R. A. Dalrymple, J. T. Kirby and P. A. Hwang, Wave diffraction due to areas of high energy dissipation, *J. Waterway, Port, Coastal and Ocean Eng.*, **110**, 67-79 (1984).
3. Z. Demirbilek, *Comparison between REFDIFS and CERC shoal laboratory study*, unpublished report, Waterways exp. station, Vicksburg, MS, 53 (1994).
4. Z. Demirbilek, B. Xu and V. Panchang, Uncertainties in the validation of harbor wave models, *Proceedings of 25th ICCE*, 1256-1267 (1996).
5. Z. Demirbilek and V. Panchang, *CGWAVE : A coastal surface water wave model of the mild slope equation*. Technical report CHL-98-xx, U.S. Army Corps of Engineer, PP. 6-11 (1998).
6. J. T. Kirby and R. A. Dalrymple, Modeling waves in surf zones and around islands, *J. of Waterway, Port, Coastal and Ocean Engineering*, **112(1)**, 78-93 (1986).
7. O. S. Madsen, Wave climate of the continental margin: Elements of its mathematical description, *Marine Sediment Transport and Environmental Management* (eds. D. Stanley and D.J.P. Swift), John Wiley, New York, 65-87 (1976).

Modeling the Dispersion of Surface Oil Slick Under Breaking Waves by Using a Multiphase SPH

X. Niu

*Department of Hydraulic Engineering, Tsinghua University,
Beijing, 100084, China
E-mail: nxj@tsinghua.edu.cn*

Waves play a very important role in the mixing process of oil slick, and greatly affect the distribution of oil pollutants in the water column near sea surface. In this study, the mixing process of spilled oil under breaking waves is simulated by using a multiphase SPH-based model. The model is an extension of our previous model. Modifications for two-phase flow are introduced to solve the instability problems caused by the high density ratio at interfaces. The surface tension between two fluids is also included in the multiphase model. The model is tested by simulating the deformation of an initially square droplet in another fluid due to surface tension, which is a classic problem usually used to test the two-phase flow model. The density ratio between the two fluids has been set to be 1/1000. The results show that the new set of SPH formulations can well deal with the large density difference at the interface of two fluids. And then it is applied to simulate wave breaking where the water surface is covered by a layer of oil. The mixing process of surface oil slick in water column is demonstrated.

Keywords: Multiphase flow; SPH; Oil slick; Breaking wave; Dispersion.

1. Introduction

The risk of oil spills has been and will continue to be a big threat to the marine environment as the growth of offshore oil production and maritime transport. In an oil spill accident, the prediction of spatial spread and concentration of the spilled oil is critical to the emergency response, which asks for a better understanding on the dynamic processes and weathering processes of oil spills. The spilled oil in marine environments undergoes a series of processes, such as surface spreading, advection, turbulent diffusion, dispersion, evaporation, dissolution, emulsification, and so on. Among those processes, the dispersion of oil slick in a wave field is the most complex processes and of great importance. Under breaking waves, surface oil slick breaks into droplets and disperses into water column. The droplets in water column rise under the effect of buoyancy and undergo continuous breakup and coalescence under the strong turbulence environment. The dispersion under waves largely affects the subsurface oil distribution and also has great influence on the weathering processes such as emulsification and dissolution. The oil dispersion under waves has been investigated experimentally by many researchers.[1,2] A series of empirical formulas have been proposed to describe the dispersion.[3] This study aims to investigate the process with numerical modeling.

It is known that smoothed particle hydrodynamics (SPH) is a very flexible Lagrangian method of flow dynamics simulation, which is easily extended to multiphase flow.[4,5] It essentially keeps the mass conservation due to its Lagrangian nature, which makes it can easily capture large interface deformation, such as breaking, merging, and splashing. So it would be very suitable for simulating the dispersion of oil slick under breaking waves.

Considering the flow dominated by the pressure gradient force, in which the computational time step is restricted by the treatment of pressure, we have developed a mesh-particle hybrid model based on SPH to improve the computational stability.[6] The basic conception is to split the pressure into a global part and a local fluctuating part. The global part is obtained by solving the pressure Poisson equation on the resolution of a background mesh, so as to keep the velocity field divergence free. And the local fluctuating pressure is related to the local particle density variation, to maintain the particles evenly distributed. The model is then extended to multiphase flows. The surface tension is invited into the two-phase modeling using the continuum surface force model.[7]

However, instability appears at the interface when we directly extend the model to multiphase flow in the cases that the density ratio of the fluids is large. This instability problem also has been found by many other researchers in multiphase SPH. The problem is caused by the discontinuity of pressure and velocity happened at the interface of two fluids, especially when the difference of density is large. Recently, some researchers proposed different ways to handle density discontinuity at the interface of two fluids correctly.[8,9,10] In this study, we proposed a new set of formula to solve stability problem due to the interface discontinuity. The model is suitable to multiphase flow with density discontinuity, surface tension, and free surface. Then, the model is preliminarily applied to simulate the dispersion of surface oil slick under breaking waves.

2. SPH-based formula for multiphase incompressible flow

The governing equations for flows under the Lagrangian framework are shown as follows.

$$\frac{d\mathbf{u}_i}{dt} = \mathbf{g} + \mathbf{f}_{p,i} + \mathbf{f}_{v,i} + \mathbf{f}_{s,i} \tag{1}$$

$$\frac{d\mathbf{x}_i}{dt} = \mathbf{u}_i \tag{2}$$

where $\mathbf{u}$ is the velocity vector, $\mathbf{x}$ is the position vector, t is time, $\mathbf{g}$ is the gravity acceleration. $\mathbf{f}_p = -\nabla p / \rho$ is the pressure force, in which p is the pressure and ρ is the density. $\mathbf{f}_v = \nu \nabla^2 \mathbf{u}$ is the viscous force, in which ν is the kinetic viscosity. The final term $\mathbf{f}_s$ is the surface tension force. The subscript i represents the i-th particle.

2.1. *Viscous force and surface tension force*

In this study, the viscous force term in the momentum equation is approximated as follows[11]

$$\mathbf{f}_{v,i} = \left(\nu \nabla^2 \mathbf{u}\right)_i = \sum_j \left(\nu_i + \nu_j\right) V_j \frac{\mathbf{u}_i - \mathbf{u}_j}{r_{ij}} \frac{\partial W}{\partial r} \tag{3}$$

where $V_j = m_j / \rho_j$ is the volume of j-th particle, in which m_j is the mass and ρ_j is the density. $W = W\left(\mathbf{r}, h\right)$ is kernel function or smoothing kernel, for which this study used the fourth-order centrosymmetric formula.[11] h is the core radius, $\mathbf{r}$ is the related position apart from $\mathbf{x}_i$ and $r = |\mathbf{r}|, r_{ij} = |\mathbf{r}_{ij}| = |\mathbf{x}_i - \mathbf{x}_j|$.

The continuum surface force model is adopted to calculate the surface tension force, which is invited into SPH by Morris (2000).[7]

$$\mathbf{f}_{s,i} = -\frac{1}{\langle \rho \rangle} \sigma \left(\nabla \cdot \hat{\mathbf{n}}_i\right) \mathbf{n}_i \tag{4}$$

where σ is the surface tension coefficient, $\langle \rho \rangle$ is the average density of the two fluids, $\mathbf{n}$ and $\hat{\mathbf{n}}$ is the normal vector and the unit normal vector of the interface. A color function c is introduced to compute the surface tension forces between different phases. In two-phase fluids flow, one phase is identified by $c=1$, and $c=0$ for the other phase. And the normal can be obtained by $\mathbf{n} = \nabla c / [c]$. Where $[c]$ is the jump in color function across the interface. The unit normal is defined as $\hat{\mathbf{n}} = \mathbf{n} / |\mathbf{n}|$.

2.2. *Treatment of the pressure force*

The pressure force term is calculated by using the pressure splitting algorithm introduced in the previous study.[6] The pressure defined at each particle is split into a spatial averaged component and a local fluctuating component, as

$$p = \tilde{p} + p' \tag{5}$$

The variable with tilde denotes the spatial averaged component defined at a coarser resolution, and the variable with prime denotes the local fluctuating component which is the difference between the variable defined at particles and the spatial averaged component.

The spatial averaged pressure is solved on the background mesh by using the pressure Poisson equation

$$\nabla^2 \tilde{p}^{n+1} = \frac{\tilde{\rho}}{\Delta t} \left(\nabla \cdot \tilde{\mathbf{u}}^* \right) \tag{6}$$

In which $\tilde{\rho}$ is the spatial averaged density; $\tilde{\mathbf{u}}^*$ is the primary estimation of spatial averaged velocity based on the known n-th time step variables without the consideration of the pressure gradient force. It can be approximately interpolated by a weighted sum of $\mathbf{u}^*$ from neighbouring particles after the prediction step. Then, the pressure force corresponding to the spatial averaged pressure on the background mesh is obtained and interpolated to particles.

The local fluctuation pressure is linked with the density deviation as

$$p_i' = C \frac{\rho_i}{\rho_0} \left(\rho_i - \rho_0 \right) \tag{7}$$

where C is a constant related to time step Δt and the particle spacing r_0; ρ_0 is the rest density of fluids. The local fluctuation pressure force is calculated based on particles using the SPH approach.

However, if we directly adopt the most widely used pressure force approach, a separation of two phases is usually observed because of the discontinuity of density on the interface. Look back to the original form of this formula. It can be seen that an essential assumption included to derive the SPH formula is that the variations of pressure and density are smooth. But on the interface with large density difference, the assumption is no longer satisfied. To overcome this problem, a slight change on the formula has been made. Considering the non-dimensional density ρ/ρ_0 are smoothly varying in all fluid phases, we can obtain

$$\mathbf{f}_{p,i} = -\left(\nabla \frac{p}{\rho} + \frac{p\rho_0}{\rho^2} \nabla \frac{\rho}{\rho_0} \right)_i = -\sum_j \left(\frac{p_j}{\rho_j^2} + \frac{p_i}{\rho_i^2} \frac{\rho_{0i}}{\rho_{0j}} \right) m_j \frac{\mathbf{x}_i - \mathbf{x}_j}{r_{ij}} \frac{\partial W}{\partial r} \tag{8}$$

In the case that interacting particles are same phase, this form reduces to the most widely used form. This form is used to ensure the pressure force is continuous, even for a discontinuous density field. This formula is then used to get pressure force corresponding to local fluctuation pressure.

Generally, a uniform background mesh for the whole computational domain is introduced to accelerate the searching for neighbouring particles in traditional particle-based models. Here, the spatial averaged pressure is computed on that background mesh. So no additional mesh is required. A prediction-correction algorithm is used, similar as the implementation procedure shown in [6].

3. Validation for the surface tension

The single phase version of this model has been validated by simulating the broken dam problem and water surface wave breaking, which showed good applicability and stability.[6] In the two-phase version, surface tension is considered. To test the added module, the surface tension driven deformation of a square droplet in another fluid is simulated.

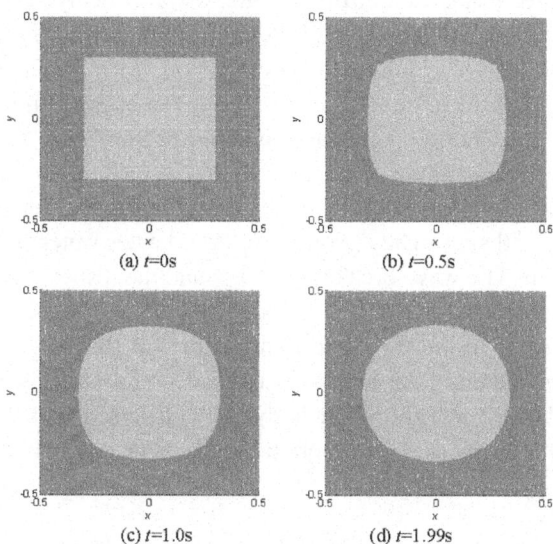

(a) t=0s (b) t=0.5s

(c) t=1.0s (d) t=1.99s

Fig. 1. Deformation of a square droplet under surface tension.

This case is a classical problem that used to test the two-phase flow model. The droplet is defined as phase 1 and the surrounding fluid is defined as phase 2. The initially square droplet with an edge length of 0.6 is located at the center of a square computational domain with an edge length of 1.0. The surface tension coefficient is set to be 1. The density ratio of phase 1 to phase 2 is set to be 1, 1/1000, 1000, respectively. Figure 1 shows the deformation of a square droplet under surface tension in another fluid with a density ratio of 1/1000. All of the three cases have been well simulated. The square droplet tends to round under the surface tension and the surface energy is partially converted to kinetic energy. Oscillation of the interface will continue for a period of time under the joint effect of surface tension and inertia. The kinetic energy will be dissipated due to viscosity, and finally the droplet reaches the static state.

4. Oil slick under breaking waves

An unsteady travelling wave with a large enough wave height will finally break. Niu and Yu (2015) has modeled the breaking wave using a modified SPH, and compared to the VOF result.[6] Here, a thin layer of oil on the water surface is considered. The wave length is set to be $L=1$, and the still water depth is $D=0.5L$. The initial wave slope $\varepsilon=0.55$ is used, which is defined as the ratio of wave amplitude to wave number. The initial water surface elevation and velocity field are calculated using the third-order Stokes wave theory. The density of oil and water are set to be 800kg/m³ and 1000kg/m³ respectively. The initial particle distance is $r_0=L/180$. The initial thickness of oil layer is set to be $3r_0$. The smoothing length is set to be $h=3r_0$ as same as the background size. The time step is $\Delta t=2*10^{-3}$s. Periodic boundary condition is applied to the left and right boundary of the computational domain.

Figure 2 shows snapshots of a time series of the wave breaking process with a layer of surface oil slick. The dark gray color indicates water and the light gray color indicates oil. The wave breaking process and the dispersion of the oil slick are reasonably simulated. As the wave skewed and its front becomes steep, the surface oil is driven to the wave crest and gathers at the wave front. After the front of the crest becomes vertical, a jet is formed and projects forward, which carries the surface oil into the water body. The dispersed oil beneath the water surface is transported by the wave motion and floats up due to buoyancy.

5. Conclusions

A SPH-based multiphase flow model is developed for modeling the dispersion of oil slick under breaking waves. The model inherits the treatment of pressure term proposed by Niu and Yu[6] to improve the computational stability. Considering the instability caused by the density difference at the interface, a new approach for the local fluctuating pressure force term for two-phase flows is presented to study the fluids even with large density difference, such as air-water interaction. The new approach is integrated into the previous SPH model, as well as the surface tension. And then the previous SPH model for viscous incompressible fluid flow is extended to two-phase fluids flows. The model is tested by the cases of square droplet deformation. The results show that the new set of SPH formulations can well deal with surface tension and the large density difference at the interface of two fluids. With the present model, the dispersion of oil slick on water surface during the process of wave breaking is preliminarily studied.

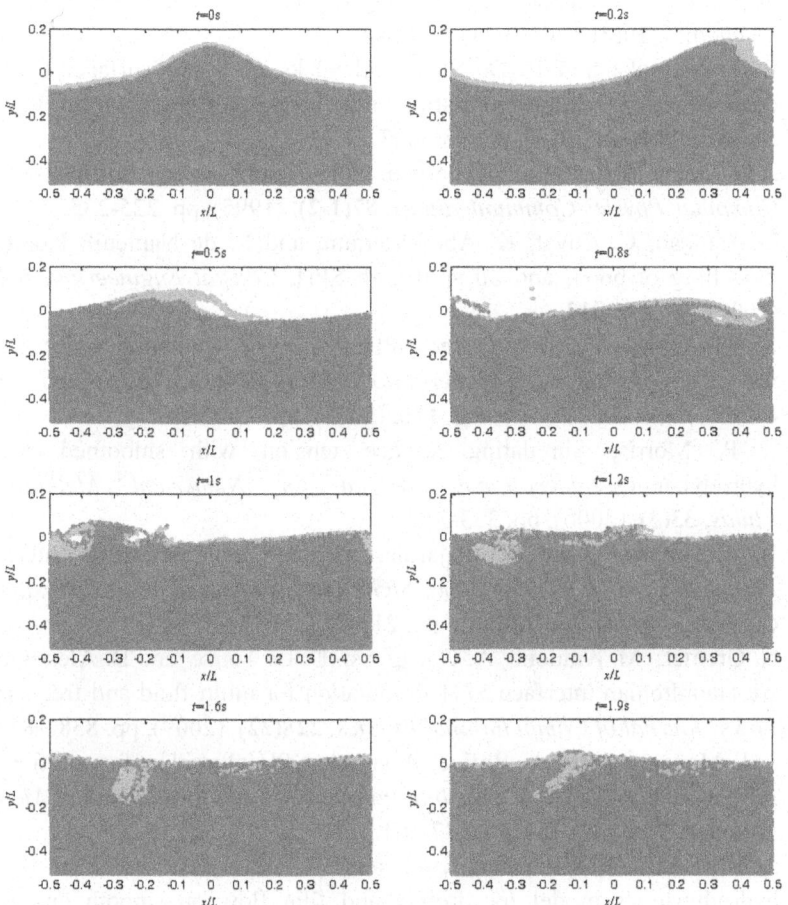

Fig. 2. Dispersion of oil slick under breaking waves.

Acknowledgments

The author would like to acknowledge the support by the National Natural Science Foundation of China under the grant No. 51479101 and Tsinghua University Initiative Scientific Research Program.

References

1. G. A. L. Delvigne, C. Sweeney, Natural dispersion of oil, *Oil and Chemical Pollution*, **4(4)**, (1988), pp. 281-310.
2. R. Parsa, M. Kolahdoozan and M. R. A. Moghaddam, Vertical oil dispersion profile under non-breaking regular waves, *Environmental Fluid Mechanics*, **16(4)**, (2016), pp. 833-844.
3. I. D. Nissanka and P. D. Yapa, Oil slicks on water surface: Breakup, coalescence, and droplet formation under breaking waves, *Marine Pollution Bulletin*, **114(1)**, (2017), pp. 480-493.
4. J. J. Monaghan and A. Kocharyan, SPH simulation of multi-phase flow, *Computer Physics Communications*, **87(1-2)**, (1995), pp. 225-235.
5. D. Violeau, C. Buvat, K. Abed-Meraïm, and E. de Nanteuil, Numerical modelling of boom and oil spill with SPH, *Coastal Engineering*, **54(12)**, (2007), pp. 895-913.
6. X. Niu and J. Yu, A modified SPH model for simulating water surface waves, *Proceeding of 8th International Conference on Asian and Pacific Coasts, Procedia Engineering*, **116**, (2015), pp. 254-261.
7. J. P. Morris, Simulating surface tension with smoothed particle hydrodynamics, *International Journal for Numerical Methods in Fluids*, **33(3)**, (2000), pp. 333-353.
8. B. Solenthaler and R. Pajarola, Density contrast SPH interfaces, *Proceedings of the 2008 ACM SIGGRAPH/Eurographics Symposium on Computer Animation* , (2008), pp. 211-218.
9. N. Grenier, M. Antuono, A. Colagrossi, D. Le Touzé and B. Alessandrini, An Hamiltonian interface SPH formulation for multi-fluid and free surface flows, *Journal of Computational Physics*, **228(22)**, (2009), pp. 8380-8393.
10. J. J. Monaghan and A. Rafiee, A simple SPH algorithm for multi‑fluid flow with high density ratios, *International Journal for Numerical Methods in Fluids*, **71(5)**, (2013), pp. 537-561.
11. J. Kordilla, A. M. Tartakovsky and T. Geyer, A smoothed particle hydrodynamics model for droplet and film flow on smooth and rough fracture surfaces, *Advances in Water Resources*, **59**, (2013), pp. 1-14.

Wave Run Up Dynamics in Jogehama Beach, Niigata Prefecture Japan

N. Inukai[†] and H. Yamamoto

Department of Civil and Environmental Engineering, Nagaoka University of Technology, Nagaoka City, Niigata Prefecture 940-2188, Japan
[†]E-mail: inu@nagaokaut.ac.jp
coastal.nagaokaut.ac.jp

K. Ogawa
East Japan Railway Company

Y. Ejiri
Penta-Ocean Construction Co., LTD

T. Ootake
Ecoh Corporation

Big wave suddenly invaded to the beach, and three children were carried off to the sea by the wave though they played on the beach. The beach characteristic topography has the cusp topography and steep slope. This study tried to comprehend the reason why this accident occurred. Firstly, this study comprehended the wave condition when the accident occurred. Secondary, this study made the survey about the geographic feature of the beach. And this study obtained the geographic data for the numerical simulation from the aerial photograph which were taken by UAV. Finally, this study comprehended the wave dynamics on the beach by the numerical simulation. This study simulated the wave dynamics by the horizontal two dimensional numerical model and the vertical two dimensional numerical model.

Keywords: Wave run up; Cusp topography; UAV; Aerial photography; Numerical simulation; Niigata Prefecture; Japan sea.

1. Introduction

The big wave suddenly arrived to the beach on 4th May 2014, and three children who played on the beach were carried off to the sea. Finally, they and 2 adult people who tried to rescue the children were drowned. The beach characteristic topography [1] has two lines of the small cusp topography. And two lines of cusp have the different phases. The beach slope is about one-tenth, therefore, the wave breaks at the beach line however the comparatively high wave height, and run up to the beach with few kinetic energy attenuated. Furthermore this slope condition can make the beach cusp easily. If the beach slope condition is one-tenth, the median diameter of the beach sand becomes to be about 1mm, and if one-eighty, the diameter becomes to be about 2mm [2]. The slope of the large beach in Niigata prefecture is about one-eighty [3], therefore the diameter of beach sand in Jogehama beach is bigger than other large beach. This study tried to comprehend the reason why this accident occurred. Firstly, this study comprehended the wave condition by using the weather chart and wave data when the accident occurred. Secondary, this study made the survey about the geographic feature of the beach. And this study made the geographic data for the numerical simulation from the aerial photograph which were taken by UAV (Unmanned Aerial Vehicle).Finally, this study comprehended the wave dynamics on the beach by the numerical simulation. This study simulated the wave dynamics by the horizontal two dimensional numerical model and the vertical two dimensional numerical model. For the horizontal two dimensional numerical model, this study used the momentum equation which used the modified Boussinesq equation, and the continuity equation. These equations were solved by the differential method. For the vertical two dimensional numerical model, this study used the CADMAS SURF-2D [4].

2. Wave and Weather Conditions When Accident Occurred

This study tried to comprehend the wave condition, and we used the observed wave data of Port of Naoetsu [5] which is 8km west from the field (Figure-1). When the accident occurred, the significant wave heights was 1.2m and the period was 7.9seconds

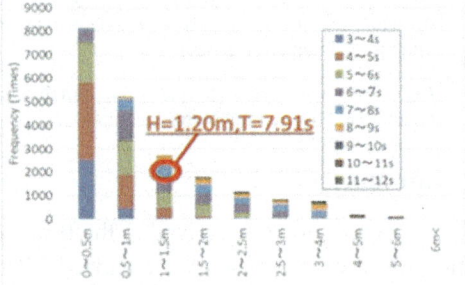

Fig. 1. Location of the field   Fig. 2. Relationship between Significant Wave Height and Period (Nowphas: Port of Naoetsu, 2007-2011)

and the wave direction was NW. When the accident occurred, the wave condition was stable, i.e. the wave condition did not change drastically.

Figure-2 shows the relationship between the significant wave heights and the periods in past 5 years. The figure shows on the day that the accident occurred, the wave heights (1.2m) was not high, however the period (7.9 seconds) was large in this wave height division. The reason why the period was large, when the accident day, the high pressure moved to east in the Japan Sea, therefore the wave heights was not large. However the pressure pattern made the long wind driven fetch. Therefore the period became long.

3. Consider The Origin of The Accident

When the accident occurred, the significant wave was not high, however the eyewitness said that the large wave occurred suddenly. Therefore this study has to explain the phenomena why the large wave occurred suddenly. To comprehend these phenomena, the study needs to use the observed raw data instead of the analyzed wave data. However, this study only has analyzed data. These data were classified under the wave period, i.e. under 6 seconds, 6-8 seconds, 8-10 seconds, 10-15seconds, 15-30 seconds, 30-60 seconds, 60-300 seconds, 300-600 seconds, over 600 seconds. This study tried to make the quasi raw data by composing these 10 data.

Figure-3 shows the quasi raw wave data for 5 minutes. This figure shows that the significant wave changes regularly while the quasi raw data changes from 0.23m to 2.08m irregularly. This means that the high wave height might reach to shore even though low height wave reaches just before. This phenomena coincides with the testimony of the eyewitness. On the other hand, the slope of this beach is steeper than the ordinary beaches. Therefore the median diameter of beach sand is bigger than the ordinary beaches. In this case, it is difficult to stand on the beach when wave arrived at the beach.

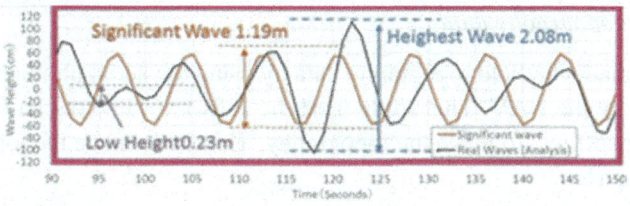

Fig. 3. Compare with Significant Wave and Waveform (Made by Analyzed Each Frequency Band)

74

According to the above results, this study considers that the victims stayed near the shoreline when the wave height was low, then the victims fell down and were carried away into the sea when the biggest wave arrived at the beach.

4. Comprehend The Topography Information

4.1. *Field survey*

This study conducted field survey to obtain the topography of the field on 4th July 2014, 29th August 2014 and 14th January 2015. The survey on 29th August was made with the Japan Society of Water Rescue and Survival Research jointly. Figure-4 shows the five coast-offing survey lines on the beach and the three coast-offing survey lines in the water. The survey in the water was made until 100m in the coast-offing direction. Furthermore, this study used the UAV to take the aerial photographs on 29th August 2014 and 14th January 2015. Especially, the topography of this field was made by the aerial photographs which was taken in 14th January 2015. Figure-5 shows the UAV which this study used. The topography was used to the numerical simulation.

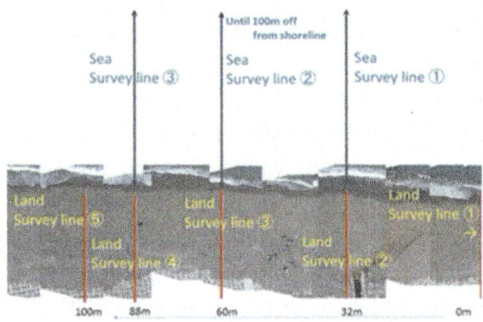
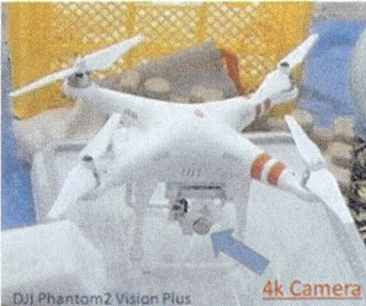

Fig. 4. Survey Line   Fig. 5. UAV

4.2. *Make topography data*

This study made the field topography data by using the aerial photographs which were taken by the UAV. This study used the software - PhotoScan Professional Edition, Agisoft Inc. to make the topography. This study made the topography in 20cm grid space.

4.3. *Verification of accuracy about topography*

This study compared the survey data and the topography data to verify the accuracy of the topography data which was made from aerial photographs. In this case, the field survey lines and the same line in topography data were compared. Figure-6 shows the two type lines. The blue narrow line is the line of the field survey, and the green thick line is the line of topography. The figure shows that the trend of two lines are nearly same. Therefore the topography which was made by UAV is duplicate with the field.

This study verified the accuracy of the data, however, this data has not underwater data, only has the beach data. Accordingly, this data made the underwater data by using the results of the survey which was conducted on 29th August 2014. Figure-7 shows the topography of the field which was combined the beach data and the underwater data. This study made the field topography data by using the aerial photographs which were taken by the UAV.

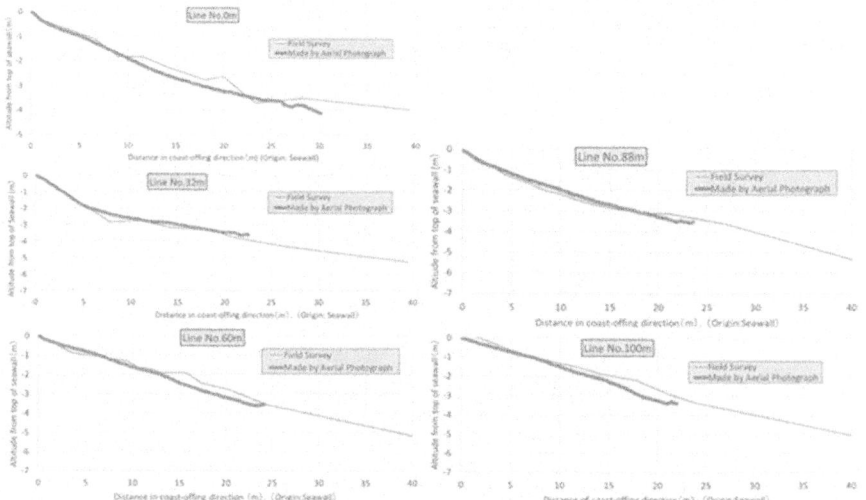

Fig. 6. Compare with Field Survey and Grid Data (Grid Data was made by Aerial Photograph)

4.4. *Comprehend feature of topography*

Figure-7 shows that the beach has the cusp topography. Especially, there are two cusp line in the beach, one cusp develops 10m from the shoreline, and another one develops 20m from the shoreline. Furthermore, the phases of these cusps are different. This result is same with Uchiyama. Figure-8 shows the heights and depth change in the coast-offing direction.

The figure shows that the slope at the front of the shoreline has the steep slope, and the slope becomes flat at the offshore. Therefore, the surf zone becomes narrow. In this case, the wave breaks at the beach line however the comparatively high wave height. And the wave run up to the beach with few attenuating the kinetic energy.

Figure-9 shows the topographical map and the depth change along two lines. The figure shows that the interval distance of the cusps are about 45m in both lines. And the differences of elevation are about 0.5m at shoreline, and about 1.0m at inner beach.

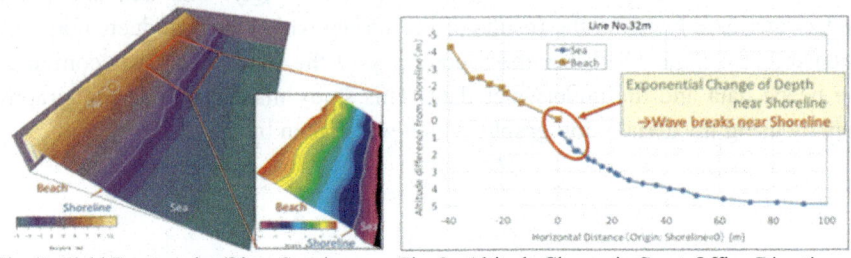

Fig. 7. Field Topography (20cm Spacing Grid Data)

Fig. 8. Altitude Change in Coast-Offing Direction

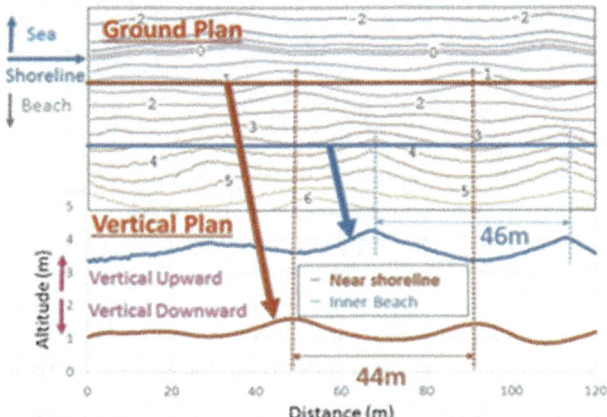

Fig. 9. Contour Map and Altitude Change in Shoreline Direction

5. Wave Condition on Survey Date

5.1. *Compare wave with accident survey*

When the accident occurred, the significant wave heights was 1.2m, the period was 7.9 seconds and the wave direction was NW. However, when the field survey was conducted, the significant wave heights was 0.96m, the period was 7.3 seconds and the wave direction was WNW. It seems that the wave condition when the field survey was smaller than the field survey. However, the both wave period were longer than the ordinary wave conditions. This study comprehended the wave dynamics by using the result of the field survey. In this case, the wave dynamics on the beach was recorded by the UAV on 14th January, 2015.

Figure-10 shows the situation of the wave run up that was recorded by the UAV. In the figure, the distance of wave run up from the shoreline when the field survey was about 24m, and the distance when the accident occurred was about 28m. Therefore the distance measured by the field survey was less than that of the accident. However, this study considers that the both motions are qualitatively same.

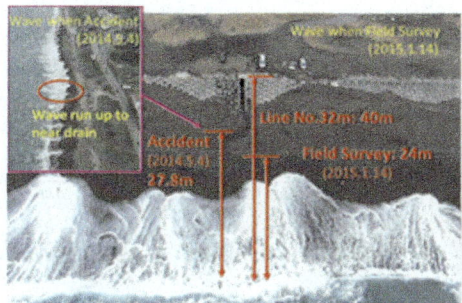 

Fig. 10. Compare Distance of Wave Run Up at Accident Date and Field Survey Date

Fig. 11. Wave Motion on Beach (Photo from Seawall, Field Survey, 14th January, 2015)

5.2. Comprehend wave condition

Figure-11 shows the wave run up dynamics which was recorded from the top of seawall. And Figure-12 shows the dynamics which was recorded by the UAV. The both figures show that the run up wave converge into the depressions of the cusps. Therefore the current on the beach makes the pectinate narrow current at the depressions of the cusps (see Figure-12).

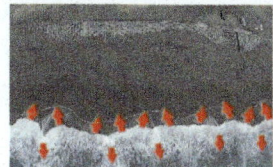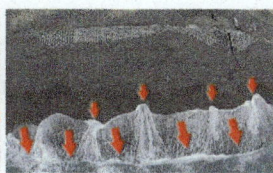

Fig. 12. Wave Dynamics (Recorded by UAV, 14th January, 2015)

6. Comprehend Wave Dynamics By Numerical Simulation

This study tried to comprehend the wave current by the numerical simulation. In this study, the horizontal two dimensional simulation and the vertical two dimensional simulation of the wave run up dynamics were pursued.

6.1. *Numerical simulation model of horizontal 2d*

This study confirmed that the current developed by the wave dynamics near the shoreline, therefore this study simulated the nearshore current in accordance with the results of field survey, and the results of simulation and filed survey ware compared.

This study use the following momentum equations which used the modified Boussinesq equation [6], and continuity equation. The wave moments calculated by the differential equations,

$$\frac{\partial Q_x}{\partial t} + \frac{\partial}{\partial x}\left(\frac{Q_x{}^2}{D}\right) + \frac{\partial}{\partial y}\left(\frac{Q_x Q_y}{D}\right) + gD\frac{\partial \eta}{\partial x} - MD_x + \tau_x = \left(B + \frac{1}{3}\right)h^2 \frac{\partial}{\partial x}\left(\frac{\partial^2 Q_x}{\partial t \partial x} + \frac{\partial^2 Q_y}{\partial t \partial y}\right) + Bgh^3\left(\frac{\partial^3 \eta}{\partial x^3} + \frac{\partial^3 \eta}{\partial x \partial y^2}\right)$$

$$\frac{\partial Q_y}{\partial t} + \frac{\partial}{\partial x}\left(\frac{Q_x Q_y}{D}\right) + \frac{\partial}{\partial y}\left(\frac{Q_y{}^2}{D}\right) + gD\frac{\partial \eta}{\partial y} - MD_y + \tau_y = \left(B + \frac{1}{3}\right)h^2 \frac{\partial}{\partial y}\left(\frac{\partial^2 Q_x}{\partial t \partial x} + \frac{\partial^2 Q_y}{\partial t \partial y}\right) + Bgh^3\left(\frac{\partial^3 \eta}{\partial y^3} + \frac{\partial^3 \eta}{\partial y \partial x^2}\right)$$

$$\frac{\partial \eta}{\partial t} + \frac{\partial Q_x}{\partial x} + \frac{\partial Q_y}{\partial y} = 0$$

where, MDxy: break water reduce term, D: depth, Qxy: current volume, h:hydrostatic pressure,η: change of water level, B:constant(1/21), g: gravitational acceleration, τxy: bottom friction of Manning roughness formula.

This study use the topography that was shown in Figure-7. The grid space was 20cm, the cusp topography were set in 44m intervals. The average bottom slope was set as about 1/10. The waves were input at offshore. The wave height was set as 1.2 m, and the Period was set 8 seconds.

6.2. *Result of numerical simulation (horizontal 2d)*

Figure-13 shows the profiles of water level in every seconds on the beach. Figure shows that the wave run up to inner 26m from the shoreline. Figure-10 shows the wave run up to inner 28m when the accident occurred, and the result of this simulation was almost same.

Therefore, this study considered that the results of numerical simulation were qualitatively same in the wave dynamics of the field. According to the result, the run up wave converges in the depressions of the cusps, and the backwash current flows in the pectinate where is the lowest place in the cusp.

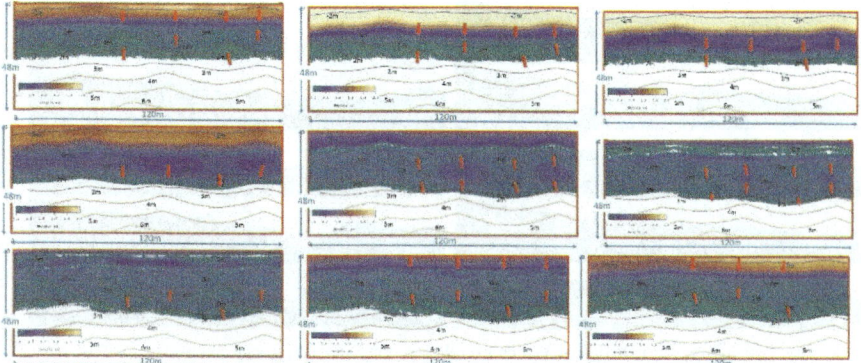

Fig. 13. Profiles of Water Level in Every Seconds on the Beach (Run up wave converge into the depressions of the cusps, and the backwash current flows in the pectinate)

6.3. *Numerical simulation model of vertical 2d*

This study simulated the vertical two dimensional wave dynamics by using CADMAS-2D [4].

For the simulation, this study used the ground-offshore topography line of 30m from the left in Figure-8, because it was conjectured that the accident occurred here. This line is the lowest depression line of the cusp. Figure-14 shows this topography. The horizontal grid space was 50cm and the vertical grid space is 5cm. The waves were inputted at offshore. The wave heights was set as 1.2 m, and the period was set as 8 seconds.

6.4. *Result of numerical simulation (vertical 2d)*

Figure-14 shows the profiles of water level over one period (eight seconds) since the incident wave arrived at the shoreline. These figures change in every second. These figures show that the wave ran up on the beach, and wave reached about 30m from the shoreline. The observed result was about 28m, and the simulation results was little bigger than the observation, therefore this study considered that the simulation results qualitatively reproduced the wave dynamics in the field.

Figure-15 shows the profiles of velocity vectors over one periods (eight seconds) since the incident wave arrived at the beach. These figures change in every second. The figure shows that when the second wave arrived at the shoreline, the backwash remains on the beach. Therefore the wave hanged over the backwash, and the current direction between the surface and the bottom became the opposite direction.

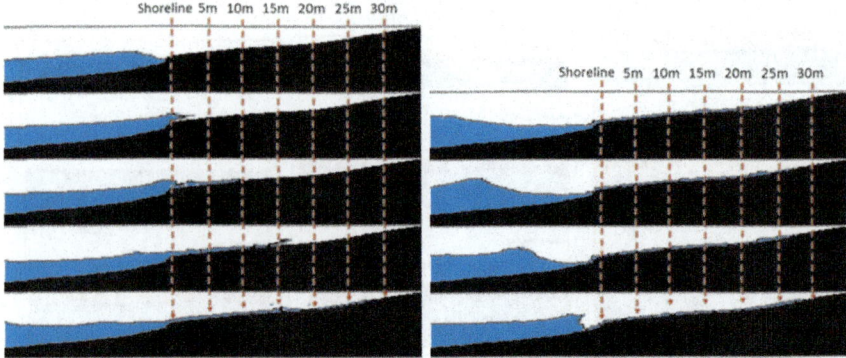

Fig. 14. Profiles of Water Level over One Period (Every Second)

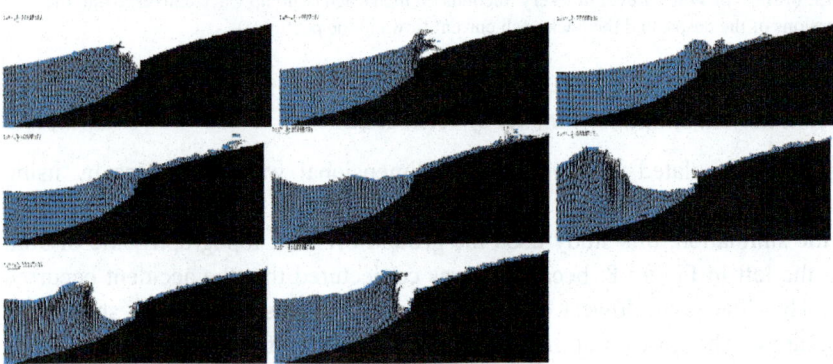

Fig. 15. Profiles of Velocity Vectors over One Period (Every Second)

Figure-16 shows the time change of the water level at every point. The location of every point are shown in Figure-14. The wave arrived at the shoreline at 32 seconds after starting the simulation. The figure shows that when the first wave arrived at the shoreline, the water depth became about 30cm until about 15m point from the shoreline. However, after the second wave, the water depth at the shoreline became over 80cm. The reason why the second water depth became higher than the first wave, there was the backwash current that hindered the running up of the incident current on the beach, therefore the depth became high due to stir the currents.

This study assumed that the slope of the beach was uniform. In actually, the mass of run up water concentrated to the depression place of the cusp topography, therefore the water depth at the depression place of the cusp became higher than the uniform slope.

Figure-17 shows the time change of the velocity at every point. The location of every point are shown in Figure-14. The wave arrived at the shoreline at 32 seconds after starting the simulation. The figure shows that when the first wave arrived at the shoreline, the velocity became about 8m/s. However, the velocity became slow after the second wave, and became fast after the fifth wave. The reason why the velocity after second wave became slower than the first wave, when the second wave arrived at the shoreline, there was the backwash current that hindered the running up of the incident current on the beach.

Furthermore, the velocity of the wave run up on the beach became about 8m/s. however, the velocity of the backwash on the beach became about 2m/s. This means, the run up wave and the backwash did not link together, i.e. the run up wave links the wave period, however, the backwash did not link the period.

When the accident occurred, the significant wave height was 1.2m, however the maximum height became about 2.1m, therefore if the maximum wave arrives at the shoreline, the water level and the velocity became bigger than the significant wave. If the maximum velocity is 10m/s, the building made by the concrete is broken by the current [7]. Therefore, when the accident occurred, the children might be able to not stand in the current.

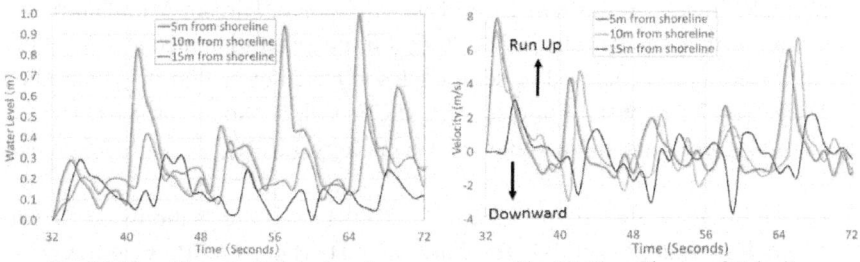

Fig. 16. Time Change of Water Level   Fig. 17. Time Change of Velocity

7. Conclusions

When the accident occurred, the significant wave had not been high, however the wave period had been larger than the ordinary period. This study consider that the victims stayed near to the shoreline when the wave height was low, furthermore the victims fell down and was carried away into the sea when the biggest wave arrived at the beach.

This study conducted field survey tree times to comprehend the topography and the wave. The survey on 29th August was conducted with the Japan Society of Water Rescue and Survival Research jointly.

This study obtained the field topography data by the aerial photographs which were taken by the UAV, and this study comprehend the feature of wave dynamics.

This study conducted the numerical simulation, and the simulation reproduce that the run up wave on the beach converge into the depressions of the cusps, and the backwash current flows in the pectinate where the lowest place in the cusp. Furthermore, the run up velocity on the beach became about 8m/s, and the water level nearby the shoreline became about 1m. According to these results, this accident occurred even though the victims were on the beach, therefore, the people on the beach need to take care not to be sudden attack of the big wave.

References

1. Kiyoshi Uchiyama (2002), "The grain size composition of the beach sediment at Jogehama's coast – The correlation between the beach cusp's characteristics and the grain size composition –", Disaster Prevention Research Institute, Kyoto University, (46), pp.637-649.
2. BEACH EROSION BOARD (1961), "Shore protection and planning", U.S. Army Beach Erosion Board, Tech. Rept., 4, 242pp.
3. Naoyuki Inukai, Yoshifumi Ejiri, Takeshi Ootake, Hiroshi Yamamoto and Tokuzo Hosoyamada (2015), "A Study for the generation mechanism of the rip current at the enclosed or semi-enclosed beach by the groin", Journal of JSCE, Ser. B2 (Coastal Engineering), JSCE, Vol.71, No.2, I_1687-I_1692.
4. Coastal Development Institute of Technology (2001), "CADMAS-SURF 2D), Taiko-sha.
5. Marin Information Group (2012). "NOWPHAS –The Nationwide Ocean Wave information network for Ports and HArbourS." Port and Airport Research Institute, online, http://www.mlit.go.jp/kowan/nowphas/. 2011.
6. Naoyuki Inukai, Takeshi Ootake, Hiroshi Yamamoto and Tokuzo Hosoyamada (2015), "A Study for the Generation Mechanism of the Rip Current at the Enclosed Beach by the Groin", Proceedings of the Annual International Offshore and Polar Engineering Conference, The International Society of Offshore and Polar Engineers, Vol. 25, pp.1359-1364.
7. Shunichi Koshimura, Yuichi Namegaya and Hideaki Yanagisawa (2009), "Fragility functions for tsunami damage estimation", Journal of JSCE, Ser. B, JSCE, Vol.65, No.4, pp.320-331.

Hydrodynamic Response Study for the Berm Breakwater Under Long Crested Random Waves

P. Neelamegam, S.A. Sannasiraj, R. Sundaravadivelu

Department of Ocean Engineering, Indian Institute of Technology Madras,
Chennai, India
bluecloud619@gmail.com
sasraj@iitm.ac.in
rsun@iitm.ac.in

S. Sakthivel

Ocean Engineering and Consultancy Pvt. Ltd.
Chennai, India
enggoecrd@gmail.com

The berm breakwater or reshaping breakwater is a special type of breakwater. The main advantage of this type of breakwater is the requirement of relatively smaller sized armour stones which accelerate the construction speed and reduce the cost of the construction. The stability of berm breakwater is strongly influenced by the weight of units used in primary layer, interlocking properties of the armour units, geometry (width and elevation) of the berm, down slope and the toe berm characteristics. In the present study, the experiments are carried out to study the structural and hydrodynamic response of trunk section in 13m water depth for the berm breakwater proposed at Gopalpur port, Orissa, India. The studies are pertaining that simulating the cyclonic wave condition of Phailin cyclone (October 12, 2013) which has crossed Gopalpur, partly damaging the berm breakwater under construction. The breakwater is designed as a non-overtopping structure with the crest level of (+) 11m CD. The berm width of 15m and the berm level of (+) 6.7m CD is adopted considering the constructional feasibility. The down slope of 1V:1.5H is considered. The armour stone gradation is 3T to 5T, 5T to 7T and 9T-12T has been adopted. For the present experiments, 1:35 scale model has been chosen and tested under long crested random waves. The stability of the structure is compared to the tests under design water level at Mean High Water Spring (MHWS) + storm surge and at MHWS. Swell and Sea wave parameters are considered for the MHWS + storm surge and MHWS respectively. The recession of the primary armour layer has been measured and which is compared with the stability and failure criteria reported in the literature and the design guidelines for berm width. It is found that recession is 2 to 3 times of the diameter of stone. The wave run-up and overtopping discharge also presented in this paper.

Keywords: Berm breakwater; Phailin cyclone; Physical model study; Armour; Recession.

1. Introduction

Conventional rouble mound breakwater is worldwide adopted to build an artificial harbour. The primary material for the construction is quarry rouble stones as armour unit. It is a well known porous structure. It can dissipate the wave energy by means of the porous medium to make a tranquil zone for safe berthing of a vessel on its sea side. The design of rouble mound breakwater is based on Hudson's formula.

$$W_{50} = \frac{w_r H_S{}^3}{K_D (S_r - 1)^3 \cot \alpha} \tag{1}$$

where, w_r is the specific weight of the armour stone/unit; H_s is the design wave height; K_D is the stability coefficient; S_r ($= w_r/w_w$) is the density; and, α is the side slope angle of armour layer. The weight of armour stones increases with increase in cubic power of design wave height. If water depth goes beyond 10 m, artificial armour units are required since larger sizes of stones are unavailable from the quarry [6]. Hence, sophisticated equipment is needed for handling the artificial armour which leads to costlier construction.

The berm breakwater is a special one and is having the capability to replace the conventional breakwaters with same efficient in wave energy dissipation. It also eliminates the large size of stone or artificial armour and heavy equipment for the construction. The berm breakwater can be built with smaller size stones to achieve the 100% quarry yield. Hence berm breakwaters are being adopted in developing countries and the places where cyclonic risk is high. The size of armour stone is based on stability number suggested by Van der Meer is given by,

$$D_{50} = H_S / \Delta N_S \tag{2}$$

where D_{50} is the size of primary armour stone, Δ is relative density and N_S is stability number [7].

The berm breakwaters are also known as reshaping breakwaters and attain equilibrium "S" profile. It is classified into four categories according to stability number.

(1) Hardly reshaping Icelandic type berm breakwater (HR-IC)
(2) Partially reshaping Icelandic type berm breakwater (PR-IC)
(3) Partially reshaping Mass Armoured berm breakwater (PR-MA)
(4) Fully reshaping Mass Armoured berm breakwater (FR-MA)

Above classification is mainly related to the stability number (N_s). The stability number and recession are inversely proportional to each other. The stones move in response to the wave action. According to this, the berm breakwater is further classified as Dynamically stable reshaping structure, Statically stable reshaping structures and Statically stable non-reshaping structures.

In the present study, a physical model study has been done to predict the hydrodynamic characteristics such as stability, recession, run-up and overtopping for the model of berm breakwater. The studies involved with simulating the natural cyclonic wave occurred off the coast of Gopalpur port, Odisha, India namely Phailin cyclone.

2. Design of Berm Breakwater

The design of berm breakwater is corresponding to respective design manuals and literature [7, 10, 12]. The berm breakwater has been designed for contour level of (-) 13 m CD. The design water level is assumed as MHWS + SS. Based on the stability number, the size of the armour stone has arrived. The stability number of 2.7 is assumed for the design, so that this would come under fully reshaping mass armoured (FR-MA) category. The berm width of 15.0 m and berm level of (+) 6.7 m CD as per the berm level is at the minimum of $0.65H_S$ above design water level has been adopted [9].

The other design parameters such as the layer thickness and toe dimensions are adopted similar to conventional breakwater as per Coastal Engineering Manual, 2006 [3]. There is a primary armour layer and three secondary layers adopted with stone sizes from 1 T to 12 T. The wide range of stones sizes up to 1 T adopted for the core layer to achieve the 100% quarry yield. The design of toe berm is according to Coastal Engineering Manual (2006) and Rock Manual (2007) [3, 8]. Projection of toe berm has made for armour stability and to avoid scour during backflow of fluid mass. Toe berm width is kept as wide to increase the stability. The location of toe is adopted as given by,

$$\frac{Hs}{\Delta D50} = \left\{ 2 + 6.2 \left(\frac{ht}{h} \right)^{2.7} \right\} Nod^{0.15} \qquad (3)$$

Other design details of prototype and model are given in Table 1 and designed cross section is shown in Figure 1.

Table 1. Design details for berm breakwater section

Description	Prototype	Model
Bed level	(-) 13.0 m CD	0.0 cm
Crest level	(+) 11.0 m CD	69.0 cm
Berm level	(+) 6.7 m CD	56.7 cm
Primary armour	9 to 12 T	210 to 280 g
First secondary layer	5 to 7 T	115 to 170 g
Second secondary layer	3 to 5 T	70 to 115 g
Third secondary layer	1 to 3 T	23 to 70 g
Toe mound	3 to 5 T	70 to 115 g
Core	Up to 1 T	Up to 23 g
Bedding layer	0.1 to 50 Kg	0.0023 to 1.16 g

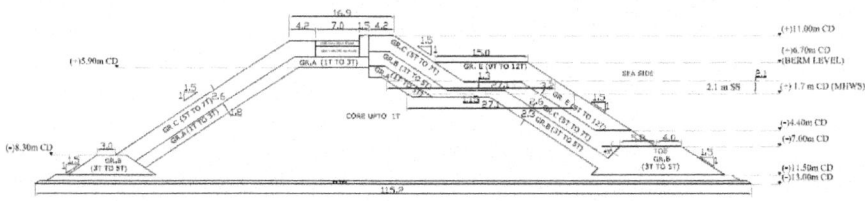

Figure 1. Cross section of prototype

3. Physical Model Study

The ascertaining the damage and recession of the berm breakwater is done by physical model study since the numerical study is ruled out. The model scale of 1:35 as per Froude model scaling law was chosen for the present study [2]. The physical model study has been done in the wave flume dimension of 72.5 m long, 2.5 m deep and 2.0 m wide at the Department of Ocean Engineering, Indian Institute of Technology Madras, India. The standard spectrum of JONSWAP is used to generating the random waves in the flume laboratory.

3.1. *Experimental model setup*

The model has been placed at 45 m from the paddle. A false bottom has been made to overcome with the feasibility in wave generation. There is one wave probe fixed at 8 m from the paddle to measure the incident wave. Run-up gauge is fixed at slope above the berm. The overtopping tray has the dimension of 2x0.2x0.2 m is fixed at lee side of the model and just after the crest of the model. The experimental setup is shown in Figure 2 and image of the model is shown in Figure 3.

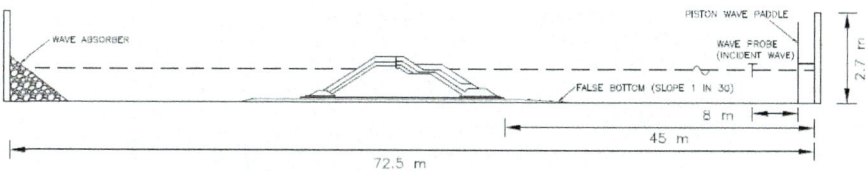

Figure 2. Experimental setup in flume laboratory

Figure 3. Image of the model

4. Design Wave

The berm breakwater was designed for the natural cyclonic wave climate. The Very Severe Cyclonic Storm (VSCS), PHAILIN occurred in Gopalpur port limited, Odisha, India on 12th October 2013.The maximum sustained wind speed of 215 kmph and lowest pressure of 940 hPa was observed during this cyclone. It caused severe damage and flooding. The cyclone track is shown in Figure 4 [1, 5, 11].

In the present study 75%, 100% and 125% of design wave were considered for the study of all hydrodynamic characteristics of the model at both water levels such as Mean High Water Spring (MHWS) and MHWS + Storm Surge. The detailed design wave parameters are listed below in Table 2.

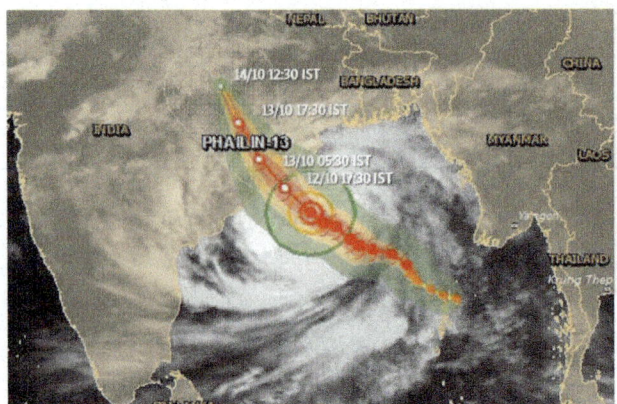

Figure 4. Track of Phailin cyclone

Table 2. Design wave parameter

Design parameters	Prototype	Model
Wave height	7.3 m	20.8 cm
Wave period for MHWS (Sea)	18 s	3 s
Wave period for MHWS + S.S (Swell)	12 s	2 s
Storm surge	2.1 m	6 cm
Duration of the cyclone	3 hrs	30 min
Water depth for MHWS	14.7 m	42 cm
Water depth for MHWS + S.S	16.8	48 cm

5. Results and Discussion

The model was tested under a long crested random wave. The percentage of damage, run-up and overtopping are estimated in each design wave conditions. The recession has been measured at the end of all the design wave conditions. The detailed results are discussed below.

5.1. *Stability and recession parameter*

The model was tested over the cyclonic duration for each design wave conditions. The 75% of design wave is to ascertain the initial settlement of stones. The 100% of design wave is to achieve the design conditions and 125% of design wave is for overloading condition as a conservative purpose.

The rocking and rolling motion of stones during the action of waves were observed. The dislodgement of stones from their initial position is accounted for the damage. The percentage of damage is predicted by using Equation 4 and plots shown in Figure 5.

$$\text{Percentage of damage} = \frac{\text{Number of dislodgement of stones}}{\text{total number of stones in berm of primary armour}} \quad (4)$$

The percentage of damage for all the models is within the 5% of damage for the 100% of design wave. This is referred as no damage condition according to Rock Manual (2007) [8]. But, the percentage of damage exceeds no damage condition for the overloading wave condition. The images of the model after the test are shown in Figure 6.

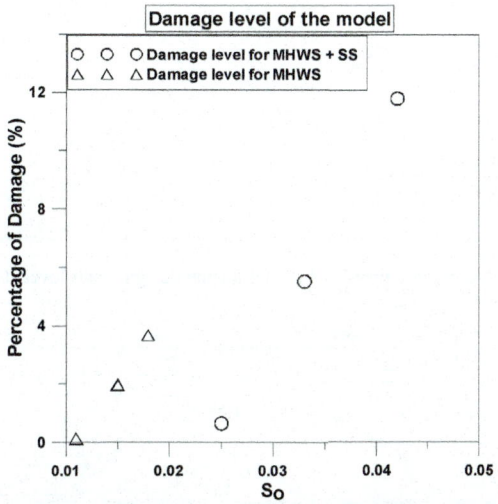

Figure 5. Damage level of the model

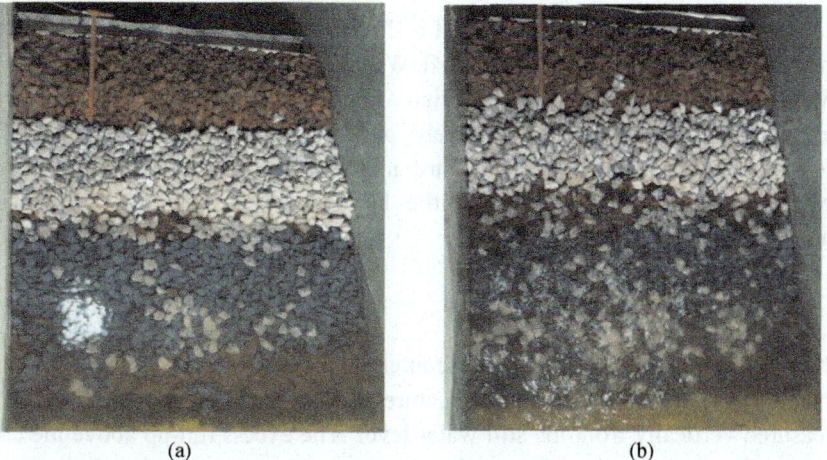

(a)　　　　　　　　　　　　　　　　　　　(b)

Figure 6. View of model after the test for design water level of (a) MHWS and (b) MHWS + SS

The erosion of stones at the seaward edge of berm under the wave action is known as the recession. The structure will attain the equilibrium "S" profile after a certain duration of the test. The eroded stones may move above the berm (or) move to the slope below the berm (or) settled down on the toe of the structure. It was measured at end of all the design wave conditions and recession profile is superimposed with the originally built cross section of models. Reshaping profiles are shown in Figure 7 and Figure 8 where red line indicates the reshaping profile.

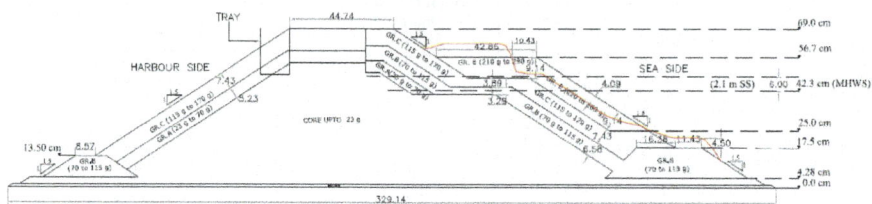
Figure 7. Recession profile of model for the design water level at MHWS

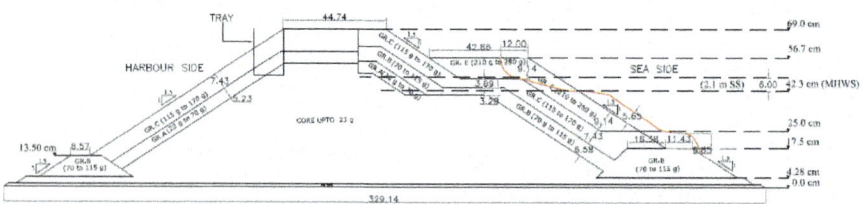
Figure 8. Recession profile of model for the design water level at MHWS + SS

The recession value of the model is 2.3 and 2.62 times of a diameter of the armour stone for the water level of MHWS and MHWS + SS respectively. The value of recession increases for design water level of MHWS+SS. The crest freeboard parameter (R_c) is inversely proportional to recession value. The recession decreases as crest freeboard increases and vice versa. Hence the design water level in relation to the berm level is the most influencing parameter.

5.2. Run-up and overtopping

The wave run-up and overtopping are interconnected parameters. The maximum up rushing of the wave on any structure is known as the wave run-up. It is measured vertically from the still water level. The excess run-up above the crest is termed as overtopping. The design wave parameter is same as the previous section. The run-up plot has been made between maximum run-up and wave steepness is shown in Figure 9. The overtopping plot has been made between average overtopping discharge and dimensionless crest freeboard is shown in Figure 10.

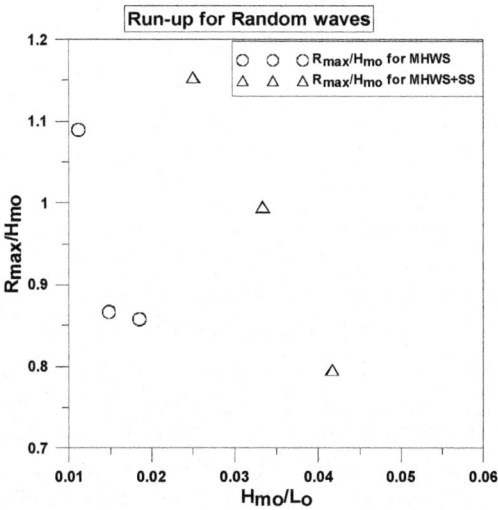

Figure 9. Wave run-up

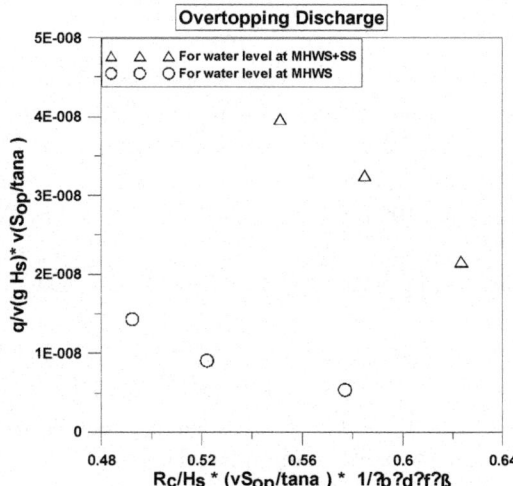

Figure 10. Overtopping discharge

It is observed from above figures that the run-up and overtopping decease with an increase of wave steepness, higher for longer waves and the water level of MHWS + SS. The observed maximum of mean overtopping discharge is 0.066 l/s/m and is termed as no overtopping as per EurOtop (2016) [4].

6. Conclusion

The detailed design for berm breakwater is described in this paper. The width of the berm, berm level and crest freeboard are main design parameters. Recession and wave overtopping decide the width of the berm and crest level respectively. Berm Breakwater utilizes the locally available stones and it will eliminate the heavy equipment for construction. The repairing will be easy since it is reshaping structure. The wastage of material which drawn from the quarry is very less and construction is more economical.

The percentage of damage of berm breakwater is within the safer limit for the design wave. The damage and recession increase with water level i.e for the MHWS+SS. The recession value is found as 2 to 3 times diameter of the primary armour stone. The run-up and overtopping increase for longer waves and is more for the water level of MHWS+SS. The wave overtopping discharge falls under the no overtopping category. The furnished sufficient berm width, crest freeboard and higher toe depth are justified that structure is safe.

References

1. Amrutha, M.M., Kumar, V.S., Anoop, T.R., Nair, B., Nherakkol, A., Jeyakumar, C., 2014. Waves off Gopalpur, the northern Bay of Bengal during Cyclone Phailin. In Annales Geophysicae 32(9), 1073-1083. European Geosciences Union.
2. Chakrabarti, S.K., 1994. Offshore Structure Modelling, World Scientific 9, 19-27.
3. Coastal Engineering Manual, 2006. Washington, D.C.: U.S. Army Corps of Engineers
4. EurOtop, 2016. Wave overtopping of sea defense and related structures–Assessment Manual. UK: Van der Meer, J.W., Allsop, N.W.H., Bruce, T., De Rouck, J., Kortenhaus, A., Pullen, T., Schüttrumpf, H., Troch, P., Zanuttigh, B., www.overtopping-manual.com.
5. Murty, P.L.N., Sandhya, K.G., Prasad, K., Bhaskaran, B., Felix Jose, R., Gayathri B., Nair, T.B., SrinivasaKumar, T., Shenoi, S.S.C., 2014. A coupled hydrodynamic modeling system for PHAILIN cyclone in the Bay of Bengal. Coastal Engineering93, 71-81.

6. Permanent International Association of Navigation Congresses, Permanent Technical Committee II. Working Group 12, 1992. Analysis of rubble mound breakwaters (No. 12). PIANC.

7. Permanent International Association of Navigation Congresses, Working Group 40, 2003. State of the art of Designing and Construction of Berm Breakwater. PIANC.

8. Rock Manual, 2007. The use of rock in hydraulic engineering. CIRIA, CUR, CETMEF, C683 CIRIA.

9. Sigurdharsan, S., Van der Meer, J.W., Burchrth, H.F., Sorensen, J.D., 2007. Optimum safety level and Design rule for Icelandic type berm breakwater. In: Proceeding of the International Conference of Coastal Structures 53-64.

10. Sigurdharsan, S., Van der Meer, J.W., 2013. The design of berm breakwater: Recession, Overtopping and reflection analysis. In: Proceedings of Coasts, Marine Structures and Breakwaters 18-20.

11. Sundaravadivelu, R., Sakthivel, S., Panigrahi, P.K., Sannasiraj, S.A., 2015. Post-Phailin Restoration of Gopalpur Port. Aquatic Procedia 4, 365-372.

12. Van der Meer, J., Sigurdharsan, S., 2014. Geometrical design of berm breakwaters. Coastal Engineering Proceedings 1(34), 25.

Research of Physical Model Test on API Simulation Spectrum

H. L. Huang, Q. L. Du, Y. R. Zhou and Q. H. Zuo

*Nanjing Hydraulic Research Institute, Hujuguan 34,
Nanjing, 210024, China
†E-mail: hlhuang@nhri.cn
www.nhri.cn*

In the nature, wind load on the ocean engineering structure is about 1/10 of the total loads under the coactions of wind, wave and current. However, for floating structures in water or ships, if the damping of the mooring system is small, and the period of the main energy of wind field is close to the oscillation period of the mooring system, the pulsating wind may generate relatively obvious oscillation on the mooring system, and the pulsating load at this situation cannot be ignored. Even though wind is one of the major loads on ships, oil platforms and offshore structures in the marine environment, uniform wind is often used to simulate in the physical model experiments of ocean engineering at present. However, wind is not uniform in nature with strong random pulsatility. It is more practical to master the simulation law of fluctuating wind field rather than to study the ship stress and motion response under random fluctuating wind field. To explore the effects of these parameters on the wind spectrum, a serial of control experiments are conducted, which provides a testing environment of the following research on dynamics of water transport engineering.

Keywords: Fluctuating wind; Wind spectrum; Harmony superposition; Physical simulation.

1. Introduction

In the nature, the wind load on the ocean engineering structure is about 1/10 of the total loads under the coactions of wind, wave and current. However, for floating structures in water or ships, if the damping of the mooring system is small, and the period of the main energy of wind field is close to the oscillation period of the mooring system, the pulsating wind may generate relatively obvious oscillation on the mooring system, and the pulsating load at this situation cannot be ignored [1].

In order to effectively study the function of wind, a number of Chinese laboratories have begun simulation research on fluctuating wind. Ji and Huang [2] presented the expression of the wind speed process based on the Longuet-Higgins stochastic model. Zhang and Yao [1] conducted a laboratory simulation of wind spectrum. Tang *et al.* [3] proposed to determine the wind speed scale by using different ship model scales through the physical model experiments. Xia *et al.* [4] explored the model similarity, scale effect and wind speed scale in combination with the simulation requirements of physical model experiment's wind field for ocean engineering, and proposed the correction coefficient of the wind speed scale. Peng *et al.* [5] gave the suggestion that when simulating the wind field, the experimental area should be 3 m away from the wind array, and its distance to the wall also should be larger than 2.5 m at the same time, and they also simulated two kinds of wind spectra. Based on the characteristics of wind spectrum, this paper conducts the wind spectrum simulation in the laboratory, and compares the results with those of the related literatures. The effects of different control parameters on the spectrum simulation are further analyzed through the experiment, which provides a testing environment for the related wind experiments.

2. Major Headings Wind Spectrum and Its Spectral Simulation

There exist many kinds of wind spectra and simulation methods. In terms of wind spectrum, there are Davenport spectrum [6], Harris spectrum, Ochi&Shin spectrum and API spectrum. In general, there are three kinds of simulation methods, such as harmonic wave superposition method, linear filtering method, and wavelet analysis method. Our paper adopts the wind spectrum simulated by the method of harmonic wave superposition. According to the theory of Shinozuka [7], the random process sample $G(t)$ can be expressed by Eq. (1).

$$G(t) = g(n \cdot \Delta t) = \sum_{j=1}^{M} \sqrt{2S_v(\omega_j)\Delta\omega} \cos(\omega_j n\Delta t + \varphi_j) + V_0, \qquad (1)$$

where, $S_v(\omega_j)$ is the wind power spectrum of the stochastic process $G(t)$, ω_j is the j-th angular frequency, $\Delta\omega$ is the increment of the angular frequency, φ_j is the random variable uniformly distributed in the interval $(0, 2\pi)$, V_0 is the average pulsating wind velocity, M is a sufficiently large number of equal parts of spectral density curve, n is the total time step of the control signal in the random process, and Δt is the time interval to generate control signals.

Taking API wind spectrum as an example, the formula of the fluctuating wind power spectrum is expressed in Eq. (2).

$$S(f) = \frac{[\sigma(z)]}{f_p} \Bigg/ \left(1 + \frac{1.5f}{f_p}\right)^{5/3}.$$ (2)

where f_p is the average frequency of the measured wind spectrum, and $f_p = 0.025 \times V(z)/z$; $\sigma(z)$ is the standard deviation of the wind speed at the

height of z, and $\sigma(z) = \begin{cases} 0.15 \times V(z) \times (z/z_s)^{-0.125} & z \le z_s \\ 0.15 \times V(z) \times (z/z_s)^{-0.275} & z > z_s \end{cases}$; $S(f)$ is random

fluctuating wind power spectrum; $V(z)$ is the average wind speed at the height of z; z_s is the standard height, usually taking 20 m as the default value.

$$S(\omega) = \left[\sigma(z)^2\right]\Big/\omega_p \cdot \left[1\Big/\left(1 + 1.5\omega/\omega_p\right)\right]^{5/3};$$
$$\omega_p = 2\pi f_p.$$

3. Laboratory Simulation System of Fluctuating Wind

The testing equipment consists of a servo driven fan, an ultrasonic anemoscope, a programmable controller DYP–16EH and a human-computer interface DOP–BO7S411. The maximum speed of the fan is 3000r/min, and the servo driven fan can change the revolution speed within a short period. Our experiments set the fastest transformation frequency of the revolution speed being 2 Hz. The measurement range of the wind speed for the ultrasonic anemoscope WindSonic is 0–60 m/s with the error of 2%, and the resolution of 0.01 m/s. And the measurement range of the wind direction is 0–359° with the error of ±3°, and the resolution of 1°. The maximum sampling frequency is 4 Hz.

Before the experiment, the anemoscope should be placed in the effective wind area of the front section of the fan, and the relationship between the revolution speed of the fan and the wind velocity should be calibrated. The fan is manipulated by the computer controlled servo driver, and different wind speeds are obtained by adjusting the computer signals to simulate unsteady wind. The whole system is shown in Figure 1.

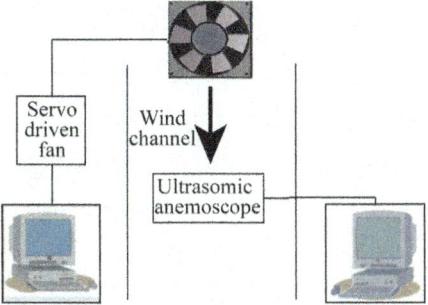

Fig. 1. Wind generating and testing system.

4. Wind Spectrum Simulation at Laboratory

The fluctuating wind simulation is a random process. The magnitude of wind can be described by the statistical characteristics of the probability theory, such as the average wind speed, the standard deviation, and the coefficient of variation, and it can also be expressed in terms of the correlation function in the time domain, or the power spectral density in the frequency domain. Therefore, as long as these two curves are consistent, we consider that the simulation process of random fluctuating wind is correct [8].

4.1. *Simulation of random fluctuating wind*

Considering the simple harmonic superposition method was adopted for the wind spectrum simulation, we have firstly carried out the blowing trial of sinusoidal wind for several times. Figure 2 shows that the result of the blowing trial is reasonable and stable. Figure 3 is the curve of measured process of the fluctuating wind, and Figure 4 depicts the correlation function of the control and measured process for the random fluctuating wind, which proves that the blowing process of the fan is conducted in accordance with the control process. Figure 5 provides the histogram of the wind speed with the wind speed as the x axis, and the event probability as the y axis. The average wind speed is about 20 m/s., and it can be found from the distribution chart that the probability obeys the hypothesis Gauss distribution.

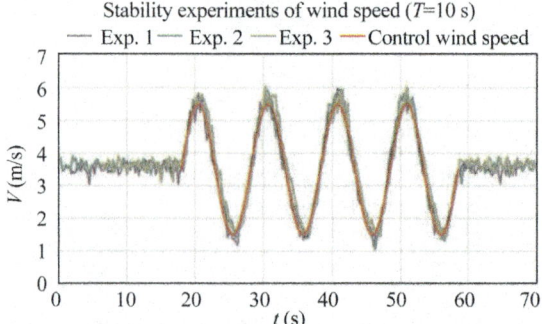

Fig. 2. Results of the stability experiments of wind speed.

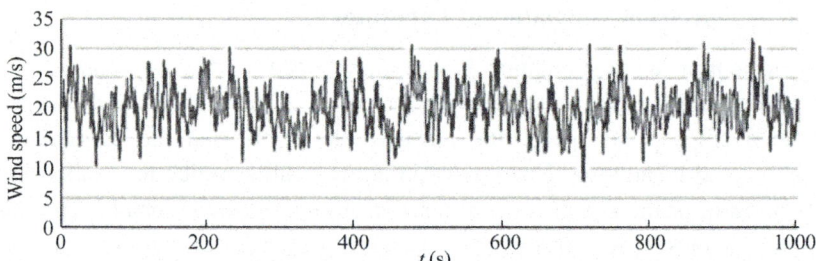

Fig. 3. Measured process of the fluctuating wind speed.

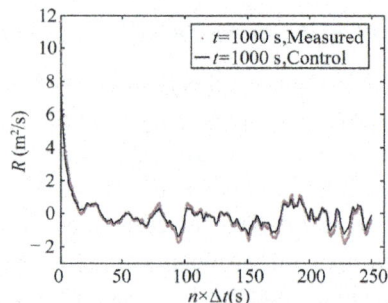

Fig. 4. Curve of the correlation function of the control and measured process for the random fluctuating wind.

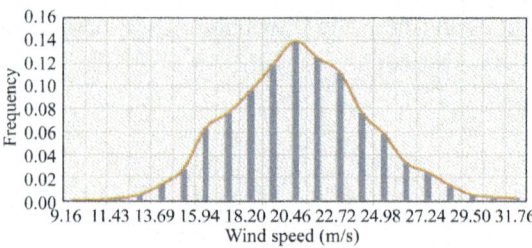

Fig. 5. Histogram of the wind speed.

4.2. *Experimental comparison of wind spectrum simulation*

In order to verify the validity of wind spectrum simulation, the experiments of Zhang and Yao [1], and Peng *et al.* [5] are repeated in our laboratory. The API wind spectrum of fluctuating wind with the wind speed of 42.69 m/s, the center height of wind pressure z=22.27 m, the computed wind frequency range of 0.005–0.7 rad/s, the model scale of 1/70, sampling time of the wind speed of 1290 s, and sampling frequency of 20 Hz, was adopted in the experiments of Zhang and Yao [1]. Whereas, the model scale of 1/140, sampling time of the wind speed of 1200 s, and the sampling frequency of 4 Hz, are utilized in our experiments. Fig. 6 is the comparison of our experimental results with those of Zhang and Yao [1].

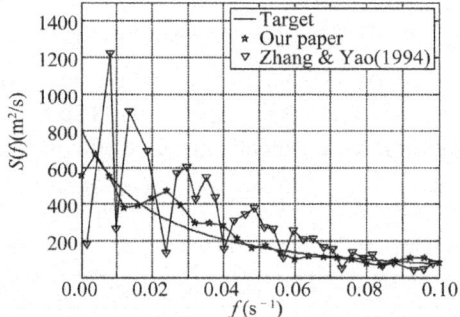

Fig. 6. Result comparison between our experiments and those of Zhang and Yao (1994).

The center height of the wind pressure is 10 m, and the average wind speed is 12 m/s in the experiments of Peng *et al.* [5]. On the contrary, the model scale is 1/40, and the sampling frequency is 4 Hz in our experiments. The comparison of our experimental results with those of Peng *et al.* [5] is shown in Figure 7.

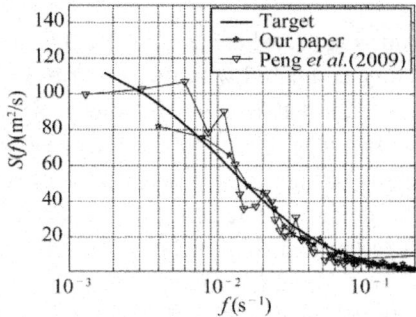

Fig. 7. Result comparison between our experiments and those of Peng et al. (2009).

Compared with the target curve, the measured spectral curve from our experiments is closer to the target spectrum, and the fluctuation and amplitude are smaller, which prove our simulation on the wind spectrum is more effective.

4.3. *Study on the main control parameters of the wind spectrum*

There are many factors which could influence the spectrum simulation in the process of the random fluctuating wind simulation, such as the time interval of the signal generation Δt, the simulation time t, and the number of equal parts of the spectral shape M. In order to investigate the influence of these parameters on the spectrum, the following experiments were carried out under specified control conditions.

4.3.1. *Study on the time interval of the signal generation, Δt*

Assuming that $t=1000$ s and $M=250$, eight scenarios were designed with $\Delta t=1.0$, 1.5, 2.0, 2.5, 2.7, 3.0, 3.5, and 4.0 s, respectively. The experimental results are depicted in Fig. 8, and the eigenvalue statistics of the measured wind spectrum are listed in Table 1.

Table 1. Eigenvalue statistics of the measured wind spectrum with different Δt.

Δt	1.0s	1.5s	2.0s	2.5s	2.7s	3.0s	3.5s	4.0s
Area of the target spectrum S_1	10.02	10.02	10.02	10.02	10.02	10.02	10.02	10.02
Area of the measured spectrum S_2	9.53	9.16	9.12	8.46	8.89	8.66	8.55	8.74
Ratio of the spectral area S_2/S_1	0.951	0.913	0.909	0.849	0.887	0.864	0.853	0.872
Variance of (S_2-S_1)	0.667	0.510	0.469	0.416	0.793	1.054	0.755	1.266
Deviation rate $\dfrac{\lvert S_1 - S_2 \rvert}{S_1}$	0.049	0.087	0.011	0.151	0.113	0.136	0.147	0.128

It can be seen from Fig. 8, when $\Delta t \in [1.0s, 2.5s]$, the simulated spectrum curve is more satisfactory compared with the scenario with $\Delta t \in [2.7s, 4.0s]$. The fluctuation is weak, and simulation is stable, which is consistent with data in Table 1. The variances between measured and target value in the scenario with $\Delta t \in [1.0s, 2.5s]$ are usually smaller than those of the scenario with $\Delta t \in [2.7s, 4.0s]$, and the same as for the deviation rate of the spectral area. Therefore, the value of Δt should be small rather than large when conducting the simulation of the wind spectrum. However, the requirement will be high for the response from the wind turbine with small Δt. Thus, it should be take the performance of the wind turbine into consideration when choosing Δt.

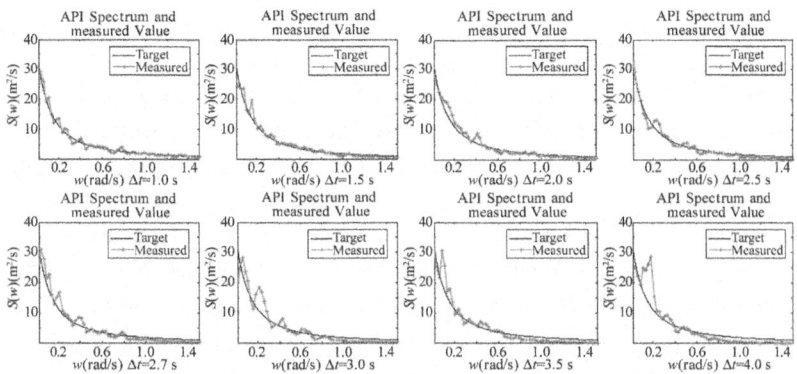

Fig. 8. Results of the measured wind spectrum with different Δt.

4.3.2. Study on the simulation time, t

Assuming that Δt=1.0 s and M=250, eight scenarios were investigated with t=150, 200, 300, 400, 500, 650, 850, and 1000 s, respectively. The experimental results are illustrated in Fig. 9, and the eigenvalue statistics of the measured wind spectrum are presented in Table 2.

Table 2. Eigenvalue statistics of the measured wind spectrum with different t.

t	150s	200s	300s	400s	500s	650s	850s	1000s
Area of the target spectrum S_1	10.03	10.03	10.03	10.03	10.03	10.03	10.03	10.03
Area of the measured spectrum S_2	13.34	10.96	9.42	10.15	12.61	8.97	8.94	9.53
Ratio of the spectral area S_2/S_1	1.330	1.094	0.940	1.011	1.257	0.894	0.891	0.951
Variance of (S_2-S_1)	4.348	1.559	1.041	0.834	1.327	1.268	0.789	0.667
Deviation rate $\dfrac{|S_1 - S_2|}{S_1}$	0.330	0.094	0.060	0.011	0.257	0.106	0.109	0.049

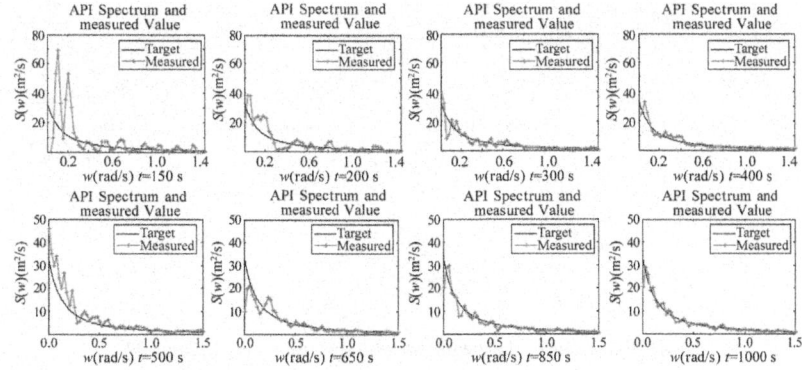

Fig. 9. Results of the measured wind spectrum with different t.

As it shown in Fig. 9, when t=150 s, the spectral fluctuation of the measured wind spectrum reaches its maximum. With the increase of t, the spectrum simulation tends to be stable, and the amplitude of the fluctuation decreases. It can be seen that the longer t, the closer the measured wind spectrum curve will be to the target one. Therefore, t is one of factors to determine the fluctuation intensity and amplitude of the measure spectrum curve. In Table 2, when t=150 s, the variance is 4.348 and the deviation rate is 0.33, which are the largest among all the scenarios. When t increases to 200 s, the variance of the measured spectrum drops to 1.559. When t>300 s, the variance and deviation rate will be varied within a small range. Thus, t must be set within a suitable range, and the shortest t should be selected from to 300 s to 400 s, however, it also should be decided according to the physical model's own characteristics.

4.3.3. *Study on the number of equal parts of the spectral shape, M*

Assuming that t=1000 s and Δt=1.0 s, eight scenarios were investigated with M=50, 100, 125, 150, 175, 200, 225, and 250, respectively. The experimental results are illustrated in Fig. 10, and the eigenvalue statistics of the measured wind spectrum are presented in Table 3.

When the number of equal parts of the target spectral shape, M=50, the fluctuation of the measured wind spectrum curve is the largest. With the increase of the value of M, the curve tends to be stable. And when M>125, the variation of the variance of the measured target becomes smaller and approaches to a steady value as shown in Table 3. It is consistent with the statement that the larger the value of M, the closer the simulation results to the target spectrum will be, proposed by Yan and Zheng [8].

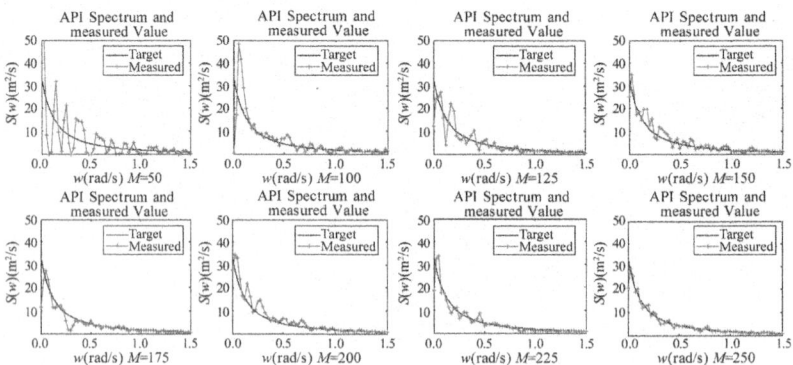

Fig. 10. Results of the measured wind spectrum with different M.

Table 3. Eigenvalue statistics of the measured wind spectrum with different t.

M	50	100	125	150	175	200	225	250
Area of the target spectrum S_1	10.02	10.02	10.02	10.02	10.02	10.02	10.02	10.02
Area of the measured spectrum S_2	10.98	11.46	9.42	11.33	8.55	11.51	9.47	9.53
Ratio of the spectral area S_2/S_1	1.095	1.140	0.939	1.129	0.852	1.147	0.944	0.951
Variance of (S_2-S_1)	3.417	2.266	1.348	0.885	1.174	0.898	0.917	0.667
Deviation rate $\dfrac{\lvert S_1 - S_2 \rvert}{S_1}$	0.095	0.140	0.061	0.129	0.148	0.147	0.056	0.049

5. Conclusions

Based on the characteristics of the fluctuating wind spectrum, the laboratory simulations of random fluctuating wind are carried out. The experimental results meet the requirements of the probability statistics on the parameters, and are better than those of the related literatures. After that, the rules of the influences of different control factors on spectral simulation are further investigated to determine the reasonable value range of each parameter, which establish a satisfactory experimental environment for the future dynamic research.

References

1. C. Y. Zhang and M. W. Yao, Simulation of wind spectrum in water basin test for offshore engineering, *China offshore Platform*, **(z1)**, 175–179 (1994). (in Chinese)
2. C. Q. Ji and X. L. Huang, Model experimental requirements and technique of ocean engineering structure, *China offshore Platform*, **11**(5), 234–237 (1996). (in Chinese)
3. X. N. Tang, Y. Q. Xia, H. Q. Yang and Y. H. Jia, A preliminary study on the determination method of wind speed scale in physical wave model test, *Coastal Engineering*, **25**(1), 1–5 (2006). (in Chinese)
4. Y. Q.Xia, H. J. Li and X. N. Tang, Wind simulation in physical model experiment of ocean engineering, *Engineering Mechanics*, **25**(1), 28–33 (2008). (in Chinese)
5. T. Peng, J. M. Yang, and J. Li, Simulation of wind field in a laboratory basin, *The Ocean Engineering*, **27**(2), 8–13 (2009). (in Chinese)
6. A. G. Davenport, The Spectrum of horizontal gustiness near the ground in high winds, *Quarterly Journal of the Royal Meteorological Society*, **87**(372), 194–211 (1961).

7. M. Shinozuka, Digital simulation of random processes and its applications, *Journal of Sound and Vibration*, **25**(1), 111–128 (1972).

8. S. Yan and W. Zheng, Wind load simulation by superposition of harmonic, *Journal of Shenyang Architectural and Civil Engineering Institute*, **21**(1), 1–4 (2005). (in Chinese)

Comparisons of Future Changes and Uncertainties of Typhoon Intensity and Storm Surge Between Super Typhoons Haiyan (2013) and Melor (2009)

M. Toyoda[†], J. Yoshino and T. Kobayashi

Graduate School of Engineering, Gifu University,
1-1 Yanagido Gifu, Gifu, Japan
[†]E-mail: w3915002@edu.gifu-u.ac.jp

To compare the future changes and their uncertainties of typhoon intensity and storm surge between Super Typhoons Haiyan (2013) and Melor (2009) due to the differences of global warming scenarios (SRESs) and general circulation models (GCMs), two kinds of ensemble pseudo-global warming experiments are conducted using a high-resolution typhoon model and a storm surge model. In the case of Typhoon Haiyan (2013), the future changes of the ensemble-averaged minimum central pressures are relatively small in both SRESs and GCMs. The uncertainty (standard deviation) of peak intensities in GCMs is greater than that in SRESs. In the case of Typhoon Melor (2009), the peak and landfalling intensities averaged tend to be intensified in both SRESs and GCMs. Because of its future intensification, the future storm surge at Mikawa Bay, Japan, is also likely to be enhanced. It is concluded that the uncertainty of typhoon intensity change of the well-mature typhoon is greater than that of the landfalling typhoon, owing to stronger vertical wind shear of a mid-latitude westerly jet.

Keywords: Typhoon Intensity; Storm Surge; Typhoon Melor (2009); Typhoon Haiyan (2013); High-Resolution Typhoon Model; Pseudo-Global Warming Experiments.

1. Introduction

The recent global warming raises concern about future changes of tropical cyclones (i.e. hurricane, typhoon, and cyclone) and their associated coastal disaster. It is thought that the increases of both the sea surface temperature and ocean heat contents by the global warming could increase typhoon intensity[1]. Many previous studies have focused on the future changes of storm surges and waves associated with future typhoons[2,3,4]. However, the recent Fifth Assessment Report (AR5) by Intergovernmental Panel on Climate Change[5] had yet to reveal the possibility whether the future climate change could influence on the future typhoon intensity, because of the large uncertainties of General Circulation Models (GCMs) and Special Report Emissions Scenarios (SRESs) used for the assessments of global warming impacts.

Under these circumstances, Toyoda et al.[6] have conducted ensemble pseudo-global warming experiments by using GCMs and SRESs to assess the future changes of typhoon intensity and storm surge for Typhoon Haiyan (2013) using a high-resolution typhoon model and a storm surge model. Typhoon Haiyan was a super typhoon that developed to a central pressure of 895 hPa in the mature stage. Haiyan made a landfall over the central part of the Philippines with keeping its intensity, and brought a severe storm surge in the coastal areas (especially at Tacloban) in Leyte Gulf[7,8]. Comparing the ensemble spreads of typhoon intensities forced by GCMs and SRESs, they pointed out that the uncertainty of minimum central pressure among GCMs was nearly 2 times larger than that among SRESs.

The uncertainty of future changes of typhoon intensity is likely to be accounted for by the uncertainties of various climate parameters in GCMs (e.g. sea surface temperature, temperature, wind velocity and relative humidity) which are used as input data. However, it remains unclear so far how each of climate parameters may affect the uncertainty of future changes of typhoon intensity. Additionally, the discussion by Toyoda et al.[6] was valid for a typhoon which occurs around the tropical ocean near the Philippines in November, but was not valid for a typhoon which develops around other oceans in other months. Further case studies should be performed to affirm the common features for the impacts of the global warming on typhoon intensity.

In order to quantitatively compare the future changes and their uncertainties of typhoon intensity and storm surge of two super typhoons, ensemble pseudo-global warming experiments forced by GCMs and SRESs are performed in this study using the high-resolution typhoon model and storm surge model. For a typhoon case which passed over the tropical ocean, Super Typhoon Haiyan (2013) is selected in common with the previous work[6]. For the other typhoon case which passed through the tropical and mid-latitude oceans, Super Typhoon Melor (2009) is chosen. Melor was a super typhoon, whose intensity reached a central pressure of 910 hPa in the mature stage, and made landfall over the central part of the Japan islands on 8 October 2009, causing a storm surge disaster at Mikawa Bay in Toyohashi City, Aichi Prefecture, Japan. The maximum sea level anomaly observed was 2.60 m, which is comparable to a maximum sea level anomaly of 2.60 m brought by Typhoon Vera (1959) (so called "Isewan Typhoon") in Japan[9]. Both super typhoons, Haiyan and Melor, are considered to show the different characteristics of future changes and their uncertainties of typhoon intensity and storm surge, because these typhoons occurred in different months over different oceans.

2. Computational settings

In this study, ensemble pseudo-global warming experiments (section 2.3) are carried out using the high-resolution typhoon model (HTM: section 2.1), empirical typhoon model and storm surge model (ETM and SSM: section 2.2) to evaluate the future change and its uncertainty of the typhoon intensity and storm surge associated with Haiyan and Melor.

2.1. *High-resolution typhoon model*

The high-resolution typhoon model (hereafter HTM) is used to simulate the intensity changes of overall life cycles of two super typhoons. HTM is based on the mesoscale meteorological model MM5[10]. MM5 is a three-dimensional, non-hydrostatic, fully-compressible, cloud-resolving atmospheric model to reproduce mesoscale and local-scale meteorological phenomena. The automatic movable nesting technique is introduced into MM5 to efficiently reproduce the high-resolution structure near the center of an intense typhoon from the genesis stage to the dissipating stage. Additionally, the several kinds of physical parameterizations (e.g. the ocean mixed layer, dissipative heating, and sea-spray processes) are newly implemented into MM5 to express realistic typhoon intensity and structure accurately. The triply nested computational domains used in this study have a horizontal grid spacing of 27-km (D1), 9-km (D2), and 3-km (D3), respectively. The movable nesting technique is applied to both D2 and D3. The four-dimensional data assimilation technique (so-called nudging) is utilized for only D1 to gradually assimilate the gridded analyses into the typhoon environment. Basically, initial and boundary conditions for HTM are provided from NCEP Final analyses FNL (temperature, geopotential height, east-west wind velocity, north-south wind velocity, relative humidity, sea-level pressure, and skin temperature).

2.2. *Empirical typhoon model and storm surge model*

Storm surge greatly changes with a slight difference of typhoon track. To avoid an error of storm surge due to the slight difference of typhoon track in HTM, the atmospheric fields (surface wind speed and surface pressure) in a typhoon are reanalyzed in this study by using an empirical typhoon model (ETM) which can be controlled by the typhoon position and intensity (central pressures) simulated by HTM. Then the typhoon position used in ETM is alternatively ingested from the best track data of Japan Meteorological Agency (JMA). Using such fixed typhoon track for the storm surge model, it is expected that global warming impacts of typhoon intensity changes on storm surge can be evaluated purely.

ETM consists of two equations in a horizontally two-dimensional plane: pressure distribution equation and gradient wind equation. The pressure distribution in a typhoon is estimated by the Myers formula. The gradient balance equation is used for the calculation of wind fields in typhoon core, considering the balance among the pressure gradient force, Coriolis force, and centrifugal force. The Blaton formula is used to consider the typhoon movement effect on wind fields. The Gold-berg-Mohn friction formula is assumed to include the surface frictional effect on the wind fields. Then the inflow angle between geostrophic wind vectors and surface wind vector is set to 10°, typical to the ocean.

For the computational settings of ETM, the radius of maximum wind speed is set to 30 km for Haiyan[7] and 50 km for Melor (based on the results of HTM). The typhoon movement vectors are defined by linear approximation by connecting two points from the JMA best track data. The vector components for Haiyan are set to 0.3375° / hr in longitude direction, 0.075° / hr in latitude direction. The vector components before landfalling of Melor are set to 0.336° / hr in the longitude direction and 0.345° / hr in the latitude direction, those after Melor's landfalling are set to 0.415° / hr in the longitude direction and 0.338° / hr in the latitude direction. The other configuration settings are followed by the previous study[6].

The storm surge model (SSM) used in this study is composed of the nonlinear longwave equations system, which predicts 2-dimensional distributions of the sea level anomaly and ocean current vectors. The typhoon atmospheric fields estimated by ETM are inputted as the boundary conditions at every 15 min. Computational domains of SSM are set with a horizontal resolution of 1 km located around the target bays: San Pedro Bay for Haiyan and Mikawa Bay for Melor. The topography data required is obtained from the global earth surface relief dataset ETOPO1 with a horizontal grid of 1 min, produced by NOAA. We use the global land use dataset with a horizontal grid of 30 sec, provided by USGS. The SSM computation for Haiyan is carried out for 2 days (48 hours) from 0000UTC on 7 November 2013 to 0000UTC on 9 November 2013. The computational period for Melor is 1 day (24 hours) from 0900UTC on 7 October to 0900UTC on 8 October 2009.

2.3. *Present climate and future climate experiments*

A number of numerical experiments are carried out according to a computational flow shown in Figure 1, using HTM, ETM, and SSM.

First, present climate experiments for Haiyan and Melor (hereafter CNTRL) are carried out with using NCEP Final Analyses (FNL) as initial and boundary conditions. Additionally, future climate experiments are also conducted by using the method of "pseudo-global warming downscaling[11]". Estimating the differences between CNTRL and future climate experiments, we can quantify the future changes of typhoon intensity and storm surge by the global warming. The initial and boundary conditions for the pseudo-global warming experiments are created by adding to FNL the differences of ten-year averaged monthly mean fields (e.g. temperature, sea surface temperature, geopotential height, east-west wind speed, north-south wind speed, and relative humidity), which are called as "global warming differences (GWDs)" obtained from different GCMs with different SRESs.

The uncertainties of the future changes of typhoon intensity and storm surge are evaluated using the differences of GWDs among SRESs or among GCMs. To conduct the ensemble pseudo-global warming experiments among SRESs, a GCM "HadCM3" in CMIP3 is selected as a reference model. Then SRESs are divided into three emission scenarios (B1, A1B, and A2) for three decadal periods (2030s, 2060s, and 2090s), as shown in Table 1(a). To perform the ensemble pseudo-global warming experiments among GCMs, a SRES "A1B in the 2090s" in CMIP3 is chosen as a reference scenario. A total of 15 GCMs are used as shown in Table 1(b). Both of two super typhoons targeted in this study show a different pattern of typhoon tracks as shown in Figure 2. Haiyan moved east-westward in the tropical ocean, nearly parallel to the latitude lines. In contrast, Melor moved along the edge of a subtropical high from the tropical ocean to the mid-latitude ocean, roughly parallel to the meridians. Comparing the results among the super typhoons simulated by HTM, ETM, and SSM, it is expected that the future

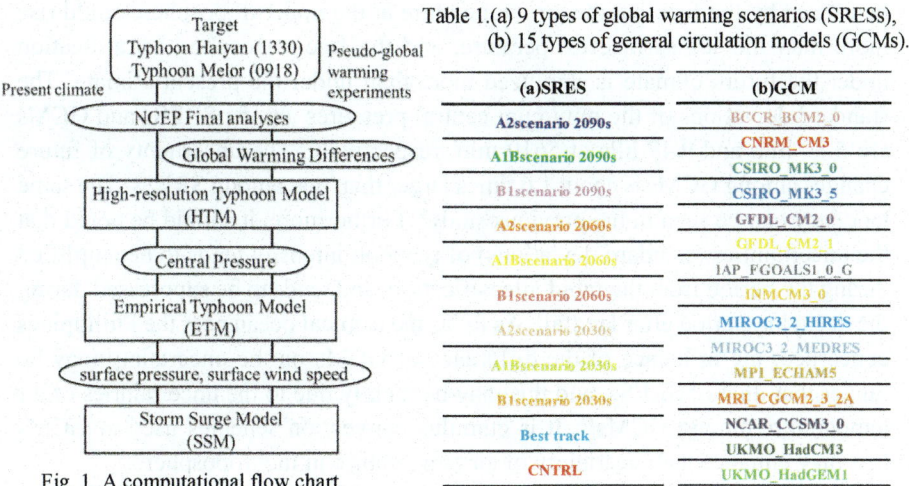

Fig. 1. A computational flow chart.

Table 1.(a) 9 types of global warming scenarios (SRESs), (b) 15 types of general circulation models (GCMs).

(a)SRES	(b)GCM
A2scenario 2090s	BCCR_BCM2_0
A1Bscenario 2090s	CNRM_CM3
	CSIRO_MK3_0
B1scenario 2090s	CSIRO_MK3_5
A2scenario 2060s	GFDL_CM2_0
A1Bscenario 2060s	GFDL_CM2_1
	IAP_FGOALS1_0_G
B1scenario 2060s	INMCM3_0
A2scenario 2030s	MIROC3_2_HIRES
	MIROC3_2_MEDRES
A1Bscenario 2030s	MPI_ECHAM5
B1scenario 2030s	MRI_CGCM2_3_2A
	NCAR_CCSM3_0
Best track	UKMO_HadCM3
CNTRL	UKMO_HadGEM1

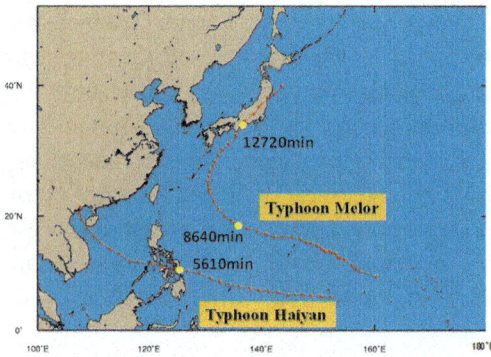

Fig 2. The 6-hourly positions of Typhoon
Haiyan (2013) and Typhoon Melor (2009)
(from the JMA best track data)

change of typhoon intensity and its uncertainty may be evaluated at different oceans at different seasons.

3. Results and discussion

3.1. *Future change and its uncertainty of Haiyan's intensity*

First, we discuss the results of ensemble pseudo-global warming experiments of Haiyan with changes of SRESs and GCMs (Figures 3, 4 and Table 2). The ensemble-averaged values of the minimum central pressures for SRESs and GCMs are 900.8 hPa and 905.0 hPa at 5610 min, respectively. The future changes from CNTRL (897.1 hPa)[12] are +3.7 hPa for SRESs and +7.9 hPa for GCMs, indicating that a well mature typhoon under the future climate tends to slightly weaken its intensity. However, at the time of landfall (5760 min), the ensemble-averaged intensities for SRESs and GCMs are 906.3 hPa and 907.5 hPa, respectively, showing a tendency to be slightly strengthened compared with CNTRL (908.8 hPa)[12]. Therefore, the resultant impact on the peak intensity of a typhoon over the tropical ocean is considered a little. As pointed out by Yoshino et al.[13], the small positive or negative changes of the typhoon intensity of Haiyan under the future climate may be caused by the fact that the air temperature at the upper troposphere could rise more than the sea surface temperature, and the free-tropospheric stratification under the future climate is stabilized more than under the present climate. The standard deviations of the minimum central pressures among SRESs and GCMs are 5.89 hPa and 9.47 hPa at 5610 min, respectively. The uncertainty of future changes among GCMs is about 1.6 times larger than that among SRESs. The same tendency can be seen in the previous study[6]. Furthermore, it should be noted that the uncertainties (standard deviations) of typhoon intensity tends to be amplified during the period from the rapid intensification, and tends to be suppressed during the decaying period after landfall. As far as the tropical ocean near the Philippines concerned, the influence of the difference of GCMs on the uncertainty may be bigger than that of SRESs, and this may be mainly due to the uncertainties of air temperature among GCMs[13]. It is cumulus convection schemes used in GCMs that may enhance the uncertainty of air temperature in the troposphere.

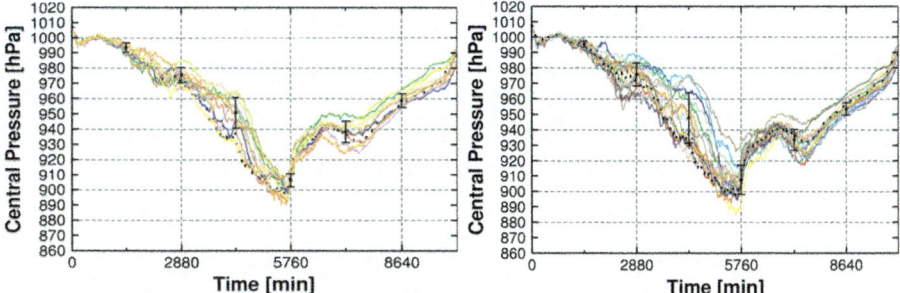

Fig. 3. Time series of the central pressure of Haiyan simulated by pseudo-global warming experiments SRESs. Error bars indicate the standard deviations of the central pressures simulated among SRESs. (dotted line: CNTRL, the colors of solid lines correspond to Table 1(a))

Fig. 4. Time series of the central pressure of Haiyan simulated by pseudo-global warming experiments GCMs. Error bars indicate the standard deviations of the central pressures simulated among GCMs. (dotted line: CNTRL, the colors of solid lines correspond to Table 1 (b))

3.2. *Future change and its uncertainty of Melor's intensity*

Next, we discuss the results of the ensemble pseudo-global warming experiments of Melor with changes of SRESs and GCMs (Figures. 5, 6 and Table 3). The ensemble-averaged values of the minimum central pressures for SRESs and GCMs are 902.3 hPa and 887.6 hPa at 8640 min, respectively. The future changes from CNTRL (907.7 hPa) are -5.4 hPa for SRESs and -20.1 hPa for GCMs, implying that future typhoons in both SRESs and GCMs are intensified in the subtropical ocean. There is a tendency that the typhoon intensities among GCMs are especially strengthened much more than SRESs. Thus future typhoons in the 2090s under the A1B scenario are likely to be intensified over the subtropical ocean. Comparing the results of Haiyan with those of Melor, there is a large difference of global warming impact on typhoon intensity between the tropical and subtropical ocean. The standard deviations of the minimum central pressure are 20.1 hPa for GCMs and 18.8 hPa for SRESs at 8640 min. The uncertainty of future changes of the peak intensity of Melor is much greater than that of Haiyan described in section 3.1, while there is a smaller difference of the uncertainty between SRESs and GCMs.

Then, the ensemble-averaged values of the landfalling central pressures of Melor for SRESs and GCMs are 948.9 hPa and 952.2 hPa, respectively. The future changes of the intensity from CNTRL (953.5 hPa) are -4.6 hPa for SRESs and -1.3 hPa for GCMs. As with the results at the peak time (8640 min), a future typhoon making landfall at the Japan islands is likely to be intensified although the future changes

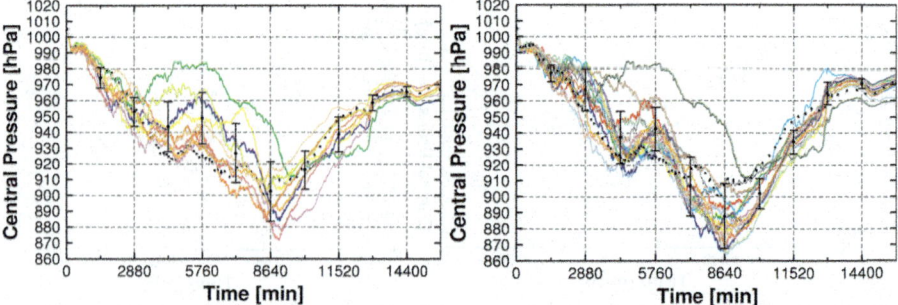

Fig. 5. Time series of the central pressure of Melor simulated by pseudo-global warming experiments SRESs. Error bars indicate the standard deviations of the central pressures simulated among SRESs. (dotted line: CNTRL, the colors of solid lines correspond to Table 1 (a))

Fig. 6. Time series of the central pressure of Melor simulated by pseudo-global warming experiments GCMs. Error bars indicate the standard deviations of the central pressures simulated among GCMs. (dotted line: CNTRL, the colors of solid lines correspond to Table 1 (b))

are relatively smaller. The standard deviations of the landfalling central pressures are 7.8 hPa for SRESs and 10.3 hPa for GCMs, meaning that the uncertainty among GCMs is greater than that among SRESs, and the uncertainty at the landfalling time is reduced more significantly than that at the peak time.

3.3. *Key factors of uncertainty of typhoon intensity change*

From the above discussion, it becomes clear that there is a large difference of the uncertainties of future changes of typhoon intensity between Haiyan and Melor. Key factors of the uncertainty of typhoon intensity changes are considered to be especially due to the uncertainties of temperature and wind speed in GWDs derived from GCMs[13]. Thus the ten-year averaged monthly mean differences of 300 hPa temperature and 300 – 850 hPa vertical wind shear between 2090s and 2000s are evaluated from a total of 15 GCMs in the A1B scenario along with the 6-hourly tracks of Haiyan and Melor (Figures. 7 and 8).

As shown in Figures. 7(a) and 8(a), the GWDs of 300 hPa temperature of Haiyan and Melor indicate that there are little variations in GWDs of temperature across the tracks of both Haiyan and Melor, and their standard deviations are approximately 1.0 K. The contribution of the uncertainties of temperature to Haiyan's intensity change is considered to be as large as that to Melor's intensity change, regardless of the differences of latitude and season.

As shown in Figures. 7(b) and 8(b), the GWDs of 300 – 850 hPa vertical wind shear of Haiyan and Melor indicate a different tendency from those of 300 across hPa temperature. It is found that the uncertainty of vertical wind changes

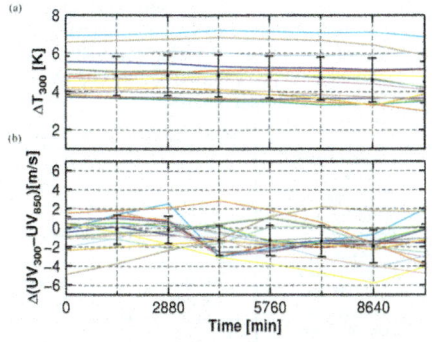
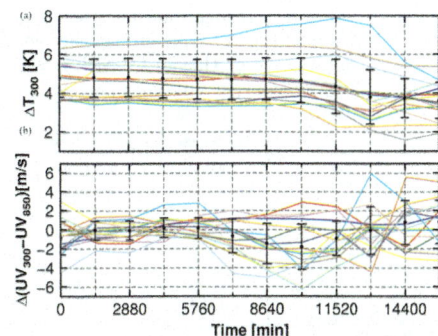

Fig. 7. Time series of ten-year averaged monthly mean differences of (a) 300hPa temperature and (b) 300 – 850hPa vertical wind shear between 2090s and 2000s (SRES A1B), across the 6-hourly positions (track) of Haiyan.

Fig. 8. Time series of ten-year averaged monthly mean differences of (a) 300hPa temperature and (b) 300 – 850 hPa vertical wind shear between 2090s and 2000s (SRES A1B), across the 6-hourly positions (track) of Melor.

shear the track of Melor is much larger than that across the track of Haiyan. In terms of an uncertainty in Melor, the standard deviations of future changes of vertical wind shear is particularly larger during the passage around the subtropical and mid-latitude oceans (during 8640 – 12960 min) than during the passage around the tropical ocean (during 0 - 8640 min). Therefore it can be considered that the high uncertainty of typhoon intensity changes of Melor during the time of peak and landfall, seen in Figure 6, are due to the uncertainty of the future changes of vertical wind shear obtained from GCMs. The upper-level westerly jet dominates in the mid-latitude ocean, and the position and intensity could be changed by the global warming, resulting in the increase of the uncertainties of future changes of typhoon intensity.

3.4. *Future change and its uncertainty of storm surge of Haiyan and Melor*

Finally, we discuss the future change and its uncertainty of storm surge induced by Haiyan (Table 2) and Melor (Table 3).

For the storm surge at the port of Tacloban by Haiyan, the ensemble averaged values of the maximum sea level height are 6.21 m for SRESs and 6.10 m for GCMs. In both experiments, the future storm surge is slightly increased from CNTRL (6.08 m). The slight increase of storm surge is accounted for by the slight increase of typhoon intensity during the time of landfall (5760 min). The standard deviations (relative standard deviations) of the maximum sea level height among SRESs and GCMs are 0.23 m (0.037) and 0.44 m (0.072), respectively, meaning that the uncertainty among GCMs is approximately two times larger than that

Table 2. The future changes and uncertainty of future change of typhoon intensity and storm surge of Haiyan for SRESs and GCMs. The values of future changes are shown in parentheses.

	SRESs (9cases)	GCMs (15cases)
Average of central pressure(5610min) (Future change)	900.8hPa (+3.7hPa)	905.0hPa (+7.9hPa)
Standard deviation of central pressure (5610min)	5.89hPa	9.47hPa
Average of central pressure(5760min) (Future change)	906.3hPa (-2.3hPa)	907.5hPa (-1.1hPa)
Standard deviation of central pressure (5760min)	4.39hPa	9.30hPa
Maximum sea level departure (Future change)	6.21m (+0.13m)	6.10m (+0.02m)
Standard deviation of maximum sea level departure (relative standard deviation)	0.23m (0.037)	0.44m (0.072)

Table 3. The future changes and uncertainty of future change of typhoon intensity and storm surge of Melor for SRESs and GCMs. The values of future changes are shown in parentheses.

	SRESs (9cases)	GCMs (15cases)
Average of central pressure(8640min) (Future change)	902.3hPa (-5.4hPa)	887.6hPa (-20.1hPa)
Standard deviation of Central Pressure (8640min)	18.8hPa	20.1hPa
Average of central pressure(12720min) (Future change)	948.9hPa (-4.6hPa)	952.2hPa (-1.3hPa)
Standard deviation of central pressure (12720min)	7.80hPa	10.3hPa
Maximum sea level departure (Future change)	2.83m (+0.20m)	2.69m (+0.06m)
Standard deviation of maximum sea level departure (relative standard deviation)	0.31m (0.11)	0.51m (0.19)

among SRESs. The result is consistent with the discussion in Figure 4.

For the storm surge at the port of Mikawa by Melor, the ensemble-averaged values of the maximum sea level anomaly are 2.83 m for SRESs and 2.69 m for GCMs. The future changes from CNTRL (2.63 m) are +0.20 m for SRESs and +0.06 m for GCMs. Therefore, the maximum sea level anomaly is more likely to increase in the mid-latitude ocean than in the tropical ocean, due to the increase of typhoon intensity by the future global warming. The standard deviations (relative standard deviations) of the maximum sea level anomalies among SRESs and GCMs are 0.31 m (0.11) and 0.51 m (0.19), respectively. As with the typhoon intensity changes of Melor (shown in Figure 6), the uncertainty of the future changes of storm surge is larger among GCMs than among SRESs. Comparing relative standard deviation at the port of Mikawa by Melor with that at the port of Tacloban by Haiyan, the uncertainty of the future change of storm surge in Japan is bigger than that in the Philippines. This larger uncertainty is considered to be affected by the uncertainty of future changes of the upper-level westerly jet seen in Figure 8.

According to the above discussion from the present climate and future climate experiments, typhoons approaching Japan will be more likely to be strengthened by the global warming, and there is a high possibility that the typhoon associated storm surge will also be intensified. It is concluded that the uncertainty of future changes of an upper-level westerly jet could play a key role in reducing the reliability of future changes of typhoon intensity in the subtropical and mid-latitude oceans.

4. Concluding remarks

In this study, ensemble pseudo-global warming experiments using the high-resolution typhoon model and storm surge model for Typhoon Haiyan (2013) and Typhoon Melor (2009) were carried out to quantify the future changes and their uncertainties of typhoon intensity and storm surge under the future climate.

The pseudo-global warming experiments for Haiyan, which was developed over the tropical ocean, showed that there was a small future change of typhoon intensities in both SRESs and GCMs. On the other hand, the uncertainty (standard deviation) of future changes of typhoon intensity was larger among GCMs than among SRESs. The pseudo-global warming experiments for Melor, which moved from the tropical ocean to the mid-latitude ocean, indicated that the typhoon intensities are strengthened at the mature stage in both SRESs and GCMs. Moreover, the standard deviations of future changes of Melor is also larger than that of Haiyan, suggesting that the uncertainty of typhoon intensity change is larger in the subtropical ocean than in the tropical ocean. Concerning the time of Melor's landfalling at Japan, the typhoon intensities in both SRESs and GCMs could be intensified under the future climate, and the uncertainty of typhoon intensity change in the mid-latitude is also increased by the global warming, due to the increase of the uncertainty of the vertical wind shear with an upper-level westerly jet.

Regarding the future changes of storm surge, both Haiyan and Melor showed that the sea level anomalies are enhanced in response to the typhoon intensity changes under the future climate. The results suggested that the expected storm surge disaster, especially in Japan, could be severer in the future climate than in the present climate. Future work will extend the method to the other coastal regions over the pan-Pacific regions.

Acknowledgements

This study was supported by JSPS Research Fellow (No.17J04771).

References

1. I-I. Lin, I.-F. Pun., and C.-C. Lien, "Category-6" super typhoon Haiyan in global warming hiatus: Contribution from subsurface ocean warming, *Geophysical Research Letters*, Vol.41, pp.8547-8553, (2014).
2. T. Yasuda, S. Nakajo, S. Kim, H. Mase, N. Mori, and K. Horsburgh, Evaluation of future storm surge risk in East Asia based on state-of-the-art climate change projection, *Coastal Engineering*, Vol. 83, pp.65-71, (2014).

3. N. Mori, T. Shimura, K. Yoshida, R. Mizuta, Y. Okada, K. Temur, M. Ishii, M. Kimito, Y. Takayabu, and E. Nakakita, Mega-ensemble projection based on 60km agcm and its application to long-term impact assessment of storm surge, *JSCE B2 (Coastal Engineering)*, Vol. 72, No. 2, pp.I_1471-I_1476, (in Japanese, 2016).

4. T. Shimura, N. Mori, T. Takemi, and R. Mizuta, Ocean wave-dependent roughness impacts on climate system by coupled atmospheric global climate-wave model, *JSCE B2 (Coastal Engineering)*, Vol.72, No.2, pp.I_1507-I_1512, (in Japanese, 2016).

5. IPCC, 2013 : *Climate Change 2013: The Physical Science Basis.*, Cambridge University Press, 1535p., (2013).

6. M. Toyoda, J. Yoshino, and T. Kobayashi, Ensemble future projections of storm surge in the Leyte gulf, Philippines, by pseudo-global warming experiments, *JSCE B2 (Coastal Engineering)*, Vol.72, No.2, pp.I_1483-I_1488, (in Japanese, 2016).

7. H. Kawai, K. Seki, and T. Fujiki, Strom surge and wave characteristics of Typhoon 1330 in central Philippines, *JSCE B2 (Coastal Engineering)*, Vol.70, No.2, pp.I_221-I_225, (in Japanese, 2014).

8. H. Kawai, T. Arikawa, K. Honda, T. Asai, S. Fujiki, and R. Kuwajima, Field survey on Typhoon 1330 wind, wave, storm surge effects on Philippine ports, *JSCE B2 (Coastal Engineering)*, Vol.70, No.2, pp.I_1436-I_1440, (in Japanese, 2014).

9. S. Aoki, and S. Kato, Strom surge in Mikawa bay caused by Typhoon No.0918, *JSCE B2 (Coastal Engineering)*, Vol.66, No.1, pp.296-300, (in Japanese, 2010).

10. J. Dudhia, A nonhydrostatic version of the Penn State-NCAR mesoscale model: Validation test and simulation of an Atlantic cyclone and cold front, *Mon. Wea. Rev.*, Vol.121, pp.1493-1513, (1993).

11. J. Yoshino, S. Arakawa, M. Toyoda, and T. Kobayashi, Intercomparison of global warming scenarios for typhoon intensity change using a High-Resolution Typhoon Model, *JSCE B2 (Coastal Engineering)*, Vol.71, No.2, pp. I_1519-I_1524, (in Japanese, 2015).

12. M. Toyoda, J. Yoshino, and T. Kobayashi, Numerical experiments of Typhoon HAIYAN (2013) and its storm surge using a high-resolution coupled typhoon-ocean model, *JSCE B2 (Coastal Engineering)*, Vol.71, No.2, pp. I_463-I_468, (in Japanese, 2015)

13. J. Yoshino, K. Shinohara, M. Toyoda, and T. Kobayashi, Sensitivity experiments on future intensity changes and uncertainties of Typhoon HAIYAN (2013) using a High-Resolution Typhoon Model, *JSCE B2 (Coastal Engineering)*, Vol.73, No.2, (in Japanese, 2017).

Sensitivity Experiments on Future Intensity Changes and Uncertainties of Typhoon Haiyan (2013) using A High-Resolution Typhoon Model

J. Yoshino[†], K. Shinohara, M. Toyoda, and T. Kobayashi

Graduate School of Engineering, Gifu University
1-1 Yanagido Gifu, Gifu, Japan
[†]E-mail: jyoshino@gifu-u.ac.jp

In order to examine the detailed mechanism of typhoon intensity changes under the future climate, a number of sensitivity experiments on typhoon intensity are made using a high-resolution typhoon model. The uncertainties of the future intensity changes are also quantified by the comparison of the sensitivity experiments among global warming differences (GWDs: sea surface temperature, air temperature, wind speed, and relative humidity) derived from general circulation models (GCMs). The comparisons of ensemble averages of simulated typhoon intensity between sensitivity experiments and a present-climate experiment indicate that sea surface temperature in GWDs is responsible for intensifying the future typhoon by about -19.6 hPa. On the other hand, air temperature in GWDs is accountable for weakening the future typhoon by about +45.5 hPa. The comparisons of standard deviations of typhoon intensity between sensitivity experiments and pseudo-global warming experiments suggest that the uncertainties of air temperature and wind speed in GWDs significantly reduce the reliability of the future projections on typhoon intensity.

Keywords: Typhoon intensity; General circulation models; High-resolution typhoon model; Pseudo-global warming experiments

1. Introduction

Sea surface temperature rise by the recent global warming has raised concern about the future intensification of typhoons over the Northwestern Pacific Ocean (Emanuel et al., 2005)[1], resulting in the increase of the risk of coastal disasters around the Asia-Pacific region. The Fifth Assessment Report (AR5) of Intergovernmental Panel on Climate Change (IPCC, 2013)[2] has yet to specify the possibility whether or not the number of intense typhoons is increased in the late 21st century. This may be due to the high uncertainties in the global warming projections.

Toyoda et al. (2015)[3] conducted a present climate experiment of Super Typhoon Haiyan (2013) using a high-resolution typhoon model (HTM, Yoshino et al., 2015)[4], and accurately reproduced a peak intensity of about 897hPa that was consistent with the best track data (895hPa). In addition, Toyoda et al. (2016)[5] estimated future intensity changes and its uncertainties, based on the ensemble pseudo-global warming experiments on Haiyan in the 2090s under the A1B scenario. They suggested that the uncertainties of typhoon intensity projections were higher in the general circulation models (GCMs) than in the global warming scenarios (GWSs). According to the theory of Maximum Potential Intensity (MPI) developed by Bister and Emanuel, (2002)[6], it should be considered that a well-matured typhoon is intensified by increasing temperature at the sea surface and is weakened by increasing temperature at the upper troposphere. Thus there is a possibility whether a typhoon under the future climate can become stronger or weaker in responding to the subtle differences between both temperatures, because many future-climate GCMs predicted that global warming may raise both temperatures in the tropics (IPCC, 2013)[2]. Yoshino et al. (2015)[4] also conducted pseudo-global warming experiments on a total of 29 typhoons in 2004 using HTM and suggested that well-mature typhoons in the present climate are hardly intensified since the rise of sea surface temperature is compensated by as the rise of upper troposphere temperature. Therefore, it is necessary to quantify how the uncertainties of several climate parameters obtained from GCMs influence on the uncertainties of the intensity changes of future typhoons.

In order to examine the detailed mechanism of typhoon intensity changes under the future climate and to quantify the uncertainties of the intensity changes of a future-climate typhoon, a number of ensemble sensitivity experiments are made in this study using HTM with changes of global warming differences (GWDs). Typhoon Haiyan (2013), which caused a huge storm surge at Tacloban in the Philippines, is selected as a well-mature disastrous typhoon case under the present climate. The GWDs used in this study are composed of sea surface temperature (SST), air temperature (T), wind speed (UV), and relative air humidity (RH), derived from the difference between the 2000s and 2090s in SRES A1B, using a total of 15 GCMs archived in the Coupled Model Intercomparison Project, phase 3 (CMIP3).

2. Computational Method

In this study, a present-climate simulation and pseudo-global warming simulations (see in section 2.2) are conducted using a high-resolution typhoon model (HTM, see in section 2.1), and a number of sensitivity experiments are also made to investigate the mechanism of typhoon intensity changes under the future climate (see in section 2.3) and to quantify the uncertainties of the future changes of Haiyan (see in section 2.4).

2.1. *High-Resolution Typhoon Model*

To reproduce the intensity and structure of Haiyan, we apply to the high-resolution typhoon model (HTM) based on the mesoscale meteorological model MM5 (Dudhia, 1993)[7], which is three-dimensional, non-hydrostatic, fully-compressible, cloud-resolving atmospheric model to predict mesoscale and local-scale phenomena (Yoshino et al., 2015)[4]. The automatic movable nesting technique is introduced into MM5 to efficiently reproduce the high-resolution structure near the center of an intense typhoon from the genesis stage to the dissipating stage. Additionally, the several kinds of physical parameterizations (e.g. the ocean mixed layer, dissipative heating, and sea-spray processes) are added into MM5 to precisely calculate boundary layer processes in a super-typhoon like Haiyan. The triply nested computational domains (D1, D2, and D3) have a grid spacing of 27-km, 9-km, and 3-km, respectively. D1 covers most of the Northwestern Pacific Ocean, including the total track of Haiyan. To resolve the internal structure of Haiyan, the movable nesting technique is applied to both D2 and D3. The four-dimensional data assimilation technique (so-called nudging) is utilized to gradually assimilate gridded analyses to the typhoon environment (nudging coefficient is set to 1.0×10^{-5}). Initial and boundary conditions for HTM are provided from NCEP Final analyses FNL. All other computational settings are identical to those used in the previous studies (Toyoda et al., 2015; Toyoda et al., 2016)[3,5].

2.2. *Present climate simulation and pseudo-global warming simulations*

Using the above-mentioned HTM, a present-climate simulation and pseudo-global warming simulations are conducted for Haiyan. These simulations are also compared with sensitivity experiments that will be described later. One of the experiments simulated by Toyoda et al., (2015)[3], of which the Haiyan's intensity showed the best accuracy with the JMA best track data, is selected as a present-climate simulation, and is used as a basic experiment (CNTRL). They showed that CNTRL accurately reproduced a peak intensity of about 897hPa that was fairly consistent with the best track data (895hPa).

As future-climate simulations, pseudo-global warming simulations are also made in this study, following the method written in Yoshino et al. (2015)[4] and Toyoda et al. (2016)[5]. The initial and boundary conditions for pseudo-global warming simulations are created by adding all of GWDs (SST, T, UV, and RH) into FNL used in CNTRL. GWDs are defined by the differences between ten-year averages during the 2000s (as the present climate) and those during the 2090s (as the future climate). The A1B scenario, in which the CO_2 concentration increases to 720 ppm by 2100, is selected as one of GWSs, because we would focus on the

Table 1. List of a total of 15 GCMs used in this study.

GCM	Research and Development Institutions	Resolution
BCCR_BCM2_0	Bjerknes Centre for Climate Research, Norway	1.9×1.9
CNRM_CM3	Météo-France/Centre National de Recherches Météorologiques, France	1.9×1.9
CSIRO_MK3_0	CSIRO Atmospheric Research,Australia	1.9×1.9
CSIRO_MK3_5		
GFDL_CM2_0	U.S. Dept. of Commerce/NOAA/Geophysical Fluid Dynamics Laboratory, USA	2.0×2.5
GFDL_CM2_1		
IAP_FGOALS1_0_G	Institute of Atmospheric Physics, Chinese Academy of Sciemces, China	2.8×2.8
INMCM3_0	Institute for Numerical Mathematics, Russia	4.0×5.0
MIROC3_2_HIRES	Center for Climate System Research (University of Tokyo), National Institute for Environmental Studies, and Frontier Research Center for Global Change (JAMSTEC), Japan	1.1×1.1
MIROC3_2_MEDRES		2.8×2.8
MPI_ECHAM5	Max Planck Institute for Meteorology, Germany	1.9×1.9
MRI_CGCM2_3_2A	Meteorological Research Institute, Japan	2.8×2.8
NCAR_CCSM3_0	National Center for Atmospheric Research, USA	1.4×1.4
UKMO_HadCM3	Hadley Centre for Climate Prediction and Research/Met Office, UK	2.75×3.75
UKMO_HadGEM1		1.25×1.875

uncertainties of GCMs rather than those of GWSs in this study. According to the above-mentioned settings, the pseudo-global warming simulations are made as the basic experiments (ALL) to investigate the physical mechanism and uncertainties of future changes of Haiyan. Table 1 shows a total of 15 GCMs used in this study.

2.3. *Sensitivity experiments for typhoon intensity changes under the future climate*

To understand the mechanism of future intensity changes of Haiyan, sensitivity experiments, in which each one of GWDs is added into the initial and boundary conditions of CNTRL (CNTRL+SST, CNTRL+T, CNTRL+UV, and CNTRL+RH), are conducted for a total of 15 GCMs. Comparing the minimum central pressure simulated by CNTRL with the ensemble average of minimum central pressures simulated by the sensitivity experiments, the key factors of intensity changes of Haiyan under the future climate could be identified. It is expected that the sensitivity experiments could estimate how different each parameter in GWDs impact on typhoon intensity changes under the future climate.

2.4. *Sensitivity experiments for uncertainties of typhoon intensity changes under the future climate*

To quantify the uncertainties of the future intensity changes of Haiyan, sensitivity experiments, in which each one of GWDs is removed from the initial and boundary conditions of ALL (ALL-SST, ALL-T, ALL-UV, and ALL-RH), are made for a total of 15 GCMs. Comparing the standard deviation of the minimum central pressures simulated by ALL with that simulated by the sensitivity experiments, the key factors of uncertainties of future intensity change of Haiyan under the future climate could be evaluated. Therefore, we could quantify how different each parameter in GWDs influence on uncertainties of future changes of typhoon intensity under the future climate.

3. Results And Discussion

3.1. *Global warming differences*

Before discussing the results of the HTM simulations, we herein examine the vertical profiles of GWDs (SST, T, UV, and RH) at the nearest grid point (10.0°N, 128.0°E), where Haiyan just made landfall in the Philippines, derived from the CO_2-warmed GCMs in the 2000s and 2090s under the A1B scenario.

Figure 1 shows the vertical profiles of future changes of sea surface temperature (SST) and air temperature (T) derived from a total of 15 GCMs. SST near the Philippines is increased by about 1~3 K at the end of 21st century and its standard deviation of the SST changes among GCMs is 0.36 K. Such a SST increase is considered to intensify future typhoons owing to the enhancement of surface energy inflows (Bister and Emanuel, 2002; Yoshino et al., 2015)[4,6]. The vertical profiles of the T changes indicates that there are obvious peaks of temperature changes at the upper troposphere and the air temperature at 250hPa are increased by +3.5~7.0 K. Its standard deviation of the T changes among GCMs is 1.15 K, resulting in larger uncertainties in T than those in SST. Such a T increase at the upper troposphere is considered to weaken typhoon intensity due to vertically stable stratification in the troposphere. Meanwhile the vertical profiles of the T changes around the lower stratosphere sharply decrease by -3.5~-0.5 K. According to the MPI theory (Bister and Emanuel, 2002)[6], such a T decrease at the lower stratosphere is thought to strengthen the typhoon intensity.

Figure 2 shows the vertical profiles of future changes of wind speed (UV) derived from a total of 15 GCMs. The UV profiles exhibit relatively larger fluctuations throughout the troposphere than the T profiles. Some GCMs show UV decreases and some GCMs do UV increases. The large deviations of UV is remarkable especially in the upper troposphere, the standard deviation of the UV changes among GCMs is 2.12 m/s at 150 hPa. The UV changes under the future climate are considered to lead to the changes of vertical wind shear in the typhoon

122

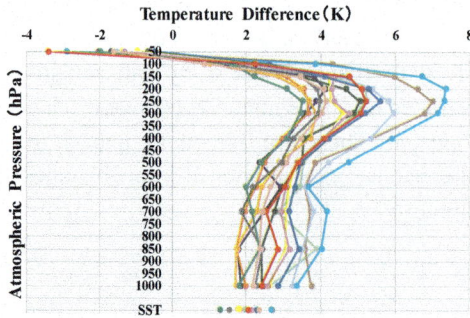

Fig. 1. Vertical profiles of the future changes of SST and T at 10.0°N, 128.0°E, derived from a total of 15 GCMs (The colors of solid lines correspond to Table 1).

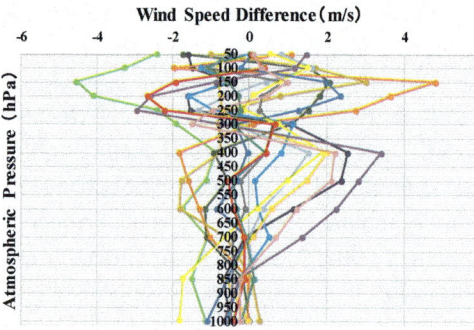

Fig. 2. Vertical profiles of the future changes of UV at 10.0°N, 128.0°E, derived from a total of 15 GCMs (The colors of solid lines correspond to Table 1).

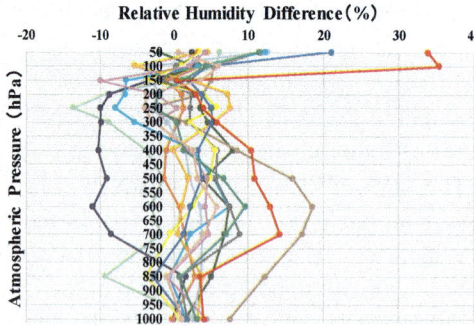

Fig. 3. Vertical profiles of the future changes of RH at 10.0°N, 128.0°E, derived from a total of 15 GCMs (The colors of solid lines correspond to Table 1).

environment, which may be a key factor to control the typhoon intensity. If the vertical wind shear is increased around a typhoon, it is more likely to reduce its intensity; on the other hand, if the vertical wind shear weaken around a typhoon, it is likely to enhance its intensity (Yoshino et al., 2009)[8]. Thus, the upper-level UV in GCMs contains the great uncertainties under the future climate, and it would affect the uncertainties of typhoon intensity projections considerably.

Figure 3 shows the vertical profiles of relative humidity (RH), derived from a total of 15 GCMs. The tropical troposphere under the future climate tends to be moisten, and the RH changes have a range of -10%~+20% at the middle troposphere. The standard deviation of the RH changes among GCMs is 6.62% at 600hPa. Such a RH increase at the middle troposphere is considered to inhibit an inflow of low equivalent potential temperature air with ventilation effect (Tang and Emanuel, 2010)[9], resulting in the enhancement of the typhoon intensity under the future climate. In addition, typhoon intensity is weakened since the increase of RH in the typhoon boundary layer decreases entropy flux from the sea surface (Bister and Emanuel, 2002)[6]. Thus, the future changes of RH is also expected to affect typhoon intensity changes and their uncertainties under the future climate.

3.2. *Present climate experiment and pseudo-global warming experiments*

We herein discuss the results of the present climate experiment (CNTRL) and pseudo-global warming experiments (ALL) which are used as the basis for comparing with the sensitivity experiments in the following sections.

Figure 4 shows the time series of the central pressure of Haiyan reproduced by HTM (CNTRL) and observed by Japan Meteorological Agency (JMA best track data). The minimum central pressure of Haiyan in CNTRL is 897.1 hPa at 5400 min from the start of the experiment, which is reasonable agreement with the observed value (895 hPa) in the JMA best track. According to the statistical verification, the mean bias error (BIAS) of the central pressures during the lifetime is 6.57 hPa, correlation coefficient (CORR) is 0.934, and root-mean-squared error (RMSE) is 13.6 hPa. As Toyoda et al. (2015)[3] suggested, it is obvious that CNTRL can realistically reproduces the intensity even of a super typhoon throughout the life cycle of Haiyan.

Figure 5 shows the time series of the central pressures of Haiyan simulated by CNTRL and ALL (based on 15 GCMs). According to the results of ALL, the ensemble-averaged value of the minimum central pressures of Haiyan under the future climate is 906.2 hPa at 5700 min from the start of the experiments. Comparing ALL with CNTRL, the averaged minimum central pressure of ALL is weakened by +11.2 hPa. Although most of GCMs in ALL indicates that the peak intensities closely match the peak intensity of CNTRL, the ensemble-averaged intensity is dragged into the extremely weakened typhoons, which are simulated by two GCMs, mpi_echam5 and miroc3_2_hires. It should be noted that the influences on typhoon intensity by the global warming strongly depends on the selection of GCM, because an extreme case in GCMs (i.e. gfdl_cm2_0) shows a

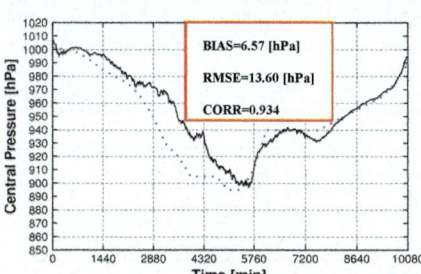

Fig. 4. Time series of the central pressures of Haiyan in CNTRL (solid line) and JMA best track (dotted line).

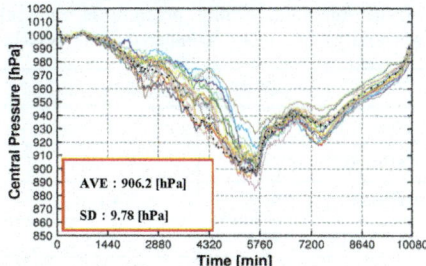

Fig. 5. Time series of the central pressures of Haiyan simulated in ALL (solid lines : the colors correspond to Table 1) and CNTRL (dotted line).

peak intensity of less than 885 hPa. In section 3.3, we will discuss which parameters of GWDs could largely alter the ensemble-averaged typhoon intensity under the future climate. Furthermore, the standard deviation of the minimum central pressure among GCMs is 9.78 hPa, which is almost twice larger than that (4.6 hPa) among GWSs shown in Toyoda et al. (2016)[5]. The standard deviation among GCMs is especially large during the mature stage of Haiyan, leading to the great uncertainties of typhoon intensity projections. In section 3.4, we will discuss which parameters of GWDs could degrade the reliability of the typhoon intensity changes estimated with GCMs.

3.3. *Mechanism of typhoon intensity change under the future climate*

We next discuss key factors of the typhoon intensity changes under the future climate, comparing the sensitivity experiments (CNTRL+SST, CNTRL+T, CNTRL+UV, and CNTRL+RH) with CNTRL. The comparisons of the ensemble-averaged typhoon intensities are summarized in Table 2.

Figure 6 shows the time series of the central pressures of Haiyan simulated by CNTRL+SST, in which the SST differences are added into the initial and boundary conditions of CNTRL. The ensemble-averaged value of the central pressure of Haiyan simulated by CNTRL+SST is 877.5 hPa at 5600 min, and the standard deviation of the central pressures among GCMs is 6.82 hPa. Comparing with the minimum central pressure (897.1hPa) in CNTRL, the ensemble-averaged minimum central pressure simulated by CNTRL+SST is intensified by about -19.6 hPa. Thus a SST rise under the future climate is responsible for increasing the typhoon intensity. It is clear from the MPI theory (Bister and Emanuel, 2002)[6] that the SST rise may increase entropy flux from the sea surface, and may accelerates to supply energy to the typhoon boundary layer.

Table 2. Contributions to the total future changes of the minimum central pressure in each of sensitivity experiments, compared with CNTRL.

	Average [hPa]	Contribution to CNTRL [hPa]
CNTRL	897.1	-
CNTRL+SST	877.5	-19.6
CNTRL+T	942.6	+ 45.5
CNTRL+UV	904.4	+ 7.3
CNTRL+RH	898.5	+ 1.4

Table 3. Contributions to the total standard deviations of the minimum central pressures in the sensitivity experiments, compared with ALL.

	SD [hPa]	Contribution to ALL [hPa]
ALL	9.78	-
ALL-SST	9.48	+ 0.30
ALL-T	7.54	+ 2.24
ALL-UV	8.99	+ 0.79
ALL-RH	15.31	- 5.53

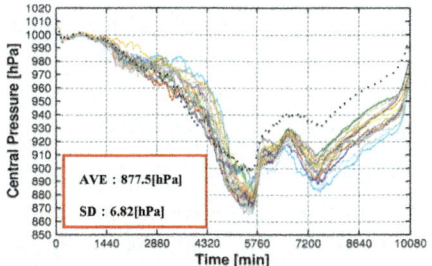

Fig. 6. Time series of the central pressures of Haiyan simulated in CNTRL+SST (solid lines : the colors correspond to Table 1) and CNTRL (dotted line).

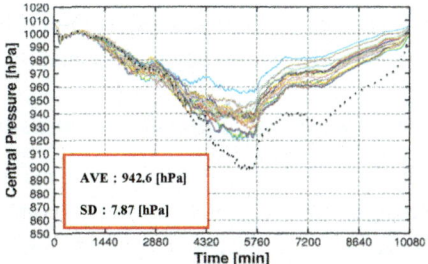

Fig. 7. Time series of the central pressures of Haiyan simulated in CNTRL+T (solid lines : the colors correspond to Table 1) and CNTRL (dotted line).

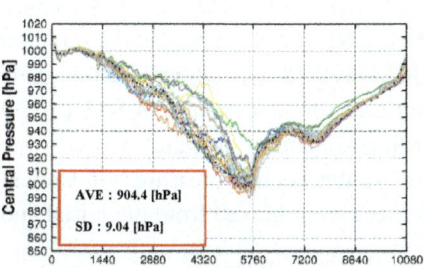

Fig. 8. Time series of the central pressures of Haiyan simulated in CNTRL+UV (solid lines : the colors correspond to Table 1) and CNTRL (dotted line).

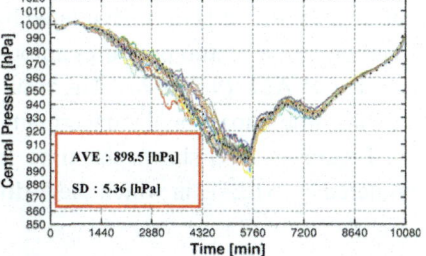

Fig. 9. Time series of the central pressures of Haiyan simulated in CNTRL+RH (solid lines: the colors correspond to Table 1) and CNTRL (dotted line).

Figure 7 shows the time series of the central pressures of Haiyan simulated by CNTRL+T, in which the T differences are added into initial and boundary conditions of CNTRL. The ensemble-averaged value of the central pressures of Haiyan simulated by CNTRL+T is 942.6 hPa at 5500 min, and the standard deviation of the central pressures among GCMs is 7.87hPa. There is a difference of the minimum central pressure of +45.5 hPa from CNTRL (897.1 hPa), causing the remarkable decay of Haiyan's peak intensity. In contrast to the results of CNTRL+SST, a T rise under the future climate is accountable for weakening typhoon intensity. According to the MPI theory (Bister and Emanuel, 2002)[6], a T rise in the upper troposphere (see in Figure 1) suppresses typhoon intensity during the mature stage. The intensification by SST (about -20 hPa) is dominated by the decay by T (about +45hPa). Therefore, it is considered that the future-climate typhoons simulated by ALL become weaker than the present-climate typhoon simulated by CNTRL.

Figures 8 and 9 show the time series of the central pressures of Haiyan simulated by CNTRL+UV and CNTRL+RH, respectively. Both CNTRL+UV and

CNTRL+RH slightly weaken the typhoon intensity by +7.3hPa and +1.4hPa, respectively, while they have relatively smaller impacts than SST and T.

3.4. *Uncertainties of typhoon intensity change under the future climate*

Finally we discuss key factors causing the uncertainties of typhoon intensity changes by the global warming, comparing the sensitivity experiments (ALL-SST, ALL-T, ALL-UV, and ALL-RH) with ALL. The comparisons of the standard deviations of the minimum central pressures are summarized in Table 3.

Figure 10 shows the time series of the central pressures of Haiyan simulated by ALL-SST, in which the SST differences are removed from the initial and boundary conditions of ALL. The ensemble-averaged value of the minimum central pressures of Haiyan simulated by ALL-SST is 939.7 hPa, and the standard deviation of the central pressures among GCMs is 9.48 hPa, which is comparable with that of ALL (9.78 hPa). The result indicates the uncertainty of SST (+0.30 hPa) has little impact on the uncertainty of future intensity change.

Figure 11 also shows the time series of the central pressures of Haiyan simulated by ALL-T, in which the T differences are removed from the initial and

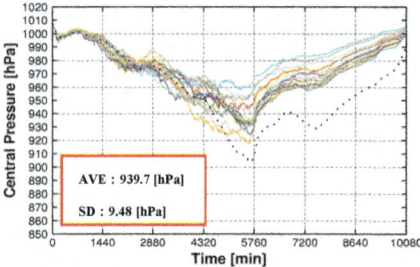

Fig. 10. Time series of the central pressures of Haiyan simulated in All-SST (solid line : the colors correspond to Table 1) and ALL (dotted line).

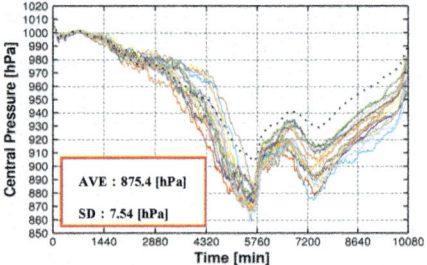

Fig. 11. Time series of the central pressures of Haiyan simulated in All-T (solid line : the colors correspond to Table 1) and ALL (dotted line).

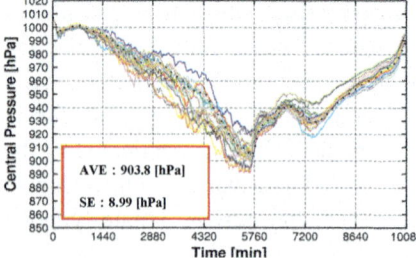

Fig. 12. Time series of the central pressures of Haiyan simulated in All-UV (solid line : the colors correspond to Table 1) and ALL (dotted line).

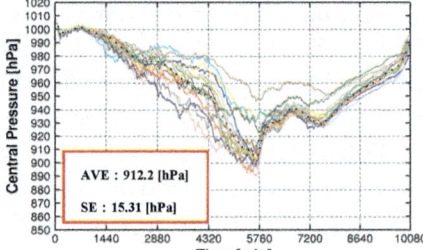

Fig. 13. Time series of the central pressures of Haiyan simulated in All-RH (solid line : the colors correspond to Table 1) and ALL (dotted line).

boundary conditions of ALL. The ensemble-averaged value of the minimum central pressures of Haiyan is 875.4 hPa, and its standard deviation of the central pressures among GCMs is 7.54 hPa. It is revealed that the uncertainty of T in the future climate contributes to the increase of the standard deviation by about +2.24 hPa, compared with ALL (9.78 hPa). According to Figure 1, the T differences especially in the upper troposphere may show the most significant contribution in all climate parameters to a reduction of the reliability for future typhoon projections.

Figure 12 shows the time series of the central pressures of Haiyan simulated by ALL-UV, in which the UV differences are removed from the initial and boundary conditions of ALL. The ensemble-averaged value of the minimum central pressures of Haiyan is 903.8 hPa, and its standard deviation among GCMs is 8.99 hPa, which is a second major contributor to that of ALL (9.78 hPa). The uncertainty of UV also accounts for +0.79 hPa of the total uncertainties on typhoon intensity change.

Figure 13 shows the time series of the central pressures of Haiyan simulated by ALL-RH, in which the RH differences are removed from the initial and boundary conditions of ALL. The ensemble-averaged value of the minimum central pressures of Haiyan is 912.2 hPa, and its standard deviation among GCMs is 15.31 hPa. It is also revealed that the uncertainty of RH has little impacts on the total uncertainty of the typhoon intensity changes (-5.53 hPa), compared with T and UV.

4. Conclusion

In this study, in order to examine the mechanism of typhoon intensity changes under the future climate and to quantify the uncertainties of the future intensity changes, a number of sensitivity experiments on typhoon intensity have been made using HTM forced by global warming differences (GWDs) derived from a total of 15 GCMs. Typhoon Haiyan (2013), which caused a huge storm surge at Tacloban in the Philippines, was selected as a well-mature typhoon case under the present climate. The GWDs used in this study were composed of sea surface temperature (SST), air temperature (T), wind speed (UV), and relative humidity (RH).

The sensitivity experiments indicated that a SST rise was responsible for intensifying future-climate typhoons by about -19.6hPa while a T rise in the upper atmosphere was accountable for weakening typhoon intensity by about +45.5hPa. Although UV and RH also weakened typhoon intensity slightly, they had relatively less impacts than SST and T. Furthermore, both T and UV tended to increase the uncertainties of typhoon intensity projections, while both SST and

RH had less effect on the total uncertainties of typhoon intensity changes than T and UV.

It is thought that these results obtained in this study are only valid when a well-matured typhoon occurs near the Philippines in November. Future work will investigate the other regions, months, and categories applying the same method.

Acknowledgments

This study was supported by JSPS KAKENHI Grant (16K00526).

References

1. K. Emanuel, Increasing destructiveness of tropical cyclones over the past 30 years, *Nature*, Vol.436, pp.686-688, (2005).
2. IPCC, 2013: *Climate Change 2013: The Physical Science Basis.*, Cambridge University Press, 1535p., (2013).
3. M. Toyoda, J. Yoshino, S. Arakawa,, and T. Kobayashi, Numerical experiments of Typhoon Haiyan (2013) and its storm surge using a high-resolution coupled typhoon-ocean model, *JSCE B2(Coastal Engineering)*, Vol. 71, No.2, pp. I_463-I_468, (in Japanese, 2015).
4. J. Yoshino, S. Arakawa, M. Toyoda, and T. Kobayashi, Intercomparison of global warming scenarios for typhoon intensity change using a high-resolution typhoon model, *JSCE B2 (Coastal Engineering)*, Vol. 71, No.2, pp. I_1519-I_1524, (in Japanese, 2015).
5. M. Toyoda, J. Yoshino, and T. Kobayashi, Ensemble future projections of storm surge in the leyte gulf, Philippines, by pseudo-global warming experiments, *JSCE B2 (Coastal Engineering)*, Vol.72, No.2, pp. I_483-I_488, (in Japanese, 2016).
6. B. Bister, and K. A. Emanuel, Low frequency variability of tropical cyclone potential intensity. 1. Interannual to interdecadal variability, *J. Geophys. Res.*, Vol.107, 4801, (2002).
7. J. Dudhia, A nonhydrostatic version of the Penn State-NCAR mesoscale model: Validation test and simulation of an Atlantic cyclone and cold front, *Mon. Wea. Rev.*, Vol.121, pp.1493-1513, (1993).
8. J. Yoshino, M. Yoshida, S. Iwamoto, T. Murakami, and T. Yasuda, Development of a high-resolution typhoon intensity forecasting model and its modification, *JSCE B2 (Coastal Engineering)*, Vol.65, No.1, pp.1261-1265, (in Japanese, 2009).
9. B. Tang and K. A. Emanuel, Midlevel ventilation's constraint on tropical cyclone intensity, *J. Atmos. Sci.*, Vol.67, pp.1817-1830, (2010).

Hindcasting of Wave Climate along the Pacific Coast of Japan in October 2014

Shinsaku Nishizaki[†], Tomoya Shibayama, and Tomoyuki Takabatake

Department of Civil and Environmental Engineering, Waseda University
4-1 Okubo 3, Shinjuku-ku, Tokyo 169-8555, JAPAN
[†]E-mail: shinsaku-nisshi@fuji.waseda.jp

Ryota Nakamura

Department of Architecture and Civil Engineering, Toyohashi University of Technology
1-1 Hibarigaoka, Tempaku, Toyohashi, Aichi 441-8580, JAPAN

One-month wave simulations were performed from Oct. 1st to Nov. 1st in 2014. The one-way numerical models composed of WRF (Weather Research and Forecasting) and SWAN (Simulating WAve Nearshore) were used. This research focused on evaluations of both a resolution of wind data and a difference between structured SWAN (STSWAN) and unstructured SWAN (UNSWAN). The results show some differences among case studies. High resolution wind data affect peak values of significant waves at typhoon periods. In addition, the overestimation of swell caused by typhoons decreased at the nearshore points, in particular at Shimoda by using UNSWAN. As a conclusion, there is little difference between STSWAN and UNSWAN on the overall simulation of wave climate. However, UNSWAN is useful for simulating nearshore waves and has possibility to predict wave climate with higher precision than STSWAN.

Keywords: one-month wave simulation; STSWAN; UNSWAN; significant wave; swell.

1. Introduction

Long term wave hindcasting plays an important role to design coastal structures. The improvement of a decision making process by coastal managers regarding coastal dykes and breakwaters is an important example.

Numerical simulations have been conducted with unstructured meshes in coastal engineering field. For example, FVCOM can simulate the phenomena of storm surge.[1] Zijlema[2] presented an unstructured SWAN, Simulating WAve Nearshore model.[3] It was installed in SWAN+ADCIRC model, simulating waves and tides caused by hurricanes and the results agree well with the observed data.[4] After this research, the application of unstructured meshes for simulating waves are gradually increasing. Some researchers compared the quality of the results between regular grids and unstructured grids and concluded that there is a little difference between them. The validity of unstructured SWAN was confirmed in the studies mentioned above, however, there are a few wave simulations containing the periods of huge waves caused by typhoons in the Pacific Ocean.

In this research, one-month wave simulation was carried out, focusing on evaluations of both a resolution of wind data and a difference between structured and unstructured SWAN; hereafter, they are called as STSWAN and UNSWAN respectively.

2. Methodology

2.1. *Target period and calculation area*

Simulations of wave climate were performed from Oct. 1st to Nov. 1st in 2014. Two typhoons, namely, Phanfone and Vongfong, passed on the Pacific Coast of Japan. The details of the typhoons are shown in Table 1. They were generated around latitude 10° N, longitude 150° E in the Pacific Ocean during the period. After passing the Pacific Ocean for northwest direction, the typhoons changed the way toward northeast direction and approached the Pacific Ocean side of Japan. When typhoon Phanfone landed on main land of Japan, it maintained the strong intensity of central pressure, 965hPa.

Table 1. Detailed information of typhoons in October 2014.

Information	PHANFONE	VONGFONG
Birth	2014-09-29 06:00:00 UTC	2014-10-03 18:00:00 UTC
Death (Latest)	2014-10-06 12:00:00 UTC	2014-10-14 00:00:00 UTC
Lifetime	174 (hours) / 7.25 (days)	246 (hours) / 10.25 (days)
Minimum Pressure	935 (hPa)	900 (hPa)

2.2. *Settings of WRF model*

SWAN requires x and y component of wind as an input for creating waves. The wind data was created by WRF, Weather Research and Forecasting[5] at 20 minutes intervals. Two domains were set: one is a parent domain of 9.9 km grids and the other is a nested domain of 3.3 km grids (Fig. 1). FNL was used as initial and boundary conditions of meteorological data. WRF Single-moment 3-class Scheme, is used for micro physics. RRTMG Shortwave and Longwave

Schemes is adopted as radiation physics of shortwave and longwave. Yonsei University Scheme is employed as planetary boundary layer physics.

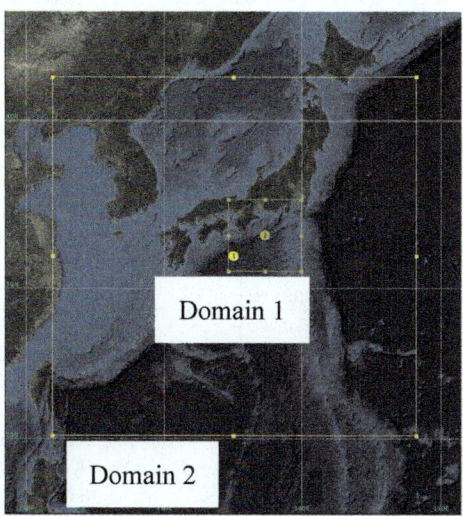

Fig. 1. Calculation domains of WRF. (Source: WRF Domain Wizard)

2.3. Settings of SWAN model

Wave simulations were performed with SWAN (Version:41.10). Calculation conditions were described in Table 2. Regular grids were available for STSWAN and unstructured grids were used for UNSWAN. The parent and nested grid resolutions of STSWAN were 0.05° and 0.02° respectively. The resolution of unstructured grid was almost 0.02° through 0.5°. Triangle elements of 0.5° resolution were set along with the water boundary. The area of triangles gradually shrimped according to the coastline and the resolution of the elements came to 0.02° around the nearshore area. The domain of UNSWAN is shown in Fig. 2.

2.4. Comparisons with NOWPHAS

The results of the wave simulation were compared with observations of NOWPHAS (Nationwide Ocean Wave information network for Ports and HArbourS[6]) on the Pacific nearshore in Japan. This network has 77 observation points around Japanese coast in 2014 and measures wave heights and periods of mean, significant, and maximum waves. Observed data are available every 2 hours at 74 points. In addition, 20 minutes interval records are obtained at 68 points, which has advantage to grasp the sudden change of significant wave heights and periods and contributes to reproduce wave climate more precisely. This research targeted the waves in Tokai area, where there are many 20 minutes interval observation points. In addition, typhoon Vongfong passed on this area. The investigated points are listed in Table 3.

Table 2. Calculation condition of UNSWAN and STSWAN.

Calculation Period (UTC)			10/01/2014 00:00 ~ 11/01/2014 00:00	
UNSWAN	Number of node		95,245	
	Number of element		188,413	
	Maximum grid resolution		0.47790 °	
	Minimum grid resolution		0.00280 °	
STSWAN	Area	domain 1	123.0E ~ 145.0E	21.0N ~ 38.0N
		domain 2	135.0E ~ 139.5E	31.5N ~ 35.3N
	resolution	domain 1	0.05 °	440 × 340
		domain 2	0.02 °	225 × 190
Calculation mode			Non-stationary / 2 dimension	
Whitecapping			Komen	
Direction division number			36 (θ=10°)	
Frequency division number			40 (0.031384 ~ 1.4204 Hz)	
maximum number of iterations			5	
time step			5 min	
Bathymetry data			GEBCO2014	

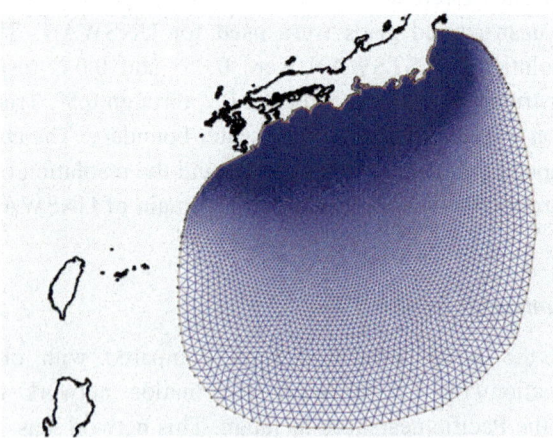

Fig. 2. Calculation area of UNSWAN

Table 3. Wave observation points of NOWPHAS.

Point name	Type	Depth	Latitude (N)	Longitude (E)
(a) Shionomisaki	Submerged	54.7	33° 25' 59"	135° 44' 50"
(b) Mie Owase offshore	GPS	210	33° 54' 08"	136° 15' 34"
(c) Ise bay mouth offshore	GPS	90	34° 22' 28"	137° 07' 29"
(d) Omaezaki	Submerged	22.8	34° 37' 17"	138° 15' 33"
(e) Shizuoka Omaezaki offshore	GPS	120	34° 24' 12"	138° 16' 30"
(f) Shimoda	Submerged	51.1	34° 38' 48"	138° 57' 11"

2.5. Case studies

The results are compared from the viewpoint of grid type, nesting availability and resolution of wind data. In this research, 4 case studies are carried out (Table 4). The simulation of Case A was performed with UNSWAN. Case B shows the results of parent domain of STSWAN. Case C and D show the results of nested domain of STSWAN with different resolution of wind data.

Case A and C show the difference between unstructured and regular grid type. The differences between Case B and C show the effect of nesting availability. The results of Case C and D show the effect of the difference of wind data resolution.

Table 4. SWAN simulation cases.

Case	Grid Type	Resolution	Nesting	Wind data
A	unstructured	0.02°~0.5°	No	9.9km
B	regular	0.05°	Parent	9.9km
C	regular	0.02°	Nest	9.9km
D	regular	0.02°	Nest	3.3km

3. Results of Calculations

One-month wave simulation was conducted at 6 points. The results of significant wave heights and periods were compared with wave observation data of NOWPHAS. In this research, time series of significant wave heights and periods are shown at two points: (b) Mie Owase offshore and (f) Shimoda in Fig. 3 and Fig. 4 respectively. (b) Mie Owase offshore is the representative of offshore points and (f) Shimoda is that of nearshore points.

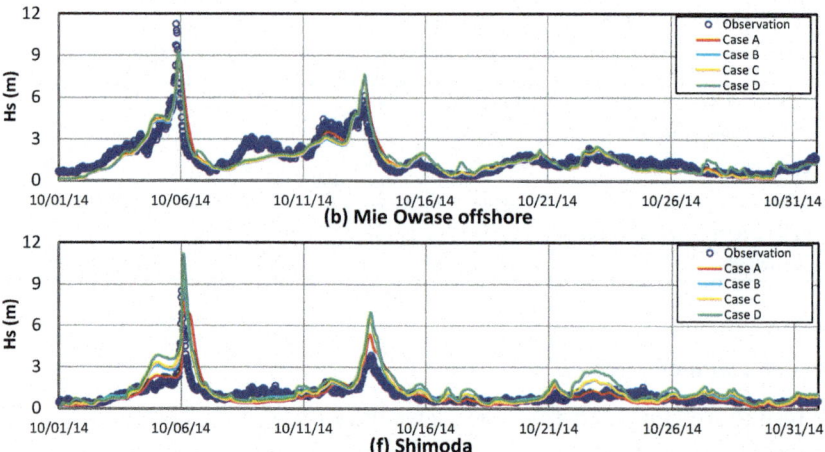

Fig. 3. Time series of significant wave heights at (b) Mie Owase offshore and (f) Shimoda.

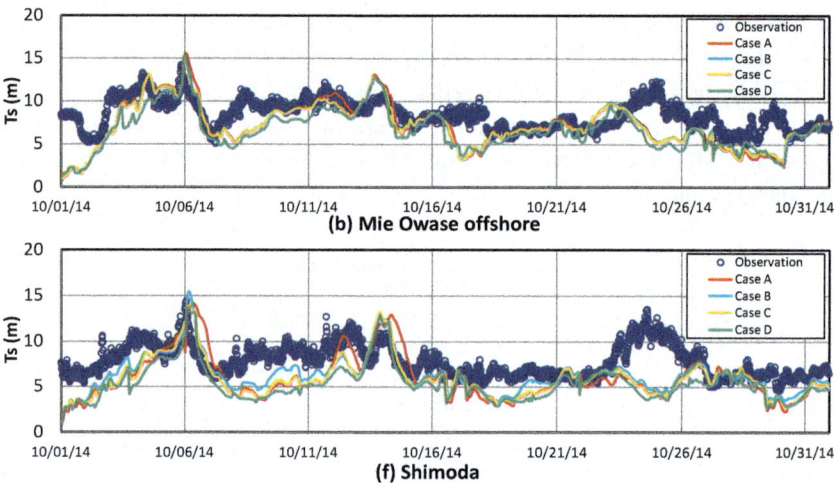

Fig. 4. Time series of significant wave heights at (b) Mie Owase offshore and (f) Shimoda.

Significant wave heights at each case were good agreement with the observation data at (b) Mie Owase offshore. Time series of significant wave heights were almost the same among 4 cases but a few overestimations were found in Case D. On the other hand, some differences at (f) Shimoda were found, in particular, during the specific events of typhoons. The simulation results suddenly increased before reaching the peak values around 4th through 5th in October. The amount of overestimation was smallest at Case A and biggest at Case D. Case B and C have middle overestimation and little difference between them. In addition, focusing on the peak values at typhoons, the results using

STSWAN of Case B, C and D get higher values compared with the results of Case A using UNSWAN. When STSWAN was used, the both of the peak values were overestimated. On the other hand, using UNSWAN, one is overestimated and the other is underestimated. The simulation results also have some overestimation periods in the latter half of October 2014. The amount of overestimations is the biggest in Case D, the second biggest in Case B and C, and the smallest in Case A. Comparing the time series of significant wave periods, the results are almost the same as each other and there is little difference among them. However, the use of high resolution 3.3 km wind data makes the periods shorter than that of wind data with 9.9 km resolution. Regarding of the reproduction of significant wave periods using STSWAN and UNSWAN, relatively longer waves appeared in typhoon periods were better agreement with the observation data. On the other hand, the periods of usual waves are somehow underestimated in all cases. The reproductions of significant wave periods are less accurate than those of significant wave heights.

4. Discussions

Comparing the simulation results around the period when the typhoon Phanfone reached Japanese coast, there are two peak values of significant wave heights (Fig. 3). One is the maximum significant wave height and the other is relatively small peak compared with the maximum height appeared later. This sometimes occurred because of the early arrival of typhoons' swell. It might lead to overestimated simulation and actually occurred due to this reason at the point of Shimoda. Focused on the amount of swell overestimation, using unstructured grids (Case A) minimize it. STSWAN simulations overestimated more than UNSWAN simulation. As high resolution wind data have the larger area of strong wind field and it makes source of wave growth increase, comparison of the results in Case C and D found that high resolution wind data caused the more overestimation. Furthermore, comparing the results at point (b) Mie Owase offshore with those at the nearshore point, (f) Shimoda, the height of the first peak increased and sudden change became milder. However, there is little difference among 4 cases investigated in this study. The offshore points are located at deep water area and the effects of the bathymetry change and bottom friction can be ignored. The same tendency is appeared at the other offshore and nearshore points. That means as the points approaching coastline, the effects become larger and they are considered well enough in UNSWAN.

The maximum significant wave heights caused by typhoons are extracted at the periods in the former half of the month (Table 5). Comparing the values at each case with observed heights, the values of Case A using UNSWAN are relatively small or underestimated. However, the values of Case B, C and D using STSWAN are greater than those of Case A and sometimes overestimated the observed values. At almost all the points, the values using high resolution wind (Case D) are the most biggest. The differences in case studies are smaller at the offshore points (b), (c) and (e).

Table 5. Maximum significant wave height caused by typhoons in October 2014.

Point name	PHANFONE					VONGFONG				
	Observation	Case A	Case B	Case C	Case D	Observation	Case A	Case B	Case C	Case D
(a) Shionomisaki	10.92	8.73	9.19	8.97	9.62	7.41	6.88	8.10	8.08	8.54
(b) Mie Owase offshore	11.27	8.69	8.59	8.92	9.30	6.14	7.31	7.19	7.25	7.66
(c) Ise bay mouth offshore	11.33	7.94	8.36	8.51	8.99	6.03	7.67	7.70	7.80	7.75
(d) Omaezaki	4.54	3.42	5.85	5.98	6.18	3.72	5.00	5.01	4.95	5.06
(e) Shizuoka Omaezaki offshore	15.85	9.10	10.67	10.65	12.01	6.89	7.13	7.65	7.63	7.67
(f) Shimoda	8.05	7.74	10.14	10.41	11.19	3.85	5.37	6.44	6.57	6.99

Except Case A at Shionomisaki, the simulations are overestimated during the period of typhoon Vongfong at every point. The gap between observation and simulations depends on the quality of wind input as well as the mesh size of the grids. Wind data quality could be poor and it makes results worse.

The relations between the observed and calculated values are investigated and the correlation coefficient (R^2) and the slope value of approximate line are compared at each case (Fig. 5). A lot of dots apart from approximate line are found in this figure and the values of correlation coefficient are less than 0.8, ranging from 0.63 through 0.78. They occurred in the high wave periods during the typhoons' approach. In the WRF simulation, the reproduced tracks of the typhoons are almost same compared with the observation data. However, the arrival times to Pacific coast of Japan were delayed. This is the reason of getting the correlation coefficient with low precision.

Focused on the values of the slope of approximate line, they are almost same between Case B and C. In addition, the differences of the maximum significant wave heights between Case B and C, ranging from 0.02 m through 0.33 m but mainly less than 0.15 m, are small enough (Table 5). This shows that the grid resolution of 0.05° is enough for the oceanic wave simulation of STSWAN. In Case D, the values of slope were the best at Mie Owase offshore but the worst at Shimoda. It suggests high resolution wind distribution doesn't necessarily contribute to the accurate calculation of wave climate.

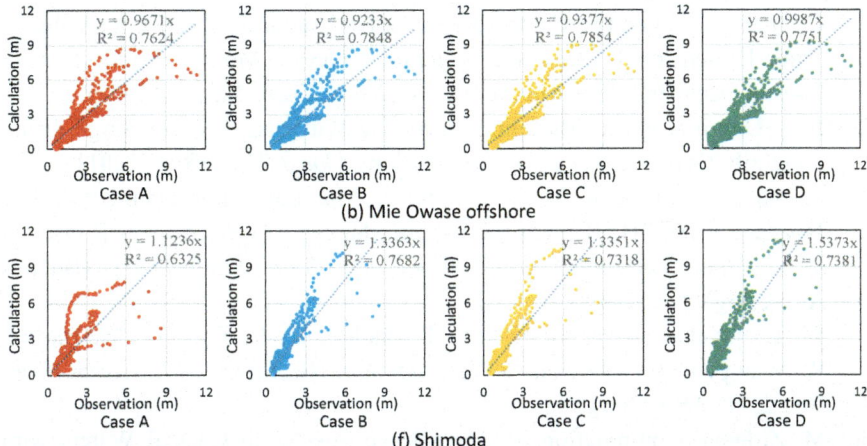

Fig. 5. Correlation coefficient and slope value of approximate line about significant wave heights.

5. Conclusions

One-month wave simulation was carried out in order to confirm the difference between UNSWAN and STSWAN by changing the grid size and the resolution of wind distribution. The results in this study show small difference among case studies because of sufficient grid resolutions as each other. However, the peak significant wave heights simulated with UNSWAN were smaller than those with STSWAN at most of the targeted points. In addition, by using UNSWAN, the overestimation of swell caused by typhoons decreased at the nearshore points, in particular at Shimoda. UNSWAN can adapt its unstructured computational meshes to the real complex coastlines and geometries. This led to the consideration of adequate energy reduction around coast. It means that UNSWAN is useful for simulating nearshore waves and it has the possibility to predict wave climate with higher precision than STSWAN.

Acknowledgments

This work was financially supported by the Strategic Research Foundation Grant-aided Project for Private Universities from Ministry of Education (Waseda University, No. S1311028).

References

1. N. Booij, R. C. Ris and L. H. Holthuijsen, A Third‐Generation Wave Model for Coastal Regions: 1. Model Description and Validation, *J. Geo. Res: Oceans* **104**, 7649 (1999).

2. C. Chen, H. Liu and R. C. Beardsley, An Unstructured Grid, Finite-Volume, Three-Dimensional, Primitive Equations Ocean Model: Application to Coastal Ocean and Estuaries, *J. Atmos. Ocean. Tech.* **20**, 159 (2003).
3. J. C. Dietrich, et al., Modeling Hurricane Waves and Storm Surge Using Integrally-Coupled, Scalable Computations, *Coast. Eng.* **58**, 45 (2011).
4. T. Nagai, K. Shimizu, M. Sasaki and A. Murakami, Improvement of The Japanese NOWPHAS Network by Introducing Advanced GPS Buoys, In *ISOPE Int. Conference on Offshore and Polar Engineering*, (Buenos Aires, Argentina, 2008).
5. W. C. Skamarock, J. B. Klemp, J. Dudhia, D. O. Gill, D. M. Barker, W. Wang and J. G. Powers, *A Description of the Advanced Research WRF*, Ver. 2, NCAR (2005).
6. M. Zijlema, Computation of Wind-Wave Spectra in Coastal Waters with SWAN on Unstructured Grids, *Coast. Eng.* **57**, 267 (2010).

Willingness to Pay for Upgrading Tsunami Co-Beneficial Structures: Example from A Railway Embankment and A Revetment in Sri Lanka

R.S.M. Samarasekara

*GPSS-GLI, the University of Tokyo,
Kashiwa, Chiba, 277-8561, Japan
E-mail: uomsameera@gmail.com*

J. Sasaki

*Department of Socio-Cultural Environmental Studies, the University of Tokyo
Kashiwa, Chiba, 277-8561, Japan
E-mail: jsasaki@k.u-tokyo.ac.jp*

M. Esteban and H. Matsuda

*GPSS-GLI, the University of Tokyo,
Kashiwa, Chiba, 277-8561, Japan
E-mail: esteban.fagan@gmail.com, matsuda@k.u-tokyo.ac.jp*

Increasing the tsunami disaster resilience of the Southwestern coastline of Sri Lanka is a pressing problem, due to the continuous presence of unprotected human settlements in tsunami-prone areas. Even though a variety of different tsunami countermeasures can be attempted, due to budgetary limitations early warning systems are typically used. Some types of coastal structures (such as coastal railway embankments, revetments etc.) have the potential to mitigate the impact of tsunami, which is often overlooked in research. The engineering resilience of these structures needs to be improved if they are to withstand a tsunami, though upgraded structures can offer a multitude of co-benefits to residents. This research assesses resident's willingness to pay (WTP) for hard defensive measures, as well as the socioeconomic factors that influence residents' WTP. WTP of residents to upgrade a coastal railway embankment and a revetment in Dimbuldooa and Wenamulla villages was measured by conducting a structured questionnaire survey of 200 residents. The results of the survey were triangulated through five expert interviews with representatives of government agencies, construction companies and academia, and two focus group discussions with residents. The findings suggest that it is necessary for disaster risk managers to pay special attention to socioeconomic factors to successfully enhance the resilience of community.

Keywords: Tsunami; WTP; Sri Lanka; Co-beneficial Structures.

1. Introduction

The South and Eastern coasts of Sri Lanka suffered heavy damage from the 2004 Indian Ocean Tsunami (IOT), with around 35,000 – 40,000 casualties [1] and 700 Million USD (United States Dollars) of property losses [2]. When a physical structure situated along the coast has the potential to mitigate the damage of a tsunami, but its main intended design function was not to act as a tsunami hard measure, it can be said to have tsunami mitigation co-benefits [3,4]. This paper investigates the Willingness to Pay (WTP) of a local community in Southern Sri Lanka for upgrading a railway embankment and a revetment as a tsunami co-beneficial structure 11 years after the IOT.

The WTP approach has already used by some developed countries to ascertain public opinion on public works [5–7]. However, such approaches are rarely used for developing regions where national budgetary allocation and public education limited. Hence, the tsunami awareness and preparedness of one community in Sri Lanka was measured, in addition to the community WTP to upgrade in monetary terms. The contribution of socio-economic features to WTP was obtained by creating a linear WTP model, with the authors finally proposing some recommendations to improve the disaster management in the area.

2. Survey design and implementation

2.1. *Study site*

Dimbuldooa and Wenamulla villages (See Figure 1) are situated within Hikkaduwa city. 740 fatalities were recorded in Hikkaduwa city due to the 2004 IOT (amounting to 2% and 4% of the inhabitants in Dimbuldooa and Wenamulla, respectively). In addition, 65% (out of 300) of the houses in Wenamulla and 68%

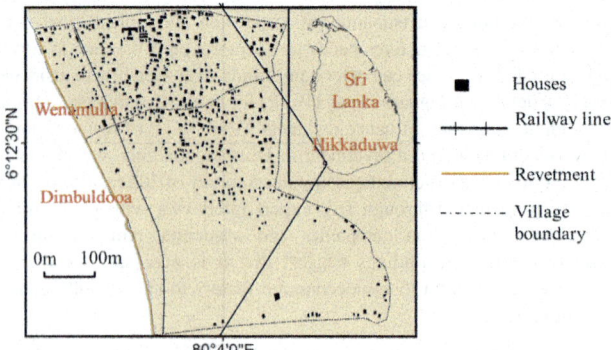

Fig. 1. Location of Dimbuldooa and Wenamulla villages and spatial extent of the railway embankment and revetment.

(out of 245) of the houses in Dimbuldooa were heavily damaged by the tsunami. The tsunami damaged the coastal railway embankment and revetment, though these were subsequently rebuilt and improved, as shown in Figure 2.

Fig. 2. (Left) Picture of an underpass through the railway embankment (2015, September), (Right) Picture of the revetment (2015, September).

2.2. *Questionnaires*

The authors administered a household level structured questionnaire survey randomly within the extent of the area inundated by the 2004 IOT to evaluate the WTP to upgrade the coastal railway embankment and revetment. This questionnaire consisted of four parts, Part 1: Questions 1 – 3: Respondent's general information, Part 2: Questions 4 – 9: Respondent's awareness of the benefits of the selected co-beneficial structure (railway embankment or revetment), Part 3: Questions 10 – 12: Respondent's WTP to upgrade the selected co-beneficial structure and Part 4: Questions 13 – 20: Respondent's awareness of tsunami risk.

The questionnaire survey differed slightly depending on whether it targeted the perception of respondents to the railway embankment or the revetment, though each house was only distributed one type of questionnaire in order to avoid any potential confusion. Drawings of one possible upgraded structure and the cost of the upgrade were shown to respondents, who were asked what percentage of the total cost of the structure that they could help to pay for in one year . The reason behind the selection of short timeframe (i.e. one year) for the question is the given flexibility to respondents to select a convenient percentage from the total upgrading cost due to economic uncertainties in Sri Lanka as a developing country. Hence authors assumed that a year was a reasonable timeframe.

The authors collected 200 valid samples (= n) (104 for the revetment and 96 for the railway embankment) during the daytime of the period between 12th and 20th September 2015. 192 samples (out of total of 364 houses that existed in September (2016) in both villages within the boundary of the inundation of 2004 IOT) were required to achieve 90% precision [8].

It is important to understand which factors influence WTP in order to suggest policy recommendations to improve tsunami preparedness. The authors assumed that WTP was dependent on nine independent parameters, and conducted a logistic regression [9] to verify their hypothesis.

The authors asked whether respondents lived (1) or not (0) in the area, whether they agreed with the government policy on disaster management (1) or not (0), whether they had witnessed the 2004 Indian Ocean Tsunami (1) or not (0), whether they trusted the Early Warning System (EWS) of Sri Lanka to save their lives (1) or not (0), whether they had participated in a tsunami drill at least once (1) or not (0), whether they were aware of safe places and safe routes to evacuate (1) or not (0) and whether they could pay more than the mean willingness to pay for public works where the government and communities work together (= μ) (1), all of which were considered as the main independent variables in logistic regression.

The residents have prior experience for such public works as water supply and road development projects. How much respondents estimated that they could pay in a month during a one year period to a given community project where both government and community invest together for a construction project was defined as the willingness to pay for public works (WTP_{PW}) in this study. The definition of mean WTP_{PW} (= μ) is given in Eq. (1). Empirical estimates prove that WTP in annual basis is 70% higher than the monthly basis [8] even though theoretical models show that the effect of time framing is negligible. Authors limited the payment duration to one year and asked monthly WTP_{PW} to minimize errors due to the economic risk in Sri Lanka. Moreover it was common to conduct WTP questionnaires in monthly basis in past water supply and sanitation projects in Sri Lanka [10–12]. The residents have prior experience for such public works as water supply and road development projects. The WTP_{PW} term is associated with respondent's willingness to pay based on their household income. Authors inquired about the WTP_{PW} before asking WTP by showing and explaining the drawings (and pictures) of proposed upgrades to capture two different decisions. One decision was WTP_{PW} which captured each respondents general attitude to contribute (in monetary terms) to public work based on their income and awareness of past similar projects (which are beneficial only for a certain community including themselves) and the other decision was WTP which captured respondents' attitude after perceiving the method of upgrading of railway embankment and the revetment.

$$\mu = \frac{\sum \text{WTP}_{\text{PW}}}{n} \qquad (1)$$

Housewives and retired respondents were not considered as having an occupation. Therefore, unemployed respondents, retired respondents and housewives were considered as 0 and other respondents were considered as 1 in the logistic regression.

Enhancement of structural resistance, such as seaward toe protection and lining against tsunami flow, can be considered as possible upgrades [4]. Figure 3 was shown to respondents to describe the proposed upgrades. Surface lining, soil improvement and introduction of a reinforced concrete block (toe protection) for railway embankment and surface lining and an interlocking block layer (instead of rock boulders) were proposed for the case of the revetment. Two photos of lining and interlocking blocks were shown to respondents (as shown in Figure 3).

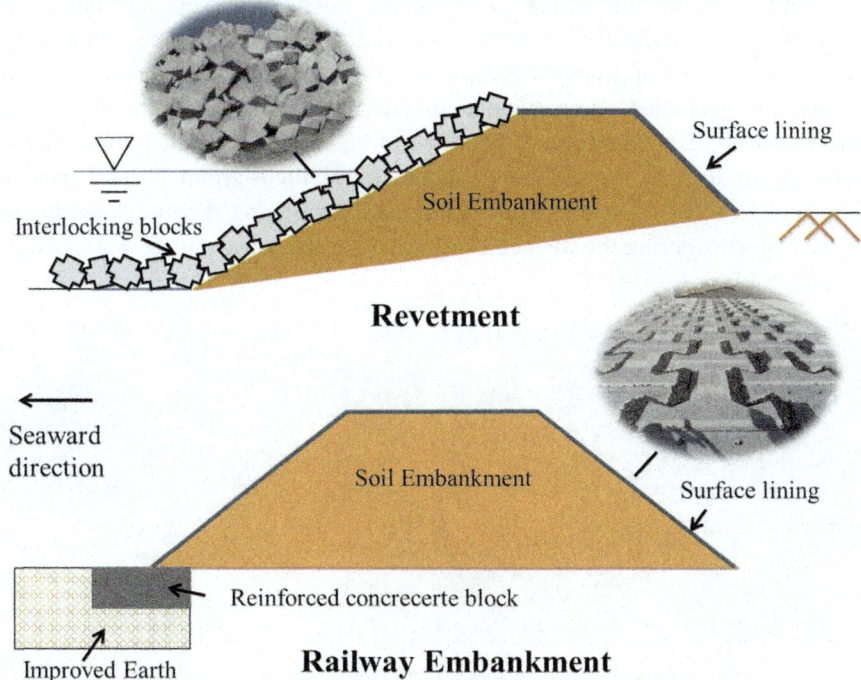

Fig. 3. Proposed upgrades for the revetment (top) and railway embankment (bottom) shown in Figure 2.

2.3. *Interviews and focus group discussions*

Key informant interviews were also conducted to clarify the results of the questionnaire survey. The experts were interviewed in September 2016 and their affiliations were shown in Table 1.

Table 1. Interviewed experts and their affiliation (in 2015, September).

Interviewed experts	Affiliation
Designers of revetments	Regional Engineers Office (Galle), Coast Conservation & Coastal Resource Management Department
Leading tsunami researchers in Sri Lanka	University of Moratuwa
Practitioners in, Early Warnings in Sri Lanka	Disaster Management Center (DMC), Ministry of Disaster Management, Sri Lanka
Designers of railway embankments	Southern Railway Development Project, Central Engineering Consulting Bureau, Sri Lanka
Practitioners of Early Warning in Indian Ocean	Indian Ocean Tsunami Early Warning System (IOTWS)

The significance of the results of the questionnaire surveys and interviews were further triangulated through the organization of two focus group discussions [13]. In the context of society in Sri Lanka, women in general are reluctant to give an opinion when men also participate in a discussion. However, the majority of questionnaire respondents were housewives and thus the women's opinion in focus group discussions cannot be ignored. Hence, focus group discussions were gender-segregated. Four men and six women (see Figure 4) participated in two separate events during the weekends in which the study was carried out [14].

Fig. 4. Focus group discussions with men (left) and with women (right).

3. Results and Discussion

3.1. *Awareness of tsunami risk and tsunami co-beneficial structures*

98% (out of 200) of respondents were residents of the case study area and the other 2% (out of 200) were local businessmen who own shops, restaurants, workshops etc. in the case study area (but did not live in the area). 98% (out of 364) of buildings were houses and rest 2% (out of 364) was commercial buildings. Authors assumed that the ratio of houses to commercial buildings was equal to the ratio of residents to nonresidents because of the total number of nonresidents was not available for the study site. The maximum possible number of nonresidents (or commercial buildings) is 6 in study site and 4 of them were included in the sample Hence the ratio of residents to non-residents in this sample quite fairly represents the study site. Approximately half of the respondents were composed of housewives and retired people.

The Government of Sri Lanka has decided not to invest in the construction of hard defensive structures due to the very high return period of events similar to the IOTs [4]. The Government had already established a unit referred to as the DMC to properly invest on a National Tsunami Early Warning System (NTEWS). 47% of residents agreed with the current Government policy. Essentially, this group of respondents believed that saving a life should be given priority over protecting physical property. However, the majority of respondents (53%) did not agree with this policy. 96% of respondents were well aware of the existence of a tsunami hazard in the area. 91% of respondents were aware of NTEWS, though 47% of them did not believe that they were safe under NTEWS.

DMC confirmed that 51% (out of 39) of the early warning towers were located in optimum places where a large number of people live. Participants of focused group discussion said that uncertainties in the dissemination of tsunami warnings and poor maintenance of the warning towers were the main reasons for a lower trust on NTEWS.

36% of respondents had never participated in a tsunami drill during the last decade, even though the DMC has organized one drill per year from 2006 onwards. The respondents who had never participated in tsunami drills provided reasons, such as illness, personal matters, their awareness of evacuation centers etc. for their absence.

The WTP_{PW} (how much respondents estimated that they could pay in a month during a one year period to a community project like this) of the 200 respondents is summarized in Table 2. The mean WTP_{PW} is 2.58 USD. The average monthly household income is 302 USD [15]. WTP_{PW} is 1% of average household income in the area.

Table 2. Community WTP_{PW} (in USD) September, 2015 (n=200).

Community WTP_{PW} (USD)	Percentage (%) of contribution
<1	22
1-2	28
2-5	24
5-10	6
10-20	3
20<	1
Not answered	16

More than 50% of respondents felt secure with the tsunami co-beneficial function (holding back a moderate tsunami and reducing the damage of a significant tsunami) of the revetment and the coastal railway embankment. Only 18% and 13% of respondents who did not feel secure with the tsunami co-beneficial function, felt secure after upgrading the present revetment and railway embankment, respectively.

3.2. *Community willingness to pay to upgrade tsunami co-beneficial measures*

Table 3 summarizes the results of the community WTP to upgrade both the railway embankment and the revetment, and the cost of upgrading up to a level that would significantly increase the protection offered by them. The cost of upgrading and the maximum tsunami mitigation co-benefit values were taken from [4].

Table 3. Cost of upgrading the revetment and railway embankment, maximum tsunami mitigation co-benefit and WTP (millions of USD).

Co-beneficial structure	Cost of upgrading	Maximum tsunami mitigation co-benefit	WTP
Revetment	0.330	0.071	0.023
Railway embankment	0.123	0.070	

53% (out of 104) of respondents were willing to pay to upgrade the revetments and 72% (out of 96) of respondents were willing to pay to upgrade the railway embankment. The reason behind this difference appears to be the relatively large upgrading cost of the revetment. The household WTP to upgrade railway embankment and revetment are respectively 250 USD and 115 USD. Both values are lower than the average household income, but higher than the WTP$_{PW}$. 86% of total respondents witnessed 2004 IOT and it might influenced the considerable difference between WTP to upgrade railway embankment (and revetment) and WTP$_{PW}$.

The stability of each regression coefficient (β) (described in section 2.2) was checked by comparing the regression coefficients of each variable. The coefficient of perception of the current level of functionality of each type of structures and the coefficient of trust on the early warning system showed instability and WTP model was reintroduced as Eq. (2). Model constant is denoted by ξ.

$$WTP = \beta 1 \text{ (Residency of the area)} + \beta 2 \text{ (Agree with Government policy)}$$
$$+ \beta 3 \text{ (Witness to 2004 IOT)} + \beta 4 \text{ (Occupation)}$$
$$+ \beta 5 \text{ (Participation to tsunami drills)} + \beta 6 \text{ (Awareness of safe places)}$$
$$+ \beta 7 \text{ (Mean WTP}_{PW}) + \xi \qquad (2)$$

Logistic regression analysis results are presented in Table 4 where $p > |z|$ denotes the z test results.

Table 4. Regression analysis result of the WTP model.

Regression coefficients (notation)	Regression coefficients (value)	Standard Error	$p > |z|$
$\beta 1$	+2.369	1.263	0.061
$\beta 2$	-0.235	0.324	0.468
$\beta 3$	-0.401	0.558	0.472
$\beta 4$	-0.180	0.323	0.576
$\beta 5$	+0.405	0.351	0.249
$\beta 6$	-0.802	0.442	0.069
$\beta 7$	+2.079	0.402	0.000
ξ	-1.682	1.313	0.200

The correctly classified value (correct classification rate) of the regression model was 69.50%. As this value was greater than 50%, the model was accepted. Residency in the area, awareness of safe places and evacuation routes and WTP$_{PW}$ had a significant contribution to WTP. A 90% confidence level was obtained regarding the coefficients of residency in the area and awareness of safe places and safe routes in the model. 99% level confidence was obtained for the coefficient of community WTP$_{PW}$. The magnitudes of regression coefficients show the contribution of each factor to the model. The positive sign (+) represents a positive correlation and the negative sigh (-) a negative correlation with regards to WTP.

Respondents who lived in the case study area, who had a high WTP_{PW} (WTP_{PW} is higher than the mean value), who had witnessed the IOT in 2004 and those who had participated in a tsunami drill at least a once showed higher WTP. Most of the people who participated in tsunami drills got exposed to deficiencies in the evacuation process, and therefore this positively correlated to the WTP. Respondents who were not satisfied with the government policy avoiding to construct hard defensive countermeasures against tsunamis, who were occupied and who were aware of safe places and safe evacuation routes were not willing to pay to upgrade the revetment and the railway embankment. However, the lowest contribution to the WTP model was from the variable of occupation.

According to the results of an interview with officials at the Disaster Management Center (DMC) in Sri Lanka, most of the people who participated in the tsunami drills were women. Most of the questionnaire respondents that declared that they did not have an occupation were housewives. This helps to explain why the correlation between those respondents who had a job and the WTP was negative.

4. Conclusions

The EWS in Sri Lanka is adversely affected by lack of regular maintenance of warning towers and average participation in tsunami drills. More than 50% respondents perceive to contribute even a small amount of money of their household income to upgrade a coastal railway embankment (or revetment). The WTP for upgrade railway embankment and revetment are respectively, 83% and 38% of their average monthly household income. Community WTP to upgrade both structures is small in magnitudes with regards to the upgrading cost. WTP for upgrade railway embankment and revetment are respectively, 40% and 7% of their upgrading costs; 71% and 32% of their tsunami mitigation co-benefit There is a substantial difference between the ways that the community perceives each structure, with WTP for the railway embankment being two times higher than that of revetment as the cost of upgrading of the railway embankment is smaller than that of the revetment, while the co-benefits are almost same. The implications of the WTP model is that population density in communities in the coastal zone might increase, though this would increase their exposure to tsunami hazards. Otherwise, it appears the WTP could be increased by encouraging residents to participate in tsunami drills and improving their income. The methodology used in this research can help disaster risk managers to understand how each individual, community and the government can manage tsunami risks and propose measures to enhance community resilience.

Acknowledgments

This study was partially funded by JSPS KAKENHI Grant Number 25303016.

References

1. J. J. Wijetunge, "Tsunami on 26 December 2004: Spatial Distribution of Tsunami Height and the Extent of Inundation in Sri Lanka," Sci. Tsunami Hazards, vol. 24, no. 3, pp. 225–239, 2006.

2. H. A. R. Ratnasooriya, S. P. Samarawickrama, and F. Imamura, "Post Tsunami Recovery Process in Sri Lanka," J. Nat. Disaster Sci., vol. 29, no. 1, pp. 21–28, 2007.

3. Y. T. J. Khew, M. P. Jarzebski, F. Dyah, R. San Carlos, J. Gu, M. Esteban, R. Aránguiz, and T. Akiyama, "Assessment of social perception on the contribution of hard-infrastructure for tsunami mitigation to coastal community resilience after the 2010 tsunami: Greater Concepcion area, Chile," Int. J. Disaster Risk Reduct., vol. 13, no. October, pp. 324–333, 2015.

4. R. S. M. Samarasekara, J. Sasaki, M. Esteban, and H. Matsuda, "Assessment of the Co-Benefits of Structures in Coastal Areas for Tsunami Mitigation and Improving Community Resilience in Sri Lanka," Int. J. Disaster Risk Reduct., vol. 23, no. April, pp. 80–92, 2017.

5. D. Graham, "Cost-Benefit Analysis Under Uncertainty," Am. Econ. Rev., vol. 71, no. 4, pp. 715–725, 1981.

6. B. O'Brien, "Cost-Benefit Analysis, Willingness to Pay," Encycl. Biostat., 2005.

7. C. Breidert, M. Hahsler, and T. Reutterer, "A review of methods for measuring willingness-to-pay," Innov. Mark., vol. 2, no. 4, pp. 8–32, 2006.

8. G. D. Israel, "Determining Sample Size," Univ. Florida, IFAS Ext., vol. PE0D6, no. April 2009, pp. 1–5, 1992.

9. O. Torres-Reyna, "Getting started in Logit and ordered logit regression," Princet. Univ., 2012.

10. H. Gunatilake, J. Yang, S. Pattanayak, and C. Van Den Berg, "ERD Technical Note Series," no. 19, 2006.

11. C. van den Berg and C. Nauges, "The willingness to pay for access to piped water: a hedonic analysis of house prices in Southwest Sri Lanka," Lett. Spat. Resour. Sci., vol. 5, no. 3, pp. 151–166, Oct. 2012.

12. S. K. P. and J.-C. Y. George Van, Houtven, Caroline van den, Berg, The Use Of Willingness To Pay Experiments : Estimating Demand For Piped Water Connections In Sri Lanka. The World Bank, 2006.

13. A. Bryman, "Encyclopedia of Social Science Research Methods," Encycl. Soc. Sci. Res. Methods, pp. 1143–1144, 2004.
14. Eliot & Associates., "Guidelines for Conducting a Focus Group," Duke Univ. Website, pp. 1–13, 2005.
15. DCS, "Household income and expenditure survey 2012/2013," Colombo, Sri Lanka, 2013.

Hindcasting of Storm Surges in Batanes during Typhoon Meranti and the Application of Sea Level Rise Scenario Simulations

Joaquin Vicente C. Ferrer

Institute of Civil Engineering, University of the Philippines-Diliman,
Quezon City, 1101, Philippines
†E-mail: jcferrer@up.edu.ph

Typhoon Meranti (locally named Ferdie) traversed the Northern Pacific Basin, affecting the Batanes Islands on September 10 to 15 of 2016. It was estimated that PhP 244.6 million in damages to houses and communities was left in the typhoon's wake. Numerically modeling the storm surges during Typhoon Meranti, and incorporating the sea level rise simulations is vital in assessing the long-term vulnerability of Batanes' coastal communities. This study aimed to simulate storm surges during Typhoon Meranti using the Advanced Multi-Dimensional Circulation (ADCIRC) model and to apply a sea-level rise scenario to the validated and calibrated model. The model was calibrated and validated versus tide station observations at Basco Port, Batan Island. The model yielded an RMSE of 14.45% versus the observations. The maximum surges simulated along Batan islands were below 0.50 meters, while it was observed that simulated storm surges in Sabtang island reached 1.50 meters. The sea-level rise scenario simulations were observed to output storm surges 93.65% higher in 2030 than those observed during Typhoon Meranti. It was also observed that the relative difference of the storm surge heights decreased from 2030 to 2100.

Keywords: Typhoon Meranti, Storm Surge, Sea-Level Rise.

1. Introduction

Typhoon Meranti (Ferdie) traversed the Northern Pacific Basin, affecting the Philippines, Taiwan, and China. Typhoon Meranti recorded drop in pressure reaching 890 hPa, a 10-minute sustained wind speed of 220 km/hour, and a 1-minute sustained wind speed of 305 km/hour[1]. On its path, the eye of the typhoon directly passed over Itbayat, Batanes. The damage caused by Typhoon Meranti in Batanes left 1 casualty, 292 destroyed houses, and 932 damaged houses[2]. It was estimated that the total damage to infrastructure cost PhP. 244.6 million. Post-disaster surveys conducted by the University of the Philippines and the University of Tokyo indicated that the sea wall in Ivana, on the island of Batan, was overtopped by the observed storm surges[3]. It was also noted that inundation in these areas were in range of 3 to 5 meters.

The damage dealt by increasing occurrence of extreme typhoons passing through the Philippines has been devastating. This trend of increased frequency of typhoons and greater intensity of passing typhoons have long been noted by the International Panel on Climate Change (IPCC) as negative side effects of climate change[4]. One particular property that has long been observed as evidence of climate change is the rise in mean sea level. The IPCC developed projections for global mean sea level (GMSL) rise through process-based models and semi-empirical models. These projections consider scenarios that account for release of emissions and its effect on melting ice sheets and glaciers. The 2013 IPCC report suggested that the range with medium confidence for the GMSL rise from 1996 to 2100 is 0.42 to 0.80 meters[5]. The rate in rise of GMSL with medium confidence is 3.5 to 8.8 mm/year.

In a study of observed mean sea level variations in Eastern Asia during 1950 to 1991, the mean sea levels on the eastern coast of the Philippines rose by 3.8-5.9 mm/year[6]. In the same report, it was indicated that this rate of mean sea level rise was much larger than that of the observed rise in GMSL at a rate of 1.05 mm/year over the same period. In a recent study of Southeast Asian seas, it was noted sea level trends are some of the highest observed in the past two decades, but will likely be less than the GMSL in the long-term[7]. It is difficult to predict the rate of sea level rise in the region, but sea level rise due to climate change and its ability to intensify the impact of storm surges is an undeniable threat to coastal communities.

The primary objective of this study is to numerically model storm surges during Typhoon Meranti using the Advanced Multi-Dimensional Circulation (ADCIRC) model. The results of the numerical model will then be validated versus recently conducted post-disaster surveys. This study also aims to incorporate the sea level rise scenarios induced by climate change to assess future vulnerability of Batanes. Using ADCIRC, storm surges due to Typhoon Meranti will be simulated after incorporating IPCC rates for GMSL rise. The long-term effects of sea level rise scenarios based on the GMSL rise projections will be considered for the years 2030, 2050 and 2100. The objective of incorporating sea level rise is to assess the community's vulnerability for future reference.

2. Methodology

Advanced Multi-Dimensional Circulation (ADCIRC) utilizes equations formulated using the traditional hydrostatic pressure and Boussinesq approximations and have been discretized in space using the finite element (FE) method and in time using the finite difference (FD) method. Surface-water Modelling Systems (SMS) is a program containing a Graphical User Interface (GUI) designed for pre-processing and post processing of the ADCIRC model. SMS was utilized for simpler and more efficient control and development of the ADCIRC model.

The typhoon track and atmospheric data used was from the Best Track data obtained from the Japan Meteorological Agency (JMA) through the Digital Typhoon website[1]. The typhoon recording taken in reference to the Coordinated Universal Time (UTC) 00:00 GMT, and was adjusted for the Philippine time zone at UTC +08:00 GMT over the period of September 10-15, 2016. The JMA Best Track data recorded the location in longitude and latitude, atmospheric pressure measurements in hPa, maximum winds in m/s and radial gust winds in m/s, in 6-hour increments. This data consisted of the typhoon track required for the ADCIRC input file: fort.22. Bathymetry data was taken from GEBCO's global grid of elevations, published in 2014 and most recently edited in March 2015[8]. The grid data's resolution is 30 arc-seconds (approximately 1 km) and follows the WGS 84 coordinate system.

The Batanes Islands is within longitude zone N51. This zone was the tide station readings at Basco Port, Batan Island, Batanes for the month of September, 2016 were obtained from the National Mapping and Resource Information Authority (NAMRIA). The tide readings were reported hourly, and in reference to mean sea level (MSL). The predicted astronomical high and low tides were taken from the NAMRIA Tide Tables for Basco Port on Batan Island, Batanes[9]. These readings provided the time of high tides and low tides for each day, referenced to the area's mean lower low water (MLLW) and were adjusted to MSL. This data recorded the water surface elevation during the typhoon event and was used to calibrate the ADCIRC model by modifying the bottom friction drag coefficient.

Shown in Figure 1 are the plots of the observed tide at Basco Port versus the predicted astronomical tide for the same port, and the difference between these two data sets. There is a noticeable difference between the predictions and observations at the tide station between 0:00 and 06:00 of 9/14/15 that indicate that the typhoon influenced the water level elevation that may have manifested in a storm surge.

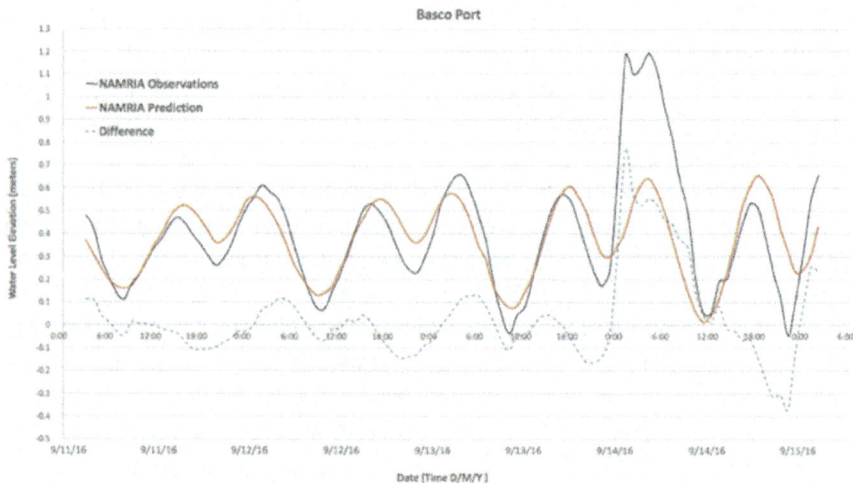

Fig. 1. NAMRIA predicted astronomic tide[9] (red) and tide station observations (black) and the difference between the data sets (broken line).

Table 1. Summary of model parameters and sources

Parameter	Source	Description
Bathymetry	GEBCO (2014)[8]	1x1 km Resolution (WGS 84 Datum)
Grid/Mesh	SMS Generated FEM	38,408.25 km² Area 64,712 Elements 33,755 Nodes
Typhoon Data	Kitamoto, A. (2016) [1]	JMA Best Track Data 9/10-15/2016
Astronomical Tide Input	NAMRIA (2016)[9]	Predicted low and high tide

The ADCIRC model parameters and data shown in Table 1. The grid was generated by the SMS program and was calibrated to a 1.00 arc degree, equivalent to a 38,408.25 km² domain. This was represented by a finite element mesh of 64,712 elements and 33,755 nodes. The depth was then taken from interpolated bathymetry data retrieved from the GEBCO 1 km resolution data, referenced from a datum of WGS 84. The astronomic tide forcing was taken from the NAMRIA tide tables with predicted high and low tides, while meteorological and atmospheric forcing was taken from JMA's best track data for the period of 9/10-15/2016.

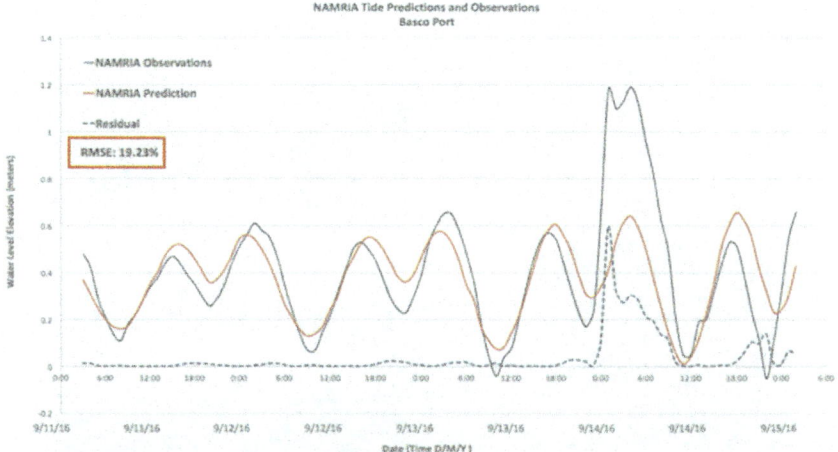

Fig. 2. Residual differences of predicted and observed tide data at Basco Port.

The predicted tides were then compared to the observations at the tide stations to determine if there was significant difference between the data sets that could constitute the occurrence of a storm surge. Shown in Figure 2 is a plot of the predicted data set and the observations at Basco Port and the residual differences between the two. The residual difference between the data set indicates that there is a significant difference between 0:00 and 06:00 of 9/14/2016 as seen in Figure 2. This indicated a storm surge occurrence and the root-mean square error (RMSE) was computed to be 19.23% between the data sets. This RMSE value was taken as the benchmark for the model calibration.

3. Results and Discussion

3.1. *Typhoon Meranti*

The model was calibrated by adjusting the bottom friction coefficient, represented by the coefficient of drag, C_d, parameter in the ADCIRC model inputs. The coefficient of drag represents the friction contributed to the materials at the boundary of the layer of the model that simulates the material at the sea bed. In order to determine the appropriate material in Batanes, pictures of the location and the coastline, taken from recent post-disaster field work, were used as references to determine the kind of material the model was going to numerically simulate.

Fig. 3. Photos of the Batanes Coastal Areas from post-disaster fieldwork observation[10].

Based on the observations of the photos shown in Figure 3, it was concluded that the expected sea bed would be made of bare rock, sand and gravel. This was then taken into account when selecting an equivalent manning's bottom friction coefficient based on Table 2. Based on Table 2, the most appropriate bottom friction coefficient was 0.040, corresponding to bare rock/sand.

Table 2. Typical Manning's roughness n coefficients for various surfaces[11].

Surface	Coefficient, n
Open Water	0.020
Mixed Forest	0.170
Commercial	0.050
Bare rock/sand	0.040
Gravel Pit	0.060
Transitional	0.100

In order to convert coefficients of drag C_d to equivalent manning's n coefficients, Equation 1 was utilized[12]. Equation 1 is given by:

$$C_d = g \cdot n^2 / H^{\frac{1}{3}} \tag{1}$$

where g is gravitational acceleration = 9.81 (*m/s*)

n is the manning's coefficient

H is the total depth of the water level (m)

The calibration runs were performed, and the analysis of results were based on the Benchmarking RMSE of 19.23%. It was also set that the conversion of the manning's coefficient n being within the range of 0.040. Table 3 summarizes the calibration runs and the evaluation parameters.

Table 3. Evaluation of calibration runs.

C_d	n	RMSE
0.030	0.039	14.86%
0.025	0.038	14.45%
Benchmarking RMSE:		19.23%

Based on the calibration runs presented in Table 3, 0.025 was the default model parameter and selected as the best parameter to represent the bottom friction for the simulation. Shown in Figure 4 are the results of the selected model calibration versus the observed NAMRIA data, with superimposed residual differences in the plot. The residual differences are considerably smaller in Figure 4 as compared to those shown in Figure 2. The corresponding RMSE indicates that the model captured factors contributing to the observed surge by the tide station. It should be noted that the model over estimates the peak prior to the surge event and underestimates the surge. This attributed to the input data not being in-phase with the observed data as well as the limitations of the model to account for wave setup that may explain the difference between the simulations and the observations.

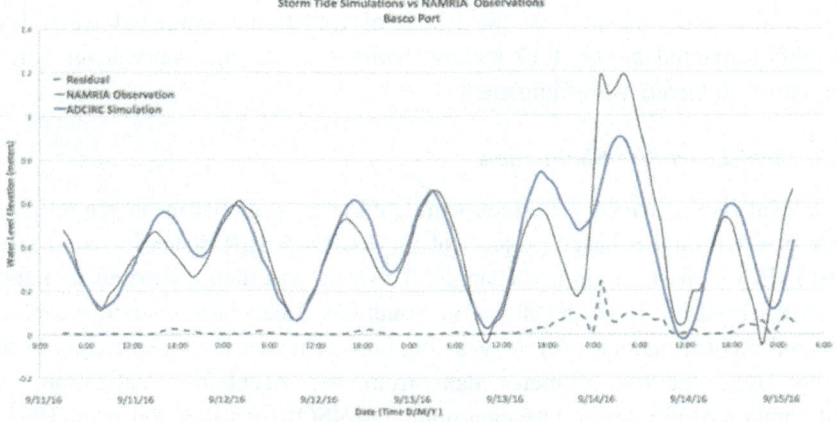

Fig. 4. ADCIRC simulation and observed data at Basco Port.

The maximum water elevations of the Batan Islands were then plotted, citing areas with settlements as seen through satellite images. The maximum storm surge levels were then plotted within the ranges of storm surge heights as shown in Figure 5.

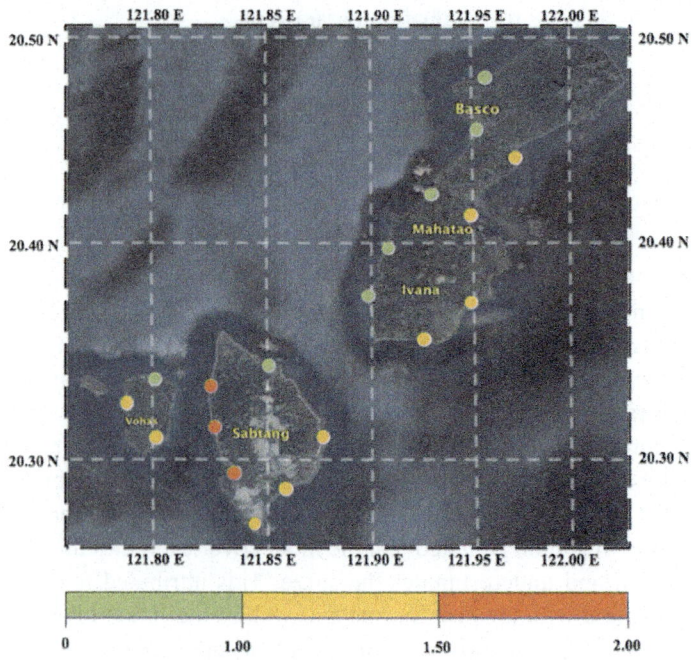

Fig. 5. Spatial representation of simulated water level heights during Typhoon Meranti[13].

It was observed that on the island of Batan, the simulated water level heights remained below 1.50 meters, while on Sabtang, water level heights beyond 1.50 meters were simulated.

3.2. *Seal-Level Rise Simulations*

Sea-level rise scenarios were then simulated using a global mean sea level rise rate of 6.1 mm/year based projects of the RCP 4.5 SLR Scenario found in the 2013 IPCC report on climate change[5]. This rate was then compared to a study that observed mean sea level rise in Southeast Asian Seas over 60 years and created reconstructions over 17-year periods[7]. The sea level rise trends in this study used satellite altimeter data from the Archiving, Validation, and Interpretation of Satellite Oceanographic (AVISO) for validation from 1993 to 2009. The average sea level trend for the waters surrounding the Batanes Islands was observed rise within the range of 4.5 to 5.5 mm per year over the time period of the available data observations. The rate of sea level rise in the areas surrounding Batanes was observed to be within the range of GMSL rise published in the 2013 IPCC report[5]. Thus, the predicted astronomic tides and the bathymetry was adjusted for the years 2030, 2050 and 2100 to run the sea-level rise scenarios. Shown in Figure 6 are the storm surge outputs from the run results of the SLR Scenario simulations at Basco Port adjusted for current sea level heights.

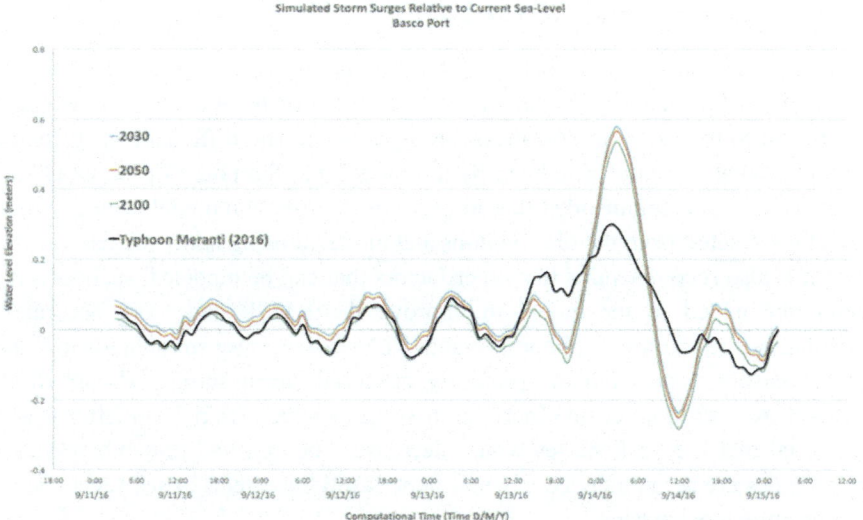

Fig. 6. Storm surge heights for SLR scenarios in 2030, 2050, 2100 versus 2016 observations.

There is a noticeable difference between the SLR storm surges simulated in 2030 and those observed at the NAMRIA tide station during Typhoon Meranti. Though the initial increase is very large, the differences between the period of 2030 to 2100 are noticeably smaller. Shown in Table 4 are the differences between the peak surges simulated and the observed peak surge.

Table 4. Relative difference of simulated SLR peak surges per year versus observed 2016 observations.

Year	Peak Surge Difference (%)
2030	93.65
2050	89.36
2100	78.73

This increase is does not follow a linear relationship with the GMSL rise value applied show in a study conducted on the effects of sea level rise in the Red River Delta in Vietnam[14]. As this study suggested, it this is attributed to the non-linear effects simulated using the ADCIRC model. It was also observed that there was a decreasing trend in the relative difference of simulated SLR water level elevations and the observations beyond 2030. This decreasing trend was attributed to the reduction in contribution of wind-induced currents to the storm surge heights.

4. Conclusion and Recommendations

This research can conclude that a validated model was developed and calibrated for Batanes Islands. The model can be concluded to have a 14.45% deviation compared to the observations, measured by RMSE. The difference is attributed to factors that were not simulated by the model and the phase of the input tide data. Thus it is recommended that in order to improve the model accuracy, tide data be extracted from the observations and modeled using harmonic analysis.

It is also recommended that other factors that can be modeled, such as wave setup, be added to the model to improve the RMSE and more accurately simulate surges during Typhoon Meranti. The sea-level rise simulations indicate an increasing trend from the presently observed storm surges. Though from 2016 there was a noticeable increase in surge height, within 2030-2100 it was observed that the peak surges would decrease. Focused and in-depth sea-level rise studies are recommended to better understand the effects of sea-level rise on storm surge mechanisms.

Acknowledgments

This research acknowledges the National Mapping and Resource Information Authority for providing tide observation data from Basco Port that made this research possible.

References

1. A. Kitamoto, Digital Typhoon: Typhoon 201614 (MERANTI) — Detailed Track Information, (National Institute of Informatics, Japan, 2016), `http://agora.ex.nii.ac.jp/digital-typhoon/summary/wnp/l/201614.html.en`.
2. National Disaster Risk Reduction and Management Council, SitRep No. 13 re Preparedness Measures and Effects of Typhoon 'FERDIE' (I.N. MERANTI), (September 2016).

`http://www.ndrrmc.gov.ph/attachments/article/2913/Si` `tRep_No_13_re_Preparedness_Measures_and_Effects_of_T` `yphoon_FERDIE_(MERANTI)_Covering_the_240600H_to_2506` `00H_September_2016.pdf`, (accessed 18 December 2016)

3. Y. Tajima, J.P. Lapidez, J. Camelo, M. Saito, Y, Matsuba, T. Shimozono, D. Bautista, M. Turiano and E. Cruz, Post-Disaster Survey of Storm Surge and Waves Along the Coast of Batanes, the Philippines, Caused by Super Typhoon Meranti/Ferdie, Coastal Engineering Journal, Vol. 59, No. 1, (2017).

4. IPCC, 2001: climate change 2001: the scientific basis. Contribution of Working Group 1 to the Third Assessment Report of the Intergovernmental Panel on Climate Change, edited by J. T. Houghton, Y. Ding, D. J. Griggs, M. Noguer, P. J. van der Linden, X. Dai, K. Maskell and C. A. Johnson (eds), (Cambridge University Press, Cambridge, UK, and New York, USA, 2001).

5. J. Church, P.U. Clark, A. Cazenave, J.M. Gregory, S. Jevrejeva, A. Levermann, M.A. Merrifield, G.A. Milne, R.S. Nerem, P.D. Nunn, A.J. Payne, W.T. Pfeffer, D. Stammer and A.S. Unnikrishnan, Sea Level Change. In: Climate Change 2013: The Physical Science Basis. Contribution of Working Group I to the Fifth Assessment Report of the Intergovernmental Panel on Climate Change [T.F.,Stocker, D. Qin, G.-K. Plattner, M. Tignor, S.K. Allen, J. Boschung, A. Nauels, Y. Xia, V. Bex and P.M. Midgley (eds.)]. (Cambridge University Press, Cambridge, United Kingdom and New York, NY, USA, 2013)

6. T. Yanagi, and T. Akaki, Sea level variation in the eastern Asia. Vol. 50, no. 6, (Journal of Oceanography, 1994), pp. 643–651.

7. M Strassburg, B. Hamlington, R. Manrung, J. Lumban-Gaol, B. Nababan, and K.-Y. Kim, Sea Level Trends in Southeast Asian Seas, (CCPO Publications, 2015), Paper 138. `http://digitalcommons.odu.edu/ccpo_pubs/138`

8. General Bathymetric Charts of the Ocean, GEBCO 2014 Grid, `http://www.gebco.net/(accessed 12 September 2016).`

9. National Mapping and Resource Information Authority, NAMRIA Tides and Currents Tables, (2016).

10. J. Camelo, Batanes Islands' Post-disaster survey photos, (2016).

11. T. Mayo, Data assimilation for parameter estimation in coastal ocean hydrodynamics modeling, Dissertation thesis, The University of Texas at Austin, (2013).

12. L. Zheng, R. H. Weisberg, Y. Huang, R. A. Luettich, J. J. Westerink, P. C. Kerr, A. S. Donahue, G. Crane, and L. Akli, Implications from the comparisons between two- and three-dimensional model simulations of the Hurricane Ike storm surge, Journal of Geophysical Research: Oceans, Vol. 118, (2013), pp 3350–3369, doi:10.1002/jgrc.20248.

13. Google Earth [Batanes, Batan, Itbayat, Sabtang, Vohas], DigitalGlobe. (accessed 8 December 2016).

14. J. Neumann, Emanuel, A. Kerry, S. Ravela, L. Ludwig, C. Verly, Risks of Coastal Storm Surge and the Effect of Sea Level Rise in the Red River Delta, Vietnam. Vol. 7, no. 6 (Sustainability, MPDI AG, Switzerland, 2015), pp. 6553-6572.

A Comparative Study of Two Different Numerical Methods on Storm Surge[*]

A. A. Mohit[†]

Department of Maritime Engineering, Faculty of Engineering,
Kyushu University, 744, Motooka, Nishiku, Fukuoka 819-0395, Japan
[†]E-mail: mohit4010@yahoo.com
www. kyushu -u.ac.jp

Y. Ide, M. Yamashiro and N. Hashimoto

Disaster Risk Reduction Research Center, Faculty of Engineering,
Kyushu University, 744, Motooka, Nishiku, Fukuoka 819-0395, Japan
E-mail: ide@civil.kyushu-u.ac.jp, yamashiro@civil.kyushu-u.ac.jp,
hashimoto-n@civil.kyushu-u.ac.jp

The present study carries out the comparison of the water level associated with the numerical methods: Finite Difference Method (FDM) and Finite Volume Method (FVM) and the simulation considered on the present climate condition along the coast of Bangladesh. The governing equations of the first model are discretized through FDM and solved by a conditionally stable semi implicit manner on an Arakawa C-grid system. For the second model, σ-coordinate is used for the irrational bottom slope representation and the mesh grid of the study domain is generated by the unstructured triangular cells. The feasible study domain with coast and island boundaries are approximated through proper stair steps for the FDM and the unstructured mesh representation for FVM. A one-way nested scheme technique is applied to the first model to include coastal intricacies as well as to preserve computational cost. Both the models are applied to extrapolate sea-surface elevation associated with the catastrophic cyclone 1991(BOB 01) along the seashore of Bangladesh. The simulation results from both the models are statistically copacetic and make a good acquiescent with some observed and reported data. In the statistical viewpoint, both the method has a good acceptance in storm surge simulation, but this study ensures the strong positive reconciliation with observed data and FVM simulation data. In Bangladesh region, it will be wise decision to use Finite Volume Methods for simulating the storm surge.

Keywords: Storm surge, FDM, FVM, Bangladesh, Cyclone, Typhoon.

1. Introduction

In south Asia, Bangladesh is a land of rivers with a cyclone prone area which is facing 5 to 6 storms every year [1]. The geographical location of the coastal region of Bangladesh makes it higher hazard region than other coastal region

---

[*] This work is supported by Ministry of Education, Culture, Sports, Science and Technology: MEXT, Japan.

around the world. An increasing number of tropical cyclones associated with surge always cause a great loss of many lives and properties along the region of interest. Mainly cyclones in November 1970, 1985, April 1991, 1997, SIDR 2007, and AILA 2009 caused a lot of the death toll It is known that, naturally, the strong circulatory wind associated with a tropical storm increases the sea level abnormally. Thus, maximum damages occurred during the surge rather than the storm itself [2], Therefore, numerical analysis is needed to assess the height of the storm surge along the coast of Bangladesh and this area still lack of adequate storm surge research due to its complexities. Some of the well known complexity factors, namely, astronomical tide, extreme bending of the coastline, offshore islands, shallow bathymetry, and huge discharge through the Meghna and other rivers which can aggravate the effect of storm surge in the Meghna estuarine region [3]. Therefore, it is necessary to have a proper knowledge of the factors that affect surge and an effective storm surge model is highly desirable for the coastal region of Bangladesh. Some analysis and prediction of storm surge associated with tide and their interaction have been investigated by many authors for the Bay of Bengal region. Worth mentioning are [1-7], where the pioneering works are due to [4] and [5]. Basically, these works focused on the coastal region of Bangladesh and its surrounding area.

As far as we know, the approximation of their physical domain was not more realistic because they were useing rectangular grid system to represent the target region for numerical storm surge analysis, which demonstrates their geometric flexibility and the capacity to solve the coastal complexities. Considering this problem of oceanographers, Finite Volume Coastal Ocean Circulation Model (FVCOM) is developed to reduce the coastal complexities and solve the governing equation numerically by flux calculation and discritized the governing equation through FVM [8]. Both the FDM and FVM are the balance numerical methods to simulate storm surge, but some few works have been done through the FVM to simulate the storm surge along the coast of Bangladesh. Notable works include the work of [9], but the combined work of FDM and FVM method is very rare. According to [10], FVM has a considerable attention to the storm surge simulation and it's widely used in computational fluid dynamics. That's why we felt that a review of two different models from a comparative figure is needed to examine the coastal complexity effect along the coast of Bangladesh .

In this study, we have analyzed the comparable figure of the two different numerical storm surge simulation method based on the dynamics of oceanography with conservation regulations. A statistical review is explained for the credence of the numerical computation which represent the accuracy of the consequence. The purpose of the study is to evaluate the operation, accuracy and espousal of the proper method for storm surge simulation so that it can be used for storm surge simulation accurately with less cost and time in the future for the complex coast.

2. Methodology

2.1 *Theoretical foundation of Nested Numerical scheme*

2.1.1 *Basic shallow water equation*

A vertically shallow water equation that was well established for the atmospheric or oceanic phenomenon with the dynamic process, including tide generating forces (TGF) written as follows [11],

$$\xi_t + \widetilde{u}_x + \widetilde{v}_y = 0 \tag{1}$$

$$\widetilde{u}_t + (u\widetilde{u})_x + (v\widetilde{u})_x - f\widetilde{v} = -g\xi_x - \Omega_x + \frac{\tau_x}{\rho(\xi+h)} - \frac{C_f u\sqrt{u^2+v^2}}{(\xi+h)} \tag{2}$$

$$\widetilde{v}_t + (u\widetilde{v})_x + (v\widetilde{v})_y - f\widetilde{u} = -g\xi_y - \Omega_y + \frac{\tau_y}{\rho(\xi+h)} - \frac{C_f v\sqrt{u^2+v^2}}{(\xi+h)} \tag{3}$$

where, (v,u) = vertically integrated components of Reynold's averaged velocity $[m/s]$; $(\widetilde{u},\widetilde{v}) = (\xi+h)u,v$; (τ_x,τ_y) =surface stress due to circulatory wind of storm along x and y direction respectively $[hPa]$; g = gravitational acceleration $[m/s^2]$; f = Coriolis parameter; C_f =friction coefficient; ρ = water density $[kg/m^3]$; h = ocean depth $[m]$; Ω = tide generating potential for equilibrium tide $[kW]$; the equilibrium surface elevation $\eta = \dfrac{\Omega}{g}$.

2.1.2 *Boundary condition*

The major dynamic equation has been used in the different nested scheme with different boundary conditions. The outer scheme if fully independent than the other inner schemes. The inner scheme boundary values of u, v and ξ are performed through the solution process in each time steps and the innermost Northwest corner, the river discharge incorporated through the equation from [3]

$$\widetilde{u} = u + \frac{Q}{(\xi+h)B} \tag{4}$$

where, $Q =$ river discharge[m^3/s]; $B =$ The breadth of the river [m]. According to [6], the open boundary condition at the west ($85^0 E$), east ($95^0 E$) and south ($15^0 N$), are

$$v + \sqrt{(\frac{g}{h})}\, \xi = 0 \tag{5}$$

$$v - \sqrt{(\frac{g}{h})}\, \xi = 0 \tag{6}$$

$$u - \sqrt{(\frac{g}{h})}\, \xi = -2\sqrt{\frac{g}{h}}\, a\sin(\frac{2\pi t}{T} + \Phi) \tag{7}$$

where, $a =$ amplitude; $\Phi =$ phase related to tidal constituent; $T =$ period.

2.1.3 Wind field generation

In this study, a famous empirical model in the Bay of Bengel region from [12] is used to simulate the surface wind field over the study region as follows

$$V_a = \begin{cases} V_0\,(\dfrac{r_a}{R})^{\frac{3}{2}}, for\ r_a \leq R \\[2mm] V_0\,(\dfrac{R}{r_a})^{\frac{1}{2}}, for\ r_a > R \end{cases} \tag{8}$$

where, V_0 =maximum sustained wind[km/h]; R =radial distance from the eye [km]; $r_a =$ some radial distance from the eye at which the wind field is described [km].

After that, the wind stress can then be parameterized under the terms and condition of the wind field associated with the storm due to [1],

$$\tau_x = C_d \rho_a u_a \sqrt{(u_a^2 + v_a^2)} \ \text{ and } \ \tau_y = C_d \rho_a v_a \sqrt{(u_a^2 + v_a^2)}$$

where, $(u_a, v_a) = x$ and y components of the surface wind [m/s]; $p_a =$ air density [kg/m^3]; $C_d =$ The surface drag coefficient that increases with wind speed. However, we set the constant value of 2.8×10^{-3} for C_d in accordance with [3].

2.2 Theoretical foundation of Finite Volume Coastal Ocean Model

Finite volume method with an unstructured grid system is used to solve the shallow water equations that consist of continuity, momentum, temperature, salinity and density equation followed by (see, [8]). The comprehensive result is carried out through the combination of three individual models, Myers model for wind and air pressure, Nao99b tide model for astronomical tide and Finite Volume Coastal Ocean Model to simulate the surge height. The external force like wind and atmospheric pressure calculated by using Myers formula and empirical reduction coefficients from the Bangladesh Meterological Department (BMD) observed best track data. The wind and atmospheric pressure distribution Myers model [13], as given in equation (9).

$$P(r) = P(c) + \Delta p.\exp(-\frac{r_0}{r})$$ (9)

where, r is the distance from the center of the cyclone; P is the pressure at a distance r; r_0 is cyclone radius, which is known as maximum sustained wind radius.

3. Numerical procedure

3.1 Nested Numerical scheme of FDM

The physical condition of our study domain is too large to compute with more accuracy because of its coastal complexities. There are some major big and small islands makes our physical domain more complex, on the other hand, a large number of grid points exercised to make more accuracy but it can significantly increase the computation time and cost. So, keeping this in mind, we referred to classify our domain in different schemes like coarse mesh scheme (CMS), fine mesh scheme (FMS), very fine mesh scheme (VFMS). The CMS covers the $15^0 N - 23^0 N$ latitude and $85^0 E - 95^0 E$ longitude with 15.8 km and 17.52 km grid resolution along x and y axis respectively, similarly, FMS and VFMS covers $21.25^0 N - 23^0 N$ and $89^0 E - 92^0 E$, $21.77^0 N - 23^0 N$ and $90.40^0 E - 92^0 E$ with 2.15 and 0.72 km along x and 3.29 km and 1.14 km along y axis. The CMS is independently calculated Based on the boundary conditions, on the other hand, the rest of the scheme depends on the outer scheme as a boundary condition for the calculation process. The calculated parameter values of ξ, u and v were linearly interpolated as a boundary condition for the FMS and the same procedure is implemented for the VFMS.

3.1.1 Model data sources and set-up

In our numerical model, we used the Meteorological, Hydrological, Geographical and Oceanographical data and this data involved with some parameters. The Meteorological input data like storm track, central pressure, maximum sustained wind radius, maximum sustained wind speed are taken from the Bangladesh Meteorological Department (BMD). The wind distribution data was generated by the empirically-based formula of [12], based on the information of BMD. The water depth data was collected from the General Bathymetry Chart of the Ocean (GEBCO). The geographic data like coastal geometry and island geometry was collected from Geophysical Data System (GEODAS) coastline shapefile and processed by the GEODAS coastline extractor software. This coastline map, island map and the other schemes CMS, FMS and VFMS were digitized through a MATLAB subroutine with a proper stair step algorithm. A conditionally stable semi implicit method used to solve the governing equation (1)-(3) as well as boundary condition (4)-(7) are discritized by finite difference (central in space and forward in time) technique. An Arakawa C system was used to considering three distinct computation points, which are (odd, odd), (even, even) and (even, odd). The equation (7) was used to compute M_2 tide, where a and Φ are calculated following [14], After that, we apply the averaging procedure to average the ξ elevation at other grid points. When the average elevation measurement of ξ was completed, then we started to measure the elevation at coastal and island boundaries. Due to this time the u velocity and v velocity are computed by the discretized equation (2) and (3). Since, the numerical model simulation has been performed through semi-implicit finite difference method, that's because to check the Courant-Friedrichs-Levy (CFL) stability criterion.

3.2 *Model description of FVM*

Boundary condition may take an important role to create the result accuracy. According to [15], boundary condition may create errors in some limited area. Due to this reason, some factors as tidal frequency, local effect, tide generating force and tidal response of solid earth are synthesized. In this study, the Meghna estuary area is affected by the large scale tide and the long term tidal gauge data is rare due to the manually operated tidal gauge. The offshore boundary at the southern node point in our computation domain (Figure 1) the ocean tidal water level has been obtained from the NAO.99b model developed by [16]. This model is the combination of empirical and numerical models, that represents 16,13 and 7 major, minor and long period tidal constituent respectively. By using Cartesian Coordinate to represent the bottom topography makes some error at the continental shelf edge where the water depth changes abruptly. So, the σ coordinates used for the vertical coordinate representation of bottom topography. We observed that the sigma layer number has an influence on surge elevation.

Fig. 1. Unstructured grid representation of Bangladesh coast

Seven layers of sigma coordinate is used in our study for expressing the sea bottom topography smoothly even at any point where the water depth rapidly changes. In our simulation, the number of mesh 81686 and nodes 42352 was used, where the size of the small network of the construction grid was 0.02 km and the size of the bigger mesh was 30 km. The calculating time was considered from 1991/04/29 2:00 (UTC) to 1991/04/30 2:00.

4. Result and discussion

4.1 *Result of FDM*

Considering with the effects of astronomical tide, the nonlinear interaction with the tide and surge is displayed in the figure 2. Our computed results are familiar with the result obtained by [17,3,7]. Our study represents the maximum surge height associated with the astronomical tide effect is 3.82-6.81m at some different locations near the coast of Bangladesh. According to the study of [3], the maximum surge height around this region is 6.9 m and another investigation of [7], the surge height is 3.82-7.29 m.

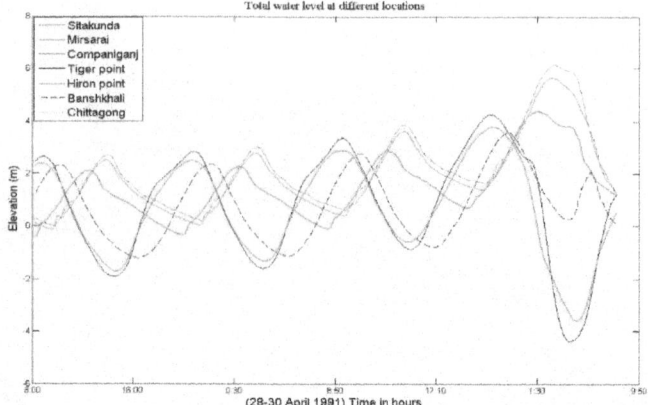

Fig. 2. Investigation of the elevation during the nonlinear interaction with the astronomical tide at that period of landfall time.

4.2 *Result of FVM*

The result obtained from the unstructured grid model FVCOM (Finite Volume Coastal Ocean Model) [8], shown in the figure 3. Figure 3 shows the height of the storm surge in different tide observation point near the coast of Bangladesh. It is also seen in the figure 3 that the high tide time and the landfall time coincide each other to make the surge height higher. It is observed from our simulation that the average water height (3.8-6.8) meter in the year 1991 cyclone, which corresponds to the result of FDM, see table 1.

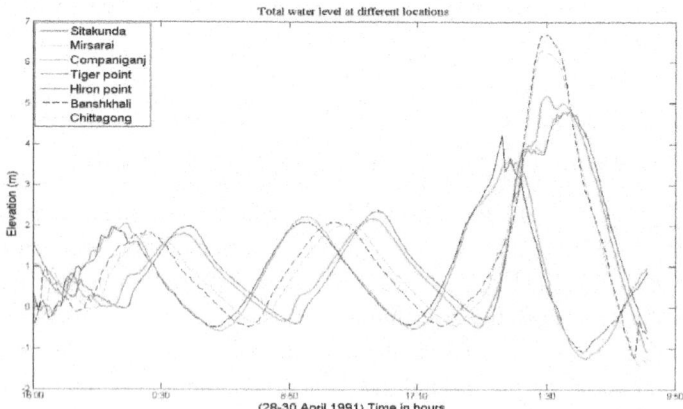

Fig. 3. Tide and storm surge simultaneous outcomes are shown at some observation point along the coast of Bangladesh.

5. Comparative results

In the study, two different methods are used to determine the height of storm surge in 1991 and a comparative study of these methods has analyzed.

Table 1. Computed water level of the different study with respect to mean sea level (in meter) associated with cyclone 1991along Bangladesh coast (wherever available)

Coastal location	Overall peak Surge level by Paul et al 2014	Simulated overall max. Surge level by Paul et al 2012a	Overall peak Surge level by Paul et al 2012b	Simulated overall max. Surge level by Roy et al. 1999	Simulated overall peak by FDM (this study)	Simulated overall peak by FVM (this study
Cox's Bazar	--	--	--	--	6.14	5.80
Moheshkhali	--	--	--	--	4.12	4.61
Banshkhali	--	--	--	--	3.58	6.81
Chittagong	6.25	4.81	5.95	5.45	4.50	6.18
Sitakunda	5.78	--	--	--	4.48	5.28
Sandwip	5.63	5.83	5.74	5.33	4.38	5.35
Mirsharai	--	--	--	--	5.66	5.05
Companiganj	7.28	7.02	6.90	--	6.15	5.52
Chital Khali	--	--	--	--	4.50	3.82
Char Jabbar	6.35	6.81	6.29	5.18	5.69	4.80
Char changa	5.81	5.49	4.95	4.31	4.12	4.50
Char Madras	5.81	--	4.97	--	4.32	4.00
Rangabali	4.50	--	3.93	--	4.07	3.92
Kuakata	3.86	3.65	3.30	--	3.96	4.80
Patharghata	--	--	--	--	4.36	4.95
Tiger point	4.57	--	--	--	4.21	4.47
Hiron point	4.01	4.52	2.69	0.70	3.80	4.30

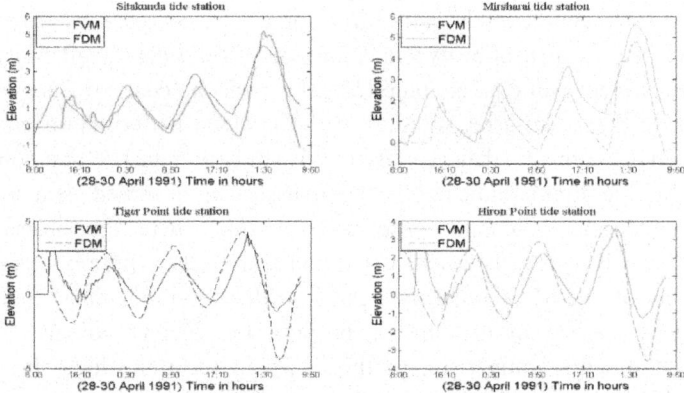

Fig. 4. The surge elevation result that obtained from the different simulation method at some observation point around the Bangladesh coastal region.

172

Figure 4 shows the result obtained from FDM and FVM at four observation point near the coast of Bangladesh, where the left column top figure shows the result obtained from FDM and FVM at the Sitakunda location near Chittagong district and the corresponding figures are displaying the result of Mirsarai, Tiger point and Hiron point respectively. The 1991 cyclone was more devastating due to the high tide during the landfall time. As a result, the death rates were much higher and more damage was occurring due to the huge flood and inundation.

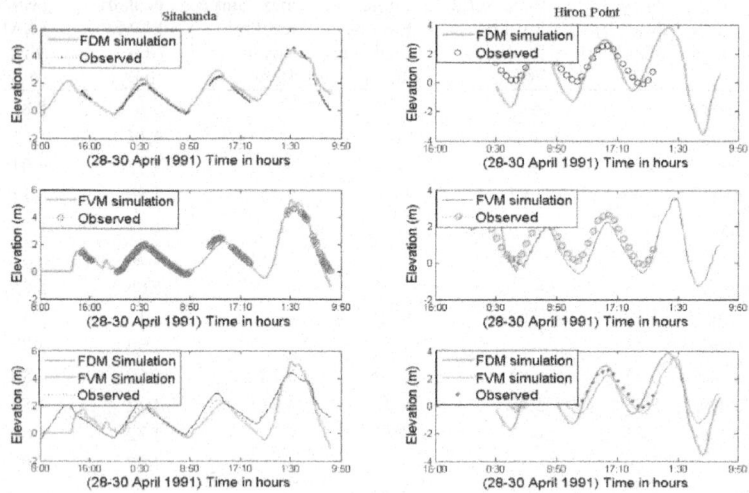

Fig. 5. The surge elevation result obtained from the different simulation method and observed result comparison. The colored line indicates the simulation result and the dotted points indicate the observed data whenever available.

Figure 5 shows the comparative representation of simulation and observed data of cyclone 1991 at Sitakunda and Hiron point. The comparative difference of results has been shown only two locations due to the data availability. In the most cases, water level data at some locations near the coast of Bangladesh are not available due to disaster weather in the cyclone period. As a result, It is usually seen the authentic data scarcity in this study region. The first column and first row of the figure indicate the comparison of observed data and FDM simulation data at Sitakunda observation point, where the observed data collected from BIWTA. The next corresponding column figures represent the comparison of FVM simulation result with observed data and their amalgamated results. In the similar fashion, the second column, first row represents the same comparison of the different location along the coast of Bangladesh. The corresponding next column shows the same description as mentioned above.

6. Conclusions

In this study, there are two numerical methods has been applied to simulate the surge height of 1991 cyclone that strike along the coast of Bangladesh. Most of the studies along the coast of Bangladesh deals only a single numerical method to investigate the surge associated with their characteristic, but we have analyzed a storm in two different numerical methods. Both the methods show the best performance and acceptable results, but FVM makes an assurance the conservation of mass, energy and heat in both in individual grid cells and over the entire domain. It can be visually perceived from the statistical analysis that the correlation coefficient of one day observed data at Hiron point is positively correlated with the correlation value 0.78, Root Mean Square Error (RSME) and Mean absolute Deviation (MAD) is 0.43 and 0.81 of the FDM, on the other hand a positive correlation with the correlated value 0.89, RMSE and MAD is 0.24 and 0.58 respectively of the FVM simulation. On the substructure of statistical viewpoint, FVM method used to simulate surge height at the complex coast will be a wise decision in the future.

Acknowledgments

One of us (A. A. Mohit) acknowledges receipt of a scholarship from the Ministry of Education, Culture, Sports, Science and Technology: MEXT, Japan with number 163618. The first author would like to express his grateful thanks to the Government of Japan for the scholarship during this study to Disaster Risk Reduction Research Center, Faculty of Engineering, Kyushu University, We would also like to thank those reviewers whose comments and suggestion improve the final results of this paper.

References

1. G. C. Paul, , A.I.M. Ismail, and M.F. Karim, Implementation of method of lines to predict water levels due to a storm along the coastal region of Bangladesh, *Journal of Oceanography.* **70**, 199–210 (2014).
2. R. A. Flather, A storm surge prediction model for the northern Bay of Bengal with application to the cyclone disaster in April 1991, *J Phys Oceanogr.* **24**:172–190 (1994).
3. G. D. Roy, A.B.M.H. Kabir, M.M. Mandal, and M.Z. Haque, Polar coordinates shallow water storm surge model for the coast of Bangladesh,*Dynamics of Atmospheres and Oceans.* **29**, 397–413(1999).
4. P. K. Das, Prediction model for storm surges in the Bay of Bengal, *Nature.* **239**, 211–213 (1972).

5. G. R. Flierl and A.R. Robinson, Deadly surges in the Bay of Bengal:dynamics and storm tide tables, *Nature.* **239**, 213–215 (1972).

6. B. Johns, A.D. Rao, S.K. Dube, and P.C. Sinha, Numerical modelling of tide-surge interaction in the Bay of Bengal, *Philosophical Transactions of the Royal Society. A* **313**, 507–535 (1985).

7. G. C. Paul and A.I.M. Ismail, Numerical modeling of storm surges with air bubble effects along the coast of Bangladesh, *Ocean Engineering* **42**, 188–194 (2012).

8. C. Chen, H. Liu, An unstructured grid, finite-volume, three-dimensional, primitive equation ocean model: application to coastal ocean and estuaries, *J Atmos Ocean Technol.* **20**,159–186 (2003).

9. K. M. Tasnim, T. Shibayama, M. Esteban, H. Takagi, K. Ohira and R. Nakamura, Field Observation and Numerical Simulation of Past and Future Storm Surges in the Bay of Bengal: Case Study of Cyclone Nargis, *Nat Hazards*, DOI 10.1007/s11069-014-1387-x (2014).

10. E. Dick, Introduction to finite volume techniques in computational fluid dynamics, *J.F. endt, ED., Springer-Verlag.* 271-297 (1994).

11. M. M. Rahman, G. C. Paul and A. Hoque, A shallow water model for computation water level due to tide and surge along the coast of Bangladesh using nested numerical schemes, *Mathematics and Computers in simulation.* **132**, 257-276 (2017).

12. C. P. Jelesnianski, A numerical calculation of storm tides inducedby a tropical storm impinging on a continental shelf, *Monthly Weather Review.* **93**, 343–358 (1965).

13. V. A. Myers and W. Malkin, *Some properties of hurricane wind field as deduced from trajectories*, National Hurricane Research Project. No. 49, U.S. Weather Bureau (Washington, D.C, 1961).

14. C . McCammon and C . Wunsch, Tidal charts of the Indian Ocean North of 15 S, *J Geophys Res.* **82**, 5993–5998 (1977).

15. M. A. Hussain and Y. Tajima, Tide-surge interaction at the Bay of Bengal along the coast of Bangladesh, *Natural Hazards.* **86(2)**,669-694. (2017).

16. K. Matsumoto, T. Takanezawa and M. Ooe, Ocean tide models developed by assimilating TOPEX/POSEIDON altimeter data into hydrodynamical model: a global model and a regional model around Japan, *J Oceanogr.* **56**, 567–581 (2000).

17. J. A. As-Salek, Negative surges in the Meghna estuary in Bangladesh, *Mon Weather Rev* **125**, 1638–1648 (1997).

Peak Flood Estimation Along Southern Coast: Kerala, India

M. Lokeshwari[†]

*Assistant Professor, Department of Civil Engineering, R. V. College of Engineering,
VTU, Bengaluru, 560059, India.*
[†]*E-mail: lokeshwarim@rvce.edu.in, www.rvce.edu.in*

Vikas Mendi

*Research Scholar, Department of Applied Mechanics and Hydraulics ,National Institute
of Technology Karnataka, Surathkal, Karnataka, 575025,India and
Assistant Professor, R. V. College of Engineering, Bengaluru, Karnataka, 560059, India
E-mail: kbl9dad@gmail.com, www.rvce.edu.in*

N Amarnatha Reddy

*Research Scholar, Department of Applied Mechanics and Hydraulics ,National Institute
of Technology Karnataka, Surathkal, Karnataka, 575025,India and
Assistant Professor, Madanapalle Institute of Technology and Science, Madanapalle ,
Andhra Pradesh, 517325, India
E-mail: amarmtech@gmail.com, www.mits.ac.in*

India being an agricultural based country depends mainly on rain water and river for the purpose of irrigation and other domestic uses. Due to deforestation and urbanization, large area of unpaved surface is converted into paved surface, which prevents percolation of rainwater into ground. As a result, surface runoff and intern the chances of flood increases. Flood causes damage to hydraulic structures, agricultural land, properties and lives. On the other hand many states of India are facing water scarcity. So it is important to carry out flood forecasting to reduce uncertainties in hydrological predictions, which balances these two problems. Some of the catchments in India are ungauged as a result, surface runoff estimation will be inaccurate and also it is very difficult to understand catchment properties and hydrological response to rainfall. Suitable methods are desirable for hydrological evaluation of watershed in the absence of runoff-rainfall data. Remote Sensing and Geographic Information System with appropriate rainfall runoff models, provides ideal tool for estimation of direct runoff, peak discharge and hydrographs. This study focuses on estimation of peak flood discharge at coastal inlets, from river outlets along the southern coast of India in Kerala. Flood estimation methods recommended by Central Water Commission (CWC) India, in report 5a and 5b is considered as reference and Soil Conservation Service Method (SCS) used for peak flood estimation.

Keywords: Flood estimation, Ungauged stations, CWC method, Soil Conservation Service method.

[†] Corresponding author.

1. Introduction

Agriculture system in India predominately depends on river and intern rainfall. Extreme rainfall patterns and/or variability become a critical production risk in this case. The rain-fed agricultural production system is vulnerable to seasonal variability which affects the livelihood outcomes of farmers and landless workers who depend on this system of agricultural production[1]. Almost 70% of earth's surface is covered by water. Ocean holds 96.5% of this water which is unfit for irrigation, domestic and industrial applications, and rest of water is found in the air as water vapour, lakes, rivers, icecaps, glaciers etc. Only about 0.3% of surface water is fresh water, found in lakes and rivers. Hence it is very essential to manage and conserve the fresh water which is coming from rain.

A sustainable water resource management should develop and implement new feasible methods of conserving water. It should properly maintain the existing sources of water along with supplying sufficient water needed for the public and develop new strategies to encounter the problems which are likely to arise due to huge demand for water during drought conditions in the country. Surface runoff is the important parameter used for water resource studies. Meteorologists use rain gauge station to measure the precipitation over a set period of time. In ungauged basins the reliable prediction of surface runoff is challenging and tedious for hydrologists. Most of the river basins in India are ungauged due to various reasons such as terrain condition, lack of financial, human resources and technology etc. Prediction of surface runoff in an ungauged catchment area is very important for various practical applications in finding out discharge to design hydraulic structures and flood protection works, forecasting the surface runoff, for water allocation and climate impact analysis. Hydrologists have developed conventional methods of finding the discharge of a given area by taking into consideration of meteorological data such as rainfall duration, intensity, area of the catchment etc. Several rainfall runoff models along with the help of tools such as Remote Sensing and Geographic Information Systems (GIS) used to estimate direct runoff and hydrographs.

In the absence of direct measurements of runoff, regional rainfall runoff relationships developed over a hydro-meteorologically similar region may be used for estimation of flood discharge at ungauged catchments by relating its basin characteristics. Many empirical methods have been developed for flood estimation in ungauged catchments. This is probably due to large differences in the catchment characteristics of small catchments, combined with the general lack of observations. Methods have been derived based on observations from single flood events or continuous flood series. Purely empirical methods range from being based solely on catchment area, to having several parameters which describe catchment characteristics in various ways. Some of the methods consider precipitation as input, others do not. However, most of the methods have been developed for flood estimation in certain region or for catchments with specific characteristics. No methods can be considered universal.

Central Water Commission of India (CWC) has well-documented reports for obtaining Unit Hydrograph (UH) & flood estimate. For the purpose of obtaining UH, CWC has divided India into 26 subzones in which the present study area comes under the subzone West Coast Region (CWC, 1998). Also the CWC has defined another synthetic method for deriving the Unit Graph which is called the dimensionless unit hydrograph approach (CWC, 2001) [3,4]. United States Soil Conservation Service (SCS) [5] as defined a dimensionless method to obtain the Unit Hydrograph for a catchment. Present study focuses on estimation of flood discharge in ungauged catchments of Kerala west coast in India. Similar kind of studies were carried out for different regions of the India by many authors to find a better way of estimating discharge [2,6,7,8].

During 1970's, various coasts have been designated as wave dominated or tide dominated or river dominated, *in order to find out the stability of the inlet*. The non-dimensional classification of tidal inlets proposed by Vu (2013) [9] also indicated that tides, waves and river flow are the three dominant forces to study the morphodynamics of tidal inlets. Since the river flow data is not available at *most of* the *costal* inlets in India, *uncertainties in the estimation of energy, calculating discharge of water for potable and recreational purposes becomes difficult. So in present study an* attempt is made to estimate the flood discharge *by river to coastal inlets using* Synthetic Unit Hydrograph (SUH) methods *and Soil Conservation Service Method (SCS)*.

Study area

Kerala, in south India is taken for the present study, which extends between 12°48' to 8°18' North Latitude and 74°52' to 77°25' East Longitude coving an area of 38,863 km² and having a coastline length of about 580km. This coastline has around 56 outlets according to the Survey of India Open Series Maps (OSM). The OSM indices covering the entire Kerala state is given in Table1.

Table 1: List of OSM Indices for the Kerala state

D43U15	C43D12	C43D14	C43E2	C43E8	C43K11	C43Q13	C43R3
D43U14	C43D11	C43D13	C43E1	C43E7	C43K10	C43K16	C43R2
C43D1	C43D10	D43V16	C43Q8	C43E6	C43E12	C43K15	C43R1
D43V4	C43D9	C43Q1	C43Q7	C43E5	C43E11	C43K14	C43L4
D43V3	D43V12	C43K4	C43Q6	C43W10	C43W15	C43K13	C43L3
D43V2	C43J15	C43K3	C43Q5	C43W9	C43W14	C43E16	C43R7
C43D6	C43J14	C43K2	C43K8	C43Q12	C43W13	C43X3	C43R6
C43D5	C43J13	C43K1	C43K7	C43Q11	C43Q16	C43X2	C43R5
D43V8	C43D16	C43E4	C43K6	C43Q10	C43Q15	C43X1	C43L8
D43V7	C43D15	C43E3	C43K5	C43Q9	C43Q14	C43R4	C43K9
D43V6							C43K12

Watershed Atlas of India provided by the Central Ground Water Board, explains that the state of Kerala comes under 3 Sub-basins of the Malabar basin like Netravathi, Varrar and Periyar, as shown in Figure 1. As the area of Netravathi sub-basin covering the Kerala state is very less compared to the other two, present work has been confined to the other two sub-basins like Varrar and the Periyar. Nine major inlets on the Malabar were identified from the OSM sheets and are shown in Figure 1. Details of the outlets are given in Table 2.

Table 2: Details of identified Outlets

Outlet ID	Nearby city	Latitude	Longitude
18	Mahe	11° 42' 14"	75° 31' 51"
19	Iringal	11° 33' 54"	75° 35' 23"
21	Elathur	11° 20' 59"	75° 43' 58"
22	Thekupuram	11° 13' 40"	75° 46' 48"
24	Balathirutti	11° 7' 25"	75° 49' 31"
26	Ponnani	10° 47' 9"	75° 54' 50"
28	Chettuva	10°30' 24"	76° 2' 15"
29	Pachallor	8° 25' 19"	76° 57' 18"
30	Poovar	8° 18' 14"	77° 4' 33"

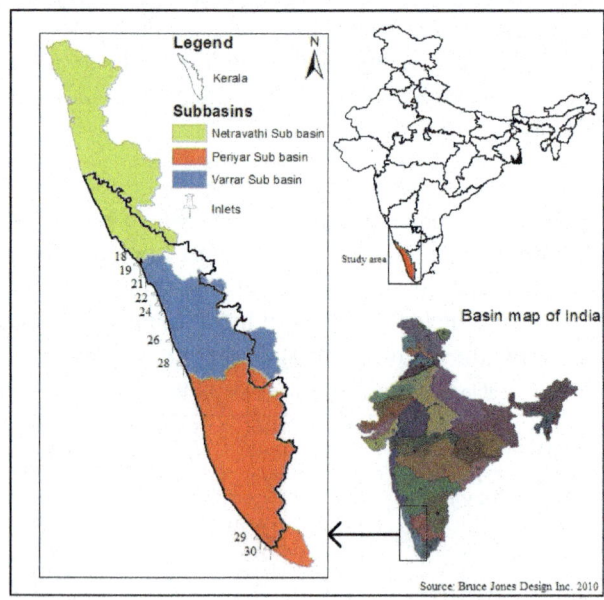

Figure 1: Study area along the coast of Kerala and the 3 sub-basins with outlets

2. Methodology

Toposheets are scanned and geo-referenced using QGIS software. Creating subset images and mosaic of all subset images is done using ERDAS software and that is given as input in ArcGIS to determine the physiographic parameters (i.e. Catchment area, Main stream length, centroid of catchment and slope of catchment). All these physiographic parameters are calculated by using the ArcGIS® software tools. Catchment Area (CA) is determined from Survey of India toposheets by marking the watershed boundary along the catchment. The length of the main stream (L) is the longest length of the main river course from the farthest watershed boundary to the point of study. The longest length of the main river course from a point opposite to the center of gravity to the point of study is denoted as L_C. Equivalent stream slope (S) of the catchment is determined by dividing the longitudinal section of the main stream into number of convenient segments representing the broad ranges of slopes of the segments and use the equation 1 to get the slope.

Where, L_i = Length of ith segment in km, L= Length of the main stream $D_{i-1}+D_i$= Elevations of river bed at intersection points of contours reckoned from the bed elevation at a point of interest considered as datum and D_{i-1} and D_i are the heights of successive bed locations at contours intersections.

Different SUH approaches used in the Study

Mostly SUH methods are used to describe the entire unit hydrograph for a gauged basin with one or two hydrograph parameters. Therefore the UH may be estimated for un-gauged basins with similar geomorphology, soils, land cover/land use, and climate similar to that for gauged basins. In present study CWC unit hydrograph method (1992) [3], CWC dimensionless UH method (2001) [4], Soil Conservation Service (SCS) method (1956) [5] are considered.

2.1. *CWC Synthetic Unit Hydrograph method*

CWC has adopted a hydro-meteorological approach for developing a regional method for flood estimation in the ungauged catchments. For developing relations between physiographic parameters and the UH parameters, the concurrent rainfall and runoff data for a number of hydro-morphologically similar catchments of different areas and sizes located in the particular subzone for minimum of 5 to 8 years during all seasons of the year are required.

Various trials of relationship between the physiographic parameters and Unit hydrograph parameters for 13 gauged catchments considered suitable for the

180

studies. The relationship between physiographic parameters and UH parameters was found to be significant. Then tp was related to unit peak discharge of the UH (q_p) and q_p was related to various UH parameters like W_{50}, W_{75}, WR_{50}, WR_{75}. The principle of least squares was used in the regression analysis to get the relationships in the form of equations to obtain the parameters of the Synthetic UH in an unbiased manner. The following relationships have been derived for estimating the 1-hr UH parameters in the subzones of Malabar and Konkan regions. Representation of CWC SUH parameters are given in Figure 2. Summary of CWC UH parameters from CWC SUH method is shown in Table 3.

Table 3: Summary of CWC UH parameters from CWC SUH method

Outlet ID	Area (km²) (10^2)	Length (km)	Slope (m/km)	q_p (m³/s /km²)	t_p (hr)	W_{50} (hr)	W_{75} (hr)	W_{R50} (hr)	W_{R75} (hr)	T_B (hr)	T_m (hr)
18	4.11	57.2	2.13	0.22	7.9	9.9	4.9	3.0	1.7	33.8	8.4
19	12.9	68.1	3.16	0.24	7.1	8.9	4.4	2.7	1.5	31.3	7.7
21	6.15	61.6	1.99	0.20	8.7	10.8	5.3	3.3	1.8	36.1	9.2
22	30.4	141	3.47	0.18	9.7	12.2	5.9	3.8	2.1	39.3	10.8
24	12.4	137	1.09	0.11	17.0	21.4	10.2	6.7	3.6	59.2	17.6
26	63.8	216	2.07	0.12	15.3	19.2	9.2	5.9	3.2	54.7	15.8
28	16.8	96.1	0.85	0.12	15.5	19.5	9.3	6.1	3.3	55.3	16.0
29	191	507	1.06	0.06	31.8	40.1	18.7	12.6	6.5	93.6	32.3
30	5.04	60.7	3.35	0.26	6.6	8.2	4.1	2.5	1.4	29.5	7.1

Where t_r = unit rainfall duration (hr), T_m = Time from the start of the rising limb to the peak of the UH (hr), Q_p = Peak discharge of Unit Hydrograph (cumec), t_p = Time from the centre of Effective rainfall duration to the UH peak (hr), W_{50} and W_{75} = Width of the UH measured at 50% and 75% of peak discharge ordinates (hr) and WR_{50} and WR_{75} = Width of rising limb of UH at 50% and 75% of peak discharge ordinates (hr), T_b = Base width of UH (hr), A = Catchment Area (km²), qp = Q_p/A (cumec/km²) respectively.

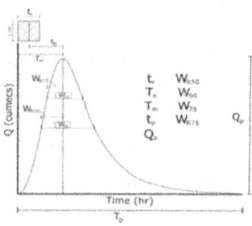

Figure 2: Representation of CWC SUH parameters

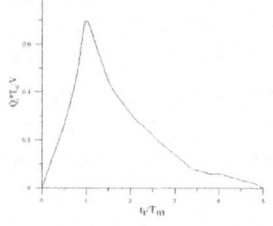

Figure 3: Dimensionless CWC unit hydrograph

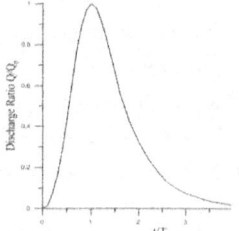

Figure 4: Dimensionless SCS unit hydrograph

2.2. *CWC dimensionless Unit Hydrograph method*

The dimensionless unit hydrograph is obtained by studying 13 measured data catchments in CWC subzones 5(a) and 5(b)[3]. The dimensionless unit hydrograph also called the non-dimensional method obtained by the CWC is as shown in Figure 3. In CWC dimensionless unit hydrograph method, the ratios used are as the ordinate, where V=0.28xCatchment area and   as the abscissa. In case of CWC non-dimensional method the value of ordinate   reaches the peak when the abscissa = 1. The summary of UH parameters are shown in Table 4. The values of T_m, and V are obtained using the relations given below

q_p= Peak discharge per unit area (cumec/km^2)

t_p= Time from the centre of Effective rainfall duration to the UH peak (hr)

T_m= Time from the start of the rise to the peak of the UH (hr)

Table 4: Summary of CWC dimensionless unit hydrograph parameters

Outlet ID	Area (km^2) (10^2)	Length (km)	Slope (m/km)	q_p (Cumee/km^2)	T_p (hr)	T_B (hr)	V (cumee) (10^3)
18	4.11	57.2	2.13	0.22	7.90	8.40	1.14
19	12.9	68.1	3.16	0.24	7.10	7.70	3.58
21	6.15	61.6	1.99	0.20	8.70	9.20	1.71
22	30.4	141	3.47	0.18	9.70	10.8	8.43
24	12.4	137	1.09	0.11	17.0	17.6	3.45
26	63.8	216	2.07	0.12	15.3	15.8	17.7
28	16.8	96.1	0.85	0.12	15.5	16.0	4.65
29	191	507	1.06	0.06	31.8	32.3	53.1
30	5.04	60.7	3.35	0.26	6.60	7.10	1.40

2.3. *Soil Conservation Service dimensionless Unit Hydrograph method*

SCS dimensionless unit hydrographs is based on a study of large number of unit hydrographs and is recommended for drawing synthetic unit hydrographs by many agencies. Figure 4 represents a typical Soil Conservation Services dimensionless unit hydrograph. In this method, the ordinate (Y-axis) is (Q/Qp) which is discharge ratio expressed with respect to peak discharge Qp, and the abscissa (X-axis) is (t/ Tp) which is the time ratio expressed with respect to time to peak Tp. By definition, (Q/Qp) =1 when (t/ Tp) = 1.

To find the values of Qp and Tp, the physiological parameters such as length of the main stream, equivalent slope of the stream and the area of the watershed are used according to the following relations. The summary of UH parameters are shown in Table 5.

t_c= Time of concentration (Min) , t_p = Lag time (hr),
T_p= Time from start of rise to peak (hr),   S = Equivalent slope (m/km),
L= Length of the main stream (km), Q_p= Peak discharge (m³/s)
A = Area of the watershed (km²)

Table 5: Summary of SCS dimensionless UH parameters

Outlet (km²) ID	Area (km²) (10^2)	Length (km)	Slope (m/km)	t_c (Min)	t_p (hr)	T_p (hr)	Q_p (m³/s)
18	4.11	57.2	2.13	15.9	9.60	10.1	85
19	12.9	68.1	3.16	15.7	9.40	9.90	271
21	6.15	61.6	1.89	17.7	10.6	11.1	115
22	30.4	141	3.37	26.8	16.1	16.6	380
24	12.4	137	0.99	42.1	25.3	25.8	100
26	63.8	216	1.97	45.8	27.5	28.0	474
28	16.8	96.1	0.84	33.9	20.4	20.9	167
29	191	507	0.96	116	69.8	70.3	565
30	5.04	60.7	3.34	14.0	8.40	8.90	117

3. Results and Discussion

CWC SUH, CWC dimensionless and Soil Conservation Service (SCS)methods of Synthetic Unit Hydrographs were applied to determine the catchment outlet discharges along the Kerala coastline of India. Results from the synthetic unit hydrograph based on these methods are presented in Table 6. Comparison of the synthetic unit hydrograph (peak discharge ordinate vs base time) are depicted in Figures 5(a) to 5(i). From results it can be observed that rising and falling limbs, Peak discharge ordinates and time to peak are closely matched for CWC SUH and CWC dimensionless method. SCS method underestimated the rising limb and overestimated the falling limb, under estimated the peak discharge ordinates and overestimated time to peak values in most of the cases, in comparison with both CWC methods. In these unit hydrograph methods, a degree of subjectivity is involved in fitting the SUH with few characteristics points. In addition, simultaneous adjustments are also required to ensure that the area under UH is unity corresponding to unit rainfall excess. However, despite their inconsistencies, these methods are still widely used in engineering applications/ problems. SCS method estimates the discharge values slightly lesser than CWC methods, since this method is applicable for small catchments ranging less than 20Km², compared to CWC SUH and CWC dimensionless methods applicable to larger areas (5000 km²). Bar chart in Figures 6 shows the variation in the estimated discharge values of all three methods at all ungauged basins outlets.

CWC dimensionless method gives an overestimated total discharge values at all basin outlets compared to CWC SUH method with a maximum of 16.12% at basin ID 29 and minimum 1.87% at basin ID 30 as shown in Table 6. At basin ID 18, SCS method underestimated by 5.62% and 9.43% in comparison with CWC SUH and CWC dimensionless methods. At basin ID 19, SCS method underestimated by 6.59% and 9.45% compared to CWC SUH and CWC dimensionless methods. At basin ID 21, SCS methods underestimated by 4.76% and 9.54% in comparison with both CWC SUH and CWC dimensionless methods. At basin ID 22, SCS method underestimated by 3.69% and 9.53% compared to CWC SUH and CWC dimensionless methods respectively. At basin ID 24, SCS method overestimated by 1.13% than that of CWC SUH method and underestimated by 9.55% than CWC dimensionless method. At basin ID 26, SCS method overestimated by 0.24% compared to CWC SUH method and underestimated by 9.58% in comparison with CWC dimensionless method. At basin ID 28, compared to CWC SUH and CWC dimensionless; SCS method over and under estimated by 0.35% and 9.60% respectively. SCS method at basin ID 29, discharge overestimated by 5.64% and underestimated by 9.58%than that of CWC SUH and CWC dimensionless methods. At basin ID 30, SCS method underestimated by 7.55% and 9.56% when compared with CWC SUH and CWC dimensionless methods. In these unit hydrograph methods, a degree of subjectivity is involved in fitting the SUH with few characteristics points. In addition, simultaneous adjustments are also required to ensure that the area under UH is unity corresponding to unit rainfall excess. However, despite their inconsistencies, these methods are still widely used in engineering applications/problems.

Table 6: Summary of total discharge values at catchments considered in the study

Outlet ID	CWC SUH method (10^2)	Total Discharge (m^3/s/hr) CWC dimensionless method (10^2)	SCS method (10^2)
18	1.22	1.26	1.15
19	3.87	3.97	3.63
21	1.81	1.89	1.73
22	8.86	9.36	8.54
24	3.45	3.82	3.49
26	17.9	19.6	17.9
28	4.70	5.17	4.71
29	50.7	58.9	53.7
30	1.52	1.55	1.42

184

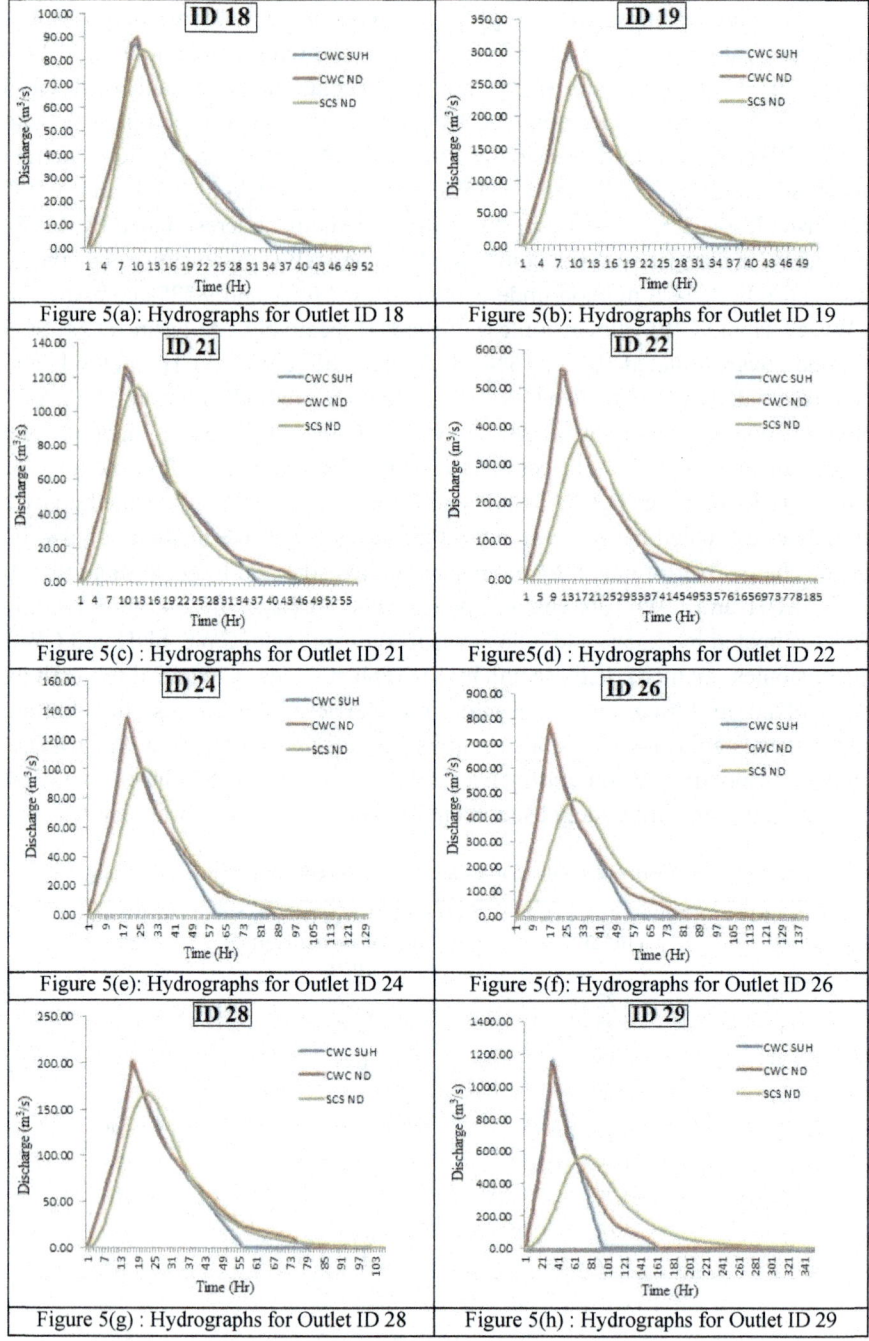

Figure 5(a): Hydrographs for Outlet ID 18

Figure 5(b): Hydrographs for Outlet ID 19

Figure 5(c) : Hydrographs for Outlet ID 21

Figure5(d) : Hydrographs for Outlet ID 22

Figure 5(e): Hydrographs for Outlet ID 24

Figure 5(f): Hydrographs for Outlet ID 26

Figure 5(g) : Hydrographs for Outlet ID 28

Figure 5(h) : Hydrographs for Outlet ID 29

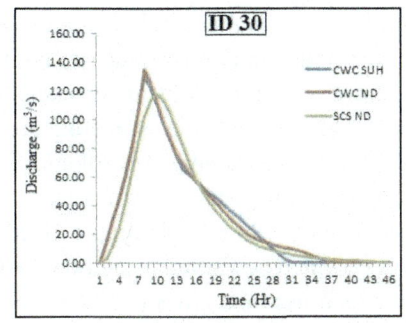

Figure 5(I) : Hydrographs for Outlet ID 30

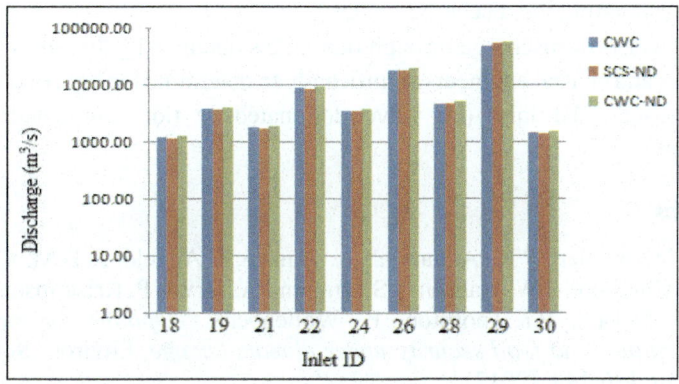

Figure 6: Comparison of total discharge for the outlets

4. Conclusion

Central water Commission (CWC) methods give higher discharge values when compared to the SCS method. CWC dimensionless method over estimates the discharge value and the Hydraulic structures which are designed based on this method will have higher safety, higher storage capacity and the structure will also be designed as an uneconomical design.

The peak discharge value obtained for the outlets with larger catchment area show larger variations. Since the larger catchment area has continuous and long stream it will have more secondary and tertiary tributaries throughout its length and these tributaries will contribute more discharge into the main stream. Therefore the longest catchment area will have longest stream and uninterrupted flow of water from the other tributaries. These factors will ultimately contribute to increase in the discharge at the outlets into the sea. Smaller catchment areas will have discontinuous and shorter streams; hence gives less discharge into the sea. Variation in the outlets with smaller catchment area is small and the values obtained from different methods are almost same. It can be concluded from the results that these methods are well suited for smaller catchments areas; For Larger catchments extra care should be taken while estimating the discharge values.

Depending upon the topography, hydrological characteristics and morphological characteristics of the catchment suitable method for the calculation of discharge has to be decided while designing a Hydraulic structure. SCS-Dimensionless method is accepted best suitable method worldwide because it uses the data pertaining to the land use land cover (LULC), ambient moisture content and also the vegetation cover. Further study concludes that, for tropical regions like India the methods given by the Central Water Commission (CWC) is best suited for estimating the discharge values.

The river flow discharge through tidal inlets obtained by this study will help researchers and coastal engineers significantly to determine the type of dominance of the coastal tidal inlet (i.e wave dominated or tide dominated or river dominated).

References

1. S.J. Vermeulen, P.K. Aggarwal, A. Ainslie, C.Angelone, B.M. Campbell, A.J. Challinor, J.W. Hansen, J.S.I.Ingram, A. Jarvis, P. Kristjanson, C. Lau, G.C. Nelson, P.K.Thornton, E. Wollenberg , *Options for support to agriculture and food security under climate change*, Environ. Sci. Policy, 15, pp. 136-144, (2012).
2. P.K. Singh, P.K. Bhunya , S.K. Mishra , U.C. Chaube, *An extended hybrid model for synthetic unit hydrograph derivation*, Journal of Hydrology, pp 336, 347– 360, (2007).
3. Central Water Commission (CWC), *Flood Estimation Report for West Coast Region Konkan and Malabar Coasts Subzones - 5a & 5b*, (1992).
4. Central Water Commission (CWC), *Manual on Estimation of Design Flood*, (2001).
5. V. Mockus, *National Engineering Handbook, U.S. Soil Conservation Service*, (1972).
6. A. Ravi and N. Sajikumar, *SUH and GIUH for Flood Estimation*, International Journal of Scientific & Engineering Research., vol. 5, no. 7, pp. 361–367, (2014).
7. C. Shu and T. B. M. J. Ouarda, *Flood frequency analysis at ungauged sites using artificial neural networks in canonical correlation analysis physiographic space*, Water Resources Research., vol. 43, no. 2 February, pp. 1–12, (2007).

8. B. Badyalina and A. Shabri, *Flood Frequency Analysis at Ungauged Site Using Group Method of Data Handling and Canonical Correlation Analysis*, Water Resources Research., vol. 9, no. 6, pp. 48–55, (2015).

9. Vu Thi Thu Thuy., *"Aspects of Inlet Geometry and Dynamics"*. Ph.D. Thesis, School of Civil Engineering, The University of Queensland, Australia. (2013).

10. *Survey of India and Department of Science & Technology, Nakshe.* [Online]. Available: http://soinakshe.uk.gov.in/. (Accessed: 26-Jan-2017).

11. *United States Geological Survey (USGS), EarthExplorer.* [Online]. Available: https://earthexplorer.usgs.gov/. (Accessed: 10 Jan 2017).

12. *Indian Space Research Organisation, "Bhuvan, Indian Geo-Platform of ISRO."* [Online]. Available: http://bhuvan.nrsc.gov.in/bhuvan_links.php. (Accessed: 15 Jan 2017).

Conceptual Solutions to Protect Shanghai Region: Design of Storm Surge Barrier System in the Yangtze Estuary

Li Yuting

Nanjing Hydraulic Research Institute,
Nanjing 210029, China
†*E-mail: ytl@nhri.cn*
www.nhri.cn

In response to the potential floods caused by typhoons to Shanghai city, China, several proposals emerged to protect the region. This paper, based on the 'System Engineering Approach', presents the design process and a feasible design for the storm surge barrier system in the Yangtze Estuary. A challenge for this 15 km long barrier system is to cross Yangtze Estuary, including two islands and three channels at the Yangtze River mouth. A storm surge barrier is required to close the coastal spine and prevent storm surges in the Yangtze Estuary, but not to obstruct shipping under normal conditions.

Keywords: Climate change; Typhoons; Storm surge barrier; Shanghai flood protection.

1. Instruction

Big coastal cities often have quite similar flooding problems: sea level rise, land subsidence and dense population. This has led to global studies in how to protect these cities in such complex water system. The Dutch and US engineers have successfully set up their own flood control systems, which always consist of multiple defense lines, including barriers, dikes, levees and various non-engineering solutions. However, Shanghai flood defense system only has sea dikes, floodwalls and urban underground drainage system (Fig. 1), which makes Shanghai a weak point to fight against the potential floods due to changing boundary conditions.

Shanghai experienced hundreds of flood disasters in 2,000 years of recorded history. The storm surge accompanying the typhoons can easily swell the tide in the Huangpu River to breaching the flood defenses along the river with inundation or large parts of the urban Shanghai area. According to the report "Analysis of Historical Flood Events in Shanghai" by the Shanghai Water Authority, the Yangtze River flood killed 145,000 people and around 28.5 million were affected in the year of 1931.

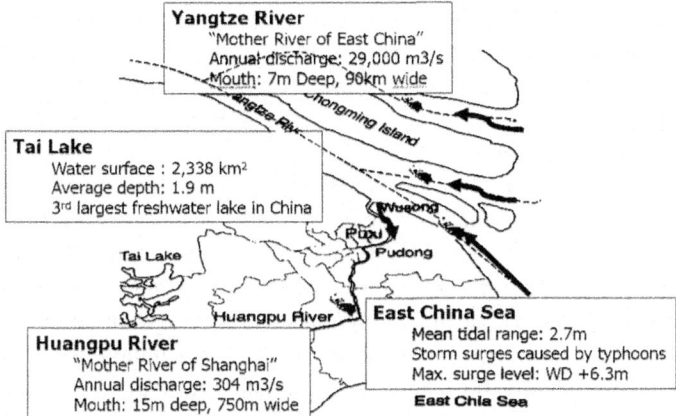

Fig. 1. Water system of Shanghai Region.

2. Philosophy of the design of coastal defenses around Shanghai

In response to the potential floods caused by typhoons to Shanghai city, China, several proposals emerged to protect the region in the first design level. By using Multi Criteria Analysis, one of them is selected as the protection system. A challenge for this 15 km long barrier system is to cross Yangtze Estuary, including two islands and three channels at the Yangtze River mouth. A storm surge barrier is required to close the coastal spine and prevent storm surges in the Yangtze Estuary.

Shanghai flood protection system is a huge and complex system to design. A design methodology applied within this research is called 'System Engineering', which gives an insight in the complexity of the object, which has to be designed. The whole design process is divided into several levels, within which a decision is made per design level. A cyclic, iterative process form large to small-scale is used in this methodology. This paper, based on a 'system engineering' approach, presents the design process and a feasible design for the storm surge barrier at this specific site.

3. Design level 1 – flood defense system in Shanghai

3.1. *Solutions description*

Four flood defense system alternatives are proposed, see Fig. 2~Fig. 5. All the possible solutions are intended to provide flood protection to Shanghai urban area as well as Chongming Island.

Solution 1 is to strengthen the floodwall and dike system. Solution 2 is to build a movable storm surge barrier in the Huangpu River. Generally, some problems will arise if the barrier is located more upstream. First, the protected area is reduced. Due to the denser population along the upstream, construction would be more difficult as less space is available. It is assumed the barrier would be constructed just at the mouth of the Huangpu River. Another option (solution 3) is to close off the tidal inlets by constructing three storm surge barriers at the mouth of the Yangtze River. An obvious advantage of this alternative is, the barriers are in the more rural opening area, resulting in a much easier construction. The drawbacks are clearly seen, the complex sand-water dynamic system in this location makes the ecosystem quite sensitive. How to maintain the salinity and to please the nature should be taken into account. The shipping is also blocked during gate closure. The only difference, between solution 3 and solution 4, is the barrier location. The construction cost of a barrier is relatively high, which is strongly related to the barrier width. So the barrier is moved more landward behind the islands.

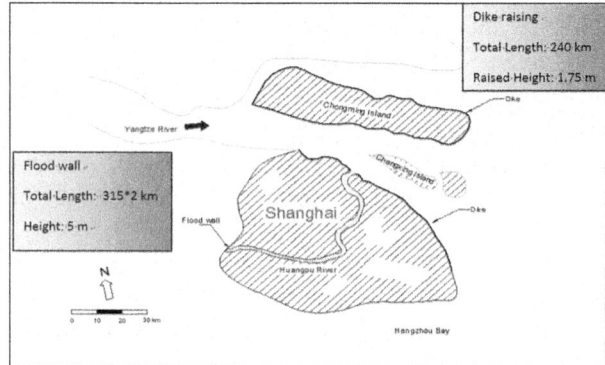

Fig. 2. Proposed Flood Defense System in Shanghai, Solution 1.

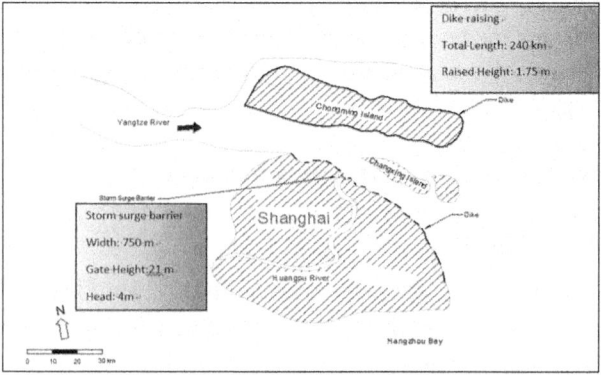

Fig. 3. Proposed Flood Defense System in Shanghai, Solution 2.

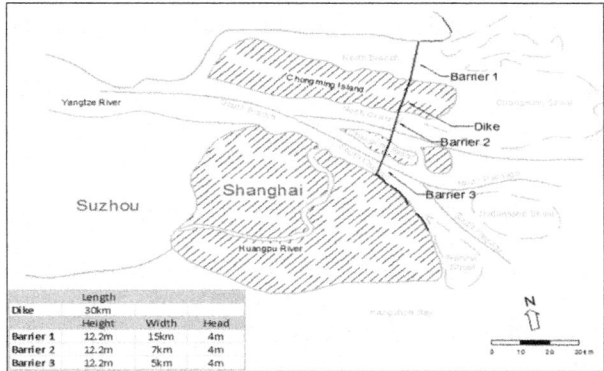

Fig 4. Proposed Flood Defense System in Shanghai, Solution 3.

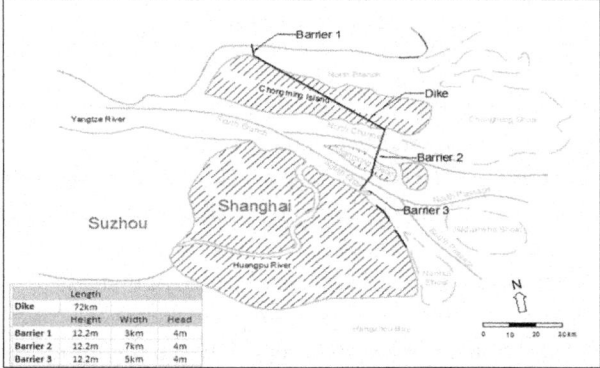

Fig. 5. Proposed Flood Defense System in Shanghai, Solution 4.

3.2. *Decision-making process*

A rough cost estimation with index numbers, as well as a Multi Criteria Analysis (MCA), is adopted as the evaluation method. According to the cost estimation, the alternative of constructing one storm surge barrier at the mouth of the Huangpu River results to be the most cost-efficient at present. It is easily understanding that, this alternative is a consequence of current standards, which seems to be set by the local government, at this moment, for short and/or mid-term solution. When considering the future climate changes, the dikes have to raise over or another defense line has to be built in the waterfronts. Including the potential extra cost, the final cost would likely double or even more.

In the Multi Criteria Analysis, the alternative of constructing three barriers behind the islands in the Yangtze River ranks the best for its excellent performance and huge additional benefits. The cost is not included in the Multi Criteria Analysis, for the reason that in some occasions it is reasonable to spend more money to realize a better performance. Except its main purpose as a flood protection structure, it also has multiple functions like prevention of salt-water intrusion, water quality improvement and road connection. In addition, the protected area (including Suzhou) and population are much larger. Furthermore, construction in the open rural area results to be a better choice, due to its less disturbance to the populated area. The problem of sedimentation in this location can also be technically solved.

To summarize, construction of three storm surge barriers behind the islands in the Yangtze River, with a better performance and lower cost, is the most flexible and cost-efficient one, and it guarantees a long-term safety to the whole region. For those reasons above, solution 4 is chosen for future study.

4. Requirements for Yangtze Estuary barrier design

This part describes the requirements for the selected barrier system in the Yangtze River Estuary. It includes the operational and functional requirements, particularly the nautical and environmental environments.

In normal conditions, the barrier is open. The discharge of the Yangtze River flows freely to the sea. Shipping must be possible through the opened barrier. When a high water level is expected, the barrier is closed. All navigation is blocked. The water level in the estuary should be kept under a safety level when the barrier is closed. During the normal condition when the barrier is opened, the barrier must be able to allow water exchange between the Yangtze River estuary and the East China Sea to maintain the local eco-system.

5. Design level 2- barrier system in the Yangtze Estuary

5.1. *Project area*

The Yangtze River Estuary is not a singer waterway to the sea. It is bifurcated into the North Branch and the South Branch. The North Branch is now suffering a severe sedimentation and the latter is the main passage for navigation and runoff. South Branch is further divided by the Changxing Island, into the North Channel and the South Channel, and the latter is then again bifurcated into the South Passage and the North Passage.

5.2. *Division of sections*

Except the barrier's fundamental function a flood defense structure, two main basic functions must be satisfied. One is to allow free shipping under normal conditions; the other one is to exchange water between the estuary and the sea so as to preserve the ecosystem. The barrier system is divided into two main functional sections: navigational section and environmental section[a] The barrier system in the Yangtze River estuary is schematized in Fig. 6.

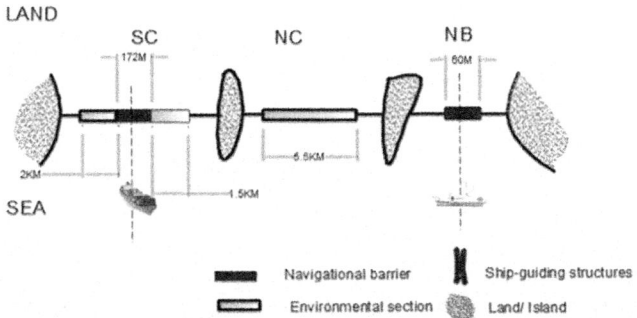

Fig. 6. Schematized barrier system in Yangtze Estuary as an entity.

6. Design level 3- barrier in the South Channel

This paper only focuses on the barrier design in the South Channel. The proposed barrier in the South Channel is composed of a navigation channel and environmental barriers, which is already illustrated. Due to the weak subsoil conditions, the foundation costs are expected to be decisive. This shortest path is chosen to the location of the barrier. The total span is 5km at this location, with 375 m navigation channel and 5km environmental barriers. The barrier is located at the head of Yangtze River mouth, connecting the Island and the urban area of Shanghai.

6.1. *Navigational section*

6.1.1. *Shipping requirements*

The navigational section has an open-closable gate, which is fully closed during storm conditions, however, opened during normal conditions. The gate should first fulfill the navigational requirements as a navigable gate.

---

[a] Environmental section is defined as the section with extra opening of the barrier to allow water exchange.

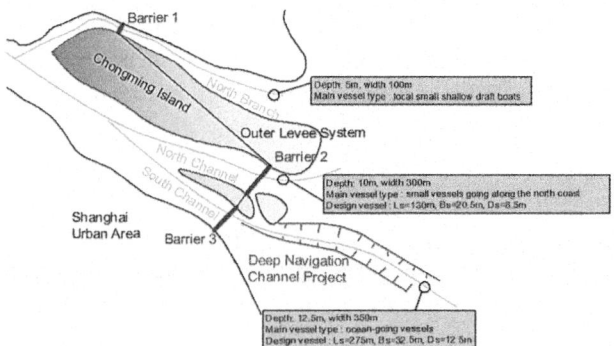

Fig. 7. Overview authorized waterways dimensions and design vessels.

The navigational requirements are set according to a report "*Approach Channel – A Guide for Design*" by PIANC (1997) and "*Waterway Guideline*" released by Rijkswaterstaat (2011).Through a 'round-and-round' calculation, the main required navigational dimensions in the South Channel are set.

Table 1 Required navigation channels dimensions

South Channel	Two-way channel	
	Width(m)	Depth(m)
	172	16.25

6.1.2. *Navigational section gate structure design*

This design level focuses on the design of navigational section of barrier system in the South Channel in the Yangtze Estuary. A large span of the barge gates is feasible and gates are suited to reverse differential head and reverse flow during operation. But the negative hydraulic head will cause some problems if the barge gate is not designed well. There is a load concentration and transfer to the hinges, which is rather a disadvantage. In terms of the construction aspect, the barge gate requires less space and the less strong foundation is acceptable. Concerning the construction aspect, the barge gate has moderate construction difficulties. It can be easier to control the project. For those reasons, the barge gate is selected. The barge gate is suitable for the wide openings, provide unlimited air draft and in the floating situation doesn't transfer too much loads to the foundations.

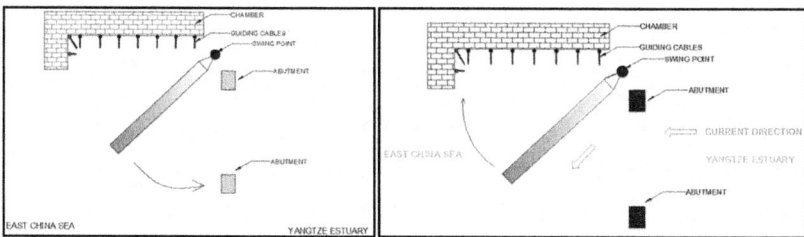

Fig. 8. Plan view of barge gate: during a storm (up); after storm condition(left). Not to scale.

According to previous computations, the navigable gate thus has a minimum 'wet profile' of 16.25 m depth and 172 m width. The gate top when the barge gate is fully opened is at WD+ 5.1m, which is 0.8m lower the design surge level, because a certain amount of overtopping is allowed in this project for economic consideration. The gate is actually a caisson, which can be ballasted with water and floated with compressed air. The initial length of the caisson is determined as the minimum required opening plus 4 m on each side to rest on the supports. It is assumed that the caisson is divided into compartments every 12m in the length direction and every 5.5 m in the width direction.

6.2. *Environmental section*

6.2.1. *Environmental requirements*

When considering the environmental aspect of the barrier, awareness in many people of the need to protect this area's natural resources and unique tidal habitat must be taken in to account. The reduction of the mouth of the estuary will cause a reduction of the tidal range. A main environmental requirement is set, as the barrier must allow the tides to enter freely, thus maintaining the tidal ecosystem.

Under non-storm conditions, all gates are opened to allow water flow. Because the Yangtze River Estuary is a sensitive hydrodynamic system, but the habitats there are not that outstanding compared to the Scheldt Estuary in the Netherlands. The criteria is set as r=80%, referring to 87% in the design of Easter Scheldt barrier. In other word, after tides pass through the barrier, the tide amplitude is reduced to 80% of their original shape, due to the effects of the constriction of the river mouth. To satisfy this criteria, the minimum flow area is obtained as $A_c = 108,000$ m^2. The maximum closure of the Yangtze River Estuary is calculated as 42.4%. As a conservative estimate, a constriction of 40% is used. In other word, 60% of the barrier system remains open during normal conditions.

Under storm conditions, the whole opening during a storm can be composed of the length of the navigational sections, as well as the potential failure length of the environmental length. For a rough estimate, the failure probability is adopted as 10%[b] during a storm. Thus, the whole opening length is: $(172+187+60)+8500$ $\times 10\%= 1,200$m. This opening of 1,200 meters will lead to a limited water level raise behind the barrier. The water level difference over the barrier (4 m) results in a maximum flow velocity $=(2gh)^{1/2}=9$m/s, resulting an inflow of 135,000 m³/s or 5.8×10^9 m³ during a 12h storm. Divided by the surface area 1800 km², this induces a 3.2m water level rise.

6.2.2. *Distribution of* environmental *sections*

Before making the decision how the sections to be located, it is of importance to introduce the hydrological conditions of the three channels. Due to the perpendicular angle between the North Branch and the main channel, the fresh water inflow is very small: less than 1% of the Yangtze River discharge in recent years; during storm conditions the salty water intrusion can be sever in the North Branch. Because of the quite small amount of water exchange, no extra opening (environmental section) is required in the North Branch. Only a small opening is left for shipping to fulfill demands from local fishery industry.

The North Channel and South Channel are responsible for almost all water exchange. The environmental section is divided in two parts according to the cross sections of the two channels. As stated above, the total length of the environmental section is 8.5 km. Thus, the environmental section in the North Channel is 8.5 km $\times (7/12) = 5$ km, while it is 3.5 km long in the South Channel.

6.2.3. *Environmental section gate design*

The environmental section is required to provide opening for water exchange between the sea and river. Thus the main requirements for the environmental sections include preventing flooding during storm conditions and supplying extra opening for water discharge during normal conditions. Due to this reason, lifting gates can be applied, because of the easy construction and rich experience. Then, the main dimensions of lifting gates are calculated. The calculations of top level of closed lifting gates lies in whether or not overtopping is allowed.

This section outlines the designed retaining height of the lifting gates (environmental section) and the associated incoming volume of water. In order to determine the retaining heights, two main requirements should be satisfied:
(1) Maximum water level rise inside the basin = 3.5 m;
(2) Maximum current velocity close to the barrier as low as possible.

---

[b] The gate failure is adopted from the design of the Eastern Scheldt Barrier. It is still sufficient to block the surge if approximately 10% of the gates fail to close during storm condition.

The cost of a barrier is mainly dependent on the main dimensions, including the maximum water level difference above the barrier, the height of the retaining construction and the barrier span. The influence of construction height (retaining height) can be minimized by optimizing the retaining heights for different sections. The retaining height for environmental section and navigational sections need not to be the same level. Peter A.L. de Vries (2014) stated, for the Bolivar Roads reduction barrier an equal retaining height over the full whole span is the least costly. The conditions in the Bolivar Roads are assumed to be similar to the system barrier in the Yangtze Estuary, thus the following calculation is based on that both the navigational section and environmental section have the same retaining height.

In this option, overtopping and/or overflow over the gates and levees occur. The barrier is modeled as a (submerged) sharp crest weir. In order to prevent large force on the backside of the structure, the maximum combined discharge over the closed gates is set at 10 l/s/m. Results show that, The maximum allowed water level rise due to overtopping of levees and gates are 3.5 m. This results in an allowed water level rise due to overtopping and/or overflow of gates of 3.5-1.5 =2.0 m. This value is used as input for the iterative calculation of the minimum required gate height. The result presents the minimum crest height of the gates, concerning the allowed water level rise in the retention are is WD 5.1m, with an overtopping discharge of 9.49 l/s/m.

Table 2 Estimated structural dimensions of the lifting gates (limited retaining)

Parameter	Value	Remarks
Opening width	80m	Experiences from reference projects shows that using gates with width of 75~100 m provides the most cost-effective solution
Number of openings	44	Based on the gate width of 80m, required total width of 3.5 km
Top level closed gate	+5.1 m	Based on the maximum overtopping and overflow requirement
Sill level	-12.5 m	Based on requirements for environment
Total height of gate	17.6 m	Based on a top level of WD +5.1m

7. Conclusions and recommendations

7.1. *Conclusions*

(1) By means of system engineering approach, presents the design process and a feasible design for the storm surge barrier system to protect Shanghai region from flooding.

(2) Considering the program of requirements of boundary conditions, several solutions have considered and evaluated for the protection system. Between all the alternatives, by using MCA and rough cost indication, the barrier system closing off the Yangtze Estuary has been selected as the best choice, which fulfills all the requirements of the project.

(3) Design level 2 discusses the operational and functional requirements for the selected barrier system. Except its fundamental function a flood defense structure, two main basic functions must be satisfied. One is to allow free shipping under all conditions; the other one is to exchange water between the estuary and the sea so as to preserve the ecosystem. Then the barrier system is divided into two main functional sections: navigational section and environmental section.

(4) In design level 3, the part of barrier in the South Channel consists of both environmental and navigational sections. A general design for the environmental section is performed, by using MCA. The vertically lifting gates are selected, because they are feasible for large span and suited to reverse differential head and reverse flow during operation. Also, with a closable barge gate in the navigational section, the minimum opening width for the free navigation is 172m.

7.2. Recommendations

(1) This report is based on design storm with the design return period of 1/1,000 [1/year]. But the design storm and required safety level should be considered in more detail, using the probabilistic design approaches. It could be useful to take the cost-benefit analysis.

(2) The Yangtze Estuary is a quite complicated system. The surrounding water system, including the Yangtze River, the Huangpu River and the Tai Lake would also influence the design philosophy. So laboratory models and tests should be realized to study more about the barrier system and operations in reality.

References

1. Gu, F. R., & Tang, Z. (2002). Shanghai: Reconnecting to the global economy. Global Networks/Linkeed Cities, New York and London: Routledge, 273-308.

2. Labeur, R. J. (2007). Open Channel Flow. Delft: TU Deft.

3. Li, C., Chen, Q., Zhang, J., Yang, S., & Fan, D. (2000). Stratigraphy and paleoenvironmental changes in the Yangtze Delta during the Late Quaternary. Journal of Asian Earth Sciences, 18(4), 453-469.
4. PIANC, I., & IMPA, I. (1997). Approach Channels A Guide for Design. Final report, PIANC Bulletin(95).
5. Qin, Z., & Duan, Y. (1992). Climatological Study of the Main Meteorological and Marine Disasters in Shanghai. Natural Hazards, 6, 161-179.
6. Ren, W., Zhong, Y., Meligrana, J., Anderson, B., Watt, W. E., Chen, J., & Leung, H.-L. (2003). Urbanization, Land Use, and Water Quality in Shanghai: 1947–1996. Environment International, 29(5), 649-659.
7. Welsink, M. W. J. (2013). Adaptation of the Hollandsche IJssel Storm Surge Barrier. (MSc Thesis), TU Delft, Rotterdam.
8. Hartsuijker, C., & Welleman, J. (2007). Engineering Mechanics: Volume 1: Equilibrium (Vol. 1): Springer.

Simulative Analysis of Storm Tide Levels in Batanes Islands during September 2016 Typhoon Meranti Using the Coupled Delft3D-SWAN Numerical Models

M.C. Turiano[†] and E.C. Cruz

*Institute of Civil Engineering, University of the Philippines,
Quezon City, Philippines
[†]E-mail: marjorie.turiano@upd.edu.ph
www.upd.edu.ph*

Super Typhoon Meranti was one of the strongest typhoon that made landfall in the Philippines in 2016. Following the typhoon event, a post-disaster survey was conducted by Tajima et al. (2017), wherein inundation levels and wave runup heights were measured at 37 locations along the islands of Batan and Sabtang, Batanes. This paper discusses the results of the numerical modeling efforts performed to simulate the storm tide levels in Batanes Islands, and the comparisons of the simulations with the observed water levels. The numerical storm tide model for the study area was developed using the two-way coupled Delft3D-SWAN hydrodynamic and wave models. The observed water levels from the tide monitoring station in Basco Port in Batan Island indicated that the peak water level during the passage of Typhoon Meranti is only 0.87 m above MSL, while the simulated peak water level is 0.72 m above MSL. Comparing the results of the simulations to the findings of the post-disaster survey conducted by Tajima et al. (2017) showed that the simulated peak water levels at the survey locations were consistently and drastically lower than the measured water levels. The observed error in the simulated peak storm tide in Basco Port is less than 20 cm, and therefore does not account for the large differences between the measured inundation levels and simulated and measured inundation levels. The large differences are attributed to other nearshore processes (e.g. surge height amplification due to the fringing reefs) that are not considered in the current storm tide numerical model setup, and shall be investigated in a future study.

Keywords: Storm surge, Storm tide, Wave setup, Numerical Modeling, Delft3D, SWAN

1. Introduction

Typhoon Meranti (Local Name: Ferdie), a Category 5 storm based on the Saffir-Simpson hurricane scale, is one of the strongest typhoons that hit the Philippines in 2016. It formed on September 9, 2016, entered the Philippine Area of Responsibility in September 11, and made landfall on Itbayat, Batanes in September 13 at about 16:00 UTC. Figure 1 shows the location of Batanes Islands and the best track of Typhoon Meranti, while Figure 2 shows the three largest islands of the Batanes group of islands. Based on the JMA typhoon bulletin, Typhoon Meranti recorded a maximum sustained wind speed of 120 knots (~222 kph) and minimum sea level pressure (MSLP) of 890 hPa. The characteristics of Typhoon Meranti were comparable to the intensity of the infamous 2013 Typhoon Haiyan that had a MSLP of 895 hPa and maximum wind of 125 knots (~232 kph). The November 2013 Typhoon Haiyan induced storm surge heights of more than 6 m in Leyte, Philippines, caused almost 90 billion pesos worth of damages, and left 6,300 deaths, 28,689 injured, and 1,061 missing [1]. The September 2016 Typhoon Meranti left the province of Batanes under a state of calamity, destroying more than a thousand houses, and left more than 55 million pesos' worth of damages on crops and infrastructure [2].

One month after the Super Typhoon Meranti event, a post-disaster survey was conducted by a joint team of the University of Tokyo and University of the Philippines in Batan and Sabtang islands. Water levels in the inundated areas or runup heights were measured at 37 locations along the coast of the islands of Batan and Sabtang. Details of the survey methodology and findings were documented by Tajima et al. [3]. The study area is located less than 50 km from the best track of the typhoon.

This paper discusses the results of the numerical modeling efforts performed to simulate the storm tide levels in Batan and Sabtang islands during 2016 Super Typhoon Meranti, and the comparisons of the simulations with the observed water levels.

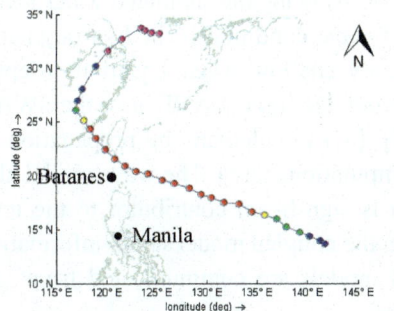

Figure 1. Location Map showing study area and track of Super Typhoon Meranti

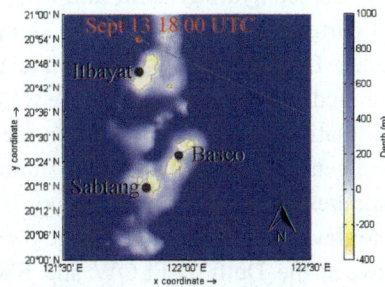

Figure 2. Identification of Batanes islands, presented with track of Super Typhoon Meranti (black line and red dot)

2. Summary of Post Disaster Survey Findings

The locations of the surveyed areas, together with the measured heights of wave runup and inundation levels, are presented in Figure 3. The elevations reported in Tajima et al. [3] were referred from the MLLW, and was translated to refer to the

MSL by applying the MTL-MLLW difference of 0.40 m obtained from the tide monitoring station in Basco Port, Batanes (NAMRIA, personal communication). The measured water surface elevations are presented in Table 1.

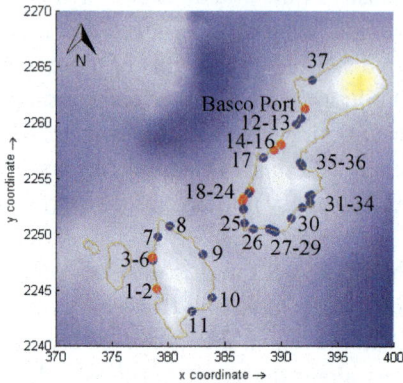

Figure 3. Locations and measured heights of wave runup (blue dots) and the water level in the inundated area (red dots)

Figure 4. Location of Basco Port, and PAGASA synoptic stations in Itbayat and Basco

3. Numerical Model Setup

The numerical storm tide model for the study area was developed using the Delft3D modeling suite ([4], [5], [6]). The storm tide model developed was a two-way coupled hydrodynamic and wave model, allowing the simulated water levels and currents interact to with the simulated wave conditions. The hydrodynamic model, Delft3D-FLOW, solves the unsteady shallow water equations (depth-averaged) in two dimensions. The wave model, Delft3D-WAVE, uses the SWAN (Simulating WAve Nearshore) model ([7], [8]) to calculate the propagation of wind generated waves and swell in the computational area. The wave model also computes the wave-induced setup, which is significant contributor to the total water level during the storm surge event. In the coupled model setup, information between the Delft3D-FLOW and SWAN models are communicated from one model to another at an interval of 30 minutes.

The storm tide numerical model consisted of three nested computational domains of varying computational extent and grid resolution. The largest domain has a uniform grid resolution of 10 km, and the smallest domain has a minimum grid size of 100 m. The applied bathymetry was based on the GEBCO-2014 dataset [9] supplemented with depth information from 1:50,000 scale topographic maps published by the National Mapping and Resource Information Authority

(NAMRIA). The outline of the fringing coral reefs was assigned a uniform depth of 1 m below MSL.

3.1. Flow Model

The open boundary conditions applied in the FLOW model was based on the TPXO 7.2 global tide model [10]. Uniform values of Manning's roughness were applied; n=0.024 for sea and n=0.035 for land areas. Due to its large extent, tidal forces were added in the physical processes of the largest computational domain.

The time and space varying surface winds and sea level pressure applied to force the storm tide model were generated following the pressure and wind distribution model by Fujita [11]. The typhoon parameters were based on the best track information obtained from the Japan Meteorological Agency (JMA) (http://www.jma.go.jp/jma/indexe.html). The average ambient pressure was set to 1010 hPa. The wind-induced surface stresses were calculated following the wind drag parameterization by Zijlema et al. [12].

3.2. WAVE Model

The WAVE model was setup to run SWAN in nonstationary mode with a time step of 10 minutes. The wave directions are divided into 36 bins separated by a $10°$ value for each bin. The wave frequencies have been set to a range lying between 0.03 to 1 Hz, divided into 37 bins.

The wave conditions at the open boundaries of the WAVE model were derived from the results of the WaveWatchIII model [13] wave hindcasts of NOAA (ftp://polar.ncep.noaa.gov/pub/history/waves) and IOWAGA-Iferemer (ftp://ftp.ifremer.fr/ifremer/ww3/HINDCAST/). The wave forces were calculated using the radiation stresses. Depth-induced breaking is computed using the model by Battjes and Jansen [14] with the breaking index of γ_B=0.80. The bottom friction was based on the JONSWAP formulation [15] with the friction coefficient set to 0.038 m^2/s^3, following the recommendation by Zijlema et al. [12] to use a smaller friction coefficient with their wind drag parameterization. Quadruplet and triad interactions were included in the model. The default whitecapping expression by Komen et al. [16] was applied. The same tropical cyclone model used in the FLOW model was applied to drive the wind generated waves. Computation of wave-induced setup is also included in the wave modeling.

4. Results and Discussions

4.1. *Tropical Cyclone Model*

A comparison of the modeled and observed surface winds and sea level pressure for the two surface observation stations in Itbayat and Basco, Batanes (**Error! Reference source not found.**) is presented in Figure 3. The radius of maximum winds was estimated following the relation given by Vickery and Wadhera [17], and the gradient winds were reduced to the near-surface level by using a reduction factor equal to 0.70 [18]. The plots show that the observed atmospheric conditions, particularly the sea level pressures in Itbayat Station, are adequately estimated by the Fujita [11] model. The modeled surface winds give an average R^2 value of 0.37, bias of 5.44 m/s, and RMSE of 6.72 m/s. The modeled sea level pressures give an average R^2 value of 0.89, bias of 2.80 hPa, and RMSE of 5.81 hPa.

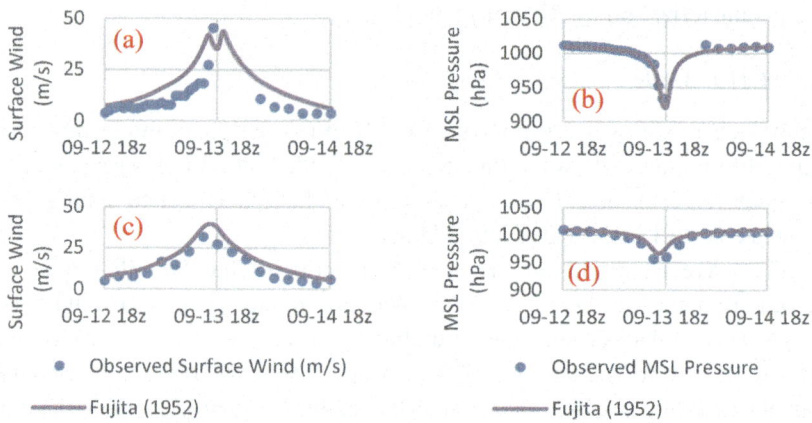

Figure 3. Comparison of observed and modeled atmospheric conditions. (a): surface winds in Itbayat; (b): sea level pressure in Itbayat; (c): surface winds in Basco; (d): sea level pressure in Basco

4.2. *Astronomic Tide Levels*

The astronomic tide levels simulated by the numerical model were first calibrated such that the simulated water levels in Basco Port, Batanes match the forecast water levels given by WXTIDE (http://www.wxtide32.com/), a free tide and current prediction program. A plot of the forecast water levels by WXTIDE and the observed water levels several days before the Super Typhoon Meranti event is presented in Figure 46. The forecast tides of WXTIDE give an R^2 value of 0.70, bias of 6 cm, and RMSE of 14 cm. It is observed, particularly during the neap tides, that the lower high tides are overestimated by WXTIDE, while the higher high tide levels were more accurately modeled. The observed low tides during the neap tides also occur above the forecast low waters. Examining the tide patterns in Figure 4, the peak of the typhoon occurred during the higher high tide, and hence the astronomic tide component of the peak storm tide level is expected to be modeled properly, while the storm tide levels during the lower high tides and low waters are expected to be overestimated.

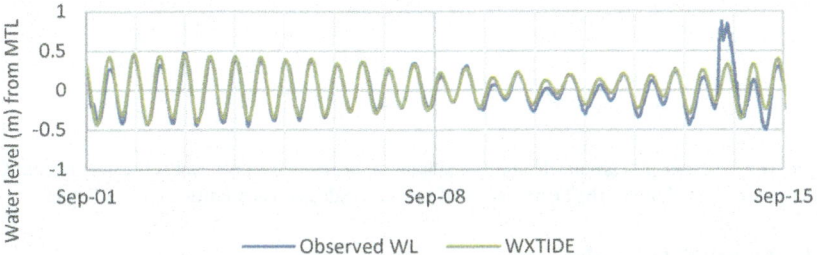

Figure 4. Comparison of forecast tide levels by WXTIDE and observed water levels in Basco Port

4.3. Storm Waves

Due to the unavailability of offshore wave observations, the results of the wave model cannot be numerically verified. To give an indication of the correctness of the results of the wave model, the wave estimates from the Coastal Engineering Manual (CEM) [19] were utilized. During the peak of the typhoon, the simulated maximum deepwater wave height 16.70 m (Figure 7a), while the deepwater significant wave height estimate given by CEM [19] is approximately 17 m (Figure 7b). Figure 6 shows similarity of the order of magnitude of the estimated and simulated values of the maximum deepwater significant wave height, which presents a good indication of the modeling of the wave processes.

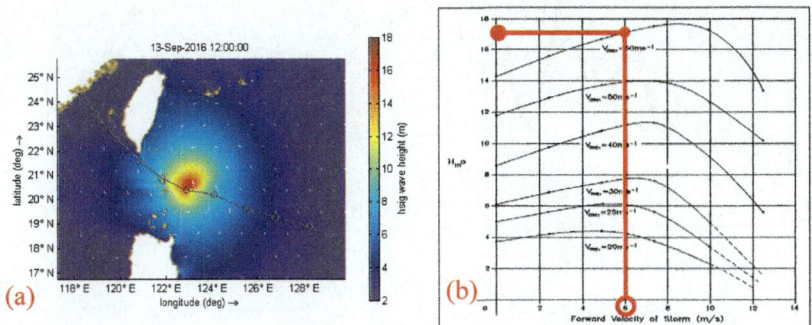

Figure 5. (a): Snapshot of simulated deepwater wave conditions. (b): Deepwater wave height estimate taken from CEM (2005)

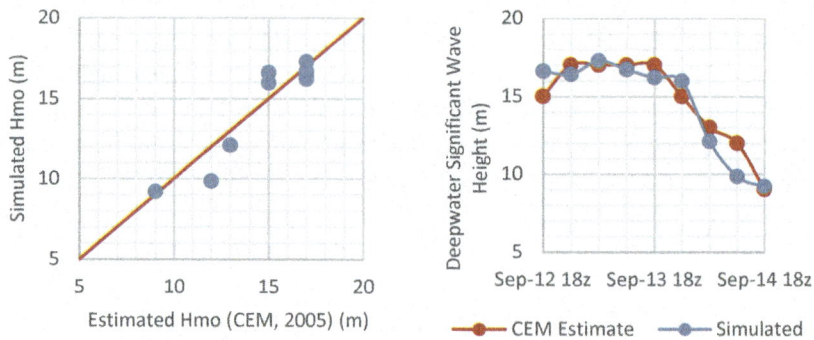

Figure 6. (a) Scatter plot around 1:1 line of maximum deepwater significant wave height from CEM (2005) versus simulations. (b) Time history of maximum deepwater significant wave heights

4.4. *Storm Tide Levels*

Several simulation cases showed that the storm surge heights in the study area were primarily due to the negative pressure of the typhoon, and was not significantly sensitive to the applied wind forcing [20]. This observation was explained by Harper et al. [21] stating that for islands with narrow continental shelves and in deep water, the main contribution of the total surge comes from the inverted barometer effect and is further increased by wave-induced setup near the coastline.

The results of the numerical storm tide model were verified against the observed water level time history measured from the tide monitoring station in Basco Port during the passage of the typhoon, and are presented in Figure 7.

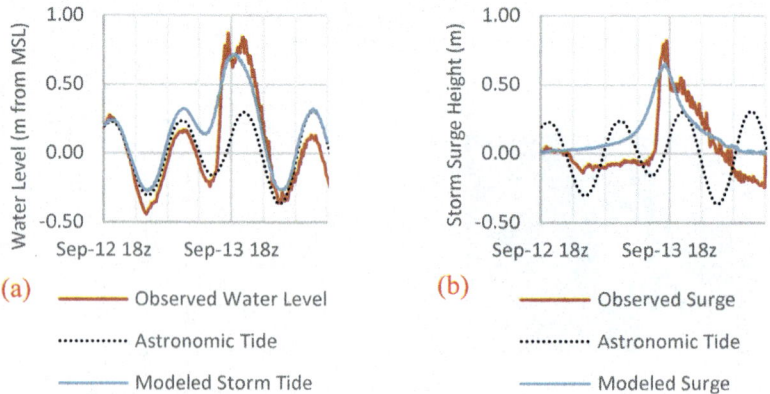

Figure 7. (a): Modeled and observed storm tide levels in Basco Port. (b): Modeled and observed storm surge heights in Basco Port.

The storm surge height is computed as the difference between the storm tide level and the astronomic tide level. The simulated peak storm surge height is 0.65 m and occurred on September 13, 17:10 UTC. This did not occur in the same instant as the simulated peak storm tide level (0.72 m above MSL) that occurred at 18:50 UTC. The calculated peak surge height occurred as the astronomic tide level is still rising from a low tide, while the simulated peak storm tide level occurred during a high astronomic tide level. The observed peak storm tide level 0.87 m above MSL (corresponding to the peak surge height of 0.82 m) and occurred at 17:30 UTC, 20 minutes later than the occurrence of the simulated peak surge height. The difference between the simulated and observed peak storm tide levels is only 15 cm.

A summary of the measured inundation and wave run up levels, together with the simulated peak water levels at the 37 survey locations of Tajima et al. [3] is presented in Table 1. It is observed that the simulated peak water levels are consistently and drastically lower than the measured water levels. The calculated differences are within the range of 1.5 to 5.5 m.

The differences in the simulated peak water levels and measured runup levels are expected and are explained by the fact that the storm tide numerical model used only gives the still water level. The wave runup elevations are more properly modeled using a nearshore phase-resolving wave propagation model, which is outside the scope of this paper. The calculated relative errors between the simulated peak water levels and the measured wave runup level therefore only give an indication of the contribution of the nearshore processes to the total wave runup elevation.

The storm tide model is expected to closely approximate the observed inundation levels, since the inundation levels give an indication of the still water level during the typhoon event. However, the simulated peak water levels are lower than the measured inundation level for all 12 inundated survey locations. Since the accuracy of the primary factors that affect storm tide levels (i.e. atmospheric forcing, astronomic tide levels, wave conditions) were properly established above, the differences between the simulated peak water levels and observed inundation levels can be attributed to other unusual nearshore processes (e.g. local surge height amplification due to fringing reefs) that are not considered in the current two-way coupled hydrodynamic model (based on nonlinear shallow water equations) and wave model (phase averaged) setup. This outcome is consistent with other numerical analyses conducted regarding storm surge height determination in other coastlines with fringing reefs ([22], [23]) wherein more detailed simulations were conducted on the nearshore using wave numerical models based on the Boussinesq wave equations to adequately model the observed storm surge heights.

Table 1. Summary of measured inundation/runup level (following Tajima et al., 2017) and simulated peak water levels. Red font: inundation observations; Blue font: wave runup observations

Point ID	Level Type (Inundation or Runup)	Measured Inundation/ Runup Level (m above MSL)	Simulated Peak Water Level (m above MSL)	Difference (m)	Percent Difference (%)
1	I	3.4	0.55	2.85	83.82
2	I	2.3	0.55	1.75	76.09
3	R	3.0	0.58	2.42	80.67
4	R	2.7	0.58	2.12	78.52
5	I	3.1	0.58	2.52	81.29
6	I	3.6	0.58	3.02	83.89
7	R	3.1	0.63	2.47	79.68
8	R	1.8	0.78	1.02	56.67
9	R	3.1	0.47	2.63	84.84
10	R	4.0	0.97	3.03	75.75
11	R	2.7	0.71	1.99	73.70
12	R	2.9	0.50	2.40	82.76
13	R	4.9	0.73	4.17	85.10
14	I	2.9	1.14	1.76	60.69
15	I	3.8	1.07	2.73	71.84
16	I	5.7	0.83	4.87	85.44
17	R	5.9	0.48	5.42	91.86
18	I	4.3	0.85	3.45	80.23
19	R	2.8	0.87	1.93	68.93
20	I	3.1	0.90	2.20	70.97
21	I	3.7	0.88	2.82	76.22
22	I	4.6	0.89	3.71	80.65
23	I	3.9	0.86	3.04	77.95
24	R	3.1	0.78	2.32	74.84
25	R	3.2	0.63	2.57	80.31
26	R	2.6	0.63	1.97	75.77
27	R	4.5	0.74	3.76	83.56
28	R	2.8	0.67	2.13	76.07
29	R	3.4	0.70	2.70	79.41
30	R	4.0	0.92	3.08	77.00
31	R	3.9	0.98	2.92	74.87
32	R	4.8	0.97	3.83	79.79
33	R	3.8	1.11	2.69	70.79
34	R	2.7	1.02	1.68	62.22
35	R	5.2	0.72	4.48	86.15
36	R	2.4	0.60	1.80	75.00
37	R	5.9	1.21	4.69	79.49

5. Conclusions

A storm tide numerical model was successfully set up and implemented to simulate the storm tide levels in Batanes islands during the September 2016

Typhoon Meranti event, using the two-way coupled Delft3D-SWAN models. The accuracy of the primary factors that affect storm tide levels (i.e. atmospheric conditions, wind-generated waves, astronomic tide levels) were established, and the contributors to errors in the simulated storm tide levels were accounted for. The ability of the model to adequately hindcast the time history of water levels in the Basco Port tide monitoring station indicates that all important processes relevant to the storm tide simulations were properly accounted for by the numerical model.

Comparing the results of the simulations to the findings of the post-disaster survey conducted by Tajima et al. [3] showed that the simulated peak water levels were consistently and drastically lower than the measured water levels. The observed differences between the simulated peak water levels and wave runup levels give an indication of the contribution of the nearshore processes to the wave runup, while the differences between the simulated and measured inundation levels can be attributed to other unusual nearshore processes (e.g. surge height amplification due to the fringing reefs) that are not considered in the current storm tide numerical model setup.

Currently, detailed inundation analyses focused on specific areas are not conducted due to the lack of more detailed information (e.g. detailed topography and bathymetry, detailed beach profile, land use) in the study area, besides the coarse 1:50,000 scale maps already used in this study. Detailed inundation and wave runup analyses are intended to be conducted using a nearshore phase-resolving wave model once this information is made available.

Acknowledgments

The authors would like to thank Professors Yoshimitsu Tajima and Takenori Shimozono of University of Tokyo for the opportunity of a joint post-disaster survey work in Batanes Islands. The authors would also like to thank John Phillip Lapidez for providing the post-disaster survey related primary information used in this paper.

References

1. National Disaster Risk Reduction and Management Council. (2014) *NDRRMC Update: Update re the Effects of Typhoon "YOLANDA" (Haiyan)*, http://www.ndrrmc.gov.ph/attachments/article/1329/U pdate_on_Effects_Typhoon_YOLANDA_(Haiyan)_17APR2014 .pdf

2.   National Disaster Risk Reduction and Management Council. (2016) NDRRMC Update: SitRep No. 13 .re Preparedness Measures and Effects of Typhoon "FERDIE" (I.N. Meranti) covering the period 240600H – 250600H September 2016, http://www.ndrrmc.gov.ph/attachments/ article/2913/SitRep_No_13_re_Preparedness_Measure and_Effects_of_Typhoon_FERDIE_(MERANTI)_Covering_th e_240600H_to_250600H_September_2016.pdf

3.   Y. Tajima, J.P. Lapidez, J. Camelo, M. Saito, Y. Matsuba, T. Shimozono, D. Bautista, M. Turiano, and E. Cruz, Post-disaster survey of storm surge and waves along the coast of Batanes, the Philippines, caused by super Typhoon Meranti/Ferdie. *Coastal Engineering Journal*, **59,** (1), 1750009, pp 11, (2017).

4.   G.R. Lesser, J.A. Roelvink, J.A.T.M. van Kester, and G.S. Stelling, Development and Validation of a three-dimensional morphological model. *Coastal Engineering*, **51** (8-9), 883, (2004).

5.   Deltares, Delft3D-FLOW: Simulation of multi-dimensional hydrodynamic flow and transport phenomena, including sediment User Manual ver 3.15, Delft:Deltares, pp 686, (2014a).

6.   Deltares, Delft3D-WAVE: Simulation of short-crested waves with SWAN User Manual ver 3.05, Delft: Deltares, pp 208, (2014b).

7.   N. Booij, R.C. Ris, and L.H. Holthuijsen, A third-generation wave model for coastal regions. 1. Model description and validation, *Journal of Geophysical Research*, **104** (C4), 7649 (1999).

8.   R.C. Ris, L.H. Holthuijsen, and N. Booij, A third-generation wave model for coastal regions. 2. Verification. *Journal of Geophysical Research*, **104** (C4), 7667, (1999).

9.   P. Weatherall, K. M. Marks, M. Jakobsson, T. Schmitt, S. Tani, J. E. Arndt, M. Rovere, D. Chayes, V. Ferrini, and R. Wigley, A new digital bathymetric model of the world's oceans, *Earth and Space Science*, **2**, 331–345, (2015).

10. G.D. Egbert and S.Y. Erofeva, Efficient inverse modeling of barotropic ocean tides, *Journal of Atmospheric and Oceanic Technology*, **19**, 183 (2002).

11. T. Fujita, Pressure Distribution Within Typhoon, *Geophysical Magazine*, **23**, 437 (1952).

12. M. Zijlema, G.Ph van Vledder, and L.H. Holthuijsen, Bottom friction and wind drag for wave models. *Coastal Engineering*, **65**, 19, (2012).

13. L.H. Tolman, A third-generation model for wind waves on slowly varying, unsteady, and inhomogeneous depths and currents, *Journal of Physical Oceanography*, **21**, 782-797, (1991).

14. J. Battjes and J. Janssen, Energy loss and set-up due to breaking of random waves, in *Proc. International Conference of Coastal Engineering*, (Hamburg, Germany, 1978).

15. K. Hasselmann, T.P. Barnett, E. Bouws, H.D. Carlson, E. Cartwright, L. Enke, J. Ewing, H. Gienapp, D.E. Hasselmann, P. Kruseman, A. Meerburg, P. Müller, D.J. Olbers, K. Richter, W. Sell, and H. Walden, Measurements of wind wave growth and swell decay during the Joint North Sea Wave Project (JONSWAP). *Deutsche Hydrographische Zeitschrift*, **8**, (12), (1973).

16. G.J. Komen, S. Hasselmann, and K. Hasselmann, On the existence of a fully developed wind-sea spectrum. *Journal of Physical Oceanography*, **14**, 1271, (1984).

17. P.J. Vickery, and D. Wadhera, Statistical models of Holland pressure profile parameter and radius to maximum wind of hurricanes from flight-level pressure and H*Wind data, *Journal of Applied Meteorology and Climatology*, **47**, 2497, (2008).

18. B.A. Harper and G.J. Holland, An updated parametric model of the tropical cyclone, in *Proc. AMS 23rd Conference on Hurricanes and Tropical Meteorology*, (Dallas, Texas, 1999).

19. D. Resio, S. Bratos, and E. Thompson, Meteorology and Wave Climate. In: Vincent, L., and Demirbilek, Z. (editors), *Coastal Engineering Manual*, Part II, Hydrodynamics, Chapter II-2, Engineer Manual 1110-2-1100, U.S. Army Corps of Engineers, Washington, DC, (2005).

20. M. Turiano, Numerical analysis of storm tide levels with wave setup in Batanes Islands during Typhoon Meranti 2016, Master's thesis, University of the Philippines Diliman (2017).

21. B.A. Harper, T.A. Hardy, L.B. Mason, L. Bode, I.R. Young, and P. Nielsen, *Queensland climate change and community vulnerability to tropical cyclones, Ocean hazards assessment, Stage 1*, Department of Natural Resources and Mines, Queensland, Brisbane, Australia, pp 368, (2001).

22. A.B. Kennedy, J.J. Westerink, J.M. Smith, M.E. Hope, M. Hartman, A.A. Taflanidis, S. Tanaka, H. Westerink, K.F. Cheung, T. Smith, M. Hamann, M. Minamide, A. Ota, and C. Dawson, Tropical cyclone inundation potential of the Hawaiian Islands of Oahu and Kauai, *Ocean Modelling*, **52-53**, 54, (2012).

23. V. Roeber, and J.D. Bricker, Destructive tsunami-like wave generated by surf beat over a coral reef during Typhoon Haiyan. *Nature Communications*, **6**, 1, (2015).

Process-Oriented Numerical Experiments of Storm-Induced Semidiurnal Surge on The South Atlantic Bight[*]

Xi Feng[†]

*College of Harbour, Coastal and Offshore Engineering, Hohai University,
Nanjing, Jiangsu Province, 210098, China;*
[†]E-mail: xifeng@hhu.edu.cn
http://en.hhu.edu.cn/

Maitaine Olabarrieta and Arnoldo Valle-Levinson
*[2]Department of Civil and Coastal Engineering, University of Florida,
Gainesville, FL, 32608, USA*

The tidal record on the East Coast of the United States reveals the frequent semidiurnal perturbation of the storm surge on the South Atlantic Bight. Tropical cyclones were examined to be one of the main triggers for these perturbation events. The peak of the storm surge is centered in the mid of the South Atlantic Bight, and radiating along the coast southward and northward. The process-oriented experiments were designed with parametric determination based on the historic events. The experiments were carried out in order to further discuss the various factors of tropical cyclones for their effect on the storm surge and tide interaction. The experimental analysis shows the storm surge and tide interaction is the most severe when a cyclone is moving orthogonal to the coastline, although the parallel-to-shore tropical cyclones are the most observed in history. The intensity of the semidiurnal perturbation to storm surge is positively correlated to the wind strength and the radial of the maximum wind speed of a cyclone, but negatively correlated to the translation velocity of a cyclone. For a cyclone moving parallel-to-shore, the corresponding storm surge perturbation lasts the longest and is the strongest, when the "fetch" of the alongshore wind with speed >17m/s (the criticality of the wind speed of the Tropical Storm based on the Saffir-Simpson category) reaches the maximum on the continental shelf of the mid of the South Atlantic Bight. High correlation is found between the tropical cyclone induced alongshore oceanic current and the semidiurnal perturbation to storm surge.

Keywords: Storm Surge, Tide, Semidiurnal Surge, Tropical Cyclone, Tide-Surge Interaction, South Atlantic Bight.

---

[*] This work is supported by NSF funding from U.S.A,
[†] Work partially supported by JSCE506 Jiangsu Key Laboratory of Hydrology and Water Resources at Hohai University, China, SOED1609, Second Institute of Oceanography, SOA, Hangzhou, China.

1. Introduction

At the places where storm often occurs, scientists found the observed tide often diverts to the time series of the astronomical tide, showing abnormal fluctuation, over-height of low tide, phase differences and other phenomena [e.g. Prandle & Wolf, 1978; Jones & Davies, 2007, 2008; Idier et al., 2012]. This can result in a semi-tidal perturbation in the signal of the residuals by subtracting the astronomical tide from the observed tide [Horsburgh & Wilson, 2007]. This perturbation influences the accuracy of the storm surge warning system and brings difficulties for predicting the peak and the time of the storm surge.

The perturbation of the storm surge is mainly due to the nonlinear interaction between the tide and the storm [Valle-Levinson et al., 2013; Feng et al., 2016]. The interaction on one side makes the energy of storm surge left on the tidal signal and results in an over-high and advanced tidal wave [Jones & Davies, 2007; 2008; Horsburgh & Wilson, 2007]; on the other hand, the energy of the tidal wave is left on the storm signal. Therefore, the residuals, which are regarded as storm surge signal through data analysis, show a strong perturbation. As storm surge is directly related to the local meteorological-condition, the storm perturbation follows the same rule. For the cyclone-induced storm surge, the perturbation is largely dependent on the wind field induced by a Tropical Cyclone (TC) [Valle-Levinson, 2013; Feng et al., 2016].

Valle-Levinson et al. [2013] revealed the tide-surge interactions in the mid-South Atlantic Bight (SAB) during the passage of Hurricane Sandy in 2012, the amplitude of the perturbation signal reached 50% of the storm surge height off the coast of Georgia State. The SAB has a semi-circular continental shelf, and the tide is predominantly semidiurnal and shows a quasi-standing wave behavior in the mid-SAB [Blanton et al., 2004]. The protruding tide-surge interaction on the mid-SAB coast, as reflected by the semidiurnal surge phenomenon, was therefore highly suspected to be related to the quasi-standing tidal behavior [Valle-Levinson et al., 2013]. In the following study, Feng et al. [2016] expand a series of investigation on the perturbation phenomenon. Through 19-years' analysis on the tidal gauge records, they found the semidiurnal perturbation signal is continuously protruding in the mid-SAB and it is associated with continuously intense alongshore winds. Through numerical simulations, they confirmed that it is the wind induced oceanic current which in scale can compare to the tidal current that enhances the Coriolis parameter and the bottom friction, which then change the mechanisms for tidal propagation. Whereas, their numerical simulation was based on constant winds while ignores the most common condition when TC occurs. On the other hand, the nonlinearity during tidal propagation movement as well as the storm surge development are all sensitive to wind effect, which reflect to different perturbation phenomena, such as amplitude, frequency, and duration [Li et al., 2003; Li & Chen,1999; Vennell, 2010; Thiebaut & Vennell, 2011].

To sum up, this study takes the SAB as an example for determining the characteristics of the semidiurnal perturbation to storm surges (simplified as 'semidiurnal surge' in the following paragraphs) induced by TCs. The following paragraphs start from each TC element and analyze the behavior of the storm-perturbation signal under different TC circumstances.

2. Methodology

Valle-Levinson et al. [2013] and Feng et al. [2016] utilized the ROMS (Regional Ocean Model System) for establishing the simulation system and successfully output the perturbation phenomenon to storm surge under the real case or a constant wind field. Therefore, the study inherits the hydrodynamic simulation system while develops the research on the sensitivity of the perturbation to TC patterns. The description of the system will not be duplicated here. The focus is on the translation speed, the path, the storm size, the intensity and the location of the eye of a TC. For each element, the experiments will take the method of controlling variables for experiment design and capturing the spatial and temporal variation of the storm perturbation.

2.1. Idealized bathymetry for mid-SAB

The SAB region sits west of the North Atlantic (NA) Ocean. It spans from Cape Canaveral in Central Florida (28 N) to Cape Hatteras in North Carolina (35 N), and is featured by a semicircle-shaped continental shelf (Figure 1A). The continental shelf varies from 30 km to 130 km in width and its widest part is located in Georgia and South Carolina. The shelf extends offshore around 100-130 km with a gentle slope and then sharply tilts at shelf break. The water depth at the shelf is generally within 100m. The mid-SAB is featured by broad and uniform topography, while it is associated with the most intense and durable signal of the semidiurnal surge. Therefore, an idealized study domain with the 2km×2km horizontal resolution is created to characterize the nearshore region between Fernandina Beach, (Florida) and Charleston (South Carolina). The domain is 800 km in length (offshore direction) and 430 km in width (alongshore direction) (Figure 1B).The continental shelf extends 130 km offshore with a constant slope of 1:1300 and the shelf break spans another 43 km eastward. The deepest water depth is 800 m in the offshore region (Figure 1C).

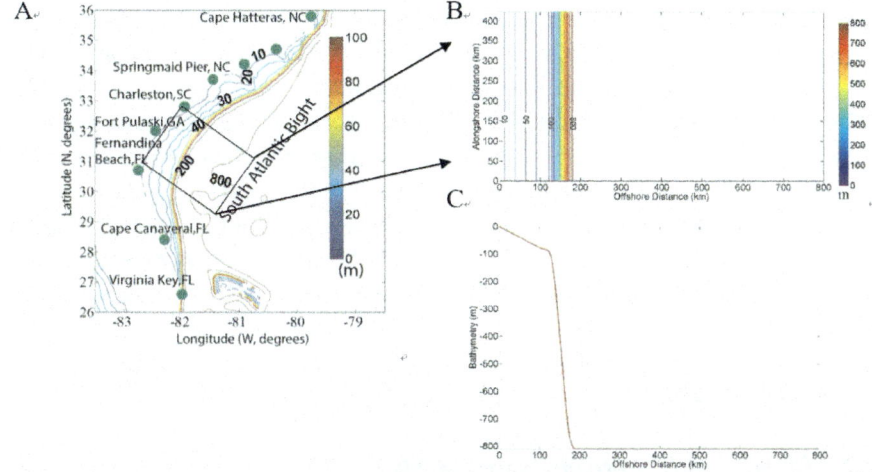

Fig. 1. A) Topography of the SAB; B) Idealized grid domain for the sensitivity tests; C) Vertical profile along the offshore direction of the Idealized grid domain.

2.2. Analytic Holland wind and pressure model

The analytical Holland model [1980] is used to derive the radial profile of winds and sea level atmospheric pressures associated with a TC. It is worth mentioning that one of the limitations of this model is that the resulting wind and pressure fields are symmetric, though real TCs demonstrate asymmetric fields. Utilizing symmetric wind and pressure fields allows us to better control the characteristics of TCs (intensity, size, translation speed, core's location and track) and thus yields a better understanding of their influence on semidiurnal surges. The wind and atmospheric pressure fields are given by the Holland model [1980] can be computed using Equations:

$$p = p_c + (p_n - p_c) \times \exp\left[-\left(\frac{RMW}{r}\right)^B\right] \tag{1}$$

$$V_g = \left[\left(\frac{RMW}{r}\right)^B \times \exp\left(1.0 - \left(\frac{RMW}{r}\right)^B\right) \times V_{max}{}^2\right]^{1/2} \tag{2}$$

$$V_{max} = \left[\frac{B \times (p_n - p_c)}{e \times \rho_a}\right]^{1/2} \tag{3}$$

where p is the sea level atmospheric pressure at radius r, p_c is the central pressure in the hurricane, p_n is the ambient atmospheric pressure and assumed constant of 1013mb, RMW is the radius of maximum winds, V_{max}, ρ_a is the air density and assumed constant at 1.15 kg m^{-3}, B is the storm scale (peakedness) parameter in Holland model, and is determined by Powell et al. [2005]'s model in the form of:

$$B = 1.881 - 0.00557RMW - 0.01097\varphi \qquad (4)$$

Where φ is latitude and is set to be 32 degrees north, a mean value for the mid-SAB. Table 1 lists the TCs that been identified for their association with semidiurnal surges through the 19-year tidal gauge water data analysis.

RMW is used to control the storm size. The gaps in Table 1 of RMW can be offset by applying empirical equation, as RMW is a function of latitude and intensity [Vickery et al., 2000; Vickery & Wadhera, 2008]. In general, RMW increases as latitude increases and wind intensity decreases [Kimball & Mulekar, 2004]. The maximum wind speed W_{max} is used to describe a tropical storm or a hurricane's wind intensity.

2.3. *Setup of process-oriented experiments*

Four sets of simulations are conducted to analyze the influence of different tropical storms (Table 2). The first set compares varied directions of the storm motion. In the second set, TCs also move parallel to the shore (0°). This set is designed to compare the influence on the interaction from different wind speeds. The third experiments sought to explore the storm size's effect on the semidiurnal surges. The design of the last set of experiments starts from the concept of the "fetch of wind" and explores the relation between storm size and offshore distance for contributing the tide-storm interaction. "Fetch" is referred to as the distance and time over which the wind blows consistently [Troitskaya et al. 2010; Mellor 1996, pp.161; Bretschneider, 1965]. For cyclones with a trajectory off the continental shelf, the near-shore region would be covered by steady and consecutive alongshore wind stress.

The parameters are described in Table 2. It should be noted that the design of the above experiments do not intend to model hurricanes or hurricane-induced storm surge in reality, but aim to provide a conceptual relation between the storm parameters and the semidiurnal surges. For example, a category 2 hurricane with RMW of 100 km is barely been seen in nature, but on the scale of the idealized domain, using a broad span of RWMs can facilitate to establish the relation between storm size and semidiurnal surge.

3. Results

The time series of the associated semidiurnal surge amplitude shows a shape as a Gaussian function. The maximum semidiurnal surge's amplitude and duration during each experiment are then taken as a proxy for quantifying the relevance or intensity of the nonlinear tide-surge interactions. A threshold of 0.05m in the semidiurnal amplitude is chosen to quantify the duration of semidiurnal surge event.

3.1. *Hurricane heading direction*

The first set of experiments shows that the semidiurnal surge is not strongly related to the direction of TC. In Figure 2A, the x-axis shows the eastward, westward and northward direction respectively. The semidiurnal surge is most intense during the landfall storm events as it is 0.2m for the land-to-ocean scenario. In the parallel-to-shore scenario, the semidiurnal surge amplitude is smaller by at least 0.1 m to the other two. The duration of the surges are proportional to the maximum amplitude.

3.2. *Hurricane wind stress*

In the second set of experiments, there are 20 experiments with a combination of different hurricane wind intensities (from 21.5m/s to 70m/s) and translation speeds (from 1.5m/s to 5 m/s). As can be seen from Figure 2B, by controlling the translation speed (not counted into the hurricane-wind stress), we found the semidiurnal surge's amplitude and duration are proportional to the category of the hurricane wind stress. Semidiurnal surge's amplitude increases at 0.1m per category on average, and its duration lengthens at 10hr per category on average. With a combination of different translation speeds, the curves are not parallel, this uncovers the combination function between the wind intensity and translation speed of a TC, which will be analyzed in the next paragraph.

3.3. *Hurricane translation speed*

The translation speed determines how long a region is affected by a hurricane. The slower a hurricane moves, the more consistent the wind stress acts on a particular region. By controlling hurricane-wind category, Figure 2C shows that the semidiurnal surge's amplitude (Max.Amp) and duration (Dur) decreases with an increase of hurricane's translation speed. The amplitude decreases linearly and the duration reduces exponentially. With stronger hurricane intensity, the semidiurnal surge's amplitude and duration increases more rapidly as the hurricane's translational speed reduces. For a category 3 hurricane, as the cyclone's translation speed decreases from 10 m/s to 1.5 m/s, the associated maximum semidiurnal-surge amplitude grows from 0 to 0.08 m and the duration changes from 0 to 50 hours. Whereas for a category 5 hurricane, as the cyclone's translation speed decreases from 10 m/s to 1.5 m/s the associated maximum semidiurnal-surge amplitude enhances from 0 to 0.4 m and the duration increases from 0 to 10 days. The rate of the semidiurnal surge's intensity and persistence is increased by a factor of 5.

3.4. *Hurricane storm size*

The storm size determines the size of the area forced by the TC induced winds and the "fetch of wind" in the near-shore region. The semidiurnal surge is enhanced linearly with the enlarged storm size (Figure 2D). Results from the third set of experiments revealed that there is no apparent interaction between storm size and translation speed, as the curves of Max.Amp and Dur. are all parallel. With every span of 10km in storm size, the semidiurnal surge's amplitude would increase by 20cm and the duration would extend by 10hrs.

3.5. *Hurricane offshore distance*

Figure 2E shows for such a TC moving from south to north, the duration and the strength of the semidiurnal surge increases as the trajectory of the cyclone came nearer to the continental shelf. However, once the cyclone's trajectory is on the continental shelf, the semidiurnal surge weakens as the TC comes closer to the shore. It is found that for a TC with RMW around 40-60 km and moving on a path near the edge of the continental shelf, the semidiurnal surge's amplitude generated on the coast is the largest. With a width similar to the mid-SAB, the average wind stress induced by the TC is the greatest over the continental shelf. Particularly, the shelf is fully covered by the field with wind velocity over 20 m/s. Within the core of a TC, the wind strength is small or zero. As the core of a cyclone came closer to the shore, the wind stress over the continental shelf becomes lower. In this case, the intensity of the semidiurnal surge goes down as well. Similarly, for a TC propagating on the shelf, the semidiurnal surge decreases as the RMW expands. The semidiurnal surge is the most persistent when the cyclone is located at an offshore distance of L around 4×RMW. In this case, the rim of a tropical depression system (atmospheric pressure <1013mb) is the closest to the open western boundary, and the "fetch of wind" with the "storm wind (17.5m/s)" status [Avila & Stewart, 2013] is the longest.

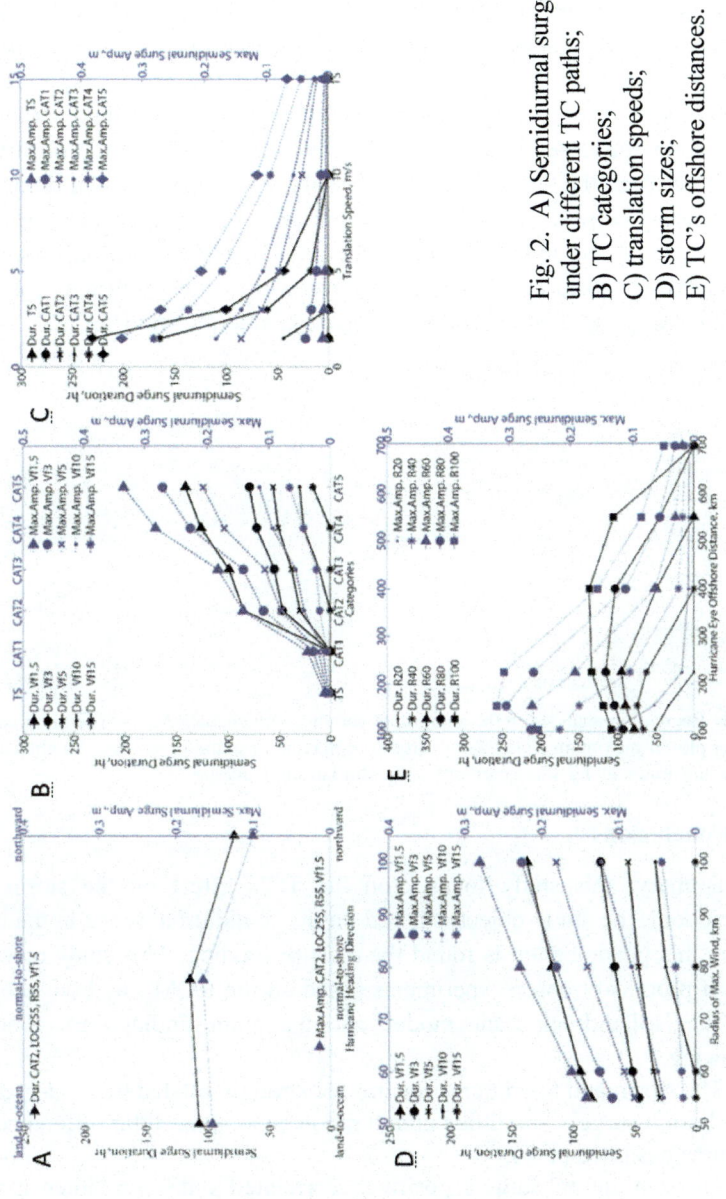

Fig. 2. A) Semidiurnal surge under different TC paths; B) TC categories; C) translation speeds; D) storm sizes; E) TC's offshore distances.

3.6. *Relation between the semidiurnal surges and alongshore currents*

It can be seen from the above experiments, in most cases, it is the alongshore winds that determine the intensity of storm and tide interaction. The correlation between the alongshore current and the semidiurnal surge's amplitude and duration is thereby explored. By summarizing the above experiments, high correlation with $R^2 = 0.89$ exists between the maximum oceanic current velocity and the maximum semidiurnal surge's amplitude (Figure 2A). A positive trend is found between the strength of the alongshore currents and the duration of the semidiurnal surges. However, the regression plot (Figure 2B) shows more scatters. It can be induced that the duration of the interaction process can be influenced by more factors rather than the wind intensity of a TC, whereas the semidiurnal surge's amplitude is heavily related to the wind speed.

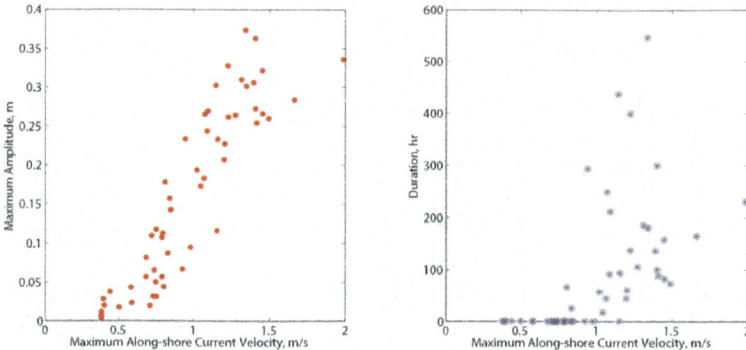

Fig. 3. Regression analysis on the alongshore currents with characteristics of semidiurnal surges. A) Scatter plot of maximum along-shore current vs. maximum semidiurnal surge amplitude; B) Scatter plot of maximum along-shore current vs. semidiurnal surge duration.

4. Conclusions

In summary, this study focuses on the TC's effect on the storm and tide interaction. The study object is based on the semidiurnal surge in the mid-SAB, where this phenomenon is found the most protruding. This study conducts four sets of process-oriented experiments based on the ROMS hydrodynamic model and the Holland hurricane model, and the main findings are concluded as following:

•   The storm and tide interaction are not strongly related to wind strength. The semidiurnal surge's amplitude and duration encompass little difference between the three designed paths;

•   The semidiurnal surge is positively correlated to the wind intensity;

•   The duration of the semidiurnal surge increases exponentially with the reduction of a hurricane's translation speed;

- The semidiurnal surge goes proportional to the RMW of TC. The amplitude and the duration grow linearly with the expansion of the storm size, RMW;
- For parallel-to-shore hurricanes of an averaged size (~55km), the semidiurnal surge is the most persistent and the peak of the semidiurnal surge is the largest, when the "fetch of wind" with storm intensity reaches the largest;
- The amplitude of the semidiurnal surge is found with a direct and positive relation to alongshore wind-induced current while the duration of which involves more influencing factors.

5. Tables

Table 1. Selected the historical record for hurricanes (with the semidiurnal surge) used to determine Holland model parameter in this study

Name	Year	Distance Offshore	Maximum Wind Speed	Forward Speed	Sea Level Pressure	The radius of Max.Wind
Range/Unit		km	m/s	m/s	mb	km
Bonnie	1998	369-545	49	6.8	963	51-55
Dennis	1999	370-410	49	14.1	969	53-56
Floyd	1999	230-480	49	16.6	933	43-47
Irene	1999	230-400	49	12.2	982	55-59
Gabrielle	2001	[a]	33	8	983	46
Frances	2004	447-578	49	11.4	960	51-54
Ophelia	2005	180-1800	49	1.5	976	52-56
Hybrid event in 2015-Oct	2005	571-1123	23	6.1	1001	56-60
Noel	2007	900-1853	49	20	980	54-58
Earl	2010	572-1005	49	4	949	48-51
Irene	2011	337-593	49	10.5	942	46-50
Sandy	2012	555-615	49	14.4	956	50-53

[a] Hurricane Gabrielle developed from the Gulf of Mexico and moved across State Florida

Table 2. Sets of experiments for studying the effect from meteorological factors

Set	Varied Parameters (Units)	Varied Values	Fixed Parameters and Values (Units)	Simulation #
1	θ, Heading Direction	1: Eastward, 2: Westward, 3: Northward L=225km	W_{max}=47m/s, MW=55km, V_f,=1.5m/s, A^a=1.0m	3
2	W_{max}, Maximum Wind Speed (m/s)	21.5, 31, 47, 52, 64, 70	RMW=55km, A =2.0m,	30
	V_f, Translation speed (m/s)	1.6, 3, 5, 10, 15	θ=Northward, L =225km	
3	RMW Radius of Max. Wind (km)	55, 60 80, 100	W_{max}=47m/s, A =2.0m,	20
	V_f, Translation speed (m/s)	1.6, 3, 5, 10, 15	θ=Northward, L=225km	
4	L, Offshore distance (km)	105, 155, 225, 400, 550, 700	W_{max},=47m/s, V_f=1.5m/s	30
	RMW, Radius of Max. Wind (km)	20,40, 60, 80,100	A=2.0m, θ=Northward	

[a]A is the astronomic tidal amplitude at the west boundary

222

Acknowledgments

This work is supported by NSF funding from U.S.A, and by JSCE506 from Jiangsu Key Laboratory of Hydrology and Water Resources at Hohai University, China, as well as partially supported by SOED1609, Second Institute of Oceanography, SOA, Hangzhou, China.

References

1. Agbley, S., 2009: Towards the efficient probabilistic characterization of tropical Cyclone-generated storm surge hazards under stationary and non-stationary conditions. Ph.D. Dissertation, Old Dominion University, pp. 151.
2. Avila, L. A., & Stewart, S. R. (2013). Atlantic Hurricane Season of 2011*. Monthly Weather Review, 141(8), 2577-2596.
3. Feng, X., Olabarrieta, M., & Valle-Levinson, A. (2016). Storm-induced semidiurnal perturbations to surges on the us eastern seaboard. Continental Shelf Research, 114, 54-71.
4. Bretschneider, C. L. (1965). Generation of waves by wind. State of the art (No. NESCO-SN-134-6). National Engineering Science Co Washington DC.
5. Holland, G. J. (1980). An analytic model of the wind and pressure profiles in hurricanes. Monthly Weather Review, 108(8), 1212-1218.
6. Horsburgh, K. J., & Wilson, C. (2007). Tide - surge interaction and its role in the distribution of surge residuals in the North Sea. Journal of Geophysical Research: Oceans, 112(C8). DOI: 10.1029/2006JC004033.
7. Idier, D., Dumas, F., & Muller, H. (2012). Tide-surge interaction in the English Channel. Natural Hazards and Earth System Sciences, 12(12), 3709-3718.
8. Jones, J. E., & Davies, A. M. (2007). Influence of non-linear effects upon surge elevations along the west coast of Britain. Ocean Dynamics, 57(4-5), 401-416.
9. Jones, J. E., and A. M. Davies (2008), On the modification of tides in shallow water regions by wind effects, J. Geophys. Res., 113, C05014, doi:10.1029/2007JC004310.
10. Kimball, Sytske K. and Madhuri S. Mulekar, (2004). A 15-Year Climatology of North Atlantic Tropical Cyclones. Part I: Size Parameters. J. Climate, 17, 3555–3575.
11. Mellor Georgia L. (1996). Introduction to physical oceanography. Woodbury, N.Y.: American Institute of Physics, pp.174.
12. Ning Lin, Kerry Emanuel. (2015) Grey swan tropical cyclones. Nature Climate Change. Online publication date: 31-Aug-2015.
13. Powell, M, G Soukup, S Cocke, S Gulati, N Morisseau-Leroy, S Hamid, L. Axe State of Florida hurricane loss projection model: Atmospheric science component. Journal of Wind Engineering and Industrial Aerodynamics, 93(8) (2005), pp. 651–674.

14. Prandle, D., & Wolf, J. (1978). The interaction of surge and tide in the North Sea and River Thames. Geophysical Journal International, 55(1), 203-216.
15. Proudman, J. (1955a). The propagation of tide and surge in an estuary. Proceedings of the Royal Society of London. Series A. Mathematical and Physical Sciences, 231(1184), 8-24.
16. Proudman, J. (1955b). The effect of friction on a progressive wave of tide and surge in an estuary. Proceedings of the Royal Society of London. Series A. Mathematical and Physical Sciences, 233(1194), 407-418.
17. Proudman, J. (1957). Oscillations of tide and surge in an estuary of finite length. Journal of Fluid Mechanics, 2(4), 371-382.
18. Rady, M. A., El-Sabh, M. I., Murty, T. S., & Backhaus, J. O. (1998). Residual circulation in the Gulf of Suez, Egypt. Estuarine, Coastal and Shelf Science, 46(2), 205-220.
19. Tang, Y. M., R. Grimshaw, B. Sanderson, and G. Holland (1996), A numerical study of storm surges and tides, with application to the north Queensland coast, J. Phys. Oceanogr., 26, 2700–2711.
20. Thiebaut, S., & Vennell, R. (2011). Resonance of long waves generated by storms obliquely crossing shelf topography in a rotating ocean. Journal of Fluid Mechanics, 682, 261-288.
21. Troitskaya, Yu. I., D. A. Sergeev, O. S. Ermakova, G. N. Balandina. (2010) Fine structure of the turbulent atmospheric boundary layer over the water surface. Izvestiya, Atmospheric and Oceanic Physics, 46, 109-120.
22. Valle-Levinson, A., Olabarrieta, M., & Valle, A. (2013). Semidiurnal perturbattions to the surge of Hurricane Sandy. Geophysical Research Letters, 40(10), 2211-2217.
23. Vennell, R. (2010). Resonance and trapping of topographic transient ocean waves generated by a moving atmospheric disturbance, J.Fluid Mech., 650, 427-442.
24. Vickery, Peter J, & D. Wadhera. (2008). Statistical Models of Holland Pressure Profile Parameter and Radius to Maximum Winds of Hurricanes from Flight-Level Pressure and H*Wind Data. Journal of Applied Meteorology & Climatology 47(10), 2497-2517.
25. Vickery, P., Wadhera, D., & Stear, J. (2010). Ss metocean / a synthetic model for gulf of mexico hurricanes. Journal of the American Statistical Association, 22(157), 75-78.
26. Zhang, W. Z., Shi, F., Hong, H. S., Shang, S. P., & Kirby, J. T. (2010). Tide-surge Interaction Intensified by the Taiwan Strait. Journal of Geophysical Research: Oceans (1978–2012), 115(C6). DOI: 10.1029/2009JC005762.
27. Li, Kunping & Zeshi Chen (1999). The characteristics of the tidal variation in the harbor of Xiaochangshan. Advances in Marine Science (in Chinese), 3, 10-15.

Observation and Numerical Investigation of Storm Wave Characteristics on the Fringing Reef along the Coast of Ivana, Batanes, Philippines under the Attack of Super Typhoon Meranti/Ferdie

John Phillip Lapidez* and Yoshimitsu Tajima

*Department of Civil Engineering, The University of Tokyo,
7-3-1 Hongo, Bunkyo-ku, Tokyo 113-8656, Japan
* E-mail: philliplapidez@coastal.t.u-tokyo.ac.jp*

Super Typhoon Meranti brought extreme waves which resulted in severe inundation in the coast of Batan on 13 September 2016. A post-disaster survey revealed the maximum wave runup height of 6.3m and the maximum inundation depth of 6.1m. This study focuses on the small coastal community called Ivana where wave amplification and wave concentration were observed. A 1D Boussinesq model is constructed to simulate the wave transformation over the fringing reefs fronting Ivana. The results show that the 1D model can capture the basic wave characteristics and how they evolve over the reef. However, the 1D model cannot account for the two-dimensional variation of the wave properties. In the case of Ivana, further investigation by a 2D model is recommended to simulate the possible occurrence of either standing or progressive edge waves.

Keywords: Meranti; Fringing Reefs; 1D Boussinesq Model

1. Introduction

Super Typhoon Meranti, with Philippine local name Ferdie, is the strongest typhoon of 2016. The Joint Typhoon Warning Center (JTWC) estimated a peak 1-minute sustained winds of 305 km/h and the Japan Meteorological Agency (JMA) estimated a peak 10-minute sustained winds of 220 km/h and a minimum pressure of 890 hPa. On 13 September 2016, while at its peak strength, Meranti passed through the province of Batanes, Philippines. Meranti generated extreme storm waves which caused coastal flooding and erosion along the coast of the province despite being protected by extensive fringing reefs. In a post-disaster survey conducted at the islands of Batan and Sabtang in Batanes, a maximum wave runup height of 6.3m and a maximum inundation depth of 6.1m were measured. The survey measurements and the track of Meranti are shown in Fig. 1. Detailed discussion of the field survey methodology and results can be found in Tajima et al.[8]. A wave hindcast using WAVEWATCH III and a wave refraction and diffraction simulation around Batan Island were conducted in Lapidez et al.[2]. It was found that the alongshore variation of the measured runup and inundation levels around Batan can be explained by refraction, diffraction, and shielding effects. However, one exception is seen in a locality called Ivana shown in Fig. 2, where wave amplification and wave concentration were observed in a very small area. Lapidez et al.[2] suggests that resonance of infragravity waves on the reef flat is the a possible reason for this phenomenon.

This study aims to investigate the a 1D wave resonance characteristics inside the reef flat using a Boussinesq-type model to simulate the wave transformation over the fringing reefs, wave breaking, and runup.

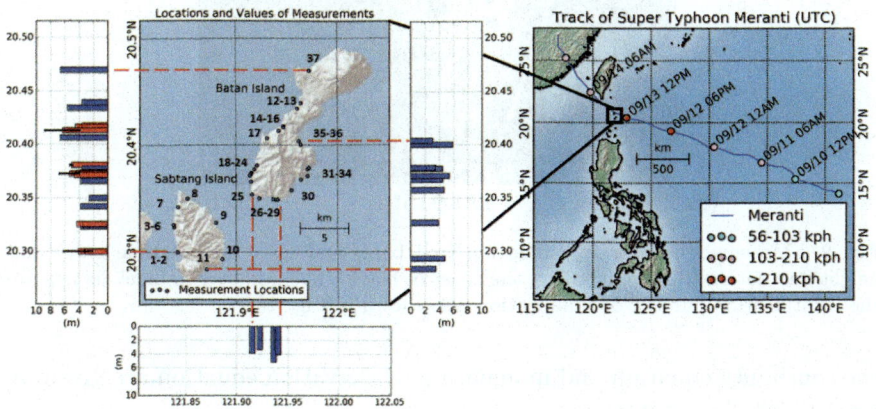

Fig. 1. Left: Locations and measured heights of wave runup (blue bars) and the water level in the inundated areas (red bars). Black bars indicate the witnessed range of fluctuating water level in the inundated area. Right: Track of Super Typhoon Meranti

2. 1D Boussinesq Model

2.1. Governing Equations

In this study, a 1D model based on Boussinesq-type equations is used to model the wave deformation and transformation as the incident wave propagates from the deep-water to the reef face and finally to the shallow reef flat. In particular, a Boussinesq model with improved linear dispersion characteristics as presented by Madsen et al.[5] is applied. For the 1D case,

226

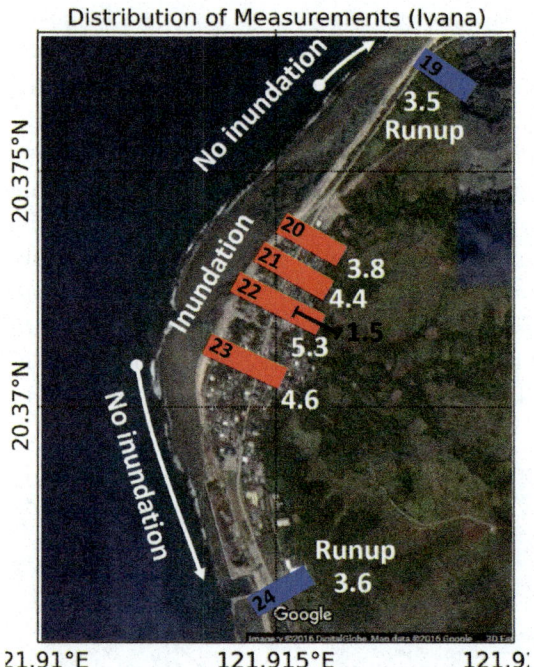

Fig. 2. Measured heights of inundation (red bars) and wave runup (blue bars) and height fluctuation (black bar) along the coast of Ivana. The numbers in black letters are the identification number of the location also shown in Fig. 1

the continuity equation and momentum conservation equation are given by Eqs. 1 and 2 respectively.

$$\eta_t + P_x = 0 \tag{1}$$

$$P_t + \left(\frac{P^2}{H}\right)_x + gH\eta_x - Bgh^3\eta_{xxx} + \left(B + \frac{1}{3}\right)h^2 P_{xxt}$$
$$-2Bgh^2 h_x \eta_{xx} - \left(\frac{h}{3}\right)h_x P_{xt} + R_x = 0 \tag{2}$$

Here, the subscripts x and t represent the space and time derivatives, P is the depth integrated volume flux, h is the still water depth, $H = h + \eta$ is the instantaneous depth, η is the instantaneous water surface elevation, B is dispersion parameter, which is set as B = 1/15 in this study and R is the vertically integrated excess momentum flux due to the wave breaking.

Equations 1 and 2 incorporate a significant improvement of the phase celerity and group velocity properties for the linear waves in water depths up to the deep water limit compared to classical Boussinesq-type model formulations. This particular set of equations is capable of simulating the propagation of irregular wave trains travelling from deep water regions (i.e., up to $h/L_0 = 0.6$, where L_0 is the deep water wavelength) to the shallow water[5].

2.2. Wave Breaking (R_x)

The natural process of waves breaking in the surf and swash zones is not inherently captured in Boussinesq-type equations alone. To address this issue, previous researches introduced empirical sub-models to add a dissipation term in the momentum equation, Eq. 2. The most common approach in published literature can be broadly categorized into two groups: (1) roller model approach (e.g., Schäffer et al.[7], Madsen et al.[4]) and (2) eddy viscosity model approach (e.g., Zelt[9], Kennedy et al.[1]). The difference in the performance of the two approaches are not significant and so both are widely used in different applications.

In this study, the eddy-viscosity type formulation proposed in Kennedy et al.[1] is adopted. This method has been shown to work well for both spilling and plunging wave breakers[1,3,6].

For a 1D, depth-integrated Boussinesq model, R_x is given by Eq. 3

$$R_x = (\nu P_x)_x \tag{3}$$

where ν is an empirical eddy viscosity term related to the total water depth, $H = h + \eta$, and to the time rate of change of the water surface level, η_t, by Eq. 4

$$\nu = \delta B H \eta_t \tag{4}$$

where δ is another empirical coefficient to account for the mixing-length and friction-velocity scales. The parameter B, whose value ranges from 0.0 to 1.0, defines the state of breaking of the waves; from non-breaking ($B = 0.0$) to fully-breaking ($B = 1.0$). For a smooth transition between the two extreme cases, the expression for B is given by

$$B = \begin{cases} 1, & \eta_t > 2\eta_t^* \\ \eta_t/\eta_t^* - 1, & \eta_t^* < \eta_t \leq 2\eta_t^* \\ 0, & \eta_t \leq \eta_t^* \end{cases} \tag{5}$$

η_t^* determines the initiation and termination of the breaking process. It is calculated by

$$\eta_t^* = \begin{cases} \eta_t^F, & t - t_0 \geq T^* \\ \eta_t^I + \frac{t-t_0}{T^*}(\eta_t^F - \eta_t^I), & 0 \leq t - t_0 < T^* \end{cases} \tag{6}$$

where η_t^I is the threshold value for breaking initiation; η_t^F is the limit for the breaking termination; t_0 is the time at which the breaking event starts; $t - t_0$ is the age of breaking event; and T^* is the duration of the breaking event.

The value of the control parameters are set as $\eta_t^I = 0.35\sqrt{gH}$, $\eta_t^F = 0.15\sqrt{gH}$, $T^* = 5\sqrt{H/g}$ and $\delta = 2.0$ for this Boussinesq model.

3. Numerical Experiment

3.1. *Numerical Model Settings*

The computational domain configuration used in the wave modeling is shown in Fig. 3. Random waves generated from a JONSWAP spectrum with significant wave height, $H_s = 11.3$ m, and wave period, $T = 8.3$ s, are introduced to an idealized reef bathymetry. The settings used in the JONSWAP spectrum are based on the results of the WAVEWATCH III wave hindcast at the locality of Ivana. The face of the reef is inclined at a fixed slope $\tan\beta_r = 1/20$ and the still water depth at the reef flat is $h_r = 1.0$ m. The width of the reef, W, is varied to investigate possible wave resonance excitation of different modes. The values used in the test cases are $W = [60, 80, 100, 120, 150, 180]$. The slope of the beach, $\tan\beta$, is also varied in different test cases. The values used are $\tan\beta = [1/10, 1/15, 1/20]$. The values set to the different parameters are chosen to simulate the conditions at the area of interest, Ivana. At each model run, the water level η is monitored at the reef end labeled as x_{end} at Fig. 3.

3.2. *Results of Numerical Experiment*

Figure 4 shows the maximum computed η at the reef end (x_{end}) for different reef widths W and beach slope $\tan\beta$. It is seen that η_{max} gradually decreases from $W = 60$ until $W = 100$, then reach an almost equilibrium value for $W > 100$. It is also seen that the influence of $\tan \beta$ on η_{max} is not significant. The tested slopes yielded almost equal η_{max} values every W value. Therefore, a representative beach slope ($\tan \beta = 1/20$) is chosen for the next analyses. Figure 5 shows the power spectra of the η timeseries

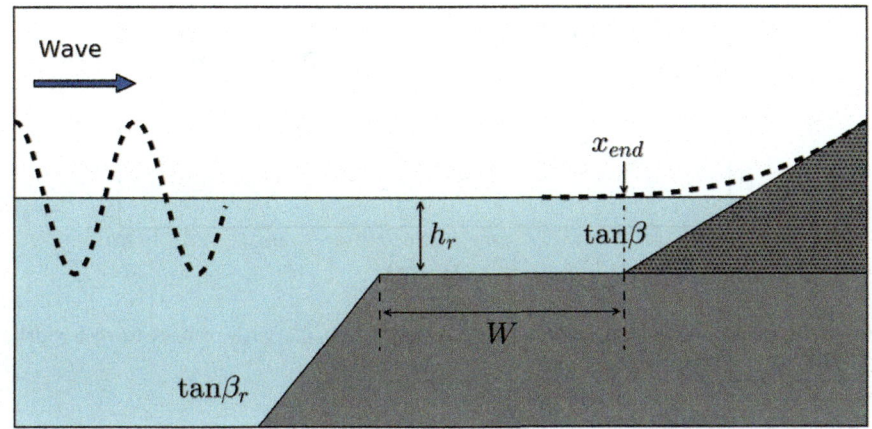

Fig. 3.   Model bathymetry for wave propagation over reefs of various widths

measured at x_{end} for each value of W. The scaled JONSWAP spectrum of the input wave is also overlain on top for comparison. It is seen in Fig. 5 that low-frequency waves or long waves are present at x_{end}. Long waves whose frequency is lower and outside the domain of the JONSWAP spectrum of the input wave were generated due to the nonlinear effects of the Boussinesq model. To get a better understanding of the different spectral wave characteristics, the spectrum is divided into two bands: long waves ($T > 20$ s), and short waves ($T < 20$ s). This division is also shown in Fig. 5. The long wave components are seen to have higher power and several peaks are also observed in the $T > 20$ s region. This shows that resonance of long wave components occurs for all the tested reef widths W.

Figure 6 shows the relationship between the mean water level, η_{avg}, significant wave height, H_s, and average period, T_{avg}, with the reef width W. First, η_{avg} is observed to increase gradually with increasing W. However, the magnitude of increase of η_{avg} is small– only about 20 cm increase from $W = 60$ to $W = 180$. The significant wave height, H_s, is divided into long wave and short wave components and is shown in the middle plot of Fig.

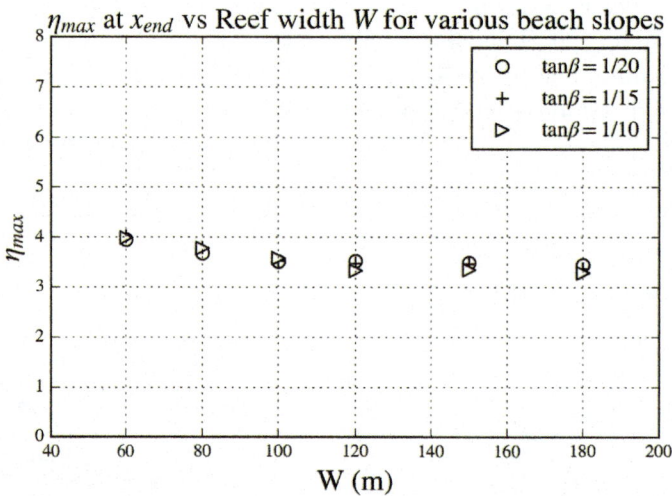

Fig. 4. Variation of maximum computed water level η_{max} with respect to reef width, W, and beach slope, $\tan \beta$

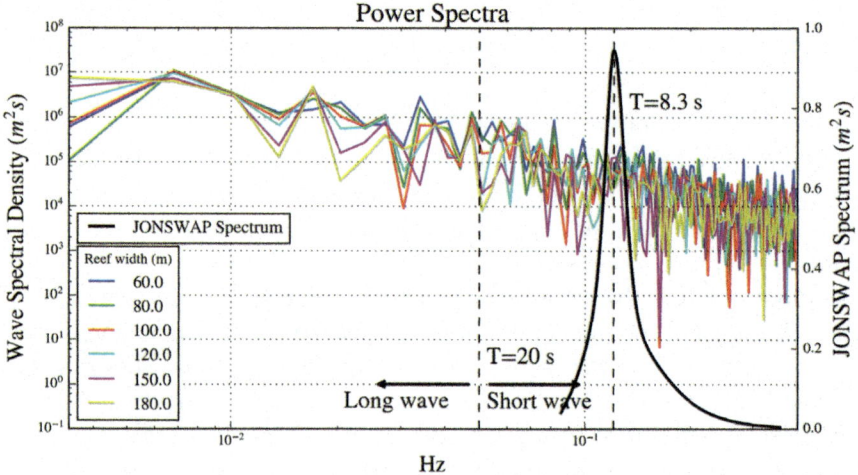

Fig. 5. Power spectrum of the computed η at x_{end} for different W's and JONSWAP spectrum of the input wave

6. For all values of W, the long wave component is consistently larger than the short wave component. This implies that the long wave components

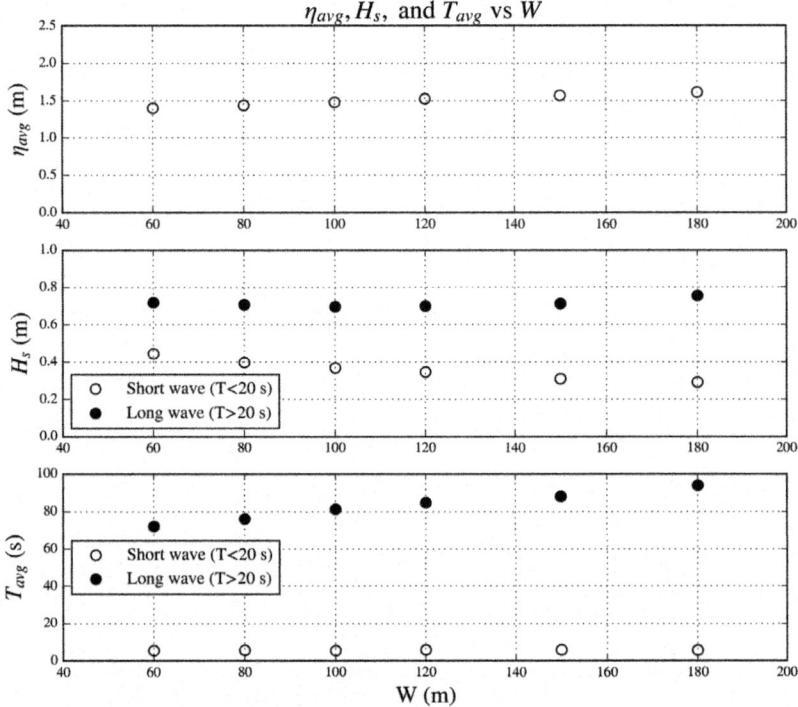

Fig. 6. Variation of wave characteristics at x_{end} with reef width W

are dominant at the region of x_{end}. Furthermore, the H_s of the short waves is seen to be gradually decreasing with increasing W. This is because a longer reef width allows more time for the short wave components to dissipate by wave breaking. Finally, the average wave period, T_{avg}, is shown at the bottom plot of Fig. 6. It is seen that the T_{avg} of the short wave components remain almost constant for all values of W while the T_{avg} of the long wave components is observed to gradually increase with increasing W.

4. Conclusion

The 1D analysis was done by a 1D Boussinesq model to simulate the effects of varying the reef width to the characteristics of the storm wave at the reef flat. It was shown that the long wave components dominate over the short waves, which were dissipated by wave breaking, inside the reef flat

232

for any of the tested reef widths. 1D resonance of the long wave components was also seen to occur for any reef width because the random wave input by the JONSWAP spectrum always generates long wave components corresponding to the resonant period of any reed width.

Fig. 7 shows the trend line of η_{max} computed by 1D Boussinesq model overlaid on top of the measured runup and inundation at Ivana (Fig. 2). It is clear that the 1D Boussinesq model cannot account for the observed wave concentration and amplification at Ivana. In fact, an opposite trend is predicted by the 1D Boussinesq model, where a wider reef results to a lower η_{max} at the reef end. This implies that the observed phenomenon at Ivana is not due to wave resonance in 1D. Further investigations using a 2D model is suggested to investigate the resonance of infragravity waves on the reef flat in two dimensions.

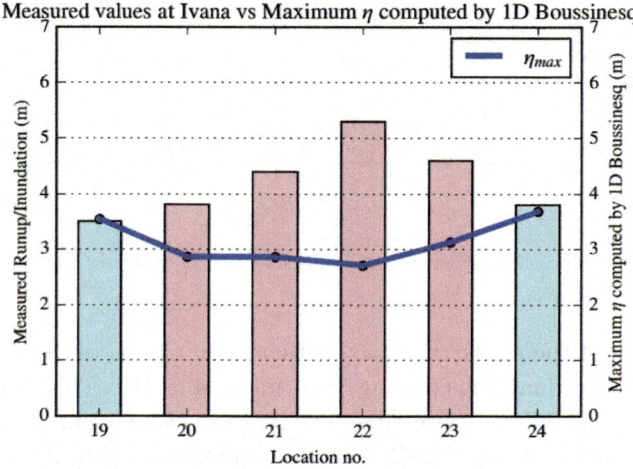

Fig. 7. Trend line of η_{max} computed by 1D Boussinesq model overlaid on top of the measured runup (blue bars) and inundation (red bars) at Ivana

References

1. Andrew B. Kennedy, Qin Chen, James T. Kirby, and Robert A. Dalrymple. Boussinesq modeling of wave transformation, breaking, and runup.i: 1d. *Journal of Waterway, Port, Coastal, and Ocean Engineering*, 126(1):39–47, 2000. doi: 10.1061/(ASCE)0733-950X(2000)126:

1(39). URL http://ascelibrary.org/doi/abs/10.1061/\%28ASCE\
%290733-950X\%282000\%29126\%3A1\%2839\%29.

2. John Phillip Lapidez, Yoshimitsu Tajima, Jeane Camelo, Mizuka Saito,
   Yoshinao Matsuba, Takenori Shimozono, Marjorie Turiano, Dominic
   Bautista, , and Eric Cruz. Locally varying inundation characteris-
   tics along coastlines of batanes induced by super typhoon meranti.
   In *Coastal Dynamics 2017 Conference Proceedings*, jun 2017. http:
   //coastaldynamics2017.dk/proceedings.html.

3. Patrick J. Lynett. Nearshore wave modeling with high-order boussinesq-
   type
   equations. *Journal of Waterway, Port, Coastal, and Ocean Engineering*, 132(5):348–357, 2006. doi: 10.1061/(ASCE)0733-950X(2006)132:
   5(348). URL http://ascelibrary.org/doi/abs/10.1061/\%28ASCE\
   %290733-950X\%282006\%29132\%3A5\%28348\%29.

4. P.A. Madsen, O.R. Srensen, and H.A. Schffer. Surf zone dynamics simulated by a boussinesq type model. part i. model description and cross-shore motion of regular waves. *Coastal Engineering*,
   32(4):255 – 287, 1997. ISSN 0378-3839. doi: http://dx.doi.org/10.
   1016/S0378-3839(97)00028-8. URL http://www.sciencedirect.com/
   science/article/pii/S0378383997000288.

5. Per A. Madsen, Russel Murray, and Ole R. Srensen. A new form of
   the boussinesq equations with improved linear dispersion characteristics. *Coastal Engineering*, 15(4):371 – 388, 1991. ISSN 0378-3839.
   doi: http://dx.doi.org/10.1016/0378-3839(91)90017-B. URL http://
   www.sciencedirect.com/science/article/pii/037838399190017B.

6. Volker Roeber, Kwok Fai Cheung, and Marcelo H. Kobayashi.
   Shock-capturing boussinesq-type model for nearshore wave processes.
   Coastal Engineering, 57(4):407 – 423, 2010. ISSN 0378-3839. doi:
   http://dx.doi.org/10.1016/j.coastaleng.2009.11.007. URL http://www.
   sciencedirect.com/science/article/pii/S0378383909001860.

7. Hemming A. Schäffer, Per A. Madsen, and Rolf Deigaard. A boussinesq model for waves breaking in shallow water. *Coastal Engineering*, 20(3):185 – 202, 1993. ISSN 0378-3839. doi: http://dx.doi.org/10.
   1016/0378-3839(93)90001-O. URL http://www.sciencedirect.com/
   science/article/pii/037838399390001O.

8. Yoshimitsu Tajima, John Phillip Lapidez, Jeane Camelo, Mizuka
   Saito, Yoshinao Matsuba, Takenori Shimozono, Dominic Bautista, Marjorie Turiano, and Eric Cruz. Post-disaster survey of storm surge
   and waves along the coast of batanes, the philippines, caused by

super typhoon meranti/ferdie. *Coastal Engineering Journal*, 59(01): 1750009, 2017. doi: 10.1142/S0578563417500097. URL http://www. worldscientific.com/doi/abs/10.1142/S0578563417500097.

9. J.A. Zelt. The run-up of nonbreaking and breaking solitary waves. *Coastal Engineering*, 15(3):205 – 246, 1991. ISSN 0378-3839. doi: http://dx.doi.org/10.1016/0378-3839(91)90003-Y. URL http://www. sciencedirect.com/science/article/pii/037838399190003Y.

Simulative Analysis of Inland Inundation Behind Roxas Boulevard Seawall Due to Storm Tide Overtopping by Historical Typhoons

J. B. Camelo, E. C. Cruz[†], L. L. B. Cruz

*Institute of Civil Engineering, University of the Philippines,
Diliman, Quezon City 1101, Philippines
[†]E-mail: eccruz@upd.edu.ph
www.up.edu.ph*

Numerical analysis of the inundation of the land behind the Roxas Boulevard Seawall is carried out to study the impacts of critical historical typhoons on coastal flooding. The numerical implementation based on ADCIRC with a coupled surge-inundation capability is discussed relative to existing conditions of the seawall. The simulative analyses are undertaken to study the implications of a "do nothing" scenario wherein the critical typhoons cause overtopping and inundate the land behind it. The results indicate the inland extent of potential inundation, reduced depth of the peak storm tide in front of the seawall, a rapid gradient of inundation depth behind the seawall, and the prominent role of Pasig River in inundation propagation inland, among other things.

Keywords: Coastal flooding; Storm tides; Roxas Boulevard seawall; Inundation; ADCIRC.

1. Introduction

In a typical year, 20 typhoons enter the Philippine Area of Responsibility (PAR), several of which make landfall along its more than 36,000 km of coastline. Strong typhoons that make landfall cause significant human casualties and tremendous economic damage. Major disasters have been caused by the strong winds, heavy rainfall intensities, and coastal inundation that accompany these strong typhoons. One catastrophic typhoon is Haiyan (local name: Yolanda) which occurred in November 2013. It generated waves as high as 10 m along the Pacific coastlines of Eastern Samar, and storm surges as high as 5.3 m along the interior coastlines of Tacloban, Leyte[1] that resulted in one of the most catastrophic natural disasters in the country in terms of human lives, property loss, and infrastructure damages.

In spite of their coastal protection, even built-up waterfronts like the Manila Bay coastline, also suffer from inland inundation caused by overtopping storm tides and waves. The city of Manila in particular has experienced recurrent coastal flooding due to strong typhoons that tracked close by. In recent years before and after Haiyan, several typhoons made landfall close to Roxas Boulevard Seawall, a reinforced-concrete gravity seawall that protects the national road Roxas Boulevard, and caused storm tides and waves to overtop it. In September 2011, at the height of typhoon Nesat (local name: Pedring), the seawall was overtopped by storm waves that caused its collapse. It was rebuilt within a year, but was overtopped several times after that. In all these overtopping cases, a vast portion of Manila that lies behind the seawall was inundated.

This study focuses on the inland inundation resulting from the overtopping of the RB seawall by carrying out simulative analyss of the storm tide-induced shallow-water motion behind the seawall. The aim of the study is to determine the spatial extent and depths of inundation after the overtopping, which will provide useful insights into post-event disaster mitigation strategies and engineering interventions for improvement and/or rehabilitation of coastal protection infrastructures.[3]

2. Critical Historical Typhoons

The national location of the study area is shown. In Figure 1 (left panel). Due to the many typhoons that passed by this area, it was necessary to shortlist the potentially critical historical typhoons that tracked the RB Seawall in the past 30 years or so. With a search circle of 150 km radius centered on the study coast, the 30 such potentially critical typhoons were identified. Their tracks, center's positions and maximum sustained wind speeds V_{max} are color-plotted in Figure 1. Table 1 also summarizes the radius of maximum winds R_{max}, central pressure p_C and their track directions relative to the study coast.

Storm tide levels (STL) induced by these 30 typhoons have been synthesized from numerical simulations based on the ADCIRC[2] model and the results for the top 6 storm tides are summarized in the last panel of Table 1. Their lifetime meteorological characteristics, including their closest distance of approach to the seawall, are also shown in the middle panel. Based on these, the highest 3 STL's were induced by typhoons Ramassun 2014, Xangsane 2006 and Fengshen 2008 with maximum storm tides of +2.36, +2.10 and +1.76 m respectively from mean tide level. By checking the time of occurrence of maximum storm tides, these values can be decomposed into the astronomic tide

(AT) and the storm surge (SS) components, which yielded maximum storm surges of +1.23 m, +1.61 m and +0.52 m respectively for Ramassun, Xangsane and Fengshen. It is clear that a high STL can occur with a nominal storm surge coinciding with a high astronomic tide, as in Ramassun's case. In contrast, a large storm surge can coincide with a low tide and yield a significant storm tide, as in Xangsane.

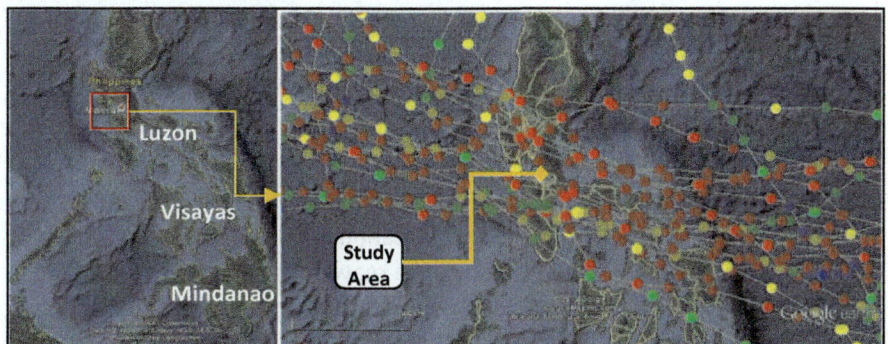

Figure 1. (left) study area; (right) track of all historical typhoons that tracked within 300 km centered at the RB Seawall

Table 1. Critical typhoons, their meteorological parameters, and simulated maximum storm tides.[2]

STL Rank	Typhoon/ Local name	Vmax (mph)	Rmax (km)	Pc (hPa)	Track rel. to Site	Closest distance (km)	AT (m)	**STL (m)**	Max. SS (m)
1	Rammasun/ Glenda 2014	103.5	130	935	S	45	1.13	**2.36**	1.23
2	Xangsane/ Milenyo 2006	98	120	940	S	25	0.47	**2.09**	1.61
3	Fengshen/ Frank 2008	103.5	90	945	E, N	20	1.24	**1.76**	0.52
4	Dot/ Saling 1985	138	330	895	N	90	1.23	**1.71**	0.43
5	Vera/ Bebeng 1983	86	130	965	S	8	1.23	**1.61**	0.38
6	Betty/ Herming 1987	126.5	190	890	S	100	1.25	**1.52**	0.27

3. Coastal Inundation Modeling

In order to analyze the inland movement of storm tide-induced inundation flow, numerical model ADCIRC (Advanced Circulation)[4] is used to compute the storm tides based on the coastal hydrodynamics under a tropical cyclone atmospheric flow boundary condition. The unstructured mesh used initially in the STL analysis is shown in Figure 2 (left panel) consisting of the marine area containing Manila Bay, coastlines of Bataan and Cavite, and a curved open wet boundary in West Philippine Sea allowing for water level specification. Details of the mesh, hydraulic properties, boundary conditions, and numerical implementation are discussed in an earlier paper.[3]

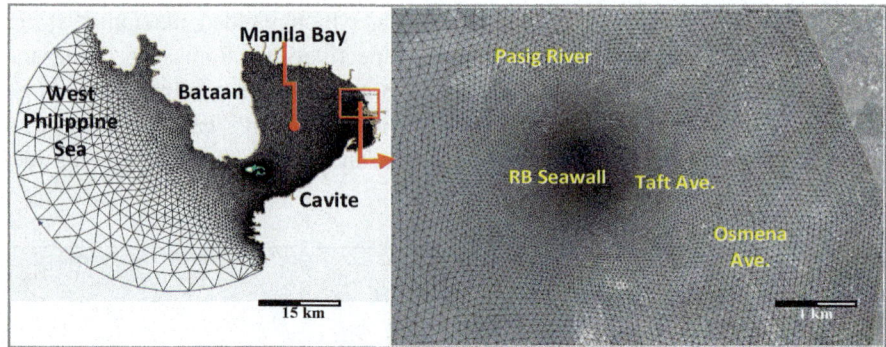

Figure 2. Storm tide and coastal inundation (partial mesh only) FEM meshes

Manila Bay was extended by 5.8 km inland to allow inundation flows behind RB Seawall (Figure 2). The delineation of the inundation domain is based on the following conditions behind the seawall: (a) alignment of important waterways such as Pasig River and the stormwater drainage channels; (b) Manila Bay coastline parallel to Roxas Boulevard; (c) a dry inland boundary that is not reached by the seawater. The dry boundary is based on the flooding hazard map published by DOST.[5] Figure 2 (right panel) shows the unstructured mesh over land that is expected to be partially inundated by storm tide-driven flows. The hydraulic parameters used in the peak STL simulations were retained except that the bottom resistance was replaced by Manning's roughness coefficient to model frictional resistance on inundated land.

3.1. Surge-inundation numerical model

A different set of model equations is implemented in ADCIRC to model the propagation of the coastal hydrodynamics into the land domain and thereby capture the coastal flooding process. ADCIRC utilizes the generalized wave continuity equations (GWCE), instead of the depth-integrated (2DDI)[4] model equations. In particular, the GWCE has been shown to yield more stable results of sea-induced inundation characteristics such as inundation depths and speeds.

ADCIRC activates the GWCE numerical model through an internal control switch. This switch implements the ADCIRC-2DDI equations in the sea domain and the GWCE on inundated land, and hence allow for a coupled simulation of storm tides and coastal flooding. Figure 2 (right panel) shows the surge-inundation mesh indicating the spatial extent and resolution of the computational domain. Finer mesh is used near the seawall where inundation is expected to commence. This unstructured mesh with varying spatial resolution aims to spatially resolve the waterways, discussed below, and simultaneously capture the dynamic effective shoreline.

Figure 3. Waterways and coastlines in inundation domain

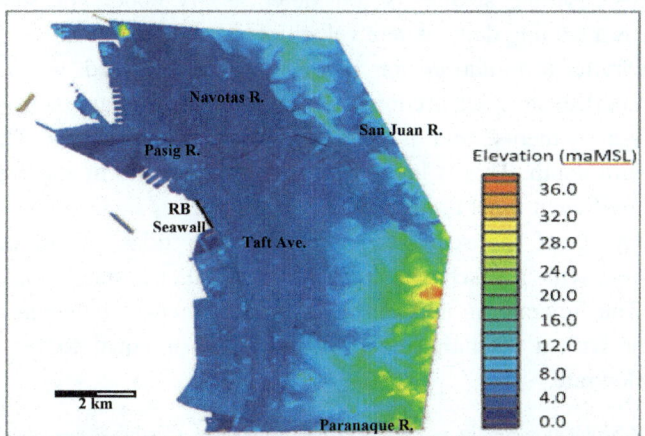

Figure 4. Digital elevation model of the inland area

Figure 3 shows the surface features of the inundation area behind the seawall. Waterways are modelled as wet boundaries with full hydraulic conveyance and roughness-dependent bottom resistance. Channels smaller than 3 m are neglected due to their small hydraulic conveyance. In order to resolve the horizontal scales of roughness within the inundation domain, digital elevation model data for the area in Figure 3 are used. Figure 4 shows the DEM map used to enter elevation data. This map was based on the optically sensed LIDAR data of the Manila Bay coastline by DOST and is able to resolve topographic features such as paved roads, paved surfaces, buildings and waterways, which is a capability needed to account for varying flow resistance in the inundation flow. However, turbulence-related features such as eddies or wakes are still not resolved in the model equations.

3.2. Numerical analysis of inland inundation

The main module computes the transient hydrodynamics (water levels and currents) which are passed onto the inundation module. In turn, the inundation flow field and depths are fed back to the main module resulting in a coupled solution of coastal hydrodynamics and inland inundation.

The parameters used in the inundation simulations are summarized in Table 2. The computational domain used 58,349 triangular elements, or an increase of 98% relative to the no-inundation case. The time marching is unchanged at 2 seconds, resulting in a maximum Courant number CFL of 0.30 for conditional stability. A minimum water depth threshold of 1.5 cm per node is considered as wet and thus anything smaller is considered as dry node. The same value is considered as a wetting depth where velocities are computed.

A quadratic formulation for bottom friction is used with Manning's roughness coefficient n set according to Table 2, which ranges from 0.017 to 0.040 on sea elements, and is 0.05 on inundation elements. The n value decreases with depth. This is in contrast with the pure storm surge simulations where the coefficient is set uniformly to 0.0025.

Overtopping of the seawall's crest by storm tides is modeled as an overflow weir with a discharge coefficient of 1.0, i.e. maximum theoretical discharge. The flowrate on an overtopped land element is modeled as a broad-crested weir with a discharge coefficient of 1.0, or equal to the theoretical maximum flowrate.

Table 2. Simulation conditions for inland inundation under existing seawall conditions

Parameters	Simulations for Inundation Flow
ADCIRC Mesh	Seabed + Land grids
No. of FEM elements	58,349
Time Step	2 sec.
Seawall grid points	Modeled as elements on grid
Friction parameter	Variable: $n = 0.05$ for depths $h < 0$m; $n = 0.017$ for $0 < h < 5$m; $n = 0.018$ for 5m $< h < 10$m; $n = 0.020$ for 10m $< h < 15$m; $n = 0.022$ for 15m $< h < 20$m ; $n = 0.025$ for 20m $< h < 30$m; $n = 0.029$ for 30m $< h < 55$m; $n = 0.040$ for 50m $< h$
Coriolis Parameter	0.0001
Drying depth	0.015 m
Wetting depth	0.015 m

4. Storm Tides and Coastal Inundation due to Critical Typhoons

The ADCIRC surge-inundation model discussed above was applied with the following data: a) uniform seawall crest elevation (=MTL +2.01m, based on DPWH 2012 construction drawings); and b) specified values of hydraulic parameters such as discharge coefficient and friction coefficient.

Figure 5 summarizes the best track positions of 4 historical typhoons that impacted the seawall's function, 2 of which (namely, Ramassun and Xangsane) yielded the highest 2 simulated storm tide levels. These 2 typhoons both tracked to the south of the seawall. The other 2 typhoons simulated are Nesat, which caused the overtopping collapse in 2011, and Saola 2012 which again overtopped the seawall barely a year after it was repaired. Nesat tracked north of the seawall while Saola tracked to the east then north past Taiwan.

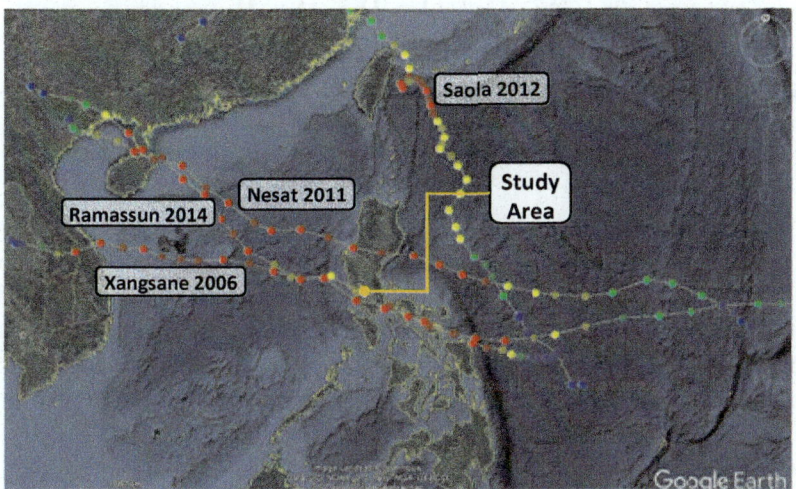

Figure 5. Tracks of critical typhoons

Figure 6 shows a comparison of the time series of recorded and computed water levels at the location of the tide gage of NAMRIA in Manila South Harbor for typhoons Nesat 2011 (top plot) and Saola 2012 (bottom plot). The surge-inundation model is able to yield the highest storm tide at the correct time (around Sept. 27 8 am UTC for Nesat, August 2 6 am UTC for Saola), and has peaks and valleys that are synchronous with those of the tide gage records. However, the amplitudes of the peaks are just fairly reproduced by the coupled model. In the case of Nesat, the model underestimates the peak storm tide by about 0.23 m, while for Saola it over-estimates the peak storm tide by about 0.17 m. The differences may be explained by the inappropriate hydraulic

242

parameters used such as the discharge coefficient and bottom resistance for both the marine and land areas. Modeling improvements are being implemented that include input of a rating curve for the seawall and better calibration of bottom resistance per typhoon case, among others.

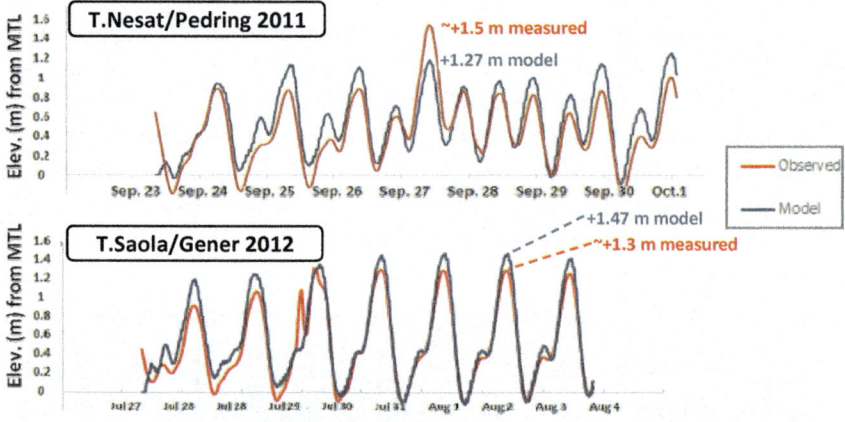

Figure 6. Comparison of simulated (blue) and actual (red) tide levels due to Typhoons Nesat 2011 and Saola 2012

One of the scenarios investigated in the RB seawall overtopping study is a "do nothing" scenario, wherein the storm tides and waves are allowed to overtop the existing seawall's crest. Figure 7 illustrates the treatment of coastal flooding depth D $(=\eta+h)$ after the time-marched solutions of the water surface displacement $\eta(x,y,t)$ from MSL are obtained. Area plots of envelopes of inundation depth D are summarized in Figure 8 for typhoons Nesat and Saola, and in Figure 9 for Ramassun and Xangsane.

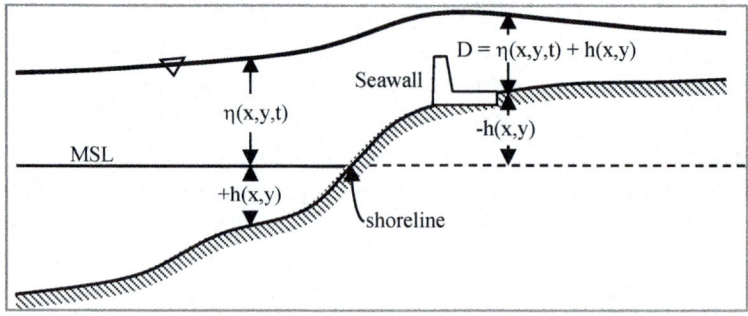

Figure 7. Definition sketch for inundation depth D

The colored areas for typhoons Nesat 2011 and Saola 2012 indicate the extent of inundation due to storm tide overtopping (note: the sea is uniformly colored to highlight the coastal flooding only). It is clear that the rear of the seawall experiences a rapid variation of inundation depth that it has decreased to 0.45 m just a few tens of meters from it. Also, the inundation occurs most inwardly along the alignment of Pasig River with the wetting front reaching as far as 3 km inland from its outfall. This result on the prominent role of channels in storm surge propagation has been reported by other studies of storm surge propagation in estuaries. Moreover, while Saola yielded almost similar inland penetration of flooding as Nesat's, the lateral extent is somewhat larger compared to Nesat, which is attributed to the higher peak storm tide predicted by the model for Saola than Nesat (see Figure 6).

Figure 9 summarizes the envelope of inundated area due to typhoons Ramassun and Xangsane. The same rapid change of inundation depth behind the seawall is seen. In contrast to Nesat and Saola (see Figure 8) however, the wetting front has advanced more inland, both behind the seawall and along the waterway Pasig River. The lateral extents of inundation is also more extensive, particularly along Pasig River, which is again attributed to the higher maximum storm tide level along the seawall.

It should be noted that the above plots involve inundation *envelopes*, not snapshots of inundation, hence, the results should be taken as inundation *potential*, not actual inundated areas.

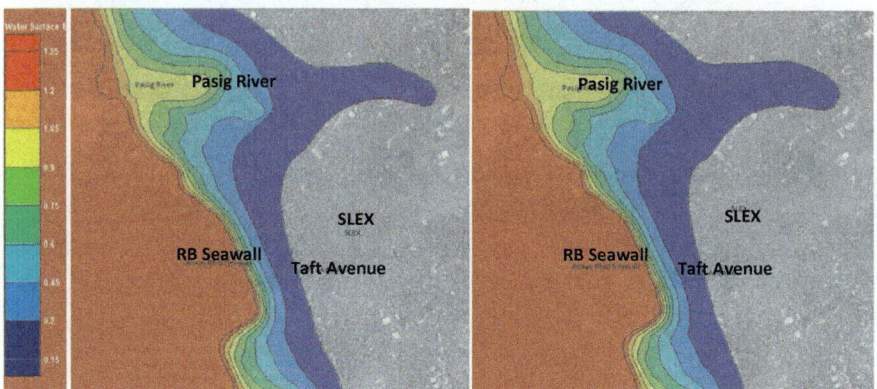

Figure 8. Simulated coastal inundation due to T.Nesat 2011(left) and Saola 2012 (right)

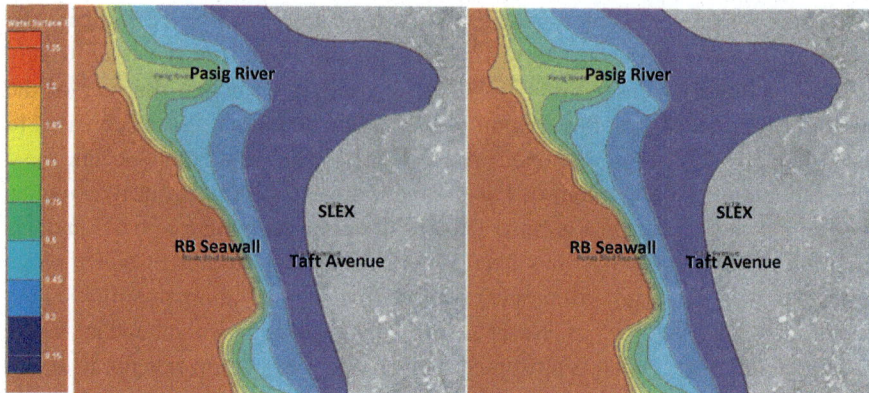

Figure 9. Simulated coastal inundation due to T.Ramassun 2014(left) and Xangsane 2008 (right)

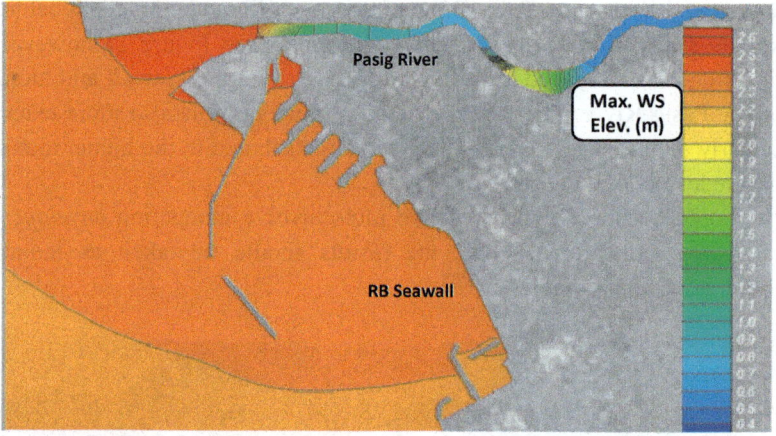

Figure 10. Maximum envelope of water surface due to T.Ramassun 2014

For Ramassun, Figure 10 shows the peak storm tides including Pasig River if the seawall and channels banks were high enough to prevent inland inundation. In this case, the maximum storm tide reached +2.36 m at the seawall, and as high as +0.5 m along Pasig River at a channel distance of 7 km upstream of the outfall.

5. Conclusions

Table 3 summarizes under existing conditions the peak depths of inundation at various inland distances from the seawall based on the synthesis of the 5 historical typhoons that impacted Manila's coastal flooding. The depths in front and behind the seawall are also summarized.

Table 3. Peak inundation depths behind RB Seawall due to 5 critical typhoons under "do nothing" scenario

Case	Typhoon/Local name	Near Seawall (40m to sea)	Near Seawall (20m inland)	Taft Ave (1km inland)	Osmeña Ave (1.85km inland)
1	Rammasun/ Glenda 2014	1.49	1.38	0.19	0.10
2	Xangsane/ Milenyo 2006	1.49	1.38	0.12	0.05
3	Fengshen/ Frank 2008	1.36	1.31	**0.37**	**0.29**
4	Saola/ Gener 2012	1.45	0.86	0.18	0.08
5	Nesat/ Pedring 2011	1.45	0.86	0.17	0.08

The largest inundated area was induced by Typhoon Fengshen 2008 with a peak depth of 0.37 m at Taft Avenue and 0.29 m at South Luzon Expressway (SLEX or Osmeña Ave.). This vast horizontal extent and high inundation depths due to Fengshen appear to be consistent with its peculiar track, which traversed east then north of the seawall, with closest distance of 26.4 km northeast from the seawall. Typhoon Ramassun 2014, which tracked 25 km at the closest distance southwest of the seawall and caused the highest storm tide level, generated a much lower inundation depth of 0.19 m at Taft Avenue.

Acknowledgment

The authors acknowledge the Department of Public Works and Highways (DPWH) for providing bathymetry data and computational resources for the study, the National Engineering Center (NEC) of the University of the Philippines-Diliman for funding the validation of tidal data, UP-DOST Dream for the Lidar data, and Engr. Eric Santos of AMH Philippines, Inc. The use of Google Earth for the aerial images is also acknowledged.

References

1. Tajima, Y; Yasuda, T; Pacheco, B; Cruz, E; Kawasaki, K; Nobuoka, H; Miyamoto, M; Asano Y; Arikawa T; Ortigas, N; Aquino, R; Mata, W; Valdez, J; Briones, F. Initial report of JSCE-PICE joint survey on the storm surge disaster caused by typhoon Haiyan. *Coast. Eng. J.* 2014, 56:1,DOI: http://dx.doi.org/10.1142/S0578563414500065.
2. U.P. National Engineering Center, *Coastal Engineering Study Report: for the Study, Preliminary Engineering and Detailed Engineering Design of the Roxas Boulevard Seawall, Manila City.* Univ. of the Phils. National Engineering Center, 2016.

3. Cruz, E.C., J.C.E.L. Santos, J.B. Camelo, M.H. Zarco, M.E.L. del Rosario, J.M.B. Gargullo, I.A.D. Inocencio, and L.L.B. Cruz, Preliminary engineering of a seawall against storm tides and waves along a built-up waterfront. *Proc., 26th International Ocean and Polar Engineering Conf.* Rhodes, Greece, 1428-1435, 2016.

4. Luettich, R.A, and J.J. Westerink, "Formulation and numerical implementation of the 2D/3D ADCIRC finite element model version 44.XX" Department of Civil Engineering and Geological Sciences, University of Notre Dame, 2004.

5. Tablazon, J., A.M.F. Lagmay, M.T. Mungcal, L. Gonzalo, L. Dasallas, J. Briones, J. Santiago, J.K. Suarez, J.P. Lapidez, C.V. Caro., C. Ladiero, V. Malano (2014). Developing an early warning system for storm surge inundation in the Philippines. DOST-Project NOAH Open-File Reports, Vol. 3 (2014), 96-111.

A Study on the Effects of Historical Typhoon Parameters on Storm Surge Generation in San Pedro Bay Using Advanced Circulation Model

I. B. O. Villalba[†], E. C. Cruz

Institute of Civil Engineering, University of the Philippines
Diliman, Quezon City 1101, Philippines
[†]Email: iovillalba@up.edu.ph
www.up.edu.ph

This study aims to determine the behavior of storm surge in San Pedro Bay generated by historical storms to understand the storm surge characteristics produced by different storm parameters. For this purpose, the Advanced Circulation Model (ADCIRC) is the hydrodynamic model used to simulate the storm surge in San Pedro Bay and the Holland 1980 Typhoon Model is used to model the pressure and wind distribution over the model domain. A total of 12 historical typhoons, including Typhoon Haiyan, are selected based on wind intensity and proximity to the study site as well as typhoons that have generated storm surges in the past. Results show that storm surge is concentrated at Basey, Samar for typhoons with tracks above or crossing the bay while typhoons with tracks south of the mouth of the bay have simulated storm surges that have almost equal rate of decrease from the inner part to the mouth of the bay. The windspeed may be an indicator of the severity of the storm surge. Typhoon Haiyan 2013 with 110 kts (204 kph) maximum sustained) wind speed has simulated storm surge peak of 4-6 meters, Typhoon Agnes 1984 with 105 kts (195 kph) wind speed has simulated storm surge peak of 2-3 meters, and Typhoon Axel 1994 and Typhoon Cecil 1979 with 75 kts (139 kph) wind speed both have simulated peak surge of 1-2 meters, inside San Pedro Bay. Moreover, the results show that Tacloban in Leyte and Basey in Samar are the most storm surge-prone areas, which may experience 1-6 meters of storm surge depending on the typhoon characteristics. Furthermore, a slow-moving typhoon has longer time to peak than fast-moving typhoons.

Keywords: Storm surge; ADCIRC; San Pedro Bay; Typhoon parameters; Disaster risk management.

1. Introduction

The Philippines is an archipelagic country sitting in the Western North Pacific, which is the most active typhoon generating area. A total of 19-20 typhoons passed through the Philippine Area of Responsibility and 8-9 typhoons make landfall annually (Cinco, 2012). In 2013, Typhoon Haiyan (local name Yolanda) struck the Visayas Region and caused catastrophic destruction, in which a total of 6,300 individuals were reported dead, 28,688 injured and 1,062 are still missing. Out of the total number of deaths, 93% came from Region VIII, where most deaths were due to drowning to storm surge and trauma (NDRRMC, 2015). According to the National Disaster Risk Reduction and Management Council, areas at risk to storm surges were identified but characteristics, behavior, and impact of the natural event were not adequately explained. As such, people were unable to imagine and visualize the impact of disaster such as that wrought by Typhoon Haiyan.

Located in Leyte Gulf, San Pedro Bay has a shallow bathymetry with a funneling coastline shape towards the San Juanico Strait. There are notable occurrences of storm surges generated by typhoons in the past, which include Typhoon Haiyan (2013), Typhoon Agnes (1984), and Typhoon 1897. The main purpose of this study is to determine the effects of historical typhoons in the generation of storm surges in San Pedro Bay using the Advanced Circulation (ADCIRC) model. The results of this study will be helpful in understanding the nature and severity of historical storm surge in San Pedro Bay with respect to typhoon parameters, such as wind speed, typhoon track, and forward speed, which could be used in disaster prevention and mitigation analysis, preliminary design of coastal protection structures, and in coastal zone management of the bay.

2. Typhoon Model

2.1. *Holland 1980 typhoon model*

The pressure and wind distribution of the typhoon are modeled using the Holland 1980 typhoon model. The Holland 1980 model for pressure and wind distribution is described as follows:

$$p(r) = p_c + (p_n - p_c)e^{-\left(r^{\frac{A}{B}}\right)}. \tag{1}$$

$$V(r) = \left[\left(\frac{R_{mw}}{r}\right)^B e^{1-\left(\frac{R_{mw}}{r}\right)^B} * V_{mw}^2 + \left(\frac{r^2 f^2}{4}\right)\right]^{0.5} - \frac{rf}{2}. \tag{2}$$

Here, r is the distance from the center of the typhoon, V is the gradient wind at radius r, p is the pressure at radius r, p_c is the central pressure, p_n is the ambient pressure, R_{mw} is the radius of maximum wind speed, V_{mw} is the maximum wind speed, f is the Coriolis parameter, and A and B are scaling parameters, given by:

$$A = R_{mw}^B. \tag{3}$$

$$B = \rho * e\frac{V_{mw}^2}{p_n - p_c}, 1 < B < 2.5. \tag{4}$$

2.2. Historical typhoons

The selection of typhoons that are simulated in this study is based on two conditions, which are: (1) proximity of typhoons to the research site and maximum windspeed and lowest pressure of the typhoon nearest to the site, and (2) historical typhoons with known and unknown storm surge occurrence as presented in the study of Soria et al. (2016). Using the data from Japan Meteorological Agency, a search radius of 150 km centered at San Pedro Bay is used to streamline the typhoons that passed through the bay and the nearest distances of the typhoon tracks with respect to San Pedro Bay are measured and the corresponding nearest central pressure and maximum windspeed are recorded. The typhoons are then ranked and selected according to maximum windspeed and nearest distance to site.

Table 1. Selected historical typhoons based on nearest windspeed, pressure and distance

	Year	Typhoon	Local Name	10-min Max. Sustained Wind speed (kt)	Central Pressure (mb)	Distance (km)	Track
1	1984	Agnes	Undang	105	940	10	North
2	2013	Haiyan	Yolanda	110	910	28	South
3	1988	Skip	Yoning	70	955	6	North
4	1978	Olive	Atang	50	985	0	Direct
5	1987	Phyllis	Trining	70	970	55	North
6	2014	Hagupit	Ruby	90	990	66	North
7	1979	Cecil	Bebeng	75	965	30	South
8	1994	Axel	Garding	75	965	30	South
9	2006	Utor	Seniang	85	975	19	North
10	2008	Fengshen	Frank	75	965	31	North

Soria et al. (2016) presented in their study 9 tropical cyclones that have generated storm surges in Leyte Gulf, from an unpublished data of PAGASA. Nine typhoons with unknown storm surge occurrences were also presented in the study. Among these identified typhoons, the typhoons that are considered in this study occurred in 1978-2016 because there are no detailed wind typhoon information data from JMA or JTWC for typhoons on or before 1977.

Table 2. Typhoons with known and unknown storm surge occurrences from [9] (maximum winds and minimum pressure are at landfall. Data are from JTWC and JMA)

	Name	Date (landfall)	Max. winds (kts)	Min. pressure (hPa)
	Typhoons with known storm surge occurrence			
1	Agnes (Undang)	4 Nov 1984	100	925
2	Skip (Yoning)	7 Nov 1988	70	955
3	Mike (Ruping)	12 Nov 1990	90	935
4	Haiyan (Yolanda)	8 Nov 2013	125	895
	Typhoons with unknown storm surge occurrence			
5	Nelson (Bising)	25 Mar 1982	105	940
6	Axel (Garding)	21 Dec 1994	95	960

Table 2 summarizes some of the typhoons that are already identified using the first criterion. As such, the total number of typhoons simulated in this study is 12, including Typhoon Haiyan (2013). In this study, the Joint Typhoon Warning Center (JTWC) typhoon best track data is the primary source for the typhoon data used in the simulation. JTWC has complete typhoon wind and pressure information from 2001-present. However, typhoons which occurred earlier than 2001 do not have data on the radius of maximum winds and central pressure. In this regard, the central pressure data from JMA is used in this study. However, data on the radius of maximum winds is also not available from JMA.To estimate the radius of maximum wind speed, the proposed method of Vickery and Wadhera (2008) is used. The radius of maximum wind speed (R_{mw}) is calculated using the following equation:

$$R_{mw} = e^{(3.015 - 6.29(10^{-5})\Delta p^2 + 0.0337\psi)}. \qquad (5)$$

where Δp is the pressure drop and ψ the latitude in degrees.

3. Hydrodynamic Model

3.1. *Advanced Circulation (ADCIRC) model*

The hydrodynamic model used in this study is the 2DDI Advanced Circulation (ADCIRC) model, which is a continuous-Galerkin, finite element model based on the depth-integrated equations of mass and momentum equations (Leuttich and Westerink, 2004). The details of ADCIRC can be found on adcirc.org.

3.2. *Hydrodynamic model set-up*

The open ocean boundaries of the model include the Pacific Ocean just beyond 128° E longitude and a portion of the Bohol Sea in the West (Figure 1a). The boundary at San Juanico Strait extends just after the San Juanico Bridge and is specified as a mainland boundary. The bathymetry of the mesh is generated from digitized bathymetric maps of the National Mapping Resource and Information Authority (NAMRIA) for shallow areas along the coastlines of Leyte Gulf merged with the deep shore bathymetry from the gridded bathymetric data of the General Bathymetric Chart of the Oceans (GEBCO) which has a spatial resolution of 30 arc seconds. Figure 1b shows the modeled bathymetry of San Pedro Bay. A depth-adaptive triangular mesh is generated with the smallest grid size of 100 m near the coast and around 12 km grid size at the Philippine Trench. The model mesh has 110,264 elements and 57,736 vertices. Also, the model mesh has a minimum elevation value of 0.3 meter and maximum elevation value of 9,887.72 meters. A time step of 5 seconds is used for the storm surge simulations.

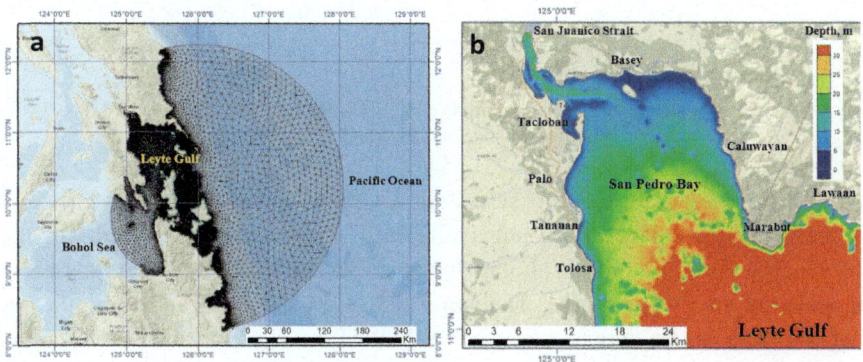

Figure 1. (a) Computational mesh of the model domain and (b) bathymetry of San Pedro Bay.

3.3. *Modeling system calibration and validation*

The tidal elevations for the Pacific Ocean boundary are extracted from the LeProvost tidal database. The constituents used are M2, S2, K1, O1, P1, and Q1. For the Bohol Sea boundary, tidal constituents are derived from harmonic analysis of WXTide tide at Nasipit Harbor and Maasin stations. The only tide station in San pedro Bay was the NAMRIA tide gaging station which was damaged by Typhoon Haiyan in 2013. Calibration of tide is done by changing the bottom roughness parameters. For this study, the manning's roughness parameter is used as the bottom shear stress is calculated using the quadratic bottom friction law. The simulated tide at Tacloban tide gaging station is compared with the observed tide (Figure 2) and a manning's roughness value of 0.025 with an RMSE of 0.1468 is used. The differences in the amplitude of the simulated tides and the actual tide may be attributed to the effect of the closing of San Juanico Strait. Other parameters used in ADCIRC are set to default values.

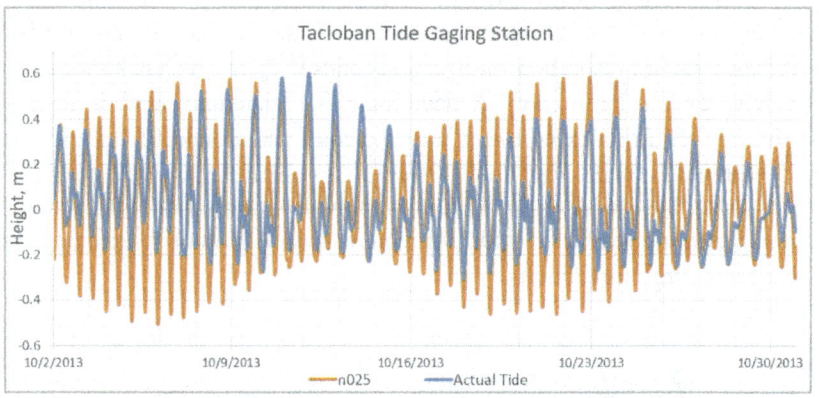

Figure 2. Simulated tides using manning's roughness of n=0.025 and observed tides at Tacloban tide gaging station from Oct. 2-30, 2013.

Observed data for the time series of the total water level during the passage of the historical typhoons are not available. For this study, the results of the storm surge simulation for Typhoon Haiyan are evaluated using observed timing of peak and high-water marks. The simulated storm surge at Tacloban indicates that the peak storm surge occurs at 0810 PHT of November 8, which is in accordance with the observed peak surge that occurred around 0800 PHT according to videos recorded by storm chaser Morgerman.[7] Observed high water marks are also compared with the results of the simulation. The simulated peak storm tide at downtown Tacloban is around 4-5 meters, which agrees well with the observed values of 4-6 meters in Tacloban (Takagi et al., 2014; Mas et al, 2015; Tajima et al., 2014). The simulated storm surge is highest at the inner part of the bay and decreases towards the mouth of the bay. The observed values of high water marks published by Tajima et al. (2014) are compared with the simulated peak storm tide levels along the coasts of San Pedro Bay, as shown in Figure 3. As expected, it is observed that measured high water marks are higher than the simulated peak storm tide levels. The discrepancy of the observed and simulated values is lower at the inner part of the Bay, however, the discrepancy at the mouth of the bay is higher. According to the study of Bricker et al. (2014), the storm surge in Tacloban is dominated by wind-driven set-up. In addition, according to interviews of Tajima et al. (2014), wave-like fluctuations were observed at Barangay #87 in Tacloban, Palo, and Tanauan. The model used in this study does not account for the effect of waves. The discrepancy in the observed and the simulated values may be attributed to the effects of waves, especially at the mouth of the bay where wave-like fluctuations were observed by witnesses.

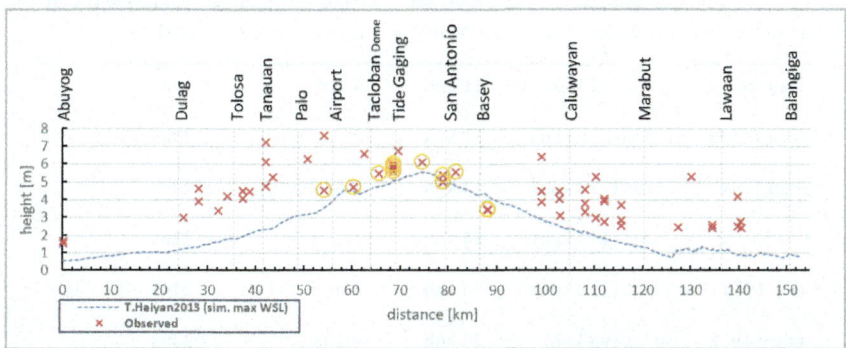

Figure 3. Simulated maximum water surface profile and observed high water inundation marks from Tajima et al. (2014) along the San Pedro Bay coasts.

4.   Results of Storm Tide Simulations

The maximum water surface elevation profiles along the coasts of San Pedro Bay and the storm surge time series for the coastal communities for the 12 historical typhoons are generated. Among these twelve typhoons, it is found that those typhoons with tracks (1) traversing the waters of San Pedro Bay, (2) north of San Pedro Bay with distance of less than 10 km and (3) south of the mouth of

San Pedro Bay, are critical to the study area, and have simulated storm surges of 1-6 meters inside the bay.

Typhoon Axel (1994) and Typhoon Haiyan (2013) which have paths traversing south of the mouth of San Pedro coming from the east-west direction have similar maximum water surface profile shape and it can be observed in Figure 4 that the maximum water surface profiles have almost equal rate of decrease from the inner to the mouth of the bay.

Figure 5 shows the maximum water surface profile of Typhoon Agnes (1984), Typhoon Cecil (1979), Typhoon Skip (1988) and Typhoon Olive (1978). It can be observed that there are concentrations of simulated maximum water surface elevation in the inner coast of Samar (Basey to Caluwayan). Typhoon Agnes (1984) has simulated storm surges above 2 meters at the inner part of the Bay, which agrees well with the observed storm surge of above 2 meters in Basey, Samar (NDRRMC, 2015). Table 3 shows the summary of the simulated storm surges and rise of peak of the critical typhoons to the study area.

Table 3. Simulated storm surge and rise of peak of historical typhoons

Typhoon	10-min max. sustained winds	Forward speed	Max. storm surge at the inner bay	Simulated rise of peak
Haiyan 2013	110 kts (204 kph)	41 kph	4-6 m	1 hr
Agnes 1984	105 kts (195 kph)	31 kph	2-3 m	2 hrs
Axel 1994	75 kts (139 kph)	14 kph	1.5 - 2m	4 hrs
Cecil 1979	75 kts (139 kph)	24 kph	1-2 m	2 hrs
Skip 1988	70 kts (130 kph)	30 kph	1 m	2 hrs
Olive 1978	50 kts (93 kph)	28 kph	< 1 m	2 hrs

The typhoons in Figure 6 that have tracks north of San Pedro Bay and with distance of more than 10 km from the head of the bay are not able to generate significant storm surge. Typhoon Utor (2006) is relatively closer to San Pedro Bay, however, the simulated storm surge is small due to its 60 kts ((111 kph) windspeed.

Typhoons following the paths below the San Pedro bay at southern portion of Leyte Gulf (Figure 7) near the Dinagat Islands are not able to generate significant storm surges in San Pedro. Although these typhoons have wind strength of 85 kts (157 kph), because of the distance of their typhoon tracks (75-85 kms south of San Pedro Bay), the typhoons are not able to generate significant storm surges in San Pedro Bay.

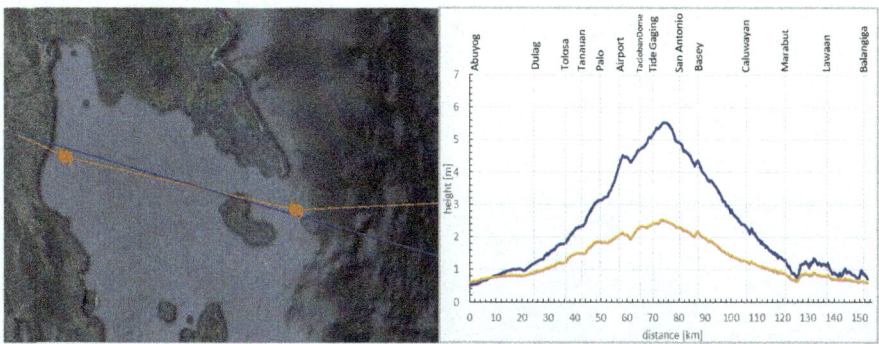

Figure 4. Typhoon tracks and simulated maximum water surface elevation profiles along San Pedro Bay coasts of Typhoons Haiyan (2013) and Axel (1994)

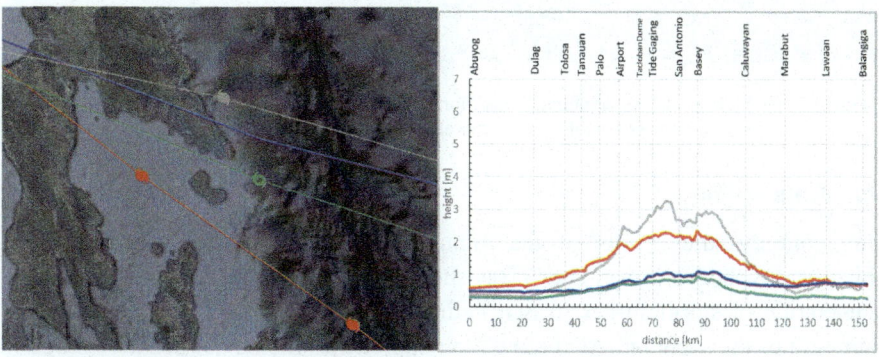

Figure 5. Typhoon tracks and simulated maximum water surface elevation profiles along San Pedro Bay coasts of Typhoons Agnes (1984), Cecil (1979), Skip (1988), and Olive (1978)

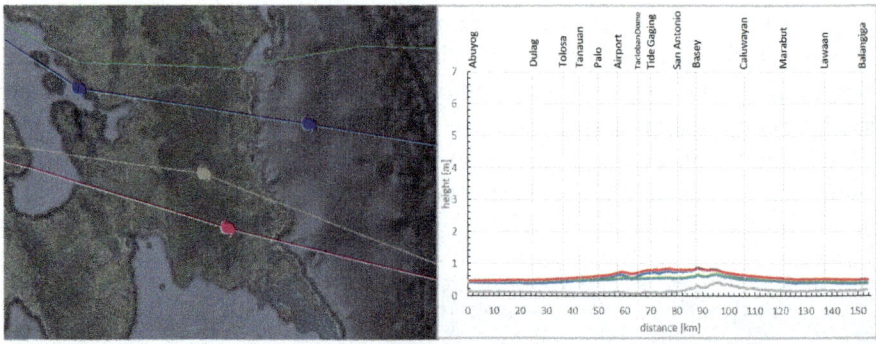

Figure 6. Typhoon tracks and simulated maximum water surface elevation profiles along San Pedro Bay coasts of Typhoons Hagupit (2014), Fengshen (2008), Phyllis (1987), and Utor (2006)

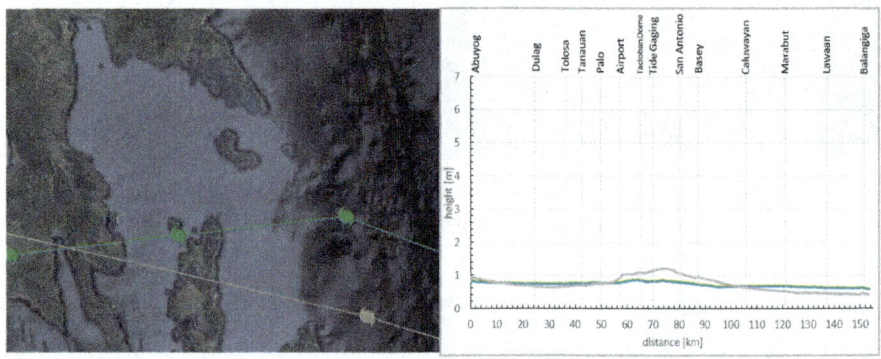

Figure 7. Typhoon tracks and simulated maximum water surface elevation profiles along San Pedro Bay coasts of Typhoons Nelson (1982) and Mike (1990)

5. Conclusions

In summary, the results of the simulations of storm surges generated by historical typhoons in San Pedro Bay show that the most important factors in the generation of storm surges in San Pedro Bay are the wind direction and wind strength of the typhoon. As the typhoon approaches the bay, a negative surge is developed, especially at the Samar coast, because the winds are coming in a direction opposing the sea. The storm surge is generated depending on the wind direction with respect to the coast, which is dependent on the track of the typhoon.

Localized concentrations of simulated storm surge along the coast of Samar are also observed for typhoons that directly crossed or passed north of San Pedro Bay with distance of less than 5 km from the head of the bay. Typhoons with tracks south of the mouth of San Pedro Bay traversing the Leyte Gulf coming from an east-west direction at the approach of the typhoon will most likely produce storm surge with almost equal rate of decrease of maximum storm surge height from the inner part of the bay to the mouth of the bay. Typhoons that follows a path that are more than 70 km from the mouth of San Pedro Bay will generate storm surge in the bay but the magnitude of the storm surge is not large.

It is observed that the typhoon forward speed influences the rate of rise of the peak surge from the minimum water level. Also, the results show that the windspeed is the most significant factor in the severity of the storm surge.

Based from the simulated storm surges generated by historical typhoons, Tacloban and Basey are the most exposed to storm surge hazard. The results of this study are mostly important in understanding the nature and severity of the generated storm surge in San Pedro Bay with respect to typhoon characteristics, such as windspeed, typhoon track, and forward speed. This study may be used for the enhancement of storm surge disaster mitigation strategies, such as storm surge hazard evaluation and in the enhancement of education and information campaigns on storm surge, especially along the coasts of Basey and Tacloban where simulated storm surges are found to be highest. This study may also be used as a benchmark study in the development of the characteristics of synthetic typhoons for future storm surge studies which can be used for detailed design and planning of coastal structures as well as in coastal zone management.

References

1. Bricker, J.D., Takagi, H., Mas, E., Kure, S., Adriano, B., Yu, C., and Roeber, V., *Spatial variation of damage due to storm surge and waves during Typhoon Haiyan in the Philippines,* Journal of Japan Society of Civil Engineers, vol. 70, no. 2, 2014, pp. I_231-I_235.
2. Cinco,T.A., *Climate Trends in the Philippines,* International workshop on the digitization of Historical Climate Data, the new SACA&D Database and Climate Analysis in the ASEAN Region, 2012.
3. Japan Meteorological Agency, Regional Specialized Meteorological Center Tokyo – Typhoon Center, Best Track Data, http://www.jma.go.jp/jma/jma-eng/jma-center/rsmc-hp-pub-eg/besttrack.html.
4. Joint Typhoon Warning Center Tropical Cyclone Best Track Data, http://www.usno.navy.mil/NOOC/nmfc-ph/RSS/jtwc/best_tracks

5. Leuttich, R.A., and Westerink, J.J., Formulation and numerical implementation of the 2D/3D ADCIRC finite element model version 44.XX, 2004.

6. Mas, E., Bricker, J., Kure, S., Adriano, B., Yi, A., Suppasri and Koshimura, S., *Field survey report and satellite image interpretation of the 2013 Super Typhoon Haiyan in the Philippines*, Natural Hazards and Earth System Sciences, vol. 15, 2015, pp 805-816.

7. Morgerman, J. iCyclone Chase Report – Preliminary: Haiyan 2013, http://icyclone.hellohelp.org/iCyclone_Chase_Report HAIYAN_2013.pdf

8. National Disaster Risk Reduction and Management Council, Y It Happened – Learning from Typhoon Yolanda, 2015, http://www.ndrrmc.gov.ph/12-others/2926-y-it-happened-learning-from-typhoon-yolanda.

9. Soria, J. L. A., Switzer, A. D., Villanoy, C. L., Fritz, H. M., Bilgera P. H. T., Cabrera, O. C., Siringan, F. P., Yacat-Sta. Maria, Y., Ramos, R. D., and Fernandez, I. Q, *Repeat storm surge disasters of Typhoon Haiyan and its 1897 predecessor in the Philippines*, Bulletin of American Meteorological Society, vol. 97, no. 1, 2016, pp. 31-48.

10. Tajima Y., Yasuda, T., Pacheco, B. M., Cruz, E. C., Kawasaki, K., Hisamichi, N., Mamoru, M., Asano, Y., Arikawa, T., Ortigas, N. M., Aquino, R., Mata, W., Valdez, J., and Briones, F., *Initial report of JSCE-PICE joint survey on the storm surge disaster caused by Typhoon Haiyan*, Coastal Engineering Journal, vol. 56, no. 1, 2014, 1450006.

11. Takagi, H., De Leon, M., Esteban, M., Mikami, T., and Nakamura, R., *storm surge due to 2013 Typhoon Yolanda (Haiyan) in Leyte Gulf, the Philippines*, Handbook of Coastal Disaster Mitigation for Engineers and Planners, edited by Esteban, M., Takagi, H., & Shibayama, T., 1st Edition, Elsevier Inc., 2015, pp. 133-144.

12. Vickery, P. J. and Wadhera, D., *Statistical models of Holland pressure profile parameter and radius to maximum winds of hurricanes from flight-level pressure and H*Wind Data*, Journal of Applied Meteorology and Climatology, vol. 47, 2008, pp. 2497-2517.

Motion of Oil Storage Tank Moored with Lines Placed in Tsunami Flow

Keisuke Murakami, Naoya Matsuki and Tsubasa Maeda

Faculty of Engineering, University of Miyazaki,
889-2192, 1-1 Gakuen Kibanadai Nishi, Miyazaki, Japan
keisuke@cc.miyazaki-u.ac.jp

Tsunami caused by Tohoku earthquake in 2011 had flooded many oil storage tanks, and these tanks brought many secondary damages on coastal area by fires. In order to reduce those secondary damages, an idea of using a mooring line to check the drift of the oil storage tank was proposed. This research examines an effective mooring system that restrains the motion of a floating tank in tsunami flow. A series of hydraulic experiments were carried out under some different mooring conditions in order to discuss the characteristics of tank motion. Furthermore, this study discusses the characteristics of a drag coefficient to obtain an equation to estimate the fluid force on a semi-immersed cylindrical tank.

Keywords: Drift of oil storage tank; Mooring system; Tsunami force.

1. Introduction

Tohoku earthquake in 2011 caused a huge tsunami, and this tsunami flooded many oil storage tanks. Some of these flooded tanks had brought the secondary damage on coastal area by fires.[1,2,3] Due to the structural reason of an oil storage tank that usually consists of a steel sheet, it is difficult to fix the tank tightly on a ground in order to check the drifting against tsunami flow. Construction of a higher oil retaining wall could be one of the measures, but it seems an unrealistic option due to its construction cost. Regarding these background, an idea of using a mooring line to check the drift of the oil storage tank was proposed, and the possibility of this idea was discussed on a numerical simulation.[4] In this idea, one end of the line was tied on the ground and the other end was bound on the side of the tank at a certain height above the bottom of the tank. In order to realize this idea, it is needed to clear the behavior of the tank and the tension of the lines under various mooring conditions.

This research examines an effective mooring system that restrains the motion of a floating tank in tsunami flow. A series of hydraulic experiments were carried out under the different mooring conditions in order to discuss the characteristics of the tank motion. Furthermore, this study discusses the characteristics of a drag coefficient to obtain an equation to estimate the fluid force on a semi-immersed cylindrical tank.

2. Experimental setup

The hydraulic experiments were carried out with using an open channel, 10m in length, 0.7m in height and 0.4m in width. This study refers a cylindrical storage tank that is located at the site of a chemical plant in Nobeoka city. Its diameter is 19.2m, and the height is 15.2m. The tank is used to store Cyclohexanol whose specific gravity is 0.941. The site of this chemical plant is assumed to have a heavy tsunami inundation due to Nankai trough earthquake which is supposed in near future. A tsunami inundation simulation revealed that the tsunami gradually inundates onto this site, and the maximum inundation depth and flow velocity at this site was simulated 10.8m and 3.5m/s, respectively. Due to this flow condition, no impulsive phenomena such as a bore flow were observed around the tank. Based on the result of this tsunami flow condition, this study measured the tank motion and the fluid force under the steady flow condition.

This study assumed a model scale as about 1/100. According to this model scale, the water depth and flow velocity in the channel was maintained as h=0.13m and u=0.34m/s, respectively. The diameter of the tank model, which was made of an acrylic cylinder, was D=0.2m. In addition to these conditions, a higher velocity case, u=0.51m/s, and a smaller diameter case, D=0.11m, were included in this experiment. The tank model was set at the center of the channel as shown in Fig. 1. It was moored from the channel floor at, a, b, c or d. The displacement of sway and heave were measured by tracing a marker set on the tank model. The submerging depth in the tsunami flow differs depending on the volume of a content in the tank. This study set some submerging conditions by changing the water volume in the tank from 0% to 80% to the tank capacity.

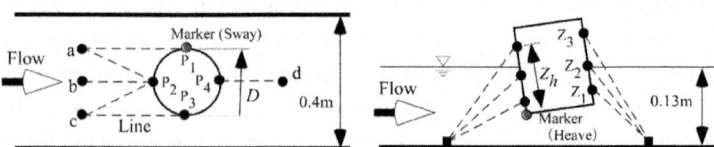

Fig. 1. Schematic of experimental setup

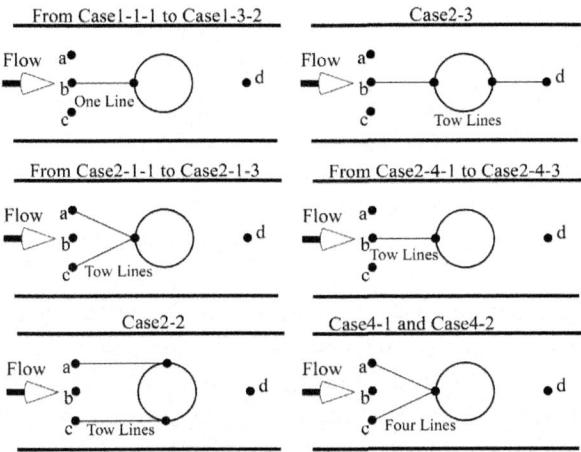

Fig. 2. Types of mooring system

Table 1. List of experimental cases

Case	Number of line	Mooring Points	e (m)	Z_h (m)	u (m/s)	D (m)
Case1-1-1	1	b-P2-Z1	0.26	0.03	0.34	0.2
Case1-1-2	1	b-P2-Z2	0.26	0.06	0.34	0.2
Case1-1-3	1	b-P2-Z3	0.26	0.09	0.34	0.2
Case1-2-1	1	b-P2-Z1	0.26	0.03	0.51	0.2
Case1-2-2	1	b-P2-Z2	0.26	0.06	0.51	0.2
Case1-2-3	1	b-P2-Z3	0.26	0.09	0.51	0.2
Case1-3-1	1	b-P2-Z1	0.26	0.03	0.34	0.11
Case1-3-2	1	b-P2-Z2	0.26	0.06	0.34	0.11
Case1-3-3	1	b-P2-Z3	0.26	0.09	0.34	0.11
Case1-4-1	1	b-P2-Z1	0.13	0.03	0.34	0.2
Case1-4-2	1	b-P2-Z2	0.13	0.06	0.34	0.2
Case1-4-3	1	b-P2-Z3	0.13	0.09	0.34	0.2
Case1-5-1	1	b-P2-Z1	0.07	0.03	0.34	0.2
Case1-5-2	1	b-P2-Z2	0.07	0.06	0.34	0.2
Case2-1-1	2	a-P2-Z1, c-P2-Z1	0.26	0.03	0.34	0.2
Case2-1-2	2	a-P2-Z2, c-P2-Z2	0.26	0.093	0.34	0.2
Case2-1-3	2	a-P2-Z3, c-P2-Z3	0.26	0.13	0.34	0.2
Case2-2	2	a-P1-Z1, c-P3-Z1	0.26	0.03	0.34	0.2
Case2-3	2	b-P2-Z1, d-P4-Z1	0.26	0.03	0.34	0.2
Case2-4-1	2	b-P2-Z1, b-P2-Z2	0.26	0.03, 0.093	0.34	0.2
Case2-4-2	2	b-P2-Z1, b-P2-Z3	0.26	0.03, 0.13	0.34	0.2
Case2-4-3	2	b-P2-Z2, b-P2-Z3	0.26	0.055, 0.13	0.34	0.2
Case4-1	4	a-P2-Z1, a-P2-Z2 c-P2-Z1, c-P2-Z2	0.26	0.03, 0.093	0.34	0.2
Case4-2	4	a-P2-Z1, a-P2-Z3 c-P2-Z1, c-P2-Z3	0.26	0.03, 0.13	0.34	0.2

Fig. 2 shows the schematics of the mooring systems. Table 1 also shows the experimental conditions in each mooring system. The cases from Case1-1-1 to Case1-5-2 examined the performance of a single mooring line in retraining the

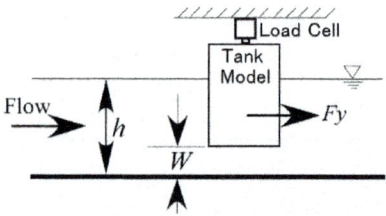

Fig. 3. Measurement system of tsunami force

sway and heave. In these cases, the length of the mooring line was changed as e=0.26m, 0.13m and 0.07m, and the height of the mooring point, Z_1, Z_2 or Z_3, was also changed from Z_h=0.03m to 0.13m. The cases from Case2-1-1 to Case2-4-3 also examined the performance of two mooring lines under the different mooring systems and mooring conditions. Case4-1 and Case4-2 examined the effect of four mooring lines in reducing the motion of sway and heave.

A tsunami force on a semi-immersed cylindrical tank, Fy, was measured in this experiment. Fig. 3 shows the schematic of a measurement system. The tank model was attached under a load cell which was fixed at the upper flange of the channel. The space below the tank model, W, was changed from 0m to 0.06m. According to the flow discharges in the channel, the water depth, h, and flow velocity, u, were ranged from h=0.05m to 0.2m and u=0.42m/s to 0.63m, respectively.

3. Results of experiments

3.1. *Motion of a cylindrical tank moored in tsunami flow*

Fig. 4(a) shows the maximum displacement of the sway motion in the case of a single mooring line, and Fig. 4(b) also shows the maximum displacement of the heave motion. The tank model was always on the channel floor in the case of 60%

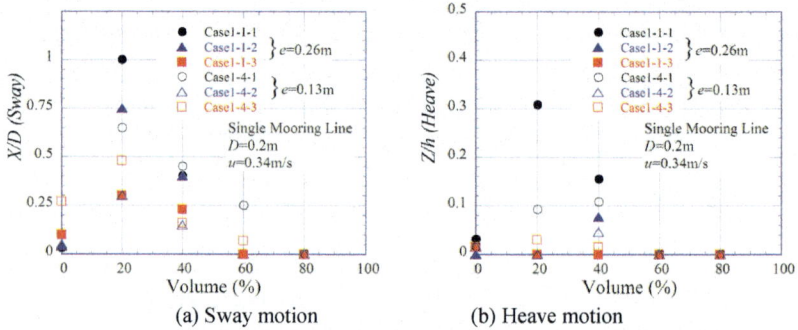

(a) Sway motion   (b) Heave motion

Fig. 4. Maximum displacement of sway and heave in the case of a single mooring line

and 80% water volume to the tank capacity. In other cases, the tank left the floor due to its buoyancy and showed the periodic motion in sway and heave. The displacement of each motion differs depending on the submerging depth, and the displacement of the sway is always larger than the heave. The sway tends to have the large displacement in the case of 20% content. These tendencies can be observed in both e=0.26m and 0.13m. An installation standard of the oil storage tank regulates the distance between adjacent tanks, and the distance should be taken at least more than the tank diameter.[5] According to this standard, this study set an allowable displacement of the sway as one quarter of the tank diameter. As shown in Fig. 4(a), the sway displacement becomes small by setting the mooring point at higher location, though the motion in 20% content never satisfies the allowable displacement.

Fig. 5 shows the maximum displacement of the sway and heave in which the tank is moored in different flow velocities with a single line. The sway and heave become smaller under the higher flow velocity, u=0.51m/s, and the large displacement observed at 20% content in sway and heave are reduced significantly. The higher flow velocity causes the increase of a drag force to the downstream direction, and the force restrains the motion in sway and heave.

Fig. 6 shows the maximum displacement of the sway motion in the case of two mooring lines. Fig. 6(a) shows the results when the tank is moored at a different height, Z_1, Z_2 or Z_3, on P_2 section. The mooring systems shown in Fig. 6(a), using two mooring lines attached on a single section, are still insufficient to restrain the sway motion within the allowable displacement. Fig. 6(b) shows the results when the tank is moored at Z_1 on P_1 and P_3 sections in Caes2-2, and at Z_1on P_2 and P_4 sections in Case2-3. These cases restrain the sway motion better than the cases in Fig. 6(a) and a single mooring system shown in Fig. 4. Especially, Case2-3 largely reduces the displacement in sway motion at 20% content, though the motion at 40% exceeds the allowable displacement slightly.

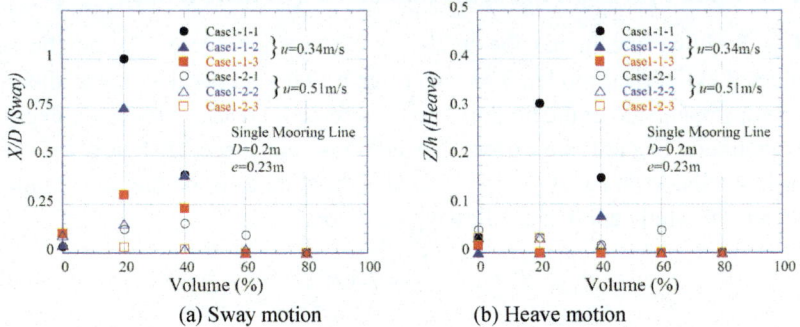

(a) Sway motion   (b) Heave motion

Fig. 5. Maximum displacement of sway and heave in the different tsunami flow velocities

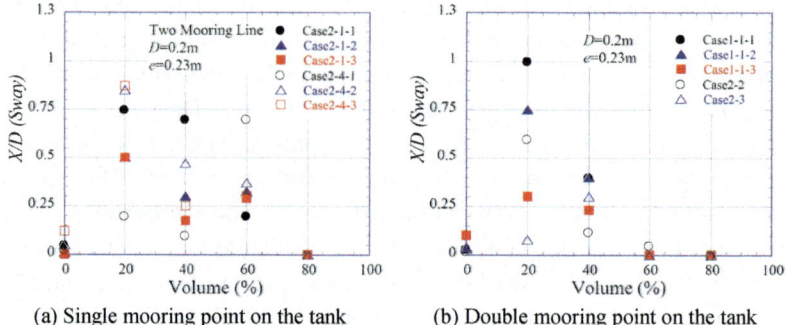

(a) Single mooring point on the tank   (b) Double mooring point on the tank

Fig. 6. Maximum displacement of sway and heave in the case of two mooring lines

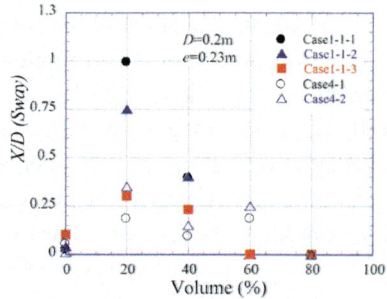

Fig. 7. Maximum displacement of sway and heave in the case of four mooring lines

Fig. 7 shows the maximum displacement of the sway motion in the case of four mooring lines. The mooring system with using four lines reduces the sway motion even in the case of attaching the lines only on a single section. Especially in Case4-2, the height of the mooring point is higher than Case4-1, and the sway motion is restrained under the allowable displacement.

3.2. Tsunami force on the semi-immersed cylindrical tank

Fig. 8 shows the increasing tendency of the fluid force, Fy, on the semi-immersed cylindrical model. The water depth, h, on a lateral axis was measured on the upstream side of the model. The fluid force increases with water depth and the magnitude of the force differs depending on the space, W, below the model. In order to obtain an equation that estimates the fluid force on the semi-immersed cylinder, this study applied a following equation.[6]

$$Fy = \frac{\rho}{2} C_d A_s u_s^2 \qquad (1)$$

where ρ is a fluid density, C_d is a drag coefficient, A_s is a reference area, and u_s is a flow velocity.

A contracted flow was formed on both side of the semi-immersed cylinder when the flow passed the object, because the tank was placed in the channel with a limited width. This experimental configuration means that the infinite numbers of semi-immersed cylindrical tanks are placed in a line normal to the tsunami flow direction. Regarding this configuration, this study employed the averaged fluid velocity, u_s in Eq. (1), at the most contracted section beside the cylinder. The reference area was applied as $A_s=D{\times}h_s$, where h_s was the depth on the same section of u_s.

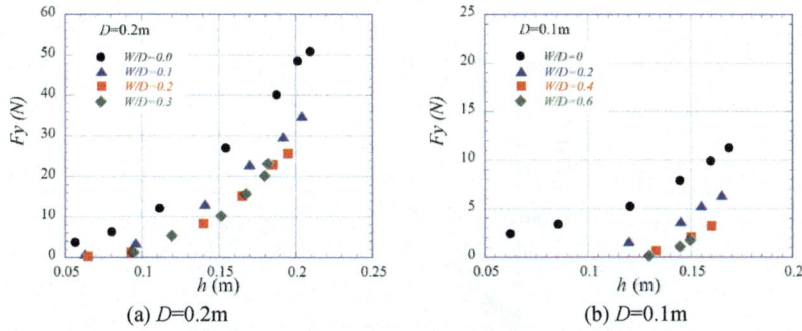

(a) D=0.2m   (b) D=0.1m

Fig. 8. Fluid force on the semi-immersed cylinder

Fig. 9 shows the distribution of drag coefficient, C_d, plotted against the space below the he semi-immersed cylinder, W/D. C_d was calculated from the measured force and Eq. (1). The drag coefficient of semi-immersed cylinder has a good correlation with W/D. The coefficient has a decreasing tendency with increase of W/D, because the submerging depth decreases with the increase of W. This study applied an exponential function as a regression equation. Eq. (2) is the regression equation and its curve is drawn in Fig. 9. Good correlation, 0.97, was obtained in this experiment. Regarding the range of water depth and tank diameter tested in this study, this equation could be applied in a limited range as $0.5{<}h_s/D{<}1.6$.

$$C_d = 1.65 \times e^{-3.32W/D} \tag{2}$$

Fig. 10 shows the correlation between measured fluid force and calculated one from Eq. (1) and Eq. (2). The results on different diameters are plotted in this figure. It is confirmed that the fluid force exerted on the semi-immersed cylinder could be estimated with good correlation, 0.95, by applying the appropriate drag coefficient calculated from Eq. (2).

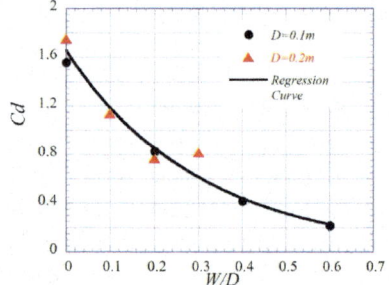

Fig. 9. Drag coefficient of the semi-immersed cylinder

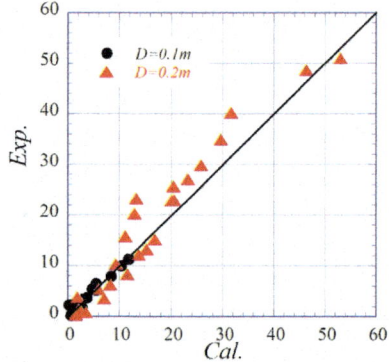

Fig. 10. Correlation between estimated force and measured one

4. Conclusions

The cylindrical tank moored with lines in tsunami flow had the periodic motion in sway and heave. Each motion in the tsunami flow depended on the submerging depth of the tank, number of mooring lines, mooring location on the tank, and mooring system. The displacement of the sway motion tended to become smaller when the mooring point was taken at a higher location, though it was difficult to restrain the motion within an allowable displacement under the mooring system with a single line. Mooring systems with two lines retrained the displacement of the sway motion better than a single mooring line, though the motion slightly exceeded the allowable displacement. The mooring system with four lines could reduce the sway motion under the allowable displacement.

The fluid force exerted on the semi-immersed cylinder showed an increasing tendency with increase of water depth and flow velocity. The magnitude of the force differed depending on the submerging depth. The drag coefficient of the semi-immersed cylinder had a good correlation with W/D, and the coefficient had a decreasing tendency with increase of W/D. This study showed a regression equation to estimate the drag coefficient, and also confirmed a good correlation between the measured force and the calculated one.

Acknowledgments

We are thankful to Asahi Kasei Corp. for supporting this research partially.

References

1. S. Zama, H. Nishi, K. Hatayama, M, Yamada, H, Yoshihara and Y. Ogawa : On damage of oil storage tanks due to the 2011 off the Pacific Coast of Tohoku Earthquake (Mw9.0), *Japan, National Research Institute of Fire and Disaster, Japan*, 15, 2012.
2. N. Fujii, M. Oomori, T. Ikegaya, S. Inagaki : Method for estimating tsunami forces and damages on oil storage tanks, *Proceedings of Coastal Engineering, JSCE*, Vol. 53, pp. 271-275. 2006.
3. S. Inagaki, T. Ikegaya, M. Oomori, N. Fujiii, K. Mukaihara and K. Hatakeyama : Method for estimating the slide and drift of tanks due to tsunami, *Proceedings of Coastal Engineering, JSCE*, Vol. 55, pp. 276-280. 2008.
4. Y. Sakamoto, K. Sugatsuki, T. Nonaka, Y. Sakamoto and T. Harada : Examination of a method of countermeasure for storage tank drifting during tsunami, *Journal of Japan Society of Civil Engineers*, B2(Coastal Engineering), Vol. 72, p.l-949-954, 2016.
5. The Fire Services Act : FDMA of the Ministry of Internal Affairs and Communications, *Ordinance on The Regulation of Hazard Materials*, Chap. 3, http://law.e-gov.go.jp/htmldata/S34/S34SE306.html, Accessed on June 6, 2017.
6. Japan Society of Civil Engineers, *The Collection of Hydraulic Formula*, pp. 218, 1999.

Study on the Risks of Tsunami Inundation Via Underground Pipelines

T. Takabatake and T. Shibayama

Department of Civil and Environmental Engineering, Waseda University,
3-4-1 Okubo, Shinjuku-ku, Tokyo, 169-8555, Japan
E-mail: tomoyuki.taka.8821@gmail.com
http://www.f.waseda.jp/shibayama

When a tsunami comes to seaside industrial areas, seawater would enter landside areas through underground pipelines and induce inundation disasters. In the present study, the authors aim to investigate the risks of this tsunami inundation via the pipelines using the numerical simulation model. The results indicated that the influence of the tsunami inundation via the pipelines on seaside areas is not small and especially it will likely to influence the initial phase of evacuation. It is therefore important for port authorities and city planners to consider such risks when developing tsunami mitigation plans.

Keywords: Tsunami; Drainage channel; Water intake/outfall.

1. Introduction

The 2011 Tohoku earthquake and tsunami devastated large parts of the northeastern Pacific coastline in Japan, highlighting the importance of considering all possible tsunami risks. Generally, when seaside areas are at risk of tsunamis, high seawalls/coastal dykes are constructed. However, tsunamis can penetrate underground pipelines (e.g. drainage channels and water intakes/outfalls) that often exist in seaside areas. Thus, seawater could enter inland areas via underground pipelines during a tsunami event, even if the protective structures function appropriately. The seawater entering inland areas could induce inundation, leading to flooding, loss of evacuation routes and damage to electrical facilities. This type of inundation, named "tsunami pipe-flow inundation" herein, has been observed in the past tsunami events. For instance, in the case of the 2004 Indian Ocean earthquake and tsunami, the seawater penetrated a submarine tunnel and flowed into the pump room at the Madras Atomic Power Station, resulting in the flooding of the seawater pump.[1] During the 2010 Chilean earthquake and tsunami, Hashimoto and Imamura[2] reported that some citizens in the Kesennuma City, Miyagi Prefecture, Japan observed seawater rising from manholes while they were evacuating. In the case of the 2011 Tohoku earthquake and tsunami, it was reported that the tsunami travelled up the discharge channel of the Kashima Thermal Power Station and entered the power station site, flooding some facilities.[3]

To evaluate the risks of tsunami pipe-flow inundation, Ito et al.[4] developed a model that could simulate the overflow volume from water channels. This model is based on a one-dimensional (1-D) pipe-flow model and thus has the advantage of low computational costs. The model was further improved by Takabatake et al.[5] so that they can be applied to a network channel and calculate the inland inundation process due to tsunami pipe-flow inundation. The model was validated through a comparison of the results of various hydraulic physical model tests. However, it has not yet been applied to real coastal industrial areas. Thus, the risks of tsunami pipe-flow inundation are not yet well understood. This study aims at investigating the risks of tsunami pipe-flow inundation for a seaside area using a simulation model.[5] Herein, the authors selected a seaside industrial area of Yokohama, Japan as a study area.

2. Methodology

The simulation model[5] was employed to investigate the risks of tsunami pipe-flow inundation. The model comprises two different simulation models: a 1-D overflow model to simulate the overflow volume from pits (manholes) of water channels and a two-dimensional inundation model to simulate the inundation flow on the ground (Fig. 1). The governing equations of the 1-D overflow simulation model are based on the following continuity and momentum equations:

$$\frac{d\eta_i}{dt} = w_i = \frac{-D_i v_i + D_{i-1}v_{i-1} + D_j v_j + D_k v_k}{A_i} \tag{1}$$

$$\frac{dv_i}{dt} = \frac{g(\eta_i - \eta_{i+1})}{L_i} + \frac{w_i^2 - w_{i+1}^2}{2L_i} + \frac{v_{i-1}^2 - v_i^2}{2L_i} - loss \tag{2}$$

where η is water surface level, D is the diameter of a pipe, A is the area of a pit, v is flow velocity inside a pipe, g is the gravity acceleration, L is the length of a pipe, w is the upward velocity of water surface, which is calculated from the continuity equation at the junction of a pit and pipes (Eq. (1)), i, j and k are the number of each pipe and pit of pipe-lineA, B and C, and the loss denotes the energy loss of head by friction and the pipe form. The inundation flow on the ground is calculated based on the shallow-water equations [Eqs. (3)-(5)].

$$\frac{\partial h}{\partial t} + \frac{\partial M}{\partial x} + \frac{\partial N}{\partial y} = \frac{Q}{\Delta x \Delta y} \tag{3}$$

$$\frac{\partial M}{\partial t} + \frac{\partial}{\partial x}\left(\frac{M^2}{h}\right) + \frac{\partial}{\partial y}\left(\frac{MN}{h}\right) + gh\frac{\partial(h+z)}{\partial x} + \frac{gn^2 M\sqrt{M^2+N^2}}{h^{7/3}} = 0 \tag{4}$$

$$\frac{\partial N}{\partial t} + \frac{\partial}{\partial x}\left(\frac{MN}{h}\right) + \frac{\partial}{\partial y}\left(\frac{N^2}{h}\right) + gh\frac{\partial(h+z)}{\partial y} + \frac{gn^2 N\sqrt{M^2+N^2}}{h^{7/3}} = 0 \tag{5}$$

where h is inundation depth, M, N are the discharge flux in x and y direction, Q is the overflow volume from a pit, Δx and Δy are mesh sizes of calculation, z is the ground height and n is the manning's coefficient. Integrating the two simulation models shown by the previous equations, it is possible to simulate the tsunami pipe-flow inundation process. The model can also be applied to water pipe networks, in which more than three pipelines are connected to a manhole.[5,6] More details of this simulation model can be found in Ito et al.,[4] Takabatake et al.[5,6]

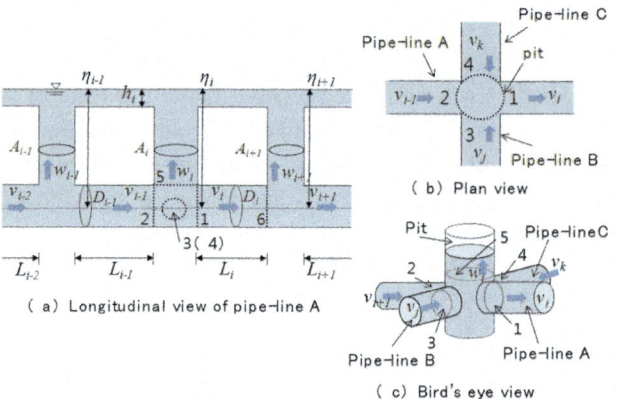

Fig. 1. A schematic view of the tsunami pipe-flow inundation simulation model.

3. Numerical conditions

A schematic view of the selected study area in a seaside industrial area in Yokohama, Japan is shown in Fig. 2. Topography data and the locations of buildings in the study area were provided by the Geospatial Information Authority of Japan, Ministry of Land, Infrastructure, Transport and Tourism.[7] The original topography data (5 m resolution data) was interpolated to obtain 1 m resolution data for use in the simulation. The black lines and orange circles in Fig. 2 denote the underground pipe-lines and manholes (The data was obtained from Yokohama City[8]). The seawater was able to flow into the water channels via the two outlets. In the simulation, hypothetical seawalls whose height can be changed were included by the pink lines in Fig. 2. A tsunami was generated by inputting a time series of water surface elevation into the easternmost meshes, as indicated in blue in Fig. 2. The head loss (friction loss and the other losses due to the changes of shape of the pipe) was accounted based on the formula book for hydraulics.[9]

In the simulation, the tsunami height (H), number of tsunamis (N), duration time (T) and height of the seawall (W) were used as variable parameters. The definitions of each parameter are shown in Fig. 3. A total of 10 different cases were simulated (Table 1). The incident tsunami waves used for each case are shown in Fig. 4. In Cases 1-6, the tsunami height was lower than that of the seawall. In these cases, inundation on the ground will be only due to tsunami pipe-flow inundation. In contrast, in Cases 7-10, the tsunami was higher than the seawall.

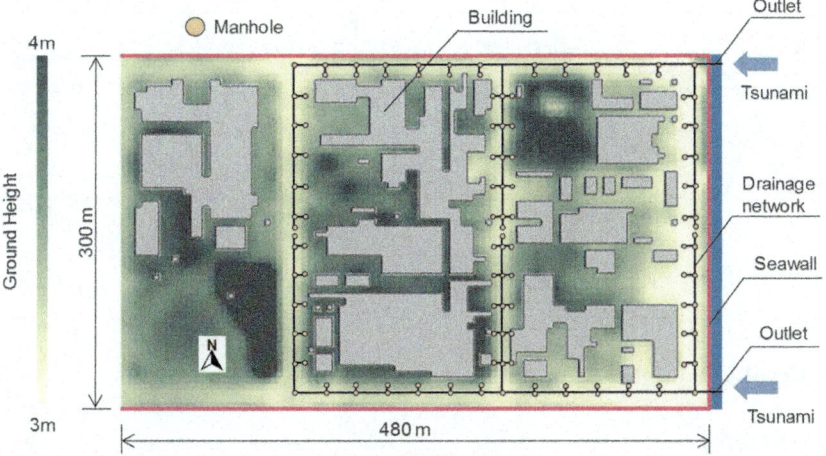

Fig. 2. A map of the study area

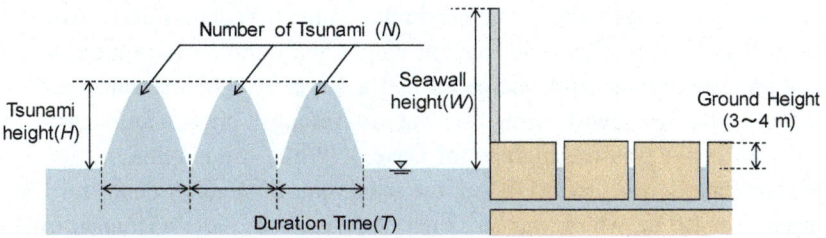

Fig. 3. Definition of the parameters used in this study

Table 1. Simulation cases.

| Case | Tsunami | | | Seawall height(W) |
	Tsunami height (H)	Duration time (T)	Number of tsunamis (N)	
Case1	7 m	10 min	1	20 m
Case2	7 m	30 min	1	20 m
Case3	14 m	10 min	1	20 m
Case4	14 m	30 min	1	20 m
Case5	7 m	10 min	3	20 m
Case6	14 m	10 min	3	20 m
Case7	14 m	10 min	1	20 m
Case8	14 m	30 min	1	10 m
Case9	14 m	30 min	1	12.5 m
Case10	7 m / 14 m	10 min	2 /1	10 m

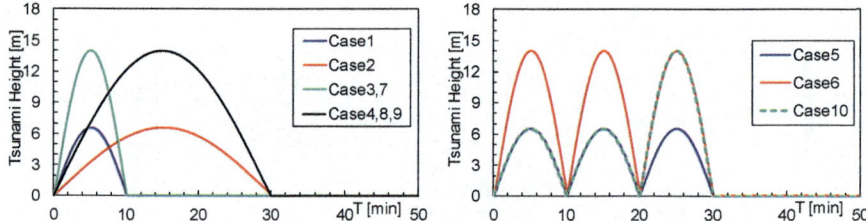

Fig. 4. Incident tsunami waves for the 10 different cases

4. Results

4.1. *Case1-4*

Fig. 5 shows the time histories of the inundation depths recorded at Point 1-5 for Cases 1-4, while Fig. 6 shows the maximum inundation depths and areas. It can be seen that, in general, as the tsunami height and duration time increased, the maximum inundation depth increased and a larger part of the study area was inundated due to overflowing seawater. Amongst these four cases, the inundation depth was the highest for Case 4. While the maximum inundation depth for Case 1 was around 0.3 m, the maximum inundation depth for Case 4 reached over 1.0 m, which was sufficient to result in the loss of human lives. In addition, it can be seen from Fig. 6(d) that the seawater reached the buildings. These results indicate that when the incident tsunami wave was higher and its duration was longer, even for cases where the tsunami does not exceed the seawall and inundation occurs only due to tsunami pipe-flow inundation, there is a risk of property damage and casualties.

Comparing Cases 2 and 3, the maximum inundation depths and area of inundation for Case 2 were larger than those of Case 3. Although the tsunami height in Case 2 was smaller than that in Case 3, the duration was longer. The results suggest that the duration would have a more significant effect on inundation depth and area of flooding due to tsunami pipe-flow inundations than the tsunami height

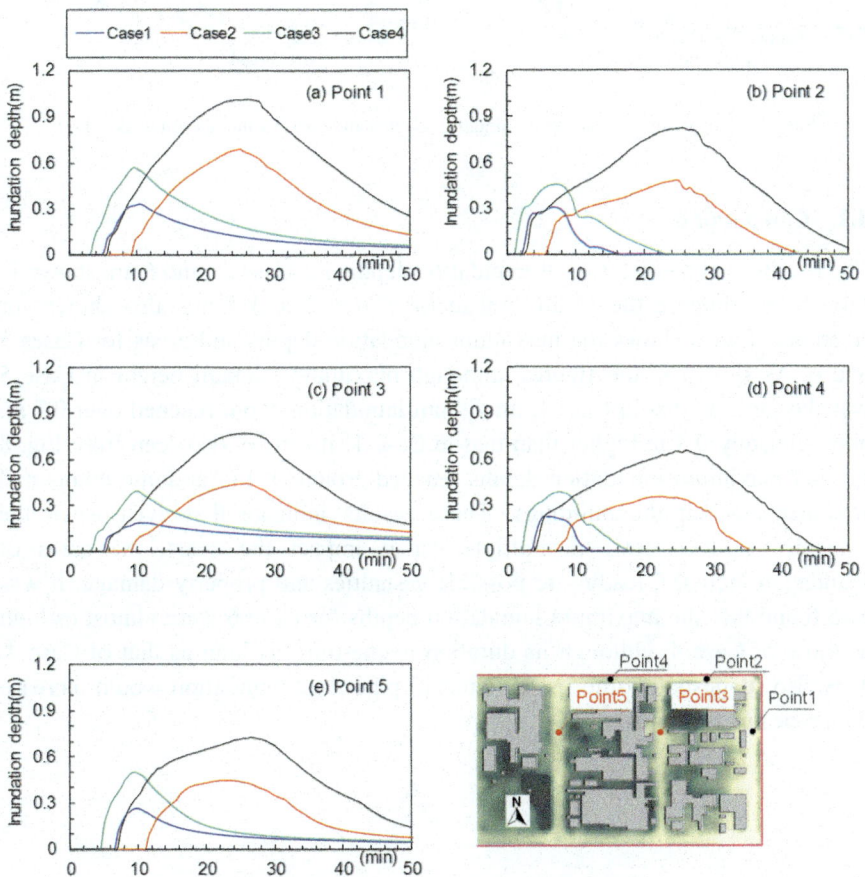

Fig. 5. Comparison of time histories of the inundation depths at Points 1-5 (Cases 1-4).

274

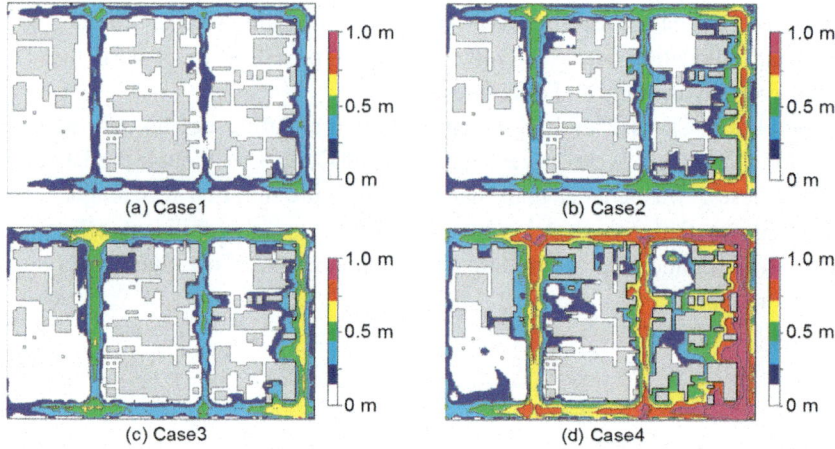

Fig. 6. Comparison of maximum inundation depths and areas of inundation (Cases 1-4).

4.2. Case5 and 6

The results of time histories of inundation depths for Cases 5 and 6 are shown in Fig. 7 To compare the results, those for Cases 2 and 4 are also shown for reference. Fig. 8 shows the maximum inundation depths and areas for Cases 5 and 6. As shown in the figures, although maximum tsunami height in Case 5 was the same as that in Case 1, maximum inundation depth reached over 0.7 m, approximately 0.4 m higher than that in Case 1. It can be also seen from Fig. 8 (a) that maximum inundation depths reached around 1.0 m at some points and seawater reached the buildings. These results indicate that even when the tsunami height is low, if multiple waves impact the coast, the areas of inundation increase, leading to possible casualties and property damage. It was also found that the maximum inundation depths for Case 6 were almost as high as those of Case 4, although its duration is one-third as long as that of Case 4. This also suggests that the risk of tsunami pipe-flow inundation would increase with repeated tsunamis.

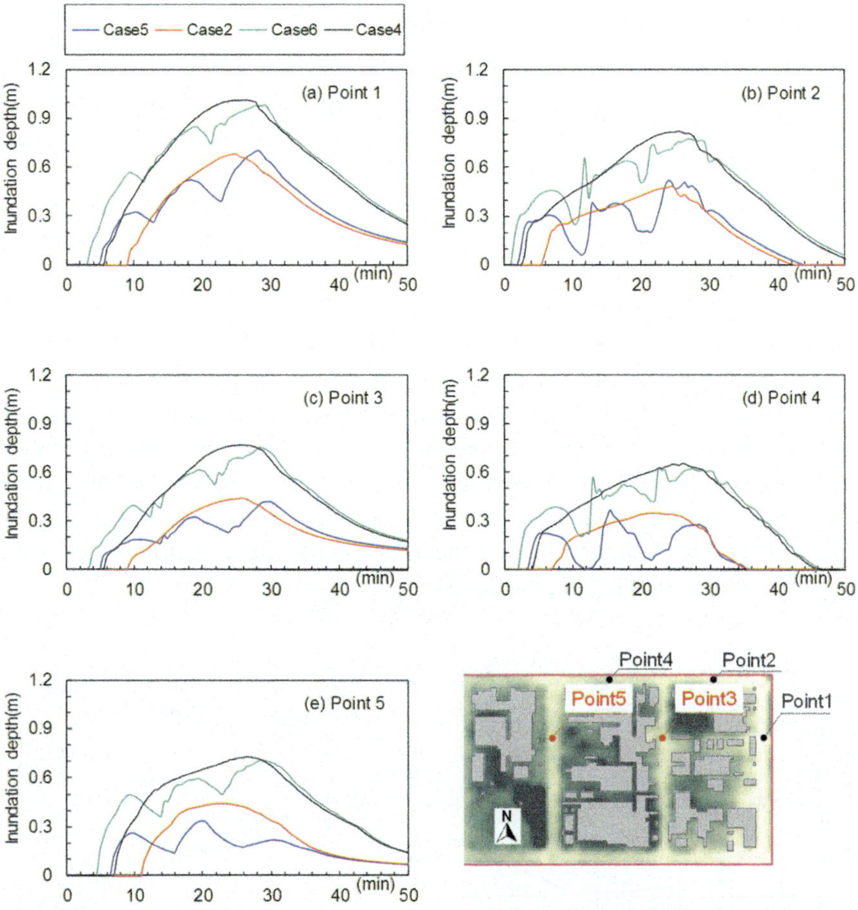

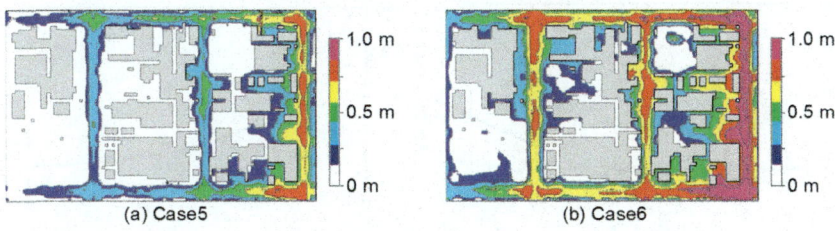

Fig. 7. Comparison of time histories of the inundation depths at Points 1 and 2 (Cases 5 and 6).

Fig. 8. Comparison of maximum inundation depths and areas of inundation (Cases 5 and 6).

4.3. *Case7-10*

For Cases 7-10, the tsunami height exceeded the seawall height. Thus, as shown in Figs. 9 and 10, the maximum inundation depth was much higher than for other cases (Cases 1-6). This means that when a tsunami exceeds a seawall, damage due to tsunami pipe-flow inundation can be neglected from the viewpoint of structural damage as most of the damage would be caused by the tsunami overflowing the seawall. However, when focusing on time when inundation begins, flooding due to tsunami pipe-flow inundation started several minutes before the overflowing tsunami arrived (see results for Cases 8 and 9). In Case 10, as shown in Figs. 10 and 11, inundation due to tsunami pipe-flow inundation occurred more than 15 minutes earlier. This means that tsunami pipe-flow inundation would influence the evacuation of population from the area. In Case 10, the inundation depth reached around 0.3 m which could influence their walking speed. It can be concluded that even when the tsunami height exceeds that of the seawall, the risks of tsunami pipe-flow inundation should not be neglected, as it is likely to influence the initial phase of evacuation, which is the most important period to save human lives.

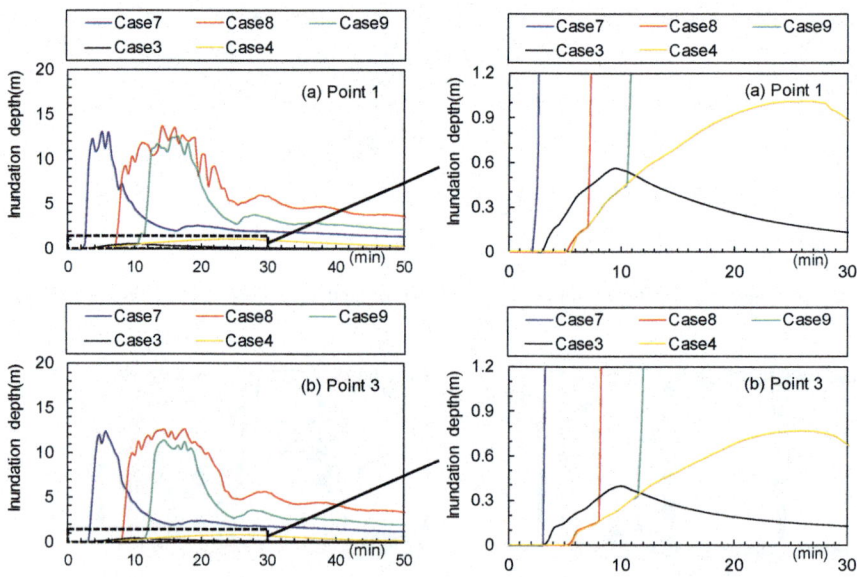

Fig. 9. Comparison of time histories of the inundation depths at Points 1 (Cases 3,4,7,8 and 9).

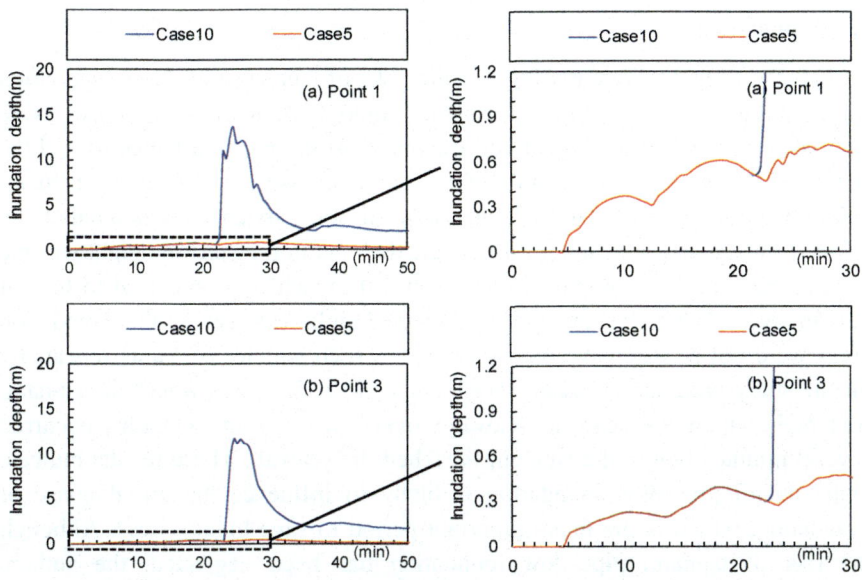

Fig. 10. Comparison of time histories of the inundation depths at Points 1 (Cases 5 and 10).

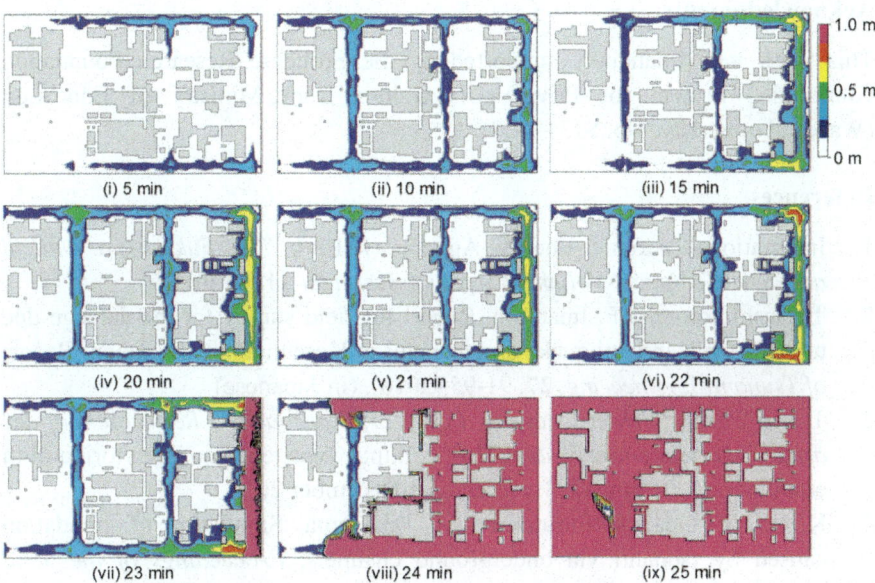

Fig. 11. Inundation process for Case 10.

5. Conclusion

In this study, the authors investigated the risks of tsunami pipe-flow inundation using a previously developed simulation model. A total of 10 cases were simulated by varying the height and frequency of the incident tsunamis and the height of the seawall. The simulation results demonstrated that as the tsunami height and duration increased, a larger area was more extensively inundated due to overflowing seawater. In the case where the maximum tsunami height was the largest, the inundation depth reached over 1.0 m, which would lead to loss of human lives. When the maximum tsunami height was small, the inundation areas were not as extensive. However, when such smaller tsunamis repeatedly hit the study area, larger areas were inundated. In the cases where the tsunami was higher than the seawall, seawater overflowing from manholes occurred several minutes before the tsunami breached the seawall. These results indicate that tsunami pipe-flow inundation is likely to influence the initial phase of evacuation, which is the most important period to save human lives. Although the risk of tsunami pipe-flow inundation has been neglected, the authors conclude that it is important for port authorities and city planners to consider such risks when developing tsunami mitigation plans.

Acknowledgments

This work was financially supported by the Strategic Research Foundation Grant-aided Project for Private Universities from Ministry of Education (Waseda University, No. S1311028).

References

1. International Atomic Energy Agency (IAEA), *The Fukushima Daiichi accident -Technical Volume 2 Safety Assessment-*, 186 (2015).
2. T. Hashimoto and F. Imamura, Report for field survey for the damage due to 2010 Chilean earthquake and tsunami in Kesennuma City, Japan. *Report of Tsunami Engineering*, **27**, 91-95 (2010). (in Japanese)
3. Tokyo Electric Power Company (TEPCO). *Approach 1: Restoration of dev astated thermal power station* (2011). http://www.tepco.co.jp/en/torikumi/th ermal/popup_01.html#s03. Accessed 1 December 2016.
4. K. Ito, Y. Oda, A. Furuta and Y. Takayama, Simulation of inundation caused by tsunami via underground channels, *Proceedings of the 32nd International Conference on Coastal Engineering*, 1530-1540 (2012). https://journals.tdl.org/icce/index.php/icce/article/view/6625/pdf_610. Accessed March 2017.

5. T. Takabatake, Y. Oda, K. Ito and T. Honda, Development of simulation model of tsunami flow coupling water channel network flow and overland inundation. *Journal of Japan Society of Civil Engineers, Ser. B3 (Ocean Engineering)*, **72**(1), 1-13 (2016). (in Japanese)

6. T. Takabatake, T. Honda, Y. Oda and K. Ito, Simulation of inundation disaster by tsunami via underground drainage channel network. *Proceedings of the 24th International Ocean and Polar Engineering Conference*, 17-24 (2014). https://www.onepetro.org/conference-paper/ISOPE-I-14-378. Accessed August 2017.

7. Geospatial Information Authority of Japan, *Download service of basic map information* (2016). https://fgd.gsi.go.jp/download/menu.php, Accessed October 2016.

8. Yokohama City, *System for providing information regarding governmental information* (2016). http://wwwm.city.yokohama.lg.jp/, Accessed October 2016.

9. Japan Society of Civil Engineering (JSCE), *Formula book for hydraulics*, 373-382. (in Japanese)

Numerical Investigation of Tsunami Run-up along the Kido River Caused by The 2011 Tohoku Earthquakes

Souki Fukazawa* and Yoshimitsu Tajima

*Coastal Engineering Lab., The University of Tokyo,
Hongo 7-3-1, Bunkyo-ku, Tokyo 113-8656, Japan
* E-mail: fukazawa@coastal.t.u-tokyo.ac.jp*

Tsunami run-up through the river is one of important features for better predictions and estimations of inundation characteristics. This study carried out numerical investigations of observed tsunami inundation around the Kido river especially focusing on the prediction of run-up speed of tsunami along the river. Based on non-linear long wave model, a sensitivity analysis was first carried out by changing several computational conditions such as tsunami profile, bottom frictions and the river discharge. It was found through this analysis that these conditions, within the range of expected uncertain variations, have certain influence on predicted run-up speed of tsunami. Second, this study also investigated the influence of the different discretization schemes of the model on predictive skills of the speed of tsunami run-up. Difference of conservative and non-conservative forms of non-linear term was investigated through numerical experiments of non-viscosity Burgers equations and it was found that the difference of these forms has significant influence on the predicted propagation speed of the bore. The same analysis was applied in the case of the Kido river and it was found that the tsunami run-up speed was increased up to 40% by selecting appropriate discretization schemes of conservative form.

Keywords: Tsunami; The 2011 Tohoku Earthquake; Run-up speed; Video analysis; Numerical schemes.

1. Introduction

The 2011 Tohoku Earthquake tsunami caused more than twenty thousands of casualties and the missing. The tsunami overflew the coastal dykes and caused significant damage even on the area protected by the coastal dykes. Development and improvement of disaster mitigation strategies in the inundated area is one of urgent and important tasks against such catastrophic tsunami.

Toward better disaster mitigation strategies, prediction of inundation characteristics is one of important factors. The tsunami may be partially blocked by coastal dykes and may propagate faster through the rivers and waterways.

It was found in several areas where the 2011 Tohoku Earthquake tsunami ran up along the river, penetrated far inland and expanded various risks caused by inundations. Lack of detailed data, however, makes it difficult to fully understand the feature of river run-up of tsunami and to investigate the validity of widely-used numerical models for predictions of such tsunami run-up along the river. Sanuki et al.[1] analyzed the video footage of tsunami run-up recorded at the Kido river in Fukushima prefecture when the 2011 Tohoku Earthquake Tsunami hit the coast and compared observed characteristics of tsunami run-up with those of their numerical simulation based on non-linear long wave model. They pointed out that widely-used tsunami run-up simulation tends to underestimate the run-up speed of tsunami along the river by about 50% while the model reasonably represents the observed area and maximum depth of inundation around the Kido river. Such underestimation of run-up speed of tsunami may result in optimistic estimation of tsunami arrival time and may have significant influence on future disaster mitigation strategies such as evacuation plans.

Based on the backgrounds discussed above, this study carried out numerical investigations of observed tsunami inundation around the Kido river especially focusing on the run-up speed of tsunami along the river. Based on widely-used model, a sensitivity analysis was first carried out by changing several computational conditions such as the incident wave conditions, river bed bathymetries and so on. Second, this study also investigated the influence of the different discretization schemes of the model on predictive skills of the speed of tsunami run-up.

2. Tsunami Inundation Characteristics around the Kido River

2.1. *Field Survey*

The Kido river mouth is located in Fukushima Prefecture. The coast on the northern side of the river mouth is covered by steep sea cliff whereas the southern side of the river has lower basin with area of around 3 square kilometers.

Sanuki et al.[1] conducted a field survey in August 2012. They found through the survey that the inundation water levels at the basin ranged from 5.4m to 8.4m above the mean sea level, i.e., Tokyo Peril. They pointed out that the inundated water levels caused by tsunami along the river were higher than those on the basin and that the tsunami overflew from the river and inundated on the basin. Besides, coastal dykes were damaged over the alongshore stretch of 300m and some part of the river dykes were also destroyed.

2.2. *Video Analysis*

Sanuki et al.[1] analyzed the video footage of tsunami run-up recorded at the Kido river in Fukushima prefecture when the 2011 Tohoku Earthquake Tsunami hit the coast. According to their analysis, the tsunami came from the southeast and started to inundate at the southern side of the Kido river mouth. The tsunami ran up through the Kido river immediately and then overflew to the basin.

They converted the video coordinate to the plane coordinate and drew the lines of tsunami front over time as shown in Fig. 1. The analysis of the tsunami run-up characteristics indicated that tsunami approached the coast with speed of around 12m/s. The propagation speed of the tsunami remained high along the Kido river while it drastically decreased in the other inundated area especially behind the coastal dykes.

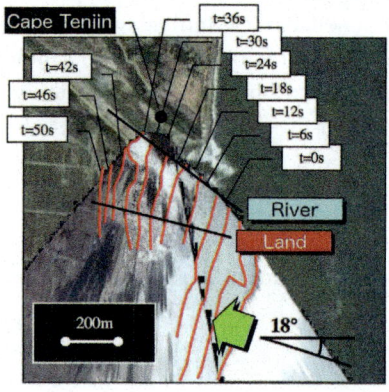

Fig. 1.   Transition of tsunami[1]

3. Characteristics of the Tsunami Run-up

In order to investigate parameters on the run-up process, we set the basic model as follows referring to Sanuki et al.[1] The governing equation is non-linear long wave model with Manning's bottom friction term. The time derivative of the governing equations were based on the leap frog scheme and the advection term is expressed in the non-conservative form and the term was discretized by the first order upwind scheme. The dry/wet boundary condition was based on the method by Kotani et al.[2] Other conditions were shown in Table 1. Collapses of the river dykes and coastal dykes, ground subsidence, and deformation of sand bars by the first wave were considered as change of bed level topography. The tsunami profile data recorded by the GPS buoy at around 19km offshore of Onahama, Fukushima Prefecture was used as the boundary condition for computation of the tsunami shoaling from the buoy to the incident boundary of the tsunami inundation computations. The tsunami profile obtained in this manner is shown by a solid red line in Fig. 2. River coordinate X with $X = 0$ at the river mouth and positive in the upstream-ward direction was set along the river for comparisons of the run-up speed of the tsunami.

According to Sanuki et al.,[1] this computation showed reasonably good predictive skills of the tsunami propagating directions and inundation area while the model tended to underestimate the run-up speed along the river. The calculated run-up speed was 6.8m/s while it was 12m/s based on the analysis of the video footage around the Kido River.

Table 1.   Calculation conditions.

Calculated Area	10km×15km
Grid Size	$\Delta x = \Delta y = 10$m
Time Step	$\Delta t = 0.2$s
Topography	Sea: M7004 series, Land: LP measurement data
Incident Wave	Waveform: GPS at Fukushima offing, Incident angle: E0°S
Bottom Friction	Sebed/River: $n = 0.025[\text{s/m}^{1/3}]$, Land: n=0.030[s/m$^{1/3}$]

In this study, seven parameters or conditions of tsunami run-up were changed from the above conditions and their influences were investigated. The calculated run-up speed results are shown in Table 2. The computational conditions and the features of computed results of each case are briefly summarized as follows.

(1) Case1: Same as Sanuki et al.[1]

(2) Case2: QUICKEST scheme was applied to the advection term.

Table 2. Computed run-up speed in different cases.

case	detail	run-up speed
case1	Base model	6.78m/s
case2	QUICKEST scheme	6.93m/s
case3	Decreased friction	6.93m/s
case4	Incident waveform	8.01m/s
case5	Incident direction	6.43m/s
case6	River bed topography	6.93m/s
case7	River Discharge	6.78m/s
case8	No flow over dykes	6.93m/s

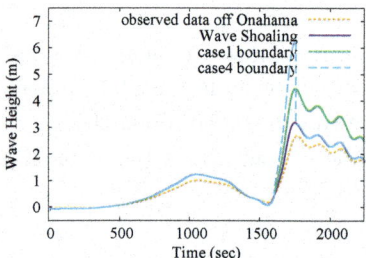

Fig. 2. Incident wave profiles

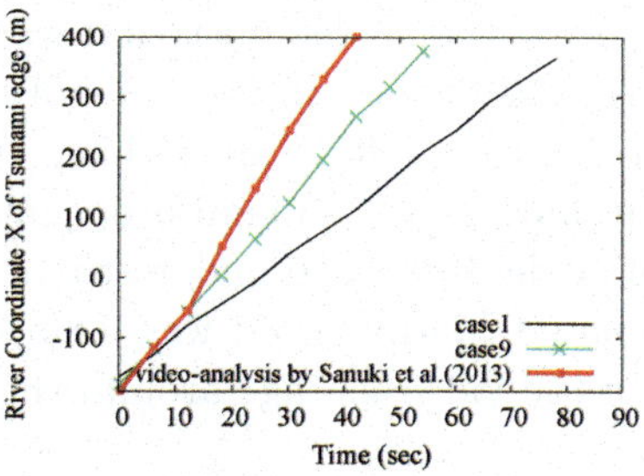

Fig. 3. Comparison of time-varying tsunami front (case1, case9 and measurement)

Compared to case 1, the run-up speed was slightly increased in case 2 at the river mouth and on the land.

(3) Case3: Manning's roughness coefficient was changed from 0.025 to $0.005[s/m^{1/3}]$. Compared to the case 1, the run-up speed was increased by up to 2.2% in the river on the relatively shallow water.

(4) Case4: Incident wave profile was changed as shown in Fig. 2. In case 4, the maximum water level of the incident tsunami profile was increased 1.4 times higher than that of case 1. As a result, the computed averaged tsunami run-up speed was 1.2 times faster than that of case 1.

(5) Case5: The incident wave angle at the offshore boundary was set to E18°S while it was set to E0°S in case1. Run-up speed in case 5 was 5% lower than that in case 1. This is because the height of tsunami inside the river was lower in case 5.

(6) Case6: The river bed levels were lowered by 0.5m from those in case 1 based on the assumption that river bed may be largely eroded by the first attack of the tsunami. The computed run-up speed in case 6 was slightly higher than that in case 1.

(7) Case7: The river discharge (20m^3/s) was induced at the upstream of the Kido river while case 1 did not account for the river discharge. The computed run-up speed in case 7 was nearly the same as that in case 1. This might be because the given discharge was too small to yield difference in this tsunami scale.

(8) Case8: All dykes were not broken. The computed run-up speed in case 8 was 0.2m/s higher than that in case 1. This is because the reflected and diffracted wave components at the coastal dyke increased the wave height in front of the river mouth.

All cases had the almost same predictive skills of the tsunami propagating directions, inundation area and maximum depth of inundation as case 1. But any of these cases still underestimated the observed run-up speed of the tsunami along the Kido river. In order to represent the observed tsunami run-up speed along the river, case 9 was set as combinations of case 2, case 4, case 6 and case 8.

Figure 3 compares the time-varying location of the tsunami front along the river. The computed results shown in the green line showed better agreement with the video-based observations shown in the red line during the first 12 seconds from the initiation of the tsunami run-up at the river mouth. Based on these analysis, it was found that the tsunami height at the river mouth is one of essential factors for estimation of the tsunami run-up speed along the river but some other model improvement may be needed for better predictions of the tsunami run-up speed along the river away from the river mouth.

4. Influence of Numerical Schemes on Predictions of Tsunami Run-up Speed

This section investigates the influence of different numerical schemes on predicted tsunami run-up speed along the river. Several numerical schemes used in this study are summarized based on the following linear advection equation.

$$\frac{\partial u}{\partial t} + c\frac{\partial u}{\partial x} = u_t + cu_x = 0 \tag{1}$$

where u is a physical quantity and ν is defined as courant number:

$$\nu = \frac{c\Delta t}{\Delta x} \tag{2}$$

(i) FTCS (Forward in Time and Central Difference in Space) scheme

$$u_i^{n+1} = u_i^n - \frac{1}{2}\nu(u_{i+1}^n - u_{i-1}^n) \tag{3}$$

(ii) 1st order upwind scheme

$$u_i^{n+1} = u_i^n - \frac{\nu}{2}(u_{i+1}^n - u_{i-1}^n) + \frac{|\nu|}{2}(u_{i+1}^n - 2u_i^n + u_{i-1}^n) \tag{4}$$

(iii) Leapfrog scheme

$$u_i^{n+1} = u_i^{n-1} - \nu(u_{i+1}^n - u_{i-1}^n) \tag{5}$$

(iv) QUICKEST scheme

QUICK (Quadratic Upstream Interpolation for Convective Kinematics) and QUICKEST (QUICK with Estimated Streaming Terms) were presented by Leonard.[3] They are 3rd order differencing schemes in space. In QUICKEST scheme, time term is also discretized. Here, discretized form in positive flow ($c > 0$) is showed.

$$u_i^{n+1} = u_i^n - \frac{\nu}{6}(2u_{i+1}^n + 3u_i^n - 6u_{i-1}^n + u_{i-2}^n) + \frac{\nu^2}{2}(u_{i+1}^n - 2u_i^n + u_{i-1}^n)$$
$$- \frac{\nu^3}{6}(u_{i+1}^n - 3u_i^n + 3u_{i-1}^n - u_{i-2}^n) \tag{6}$$

(v) MacCormack scheme

MacCormack scheme is one of the predictor-corrector methods.

$$\bar{u}_i = u_i^n - \nu(u_i^n - u_{i-1}^n) \tag{7}$$

$$u_i^{n+1} = \frac{1}{2}(u_i^n + \bar{u}_i) - \frac{\nu}{2}(\bar{u}_{i+1} - \bar{u}_i) \tag{8}$$

(vi) TVD-MacCormack scheme

In TVD (Total Variation Diminishing)-MacCormack scheme, artificial viscosity term is added to corrector step in MacCormack scheme.

$$TVD_i = \{G^+(r_i^+) + G^-(r_{i+1}^-)\}\Delta U_{i+1/2}$$
$$- \{G^+(r_{i-1}^+) + G^-(r_i^-)\}\Delta U_{i-1/2} \tag{9}$$

$$\Delta U_{i+1/2} = u_{i+1} - u_i, \tag{10}$$

$$r_i^+ = \frac{1}{r_i^-} = \frac{\Delta U_{i-1/2}}{\Delta U_{i+1/2}}, \tag{11}$$

$$G^\pm(r_i^\pm) = 0.5|\nu|(1 - |\nu|)\{1 - \phi(r_i^\pm)\}, \tag{12}$$

$$\nu = \frac{c\Delta t}{\Delta x}, \tag{13}$$

$$\phi(r) = \begin{cases} \min(2r, 1) & (r > 0) \\ 0 & (r \le 0) \end{cases} \tag{14}$$

(vii) CIP scheme

CIP (Constrained Interpolation Profile) scheme was presented by Yabe et al.[4] This is one of the semi-Lagrange methods. In this scheme, a special profile within each grid is interpolated with a cubic polynomial and both the value and its spatial derivative are calculated in advection.

The following three subsections investigates the difference of these numerical schemes in different case problems.

4.1. Shock wave problem

Non-viscosity Burgers equation is expressed as follows:

$$\frac{\partial u}{\partial t} + u\frac{\partial u}{\partial x} = \frac{\partial u}{\partial t} + \frac{1}{2}\frac{\partial u^2}{\partial x} = 0 \tag{15}$$

In order to investigate the influence of non-linear term, numerical schemes were applied to this equation. Under the assumption of the Shock wave problem, the initial condition was set to $u = 1(x < 1.5)$ and $u = 0(x \ge 1.5)$. Analytical solution of this problem indicates that the front face of the shock wave should propagate with speed of 0.5m/s. For comparisons of each numerical scheme, either MacCormack scheme or TVD-MacCormack scheme was applied for discretization of the conservative form of the advection term while all the other schemes listed above were respectively applied to discretization of the non-conservative form. Figure 4 shows the comparison of the theoretical solution and the result of each numerical scheme at $x = 0.05$ and $t = 0.01$.

As seen in Fig. 4, the discretization of conservative form reasonably expresses the theoretically-derived propagation of the shock wave while the discretization of non-conservative form tends to fail predictions of the shock wave propagation. This is because, in the case of non-conservative form, the advection term is always equal to zero if $u = 0$ and thus the value of u never changes in the area where $u = 0$ is set as initial conditions. In the case of conservative form, on the other hand, the advection term yields non-zero value in the vicinity of the shock wave and thus yields reasonable predictive skills of the wave propagation. Besides the problem in the prediction of the shockwave propagation, the cases in FTCS and Leapfrog scheme show unstable fluctuation in the vicinity of the shock wave.

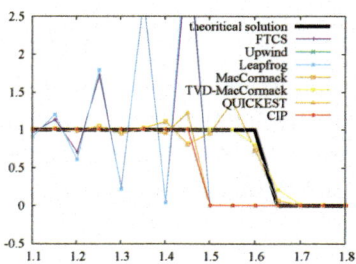

Fig. 4.   Difference of solution of Burgers equation in each numerical schemes

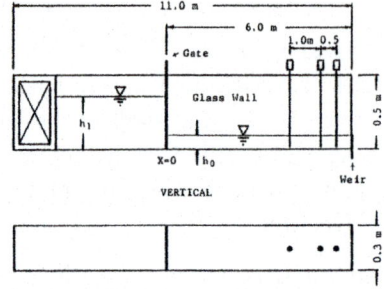

Fig. 5.   Experimental setup[6]

4.2. Bore problem

This section carries out the sensitivity analysis similar to the previous section but in the case of predictions of tsunami bore. The governing equations of non-linear shallow water wave are expressed as follows:

$$\frac{\partial D}{\partial t} + \frac{\partial M}{\partial x} = 0 \tag{16}$$

$$\frac{\partial M}{\partial t} + u\frac{\partial M}{\partial x} + gD\frac{\partial}{\partial x}(D + z_b) + \frac{gn^2}{D^{7/3}}M|M| = 0 \tag{17}$$

where D is the total water depth[m]; M is volume flux[m^2/s]; u is depth-averaged velocity[m/s]; g is gravitational acceleration[m/s^2]; z_b is the bed level[m]; and n is Manning's roughness factor[s/m$^{1/3}$].

In this analysis, we applied and compared the following numerical schemes: (i) CIP scheme with characteristic curve method;[4] (ii) First upwind scheme discretized in non-conservative form; (iii) QUICKEST scheme discretized in non-conservative form; (iv) First upwind scheme discretized in conservative form; [5] (v) TVD-MacCormack scheme; and (vi) MacCormack scheme.

The developed simulator was applied to the experiment of moving hydraulic jump and undular bore conducted by Matsutomi. [6] Figure 5 shows the experimental setup. In the experiment, h_0, shown in Fig. 5, is set to 0.05m or 0.06m, and h_1 was changed from 0.10m to 0.17m for every 0.005m. Time waveform at $x = 5$ with a wave gauge and space waveform by a camera were recorded.

Among the results of twenty cases presented by Matsutomi, [6] this study applied the model only to the case 20, case 20 with $h_0 = 0.06$[m] and $h_1 = 0.165$[m] since the observed results were similar to each other in all the other cases. Six different numerical schemes as mentioned above were used in our simulation with $x = 0.02$[m], $t = 0.005$[s], $n = 0.01$[sm$^{-1/3}$].

The result is shown in Fig. 6. It suggests that six solutions can be classified into two types: (i) the first type yielded the wave height of around 0.045 meter and the wave arrival time at the wave gauge from gate opening of around 4.1 seconds; (ii) the other type yielded the wave height of around 0.054 meter and the wave arrival time of around 5.0 seconds. The former is based on conservative scheme while the latter is based on the non-conservative form. Matsutomi[6] reported the wave height in the experiment case 20 was 4.5cm, which is equivalent to the predictions based on conservative form.

Figure 7 is the comparison of non-dimensional bore speed between the present computations and Matsutomi's experimental results. This figure also shows solutions can be classified into two types: (i) conservative schemes and (ii) non-conservative schemes. Non-conservative schemes underestimated the speed. Conservative schemes got better solution.

It was found through comparison with the bore experiments that difference of conservative and non-conservative discretization scheme has significant influence on predicted propagation speed the bore.

4.3. *Tsunami propagation along the river*

The previous analysis in Section 3 was based on the non-conservative form, observed underestimation of the tsunami run-up speed along the river may be significantly improved by the use of conservative form of the advection term. This subsection also applied different discretization schemes to the 1-D tsunami propagation along the Kido river. We applied first upwind scheme discretized in conservative and non-conservative form to the 1-D

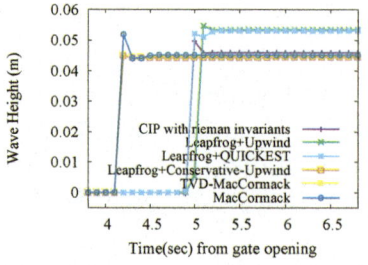

Fig. 6.   Time profile of the computed bore

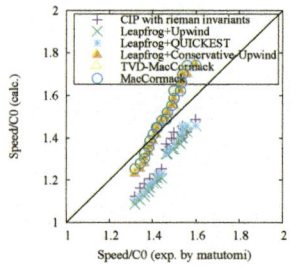

Fig. 7.   Ratio of bore speed of experiment and simulation

model. These were the same except for the discretization scheme of the advection term. The condition was the same as case 1 in the preceding chapter.

Both models reasonably represented maximum depth of inundation along the Kido river. However, there was large difference in the computation of the run-up speed.

Figure 8 is the comparison of simulations (green and blue line) and video-analysis (black line). In the figure, the time of all the computed results were synchronized at $X = -961$ m. It shows that run-up speed near the river mouth is higher in conservative scheme than in non-conservative scheme. It was found that the tsunami run-up speed was increased up to 40% by selecting appropriate discretization schemes of conservative form. Run-up speed with conservative scheme was not decreased in the process of tsunami propagation and run-up. This is similar trend with the video footage. Figure 9 is the comparison of space waveform at $t = 2804, 2876$ and 2932[s]. It shows the solution of non-conservative scheme is lower and faster than conservative one. This is similar to the case study of Matsutomi's experiment.

5. Conclusion

This study investigated the influence of various conditions on the predictive skills of the tsunami run-up speed along the river. It was found through the analysis that the wave form and the wave height at the river mouth has significant impact on the tsunami run-up speed especially in the vicinity of the river mouth while the other uncertain factors such as bottom friction and erosion of the river bed, showed relatively less impact on the predictive skills of the tsunami run-up speed. It was also found that conservative

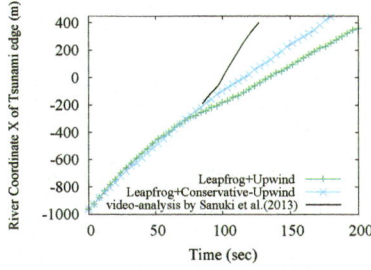 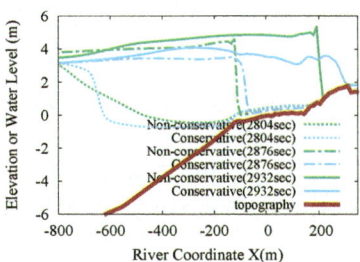

Fig. 8. 1D Run-up speed depended on numerical discretization schemes

Fig. 9. Spatial tsunami profile around the river mouth in 1D run-up model

and non-conservative forms of the advection term had significant impact on the predictions of the tsunami run-up speed and non-conservative form is recommended for the prediction of the tsunami run-up especially in the vicinity of the bore face.

Acknowledgments

This work was supported by JSPS KAKENHI Grant Number 16J06386.

References

1. Sanuki, H., Takemori, R., Tajima, Y., and Sato, S.: Study on tsunami flooding in river based on video images and numerical simulation, Journal of Japan Society of Civil Engineers, Ser. B2 (Coastal Engineering), Vol. 69 No. 2 pp. I_196-I_200 (2013).
2. Kotani, M., Imamura, F., Shuto, N.: Tsunami run-up simulation and damage estimation by using GIS. Proceedings of the Coastal Engineering JSCE 45, 356-360 (in Japanese) (1998).
3. Leonard, B.P.: A Stable and Accurate Convective Modelling Procedure Based on Quandratic Upstream Interpolation, Computer Methods in Applied Mechanics and Engineering, No. 19, pp. 59-98 (1979).
4. Yabe, T. and Aoki, T.: A universal solver for hyperbolic equations by cubic-polynomial interpolation. I. One-dimensional solver, Comput. Phys. Commun., 66, 219 (1991).
5. UNESCO: IUCG/IOC Time project numerical method of tsunami simulation with the leap-frog scheme,(1997).
6. Matsutomi, H.: An Examination of Moving Hydraulic Jump Condition, Proceedings of the japanese conference on hydraulics, Vol. 33 pp. 271-276 (1989).

Basic Study on Real-time Prediction of Meteotsunami Using Artificial Neural Network Model Calibrated by Stationary Measurement Data

S. Nakajo

Department of Engineering, Osaka City University,
Osaka, Sumiyoshi-ku, Sugimoto 3-3-138, Japan
†E-mail: nakajo@eng.osaka-cu.ac.jp

R. Yamaguchi

Kumamoto City Office,
Kumamoto, Chuo-ku, Tedori-Honcho, 1-1, Japan
E-mail: yamaguchi.ryuta@city.kumamoto.lg.jp

S. Kim

Department of Engineering, Tottori University,
Tottori, Koyama-minami, 4-101, Japan
E-mail: sooyoul.kim@sse.tottori-u.ac.jp

G. Tsujimoto

Faculty of Advanced Science and Technology, Kumamoto University
Kumamoto, Chuo-ku, Kurokami, 2-39-1, Japan
E-mail: tgozo@kumamoto-u.ac.jp

The real-time prediction of the meteotsunami is very difficult because the micro-pressure wave is too small and quick to identify from numerical weather forecasting results. In this study, we analyzed the meteotsunami from high time resolution measurement data and examined predictability of it by an artificial neural network model (ANN model). The appropriate combination of input data is very important factor in order to make a high-reproducible model. In addition, the number of units of intermediate layer, training epochs and a lead-time are also notable model parameters. We discussed about these effects based on sensitivity analysis of an ANN model.

Keywords: Meteotsunami; Neural network; Real-time prediction.

1. Background and research purpose

The long wave caused by subtle atmospheric pressure waves has been called as 'Abiki' in west-coastal area of Kyushu Island in Japan conventionally. The wave is a kind of meteolorogical-tsunami which has a maximum wave height is approximately 3.0 m (March, 1973 in Nagasaki, Hibiya and Kajiura [1]). When this long wave occurs at high tide, the inundation damage has been brought. The period of this wave is approximately 5 to 30 minutes. The rapid change of water level in harbor caused the overturning and the outflow of small crafts and damage of mooring structures such as a pontoon. This wave often observed in spring in west-coastal area of Kyushu Island.

For example, according to Hibiya and Kajiura [1] or Monserrat et al. [2], micro-barometric wave is considered as a primary factor of this phenomena. The order of the amplitude of this micro-barometric wave is too small to detect from some weather forecast information and it was estimated about 3 hPa according to Hibiya and Kajiura [1]. They explained that the major mechanism of this long wave is a Proudman resonance and a wave resonance in closed harbor. Therefore, the propagation speed of this small amplitude wave has been major subject of mechanism of meteotsunami development. For example, Matsuo and Asano [3] discussed the effect of propagation speed to amplitude of meteotsunami based on numerical simulation results. They concluded the propagation speed of micro-barometric wave has to be about 100 km/h. This propagation speed is very fast and it is not realistic to calculate it by real-time weather forecasting.

The difficulty of forecasting of meteotsunami depends on these features of micro-barometric wave. In the case of real-time forecast of storm surge, we could predict it by using real-time numerical weather forecasting data (or its downscaling information created from regional scale model like WRF) and long-wave propagation model. However, a simulation of micro-barometric wave with enough spatial and temporal resolution is difficult relatively. Therefore, in previous studies, some researchers have been trying approximate simulation of long-wave using external force of artificial micro-barometric wave (Hibiya and Kajiura [1], Sakamoto et al. [4], Matsuo and Asano [5]).

The purpose of this study is to investigate a possibility of application of a non-physical empirical model for real-time prediction of meteotsunami. We could use high temporal resolution meteorological and tide data at observatory stations although the usage of high resolution data for input of two-dimensional numerical simulation of long-wave is difficult. By using these data, we constructed an artificial neural network model (ANN model) for prediction of meteotsunami.

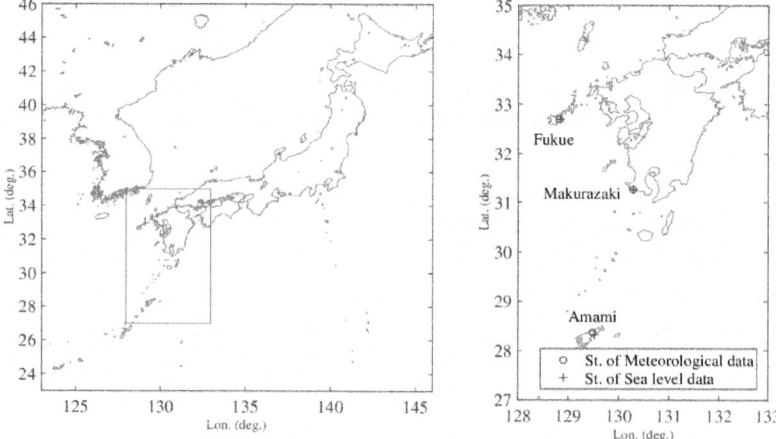

Fig. 1. Location of observatory stations

2. Basic analysis of meteotsunami, Abiki

2.1. *Observation data*

The micro-barometric wave which is a primary factor of meteotsunami is very small and rapid phenomena. Therefore we have to pay attention the temporal resolution of observation data when we analyze the indication of meteotsunami. Around Kyushu Island, Japan Meteorological Agency has provided about a dozen constant monitoring stations of meteorological data and tide data. However, they are distributed at only major populated area. According to Tanaka et al. [6], significant micro-barometric waves generated around the center of the East China Sea. Therefore we could not use the data around the meteotsunami generation area. Figure 1 shows location of observatory stations used in this study. In these stations, Makurazaki is a typical harbor where we observed significant meteotsunami almost every year. Although the past observation data was open access in web site of JMA, this public data was not a raw data with high temporal resolution. We requested this raw data during particular period to JMA individually. We analyzed the data of three period; (A) 2009 Feb. 21st-28th, (B) 2009 Jul. 14th-16th, (C) 2010 Jan. 31st-Feb. 2nd. In these period, significant meteotsunami has been observed in tide data. The temporal resolution of meteorological data is 10 seconds, and that of tide data is 15 seconds. The meteorological data includes atmospheric pressure P, wind speed W, wind direction θ, temperature T and relative humidity near the ground surface R.

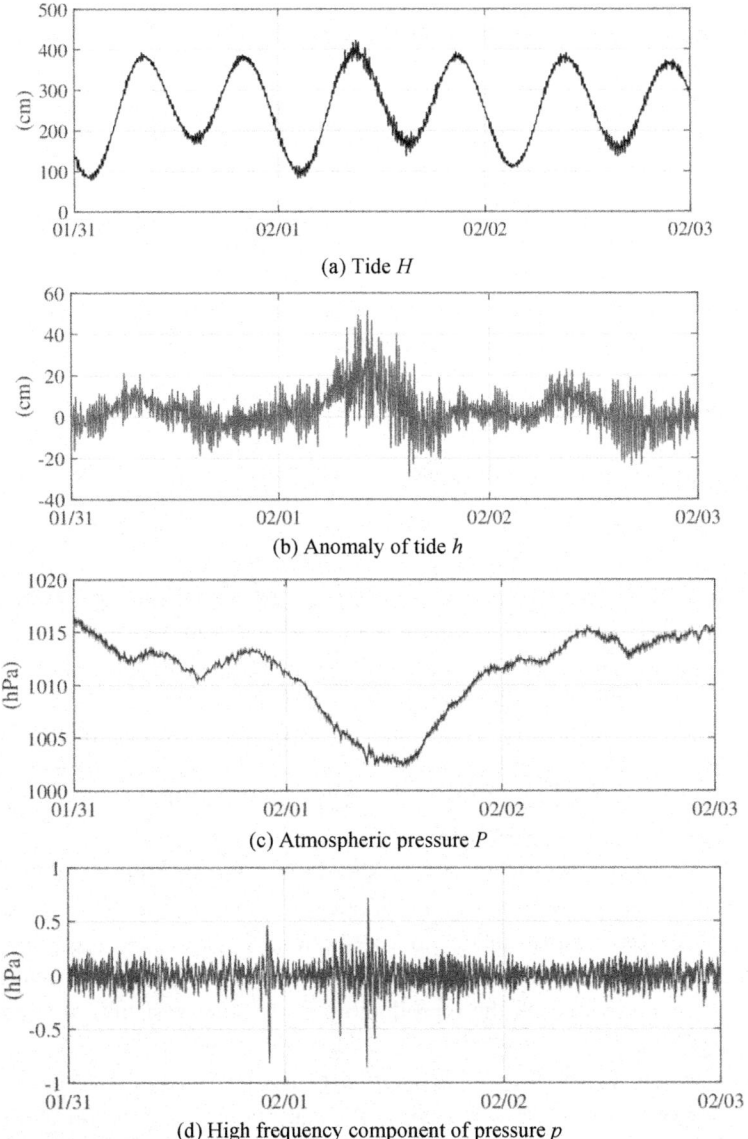

Fig. 2. Time series of observations (tide and atmospheric pressure) and analyzed secondary data (anomaly of tide and high frequency component of pressure) at Makurazaki stations during 2010 Jan. 31st-Feb. 2nd

Figure 2 shows a sample of time series of observation data and secondary data. Anomaly of tide h was calculated from difference between observation and astronomical tide calculated by empirical tidal components. Then a high frequency component of atmospheric pressure p was calculated by difference between observation and moving average of pressure. The meteotsunami was observed around high tide phase and its magnitude was about 0.4 m. The macroscopic drop of atmospheric pressure shows a passing of low pressure. Then time series of high frequency component of pressure shows some spike signals occurred close on the timing of the trough of pressure and the maximum meteotsunami. This spike signal would correspond to a passing of micro-barometric wave. However, we also have to pay attention to the existence of the similar spike pressure observed before low pressure approach. It is assumed that this spike signal could not bring development of significant meteotsunami.

3. Methodology

3.1. *Annual neural network model*

ANN model has been used in some coastal engineering field in the past (For example, Mase et al. [7], Huang et al. [8], Kim et al. [9]). An ANN model can correlate multiple input variables with the output signal through many nodes. Mase et al. [7] adopted ANN to simulation of the stability of armor unit and rubble mound breakwater. Huang et al. [8] studied ANN model for prediction of water level at objective point from other water level stations to get long-term water level data. Kim et al. [9] developed ANN model for prediction of time series of water level caused by storm surge. A storm surge is similar physical phenomena to meteotsunami. However, generally a storm surge is caused by remarkable meteorological factor such as tropical cyclone and subtropical cyclone. For example, Kim et al. selected typhoon location and central atmospheric pressure as ANN model input. In the case of a meteotsunami, we could not define distinct factor for prediction. Therefore we selected many meteorological observation data as candidate of factor first, and we investigated important factors based on the reproducibility of actual variation of water level inductively.

In this research, we studied the possibility of the feedforward type hierarchical neural network model with three-layer of input, hidden neuron and output. The learning algorithm was selected the back-propagation technique. Then the Levenberg-Marquardt method which is an optimization algorithm for minimization of an error was employed. This basic structure is same to that of previous study for storm surge prediction (Kim et al. [9]). The simulation result depends on the combination of input variables, number of hidden layer, number of iteration for learning and lead-time. We conducted the sensitivity analysis for these factors.

3.2. *Input variables and prediction target*

We selected all observation data as input of ANN model. However, wind speed, temperature and relative humidity were not good factors for prediction. We have no space to show all results, therefore in this paper, limited results are shown.

Table 1. Experimental conditions

Case No.	Validation data	Learning data	Input variables	Number of hidden layer	Training epochs	Lead-time (hour)
A-0			$h(0), P, \theta, p(0)$			
A-1			$h(0), P, \theta, p(0,1)$			
A-2			$h(0,1), P, \theta, p(0)$			
A-3			$h(0,1), P, \theta, p(0,1)$			
A-4			$h(0), P, \theta, p(0,2)$	30	100	1.0
A-5			$h(0,2), P, \theta, p(0)$			
A-6			$h(0,2), P, \theta, p(0,2)$			
A-7			$h(0,1,2), P, \theta, p(0,1,2)$			
A-8			$h(0,2), P, \theta, p(0,1)$			
B-1	Anomaly of tide at Makurazaki 2010 Jan. 31st ~ Feb. 2nd	Anomaly of tide at Makurazaki 2009 Feb. 21st ~ Feb. 28th		5		
B-2				10		
B-3				30	100	1.0
B-4				50		
B-5				100		
C-1					25	
C-2			$h(0,2), P, \theta, p(0)$	30	50	1.0
C-3					100	
C-4					200	
D-1						0.5
D-2						1.0
D-3				30	100	2.0
D-4						3.0
D-5						5.0

h: anomary of tide, P: atmospheric pressure, θ: wind direction, p: high frequency component of pressure
Subscripts of input variables mean the station number; 0: Makurazaki, 1: Amami, 2: Fukue

Finally, we selected four type data for input variables, anomaly of tide h, atmospheric pressure P, wind direction θ and high frequency component of pressure p. These input data were normalized by maximum value of them. Prediction target of ANN model is a meteotsunami, therefore here we planned to predict anomaly of tide h. However, h is very fluctuated and noisy unlike the storm surge or tide. First, we tried to predict it directly, but it is very difficult. Therefore we decided to change a prediction target to moving average of h. Table 1 shows experimental conditions of ANN model.

4. Results

4.1. *Effect of input variables*

We validated prediction results from 20 ensemble trial results for every experimental conditions. Every ensemble member has used different learning data sampling randomly. The RMSE and the correlation coefficient between simulation and actual data were used as macroscopic validation indices. Then the error of maximum peak of h and maximum amplitude of h were used as microscopic validation indices. Figure 3 shows an example of comparison results of different experiments (A-series). In these figures, circle and line length show an average and standard deviation between 20 ensembles, respectively. First, the observation data near the target point (A-0) is an essential for good prediction. Then, the use of observation data on different stations has a potential of improvement of prediction result, although it depends on the combination of input data. For example, in this case, we could get a better result (lower error and higher correlation) when we used the data at Fukue (A-5 and A-6). Comprehensively, the result of A-5 showed the best prediction result on average. The excessive input variables were not helpful for good prediction.

4.2. *Effects of number of hidden layer and training epochs*

The optimum number of hidden layer would depends on the combination of input variables. We defined this number as 30 based on sensitive analysis results (see Table 1, B1-B5). If this number was larger than this value, the RMSE tended to become large. The effect of a training epochs is shown in Figure 4(a). The reproducibility of ANN model did not much depend on the training epochs in the range of this investigation, although the standard deviation of the error slightly tended to be small when the training epochs are relatively large.

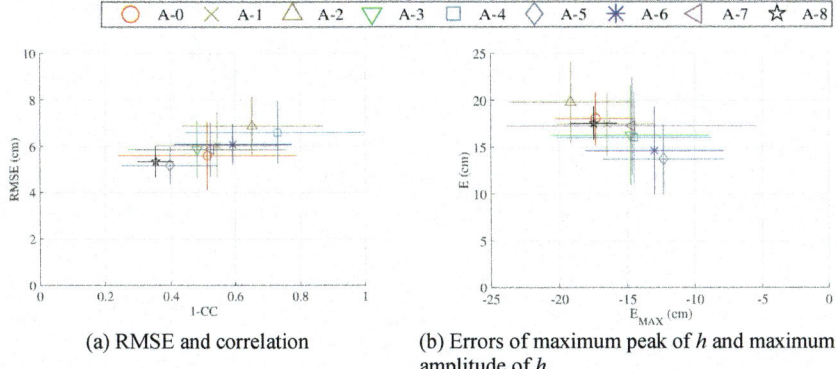

(a) RMSE and correlation

(b) Errors of maximum peak of *h* and maximum amplitude of *h*

Fig. 3. Samples of validation results between observation and prediction of anomaly of tide *h* at Makurazaki, 2010 Jan. 31st-Feb. 2nd in case of A-series.

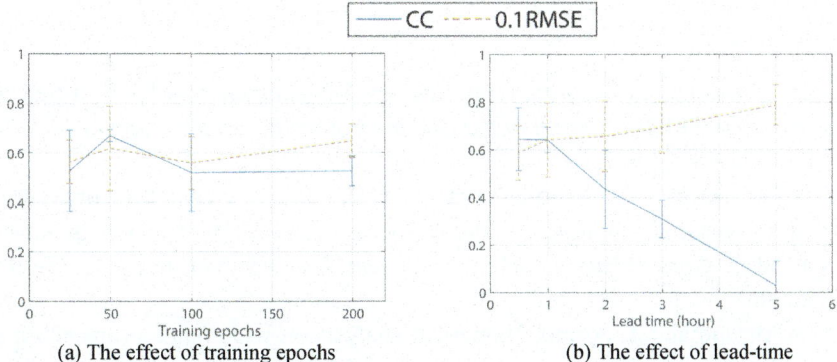

(a) The effect of training epochs

(b) The effect of lead-time

Fig. 4. The effect of training epochs and lead-time to the accuracy of ANN model (the unit of RMSE is [10 cm])

4.3. *Effect of a lead-time of prediction*

Figure 4(b) shows the effect of a lead-time to the reproducibility of ANN model. The lead-time becomes longer, the RMSE value becomes larger and the correlation becomes smaller, respectively. However, the RMSE and correlation were not so much changed when the lead-time is smaller than 1 or 2 hours. This result shows present model has a potential to predict the meteotsunami before about 1 hour.

5. Conclusion

Relation between anomaly of tide and micro-pressure wave has been confirmed from high time resolution observation data. However, it is difficult to predict the meteotsunami from observation of atmospheric pressure at fixed point because not all micro-pressure caused the meteotsunami. Some standards of appropriate parameters of ANN model have been estimated from sensitivity analysis, although model results are depending on the number of intermediate units and training epochs. In a present basic model, the appropriate longest lead-time was approximately 1 or 2 hours. The combination of learning data and validation data was very important for model accuracy. It would be efficient to use some ensemble models reflecting the effect of different factors for actual prediction. Prediction result has to be considered as one sample of statistical trial.

Acknowledgments

The observation data analyzed in this research was supported JMA. We are deeply grateful to their kind intentions.

References

1. T. Hibiya and K. Kajiura, Origin of Abiki Phenomenon (a Kind of Seiche) in Nagasaki Bay, *Journal of the Oceanographical Society of Japan*, Vol. **38**, pp. 172-182 (1982).

2. S. Monserrat, I. Vilibic and A. B. Rabinovich, Meteotsunamis: atmospherically induced destructive ocean waves in the tsunami frequency band, *Natural Hazards and Earth System Sciences*, Vol. **6**, pp. 1035-1051 (2006).

3. S. Matsuo and T. Asano, Numerical analysis on wide range propagation of meteo-tsunami causing secondary undulation on the coasts of Kyushu along the East China Sea, *Journal of Japan Society of Civil Engineers, Ser. B2 (Coastal Engineering)*, Vol. **71**, No. 2, pp. I_133-I_138 (2015) (in Japanese).

4. K. Sakamoto, G. Yamanaka, H. Tsujino, H. Nakano and M. Hirabara, Large Predictability of the Abiki phenomenon using a next-generation Japanese-coastal-seas model, MRI.COM-JPN, *UMI TO SORA*, Vol. **88**, No. 3, pp. 85-98 (2013) (in Japanese).

5. K. Tanaka and T. Asano, Meteorological conditions inducing meteotsunami at Kamikoshiki Island in February 2009, *Journal of Japan Society of Civil Engineers, Ser. B2 (Coastal Engineering)*, Vol. **66**, No. 1, pp. 181-185 (2010) (in Japanese).

6. H. Mase, M. Sakamoto and T. Sakai, Neural network for stability analysis of rubble-mound breakwaters, *Journal of Waterway, Port, Coastal, and Ocean Engineering*, Vol. **121**, 6, pp. 294-299 (1995).

7. W. Huang, C. Murray, N. Kraus and J. Rosati, Development of a regional neural network for coastal water level predictions, *Ocean Engineering*, Vol. **30**, 17, pp. 2275-2295 (2003).

8. S. Kim, Y. Matsumi, S. Pan and H. Mase, A real-time forecast model using artificial neural network for after-runner storm surges on the Tottori coast, Japan, *Ocean Engineering*, Vol. **122**, pp. 44-53 (2016).

Evaluation of Tsunami Force with Overflow on Breakwater

N. Tsuruta[†], K. Suzuki and T. Kita

Coastal and Ocean Engineering Research Department, Port & Airport Research Institute,
Yokosuka, Kanagawa 239-0826, Japan
[†]*E-mail: tsuruta-n@pari.go.jp*
www.pari.go.jp/en

M. Miyata and M. Takenobu

Port and Harbor Department, National Institute for Land and Infrastructure Management, 3-1-1 Nagase,
Yokosuka, Kanagawa 239-0826, Japan
www.nilim.go.jp/english/eindex.htm

A tsunami force has unknown features due to its complexity, and makes it difficult to be predicted accurately. As a typical example, in the 2011 off the Pacific coast of Tohoku earthquake with huge tsunami, many breakwaters slid or fell down by the unexpectedly terrible damages caused by the tsunami. For effective design of coastal structures, achieving a correct prediction of the tsunami pressure should be an urgent subject. In the existing studies, the tsunami force F with an overflow is estimated by using a compensation coefficient α based on the hydrostatic pressure as: $F = \alpha \rho g h$. The compensation coefficient α is usually set as $\alpha = 1.1$ for the front of the caisson and 0.9 for the rear of that constantly. However, in recent studies, it was found that Parameter α can be easily and randomly changed depending on the boundary condition, and unfortunately, estimation method of its effective has not been developed yet. To resolve this problem, in this study, hydraulic experiments targeting on tsunamis with their overflows on a breakwater are implemented to examine the variability of the compensation coefficient α. And from the experimental results, a simple and effective estimation method for the compensation coefficient α is newly proposed.

Keywords: Tsunami; Wave force; Overflow; Breakwater; Hydraulic experiment.

1. Introduction

In the 2011 off the Pacific coast of Tohoku earthquake with huge tsunami, some breakwaters had unexpectedly terrible damages resulting in falling down or sliding caused by the tsunami forces directly. Its prime example can be shown, e.g. in Soma and Kamaishi ports as Figure 1. Reflecting on this fact, requirements of breakwaters in design have been changed so as to keep their performance even against unprecedentedly huge tsunami, i.e. the largest tsunami in history (around this 1,000 years) corresponding to the so-called "Level 2 Tsunami" in Japan. Nevertheless, achieving complete protection from such an immense tsunami is actually unrealistic, thus, the design concept should aim to keep their resilience even during their collapse as long and as effectively as possible to gain time for evacuation of people. Following this concept, the Japan design guideline of breakwater against tsunami has been revised in 2013.[1] However, for the new considered situation targeting the huge tsunami, it requires us to comprehend not only a steady condition of structures, but also more flexible situations with its dynamically changeable performance in collapse through 0 to 100% process. Unfortunately, details of the damage mechanisms and even the tsunami force are still not studied sufficiently due to their complexity. As an urgent issue with the highest priority, more appropriate design criterions should be built into the guideline with accurate prediction of tsunami forces.

Fig. 1. Snapshot of broken breakwaters by tsunami in Kamaishi port.

Targeting a significantly large tsunami, it is presupposed that a concomitant overflow attacks the target breakwater as presented in Figure 2.

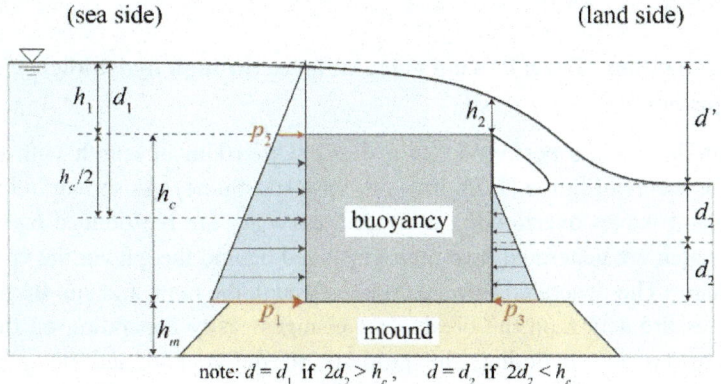

note: $d = d_1$ if $2d_2 > h_c$, $d = d_2$ if $2d_2 < h_c$

Fig. 2. Graphical presentation of the existing formula of overflowing tsunami force.

In the guideline, the existing formula is based on the hydrostatic pressure of foreside and backside of the target caisson as:

$$p_1 = \alpha_f \rho_0 g (h_c + h_1)$$

$$p_2 = \frac{h_1}{h_c + h_1} p_1 \tag{1}$$

$$p_3 = 2\alpha_r \rho_0 g d_2.$$

Where p_1, p_2, p_3: wave forces at the height of the free surface, the bottom of the caisson at the front and rear, respectively, α_f, α_r: compensation coefficients for the front and rear, respectively, ρ: fluid density, g: vector of the gravitational acceleration, h_c: height of the caisson from the mound, h_1: the foreside height of the target tsunami from the top of the caisson and d_2: half of the backside height of the target tsunami from the mound behind the caisson. In accordance with an experimental investigation by Arikawa et al.,[2] the arbitrary parameters corresponding to the hydrostatic-pressure calibration coefficients α_r and α_f (hereinafter abbreviated as "Parameter α") are set with 0.9 and 1.05, respectively, as recommendation values. However, the applicability of the recommendation values has not been proved in a wide range because of their limited experimental setups. Subsequently, Arikawa et al.[3] and Miyata et al.[4] later, implemented experiments with various flows, caissons and parapets, and arranged Parameter α_r with an original method based on d^* ($=d'/d$). In [3,4], it was unfortunately found that the arranged Parameter α_r varies, and thus, it has difficulty in being predicted correctly.

In order to resolve this problem, this study presents a simple and effective estimation method of Parameter α_r, which is a key to decide the tsunami forces. For the goal, hydraulic experiments targeting a breakwater under overflows are implemented.

2. Investigation of overflowing tsunami force through hydraulic experiment

The experiments are performed by a flume with 105m in length with 0.8m in width for the main flume (2.2m in width for a sub-flume). As shown in Figure 3, the tsunami waves overflowing a target breakwater are reproduced by uniform flows, which are generated by a pump installed behind the gate at the end of the main flume. The generated flows circulate through the main and sub flumes. The wave pressure acting on the breakwater changes easily depending on the types of the breakwater including the property of the mound and the additional construction, e.g. parapet, additional rubble mound and etc. In this study, the changeable property of the tsunami wave force is examined by targeting the plural breakwaters. The first target is a breakwater above an impermeable mound to examine the influence of the mound. The second is a breakwater above a permeable mound corresponding to the standard breakwater. The third and fourth models are a breakwater above the permeable mound with a parapet and an additional rubble mound, respectively. Regarding settings of the water heights at the front and rear of the breakwater, their relations are shown in Figure 4. Here, the both water heights are normalized by the height of the caisson.

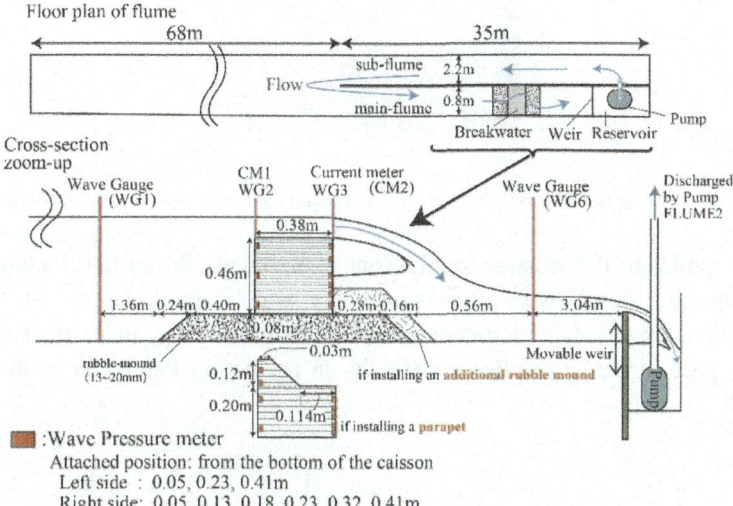

Fig. 3. Experimental condition.

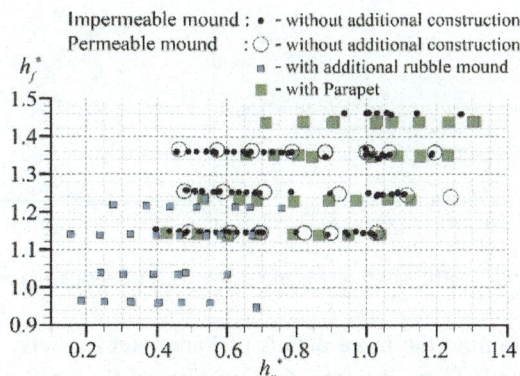

Fig. 4. Settings of the water height ($h_r{}^*$ vs. $h_f{}^*$).

In Figure 5, a snapshot of our present experiment is shown with the vertical distribution of the tsunami pressure acting on the caisson. It is found that the tsunami pressure on the rear side of the caisson has a significant gap from the hydrostatic pressure.

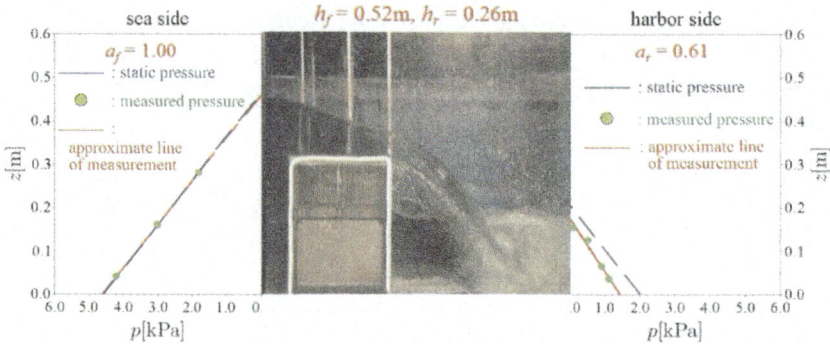

Fig. 5. Snapshot of the experiment with its vertical distribution of tsunami pressure.

In similar to the existing studies, our experiments shows that Parameter α_r goes up and down with dispersion in a wide range (Figure 6) by the conventional method.[2,3,4] Parameter α_r varies significantly, in particular, as d^* (=d'/d) goes larger. It implies a difficulty in predicting Parameter α_r from this method.

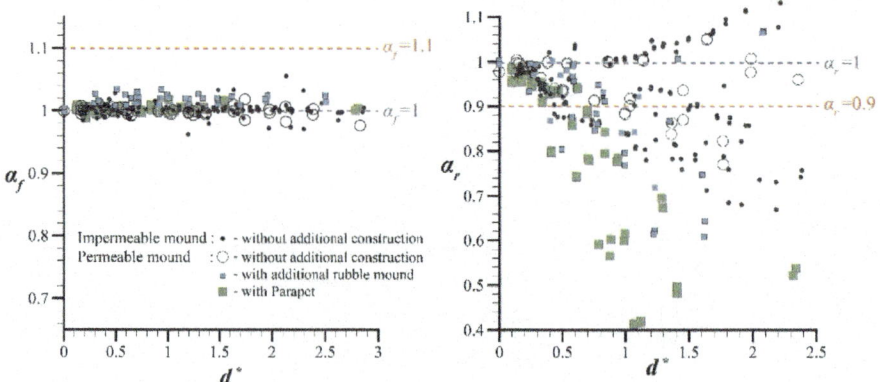

Fig. 6. Compensation coefficients α_f and α_r arranged by existing formula in our experiments.

In order to examine the more details of Parameter α_r, here, it is rearranged with another method. Then, the total force acting on the back of the caisson is normalized by the hydrostatic pressure of the forward tsunami force, which is defined with the water height h_f at the front of the breakwater through WG1 as:

$$\frac{\alpha_r \rho g h_r}{\rho g h_f} = \alpha_r \frac{h_r}{h_f} = \alpha_r h^*. \tag{2}$$

Where h^*: normalized water height $(=h_r/h_f)$. Figure 7 shows a relation between the normalized tsunami force corresponding to Eq. (2) and the normalized water height h^*. From the figure, it is recognized that the normalized tsunami force significantly disperses as the normalized water height h^* decreases.

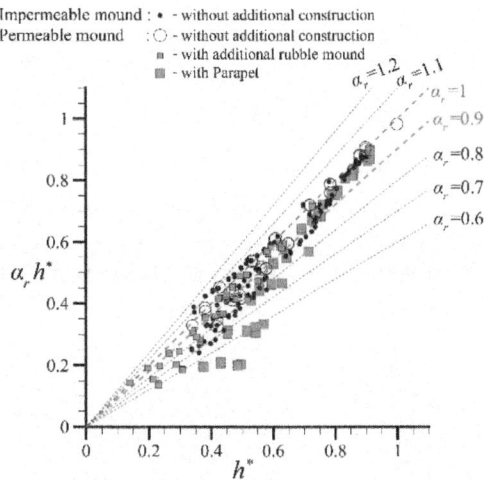

Fig. 7. A relation between the normalized tsunami force acting on the back of the caisson corresponding to Eq. (2) and the normalized water height h^*.

The target water level behind a structure is generally represented by a water height at a position adequately apart from the structure beyond the falling water-stream in an overflowing situation. Figure 5 shows that the local displacement of the water level, which is caused by the falling water-stream, in the vicinity of the structure. It would give a noteworthy hint to detect the appropriate Parameter α_r. From this viewpoint, the following normalized waterfall length L^*, which corresponds to a kind of a relation between the mound length B_m and the waterfall length L_w on the mound from the rear edge of the caisson (Figure 8), is extracted in each case as:

$$L^* = \frac{L_w - B_m}{B_m}. \tag{3}$$

308

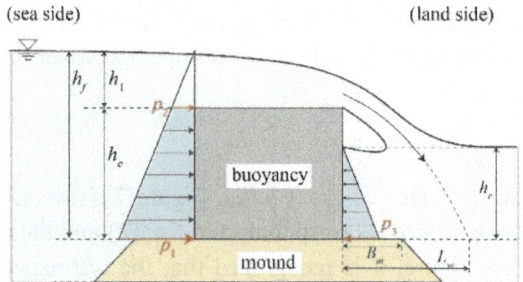

(sea side)   (land side)

Fig. 8. Graphical presentation of presented formula of overflowing tsunami force.

And, this normalized waterfall length L^* is introduced into the normalized water height h^* on a trial basis as:

$$h^{**} = h^* + \beta L^*. \tag{4}$$

Where β: empirical parameter (β=0.1). Here, the waterfall length L_w is estimated from the firing speed of the water-stream at the edge of the caisson by regarding the state of the waterfall as free falling.

By using Eq. (4), Parameter α_r is rearranged in Figure 9 by introducing the physical feature of the overflow, which would be a key to change Parameter α_r.

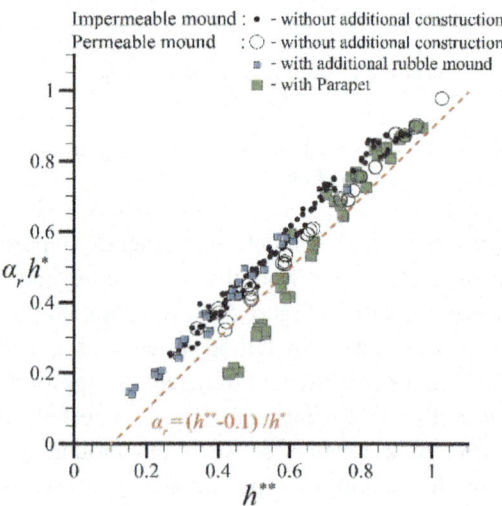

Fig. 9. Compensation coefficients α_f and α_r arranged by the proposed method.

From the figure, the dispersion of the normalized tsunami forces at the relative small x ($=h^{**}$) is clearly suppressed comparing with Figure 7. From Table 1 showing the dispersion of each case in Figure 7 and Figure 9, it can be recognized that the dispersion decreases by the rearrangement as Figure 9 except the cases with the parapet. In the cases with the parapet, the rearrangement deteriorates the dispersion of the plots where x $(=h^{**})$ is relative high despite of the improvement at the small x. Therefore, the totally averaged dispersion in the Parapet-cases does not change much.

Table 1. Dispersion of the plots of the compensation coefficient α_r in the present experiments by existing method (Figure 7) and proposed method (Figure 9).

Model of Breakwater	Fig.7	Fig.9
Impermeable mound without additional construction	1.99×10^{-3}	5.02×10^{-4}
Permeable mound without additional construction	9.09×10^{-4}	3.67×10^{-4}
Permeable mound with additional rubble mound	1.27×10^{-3}	6.63×10^{-4}
Permeable mound with parapet	2.30×10^{-3}	2.18×10^{-3}

From Figure 9, an approximation of Parameter α_r for design is gotten from an asymptote of the plots of the cases with impermeable mound and permeable mound and with/without additional rubble mound (except of the cases with parapet) as:

$$\alpha_r = \frac{\left(h^{**} - 0.1\right)}{h^*} = \frac{\left(h^* + 0.1L^* - 0.1\right)}{h^*} = 1 + 0.1\frac{L^* - 1}{h^*}. \tag{5}$$

3. Concluding remarks

The overflowing tsunami forces have unknown features due to their complexity. For more effective design of resilient breakwaters, the accuracy of the estimation formula of the forces should be enhanced as one of the most urgent subject. The problem of the existing formula is its inaccuracy of the compensation coefficient α_r to estimate the overflowing tsunami force acting on the rear side of the breakwaters. Even the tendency of the variability is not sufficiently comprehend. In this study, hydraulic experiments are performed with targeting several types of breakwaters including additional constructions, e.g. parapet, additional rubble mound with considering the permeability of the mound under the caisson in order to examine the compensation coefficient α_r.

Firstly, we confirmed the random variability of the compensation coefficient α_r, which was reported by Arikawa et al.[2,3] and Miyata et al.[4] as previous studies. Secondly, focusing on the hydraulic condition behind the breakwaters, the variability of the compensation coefficient α_r was rearranged with a parameter related to the waterfall length of the overflows. From the result of the rearranged plots of the compensation coefficient α_r, we newly constructed an estimation formula for the compensation coefficient α_r for comprehensive types of breakwaters including an additional rubble mound. On the other hand, if a parapet is installed on the target caisson, its plots still scatter randomly despite of the improvement of the dispersion by the proposed method.

For our future works, the detail of such an influence by the additional parapet should be examined. Moreover, the scaling effect also should be examined.

Acknowledgments

We would like to express our sincere gratitude to Kenichiro Shimosako, Senior Director for Research of Port and Airport Research Institute, for his continuous support and insightful guidance. We also want to thank Sogo Zikuhara, Subsection Chief of National Institute for Land and Infrastructure Management, for his kind support for our hydraulic experiments.

References

1. The Japan design guideline of breakwater against tsunami (2013), http://www.mlit.go.jp/kowan/kowan_tk5_000018.html.
2. T. Arikawa, S. Satoh, K. Shimosako, K. Tomita, D. Tatsumi, G-S. Yeom and K. Takahashi, Investigation of the Failure Mechanism of Kamaishi Breakwaters due to Tsunami – Initial Report Focusing on Hydraulic Characteristics –, *Technical note of the port and airport research institute*, **1251** (2012).
3. T. Arikawa, S. Satoh, K. Shimosako, K. Tomita, D. Tatsumi, G-S. Yeom and T. Niwa, Failure Mechanism and Resiliency of Breakwaters under Tsunami, *Technical note of the port and airport research institute*, **1269** (2013).
4. M. Miyata, Y. Kotake, M. Takenobu, T. Nakamura, N. Mizutani and S. Asai: Experimental study on hydraulic characteristics of tsunami overtopping flow over a caisson-type breakwater, *Journal of Japan Society of Civil Engineers, Ser. B3 (Ocean Engineering)*, **70**(2), I_504–I_509 (2014).

Characteristic of Tsunami Flow Over The Breakwater with Three-Dimensional Configuration

A. A. Sulianto[†] and N. Lusiana
Department of Environmental Engineering, Universitas Brawijaya,
Veteran Street Malang 65145, East Java, Indonesia
[†]E-mail: adi_sulianto@ub.ac.id
www.ub.ac.id

K. Murakami
Dept. of Civil and Environmental Engineering, University of Miyazaki,
1-1 Gakuen Kibanadai Nishi, Miyazaki, 889-2192, Japan
E-mail: keisuke@cc.miyazaki-u.ac.jp

After field surveys on the Tohoku earthquake tsunami in 2011, coastal engineers recognized the importance of persistent coastal structure to reduce the tsunami damages. It is important to understand the characteristics of tsunami flow formed around the breakwater and rubble mound with three-dimensional configuration. The applicability of CADMAS-SURF has been investigated mainly in two-dimensional wave deformation problems and few investigation has been done in the case of three-dimensional tsunami flow problems. CADMAS-SURF/3D was employed to simulate tsunami flow over the breakwater. In this study, a numerical half-basin domain was considered in 3D simulations. Hososhima Yojima breakwater was selected as the prototype structure. This is one of the breakwaters that are needed to be reinforced to realize a persistent breakwater against a supposed L2 tsunami attack on the eastern side of Kyushu Island. This study firstly discusses the applicability of CADMAS-SURF to the three-dimensional tsunami flow problem. Some hydraulic experiments were conducted to verify the validity of numerical results. This study also discussed some characteristics of tsunami flow over the breakwater such as tsunami flow passes through the gap between caissons, the effect of the partial subsidence of caissons on tsunami flow above the rubble mound, and the characteristics of tsunami flow in around a breakwater..

Keywords: Tsunami flow; breakwater; CADMAS-SURF/3D.

1. Introduction

Field surveys after the Tohoku earthquake tsunami in 2011 reported many types of tsunami flows that caused damage of breakwaters. In order to design the persistent breakwater against the supposed L2 tsunami, it is important to understand the characteristics of tsunami flow formed around the breakwater and rubble mound with three-dimensional configuration. CADMAS-SURF/3D (Coastal Development Institute of Technology, 2010) is employed to simulate tsunami flow over the breakwater.[1] The applicability of CADMAS-SURF has been investigated mainly in two-dimensional wave deformation problems,[2,3] and few investigation has been done in the case of three-dimensional tsunami flow problems. Based on above background, this study is intended to investigate the characteristic of tsunami flow over the breakwater in case of Hososhima Yojima breakwater. This is one of the breakwaters that are needed to be reinforced to realize a persistent breakwater against a supposed L2 tsunami attack on the eastern side of Kyushu Island. The numerical simulation is based on CADMAS-SURF/3D model.

2. Model Setup

2.1. *Hydraulic Experiment*

Some hydraulic experiments were conducted to verify the validity of numerical results obtained from CADMAS-SURF/3D. Figure 1 shows the configuration of the hydraulic experiment. The model scale of experiment was 1:20. The width of the channel was 0.4m. The width of the caisson model was 0.25m, and gap between each caissons were set as 0.05m. Three test lines, *U1*, *U2* and *U3* in Figure 1, were set along open channel to measure the flow velocity. Three kinds of initial impoundment (W_0), 0.1875m, 0.125m and 0.085m, were set on the upstream side of the open channel to generate the bore type tsunami.

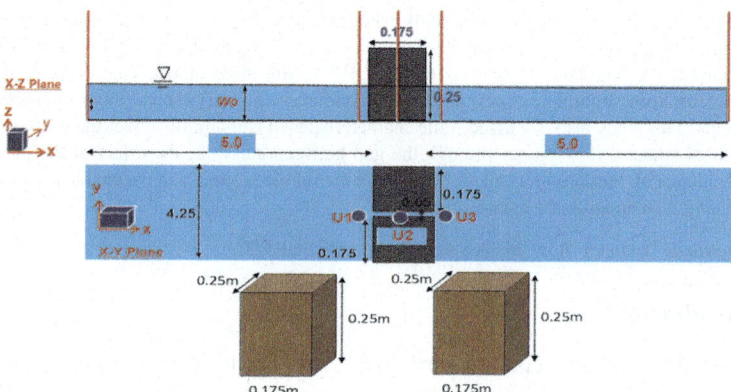

Fig. 1. Configuration of hydraulic experiment (not scale; unit in m)

2.2. *Numerical Model*

In generating a bore type tsunami using CADMAS-SURF, time histories of water surface elevation and fluid velocity were prescribed as inflow properties on the face of the input boundary. These values were uniformly assigned on the ghost cells, which were set outside the upstream cells. The fluid velocity was estimated by applying an analytical formula derived by Fukui,[4] and previously this bore type tsunami using CADMAS SURF was used in numerical simulation of tsunami bore pressure on cylindrical structure by Wijatmiko and Murakami.[5]

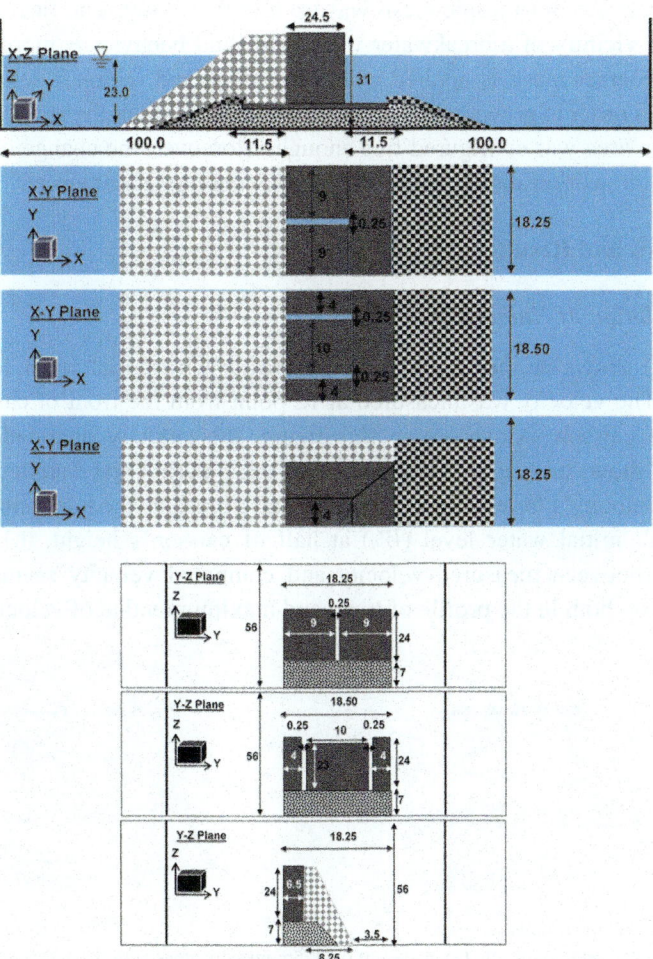

Fig. 2. Configuration of numerical simulation (not scale; unit in m)

314

In this study, a numerical half-basin domain was considered in 3D simulations in order to save computational time. This half-basin domain employs half the channel width by considering a symmetry axis of the structure model. Therefore the total number of elements is about a half those of the full-basin model. In the case of a tsunami wave impact on a cylindrical structure, the results obtained from the half-basin model are not significantly different in comparison with the results obtained from the full-basin model.[6] The configuration of numerical wave tank can be seen in Figure 2. In order to conduct an effective computation, the mesh optimization is important for a big simulation domain, which requires a fine grid configuration. An anisotropic mesh was used in this study. The finer mesh was used at the vicinity of a breakwater where the fluid behavior is more complex, while the coarser mesh is applied where the numerical region is less intensely observed. In order to prevent the excessive wave dissipation, the mesh at the bore propagation area was configured fine enough. Moreover, the change of grid size in the mesh transition should not be abrupt to avoid numerical errors.[7]

3. Analysis and Result

3.1. Validation of Numerical Model

Figure 3(a) shows the horizontal velocity profile with initial water level (W_0), 0.1875m. The velocity was measured at 12 point from the front of caisson until behind the caisson. As shown in this figure, the velocity obtained from 3D simulation shows quite good agreement with the experimental data in case of the profile of velocity's trend value. Figure 3(b) also presents the horizontal velocity profile with initial water level (W_0) at half of caisson's height, 0.125m. The agreement between measured velocity and computed velocity seems in good agreement on both in the profile of trend and maximum value of velocity.

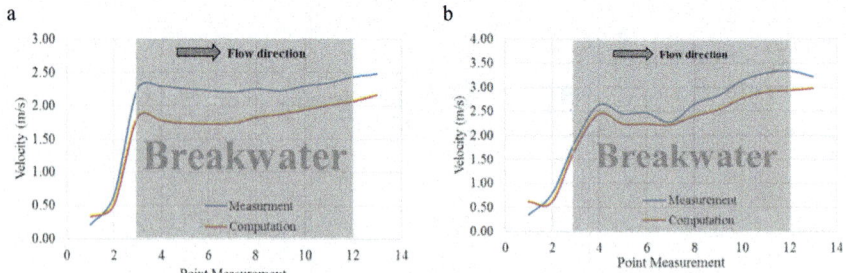

Fig. 3. Horizontal velocity profile (a) in case of 0.1875m initial water depth (b) in case of 0.125m initial water depth

Observation of water surface elevations in front and behind of the breakwater is to identify steady overflow in time history during tsunami flow over the breakwater. Figure 4 shows water surface elevation profiles in front of and behind the breakwater. The figure shows that tsunami hit the breakwater at 7second then flowed over the breakwater, and the elevation increased irregularly due to wave reflection. The steady overflow occurred during time history from 10-second until 16-second with the average elevation of 10m from the initial water level or 2m above breakwater. Flow characteristic and hydrodynamics process will be discussed based on the steady overflow time history.

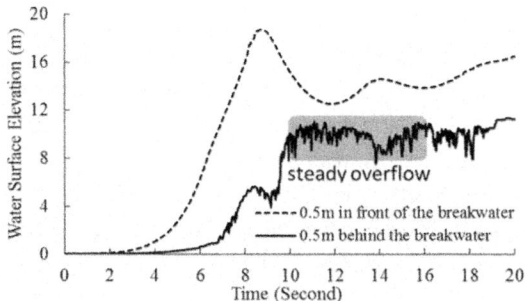

Fig. 4. Water surface elevation profile.

3.2. Effect of gap between caisson breakwaters on wave velocity

The investigation about horizontal velocity profile due to the gap of caisson can be seen in Figure 5. The figure shows the average velocity profile during steady overflow in the gap of caisson breakwaters. The velocity in front of the breakwater tends to be a negative value. It means that the velocity tends to be the opposite of the flow direction (to the left side). It occurs due to the existence of tetrapod in front of the caisson. This velocity increases steeply when reached the gap of the caisson and becomes constant while reaching the middle of the gap, and slightly increases when the flow reaches outside of the breakwater. The increasing velocity occurs due to the principle of continuity and conservation mass. On the other hand, the velocity above the breakwater gradually increases and it reaches to a maximum point, three times larger than the initial velocity when the flow reached outside of the breakwater. The velocity above the breakwater accelerates when tsunami flow over the breakwater.

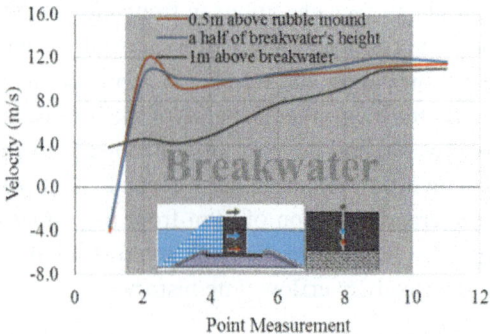

Fig. 5. Velocity profile in the gap of the caisson.

Tsunami wave hit and flow over the breakwater forms complex phenomena including turbulence flow, supercritical flow, and also vortex in the behind of breakwater. The complex flow was also affected by seepage flow from rubble mound. The vertical velocity from above rubble mound infiltrates the inner part of the rubble mound. These phenomena can be seen in the Figure 6(a). Furthermore, because the existence of the gap between caissons, the contracted flow forming affects the velocity at the area behind the breakwater. The flow pattern from a plane view is presented in Figure 6(b).

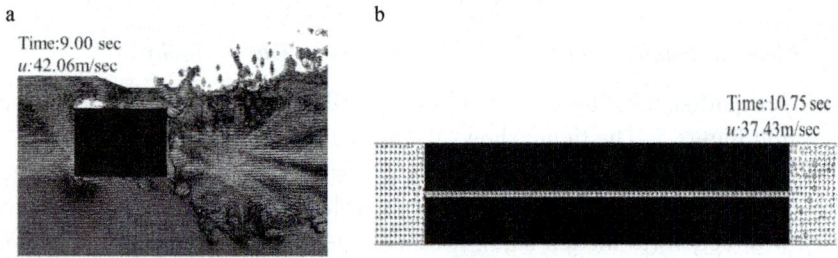

Fig. 6. Flow pattern; (a) from the side view (b) from top view

3.3. *Effect of the subsidence of the caisson-breakwater*

In some natural disaster such as earthquake and tsunami, the disaster may cause subsidence to the breakwater construction. This study investigated the effect of subsidence to the characteristic of flow in around the structure. Caisson breakwaters typically have vertical sides and are usually erected where it is desirable to berth one or more vessels on the inner face of the breakwater. They used the mass of the caisson and the fill within it to resist the overturning forces applied by waves hitting them. In this study, CADMAS-SURF/3D will simulate the flow patterns and flow characteristics around the breakwater.

The relation and the effect of caisson's subsidence to the scouring of rubble mound were investigated by measuring the velocity profile behind of breakwater. The point measurement located at x-axis 1.5m behind breakwater and z-axis at several locations start from one grid above the rubble mound (9m) until under the initial water level (22.5m). Figure 7(a) shows the velocity profile behind the gaps. There are two gaps in y-axis, that is y=4.24m and y=14.5m. The velocity profile behind the two gaps have similar pattern. The rapid increase can be seen in the results of velocity profiles when tsunami hit the breakwater. After these increasing phases, velocity profiles maintained a fluctuation profile as a continuous overflow phase between 10-second and 16-second. The maximum value of velocity reached 14m/s, the minimum of about 3.8m/s, and the average velocity at around 10.8m/s. This fluctuation profile occurred due to the vortex flow formation behind the breakwater.

Figure 7(b) presents the velocity profile behind the subsidence caisson. The figure shows that the velocity in the behind of caisson tends to be a negative value. The negative value was caused by the formation of vortex. Its velocity has fluctuation profile of both positive and negative value due to the vortex flow. The fluctuation of velocity describes that the complex flow occurred during tsunami flow over the breakwater. The positive velocity was about 1.5m/s and the negative velocity was about -5.37m/s. This indicates that the horizontal velocity tends to lead to the left direction or in the opposite direction with waves flow. In other words, the vortex formed during tsunami overflow on the breakwater. The fluctuation of velocity value behind the breakwater is presented in Table 1.

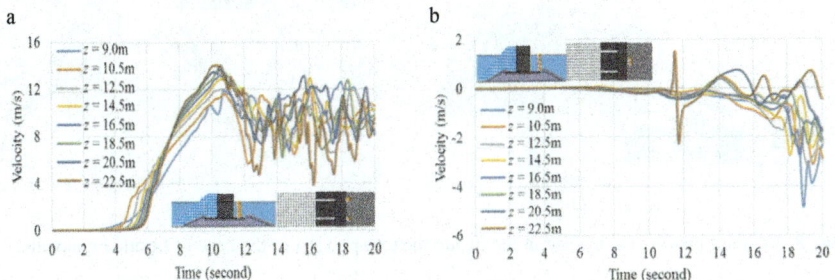

Fig. 7. Velocity profile; (a) behind gap of the breakwater (b) behind the subsidence breakwater

318

Table 1. The range of wave velocity behind the breakwater

Velocity	Behind 1st Caisson (m/s)	Behind 1st Gap (m/s)	Behind subsidence Caisson (2m from left side) (m/s)	Behind subsidence Caisson (Middle) (m/s)	Behind subsidence Caisson (2m from right side) (m/s)	Behind 2nd Gap (m/s)	Behind 2nd Caisson (m/s)
Maximum	5.62	18.29	6.39	5.93	4.90	20.17	5.82
Minimum	-1.37	4.73	-0.19	-0.30	-0.82	6.42	-0.88
Average	2.60	12.80	3.45	2.73	1.99	12.62	3.50

3.4. Flow characteristics beside breakwater

Figure 8(a) presents the velocity profile beside the breakwater. The measurement point of the x-axis located at the center point of the breakwater, and the y-axis located inside the tetrapod, 7.5m or 1m beside the caisson structure, there are eight measurement point in the z-axis, that is 7.5m (above mound foundation), 10m, 12.5m, 15m, 17.5m, 20m, 22.5m, and 25m. The velocity reached a maximum value around 2.3m/s, minimum value -0.2m/s, and the average velocity ranged from 0.5ms to 2m/s while overflowing on the breakwater. As shown in the figure, the profile of velocity inside the tetrapod (beside the caisson) have fluctuation velocity. This fluctuation was affected by the seepage flow both from the front side of the breakwater and also from beside of the breakwater.

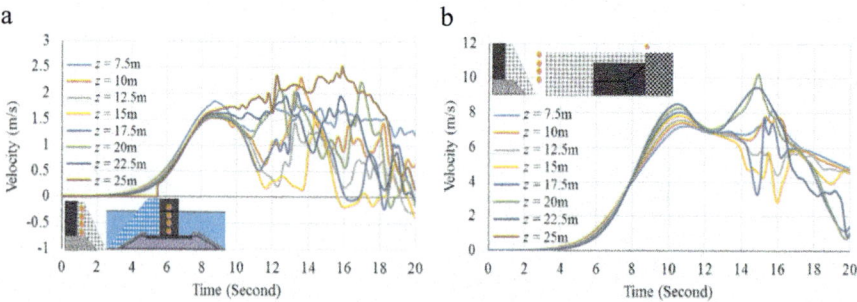

Fig. 8. Velocity profile; (a) in case in the inside the tetrapod (b) in the case of beside breakwater

Figure 8(b) shows the velocity profile in the rear corner of the breakwater. The measurement point of x-axis 126m was located at the rear corner of the structure, y-axis located beside the structure, 15m or 0.5m beside the tetrapod, and there were several measurement points in the z-axis, that is 7.5m (above mound foundation), 10m, 12.5m, 15m, 17.5m, 20m, 22.5m, and 25m. As shown in the figure that the fluctuation velocity occurs during tsunami overflowing at on the breakwater. It can be seen from the velocity profile that tsunami flowed over the breakwater forming a complex flow including supercritical flow due to high-speed flow and vortex flow. Furthermore, the flow characteristic beside the breakwater was presented in more detail from the flow pattern.

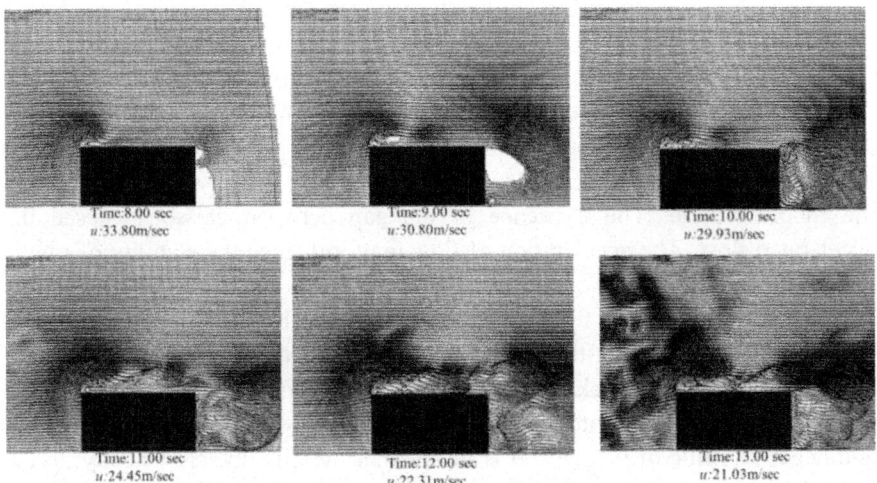

Fig. 9. Flow pattern from top view

Figure 9 presents the flow pattern and characteristic beside a breakwater from the top view. As shown in the figure, tsunami flowed from the front side of the breakwater and formed a vortex in the front side of the obstacle. Furthermore, the flow came around into the side and behind the breakwater, and vortex flow also formed at the back-side of the obstacle. The figure shows that vortex phenomena occurred during wave overtopping of the breakwater. The vortex indicated two trends of velocity profile, positive value and negative value of velocity. It means there is reverse velocity, to the right direction and back to left direction. When the velocity shear was imposed on the free surface, a mixing layer seemed to appear at the interface due to the sharp velocity gradient. Therefore, when the initial bore pressure acted on the breakwater, the safety factor of the breakwater bearing capacity rapidly decreased. In the case where the excess pore water pressure in the ground was considered, the safety factor of the breakwater bearing capacity markedly decreased when the continuous wave pressure was acted. During dilatational waves, the safety factor also decreased.

4. Conclusion

The characteristics of three-dimensional flow around a breakwater in the prototype scale of Hososhima Yojima breakwater was investigated to identify the characteristics of tsunami flow in and around the breakwater. In this study, a numerical half-basin domain was considered in 3D simulations in order to save computational time. This half-basin domain employed half the channel width by considering a symmetry axis of the structure model. The steady overflow occurred during time history from 10 sec until 16 sec with the average elevation 10m from initial water level or 2m above breakwater.

Tsunami wave hit and flowed over the breakwater forming a complex phenomenon including turbulence flow, supercritical flow, and also vortex in the behind of breakwater. The complex flow was also affected by seepage flow from the rubble mound. The existence of the gap between caissons caused the accelerated flow. The acceleration of flow may influence the stability of rubble mound foundation.

The tsunami flowed from the front side of the breakwater, and formed a vortex in the front side of the obstacle. Furthermore, the flow came around into side and behind of the breakwater. Vortex flow also formed at the back side of the obstacle. The vortex indicated two trends of velocity profile that are positive value and negative value of velocity. It is means that there is reverse velocity, to the right direction and back to left direction. As such, when the velocity shear is imposed on the free surface, a mixing layer seems to appear at the interface due to the sharp velocity gradient thereby affecting the stability of the breakwater.

References

1. Coastal Development Institute of Technology: CADMAS-SURF/3D Research and development of numerical wave tank, Coastal Technology Library, No. 39, pp. 1-235, 2002.
2. K. Kawasaki, S. Yamaguchi, N. Hakamada, N. Mizutani, S. Miyajima, Wave Pressure Acting on Drifting Body after collision with Bore, *Annual J. of Civil Engineering in the Ocean, JSCE*, Vol. 53, pp. 786-790. 2006.
3. T. Suzuki, A. Okayasu, T. Shibayama, A numerical study intermittent sediment concentration under breaking waves in the surf zone. *Coastal Engineering.* Vol. 54(5), pp. 433-444, 2007.
4. Y.S. Fukui, M.N. Hidehiko, Y. Sasaki, Study of tsunami -Investigation of wave velocities in case of bore type tsunami-, *Annual J. of Coastal Engineering in Japan*, Vol. 9, pp. 44-49, 1962 (in Japanese).

5. I. Wijatmiko, K. Murakami, Numerical simulation of tsunami bore pressure on cylindrical structure", *Annual J. of Coastal Engineering in Japan, JSCE*, Vol. 26, pp. 273-278, 2010a.

6. M. Bozorgnia, J.J. Lee, Computational Fluid Dynamic Analysis of Highway Bridges Exposed to Hurricane Waves, Coastal Engineering Journal, Vol. 33, pp. 1-14, 2012.

7. S.C. Yim, W. Zhang, A Multiphase Multi scale 3-D Computational Wave Basin Model for Wave Impact Load on a Cylindrical Structure, *J. of Disaster Research*, Vol. 4, No. 6, pp. 450-461, 2009.

An Experimental Study of the Maximum Run-up Height Under Dam-break Flow on the Initial Dry-bed

Senxun Lu

Ocean College, Zhejiang University, 866 Yuhangtang Road,
Hangzhou, Zhejiang 310058, China
jimlu@zju.edu.cn

Haijiang Liu

College of Civil Engineering and Architecture, Zhejiang University,
866 Yuhangtang Road, Hangzhou, Zhejiang 310058, China
haijiangliu@zju.edu.cn

Hydrodynamic properties of the breaking bore run-up process were investigated under a sloping dry bed condition. A series of laboratory experiments were performed using a 1:7.5 stainless steel slope and a dam-break type PVC-made flume. Under the dry bed condition, different flow velocities were generated by adjusting initial water levels from 8cm to 24cm in the upstream reservoir. The bore propagation velocity and the maximum run-up height were obtained from images recorded using a high speed video camera. Existing models underestimate the present measured maximum run-up height, which is ascribed to the phenomenon that multiple ensuing bores generated by the water mass compression overpass the preceding bore front during the run-up process. Nevertheless, water depth of run-up front under the dry bed condition is much smaller than that of the wet bed condition which contributes to the generation of multiple ensuing bores.

Keywords: Dam-break; Dry bed; Run-up height; Ensuing bores.

1. Introduction

Studies of the wave/bore run-up process are of great importance for the design of coastal structures, sediment transport and estimation of wave overtopping rate. Precise prediction of the maximum wave/bore run-up height is crucial for the coastal disaster prevention and mitigation efforts.

Previous experimental studies have been designed to investigate the run-up process under the initial wet bed condition with a certain downstream water level. Accordingly, two types of bore (breaking bore and undular bore) are observed under different ratios of upstream to downstream water levels.[2,7] In terms of initial dry bed condition (downstream water level is zero), the hydrodynamic features of dam-break flow on horizontal plane have also been studied. Detailed results of wave profiles and distributions of flow velocity in the water body have been described in [4]. However, few studies have been conducted on the dam-break induced run-up process over the initial dry bed condition.

In this study, hydrodynamic properties of the maximum run-up height under the dry bed condition were investigated. The bore propagation velocity and the maximum run-up height were obtained from video records using a high speed video camera (HSVC). Present run-up data were compared with theoretical run-up heights which are only governed by the instantaneous bore velocity when reaching the beach toe for the dry bed velocity and initial shoreline for the wet bed condition. Besides, following the first bore front, multiple ensuing bores were generated to overpass the preceding bore front in the run-up process, contributing to a higher run-up height. Further discussions on the maximum run-up height, multiple ensuing bores and water depth of run-up process were made to explain the physical insight of run-up properties under the dry bed condition.

2. Experimental set-up

A series of laboratory experiments were performed using a dam-break flume with a dimension of 6.5m long, 0.4m wide and 0.4m high (as seen in Figure 1). The horizontal flume made of transparent PVC plates consists of three main parts: the upstream zone (the dam-break bore generating area), the bore propagation zone and the swash zone, whose horizontal lengths are 1.5m, 1.8m and 3.0m respectively. For the upstream zone, the upstream dam-break reservoir is separated from the downstream channel by a 1mm-thick stainless steel gate which could be uplifted quickly through a pulley system using a heavy mass. The gate fully releasing time satisfies the sudden opening criterion of generating the dam-break flow.[3,5] The length of bore propagation zone enabled the dam-break flow to be fully developed into a steady form before the flow entering the swash zone. Based on the laboratory measurements, [4] confirmed that hydrodynamic features of dam-break flow, e.g., the bore height and bore propagation velocity, become stable only after a certain distance away from the gate location. With respect to the swash zone, a 3 mm thick stainless steel plate is deployed to make a 7.6° sloping beach (3.0m long in horizontal direction, 0.4m high in vertical direction) in order to simulate the run-up process along an impermeable beach.

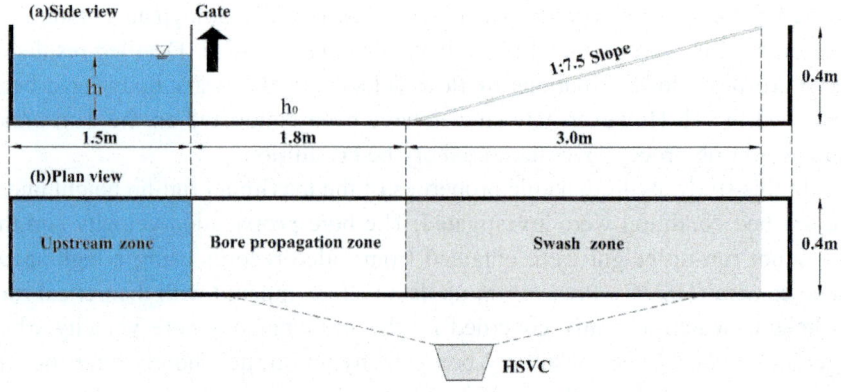

Fig. 1. Schematic diagram of the experimental set-up.

As shown in Table 1, all experiments were conducted under the dry bed condition where the downstream water level h_0 is 0 cm and the upstream water level h_1 ranges from 8 cm to 24 cm. For comparison, a wet bed test "24-8" (h_1=24cm, h_0=8cm) was also conducted. The bore propagation velocity and the maximum run-up height were recorded using a HSVC (recording frequency of 150 Hz and image resolution of 2048×1088 pixels). Experimental repeatability was confirmed by conducting each test three times and the averaged results of three repeated experiments were provided. The bore front velocity and the maximum run-up height were estimated on the basis of the image analysis of the video recording, same as [4]. For a clearer video recording, two spotlights setting at the same side of HSVC and two LED lights standing on the top of flume have been utilized to illuminate the flow process. In this study, the bore propagation velocity was obtained at the time when the flow is reaching the beach toe. Based on the experimental observation, it is confirmed that dam-break type induced bores are all in breaking bore form under an initial dry bed condition.

Table 1. Experimental set-up conditions.

Tests	h_1 (cm)								
	8	10	12	14	16	18	20	22	24

3. Results and discussions

3.1 *The maximum run-up*

Table 2 shows the instantaneous incident bore front velocity when reaching the beach toe. Since the moving distance of bore front and corresponding moving time are both available to be extracted from the video records, the flow velocity can be obtained through this image analysis method. The flow velocity increases as the upstream water level becomes higher. Based on the frictionless nonlinear shallow water wave equation and using the method of the characteristics, [6] derived the maximum run-up height along a uniform slope,

$$R = \frac{U^2}{2g} \tag{1}$$

where R is the maximum run-up height, U is the incident bore front velocity when arriving at the beach toe (as listed in Table 2), g is gravitational acceleration. Eq. (1) demonstrates the energy conservation during the run-up process, i.e., changing from kinematic energy at the beach toe to the potential energy completely at the run-up limit as bottom resistance is neglected here.

Table 2. The instantaneous incident bore front velocity at the beach toe

Tests	8-0	10-0	12-0	14-0	16-0	18-0	20-0	22-0	24-0
Velocity (m/s)	1.00	1.15	1.30	1.78	1.90	1.98	2.16	2.27	2.49

Later, taking into account the effect of bottom friction, [8] obtained an analytical expression of breaking bore run-up height over a uniform slope. To achieve the dimensional form of their expression, the frictionless run-up height needs to be obtained first, which could be estimated from Eq. (1). Accordingly, the dimensional run-up height after considering the bottom resistance can be estimated as,

$$R_f = \frac{R}{9.66} \ln \frac{1.08 \tan \gamma}{C_f} \tag{2}$$

where R_f is the maximum run-up height with the bottom friction effect, γ is the slope of the beach; C_f is a dimensionless resistance coefficient, which is correlated to the Darcy–Weisbach resistance coefficient f as [1],

$$C_f = \frac{1}{8} f = \frac{gn^2}{r^{1/3}} \tag{3}$$

where n is Manning's coefficient ($n=0.012$ for the stainless steel), r is hydraulic radius of the flume.

326

Figure 2 presents the comparison between the measured maximum run-up height and theoretical ones which are derived by [6] and [8] respectively. As shown in Figure 2, in general, all the measured data are larger than theoretical ones. Comparing between [6] and [8], it is obvious that with bottom friction, the estimated theoretical maximum run-up heights are significantly reduced. With the increase of upstream water level, the influence of bottom resistance to the maximum run-up height becomes obvious leading to a significant reduction of the run-up height with respect to [6] and [8]. Nevertheless, the present experimental data, although measured under a frictional bottom condition, are similar to or even larger than the frictionless run-up heights estimated using [6] under the same initial hydrodynamic conditions. Meanwhile if comparing with the theoretical run-up heights with bottom friction using [8], the present experimental data are much larger, indicating there may exist other mechanisms which enhance the run-up process in the present experiments.

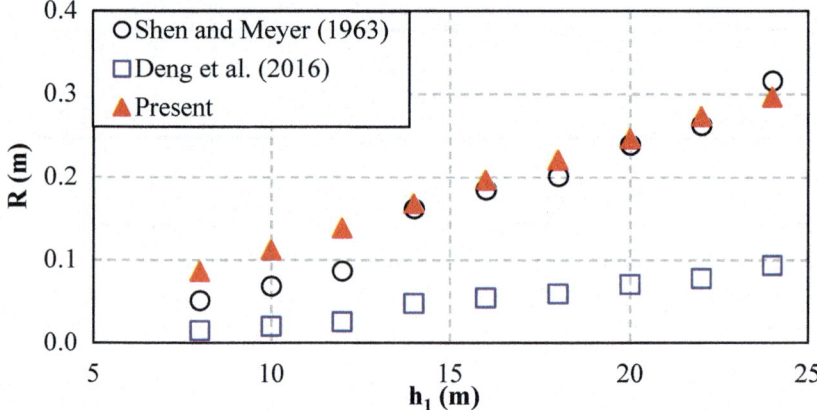

Fig. 2. Comparison between the present maximum run-up height and theoretical estimations.

3.2 *Multiple ensuing bores*

Figure 3 shows four snapshots during the breaking bore run-up process in case of "24-0". When the bore swashes up as Figure 3(a) shows, the velocity of the initial bore front begins to decelerate quickly owing to the bottom friction. Meanwhile the ensuing water body swashes up faster than the initial bore front, which makes water mass compressed, generating a new crest indicated using a white arrow in Figure 3(b). Subsequently, this newly born bore moves forward, overpassing the initial bore front and forming a new bore front as indicated in Figure 3(c). Similar phenomenon happens in Figure 3(c) and Figure 3(d) until the ensuing new bore in Figure 3(d) becomes the final bore front swashing up to the run-up limit. According to experimental observation, three new bores could be generated till the swash end for most experimental tests.

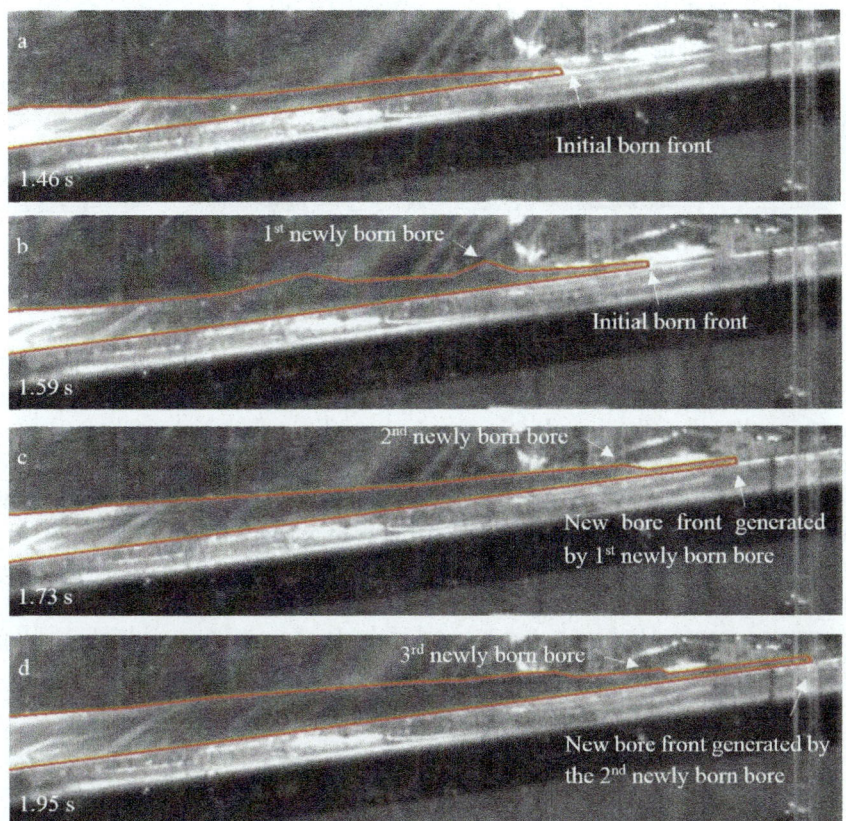

Fig. 3. Snapshots showing bore run-up process of case "24-0" (red lines envelope the water body, numbers at the bottom left corner indicate the time since the bore reaches the beach toe).

3.3 *Run-up comparison between dry and wet bed conditions*

Since the phenomenon of multiple ensuing bores run-up was not apparent under the wet bed condition in previous studies,[2,7] experiments under both dry and wet bed conditions were conducted to further reveal the physical insight of such difference. Here, cases of "16-0" and "24-8" with the same initial water head were considered for inter-comparison.

Figure 4 shows the image recorded at the instant when the bore front arrives at the beach toe or the initial shoreline. It is obvious that the bore height of case "16-0" is much smaller than those of case "24-8", similar to [4]. Such difference in incident bore water depths leads to different bore front water depths during the run-up process. Figure 5 shows the comparison of water depths between cases of "16-0" and "24-8" near the bore front at 0.67 sec after bore front passing the beach toe or the initial shoreline. It is clear that the water depth of bore front in Figure 5(a) is much smaller than that in Figure 5(b).

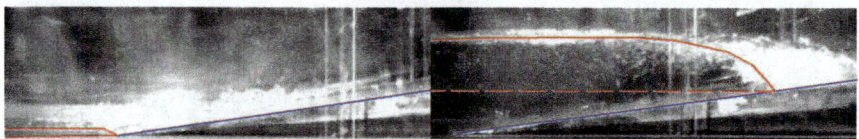

Fig. 4. Snapshots of incident bore profiles (red lines envelope the incident water body, blue lines represent the side view profile of slope, dashed line represents the still water level) at the beach toe or the initial shoreline of case "16-0" (left panel) and "24-8" (right panel).

According to Eq. (3), smaller water depth results in a larger bottom friction. Since the water depth under the dry bed condition (approximately 0.005m) is much smaller than those under the wet bed condition (approximately 0.01m), a larger bottom friction exerting on the bore front is expected for the dry bed case. This leads to a significant velocity deceleration and water mass compression, making ensuing bores easily overpass the preceding bore front. While under the wet bed condition, a relative large water depth of bore front reduces the retarding effect from bottom friction, thus no apparent multiple ensuing bores occur during its run-up process. However, the physical insights of generating multiple ensuing bores remain unclear.

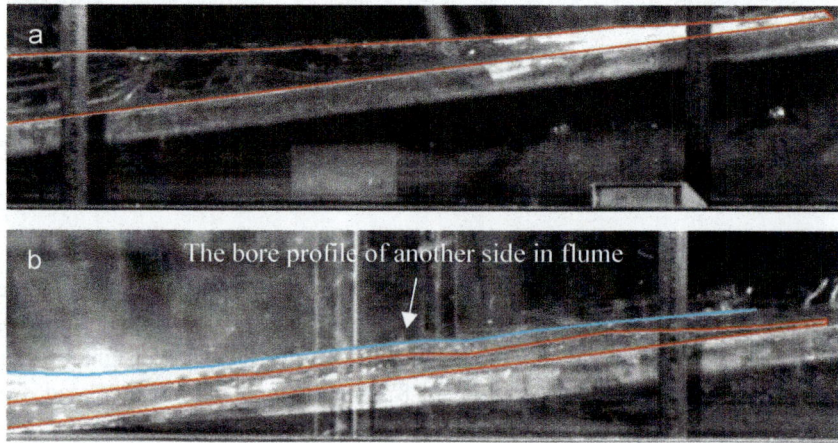

Fig. 5. Snapshots of bore front profiles during run-up of case "16-0" (a) and "24-8" (b).

4. Conclusions

In this study, a series of experiments have been performed to investigate the breaking bore run-up process and the corresponding maximum run-up height over an initial dry bed condition using a dam-break flume. High speed video camera was used to estimate the bore propagation velocity and the maximum run-up height.

As the nonlinear shallow water equations only solve for the run-up process of the single bore, both [6] and [8] underestimate the present measured maximum run-up heights. Based on the laboratory observation, it is ascribed to the multiple ensuing bores presented in the present experiments. Due to the smaller water depth at the bore front, the larger bottom friction under the dry bed condition contributes to the succeeding water mass compression during the run-up process, generating multiple ensuing bores. Subsequently, these newly born bores overpass the preceding bore front till the swash end, which elongates the run-up process and enlarges the corresponding breaking bore run-up height. Nevertheless, compared with the dry bed condition, such ensuing bores phenomenon does not appear in the bore run-up process over the initial wet bed condition.

Acknowledgements

This study was financially supported by the Natural Science Foundation of Zhejiang Province, China (No. LR14E090002), the Natural Science Foundation of China (No. 11632012) and the Open Research Fund Program of State key Laboratory of Hydroscience and Engineering (No. sklhse-2016-B-02).

References

1. B. C. Yen, Open channel flow resistance. *J. Hydraul. Eng*, 128(1), 20-39 (2002).
2. H. H. Yeh, A. Ghazali and I. Marton, Experimental study of bore run-up. *J. Fluid. Mech*, 206(206), 563-578 (1989).
3. H. Liu, B. Ying and H. Liu, An experimental study of the influence of gate releasing time to the dam-break hydrodynamics. *Proc. 2nd Conf. Global Chinese Scholars on Hydrodynamics*, Wuxi, 251-256 (2016).
4. H. Liu and H. Liu, Experimental study on dam-break hydrodynamic characteristics under different conditions. *J. Disaster. Res*, 12(1), 198-207 (2017).
5. G. Lauber and W. H. Hager, Experiments to dambreak wave: horizontal channel. *J. Hydraul. Res*, 36(3), 291-307 (1998).

6. M. C. Shen and R. E. Meyer, Climb of a bore on a beach. Part3. Run-up. *J. Fluid. Mech*, 16 (1), 113–125 (1963).

7. R. L. Miller, Experimental determination of run-up of undular and fully developed bores. *J. Geophys. Res*, 73(14), 4497-4510 (1968).

8. X. Deng, H. Liu, Z. Jiang and T. E. Baldock, Swash flow properties with bottom resistance based on the method of characteristics. *Coast. Eng*, 114, 25-34 (2016).

Analysis Current Mechanism of Cambodian South Coastal Region

N. Inukai[†]

Department of Civil and Environmental Engineering, Nagaoka University of Technology,
Magaoka city, Niigata prefecture 940-2188, Japan
[†]E-mail: inu@nagaokaut.ac.jp

N. Kopy and C. EM[†]

CES Co., Ltd.
Phnom Penh city, Cambodia
[†]E-mail: chamnab@ces.com.kh

The authors have concerned in the assessment about Cambodian shoreline since 2008, and made many observations and made the geographical map, water flow mechanism, weather and etc. This study reports the observation result and feature of geography, tidal change and weather. And furthermore, this study simulate the tidal flow and wind-driven flow. Ultimately, this study try to know the flow mechanism of this region. Firstly, this study performed the field survey and got the information about the feature of the current, topography, the tidal level and the wind. Secondary, this study made the geographic data by the sonar data. Finally, this study did the numerical simulation about the tidal current and the wind-driven current, and tried to comprehend about the current mechanism in this region.

Keywords: Cambodia; Tidal current; Wind-driven current; Objective analysis; Numerical simulation.

1. Introduction

The total extension of the Cambodian shoreline is approximately 500km (see Figure 1). However this region has the abundance of natural piscatorial resources by reason of the semi-enclosed, calm, shallow sea and the enormous mangrove grown coast. The Cambodian Environmental Impact Assessment Law was put a law into force, recently. For perform the environmental impact assessment, the flow mechanism, the water quality and another physical quantities need to be known, however the observed data in this region does not supply in the present.

The authors have concerned in the assessment about this region since 2008, and made many observations, for example, the geographical map, the water flow mechanism, weather and etc. This study reports the observation result and feature of geography, tidal change and weather. Law. And furthermore, this study simulates the tidal flow and wind driven flow. Ultimately, this study try to know the flow mechanism of this region.

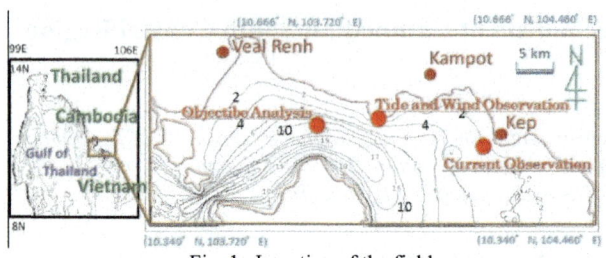 

Fig. 1. Location of the field   Photo 1. Mangrove

Firstly, this study did field survey and comprehended the feature of the field about the current, topography and etc. After the field survey, this study observed the time history of the tidal level and the wind. Furthermore, the tidal harmonic constant, the prevailed wind speed and the prevailed wind direction in this region were decided by the observation data and by the prediction model.

Secondary, this study made the geographic data. In this case, the observed data by the sonar that were set on the boat were mainly used, and the digitized data from the satellite image were used to make the shoreline. Furthermore, this study used Etopo1 data set [1] to complement the survey blank area. This data was used for the numerical simulation about the tidal current and the wind-driven current. Finally, this study did the numerical simulation about the tidal current and the wind-driven current, and tried to comprehend about the current mechanism in this region.

2. Field Survey

Firstly, this study did field survey and comprehended the feature of the field about the current, topography and weather. The survey terms are since 20th to 22th May 2008 and 6th January 2013. Photo 1, Photo 2 and Photo 3 show the field. Photo 3 shows that the topography has the gentle slope and the shelving bottom. And Photo 1 shows that Rhizophora which is one of mangrove vegetates at almost coastline, and the piled up bottom material near the mangrove is the bottom sludge which changed from the withered leaves and the animal waste. Furthermore, Photo 2 shows that many eelgrass vegetates at the 2-3m depth, therefore the small fish and plankton can live easily in this conditions. According to these results, there is the good fishing ground. Photo 3 shows that the tidal water level range is large in this area, and the shoreline move to 200m

Photo 2. Eelgrass   Photo 3. Left: High Tide (15:48, 22nd May, 2008), Right Low Tide
(6th Jan, 2013)   (08:29, 20th May, 2008, UTC+7)

offshore when the low tide. When the water level is the lowest, the depth of 500m offshore is about 1m.

The sea area is dotted with islands, and the fetch of the sea wind is short. Therefore, the no big wave occur in this area. This fact can inferred from the results of the hearing investigation at there and many house are built at the seaside.

The SE and W sea wind blows hard because the straits are locate to these direction. It conceivable that the strong wind driven current will occur when the wind blow from these direction.

3. Hearing Investigation

3.1. *Particle of hearing investigation*

This study interviewed a delegate of the fishing village on 20th May 2008, to comprehend the feature of oceanographic phenomena around this area. Table 1 shows the particle of interview to the delegate of the fishing village.

Table 1. Particle of hearing investigation

a. Occur or not occur ocean wave
b. Wind direction every season, and strength
c. Difference of Tidal Wave between High and Low level
d. Occur or not occur storm surge, storm wave and tsunami

3.2. *Result of investigation*

The results of the hearing investigation are shown below.
(1) Breaking wave does not occur in this area.
(2) Breaking wave does not occur in this area.
(3) There is a dry season and the rainy season. Wind blows from NE in the dry season, and from W in the rainy season. Especially, the strong wind blows from W in the rainy season.
(4) Tidal level difference is about 1m between the high and low tide level. The sea level is the lowest at about 7 am, and the highest in the evening. When the tide level is the lowest, the shoreline move to 200m offshore when the low tide.

334

(5) There is no damage of the storm surge, high wave and tsunami.

According to the above results, the tidal current is prevalent in this area. And there are the prevailing wind directions, and these directions accord to channel open directions. Therefore this is considered that the wind driven current will be strong when these winds blow from these directions.

4. Make Topography Information

This study tried to comprehend the current condition of this area by the numerical simulation. In this case, the topographic information is needed for the numerical simulation. However, Cambodia government does not have the information. Therefore, this study did field survey to get and make the topography information. When this study did the field survey, the memory type sonar was set at the boat side. In this area, there is the border between Vietnam and Cambodia, and the Cambodian boat cannot go into Vietnam. Therefore, this study used the information of Etopo1 [1] at these area.

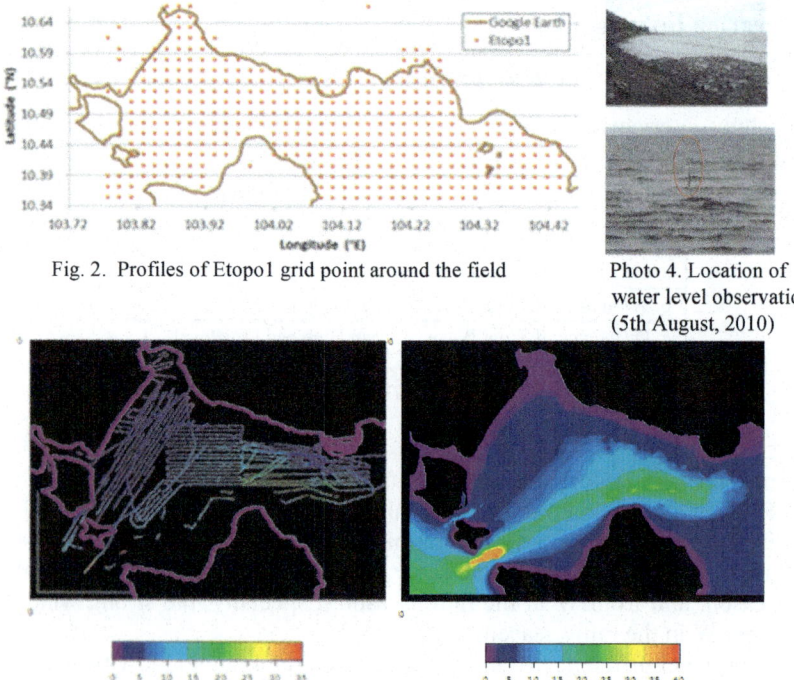

Fig. 2. Profiles of Etopo1 grid point around the field

Photo 4. Location of water level observation (5th August, 2010)

Fig. 3. Depth information by survey

Fig. 4. Topography Information

Figure 2 shows the profiles of Etopo1 grids. The coordinates of shoreline and estuary were gotten from Google Earth.

The time change of tidal level was observed at the same time as the field survey, and the tidal data was deleted in the observed depth information.

Figure 3 shows the boat trail and the depth at every point. Figure 4 shows the topography information which was made by the field survey and Etopo1.

5. Comprehend Tide Information

5.1. *Observation*

This study made field survey to get the tidal information.

Firstly, this study measured the difference between the tidal high level and the tidal low level by the visual observation in 20th and 22nd May, 2008. The point was located the 5km from Kampot where was shown in Figure 1. The photos at the tidal high water level and the low water level are shown in Photo 3. The left photo shows the high water level. The piles of the house on the sea sank out of sight, however the shoreline went back to 200m offshore in the right photo when the low water level. The difference between the high water level and the low water level was about 1m when the observation.

Secondary, this study observed the time change of the tidal water level since 7 am 5th to 2pm 13th August, 2010. Photo 4 shows the observation point and the measure which was set in the sea water. The water level was observed at every one hour by observing the scale of the measure.

5.2. *Numerical model for tide level*

This study calculated the tidal water level by NAO.99b [2], and the result was compared with the result of the observation.

NAO.99b is a calculate model which calculates a tidal water level at decided point by the observed data of TOPEX/POSEIDON satellite. Figure 5 shows the result of the calculation. The figure shows that the daily tide is prevalent in this

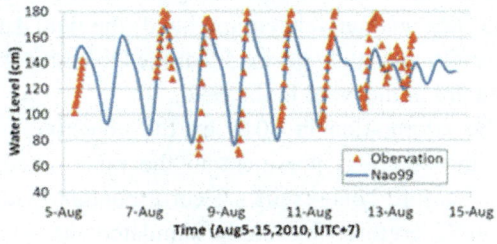

Fig. 5. Compare with NAO.99b and Observation

area, and the tidal amplitude was 60cm-100cm. And the periods of calculation were almost same with the observed results, however, amplitude of the observation was slightly larger than calculate. The reason is conceivable that the observed point locates in a semi enclosed bay as shown in Figure 1, and the amplitude increases due to the effect of the complex topography. However, this study considers that the tide model qualitatively calculates and comprehend the tide of the field.

6. Comprehend Seasonal Wind

According to the results of the hearing observation, the sea wind blows in this area, and this study have to comprehend the effect of the wind driven current. Therefore, this study analyze the objective analysis data and the observation data to comprehend the feature of the seasonal wind.

6.1. *Field survey and objective analysis data*

This study observed the wind velocity and wind direction in the dry season since 7:28 16th to 7:20 17th March, 2010, totally 25 hours. In rainy season, since 15:40 13th to 14:00 15th June, 2010, totally 46 hours. The observed point locates 5km west from Kampot in Figure 1.

The Objective analysis Data is supplied by Japan Meteorological Business Support Center [3]. In this case we use GSM Objective Analysis data. The data term is since January to December, 2010. Locate of data was same with the observed data, and the term of data was 1 year of 2010.

6.2. *Compare observed data and objective analysis data*

Figure 6 and Figure 7 show the results of the observation and analyze. According to these figures, there are the difference of angle between the both figures. This reason is considered that the grid space of GSM is 1 degree, and GSM could not reflect the effect of the complex field topography. However, the both wind velocities are almost same. And this study considered that the Objective Analysis Data qualitatively comprehends the field data.

According to this result, this study decided that the Objective Analysis is used to comprehend the feature of the wind in the field.

Figure 8 shows the information in 2010 by the objective analysis data. The figure shows that the WSW is prevalent in the rainy season and the SE is prevalent in the dry season. This results are corresponding with the result of the hearing observation. Therefore, this study simulated the wind driven current when the WSW and SE wind blows. The wind velocity was set as 10 m/s.

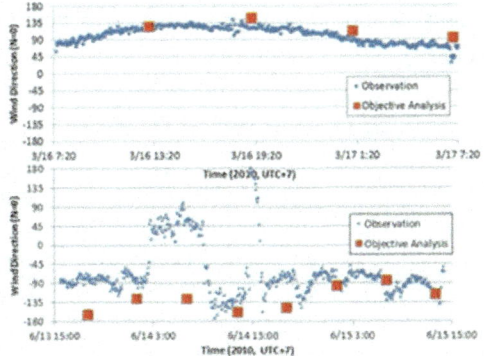

Fig. 6. Wind observation and objective analysis. (Wind Direction)

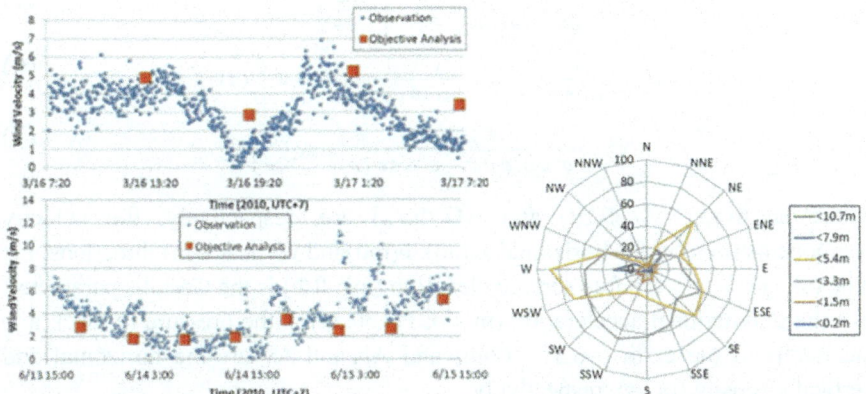

Fig. 7. Wind observation and objective analysis (Wind Velocity)

Fig. 8. Wind direction by appearance frequency table of objective analysis (2010)

7. Numerical Simulation of Tidal Current and Wind Driven Current

7.1. *Numerical model for current simulation*

(1) A Spherical Coordinate Model

The current in this domain is expected to be complicate involving the large region.

Therefore, the study has to take the large domain, so that numerical hydrodynamic model used for such case is the three dimensional spherical coordinate model following the earth's geometry [4][5]. The governing three-dimensional equations under the spherical coordinate describing constant density, free surface flow can be derived from the Navier-Stokes equations. After turbulent averaging, and applying the hydrostatic and Boussinesq approximations, λ and Φ momentum equations and continuity equation are given as equations (1)-(7).

$$\frac{1}{a\cos\varphi}\frac{\partial u}{\partial\lambda}+\frac{1}{a\cos\varphi}\frac{\partial}{\partial\varphi}(v\cos\varphi)+\frac{\partial w}{\partial z}=0 \tag{1}$$

$$\frac{\partial u}{\partial t}+L(u)-\frac{uv\tan\phi}{a}-fv=-\frac{1}{\rho_w}\frac{1}{a\cos\phi}\frac{\partial P}{\partial\lambda}+A_h\left\{\nabla^2 u+\frac{(1-\tan^2\phi)u}{a^2}-\frac{2\sin\phi}{a^2\cos^2\phi}\frac{\partial v}{\partial\lambda}\right\}+A_v\frac{\partial^2 u}{\partial z^2} \tag{2}$$

$$\frac{\partial v}{\partial t}+L(v)-\frac{u^2\tan\phi}{a}+fu=-\frac{1}{\rho_w}\frac{1}{a}\frac{\partial P}{\partial\phi}+A_h\left\{\nabla^2 v+\frac{(1-\tan^2\phi)v}{a^2}-\frac{2\sin\phi}{a^2\cos^2\phi}\frac{\partial u}{\partial\lambda}\right\}+A_v\frac{\partial^2 v}{\partial z^2} \tag{3}$$

$$-\rho_w g-\frac{\partial P}{\partial z}=0 \tag{4}$$

$$\frac{\partial\zeta}{\partial t}=-\frac{1}{a\cos\phi}u_s\frac{\partial\zeta}{\partial\lambda}-\frac{1}{a}v_s\frac{\partial\zeta}{\partial\phi}+w_s \tag{5}$$

$$L(a)=\frac{u}{a\cos\phi}\frac{\partial}{\partial\lambda}(\alpha)+\frac{v}{a\cos\phi}\frac{\partial}{\partial\phi}(\cos\phi\alpha)+w\frac{\partial}{\partial z}(\alpha) \tag{6}$$

$$\nabla^2\alpha=\frac{1}{a^2\cos^2\phi}\frac{\partial^2\alpha}{\partial\lambda^2}+\frac{1}{a^2\cos\phi}\frac{\partial}{\partial\phi}\left(\cos\phi\frac{\partial\alpha}{\partial\phi}\right) \tag{7}$$

where $u(\lambda,\Phi,z,t)$, $v(\lambda,\Phi,z,t)$ and $w(\lambda,\Phi,z,t)$ are, respectively, the velocity components (m/s) in the horizontal λ(rad),Φ(rad)and vertical z(m) directions; t is time (s); $\zeta(\lambda,\Phi,t)$ is the free surface elevation (m); $f(\Phi)$ is the Coriolis parameter; g is the gravitational acceleration (m/s2); $P(\lambda,\Phi,z,t)$ is the pressure (N/m2); a is the radius of the earth (=6.37×106m) and Ah and Av are the horizontal and vertical viscosity (m2/s), respectively.

In this study, the wetting and drying scheme is implemented on the only surface layer.

(2) Vertical Boundary Conditions in Spherical Coordinate

The boundary conditions at the free surface are specified as the wind stresses:

$$\left(\rho A_v\frac{\partial u}{\partial z},\rho A_v\frac{\partial v}{\partial z}\right)=\left(\tau_\lambda^w,\tau_\Phi^w\right)=C_{da}\left(u_w,v_w\right)\sqrt{u_w^2+v_w^2} \tag{8}$$

where τw is the wind stress at the free surface; uw and vw are the components of wind speed measured at some distance above the free surface and Cda is the drag coefficient at the water surface.

The drag coefficient is normally a function of the roughness of the sea surface and the wind speed at some height above the water surface. For this study, a empirical relationship developed is used. Garrat [6] defined the drag coefficient as a linear function of wind speed measured at 10 meters above the water surface:

$$C_{da} = 0.001(0.75 + 0.067W_s) \tag{9}$$

where Ws is the wind speed in meters per second.

The bottom boundary conditions for the three dimensional model satisfy the quadratic stress law:

$$(\tau_{b\lambda}, \tau_{b\Phi}) = C_d(u_b, v_b)\sqrt{u_b^2 + v_b^2} \tag{10}$$

where ub and vb are the near bottom velocities, ρis the density of water and Cd is the drag coefficient. The drag coefficient is defined as

$$C_d = \frac{\kappa^2}{\ln^2\left(\dfrac{z_1}{Z_0}\right)} \tag{11}$$

where Z0 is the size of the bottom roughness elements and Z1 is the height at which the velocity is measured. This formulation is appropriate within the constant flux layer above the bottom. In very shallow water the drag coefficient is set as a constant value which linearly varies to the value of the above equation at a certain depth.

7.2. *Calculating area and calculation condition*

Figure 9 shows the calculating area. This study set the grid space as 50m for the numerical simulation of the tidal current and the wind driven current. The boundary condition for the simulation of the tidal current is shown in Figure 9. The amplitude of the major 4 tidal components and the period of K1 component were set at the boundary. These condition were calculated by NAO.99b. The ocean wind speed for the numerical simulation of the wind driven current was set as 10 m/s. Furthermore, the wind direction of the simulation were set as WSW and SE of the prevalent directions. The tidal current simulation calculated in one layer, and wind driven current simulation calculated in 3 layers (0-2 m, 2-5 m, 5 m-).

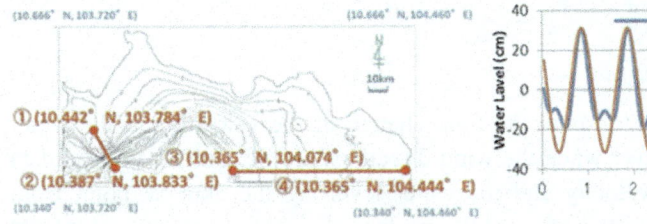 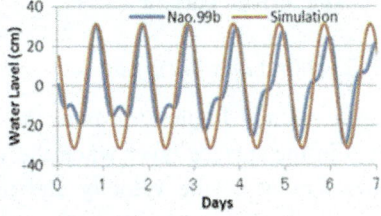

Fig. 9. Simulation Area and Boundary Condition   Fig. 10. Compare with NAO.99b and simulation (Tidal Water Level)

7.3. *Result of simulation*

7.3.1. *Tidal current*

Figure 10 shows the comparative the result of tidal model NAO.99b and simulation about time change of tidal water level. The result of tidal model changes under the influence of the diurnal tide and the semidiurnal tide. According to the result of the tidal model, when the two phases get consistent, the amplitude become the maximum. And, according to the simulation, the amplitude always become the maximum due to the simple amplitude. However, the maximum amplitude and both phase get consistent. Furthermore, this study confirmed that NAO.99b could calculate accurately the tidal water change.

According to these results, this study considers that the result of the numerical simulation can calculate the maximum tidal water change.

Figure 11 and Figure 12 show the profiles of the tidal current when the point where is 5km W from Kampot was ebb (Figure 11) and flood (Figure 12). Figure 11 shows that the current direction becomes S when the tidal ebb, and becomes NW when the tidal flood. The velocity becomes 30 cm/s at the deep area, and 5 cm/s at shallower than 5m area. The scale of tidal current depends on the depth, therefore the velocity becomes large at the deep area.

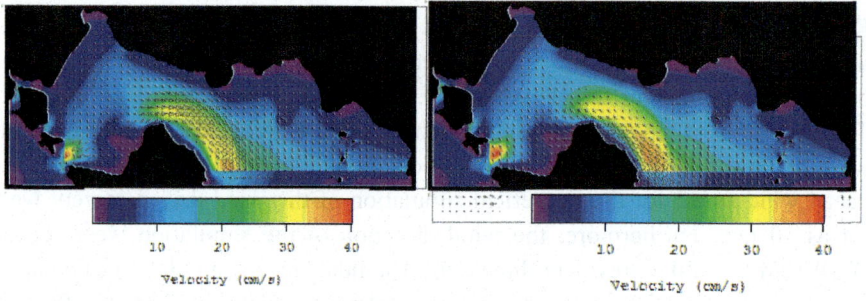

Fig. 11. Velocity and Direction (Ebb)   Fig. 12. Velocity and Direction (Flood)

(1) Wind driven current

7.3.2. *Wind driven current*

Figure 13 and Figure 15 show the time change of the velocity and the water level in the surface layer when the wind directions are SE and WSW. The data extract point of the velocity and the water level are the 5km W point from Kampot in Figure 1. Both figure shows that the water velocity and level drastically changes when the wind start blowing, however the change become stable after a few hours.

Figure 14 and Figure 16 show the profile of current velocity when the wind direction are SE and WSW. These figures show that the maximum water velocity in the surface layer becomes about 40 cm/s, and this is faster than the tidal current.

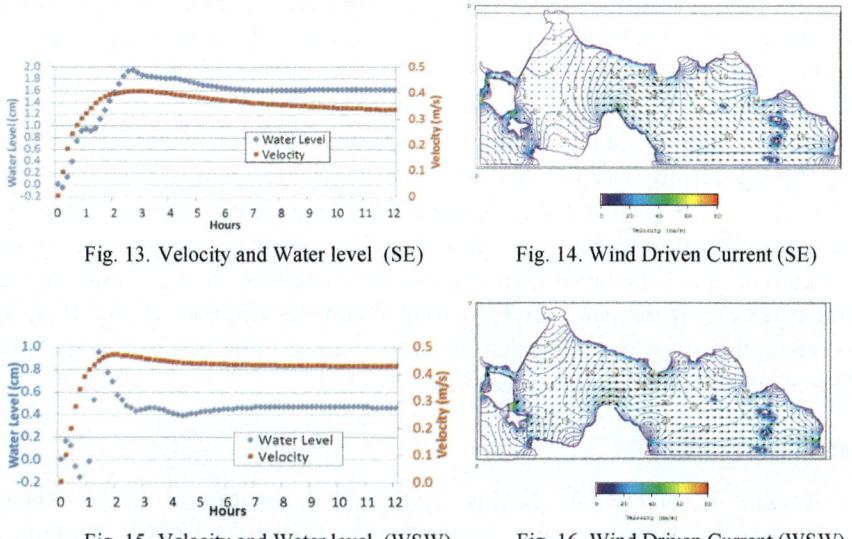

Fig. 13. Velocity and Water level (SE)

Fig. 14. Wind Driven Current (SE)

Fig. 15. Velocity and Water level (WSW)

Fig. 16. Wind Driven Current (WSW)

7.3.3. *Compare simulation and observation*

This study observed the wind velocity and the current velocity at 11 am 8th August, 2010. The location was the offshore from KEP where is shown Figure 1. The depth of the field was 4m and wind velocity was 3.5 m/s and the direction was WSW. The current velocity was observed by the propeller anemometer. The velocity which was shallower than about 40cm was 7cm/s, and the current direction was WSW. The current direction was almost same direction with the wind. And the current velocity which was deeper than about 40cm was 10 cm/s and the current direction was S. The current direction was almost same with the tidal current. Therefore, this study considered that the current formed the 2 layer, and the surface layer affected by the wind driven current and the deeper layer affected by the tidal current. Furthermore, according results are almost same with the simulation of tidal current ant wind driven current, therefore, this study considers that the numerical simulation can calculate the tidal current and the wind driven current in the field.

8. Conclusion

This study confirmed that the tidal current prevents in this area by the observation and calculation. This study comprehend the feature of the ocean wind over one year by the observation and the objective analysis.

According to the results, the WSW wind prevalent in the rainy season and the wind velocity becomes maximum.

This study simulated the tidal current and the wind driven current in Cambodian coastal area, and comprehend the current profiles, the maximum velocity and etc. The topography data which was used for the simulation was made by the observation and the satellite data. The diurnal tide prevents in this area, and the difference of tidal water level becomes about 1m. The maximum velocity of the tidal current in the shallower than 5m area becomes about 5 cm/s, however the surface velocity if the wind driven current sometimes becomes about 40 cm/s. Therefore, this study considers that the current around this area was sometimes affected by the seasonal wind driven current.

References

1. Amante, C. and B. W. Eakins : ETOPO1 1 Arc-Minute Global Relief Model: Procedures, Data Sources and Analysis., NOAA Technical Memorandum NESDIS NGDC-24, 19 pp, 2009.
2. Matsumoto, K., T. Takanezawa, and M. Ooe : Ocean Tide Models Developed by Assimilating TOPEX/POSEIDON Altimeter Data into Hydrodynamical Model, A Global Model and a Regional Model Around Japan, Journal of Oceanography, Vol. 56, pp. 567-581, 2000.
3. Japan Meteorological Business Support Center : GSM Objective Analysis Since January to December, 2010.
4. Inukai N., N. Hayakawa, Y. Fukushima and T. Hosoyamada : Numerical Simulation of Drift Current by Tracing the Model of Surface Particles, Proceedings of Coastal Engineering, Japan Society of Civil Engineers, Vol. 44, pp. 1046-1050, 1997.
5. Inukai N., N. Hayakawa and Y. Fukushima : Estimation in Predict Method of Sea Surface Wind and Wind-Driven Flow by Weather Chart, Proceedings of Coastal Engineering, Japan Society of Civil Engineers, Vol. 49, pp. 316-320, 2002.
6. Garrat, J.R. : Review of drag coefficients over ocean and continents, Monthly Weather Review, Vol. 7, pp. 915-929, 1997.

Estimation of Potential Tidal Energy Along the West Coast of India[*]

Vikas Mendi[†]

Research Scholar, Department of Applied Mechanics and Hydraulics, National Institute of Technology Karnataka, Surathkal, Karnataka, 575025, India and Assistant Professor, R V College of Engineering, Bengaluru, Karnataka, 560059, India
[†]*E-mail: kbl9dad@gmail.com*
www.rvce.edu.in

Jaya Kumar Seelam

Principal Scientist, National Institute of Oceanography, CSIR-NIO, Goa, India, 403004
E-mail: jay@nio.org

Subba Rao

Professor, Department of Applied Mechanics and Hydraulics, National Institute of Technology Karnataka, Surathkal, Karnataka, 575025, India
E-mail: sura@nitk.ac.in

Tidal energy is one of the clean and non-depleting renewable energy sources. In contrast to other clean sources, such as wind, solar, geothermal etc., tidal energy can be predicted for years ahead. Also, the medium, seawater, is more than 800 times denser than air and the astronomic nature of the underlying driving mechanism results in an essentially predictable resource, although subject to weather-related fluctuations. These features make it an important energy source for global power production in the near future. There are various types of tidal power plants across the world with varying tidal elevation. Also the method of conversion of the tidal energy into electrical energy is site specific. For example, we can adopt conventional method to extract energy in high tidal regions. But when it comes to low tidal regions like the southern India where tidal elevation measurements does not exceed 2.5m, there is a need of low flow turbines which can extract higher energy from lower head. An important factor that is responsible for the velocity of tidal stream is the tidal inlet dimensions. In this paper, an attempt has been made to identify the feasible locations for extraction of potential tidal energy along the Indian Coast.

Keywords: Tidal energy; Extraction; Renewable energy.

---

[*] This work is supported by NITK, Surathkal and R V College of Engineering, Bengaluru.
[†] Corresponding author.

1. Introduction

It's a well-known fact that the density of sa water is 800 times that of the density of air. This helps the driving mechanism of turbine blades more efficiently resulting in an effectively predictable resource, subject to changes in weather conditions[1]. The tides can be forecasted well in advance and hence can be considered as a major contributor in assessing tidal energy potential of any location[2]. While using tides for energy extraction, it is also important to understand its impacts on the local hydrodynamics[1]. Most of the existing technology used for tidal energy conversion is from the wind power industry[9]. The tides that are generated along some parts of the Indian coastline have the potential to extract energy from the turbines. The tidal elevation in India is as high as 8.5m at Bhavnagar, Gujarat and as low as 0.5m at the Southern part of India. Survey of India forecasts tidal elevations at some locations along the coastline of India.

Tidal prism is another important parameter in predicting potential energy from the tides. It can be defined as the amount of water necessary to fill up the tidal basin between high tide level and low tide level and is quantified by multiplying the tidal range with the basin area and removing the volume of sandy shoals[3]. Most of the tidal power stations existing today generate energy by the stored head of water, which is mostly done by constructing a barrage across the stream. During the high tide, the water flows into the basin and that water is stored in order to release during the low tide while the turbines are activated for power generation.

The quantity of water that flows into the tidal basin is also influenced by the width of the inlet (referred as throat width in this paper). The throat width of the tidal inlet can be directly responsible for the energy extraction at that inlet. The relation between the inlet width, cross-sectional area and tidal prism is mentioned by de Bok et al.[5]. Tidal period being 12 hours and 25 minutes for semi diurnal tides, the tidal basin has to be filled up during the high tide and emptied during the low tide for efficient energy extraction. Also, narrow throat width accelerates the flow of water thereby producing a venturi effect which helps in faster rotation of the rotor. However, if the throat width is too small, the installation of turbines cannot be done and the tidal basin is not fully filled during high tide condition.

In India, tidal energy estimation studies are mostly carried out for the Gulf of Kachchh, Gulf of Khambhat and the Sundarbans. The bay areas of the Gulf of Kachchh, Gulf of Khambhat and in the Sundarbans are approximately 4000km^2, 2350km^2 and 600km^2 respectively and the respective throat widths are approximately 40km, 45km and 18km. The energy extraction at all these places is different. Similar locations are identified along the Indian coast and the feasibility of energy extraction at those places is discussed in this paper.

2. Study Area

This study is carried out for the coastal states along the west coast of India. The western coastal plain of India consists of a narrow strip of land (i.e. width varies from 50-80 km), extended from Gujarat (GJ) in North to Kerala (KL) in the South. Rivers originating from the Western Ghats are mostly perennial and fast flowing. West coast of India has approximately 3200 km of coastline excluding southern most part on Tamil Nadu. A small portion of Tamil Nadu also belongs to the west coast at the southern part of India which is not considered in this study.

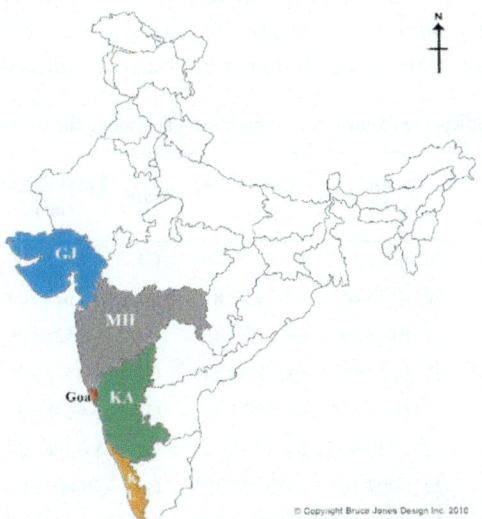

Figure 1: Study area showing maritime states of west India

3. Methodology

Table 1 in the appendix shows the tidal inlets identified in 5 maritime states along the west coast of India using Google Earth®. Lagoon area or basin area for the identified basins is calculated from high-resolution data obtained from Landsat 8 OLI (Operational Land Imager) and TIRS (Thermal Infrared Sensor) from U.S. Geological Survey Department. This data consists of various high definition images of different bands and these are overlaid and the final image is obtained using ArcMap® 10.1. From the composed image, the lagoon or bay area is calculated by creating new shape files in ArcMap® and further verifying the same using area calculation tool from Draftlogic® software.

The tidal elevation measurements were obtained by simulating the tidal variation along the coast using MIKE21 flow model (FM). The boundaries for the MIKE21 FM tidal model are taken from the global tidal constituents available through the MIKE21 tidal prediction toolbox. Some of the tidal elevation values are given in Sanil Kumar et al.[4]. The results obtained are cross-checked with the tidal stream flow in hydrographic charts, published by the National Hydrographic Office, Dehra Dun, India.

The basin area, tidal elevation and tidal prism recalculated for these basins. The basin area, tidal prism and throat widths at all identified tidal inlets along the west coast are tabulated in Table 1. It may be noted that the calculation of basin area has been restricted approximately to 15 km for some of the basins whose extent is very large (considering the extent of tidal effect) and for those basins which are less than 10 km, full area has been considered.

Table 1. Tidal inlets are identified in 5 maritime states along the west coast of India.

No	Name	Latitude (°N)	Longitude (°E)	State	Tidal Prism (m³)	Basin Area (km²)	Throat Width (m)
1	Gandhiya	22°22'24.74"	69°31'00.26"	GJ	1.865E+05	0.09	18
2	Kalawad	22°20'15.65"	69°29'22.81"	GJ	3.167E+04	0.02	68
3	Panero	22°19'58.58"	69°27'27.42"	GJ	1.252E+06	0.54	-
4	Mithapur	22°24'12.82"	68°58'39.24"	GJ	1.457E+05	0.07	27
5	Makanpur	22°21'11.88"	68°57'34.30"	GJ	5.964E+05	0.25	116.5
6	Dwaraka	22°15'20.92"	68°57'33.43"	GJ	8.939E+04	0.04	23
7	Hari kund	22°14'07.26"	68°57'55.39"	GJ	9.680E+05	0.38	43
8	Goji	21°58'58.62"	69°11'44.70"	GJ	3.640E+05	0.14	35
9	Navadra	21°56'02.00"	69°14'42.80"	GJ	2.510E+05	0.1	30
10	Miyani	21°49'52.61"	69°22'14.90"	GJ	5.552E+07	22.88	142
11	Porbandar	21°38'20.22"	69°35'28.17"	GJ	8.114E+05	0.37	104
12	Tukada	21°32'12.73"	69°42'21.71"	GJ	5.057E+06	2.36	-
13	Chikasa	21°27'37.87"	69°46'35.66"	GJ	2.091E+06	1.01	118
14	Shil	21°10'38.96"	70°01'47.76"	GJ	1.119E+06	0.57	330
15	Ravon	20°54'58.10"	70°20'45.12"	GJ	2.437E+05	0.14	-
16	Prabhas	20°52'51.13"	70°24'39.02"	GJ	1.234E+05	0.07	1865
17	MulDwarka	20°45'42.59"	70°40'22.49"	GJ	2.109E+05	0.16	2705

18	Sarkhadi	20°42'32.80"	70°52'16.25"	GJ	1.112E+07	5.13	18
19	Diu	20°43'11.68"	70°59'13.56"	GJ	1.514E+07	6.55	1000
20	Bandar	20°44'35.99"	71°04'33.79"	GJ	2.714E+05	0.12	-
21	Rajput	20°45'22.00"	71°05'07.98"	GJ	2.086E+05	0.09	355
22	Rajpara	20°47'52.91"	71°12'14.39"	GJ	1.138E+06	0.43	113
23	Jafrabad	20°52'03.14"	71°22'18.34"	GJ	7.354E+06	2.4	234
24	Chanch	20°57'08.10"	71°32'18.40"	GJ	7.672E+06	2.24	110
25	Tena river	21°13'40.12"	72°36'29.15"	GJ	8.658E+06	1.36	22
26	Mindhola river	21°03'59.65"	72°41'24.75"	GJ	1.469E+07	2.38	1150
27	Puma river	20°54'32.08"	72°46'42.98"	GJ	6.264E+07	10.52	72
28	Kavai creek	20°48'16.88"	72°50'02.88"	GJ	4.551E+06	0.75	215
29	Ambika river	20°45'09.94"	72°51'06.84"	GJ	2.674E+07	4.91	283
30	Auranga river	20°37'59.02"	72°53'19.71"	GJ	5.862E+06	1.02	144
31	Umarsadi	20°31'57.29"	72°53'19.62"	GJ	5.767E+06	1.03	167
32	Kolak river	20°28'01.27"	72°51'27.38"	GJ	1.487E+06	0.27	91
33	Daman	20°24'41.47"	72°49'50.96"	GJ	6.231E+06	1.15	143
34	Bhathaiya	20°23'06.00"	72°49'35.44"	GJ	2.982E+05	0.06	17
35	Kalu river	20°22'05.88"	72°49'16.63"	GJ	2.052E+06	0.38	133
36	Kalai	20°21'17.96"	72°48'41.14"	GJ	7.366E+04	0.02	33
37	Kalgam 1	20°19'27.77"	72°46'31.09"	GJ	9.649E+04	0.02	20
38	Kalgam 2	20°20'02.08"	72°46'36.88"	GJ	4.211E+04	0.01	13
39	Maroli	20°18'08.68"	72°46'23.57"	GJ	6.293E+04	0.02	47
40	Varoli river	20°15'40.21"	72°45'08.94"	GJ	9.112E+04	0.02	36
41	Nargol	20°12'09.18"	72°44'51.00"	GJ	5.767E+06	1.3	121
42	Deheri	20°09'54.97"	72°44'31.92"	GJ	1.082E+05	0.03	20
43	Zai	20°07'42.89"	72°44'18.81"	MH	2.099E+05	0.05	90
44	Gholvad	20°04'55.60"	72°43'34.53"	MH	2.661E+04	0.01	18
45	Dahanu	19°58'09.91"	72°43'33.52"	MH	9.796E+06	1.98	383
46	Navapur	19°47'32.14"	72°41'47.30"	MH	3.167E+06	0.66	80
47	Dudh river	19°43'47.60"	72°43'07.00"	MH	7.919E+06	1.67	500
48	Darapada	19°37'13.80"	72°43'44.81"	MH	2.632E+05	0.06	66
49	Kelwa	19°35'53.92"	72°45'35.45"	MH	2.988E+06	0.64	256
50	Bhadve	19°29'35.95"	72°47'40.71"	MH	1.516E+08	32.6	1921
51	Rajodi	19°25'01.74"	72°45'41.27"	MH	3.307E+04	0.01	19
52	Bhuigaon Kh.	19°23'34.91"	72°46'03.57"	MH	1.080E+06	0.24	92
53	Vasai	19°19'01.34"	72°48'12.01"	MH	2.029E+08	43.91	2291

54	Malad	19°11'52.98"	72°48'01.66"	MH	2.134E+07	4.66	304
55	Versova	19°08'40.27"	72°50'02.88"	MH	5.876E+06	1.29	130
56	Revas	18°53'43.55"	72°53'00.05"	MH	1.059E+09	249.34	7670
57	Mandve	18°50'09.28"	72°57'33.40"	MH	1.298E+08	30.58	1684
58	Awas	18°46'10.60"	72°51'52.17"	MH	2.648E+04	0.01	63
59	Surekhar	18°45'27.00"	72°52'01.58"	MH	3.430E+05	0.1	33
60	Navgaon	18°42'04.28"	72°51'48.01"	MH	1.912E+05	0.05	44
61	Navapada	18°40'07.18"	72°52'00.98"	MH	4.018E+05	0.1	110
62	Akshi	18°38'18.24"	72°53'19.62"	MH	3.195E+06	0.8	137
63	Bagmala	18°34'42.10"	72°55'29.89"	MH	2.640E+05	0.08	94
64	Revdanda	18°32'18.35"	72°55'58.00"	MH	4.413E+07	11.97	549
65	Wandeli	18°30'12.20"	72°54'37.73"	MH	3.742E+05	0.1	124
66	Surulpeth	18°22'23.88"	72°55'46.98"	MH	4.809E+05	0.13	97
67	Eakdara	18°19'12.47"	72°58'23.06"	MH	1.040E+06	0.29	103
68	Rajapuri	18°16'53.15"	72°58'48.79"	MH	2.052E+08	56.9	1637
69	Velas Agar	18°11'44.52"	72°59'18.88"	MH	6.784E+05	0.2	189
70	Diveagar	18°09'32.72"	73°00'00.61"	MH	2.087E+05	0.06	30
71	Walavatikh	18°04'20.10"	73°01'00.09"	MH	4.647E+05	0.14	43
72	Shrivardhan	18°01'34.86"	73°02'31.35"	MH	4.684E+06	1.4	134
73	Bankot	17°58'59.52"	73°02'59.86"	MH	6.483E+07	19.78	701
74	Sakhari	17°55'54.52"	73°04'08.17"	MH	2.973E+06	0.93	163
75	Koliwada	17°53'25.37"	73°05'27.49"	MH	5.167E+05	0.17	83
76	Jaikar	17°50'30.62"	73°10'06.76"	MH	2.656E+06	0.85	108
77	Dabhol	17°34'47.14"	73°10'36.96"	MH	5.272E+07	18.37	664
78	Mouje	17°30'54.72"	73°11'51.31"	MH	3.074E+05	0.11	27
79	Palshet	17°26'15.97"	73°13'15.74"	MH	9.637E+04	0.04	38
80	Muslondi	17°21'16.13"	73°13'37.22"	MH	1.009E+05	0.04	12
81	Jaigad	17°17'32.78"	73°14'24.96"	MH	4.891E+07	18.33	931
82	Bhandrawada	17°11'58.02"	73°15'42.85"	MH	7.798E+05	0.3	32
83	Bhandarpule	17°09'30.49"	73°16'29.11"	MH	2.951E+05	0.12	63
84	Kajitbhati	17°06'00.40"	73°17'17.98"	MH	4.436E+05	0.18	67
85	Sadye	17°04'40.84"	73°17'33.98"	MH	8.289E+05	0.33	47
86	Sadamirya	17°02'17.30"	73°17'03.86"	MH	6.065E+06	2.38	314
87	Karle	16°59'09.17"	73°18'02.16"	MH	7.374E+06	2.95	215
88	Bhatiwadi	16°53'27.60"	73°18'59.66"	MH	6.606E+06	2.69	79
89	Khsheli	16°48'23.26"	73°19'38.43"	MH	1.128E+07	4.66	219

90	Wadativare	16°41'13.74"	73°21'26.23"	MH	7.087E+05	0.3	105
91	Vijayadurg	16°34'14.38"	73°19'47.99"	MH	4.371E+07	19	1639
92	Phanase	16°26'36.31"	73°22'21.04"	MH	7.575E+05	0.34	36
93	Padavne	16°23'34.26"	73°23'10.95"	MH	1.772E+07	7.93	730
94	Taramumbari	16°21'36.50"	73°24'40.81"	MH	2.847E+06	1.29	100
95	Morve	16°16'28.34"	73°26'03.17"	MH	2.037E+06	0.94	53
96	Pirawadi	16°12'04.03"	73°27'46.55"	MH	3.372E+06	1.57	117
97	Sariekot	16°05'16.91"	73°27'56.98"	MH	1.481E+07	6.94	183
98	Malvan	16°04'15.67"	73°29'26.36"	MH	5.286E+05	0.25	33
99	Kalethar	16°00'36.72"	73°30'13.56"	MH	2.468E+04	0.02	10
100	Bhogwa	15°58'01.20"	73°32'02.41"	MH	1.159E+07	5.51	392
101	Shriramwadi	15°56'17.45"	73°34'03.09"	MH	2.948E+05	0.15	27
102	Khalchiwadi	15°55'08.80"	73°35'11.19"	MH	9.719E+04	0.05	-
103	Kelus 1	15°54'37.76"	73°35'48.19"	MH	3.714E+05	0.18	40
104	Kelus 2	15°53'42.32"	73°36'40.63"	MH	7.977E+04	0.04	-
105	Dabholi	15°51'56.74"	73°39'12.07"	MH	2.333E+05	0.12	27
106	Tank	15°47'37.82"	73°39'49.45"	MH	1.314E+06	0.64	77
107	Khalchikar	15°45'12.49"	73°41'23.10"	MH	1.357E+06	0.66	120
108	Tiracol	15°43'16.82"	73°44'10.86"	Goa	8.801E+06	4.25	270
109	Siolim	15°36'38.81"	73°44'53.07"	Goa	1.874E+07	9.08	550
110	Baga	15°33'46.08"	73°47'24.54"	Goa	1.927E+05	0.1	25
111	Mandovi river	15°29'08.81"	73°48'56.41"	Goa	1.180E+08	57.65	3235
112	Zuari estuary	15°25'36.88"	73°57'03.30"	Goa	6.692E+07	32.7	4229
113	Mobor	15°08'30.41"	73°58'47.73"	Goa	3.897E+06	1.91	166
114	Cola	15°03'15.66"	74°00'56.30"	Goa	4.361E+05	0.22	38
115	Anjadip	15°00'48.28"	74°02'07.85"	Goa	8.339E+04	0.05	13
116	Canacona 2	14°59'36.53"	74°02'19.25"	Goa	1.992E+06	0.98	8
117	Canacona 1	14°59'02.65"	74°02'59.10"	Goa	1.545E+05	0.08	46
118	Mashem	14°57'27.72"	74°07'19.97"	Goa	6.093E+05	0.3	60
119	Karwar	14°50'29.72"	74°09'31.59"	KA	2.507E+07	12.38	680
120	Chendia	14°45'08.89"	74°13'41.79"	KA	1.153E+06	0.57	30
121	Todur	14°44'41.14"	74°15'51.25"	KA	2.348E+05	0.12	75
122	Belekeri	14°42'42.52"	74°16'45.48"	KA	1.358E+06	0.67	59
123	Ankola	14°39'41.98"	74°17'02.63"	KA	4.444E+05	0.22	24
124	Belambar	14°38'49.38"	74°17'32.99"	KA	5.553E+05	0.28	7
125	Manjaguni	14°36'00.86"	74°21'28.42"	KA	8.920E+06	4.42	75

126	Tadadi Port	14°31'11.78"	74°23'34.51"	KA	3.162E+07	15.7	595
127	Alvekodi 2	14°25'03.94"	74°25'27.75"	KA	6.802E+05	0.34	33
128	Karki	14°17'56.65"	74°28'11.32"	KA	2.636E+07	13.31	155
129	Nakhuda	14°11'20.62"	74°30'05.06"	KA	7.838E+04	0.04	6
130	Alvekodi 1	14°01'36.62"	74°31'04.58"	KA	2.975E+06	1.55	60
131	Jali	13°59'05.53"	74°32'05.18"	KA	9.137E+04	0.05	15
132	Mavakurve	13°58'01.81"	74°33'59.88"	KA	5.020E+05	0.27	159
133	Hadin	13°56'59.86"	74°35'08.18"	KA	6.787E+04	0.04	55
134	Alivey Gadde	13°55'20.78"	74°36'25.66"	KA	5.814E+05	0.31	17
135	Paduvari	13°52'05.52"	74°37'27.52"	KA	2.225E+06	1.2	100
136	Koderi	13°47'38.69"	74°40'11.33"	KA	1.285E+06	0.7	85
137	Gangoli	13°38'02.83"	74°41'43.50"	KA	3.707E+07	21	203
138	Kodithale	13°27'00.94"	74°44'11.12"	KA	2.581E+07	15.4	183
139	Malpe	13°20'50.24"	74°41'44.20"	KA	8.755E+06	5.4	98
140	Kaup	13°13'26.04"	74°46'01.08"	KA	3.799E+04	0.03	8
141	Nadsal	13°06'42.77"	74°46'36.57"	KA	4.352E+05	0.28	12
142	Hejamadi	13°04'31.04"	74°48'22.09"	KA	5.906E+06	3.82	178
143	NMPT	12°55'37.45"	74°49'40.85"	KA	2.333E+06	1.54	505
144	Bunder	12°50'43.55"	74°51'49.84"	KA	2.815E+07	18.82	332
145	Kanwatheertha	12°45'38.70"	74°53'15.69"	KA	2.950E+05	0.2	35
146	Hosabettu	12°42'28.76"	74°55'13.85"	KL	7.261E+05	0.5	31
147	Bandiyod	12°37'55.99"	74°55'54.92"	KL	1.007E+05	0.07	-
148	Shiriya	12°36'25.02"	74°57'30.22"	KL	2.306E+06	1.6	86
149	Puthur	12°32'28.03"	74°59'15.77"	KL	8.111E+05	0.57	36
150	Thalangara	12°28'30.68"	75°00'16.70"	KL	6.595E+06	4.61	81
151	Chembirika	12°26'26.99"	75°00'48.79"	KL	2.539E+05	0.18	22
152	Thekkekara	12°25'15.06"	75°01'40.95"	KL	4.264E+04	0.03	-
153	Tharavadu	12°23'56.69"	75°03'41.29"	KL	5.146E+05	0.37	-
154	Kadapuram	12°20'39.37"	75°07'12.98"	KL	9.750E+05	0.7	28
155	Kaithakkad	12°11'59.86"	75°13'37.10"	KL	2.785E+07	19.97	260
156	Madayi	12°01'18.30"	75°17'46.82"	KL	4.322E+06	3.18	86
157	Azhikkal	11°56'37.00"	75°24'16.63"	KL	3.358E+07	24.76	366
158	Thottada	11°50'17.81"	75°25'56.01"	KL	5.459E+04	0.05	23
159	Nadal	11°48'30.71"	75°27'18.12"	KL	3.174E+04	0.03	25
160	Dharmadom	11°46'42.28"	75°28'17.94"	KL	1.840E+06	1.39	118
161	Koduvalli	11°45'55.98"	75°31'53.97"	KL	1.398E+06	1.06	383

162	Mahe	11°42'13.75"	75°32'36.54"	KL	1.549E+06	1.18	110
163	Olavilam	11°40'46.16"	75°32'51.21"	KL	2.818E+03	0.01	14
164	Chombala	11°40'14.70"	75°33'31.73"	KL	6.880E+03	0.01	-
165	Madappallly	11°38'49.13"	75°35'22.90"	KL	7.939E+03	0.01	-
166	Iringal	11°34'06.64"	75°37'50.48"	KL	4.468E+06	3.43	146
167	Nandi	11°28'06.82"	75°42'07.03"	KL	1.111E+04	0.01	24
168	Puthiyapurayil	11°25'02.50"	75°44'08.04"	KL	4.106E+03	0.01	8
169	Elathur	11°20'46.21"	75°46'46.46"	KL	9.295E+06	7.33	82
170	Thekepuram	11°13'39.50"	75°48'13.03"	KL	4.530E+05	0.37	138
171	Beypore	11°09'42.91"	75°49'32.80"	KL	1.146E+07	9.22	272
172	Kudalundi	11°07'27.70"	75°51'26.16"	KL	1.746E+06	1.42	222
173	Chiramangalam	11°01'11.50"	75°54'42.27"	KL	8.153E+05	0.68	82
174	Malappuram	10°47'13.70"	75°56'10.66"	KL	1.590E+07	13.38	229
175	Veliancode	10°43'56.06"	75°58'48.42"	KL	1.831E+06	1.56	81
176	Nallamkallu 2	10°37'57.94"	75°58'56.33"	KL	2.449E+03	0.01	14
177	Nallamkallu 1	10°37'39.72"	76°02'18.55"	KL	3.528E+03	0.01	26
178	Chettuva	10°30'30.17"	76°04'56.79"	KL	9.681E+06	8.36	330
179	Nattika	10°25'09.77"	76°05'54.13"	KL	5.644E+03	0.01	-
180	Muriyamthodu	10°22'45.08"	76°06'34.72"	KL	2.000E+04	0.02	15
181	Palapetty	10°20'59.10"	76°06'44.76"	KL	1.182E+04	0.02	15
182	Chamakkala	10°20'34.04"	76°07'24.18"	KL	1.493E+04	0.02	19
183	Vazhiyambalam	10°18'41.40"	76°07'39.62"	KL	1.231E+04	0.02	10
184	Perinjanam	10°17'54.49"	76°08'12.04"	KL	6.405E+03	0.01	25
185	Ambalanada	10°15'47.63"	76°09'51.08"	KL	2.603E+03	0.01	15
186	Munambam	10°10'41.70"	76°14'03.21"	KL	1.212E+07	10.76	196
187	Kochi	9°58'10.93"	76°17'04.09"	KL	2.195E+08	198.26	479
188	Andhakaranazhy	9°44'57.35"	76°17'31.71"	KL	9.316E+05	0.89	44
189	Arthunkal	9°39'45.00"	76°17'35.82"	KL	1.381E+04	0.02	-
190	Kackary	9°39'04.59"	76°17'46.46"	KL	4.327E+04	0.05	-
191	Perunneermangal am	9°37'23.30"	76°17'55.12"	KL	5.272E+04	0.06	85
192	Janakshemam	9°36'06.25"	76°18'01.46"	KL	6.993E+03	0.01	-
193	Valavanadu 2	9°35'18.41"	76°18'04.11"	KL	1.112E+04	0.02	-
194	Valavanadu 1	9°34'58.61"	76°18'09.90"	KL	2.012E+04	0.02	-
195	Pollethai	9°34'20.14"	76°18'19.63"	KL	1.921E+04	0.02	14
196	Kattoor	9°33'25.21"	76°18'28.43"	KL	2.582E+04	0.03	10

197	Omanapuzha	9°32'37.53"	76°18'45.79"	KL	2.219E+04	0.03	-
198	Poomkavu	9°31'04.68"	76°18'52.78"	KL	1.880E+04	0.02	-
199	Padinjare	9°30'25.27"	76°19'26.59"	KL	9.479E+03	0.01	13
200	Eravukadu	9°27'51.94"	76°19'34.26"	KL	2.288E+04	0.03	-
201	Punnapra 2	9°27'25.22"	76°19'55.53"	KL	2.190E+04	0.03	-
202	Punnapra 1	9°26'06.78"	76°22'59.14"	KL	1.915E+04	0.02	13
203	Thottapally lake	9°18'42.42"	76°27'46.41"	KL	1.349E+06	1.36	38
204	Azheekkal	9°08'12.89"	76°31'10.27"	KL	1.078E+07	11.12	200
205	Kandathil	9°01'04.13"	76°32'18.94"	KL	3.573E+06	3.76	-
206	Neendakara	8°56'03.60"	76°38'55.56"	KL	4.839E+07	51.4	225
207	Paravur lake	8°48'44.33"	76°40'33.49"	KL	6.844E+06	7.43	41
208	Kappil	8°46'38.49"	76°47'11.70"	KL	3.992E+06	4.35	19
209	Madanvila	8°38'00.09"	76°53'10.64"	KL	3.409E+06	3.75	131
210	Veli	8°30'30.05"	76°57'29.16"	KL	4.825E+05	0.54	-
211	Pachalloor	8°25'29.72"	77°01'43.04"	KL	1.895E+05	0.22	15
212	Kochupally	8°20'45.56"	77°04'45.62"	KL	2.337E+04	0.03	23
213	Paruthiyoor	8°18'20.74"	77°09'59.76"	KL	3.782E+05	0.44	26

3.1 Calculation of potential tidal energy

The potential tidal energy is the energy due to release of the stored water in the basin. It is also a fact that the increase in tidal variation results in increase of energy extraction to a large extent[10]. The potential energy mainly depends on the tidal prism of the basin. Potential energy obtained due to the stored water can be calculated as shown Eq. (1)[11].

$$E = \tfrac{1}{2} A \rho g h^2 \qquad (1)$$

where, h is the spring tidal range, is the area of the tidal basin, is the density of water = 1025 kg/m³, and g is the gravitational force = 9.81 m/s².

From the above equation, it can be seen that the potential energy varies with square of tidal range. So, a barrage should be placed in such a location where it is possible to achieve maximum storage head. The inlets that are between islands having large basin area are considered to have a greater amplification effect because of the reduction in the throat area and the water depth relative to the surroundings, producing a venturi effect. This accelerates the water as it is forced through a channel with a smaller cross-sectional area[12].

Sample calculation of potential tidal energy at Kavai creek, Gujarat 20°48'16.88" 72°50'02.88"

The tidal range of tide at Kavai creek, Gujarat = 6.11 m

The surface of the tidal energy harnessing plant = 744537.83 m^2

Density of sea water = 1025.18 kg/m^3

Mass of the sea water = volume of sea water × density of sea water =
(area × tidal range) of water × mass density
= (744537.83 m^2 × 6.11 m) × 1025.18 kg/m^3 = 4.6 × 10^9 kg (approx.)

Potential energy content of the water in the basin at high tide = $E = \frac{1}{2}A\rho gh^2$
= ½ × 744537.83 m^2 × 1025 kg/m^3 × 9.81 m/s^2 × (6.11 m)^2 = 13.97 × 10^{10} J
(approx)

Now we have 2 high tides and 2 low tides every day. At low tide the potential energy is zero.

Therefore the total energy potential per day = Energy for a single high tide × 2
= 13.97 × 10^{10} J × 2= 27.94 × 10^{10} J

Therefore, the mean power generation potential
= Energy generation potential/time in 1 day = 27.94 × 10^{10} J/86400 s = 323.37
MW

4. Results and Discussion

A threshold of 15 km^2 basin area is considered for possible power extraction from basins along the west coast of India. The total number of inlets found suitable for energy extraction are 20 considering the basin area. From the tidal basins with less than 15 km^2 the extracted energy may not provide sufficient power output.

The tidal prism varies proportionally with the basin area and those common inlets which have large basin area and greater tidal prism values are favourable for energy extraction. Another important parameter considered in the selection of tidal basin with respect to tidal prism is tidal range. Obviously, if the tidal range and basin area are larger, the tidal prism is higher. It can be said that, the consideration of tidal inlets for energy extraction based on tidal prism is dependent on the tidal range and basin area. A threshold of 10×10^6 m^3 is considered to extract power. The total number of inlets found suitable for energy extraction are 38 considering the threshold tidal prism. By definition, tidal prism is the volume of water filled in the tidal basin. The volume of water filled in the tidal basin also depends on the width of the inlet. The width of the tidal inlet should be such that it allows enough water to fill the Basin during the high tide condition so that the required head is obtained in order to extract energy. A threshold of 200 m is considered for the throat width to extract power. The total number of inlets found suitable for energy extraction are 46 considering the throat width.

Table 2. State wise potential energy extraction based on basin area, tidal prism and throat width.

	Estimated potential energy considering Basin area greater than 15 km²		Estimated potential energy considering Tidal prism greater than 10⁷ m³		Estimated potential energy considering Throat width greater than 200 m	
	No. of inlets	Energy generated (MW)	No. of inlets	Energy generated (MW)	No. of inlets	Energy generated (MW)
Gujarat	1	646.1	6	2126.9	9	819.43
Maharashtra	9	22792	15	24199	19	24596.4
Goa	2	2151	03	2472.42	4	2472.42
Karnataka	4	1427.45	06	2026.05	5	2053.21
Kerala	4	3832.85	08	4417.75	9	4567.01
Total gross energy	**20**	**30849.40**	**38**	**35242.12**	**46**	**34508.47**

The above threshold values are set by considering the operating/proposed tidal power generating stations around the globe. Hence, it can be recommended that the selection of site for energy extraction can be a combination of basins having large area, greater tidal prism and wide throat width.

5. Conclusions

A total number of 213 locations are identified for potential tidal energy extraction, of which 20 are found favourable for energy extraction based on basin area, 38 are found favourable based on tidal prism and 46 are found suitable based on inlet throat width. Out of all the potential locations identified based on basin area, tidal prism and throat width, 18 number of inlets can be considered suitable for energy extraction considering the potential in all the three considerations mentioned above along the west coast. This is a preliminary study on the identification of locations for energy extraction. Many other parameters like depth of the channel, velocity of the stream, required head for energy extraction etc., are also to be considered in order to establish a tidal power plant. This will be considered in our further studies.

References

1. I.G. Brydena, T. Grinstedb, G.T. Melvillea, *Assessing the potential of a simple tidal channel to deliver useful energy*, Journal of Applied Ocean Research, Volume 26, Issue 5, July 2004, pp 198–204.
2. I. Bryden and G.T. Melville, *Choosing and evaluating sites for tidal current development*, Proceedings of the Institution of Mechanical Engineers - Part A, Journal of Power and Energy, December 2004.
3. N. Amaranatha Reddy, M. Vikas, Subba Rao, Jaya Kumar Seelam *Classification of Tidal inlets along the Central East Coast of India*, 8th International Conference on Asian and Pacific Coasts (APAC 2015), Volume 116, 2015, pp 922-931.
4. V. Sanil Kumar, G. Udhaba Dora, Sajive Philip, P. Pednekar, and Jai Singh, *Variations in Tidal Constituents along the Nearshore Waters of Karnataka*, West Coast of India, Journal of Coastal Research, Volume 27, Issue 5: pp 824–829, 2011.
5. Fraenkel P.L., *Power from marine currents,* Proc. Inst. Mech. Eng, Part A, J Power and Energy, 216(A1): pp 1–14, 2002.
6. Bahaj, A.S., Myers, L.E. and Thompson, G., *Characterising the wake of horizontal axis marine current turbines*, In: Proceedings of the Seventh European Wave and Tidal Energy Conference. Porto, Portugal, 11-14 September 2007.
7. Batten, W.M.J., Bahaj, A.S., Molland, A.F., Chaplin, J.R., (2007), *Experimentally validated numerical method for the hydrodynamic design of horizontal axis tidal turbines*, Ocean Eng. 34 (7), pp 1013–1020.
8. Shaikh Md. Rubayiat Tousif, Shaiyek Md. Buland Taslim, *Tidal Power: An Effective Method of Generating Power*, International Journal of Scientific & Engineering Research Volume 2, Issue 5, May-2011.
9. Gorlov, A.M., *Tidal energy*, North eastern University, Boston Massachusetts, USA pp 2955–2960, 2001.
10. Bryden, I.G. and Melville G.T., *Choosing and evaluating sites for tidal current development*, Proceedings of the Institution of Mechanical Engineers Part A Journal of Power and Energy 219(A-3) 235–247, 2004.
11. Frost C., C.E. Morris, A. Mason-Jones, D.M. O'Doherty, T. O'Doherty, *The effect of tidal flow directionality on tidal turbine performance characteristics*, Renewable Energy, Volume 78, pp 609–620, 2015.

Rip Current Observation
with a X-band Radar at Uchinada Beach, Japan

S. Takewaka[†] and M. Kojima

*Department of Engineering Mechanics and Energy, University of Tsukuba,
Tsukuba, Ibaraki 305-8573, Japan
[†]E-mail: takewaka@kz.tsukuba.ac.jp*

Rip current observation has been conducted with a X-band radar at a swimming beach in Japan for 40 days in a summer. The radar set is on the beach, approximately 300 m backwards from the water line, to observe the sea surface states over 1500 m in the longshore. Instantaneous radar images are processed every 2 minutes to yield time averaged images. From the time averaged images, features like distribution of waterlines, breaker zone, locations of rip channels, rip heads and currents can be interpreted. Rip channels and occurrence of rip currents are manually digitized from the time averaged images. In the first half of the observation period, rip currents seemed to occur randomly, and were not connected with rip channels. On the contrary, after an attack of high waves in the middle of the observation period, most of the rip currents were detected in the vicinities of rip channels. During this period, rip channel migrations were observed. Occurrences of rip currents have been detected in the whole period when the incident wave height exceeded 0.3 m.

Keywords: Rip current; X-band radar; Remote sensing.

1. Introduction

Rip currents are narrow and transient seaward flows extending over the breaker zone. They play important roles in offshore sediment transport, shoreline and bar system evolutions, and to keep a swimming beach safe, to know their behaviors are key factor. Rip current observation with instruments requires extensive efforts, since it is unsteady and sometimes migrates rapidly in the longshore. Remote sensing using X-band radar [1], providing both spatial and temporal data, has the potential to observe the entire rip current system and is already tested at Hasaki Beach facing Pacific Ocean [2] in this context. In this work, same method is applied at a swimming beach, Uchinada Beach facing Sea of Japan, to check detectability of rip current occurrence to help keep safety for swimmers. The observation was conducted continuously for 40 days in summer of 2015.

2. Site and X-band Radar Observation

The observation was conducted at Uchinada Beach in Kanazawa Prefecture of Japan, facing Sea of Japan (Figure 1). The beach is partially sheltered by the breakwaters of Port of Kanazawa. The radar used in the study has a small antenna which is enclosed in a radome of 60 cm diameter (Figure 2). Figure 3 shows the close-up of the site and longshore coordinate system used in this study. The area is a micro-tidal coast with tidal range of approximately 0.3 m.

Fig. 1. Observation site. ● Radar location (Lat/Lon = 36.642/136.624 deg.), ● NOWPHAS wave gauge (Depth = 20 m, Lat/Lon = 36.614/136.568 deg.).

Fig. 2. Radar (JRC, JMA-5104, diameter ~ 0.6 m).

358

X-band radar is an imaging device that provides instantaneous distributions of wave crests and shorelines. Ensembles of original radar images enables us to estimate the intertidal morphology and occurrence of rip currents which is already verified in the previous study [2].

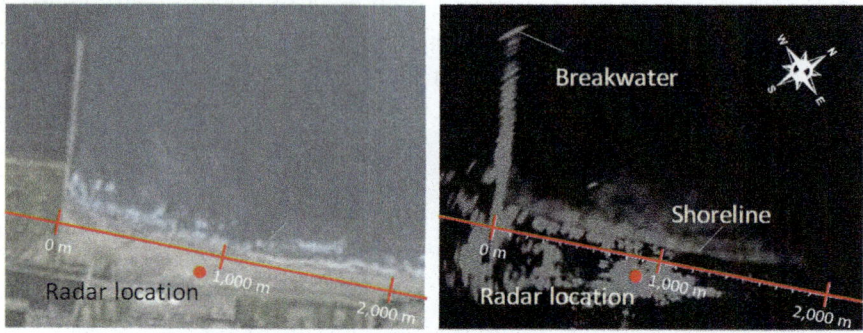

Fig. 3. (Left) Longshore coordinate and (Right) time averaged radar image.

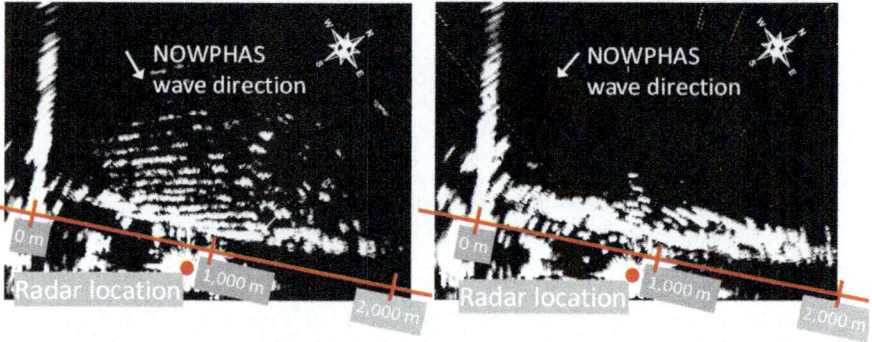

Fig. 4. Examples of radar images. (Left) Western wave incidence. (Right) Northern wave incidence. Wave crests which propagate to the shore are visible. Black dot in the center denotes radar location.

The observation started from end of August and continued until beginning of October in year 2015. Original radar images (see Figure 4) were collected every 2 seconds, and time averaged images were processed at every 2 minutes. An example of time average image is shown in Figure 5, which covers approximately 1500 m of the beach. From image interpretation of the time averaged images, we can identify the locations of shoreline, breaker zone, rip channels and rip heads, and locations of these items were digitized by manual inspection. The breaker zone appears as bright belt extending in the longshore, since surface turbulence yields higher backscatter of the emitted radar beam. On the contrary, above a rip channel, a relatively deeper portion compared to the surroundings, wave breaking occurs less, which resulting smaller backscatter, and this area appears as a dark strip in the breaker zone, or can be identified as dark gaps in the bright belt extending into the horizontal direction in the image.

During the observation period, bathymetric surveys and floater releases at the rip channels have been also done to validate the results of interpretation of radar images. Figure 6 are comparisons between survey results and time averaged radar images. Rip channels can be identified in the survey results as deep valleys close to the shoreline, and relevantly, rip channels can be found at similar locations in the time averaged radar images. Locations of the channels were not fixed during the observation, which will be discussed in the next.

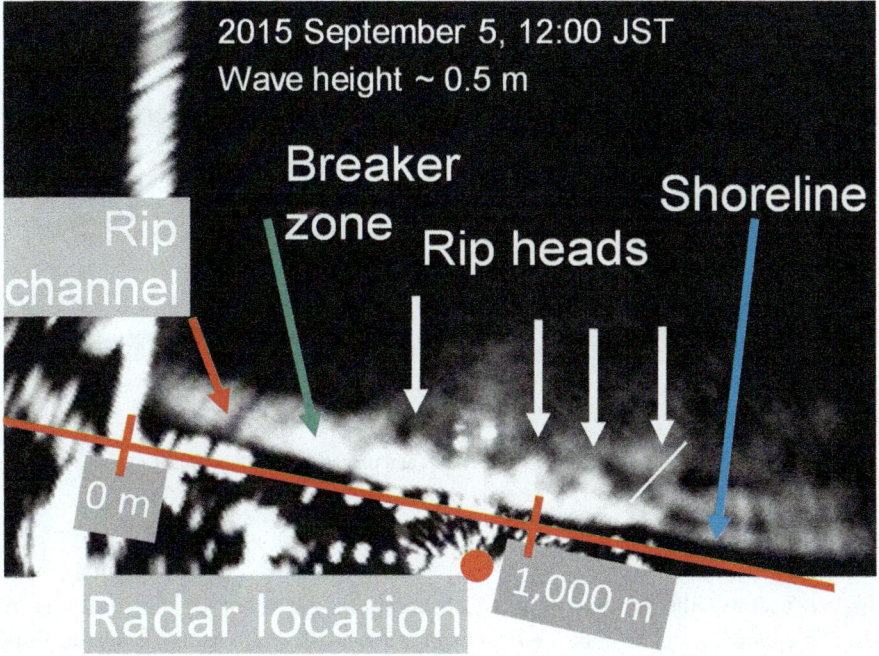

Fig. 5. Example of time averaged radar image. Several rip currents and heads are captured in the image.

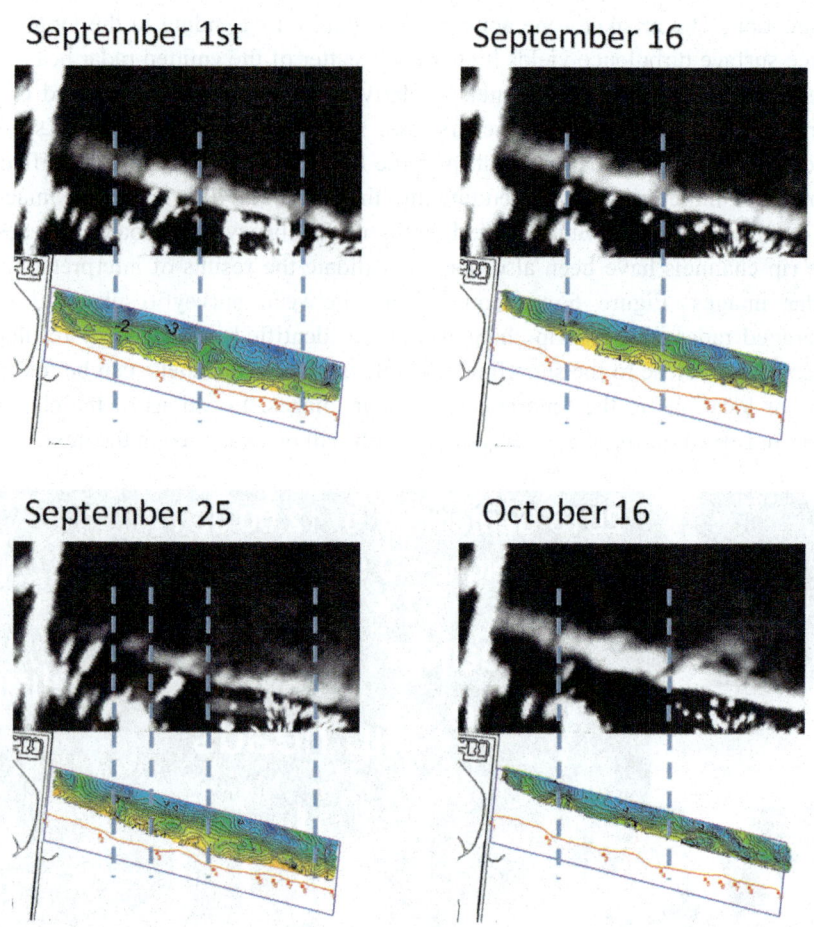

Fig. 6. Comparisons of survey results and time averaged radar images.

3. Results of the Observation

Figure 7 shows the summary of the observation: variations of the locations of the rip channels, occurrence of rip currents (identified as rip heads in the time averaged images), and wave height and direction measured at NOWPHAS wave gauge (see Fig. 1). Basically, occurrences of rip currents have been detected in the whole period when the incoming wave height exceeded 0.3 m.

In the first half of the observation, it seems that rip currents occurred randomly and are not connected with rip channels. During and after an attack of high waves in the middle of the observation period, two rip channels initially found at longshore position approximately 600 m and 1,100 m, shifted closer by 80 m, as shown in Figure 8. This is indicating complex nearshore circulation must be existent during this period which resulted longshore migration of rip channels into opposing directions. After this period, most of the rip currents were detected in the vicinities of rip channels, which suggesting that rip currents are connected to rip channels. Mechanism, however, for these different behaviors in the first and latter periods are not understood.

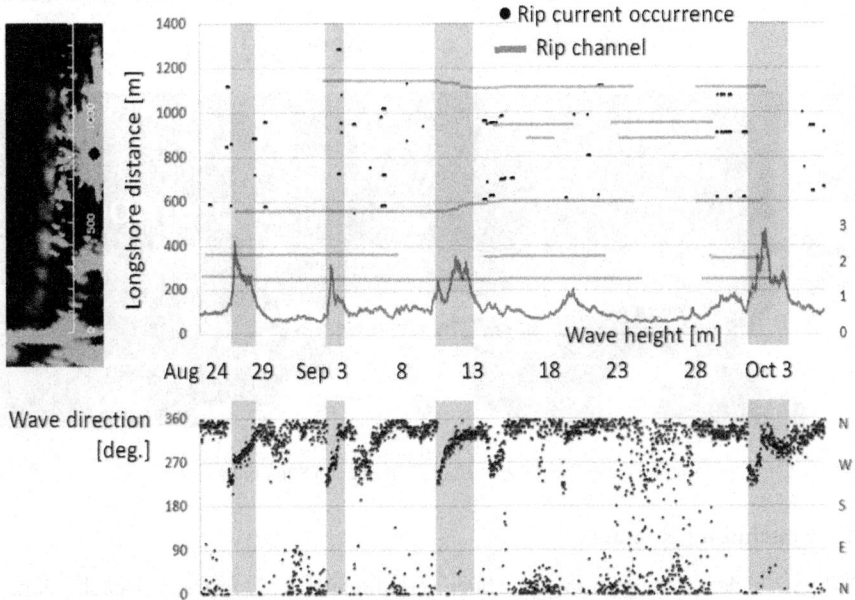

Fig. 7. Variation of locations of rip channels and rip current heads. Lateral axis is the elapsed time, and the vertical, longshore extent. Main panel shows locations of rip channel (grey solid line) and rip heads detected (black dots), and variation of wave height measured at NOWPHAS wave station. Bottom panel shows variation of wave incidence angle. Grey belts extending in the vertical denote stormy periods accompanying sharp changes of incident wave direction at the beginning of the periods..

362

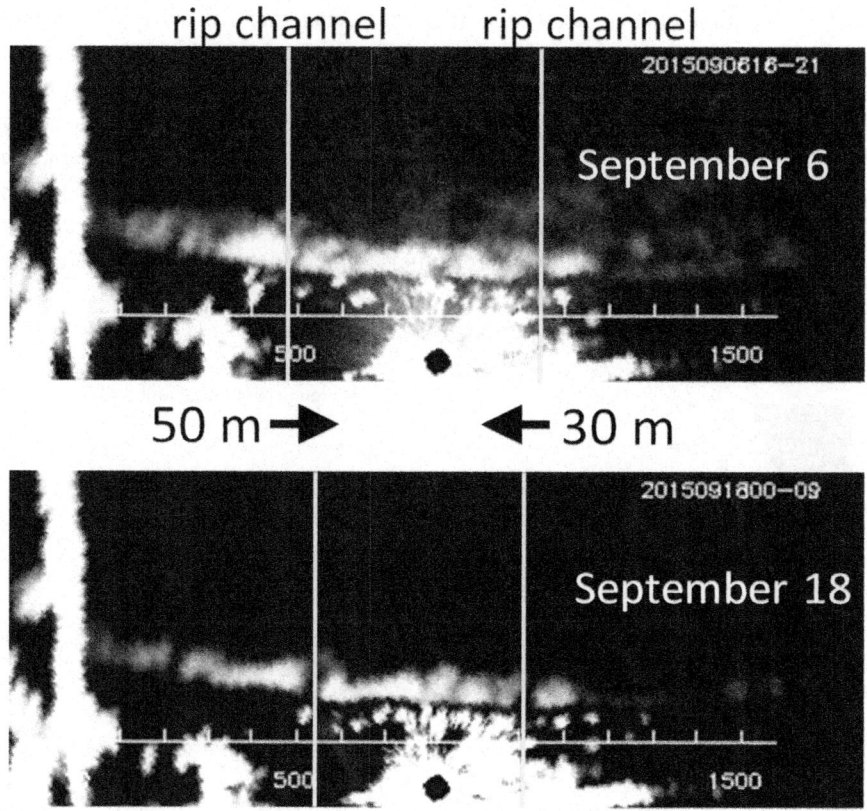

Fig. 8. Migrations of rip channels from 2015 September 6 to 18.

4. Concluding Remarks

X-band radar observation has been conducted at Uchinada Beach, Japan. Occurrence of rip currents were identified from the time averaged radar images: Rip currents developing from the breaker zone to the offshore with rip heads appear as bright streaky patterns in the time averaged radar images. At the same time, rip channels at low wave condition were identified from the time averaged images. Variations of the locations of the rip channels, occurrence of rip currents (~ rip heads), and wave height and direction were summarized for the observation period. In the first half, rip currents seemed to occur randomly and were not connected with rip channels. During and after an attack of high waves in the middle of the observation period, two rip channels shifted closer by 80 m, and locations of the rip currents were fixed more to the rip channels. Mechanism, however, for these different behaviors are not understood.

Acknowledgments

The authors are grateful to Dr. Ono, Dr. Sakai and Dr. Kuroki of ECOH Corporation, who assisted the field observation.

References

1. S. Takewaka, Measurements of Shoreline Positions and Intertidal Foreshore Slopes with X-band Marine Radar System, *Coastal Engineering Journal*, **47**, 2005.
2. S. Takewaka and T. Yamakawa, Rip Current Observation with X-band Radar, in *Proceedings of International Conference of Coastal Engineering 2010*, (Shanghai, China, 2010).

Numerical Analysis of Sea Level Rise Effect on Design Water Levels for the Roxas Boulevard Seawall Rehabilitation[*]

E. C. Cruz[†], J. B. Camelo, L. L. B. Cruz, J. C. E. L. Santos

Institute of Civil Engineering, University of the Philippines,
Diliman, Quezon City 1101, Philippines
[†]*E-mail: eccruz@upd.edu.ph*
www.upd.edu.ph

Simulative analyses were carried out to study the effects of sea level rise (SLR) and global climate change scenarios on storm tide levels fronting the Roxas Boulevard seawall. Storm tides under historical typhoons and various periods of SLR were computed using ADCIRC storm surge model. The contribution of storm waves to the non-overtopping crest elevation was also studied using a nearshore wave model. Return periods of various design water levels were also associated with the various SLR and GCC scenarios to establish bases for the rehabilitation design of the RB seawall.

Keywords: Sea level rise; Storm tide; Seawall; Roxas Boulevard; Global climate change.

1. Introduction

Roxas Boulevard is an important component of the transport infrastructure of Metro Manila and also serves as a tourism destination for its renowned unique view of the sunset in Manila Bay. Along a 1.34-kilometer long stretch, a seawall protects the road pavement, pedestrian promenade and the high-value commercial properties behind it (Figure 1). However, during the height of typhoon Pedring in September 2011, the RB seawall was overtopped by the storm tides and waves that caused its collapse, and tremendous flooding inland. It was quickly restored after that but subsequent typhoons still caused wave overtopping, with the ensuing inundation of the roads and properties behind it. The national public works agency DPWH commissioned a coastal engineering study in order to understand the problem and formulate a more permanent solution to the perennial coastal flooding of Manila City.

One of the possible overtopping mitigation schemes based on the study was the redesign of the RB seawall in order to contain the historically highest storm

[*] This study was supported by a UP-DPWH Memorandum of Agreement in 2014 on the coastal engineering study and detailed engineering design of Roxas Boulevard Seawall.

tides and storm wave runup. Part of the design consideration is the inclusion of effects of sea level rise and global climate change in the rehabilitation design. This paper discusses the numerical analysis of the implications of sea level rise and global climate change on the historical storm tides and the synthesis of design water levels based on the results.

Figure 1. Location of study area Roxas Boulevard Seawall

2. Sea Level Rise Scenarios

Several studies have yielded various magnitudes of sea level rise (SLR) based on a number of causative processes, including global increases of atmospheric temperatures, melting of the glaciers and ice caps in the poles and the tropics, and extreme terrestrial events (flooding). Table 1 summarizes estimates of rates of the sea-level rise for the Manila Bay area. Some of these also presented estimates of the annual rates of land subsidence.[4]

Table 1. Summary of Sea Level Rise estimates for the Manila Bay coastline

Source	SLR rate (mm/yr)	Land subsidence (mm)	SLR for next 15-yr (mm)	Note
IPCC (2007)	3.8	0	57	Based on global estimates by IPCC
IPCC (2013)	5.9	0	88.5	IPCC (worst case scenario)
Rodolfo and Siringan, 2006	6.6	Included but at reduced rates	99	Unknown contribution of subsidence to recorded sea level rise
Clavano (2012)	6.25	0	93.75	-
NASA (2015)	3.39	0	50.85	Based on 1993-2015 data

As the base times of these studies are varied, the absolute sea-level at the end of the design return period of the seawall may be estimated by multiplying this annual rate of rise by the design life. Only results that pertain to pure sea-level rise are considered. Based on these conditions, the most relevant site-specific estimate is due to Clavano (2012)[4] at 6.25 mm/year, which yields an estimated SLR of 93.76 mm at the end of 15 years.

3. Analysis of Storm Tides under SLR Scenarios

In order to account for the impact of sea level rise on the seawall design, magnitudes of SLR are estimated for 3 scenarios: 1) Arbitrary design life of 15 years; 2) period of 23.76 years of existing seawall with crest elevation at MTL+2.01 m; and 3) period of 50 years.

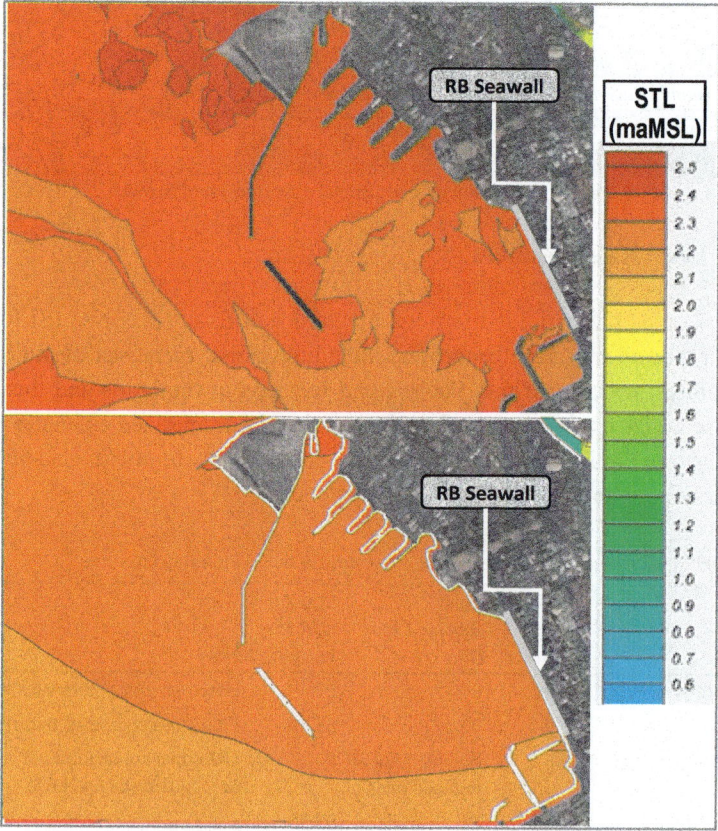

Figure 2. Simulated storm tide levels due to Typhoon Ramassun 2014 under existing conditions (top), and under 50-year SLR (bottom).

The periods in scenarios 2 and 3 are based on the results of the frequency analysis of storm tides generated by the strongest 30 typhoons in the most recent past. Scenario 2 corresponds to a return period of 23.76 years which is associated with the crest elevation (= MTL+2.01m) of the existing seawall, and thus represents the SLR at exceedance of the present seawall crest elevation. Scenario 3 is the SLR attributed to the historically highest storm tide due to Typhoon Ramassun 2014.

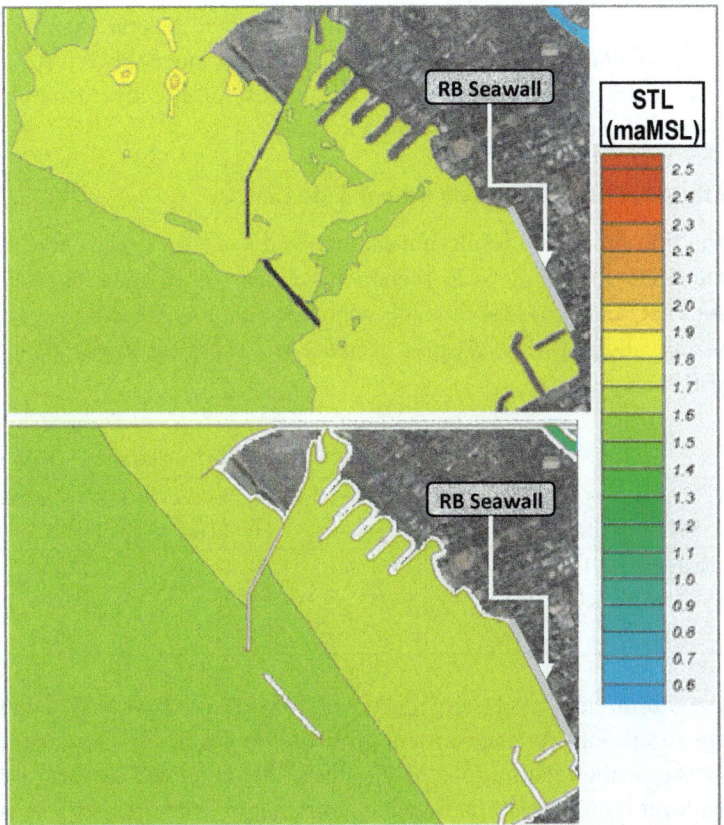

Figure 3. Simulated storm tide levels due to Typhoon Fengshen 2008 under existing conditions (top), and under 50-year SLR (bottom).

The estimated SLR magnitudes for the 3 scenarios above are 94 mm, 149 mm and 313 mm respectively. Following the base study for the typhoon hydrodynamics under existing conditions[6], storm tide simulations based on the ADCIRC hydrodynamics model[7] were carried out by adding these SLR magnitudes to the input bathymetry data, effectively increasing the still-water depths of the computational domain by the same amount. In the simulations for the 3 most critical typhoons, the astronomic tides along the boundaries and meteorological forcing were unchanged. The results under scenarios 1 and 3 are shown in Figure 2 for Typhoon Ramassun, and in Figure 3 for Typhoon Fengshen, which are among the critical typhoons that yielded the highest 3 storm tides for the study area[6] (see Table 2).

It is seen that these sea level rises result in decrease of the storm tide levels along the seawall. This is consistent with the known general characteristic of long-period oscillations, of which storm surge is one, since the return flow from the flow-impeding wall is enhanced by an increase of effective depth, resulting in alleviation of the storm surge on the surface.

4. Synthesis of SLR Effects on Storm Tide Levels

The storm tide simulation results are summarized in Table 2. The decrease of the simulated STL for all 3 SLR scenarios relative to existing condition (see "Without SLR" column) is seen in all typhoon cases. It is noted however that the decreases are slight, i.e. the greatest decrease is 22 mm for Xangsane under the 50-yr SLR scenario.

Table 2. Storm tide levels due to SLR in Manila Bay

Critical Typhoon	Without SLR	With 15-yr SLR	With 23.76-yr SLR	With 50-yr SLR
Sea Level Rise:	0 m	0.094 m	0.149 m	0.313 m
Ramassun/ Glenda 2014	2.358	2.351	2.351	2.348
Xangsane/ Milenyo 2006	2.088	2.075	2.075	2.066
Fengshen/ Frank 2008	1.763	1.756	1.751	1.747

Since the storm tide levels are reckoned from a fixed vertical datum, it is necessary to account for the translation of the datum in the presence of SLR. Figure 4 shows the relationships of the mean tide level MTL, storm tide level STL, sea level rise, astronomic tide and storm surge. Under existing or baseline condition, i.e. no SLR, the storm tide level is given by

$$STL_{No\ SLR} = MTL_{present} + AT + SS_{No\ SLR} \qquad (1)$$

where the sum of the last 2 terms are obtained from ADCIRC simulations (see Table 2). With SLR, the storm tide level is

$$STL_{With\ SLR} = MTL_{present} + SLR + AT + SS_{With\ SLR} \qquad (2)$$

Thus the change in the storm tide level Δ_{SLR}, or net meteorological tide, reckoned from the present MTL is as given by

$$\Delta_{SLR} = SS_{SLR} - SS_{No\ SLR} + SLR \tag{3}$$

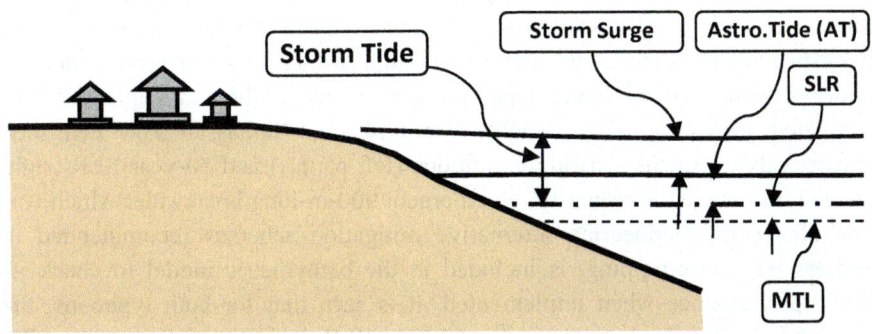

Figure 4. Definition sketch of sea level rise and storm tide levels

Table 3 summarizes the results of applying Eq.(3) based on the simulated STL's under the 3 SLR scenarios of 15, 23.76 and 50 years. It is clear that the net change of the STL when reckoned from the MTL, is an *increase* in the mean sea surface. This increase follows the SLR value, but is slightly lower by about 10 mm due to the decrease of the storm tide level (see Table 2). Thus, for the RB seawall study area, the effect of the sea level rise is a consistent *reduction* of the net meteorological tide Δ_{SLR}.

Table 3. Changes in storm tide levels due to sea level rises in Manila Bay

Critical typhoon	Changes in storm tide level (m above present MTL)			
	Without SLR	With 15-yr SLR	With 23.76-yr SLR	With 50-yr SLR
Sea Level Rise (m):	0	0.094	0.149	0.313
Ramassun/ Glenda (2014)	0	0.087	0.141	0.302
Xangsane/ Milenyo (2006)	0	0.081	0.135	0.290
Fengshen/ Frank (2008)	0	0.087	0.136	0.296

5. Analysis of Storm Waves under Sea Level Rise Scenarios

Waves generated by typhoons interact with the coasts, further elevating the sea surface locally along the coastlines reached by the ensuing wave runup. In the shallow waters of the foreshore, the wave heights are affected by wave shoaling, refraction and wave breaking. However, an uplifted sea surface resulting from sea level rise generally modulates these shallow water wave transformations. To determine the final water surface elevation effectively reached by the water surface during historical typhoons, it is thus necessary to determine the storm wave heights along the coastline.

Numerical simulations of storm waves based on a Boussinesq-type wave model[6] under the 3 SLR scenarios were carried out for typhoons Ramassun and Fengshen which respectively yielded the highest STL and storm wave heights under existing seawall conditions. Figures 5 and 6 show a comparison of simulated nearshore wave heights for typhoons Ramassun and Fengshen respectively under the existing conditions (left panels) and 50-year SLR (right panels). In these SLR scenarios, an emergent 900-m-long breakwater which was one of several engineering alternative mitigation schemes recommended to address wave overtopping, is included in the bathymetric model to check its likely performance when implemented. It is seen that for both typhoons, the wave height distribution under the 50-year SLR is more uniform along RB seawall relative to existing conditions. Under both scenarios, the storm tides exceed the crest elevations of the breakwaters even under existing conditions. Consequently, the combined SLR and storm tide level lead to an increased penetration of storm waves not only towards the seawall but into the MYC marina as well. As a result, the maximum wave heights along the seawall tend to be *lower* for both cases. Moreover, in both cases, the non-overtopping crest elevation, or NOCE, the redistribution of the wave heights along the seawall will likely not result in pronounced increase in the NOCE that was based on historical conditions.

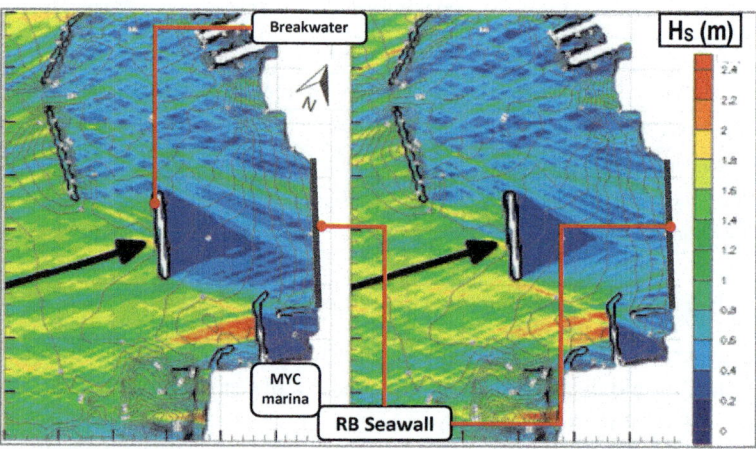

Figure 5. Nearshore wave heights due to Typhoon Ramassun 2014 under

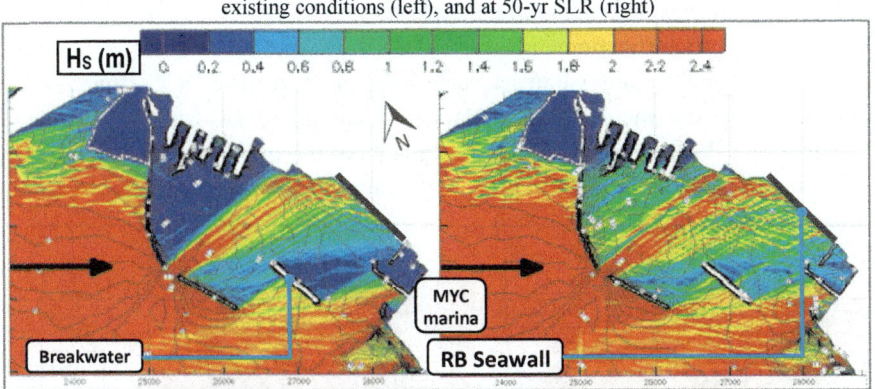

Figure 6. Nearshore wave heights due to Typhoon Fengshen 2008 under
existing conditions (left), and under 50-yr SLR (right)

6. Analysis of Storm Tides under Global Climate Change Scenarios

Global Climate Change (GCC) scenario includes the increase of atmospheric temperature as a result of increase in greenhouse gases' emission. The higher temperature warms up the sea surface which initiates and intensifies tropical cyclones. The correlation of higher atmospheric temperature and intensity of tropical storms is well established. Emanuel et al. (2008) reported on progress in identifying and predicting the effects of climate change on tropical cyclone activity in the Atlantic and Pacific Oceans. Carey (2005) reported that major storms in the Atlantic and Pacific Oceans since the 1970's have increased by about 50 percent in both duration and intensity. Using historical tropical cyclone data archives from 1980 to 2010, however David et al. (2013) reported that the general trend in historical tropical cyclone activity in the western Pacific basin, which includes the eastern seaboard of the Philippines, shows *no* increase in annual mean value of the maximum wind speed.

In order to reasonably account for the effects of GCC on the rehabilitation design of RB seawall, the meteorological properties of the 3 most critical typhoons (see Table 2) were modified to reflect the effect of a sea surface temperature rise on typhoon intensity, keeping the same their track and size, based on the best-track data[11]. In this study, this increase of typhoon strength is accounted for as follows:

a) Apply a magnification factor r to the maximum sustained wind speed V_{max}.
b) Apply a magnification factor of r^2 to the central pressure difference, using the non-storm atmospheric pressure as ambient pressure.
c) Retain the radius of maximum wind speed R_{max}.

d) Retain the best track data of the typhoon.

An arbitrary magnification factor $r = 1.10$ is applied, which is assumed compatible to a 2-degree rise in sea surface temperature. The GCC effects above are applied to the meteorological data at each of the known storm center's locations. As illustration for Typhoon Ramassun, the maximum lifetime V_{max} is increased from 166 kph to 182 kph, and the minimum central pressure decreased from 935 hPa to 919 hPa.

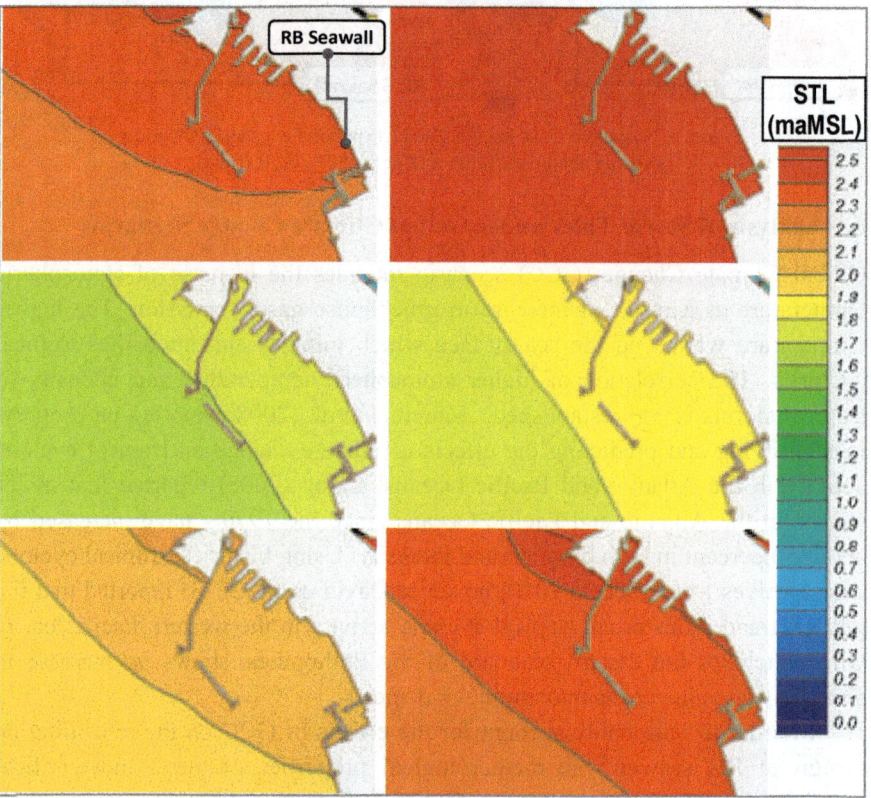

Figure 7. Simulated storm tide levels under existing conditions (left panels) and under 10% -increase GCC scenario (right panels), due to typhoons Ramassun 2014 (top), Xangsane 2006 (middle), and Fengshen 2008 (bottom)

With the intensified meteorological data of the historical typhoons, ADCIRC was again used to determine the envelopes of storm tide levels. The resulting STL's are summarized in the right-side plots in Figure 7 for the 3 critical typhoons Ramassun, Xangsane and Fengshen. For comparison, the

STL's under existing condition are shown in the left plots. It is seen that the STL's have all increased under the GCC scenario. In the case of Ramassun, the maximum STL also covered a much larger area of the Manila Bay basin, since it tracked very close to the RB seawall.

Table 4 compares the storm tide levels under the existing condition and under +10% GCC scenario. It is clear that all STL's have materially increased relative to the historical cases. The highest increases are in Ramassun and Xangsane, which both tracked south of RB seawall, in contrast to Fengshen which tracked north. The closest points of approach of the 3 typhoons are all within 45 km from the seawall. While this observation is not conclusive due to the limited number of simulated GCC cases, it appears that a 10% increase in the wind speeds will translate to higher increases in storm tide levels for south-tracking typhoons.

Table 4. Changes in Storm Tide Levels due to GCC scenario

Historical Typhoon	Storm Tide Levels (m above present MTL)		
	No GCC	With +10% GCC	% change
Ramassun/ Glenda (2014)	2.358	2.644	+12
Xangsane/ Milenyo (2006)	2.087	2.421	+16
Fengshen/ Frank (2008)	1.762	1.897	+8

7. Return Periods of SLR and GCC Scenarios

By combining the SLR and the simulated STL reported above, various design water levels corresponding to various scenarios of sea level rise and global climate change can be considered in the rehabilitation of the seawall. By applying the best-fitting probability density function through the simulated storm tide levels under existing conditions, a return period for each SLR and GCC scenario can be determined.

Table 5. Summary of return periods of SLR and GCC scenarios

Case	Scenario	Design Water Level	Return Period (years)
1	Existing (MTL +2.01m)	MTL +2.01	23.76
2	Historical Highest	MTL +2.358	57
3	SLR Scenario 23.76-yr STL	MTL +2.499	78
4	SLR Scenario 50-yr STL	MTL +2.660	109
5	GCC Scenario ($r = 1.1$)	MTL +2.644	106

Table 5 summarizes the return periods for these design water levels. It is seen that the 23.76-yr and 50-yr SLR scenarios have estimated recurrence of 78 and 109 years respectively, which are longer than the return period (57 years) of the historically highest STL due to Ramassun 2014. The 10%-increase typhoon intensity GCC scenario has an associated return period of 106 years, which is also longer than the historically highest storm tide level. Since the historical STL is associated with a return period of 57 years, the SLR and GCC scenarios considered in this study have recurrences of almost twice as long as the existing climate.

Conclusions

This study on SLR and GCC has found out that a sea level rise of 6.25 mm/year is indicative of the rate of mean sea level for the RB seawall area.

Simulative analyses of storm tide levels for the critical typhoons that tracked the seawall under 3 scenarios of SLR corresponding to return periods of 15, 23.76 and 50 years consistently showed a *decrease* in the net meteorological tide. This is attributed to the modulating effect of an increased effective depth on long-wave motion such as the storm surge responsible for the STL.

Simulative analyses of storm wave heights, which are used to determine the non-overtopping crest elevation of the seawall, also indicated a more uniform distribution of wave heights along the seawall under a 50-year SLR scenario. This is also attributed to the modulating effect of an increased effective depth on short-wave motion such as storm waves.

Storm tide simulations intended to account for a 10% increase in typhoon wind intensity due to global climate change showed a consistent rise of the storm tide levels for the 3 critical typhoons of RB seawall. Likely a site-specific conclusion for the RB seawall, a higher percentage increase in storm tide level is obtained for the south-tracking critical typhoons.

Based on frequency analysis of typhoon-induced storm tides, a return period of 109 year is associated with a 50-year sea-level rise, while an arbitrary 10% increase in typhoon intensity under global climate change yielded a return period of 106 years for the RB seawall area.

References

1. IPCC, Climate Change 2007: The Physical Science Basis. Contribution of Working Group I to the Fourth Assessment Report of the Intergovernmental Panel on Climate Change. *Cambridge University Press*, Cambridge, United Kingdom and New York, NY, USA.

2. IPCC, Climate Change 2013: The Physical Science Basis. Contribution of Working Group I to the Fifth Assessment Report of the Intergovernmental Panel on Climate Change. *Cambridge University Press*, Cambridge, United Kingdom and New York, NY, USA.

3. Rodolfo, K.S., F.P. Siringan, Global sea level rise is recognised, but flooding from anthropogenic land subsidence is ignored around northern Manila Bay, Philippines. *Disasters. Special Issue on Climate Change and Disasters*. 30 (1), 118-139, 2006

4. Clavano, W, Changing sea levels: The global context and Philippine coastal vulnerability (2012). http://www.ecojesuit.com/changing-sea-levels-the-global-context-and-philippine-coastal-vulnerability/2521/

5. NASA, Climate Change: Vital Signs of the Planet, "Vital Signs: Sea Level". N.P, 2015. http://climate.nasa.gov/vital-signs/sea-level/

6. Cruz, E.C., J.C.E.L. Santos, J.B. Camelo, M.H. Zarco, M.E.L. del Rosario, J.M.B. Gargullo, I.A.D. Inocencio, and L.L.B. Cruz, Preliminary engineering of a seawall against storm tides and waves along a built-up waterfront. Proc., 26th *International Ocean and Polar Engineering Conf.* Rhodes, Greece, 1428-1435, 2016.

7. Luettich, R.A, and J.J. Westerink, "Formulation and numerical implementation of the 2D/3D ADCIRC finite element model version 44.XX" Department of Civil Engineering and Geological Sciences, University of Notre Dame, 2004.

8. Emanuel, K., Sundararajan, R., and Williams, J, Hurricanes and global warming: Results from downscaling IPCC AR4 simulations. *Bull. Amer. Meteor. Soc.*, 89, 347-367, 2008.

9. David, C.P., Racoma, B.A., Gonzales, J., Clutario M.V., A manifestation of climate change? a look at Typhoon Yolanda in relation to the historical tropical cyclone archive. *Science Diliman* Vol. 25, No. 2, 2014.

10. Carey, B., Increase In Major Hurricanes Linked To Warmer Seas. LiveScience.com. N.P., 2015. Web. 4 Feb. 2016.

11. Digital Typhoon: Typhoon Images and Information. (2001-2011) Kitamoto Asanobu/National Institute of Informatics (NII). http://agora.ex.nii.ac.jp/~kitamoto/.

Simulation of Hydrodynamic and Associated Sediment Transport for Relocation of Sand Trap in Chennai Coastal Region

R. Manivanan[†] and J. D. Agrawal

Ports and Harbours, Central Water and Power Research Station,
Pune, Maharastra 411024/, India
[†]E-mail: vananrmani@rediffmail.com
www.cwprs@gov.in

T. Nagendra

Coastal Engineering Group, Central Water and Power Research Station
Pune, Maharastra 411024/, India
E-mail: nagendra_t@cwprs.gov.in

Sediment transport is the mechanism by which coastal erosion and accretion proceeds. Siltation in harbours and their approach channels is one of the major concern connected with the expansion of existing and development of new harbours in the coastal ecosystem. The economic of harbours is directly related to their annual maintenance dredging, and as such a proper assessment of the quantity of siltation and provision of adequate measures for the maintenance of approach channel depths would form an important integral part of planning and maintenance of the port. Model studies on littoral drift and sediment transport along coastline south of Chennai port is carried out using LITPACK and integrated Coupled Model MIKE21 FM with Hydrodynamics (HD) model, Spectral Wave (SW) model and Sediment Transport (ST) model. Two alternative locations for the sand trap were considered and these sand traps were examined in detail for the trapping efficiency and safe dredging operations. The annual sediment deposition in the proposed Sand Trap near the approach channel is of the order of 0.58 Million m^3. The model simulations indicated accumulation of sediments immediately on the up drift side of the sand trap with some siltation near the junction of trap and the approach channel. A minimum gap of 75 m to 150 m has to be provided between the breakwater toe and the top edge of the side slope of sand trap

Keywords: Mathematical model; Hydrodynamics; Sediment Transport; Wave Transformation; Integrated coupled model; Dispersion coefficients; Relocation sand trap and East coast of India.

1. Introduction

Siltation in harbour and their approach channels is one of the major concern connected with the expansion of existing and development of new ports and harbours in the coastal ecosystem. Siltation could occur due to various reasons viz. deposition of littoral drift which is interrupted by the approach channel, break water, groins, deposition of sediments brought into suspension by wave action, Whenever the alongshore drift is large, wave action obviously is quite substantial which renders the maintenance of depths during this period by dredging difficult. In such cases, it would be necessary to make adequate provision to ensure that the depths are not deteriorated to any substantial extent by the movement of the sand drift. The design of the sand trap would be governed by a number of factors such as the extent over which a major part of the drift takes place, quantity of material transported, size distribution of sediments, velocity of currents and mode of dredging. There are two monsoons along the east coast of India from south and south west (SSW)direction during south west monsoon period from May to September and from North Easterly (NE) direction during North East monsoon beginning from November. In view of this, the direction of drift along the shore changes with monsoon. The Chennai Port is located on the open coastline and port developed in stages in a span of more than 100 years. A major impact of the port has been on the littoral drift along the coastline towards north resulting in phenomenal changes to the coast.

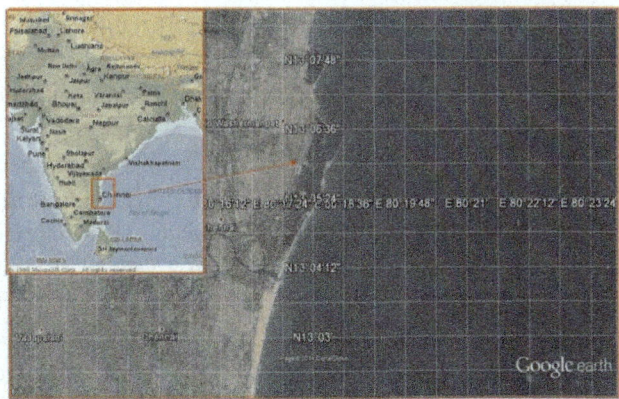

Fig. 1. Location Plan of Chennai Port

The coastline in the vicinity of Chennai Port experiences large movement of sediment in the littoral drift zone moved by the waves and currents during southwest and northeast monsoon. It is estimated that a net sediment transport of the order of 0.5 Million m³/yr takes place from south to north at Chennai. The accretion of the sand on the south side of the port has resulted in the formation of the 'Marina Beach' with advancement of shoreline being of the order of 750 m. As a result of blocking the movement of littoral drift by south arm of

Chennai Port, erosion of similar magnitude of the coastline on the north of the existing fishing harbour has taken place. The accretion on the southern side of the port has now extended almost up to the tip of the breakwater as shown in Figure 1. This present study would be assessed the existing sand trap and relocated sand trap using mathematical modelling techniques.

2. Environmental Site Conditions

The sediment transport in deeper water due to the combined effects of currents and waves was also estimated as reported by Van Rijn, 1989. Thus, the sediment carrying capacity of the currents is assessed, leading to estimates of the sediment transports at the dump-site and in the "bar" area. Analysis and interpretation of all of the above lead to a relatively complicated sediment balance for the littoral sub cell at the Port of East London (CSIR, 1994a, 1995).Sediment transport modelling must resolve the following site-specific processes. Sedimentation in dredging areas is due in most part to marine silt carried into the harbour by the bottom density current. Mathematical model studies is carried out using an Integrated Coupled Mathematical Model of Hydrodynamics (HD), Wave Transformation (Spectral Wave) and Sediment Transport (ST) in the high littoral drift zone of Chennai port area. The model studies would be simulated the long-shore sediment transport along the coastline south of the harbour entrance and assess the trapping efficiency of the proposed sand trap. The present paper describes the details of the studies carried out and the recommendation are made for feasible sand trap relocation in this region.

The littoral drift rates were computed by various researchers in Indian coastal region particularly east coast of India. Monthly littoral drift along the Madras coast was computed by Prasad and Reddy (1988) by energy flux method. The total northerly and southerly drifts in a year were estimated to be 1,550,000 m^3 and 1,010,000 m^3 respectively. Thus the annual gross and net (towards north) drifts were of the order of 2.56 million m^3 and 0.54 million m^3 . Using deep water wave data, Chandramohan et al. (1990) estimated northward drift of $1.03x\ 10^6\ m^3$ and southward drift of $0.68\ x10^6\ m^3$ with net drift(towards north) to be $0.34x\ 10^6\ m^3$ along Madras coast. Shoreline stability studies were carried out by National Institute of Oceanography (NIO), Goa. On an average, for the whole Kalpakkam coast, the annual gross northerly transport was about 1.44 million m^3 and southerly transport was around 0.35 million m^3. The average net annual transport towards north was about 1.09 million m^3. Sunder (2001) estimated littoral drift rates in madras region which the average northward and southward transport is 2.44 million m^3 and 1.37 million m^3 with the net northward transport to be 1.06 million m^3 per year. Marine EIA studies were carried out by INDOMER Coastal Hydraulics (P) Ltd. for desalination plant at Minjur. They estimated northward drift of 0.977313×10^6 m^3 and southward drift of 0.52×10^6 m^3 with net drift (towards north) to be 0.47×10^6 m^3 along Minjur coast. Shoreline studies were carried out by ICMAM in which the net transport (towards north) at Ennore coast was assumed to be 0.35×10^6 m^3. Studies for simulation of littoral drift and shoreline changes at Ennore port Tamilnadu carried out by CWPRS during 2009 and 2014 in which the annual net northward transport 0.52 million m^3 and annual gross transport is 0.70 million m^3 is computed. The estimated northward and southward transport is about 0.61 and 0.09 respectively. Based on the above data, the average northward and southward transports for Chennai port area coastline are assumed to be 0.62 and 0.31 million m^3 so that net and gross transports are 0.31 and 0.94 million m^3. To define the littoral regime at a specific site, certain environmental data and an understanding of the relevant coastal processes, are required (CSIR, 1994a, 1995). The narrow near shore zone is subjected to a high energy swell regime, and is covered by a thin wedge of sandy sediment. This wedge appears to have achieved dynamic equilibrium with the prevailing energy regime, and additional sediment is rapidly dispersed and fed into a sand stream (Martin and Flemming,1986).

2.1. *Tidal levels and currents*
The Chennai port is situated in east coast of India. Tidal levels as per admiralty chart No. 3004 at Chennai are given in Table 1 below.

Table 1. Details of Tidal Level at Chennai Port

Sl No.	Types Of Tide		Tidal Levels(m)
1	MHWS	:	1.1
2	MHWN	:	0.8
3	MLWN	:	0.4
4	MLWS	:	0.1
5	MSL	:	0.6

The observed tidal levels near Chennai port region is considered for the present studies as reported by Kankara et al. (2013) for a period of 15 days during northward current used for model development. The tidal levels of 15 days duration typically cover spring and neap tides prevailing at the site. In general, the tides are semi-diurnal with low amplitude. The maximum spring tidal range during the period is 1.3 m while the minimum neap tidal range is 0.2 m. The average U- V velocities were used for calibration of the model. It is observed that the value is of the order of 0.03 m/s and 0.11 m/s during south west monsoon and the average U- V velocities were of the order of 0.05 m/s and 0.30 m/s during north east monsoon.

2.2. *Waves*

The maximum value of significant wave height and maximum wave height were found to be 3.00 m and 4.5 m respectively with predominant wave directions are South West and East of North East, waves from East and East of South East wave for a smallest duration.

2.3. *Cross shore profiles*

The data on the cross shore profile (Figure 2)   was extracted   from the hydrographic surveys of the year 2006-2007 and 2014. Cross shore profile data for coastline north and south of the Sand Screen have been analysed.

3. Littoral Drift Simulation Studies

LITPACK model is used to estimate seasonal/annual littoral drift rates and its distribution on the profiles normal to shoreline. Six (6) representative cross-shore bed profiles of North side from sand screen S1, S2, S3, S4, S5 and S6. 3 representative cross-shore and three bed profiles south side from sand screen SS1, SS2 and SS3) were considered.

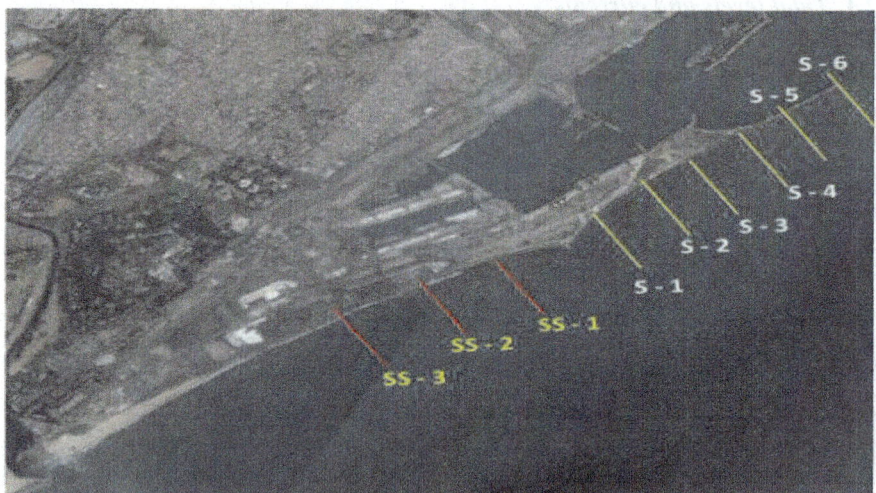

Fig. 2. Locations of Cross shore Profiles

Each cross shore profile covers a distance of 500 m extending up to about -12 m depth contour. Each profile is discretised into 50 grid points with grid size of

10 m. Besides the inshore wave climate and bathymetry along a cross-shore profile, grain size distribution over the profile is also used. The grain size is observed to be in the range of 0.15-0.35 mm. Hence D_{50} over the profiles was assumed to be in the range 0.1-0.4 mm. The bed roughness was adjusted to get the annual net and gross transports of the order of 0.31 and 0.92 million m^3. On the basis of annual wave climate the model computes annual transport rates. The model was run for the nine representative profiles. Using the near shore annual wave climate as input for each cross shore profile, northward, southward, net and gross transport rates were computed. The littoral drift rates of these profiles are tabulated in the following Table 2.

Table 2. Annual Transport Rates (Million M^3)

Profile	Northward (million m^3)	Southward (million m^3)	Net (million m^3)	Gross (million m^3)
Profile S-1	0.62	0.12	-0.50	0.74
Profile S-2	0.59	0.32	-0.27	0.90
Profile S-3	0.62	0.29	-0.33	0.91
Profile S-4	0.61	0.31	-0.30	0.92
Profile S-5	0.55	0.31	-0.25	0.86
Profile S-6	0.62	0.31	-0.31	0.92
Profile SS-1	0.59	0.35	-0.23	0.94
Profile SS-2	0.54	0.35	-0.19	0.88
Profile SS-3	0.61	0.27	-0.34	0.87

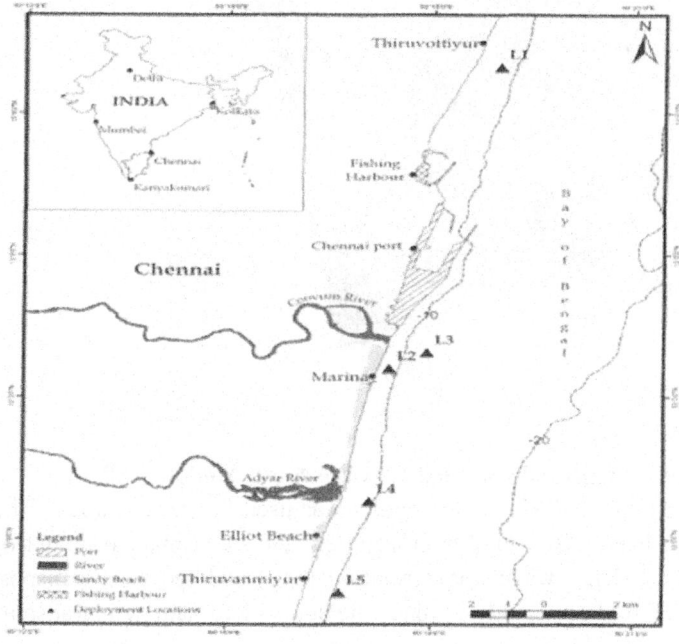

Fig. 3. Sampling Locations

382

The depth of the Sand trap, considering the bathymetry in the region and maximum trapping efficiency, should be atleast (-) 22.0 m (CD) for Alternative – I and (-) 18.0 m (CD) for Alternative – II. The length of the Sand trap based on twice the volume of annual littoral drift volume would work out to a little more than 200 m. Since the Trap will be exposed to the influence of waves and currents, safe operation of dredging will need that the Sand trap length is at least 500 m in length. This will provide for economical dredging cycle as well.

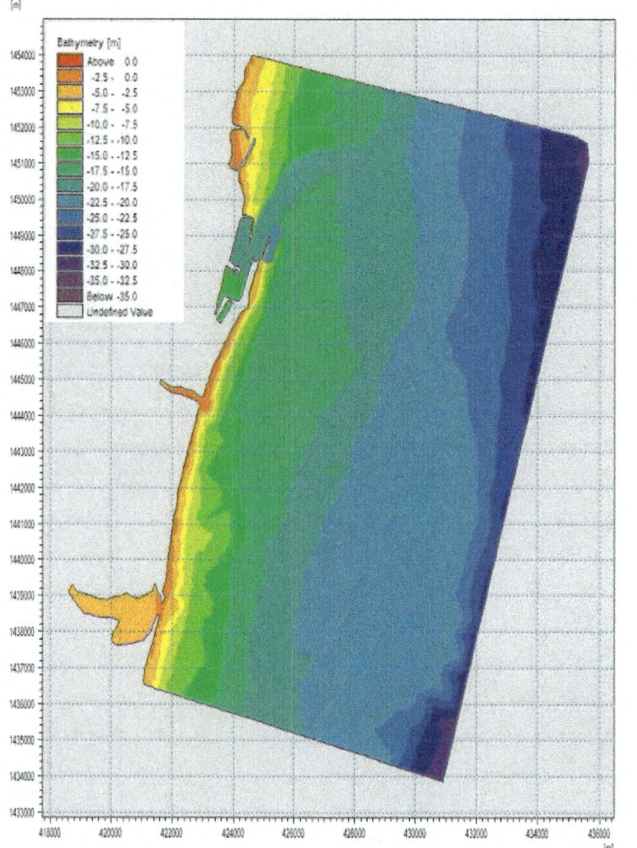

Fig. 4.
Bathymetry
(Model)

4. Simulation in an Integrated Coupled Flexible Mesh Model

As stated earlier, MIKE 21 FM coupled model was used for these studies. The Alternative - I has been tested further in an integrated mathematical model of tidal hydrodynamics (HD), wave transformation (SW) and sediment transport (ST). These model studies would simulate the long-shore sediment transport along the coastline south of the harbour entrance and assess the trapping efficiency of the proposed Sand trap.

4.1. *Model calibration*

Several trial runs of hydrodynamic model were made by varying the northern boundary i.e. U and V velocities as well as by varying Mannings number 'N' for bed roughness to get the observed flow field in the model region. With Mannings bed roughness of 0.0312 and Smagorinsky eddy viscosity factor 0.28,the observed flow of May 2009 was simulated. The model was run for a period of 10 days. The sampling location (L3) at the site is shown in figure 3. The model domain is shown in Figure 4. The time history of the observed and computed surface elevation is shown in Figure 5. it is considered as good matching in surface elevation values at L3 location. Similarly it is tried to match for the velocities also but there is no observed data in the selected location. hence it is noted that the published data on U velocity and V velocity variation are matched with the simulated values near the L3 location is shown in Figures 6(A) and 6(B) respectively. Calibration of the U velocities and V velocities is considered to be satisfactory with the published values (R.S. Kankara et al., 2013)

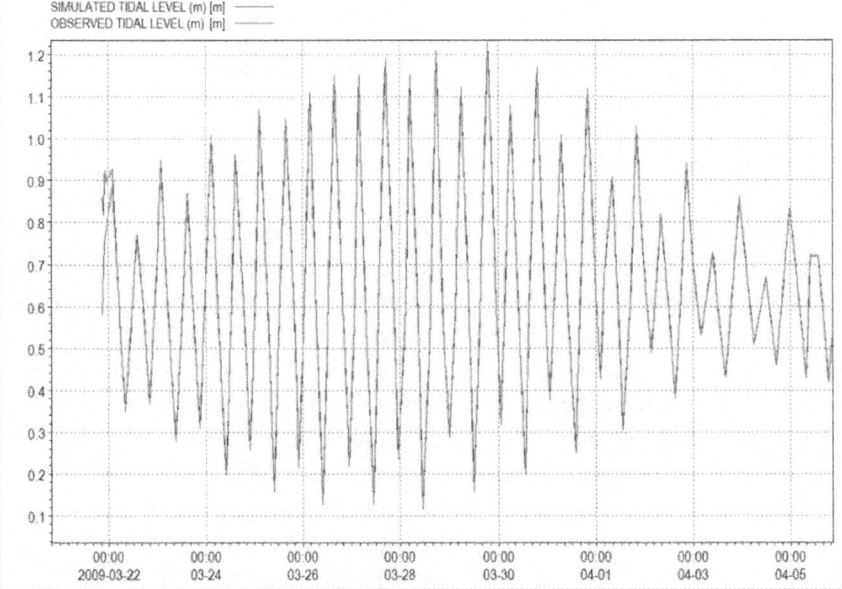

Fig. 5. Observed & Simulated Surface Elevation at L3 Location

384

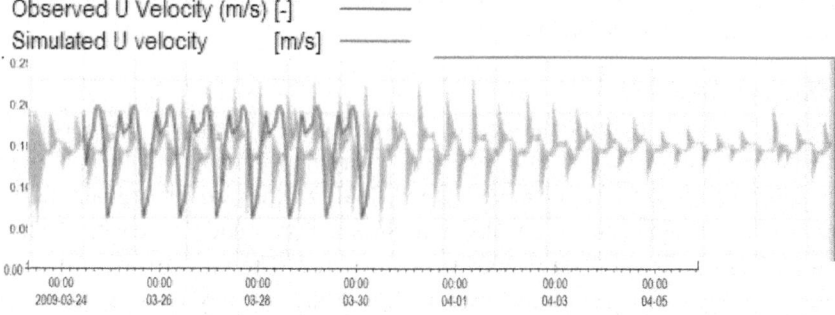

Fig. 6(A). Observed & Simulated U Velocities at L3 Location

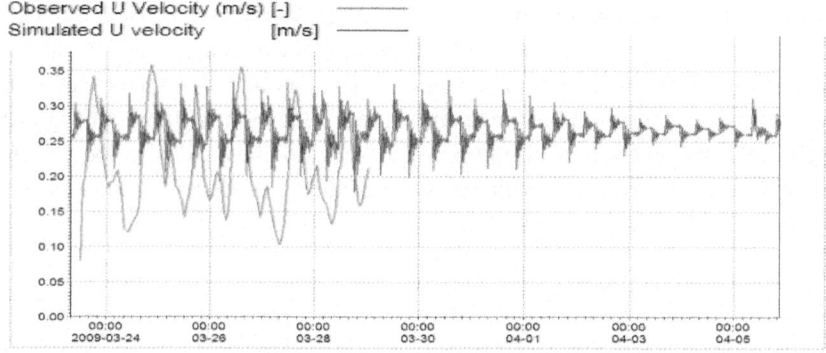

Fig. 6(B). Observed and Simulated V Velocities at L3 Location

4.2. Predictive model studies

The model is simulated for the northerly drift accordingly waves approaching from South–East (SE) Quadrant is simulated in the model. The predominant directions for the waves approaching Chennai Port are ENE, East, ESE and SE. The occurrence from SE is 60% with maximum wave height of 2.5 m. Wave heights of 3.5 m occur from ESE direction and the percentage occurrence is very marginal. The wave induced currents are active just before the sand trap as wave breaking will not occur in the trap due to large depth. The tide and wave induced currents in the whole region and surf zone for alternative I are shown in Figures 7. The tide and wave induced currents for alternative II identical to Alternative I. There is no circulations and abnormal current pattern observed in this region. The active zone of sediment transport in the surf zone indicated to be in the south of Cooum River (Marina Beach) and in the coastline north of the Sand Screen. Therefore the location of Sand trap close to the Harbour entrance i.e. Alternative-1 (Figure 9) is more efficiency than Alternative II. Based on the experiences of the authors and references, it is clear that the sand trap location should have a minimum gap of 75 to 150 m has to be provided between the breakwater toe and the edge of the side slopes of Sand trap.

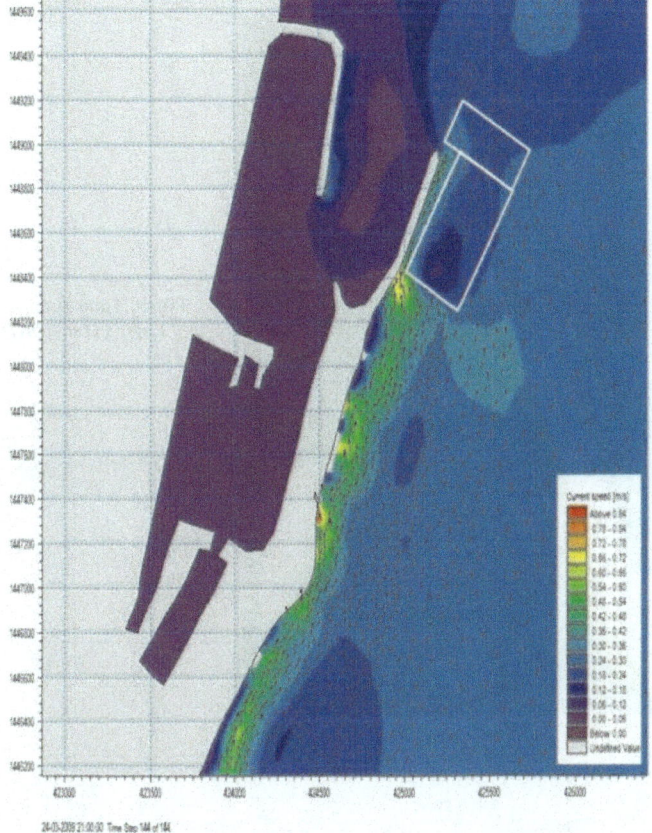

Fig. 7. Tide and Wave Induced Currents

5. Conclusion
The simulations clearly indicate the qualitative trends for the trapping efficiency of proposed sand trap. Accumulation of sediments is evident immediately on the up drift side of the sand trap with some siltation near the junction of sand trap and the approach channel. In the gap between the proposed sand trap (Figure 9) and the south breakwater concentration of currents and some erosion is indicated by the model. A minimum gap of 75 to 150 m has to be provided between the breakwater toe and the edge of the side slopes of sand trap.

386

Acknowledgments: Authors are greatful to Dr. V. V. Bhosekar, Director, Additional Charge, Central Water and Power Research Station, Pune for his kind consent for publishing this paper.

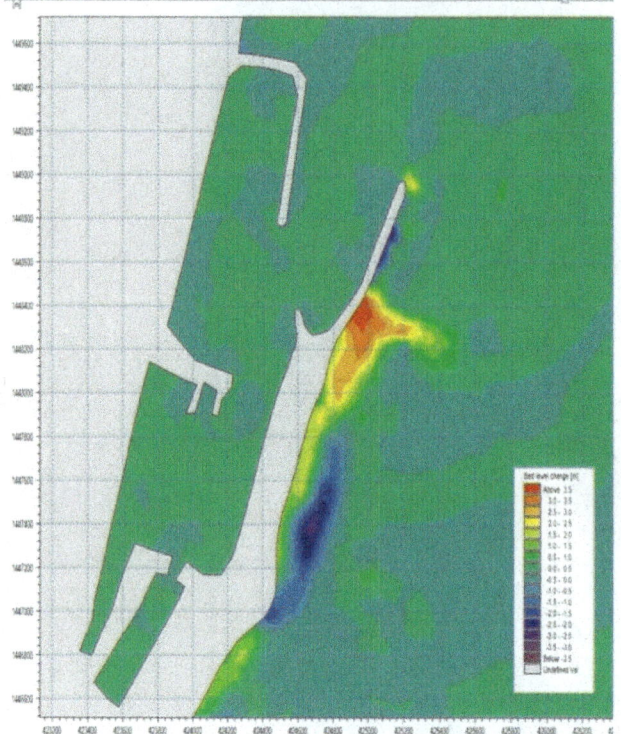

Fig. 8: Total Sediment Load (Transport)

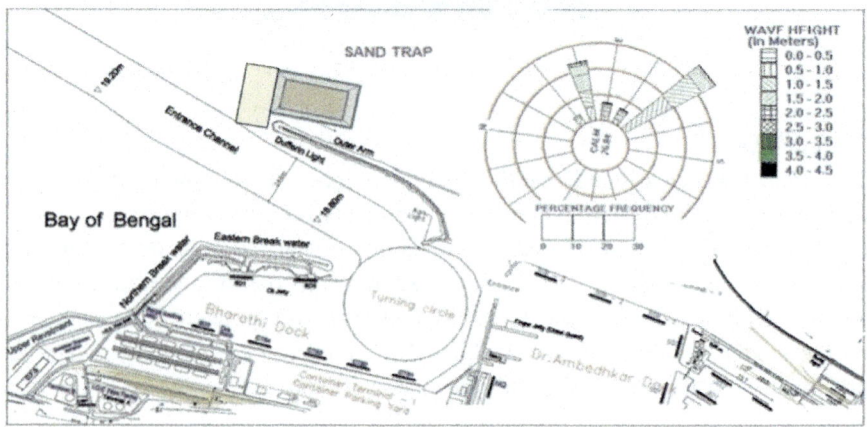

Fig. 9. Relocated Sand trap

References

1. Chandramohan P., Nayak B. U. and Raju V. S., (1990), 'Longshore sediment model for south Indian and Shri Lankan coasts', ASCE Jl. Of Waterway, Port, Coastal, and Ocean Engineering, Vol. 116, pp. 408-424.

2. CWPRS Technical Report (2009) 'Mathematical model studies for simulation of littoral drift and shoreline changes for the proposed phase II development at Ennore port, Tamil nadu. Technical Report No. 4657.

3. CWPRS Technical Report (2014) 'Mathematical model studies for integrated Morphological changes in the coastline between Ennore creek and L&T port including north of L&T port at Ennore, Tamilnadu Technical Report No. 5142.

4. CSIR (1994a). Oos-Londenhawe: Sedimentasie, sandvangput en storting. CSIR Report EMAS-C 94020, Stellenbosch.

5. CSIR (1995). Port of East London: Sedimentation on the inside of the southern breakwater. CSIR Report EMAS-C 95034, Stellenbosch.

6. Kankara R S, R.Mohan, and R Venkatachalapathy (2013) 'Hydrodynamic modelling of Chennai coast from a coastal zone management perspective' Jr. of Coastal Research, Vol. 29, No. 2, pp. 347-357

7. LITPACK 2000, An Integrated Modelling System for Littoral Processes and Coastal Kinetics, Danish Hydraulic Institute, 2000.

8. Martin A K and Flemming B W (1986). The Holocene shelf sediment wedge off the south and east coast of South Africa. Knight R J and McLean J R (Eds), Shelf Sands and Sandstones. Canadian Society of Petroleum Geologists, Memoir II, pp. 27-44.

9. NIO, (2001) Goa, Shoreline stability studies for 500 MWe PFBR project at Kalpakkam,

10. Prasad K.V.S. R. and Reddy B.S.R., (1988) 'Near-shore sediment dynamics around Madras port, India', ASCE Jl. of Waterway, Port, Coastal, and Ocean Engineering, Vol. 114, No. 2, March, pp. 206-219.

11. Rapid Marine EIA for CWDL Desalination Plant at Minjur, (2005), Indomer Coastal Hydraulics(P) Ltd.

12. Shoreline Management Plan for Ennore Coast, TamilNadu, (2007) ICMAM,

13. Sunder V., (2002) Chapter 9 on Littoral Drift in Harbour and Coastal Engineering (Indian Scenario), Vol II, Ed. S. Narasimhan and S. Kathiroli, NIOT Chennai, 2002.

14. Van Rijn L C (1989). Sediment transport by currents and waves - Handbook. Delft Hydraulics, Report H461, Volume 2, page 10.15-10.21.

Effects of Soil Resistance Damping on Wave-Induced Pore Pressure Accumulation in a Sandy Seabed[*]

Linlong Tong, Jisheng Zhang[†], Jinhai Zheng and Rui He

State Key Laboratory of Hydrology-Water Resources and Hydraulic Engineering, Nanjing, Jiangsu, China
College of Harbor, Coastal and Offshore Engineering, Hohai University, Nanjing, Jiangsu, China
[†]E-mail: jszhang@hhu.edu.cn
www.hhu.edu.cn

A two-dimensional porous-elastoplastic model, in which the influence of the reduction of the effective stress on the soil strength has been considered, is proposed to investigate the accumulation of pore water pressure under standing wave. The simulation results show that the liquefaction is likely to occur around the node due to the accumulation of pore water pressure. The liquefaction leads to the decrease of soil resistance, which has great effect on the development of the residual pore pressure. The simulation demonstrates that if the decrease of soil resistance is not considered, the soil liquefaction depth will be overestimated.

Keywords: Residual pore pressure; Shear modulus; Liquefaction; Standing wave.

1. Introduction

It is well known that wave propagating over the porous seabed will generate oscillatory water pressure on the sea floor. The soil skeleton within the seabed is compressed under the wave crest and expanded under the wave trough, resulting in the changes of the pore pressure and effective stresses in the seabed. When the incident wave is large, the excess pore pressure can be great and comes to the initial vertical effective stress. In this situation, the effective stress vanishes and the granular material will transform from a solid state to a non-Newtonian fluid, which has little resistance to any shear loadings. According to Seed and Rahman[1], this phenomenon is the so-called soil liquefaction. Soil liquefaction

---

[*] Work partially supported by the National Natural Science Foundation of China (Grant No. 51425901), the Natural Science Foundation of Jiangsu province (Grant No. BK20150804), the 111 Project (Grant No. B12032), the marine renewable energy research project of State Oceanic Administration (GHME2015GC01), and Colleges and Universities in Jiangsu Province Plans for Graduate Research and Innovation Projects (Grant No. B1504708).

hazards to the stability of the harbor, coastal and offshore engineering, as it may give rise to an incline of offshore oil platforms and serious subsidence of breakwaters. Due to the practical engineering application, many studies have been carried out to investigate the wave-induced soil dynamic response.

Two mechanisms of wave-induced liquefaction have been observed in the field and laboratory[2,3]. Suggested by Sumer and Fredsøe[4], the first one is termed as the momentary liquefaction and the other one is termed as residual liquefaction. Momentary liquefaction is induced by the oscillatory pore pressure, while residual liquefaction is mainly induced by the residual pore pressure. As to momentary liquefaction, it generally occurs within unsaturated seabed and has been extensively investigated using both the physical method[5] and theoretical method[6,7]. Those studies indicate that the amplitude of the pore pressure decreases with the increasing of soil depth and phase lag of excess pore pressure were observed along the vertical direction of the seabed. The Biot's consolidation equations[8] and its derivative models[9] were widely used to study the soil dynamic response in this mechanism.

Compared with the momentary liquefaction, wave-induced residual liquefaction has received relatively less attention and it is mainly focused in this study. Based on many experimental measurements, Seed and Rahman[1] suggested that the cyclic shear stress induced by the periodic wave loading caused the pore pressure accumulation in seabed. Therefore, a source term related to the cyclic shear stress has been introduced into the diffusion equation to govern the pore pressure accumulation. Following their study, McDougal et al.[10] gave a simple analytic solution of this model to evaluate the liquefaction potential of soils. However, all these studies were limited to one-dimension (1D). Recently, this model has been further extended to two-dimension (2D) by Jeng and Zhao[11]. It should be noted that all these models can only investigate the pore pressure accumulation before liquefaction. Sekiguchi et al.[12] studied the pore pressure accumulation in a sandy bed using a centrifuge and they thought pore pressure accumulation was induced by plastic deformation. Sassa et al. [13] further gave a formula to calculate the plastic deformation of soil and proposed a 1D two-layer fluid model which is used to investigate the progressive liquefaction and densification of soils. Recently, this model also has been extended to 2D by Liao et al.[14]. Their study shows that the residual pore pressure has an obvious 2D distribution under the standing wave loadings. However, it should be noted that they treated the soil skeleton as an invariant medium similarly to other studies, as the damping of the soil resistance due to the reduction of the effective stress is ignored. As such, the simulated residual pore pressure can greatly exceed the initial vertical effective stress in the liquefied areas. Other studies on the wave-induced pore pressure accumulation are those by Sassa and Sekiguchi[15] and Jeng and Ou[16] who used distinctive models including the sophisticated elastic-plasticity constitutive relation.

Though those studies demonstrated some characteristics of the soil dynamic response to wave loading, most studies treat the soil skeleton as an invariant medium which ignores the damping of the soil resistance due to the reduction of effective stress. The classic formula stating the relationship between the effective stresses and the shear modulus (G) of a sandy soil can be expressed as[17]

$$G = C_g p_a \frac{(e_r - e)^2}{1 + e} \left(\frac{p}{p_a} \right)^{n_g} \tag{1}$$

where C_g and n_g are non-dimensional soil parameters, e_r=2.97 for angular sands, ng is about 0.5 for most granular soils, p_a is air pressure, e is the void ratio of the sand and p is the mean effective stress. This widely used formula[18] indicates that the damping of the soil resistance should be considered when the residual pore pressure is non-ignorable compared with the initial effective stresses. In this study, a 2D poro-elastoplastic model in which the influence of the damping of the soil resistance is considered is proposed to investigate the residual pore pressure within a sandy seabed. An integrated poro-elastic model is used to calculate the shear stress ratio in the seabed, in which the Reynold-averaged Navier-Stokes (RANS) equations are used for describing the wave motion, and a modified 2D accumulation model is adopted to describe the residual pore pressure in the porous seabed.

2. Formulations

2.1. *Seabed model*

It is well known that the soil in seabed consists of soil particles, water and trapped air. The soil particles form the skeleton of the porous structure, and the water and air fill the void between the soil particles. The storage equation of the porous medium can be expressed as

$$\frac{\partial \varepsilon_v}{\partial t} - n\beta \frac{\partial u_e}{\partial t} = -\frac{k}{\gamma_w} \left(\frac{\partial^2 u_e}{\partial x^2} + \frac{\partial^2 u_e}{\partial z^2} \right) \tag{2}$$

where n is the porosity of soil, k is the permeability of soil, γ_w represents the unit weight of pore water, u_e represents the total excess pore pressure in the seabed, β is the compressibility of pore fluid (the mixture of pore water and air), which is defined as

$$\beta = \frac{1}{K_f} + \frac{1-S_r}{p_{w0}} \tag{3}$$

where K_f is the bulk modulus of pore water, S_r is the saturated degree of the seabed and p_{w0} is the absolute static pressure.

The total excess pore pressure in seabed can be divided into two parts:

$$u_e = u_e^{(1)} + u_e^{(2)} \tag{4}$$

where $u_e^{(1)}$ indicates the transient or oscillatory excess pore pressure, whose temporal average over any wave cycle is zero by definition; $u_e^{(2)}$ is the residual pore pressure which is induced by the plastic volumetric deformation of the seabed. The primary focus of this study is to develop the residual pore pressure. ε_v in Eq. 2 is the total volumetric strain of the soil, which consists of two parts: an elastic component (ε_e) and a plastic component (ε_p):

$$\varepsilon_v = \varepsilon_e + \varepsilon_p \tag{5}$$

The elastic component (ε_e) is recoverable during each wave cycle and is mainly related to the spherical stress in the form of

$$d\varepsilon_e = \frac{1}{K}dp = \frac{1}{K}d\left[\frac{1+2K_0}{3}(\sigma_v - u_e)\right] \tag{6}$$

where σ_v is the wave-induced total normal stress in vertical direction, K_0 is the ratio between the horizontal and the vertical effective stress. K is the bulk modulus of the soil and can be determined as

$$K = \frac{1}{1-2\mu}G \tag{7}$$

where μ is the Poisson's ratio. As the period-averaged total normal stress σ_v and oscillatory pore pressure $u_e^{(1)}$ are negligible, the governing equation of residual pore pressure can be written as

$$\frac{\partial u_e^{(2)}}{\partial t} = \frac{K}{1+Kn\beta}\frac{k}{\gamma_w}\left(\frac{\partial^2 u_e^{(2)}}{\partial x^2} + \frac{\partial^2 u_e^{(2)}}{\partial z^2}\right) + \frac{K}{1+Kn\beta}\frac{\partial \varepsilon_v}{\partial t} \tag{8}$$

Combining Eq. 1 with the effective stress principle of soil yields:

$$K = \frac{C_g P_a}{1-2\mu}\frac{(e_r - e)^2}{1+e}\left(\frac{1+2K_0}{3}\frac{\sigma'_{v0} - u_e^{(2)}}{P_a}\right)^{n_g} \tag{9}$$

where σ'_{v0} is the initial vertical effective stress. In this study, the decrease of the void ratio due to the compression of soil is neglected as it is very small compared with its initial value. ε_p is generally considered as a function of the cyclic stress ratio in wave-induced soil dynamics[12]. In this study, the empirical formula proposed by Sassa et al.[15] is used to calculate the plastic strain, viz

$$\varepsilon_v^p = \left(1 - \exp(-\theta\xi)\right)\varepsilon_\infty^p \tag{10}$$

$$\varepsilon_\infty^p = R\left[\exp\left(\alpha\chi\right)-1\right] \tag{11}$$

where θ, R and α are material parameters, ξ represents the number of wave repetitions ($\omega t/2\pi$), t is time, ω is angular frequency of waves, and ε_∞^p denotes the plastic volumetric strain that is attained ultimately with ξ approaching infinity. In Eq. 8, the cyclic stress ratio is defined as

$$\chi = \frac{|\tau_{xz}|_{max}}{\sigma_{v0}'} \tag{12}$$

in which $|\tau_{xz}|_{max}$ denotes the amplitude of the shear stress. $|\tau_{xz}|_{max}$ and σ_{v0}' can be obtained by solving the Biot's consolidation equation:

$$\frac{\partial\sigma_x'}{\partial x} + \frac{\partial\tau_{zx}}{\partial z} = -\frac{\partial u_e^{(1)}}{\partial x} \tag{13-a}$$

$$\frac{\partial\tau_{xz}}{\partial x} + \frac{\partial\sigma_z'}{\partial z} = -\frac{\partial u_e^{(1)}}{\partial z} \tag{13-b}$$

$$\frac{k}{\gamma_w}\Delta u_e^{(1)} = n\beta\frac{\partial u_e^{(1)}}{\partial t} - \frac{\partial\varepsilon_e}{\partial t} \tag{13-c}$$

where σ_x', σ_z' and τ_{zx} indicate the horizontal effective stress, the vertical effective stress and the shear stress, respectively. Under the condition of plane strain, the linear stress-strain relationship is given by

$$\sigma_x' = 2G\left(\frac{\partial u}{\partial x} + \frac{\mu}{1-2\mu}\varepsilon_e\right) \tag{14-a}$$

$$\sigma_z' = 2G\left(\frac{\partial w}{\partial z} + \frac{\mu}{1-2\mu}\varepsilon_e\right) \tag{14-b}$$

$$\tau_{zx} = \tau_{xz} = G\left(\frac{\partial u}{\partial z} + \frac{\partial w}{\partial x}\right) \tag{14-c}$$

where u and w are the horizontal and vertical displacement of soil particles. As the damping of soil resistance has been considered in this study, $|\tau_{xz}|_{max}$ is not a constant any more but decreases with time. It should be noted that it is difficult to calculate $|\tau_{xz}|_{max}$ at certain time. Thus, the value of $|\tau_{xz}|_{max}$ under the initial condition is used to approximate the cyclic stress ratio in the simulation. In this study, the seabed model is solved by finite element method (FEM).

2.2. Seabed model

Wave motion and seepage flow in the porous structures are determined by solving the RANS equations[19]. The mass and momentum conservation equations can be written as

$$\frac{\partial \overline{u}_{fi}}{\partial x_i} = 0 \tag{15}$$

$$\frac{\partial \overline{u}_{fi}}{\partial t} + \overline{u}_{fj}\frac{\partial \overline{u}_{fi}}{\partial x_j} = -\frac{1}{\rho_f}\frac{\partial \overline{p}_f}{\partial x_i} - \frac{\overline{u'_{fi}u'_{fj}}}{\partial x_j} + \frac{1}{\rho_f}\frac{\overline{\tau}_{ij}}{\partial x_j} + g_i \tag{16}$$

where $\overline{u}_{fi}$ is the Reynold-averaged flow velocity, x_i is the Cartesian coordinate, ρ_f is water density, $\overline{p}_f$ is the Reynold-averaged water pressure, $\overline{\tau}_{ij}$ is the Reynold-averaged viscous stress tensor of the mean flow, g_i is the acceleration due to gravity. The effect of turbulence fluctuation on the mean flow, denoted as $\overline{u'_{fi}u'_{fj}}$, is obtained by solving the modified k-ε turbulence model where k is the kinetic energy and ε is the dissipation rate of the kinetic energy. In above description, the prime denotes turbulence fluctuation with respect to the ensemble mean.

The wave model COBRAS is solved by finite-difference method (FDM). Two-step projection method is used to decouple the RANS equations. With high-order finite difference scheme of staggered grid, highly precise solutions in space can be reached. A modified VOF method is employed to track the free surface of wave. For more details, readers are referred to Lin and Liu[19].

3. Validation

In this study, the proposed model is validated by comparing with the experiments by Sekiguchi et al.[12] in a centrifuge and Tzang et al.[20] in a wave flume. In the experiment of Sekiguchi et al.[12], a balanced-beam centrifuge was used to investigate the wave-induced instability of sand bed. During the experiment, the centrifugal acceleration was 50 times of the ordinary gravity. A small flume was installed in the centrifuge. In the flume, the depth of the sand layer and the fluid layer were 0.047 m and 0.044 m, respectively. Silicone oil with a viscosity of 50 cSt was used as the pore fluid as well as the exterior fluid over the seabed. The gravity of silicone oil was set as $0.96\gamma_w g$ in this study. Quasi-standing waves of the first mode were formed when the fluid was excited at a frequency of 8.8 Hz. The amplitude of the input pressure fluctuation was 1.7 kPa. In the vertical direction, two pore pressure transducers were used to measure the pore pressure in the seabed respectively at depth of 0.011 m and at the seabed bottom. The porosity of soil was 0.5 and the specific gravity was 2.65. In computation, α is set as 55, the plastic parameter θ is 1.4 and R is given as 0.000018, and C_g is assumed to be 265.

Two cases, with and without shear modulus damping, are simulated in this validation. Fig. 6 is the comparison of simulation and measurement in which red

394

solid line is the simulation with soil resistance damping while dashed line is without. In general, the comparison shows that the numerical simulation with soil resistance damping agrees better with the measurement. It should be noted that the residual pore pressure is overestimated when soil resistance damping is neglected. As it is seen in Fig. 1 that the maximum residual pore pressure at depth of 0.011 m excesses its initial vertical effective stress (σ'_{v0}) 4.61 kPa when the soil resistance damping is ignored.

In the wave flume experiment, a sediment basin (3 m (length) $\times$ 0.5 m (height) $\times$ 1 m (width)) was located in a wave flume (37 m (length) $\times$ 1.2 m (height) $\times$ 1 m (width)). During the experiment, the basin was full of soil and covered by water layer with depth of 0.45 m. Five pore pressure transducers were installed in the seabed to measure the pore pressure under progressive wave loadings. The wave conditions and soil properties are listed in Table 1. Readers are referred to Tzang et al (2011) for more details of the experiments.

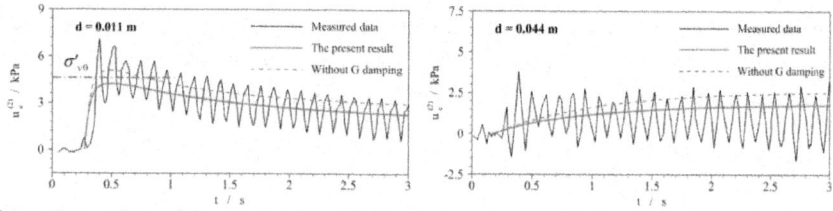

Fig. 1. Comparison of the simulated residual pore pressure and the experimental data measured by Sekiguchi et al.[12] in centrifuge.

Table 1. Soil properties and wave characteristics used in the experiment

Medium	Experiment parameter	Value
Wave	Wave height	0.091 m
	Wave period	1.5 s
	Soil porosity	0.48
	Poisson's ratio	0.49
	Permeability	1.5×10^{-9} m/s
	Specific gravity	2.64
Seabed	C_g	12.2
	R	0.0018
	α	65
	θ	0.11
	K_0	0.5

Fig. 2 shows the simulated and measured variation of the accumulation of residual pore pressure with time. Both the measurement and simulation shows that the residual pore pressure has a sharp increase with time at the beginning when wave arrives. The residual pore pressure then reaches a maximum value and starts to decrease with time. This may be ascribed to the fact that the residual pore pressure gradient becomes upwards as the residual pore pressure can only reach the initial vertical effective stress which is smaller over this point. Fig. 2 also shows that the simulated residual pore pressure is in reasonably good agreement with measurements.

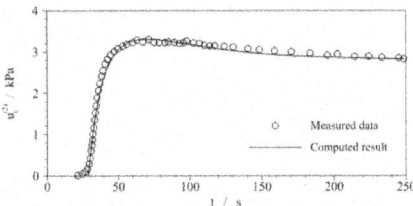

Fig. 2. Comparison of the simulated residual pore pressure and experimental data measured by Tzang et al.[20] in a wave flume.

4. Result and Discussion

In this section, the modified model is used to study the accumulation of pore pressure under standing wave, which is often generated in front of the marine structures, such as breakwaters and sea walls. The computational domain and dimension sizes of this numerical study are shown in Fig. 3. The incident wave is reflected by a vertical wall which is located at the right side of the computational domain, then standing wave is generated in front of the wall. The properties of the simulation system, including the incident wave and the seabed, are listed in Table 2. In simulation, an internal wave maker is placed at x=0 m on the left-hand side of seabed. The wave maker of the second-order Stokes wave is adopted to simulate the generation of wave.

For the mean flow field, no-slip boundary condition is applied on the seabed surface where the kinetic energy k and kinetic energy dissipation rate ε are zero. On the free surface, the atmospheric pressure, the normal stress, tangential stress, the normal gradient of k and ε are all set to be zero. At the left side, a combination of weakly reflecting boundary condition and numerical sponge layer are used to avoid wave reflection. On the seabed surface, the vertical effective stress and shear stress is ignored. The oscillatory pore pressure $u_e^{(1)}$ equals to the dynamic wave pressure, which is calculated by the RANS equations. As the vertical effective stress is always zero, the residual pore pressure $u_e^{(2)}$ is assumed to be zero in the surface. In this way, the pore pressure is continuous at the interfaces between the porous seabed and the sea water. The bottom of the porous seabed is treated as impermeable and rigid with zero displacement and no vertical flow. Both lateral boundaries of seabed are fixed in horizontal direction but free in vertical direction.

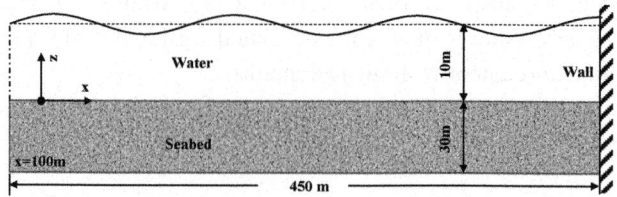

Fig. 3. Sketch of the computational domain for the development of the residual pore pressure under standing wave.

Table 2. Properties of the incident wave and the seabed

Medium	Experiment parameter	Value
Seabed	Wave height	2 m
	Wave period	10 s
	Permeability	5×10^{-5} m/s
	Porosity	0.35
	Poisson's ratio	0.33
	S_r	1.0
	C_g	600
	α	4.6×10^{-5}
	θ	0.1
	R	55
	K_0	0.5

In order to solve Biot's equation, the initial distribution of the shear modulus is determined by Eq. 9, as the initial vertical effective stress is defined as $\sigma'_{v0} = \gamma' z$, where γ' is the buoyant weight density of the soil. Then the elastic response of the seabed is studied by solving the Biot's consolidation equation, and in this stage, the dynamic water pressure calculated by COBRAS is imposed on the surface of the seabed. Fig. 4 shows the distribution of the shear stress at the wave crest (Fig. 4a), trough (Fig. 4b) and the amplitude of the shear stress (Fig. 4c). It can be found that the shear stress is significantly affected by the nonlinear wave, as the contours lean to the wave crest in the horizontal direction, and the max shear stress has two local maximum value points near the node (see Fig. 4c).

4.1. Development of the residual pore pressure

When the maximum shear stress and the initial vertical effective stress are obtained, the cyclic stress ratio in the seabed is determined by Eq. 12. Then the residual pore pressure can be simulated by afore validated accumulation model.

Fig. 5 shows the distribution of residual pore pressure within the seabed at some different wave cycles. It can be clearly seen that the residual pore pressure accumulates rapidly at the node and slowly at the antinode, which is consistent with the observation of the laboratory experiments[21]. Due to this characteristic, a gradient of wave-induced residual pore pressure is generated in horizontal direction and the pore water is forced to flow from the node to the antinode, which promotes the pore pressure accumulation at the antinode. Secondly, the pore pressure accumulates rapidly in the surface layer (see Fig. 5a and Fig. 5b) but slowly in the bottom layer (see Fig. 5c). However, the residual pore pressure in the surface cannot accumulate sustainably and it will stop growth when it reaches a large value (see Fig. 5b and Fig. 5c). As Fig. 5 only shows the residual pore pressure at some different wave cycles, Fig. 6a further shows the development of that at different depths of the vertical profile x=435 m. In order to study the effects of the soil resistance damping on the pore pressure accumulation, a contrast case is simulated in which the shear modulus of the soil remains unchanged during the simulation and the corresponding results is displayed in Fig. 6b. Fig. 6a shows that the residual pore pressure in the surface layer built-up rapidly at the beginning of wave loading. After this process, it

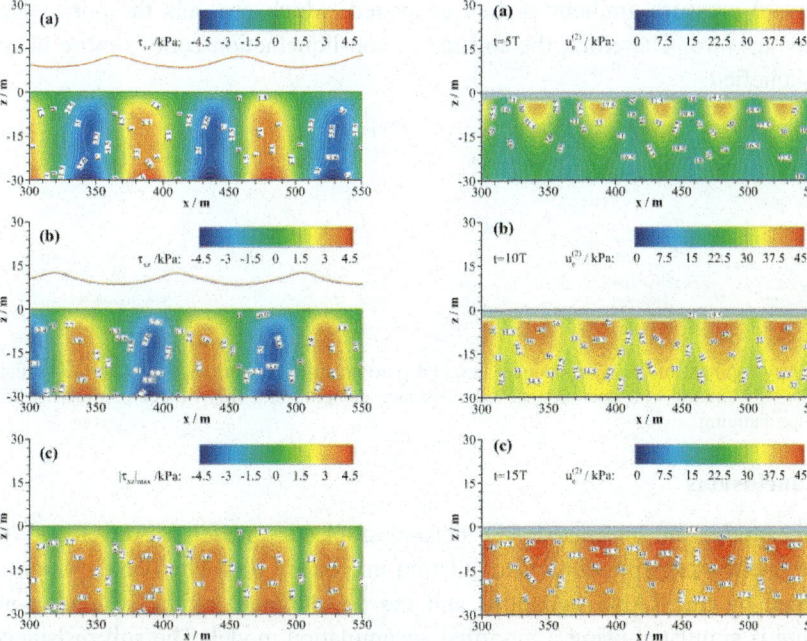

Fig. 4. The distribution of shear stress (a, b) and the amplitude of shear stress (c) in the seabed.

Fig. 5. The distribution of residual pore pressure within the seabed at different wave cycles.

398

keeps unchanged within a period of time and the value of the residual pore pressure in this stage equals to the initial vertical effective stress. Based on the liquefaction criterion proposed by Seed and Rahman[1], the soil is liquefied when this process appears. This is consistent with the observation of the experiments[22]. However, if the soil resistance damping is not considered, the accumulation of the residual pore pressure can exceed the local initial vertical effective stress, which is never observed in the field and laboratory.

4.2. *Assessment of liquefaction zones*

The assessment of the liquefaction zone is of importance in offshore engineering. Based on afore mentioned liquefaction criterion, Fig. 7 shows the liquefaction depth with (Fig. 7a) and without (Fig. 7b) soil resistance damping. It can be found that the maximum liquefaction depths at t=15T in the two cases are about 4.5 m and 8.5 m, respectively. Thus, the liquefaction zone is overestimated when soil resistance damping is not considered. This can be explained by the vertical distribution of wave-induced residual pore pressure (see Fig. 6b). When the soil resistance damping is not considered, the residual pore pressure in the liquefaction zone can further accumulate to a large value. A downward pressure gradient is then generated, which prevents the pore water seepage upward. Therefore, the soil in the non-liquefaction zone is more likely to be liquefied.

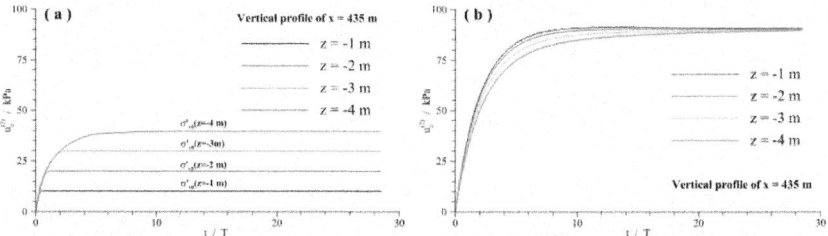

Fig. 6. Development of the wave-induced residual pore pressure for various depths in the vertical profile x=435 m: (a) with considering soil resistance damping and (b) without considering soil resistance damping.

5. Conclusions

In this study, an integrated 2D poro-elastoplastic model is developed to investigate the pore pressure accumulation in a sandy bed. The wave motion is governed by the RANS equations and the development of the residual pore pressure is simulated using a modified accumulation model. The soil resistance damping has been considered in the model which has been validated using available laboratory experiments. The validated model is then applied to investigate the standing wave-induced pore pressure accumulation in a sandy bed. The results indicate that the pore pressure accumulates rapidly at the node and this area is more likely to liquefy. The damping of soil resistance has significant effects on the pore pressure accumulation in the seabed. Without considering this effect, the liquefaction depth is overestimated.

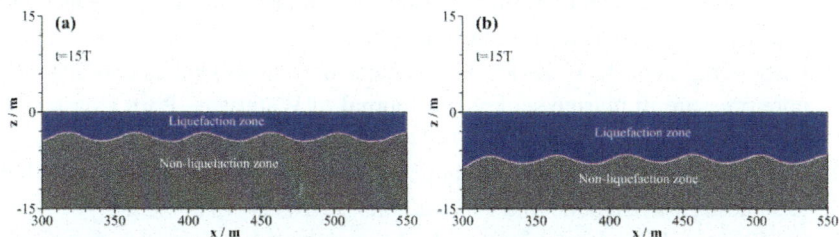

Fig. 7. Liquefaction zones at t=15T: (a) prediction with considering soil resistance damping and (b) prediction without considering soil resistance damping.

References

1. H. B. Seed, and M. S. Rahman, Wave-induced pore pressure in relation to ocean floor stability of cohesionless soils, Marine Geotechnology, **3(2)**, 123-150 (1978).
2. J. S. Zhang, JS, Q. Z. Li, C. Ding, J. H. Zheng, and T. T. Zhang, Experimental investigation of wave-driven pore-water pressure and wave attenuation in a sandy seabed, Advances in Mechanical Engineering, **8(6)**, 1-10 (2016).
3. B. M. Sumer, J. Fredsøe, S. Christensen, and M. T. Lind, Sinking/floatation of pipelines and other objects in liquefied soil under waves, Coastal Engineering, **38(2)**, 53-90 (1999).
4. B. M. Sumer, and J. Fredsøe, The mechanism of scour in the marine environment (World Scientific, Hackensack, NJ, 2002).
5. B. Liu, D. S. Jeng, G. L. Ye and B. Yang, Laboratory study for pore pressures in sandy deposit under wave loading, Ocean Engineering, **106**, 207-219 (2015).
6. J. S. Zhang, D. S. Jeng, and Liu, PL-F, Numerical study for waves propagating over a porous seabed around a submerged permeable breakwater: PORO-WSSI II model, Ocean Engineering, **38(7)**, 954-966 (2011).
7. J. S. Zhang, D. S. Jeng, Liu, PL-F and C. Zhang, Response of porous seabed to water waves over permeable submerged breakwaters with Bragg reflection, Ocean Engineering, **43(2)**, 1-12 (2012).

8. M. A. Biot, General theory of three-dimensional consolidation, Journal of Applied Physics, **12(2)**, 155-164 (1941).
9. O. Zienkiewicz, C. Chang and P. Bettess, Drained, undrained, consolidating and dynamic behavior assumptions in soils, Géotechnique, **30(4)**, 385-395 (1980).
10. W. G. McDougal, Y.T. Tsai, Liu, PL-F and E. Clukey, Wave-induced pore water pressure accumulation in marine soils, Journal of Offshore Mechanics and Arctic Engineering, **111(1)**, 1-11 (1989).
11. D. S. Jeng, and H. Y. Zhao, Two-dimensional model for accumulation of pore pressure in marine sediments, Journal of Waterway, Port, Coastal, and Ocean Engineering, **141(3)**, 04014042 (2015).
12. H. Sekiguchi, K. Kita and O. Okamoto, Response of poro-elastoplastic beds to standing waves, *Soils and Foundations*, **35(3)**, 31-42 (1995).
13. S. Sassa, H. Sekiguchi and J. Miyamoto, Analysis of progressive liquefaction as a moving-boundary problem, Géotechnique, **51(10)**, 847-857 (2001).
14. C. C. Liao, H. Y. Zhao and D. S. Jeng, Poro-elastoplastic model for the wave-induced liquefaction, *Journal of Offshore Mechanics and Arctic Engineering*, **137(2)**, 042001 (2015).
15. S. Sassa and H. Sekiguchi, Analysis of wave-induced liquefaction of sand beds, Géotechnique, **51(2)**, 115-126 (2001).
16. D. S. Jeng and J. H. Ou, 3D models for wave-induced pore pressures near breakwater heads, *Acta Mechanica*, **215(1)**, 85-104 (2010). "
17. F. E. Richart, J. R. Hall and R. D. Woods, *Vibrations of soils and Foundations*, (Prentice Hall, Englewood Cliffs, New Jersey, 1970).
18. M. Rahman, S. Lo and Y. F. Dafalias, Modelling the static liquefaction of sand with low plasticity fines, *Géotechnique*, **64(11)**, 881-894 (2014).
19. P. Z. Lin and Liu, PL-F, A numerical study of breaking waves in the surf zone, Journal of Fluid Mechanics, **359(1)**, 239-264 (1998).
20. S. Y. Tzang, Y. L. Chen and S. H. Ou, Experimental investigations on developments of velocity field near above a sandy bed during regular wave-induced fluidized responses, *Ocean Engineering*, **38(7)**, 868-877 (2011).
21. V. O. Kirca, B. M. Sumer and J. Fredsøe, Residual liquefaction of seabed under standing waves, Journal of Waterway, Port, Coastal, and Ocean Engineering, **139(6)**, 489-501 (2013).
22. J. Miyamoto, S. Sassa and H. Sekiguchi, Progressive solidification of a liquefied sand layer during continued wave loading, *Géotechnique*, **54(10)**, 617-629 (2004).

Influence of Adsorbed Ions and Ionic Strength in Sediment on Liquid Limit

T. Suzuki[§], Y. Morimoto, T. Hibino, S. Nakashita

Hiroshima University, 3-9-1, Kagamiyama, Higashihiroshima, 739-0046, Japan
[§] *E-mail: m174528@hiroshima-u.ac.jp*

Transport of sediment discuss with physical phenomena including bottom shear stress. However, soil particle surface has negatively charge and adsorb cations. Understanding of transport and deposition of sediment, we need to consider not only physical phenomena but also chemical phenomena including adsorbed ion and ion concentration. In this paper, we focused on variation on property of sediment by ratio of adsorbed cation, pore water ion concentration. To consider the resuspension of sediment, it is important to clarify the factor of critical shear strength change. In this study, we focused on liquid limit related with shear strength. Considering cohesion and repulsion force of sediment based on DLVO theory, ionic strength of pore water and adsorbed ion are important parameters determining electric double layer thickness and surface potential. It is expected that when the ionic strength becomes higher, the electric double layer becomes thinner and sediment tends to aggregate easily. In this study, to understand the factor of shear strength change, we discussed relationship among liquid limit, adsorbed cation of sediment, ion concentration of pore water. The results revealed that amount of adsorbed ion have a little effect for liquid limit. However, strong correlation exists between the ionic strength of pore water and the liquid limit.

Keywords: Adsorbed ion; Ionic strength; Liquid limit; Coastal sediment.

1. Introduction

Excess deposition of sediment aggravates tidal flat and sea bottom environment. Explanation of sediment circulation solves excess deposition of sediment. Especially considering the sediment resuspension of sediment is necessary. Critical shear stress of sediment is important factor for sediment resuspension. In the past, water content and organic matter content have been considered as parameters of critical shear stress. On the other hand, it is expected that electrochemical phenomenon is important due to cohesion of sediment. Cohesion of sediment is defined by inter particle force balance of attraction and repulsion. Thus, ion concentration of pore water and adsorbed ions of soil particles are important for the cohesion. The cohesion of minerals is many researches from the electrochemical phenomenon. However, few studies have been conducted on sediment which organic matter is adsorbed on clay minerals.

It has been reported the penetration depth obtained from the fall cone test related to the shear stress of sediment. In this study, we focused on the liquid limit obtained from the fall cone test as an index of sediment resuspension.

Warkentin (1961) reported that liquid limit of clay minerals decreased with increasing salt concentration. Nakashita et al (2016) reported that the liquid limit related to the ion adsorbed on sediment. However there is no research which clarified simultaneously the influence of ion concentration and adsorbed ions on the properties of sediment. In this study, we focused on the liquid limit correlated with the critical shear stress of the sediment. We researched the liquid limit of the sediment due to adsorbed ion sediment and ionic strength of pore water.

2. Relationship between Ions and Physical Properties in Sediment

We refer to the studies of Michaels et al (1962) and Adachi et al (1998). It is assumed that the main cause of the critical shear stress of sediment is inter particle interaction due to surface force. Dispersion and cohesion force of sediment are based on DLVO theory. According to the DLVO theory, the potential energy U per unit area around the soil particle is expressed by the equation (1) as the summation of the energy U_e(the electrostatic repulsion) and the energy U_v (the intermolecular force).

$$U = U_e + U_v = \frac{64k_BTv_\infty\gamma^2}{\kappa}\exp(-kL) - \frac{\alpha}{12\pi L^2} \tag{1}$$

$$\kappa = \sqrt{\frac{2e^2}{\varepsilon k_BT} \times \frac{1}{2}\Sigma v_i z_i^2} \tag{2}$$

k_B [J/K] is the Boltzmann constant, T [K] is the absolute temperature, v_∞[mol/L] is the total ion concentration of the solution, γ [V] is the surface potential, κ [1/m] is the reciprocal of Debye length, v_i [mol/L] is the total ion concentration of the solution of i type, z_ie [C] is the ion charge number of type i, ε [C²/J · m] is the dielectric constant. κ is expressed as equation (2) using the Debye length or the reciprocal number of the electric double layer, and $\frac{1}{2}\Sigma v_i z_i^2$ [mol/L] is the ionic strength in the solution.

The adsorbed ion is important to determine the surface potential. Increasing the amount of adsorbed ions expects to increase the surface potential. The electric double layer become thick when the ion concentration of pore water is low. Fig. 1 shows the dispersion and cohesion force of inter particles by electrochemical factors. The repulsive force becomes large when the surface potential is high and the electric double layer is thick (Fig. 1a). it means the soil particles disperse easily. On the other hand, the repulsive force between the soil

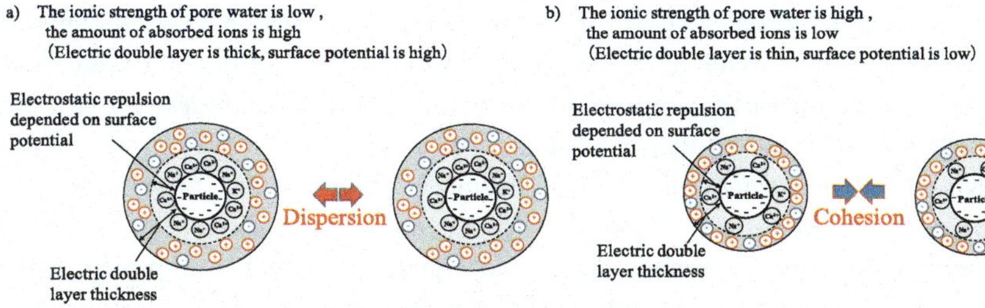

Fig. 1. Dispersion and cohesion force of inter particles by electrochemical factors

particles becomes small when the surface potential is low and the electric double layer is thin (Fig. 1b). It means the soil particles cohere easily. In this study, it is assumed that the main cause of the occurrence of the critical shear stress is interaction between particles due to surface force in sediment.

3. Materials and Method

3.1. *Measurement of properties and liquid limit of sediment*

We used the marine sediment collected on May 23, 2016 at Kaita Bay, Kure Bay in the southern part of Hiroshima Prefecture in Japan as shown in Fig. 2. The water content at the time of collection of these sediments were 222.8% and 388.7%, respectively.

Liquid limit tests conducted based on JGS0142 and this examination measured the liquid limits of the samples using a fall cone tester (DH-22NM, SEIKEN, INC.). To make flow curves (relationship between cone penetration depth and water content), we need more than 5 data in total. These data use two or more cone penetration depth in the range of 7 to 11 mm and 11 to 15 mm respectively. The water content of 11.5 mm of the flow curve defined as liquid limit.

3.2. *Method and measurement for changing adsorbed ions (Case 1)*

Considering the ions exchange order ($Ca^{2+}>Mg^{2+}>K^{+}>Na^{+}$), $CaCl_2$ solution, which high ion exchange order expects to change adsorbed ions on sediment. In this experiment, we changed adsorbed ion consider with ion exchange order. Case 1-1 was sediment collected in Kaita Bay. Case 1-2 was sediment replaced pore water with deionized water. Case 1-3 was the sediment replaced the pore water with deionized water after stirring twice with 0.5 mol/L $CaCl_2$ solution. Adsorbed ions and liquid limits of each samples was measured. To replace pore water with deionized water, added 500 mL of deionized water to 300 mL of organic mud and stirred at 200 rpm for 5 hours. After putting the samples for 10 hours, the electric conductivity of the supernatant water was measured using an electric conductivity meter (ES-51, HORIBA), and the supernatant water was removed. This operation repeated until the electric conductivity of the supernatant water was less than 0.1 S/m.

Measuring adsorbed ions in the marine sediment needs to remove pore water ions from seawater. 2.5 g of the dried sample was shaken with 30 mL of deionized water for 1 hour. Then, the supernatant water removed after centrifugation. To remove the influence of pore water ions from seawater, this operation repeated three times. After removing pore water ions, the same operation was conducted using $BaCl_2$ solution and collected the supernatant water. The cations (Na^+, Ca^{2+}, K^+) contained in the supernatant water measured using water quality analyzer (LAQUA Twin, HORIBA). This operation replaces all adsorbed ions on the soil particles with Ba^{2+} ions. Ba^{2+} ion is easy to adsorbed on the soil particle because of exchange order ($Ba^{2+} > Ca^{2+} > K^+ > Na^+$). By replacing the adsorbed ions on the soil particles with Ba^{2+}, adsorbed other cations are released into the supernatant water.

3.3. *Method of changing ionic strength in pore water (Case 2)*

It is expected the ionic strength (summation of $0.5 \times$ ion concentration $\times$ square of the ion valence number) is related to the liquid limit according to DLVO theory. The pore water of Case 2-1 was replaced with $CaCl_2$, and the pore water of Case 2-2 was replaced with NaCl solution. Case 2-1 and Case 2-2 were the same ion concentration at 0.5 mol / L, however ionic strength was different. The pore water of Case 2-3 was replaced with $CaCl_2$, and the pore water of Case 2-4 was replaced with NaCl solution. The ion concentration of Case 2-3 was 0.2 mol/L and the ion concentration of Case 2-4 was different at 0.6 mol /L, however the ionic strength are the same. To exchange pore water, 300 mL of sediments and 500 mL of the solution to be exchanged were added and stirring the samples at 200 rpm for 5 hours. After putting the samples for 10 hours, the electric conductivity of the supernatant water was measured and the supernatant water was removed. This operation repeated until the electric conductivity of the supernatant water was same as each solution. Liquid limit of these samples was measured.

4. Result and Discussion

4.1. *Ion exchange with ions adsorbed in sediments*

Fig. 3 shows the change in the amounts of adsorbed ions after exchanging the pore water of sediments. Comparing Case 1-1 and Case 1-3, 74% of the adsorbed ions were Ca^{2+} ions in Case 1-1, and more than 95% of the adsorbed ions were Ca^{2+} ions in Case 1-3. In Case 1-2, the amount of Na^+ and K^+ ions decreased and Ca^{2+} ions increased by exchanging the pore water of the sample with deionized water. This is because the pore water of the collected sediment was seawater in the initial condition.

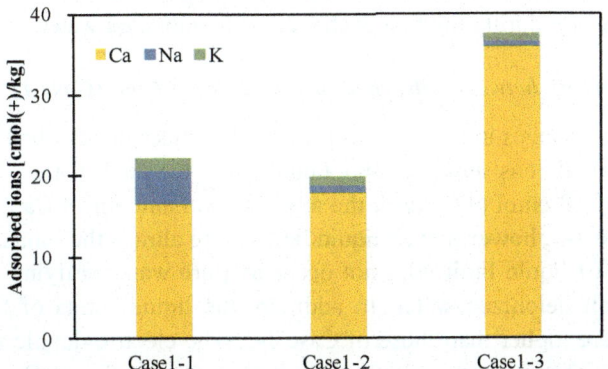

Fig. 3. Changes in adsorbed ions due to exchange of pore water

To clarify the speed of ion exchange, we examined the change of stirring time and amounts of adsorbed ions of sediments collected in Kure Bay. The types and amounts of adsorbed ions of collected sediments in Kure Bay and Kaita Bay were different. Because the adsorbed ions depend on the fine particle fraction, the organic matter content, the salt concentration in marine area. The pore water of the sediment was extracted by centrifugation, and 30 mL of pore water was added to 2.5 g of dried sediments in a 60 mL centrifuge tube.

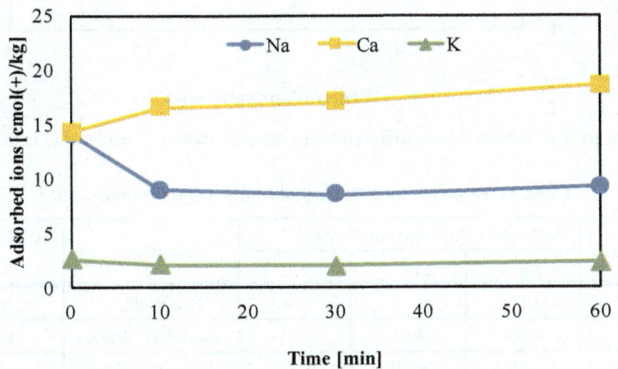

Fig. 4. Changes of adsorbed ions by stirring sediment with pore water

After stirring for 10, 30 and 60 minutes, we measured adsorbed ions. Fig. 4 shows the change in adsorbed ions.

The Ca^{2+} ion in the seawater exchanged with the adsorbed Na^+ ion in the sediments at the time of stirring for 10 minutes. In Case 1-2, the amount of adsorbed Ca^{2+} ions increased due to the high ion exchange rate. The adsorbed Na^+ and K^+ ions in the sediments were exchanged with the Ca^{2+} ions in the initial pore water during the first exchange with deionized water.

4.2. Relationship between adsorbed ions and liquid limit (Case 1)

Fig. 5 shows changes in flow curves due to differences in adsorbed ions, Table 1 shows adsorbed ions amounts and liquid limits of each Case . There was a difference of 18 cmol (+) / kg in the adsorbed ion amount of Ca^{2+} ions of Case 1-2 and Case 1-3, however their liquid limits were almost the same. It was found that change of liquid limit does not occur by pore water unifying the same ion intensity with deionized water. In addition, the liquid limits of Case 1-2 and Case 1-3 were higher than those of Case 1-1. The electric double layer of Case 1-1 becomes thinner and easier to cohere than Case 1-2 and Case 1-3, where pore water was deionized water, because pore water in Case 1-1 was seawater.

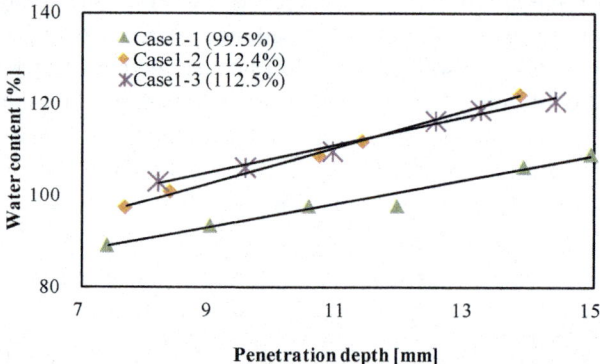

Fig. 5. Changes in flow curves due to differences in adsorbed ions (Inside of () is the liquid limit)

Table 1. Adsorbed ions and liquid limits in experiment Case 1

Sample	Main absorbed ions [cmol(+)/kg]			Pore Water	Liquid limit [%]
	Na^+	Ca^{2+}	K^+		
Case1-1	4.17	16.50	1.64	Sea Water	99.4
Case1-2	0.80	18.05	1.13	Deionized Water	112.4
Case1-3	0.61	36.00	0.90	Deionized Water	112.5

4.3. *Changes in liquid limit by changing ionic strength of pore water (Case 2)*

Fig. 6 shows changes in flow curves due to differences in ionic strength of pore water, Table 2 shows the ion concentration and ionic strength of pore water and liquid limits. Case 2-1 and Case 2-2 were the same ion concentration of pore water, however ionic strength was different. The liquid limit of Case 2-1 with 1.5 mol/L ionic strength was 20% lower than Case 2-2 with 0.5mol/L ionic strength. Case 2-3 and Case 2-4 were the same ionic strength of pore water, however ion concentration was different. The liquid limit of Case 2-3 and Case 2-4 were only about 7% difference.

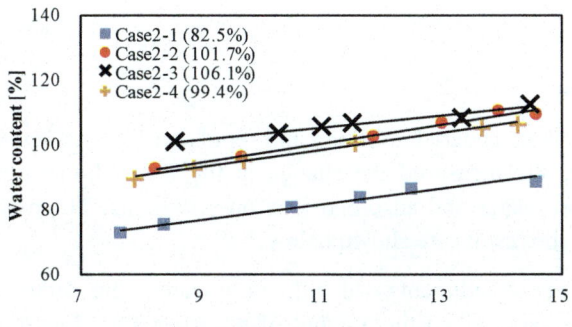

Penetration depth [mm]

Fig. 6. Changes in flow curves due to difference in ionic strength of pore water

Table 2. Ion concentration of pore water and ionic strength, liquid limit of sediment in experiment Case 2

Sample	Pore Water	Ion concentration [mol/L]	Ionic strength [mol/L]	Liquid limit [%]
Case2-1	$CaCl_2$	0.5	1.5	82.5
Case2-2	NaCl	0.5	0.5	101.7
Case2-3	$CaCl_2$	0.2	0.6	106.1
Case2-4	NaCl	0.6	0.6	99.4

Fig. 7 shows the relationship between the ionic strength of the pore water and the liquid limits including experiment Case 1 and Case 2. There was a strong correlation between the ionic strength and the liquid limit. As the ionic strength of pore water is larger, the liquid limit becomes lower.

408

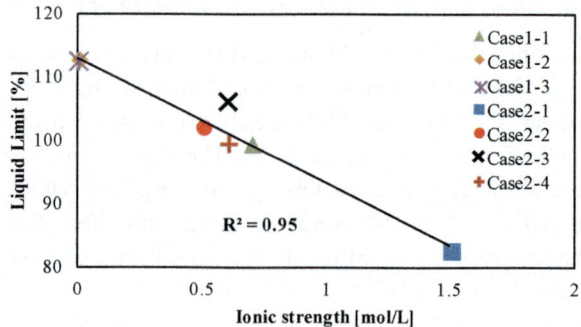

Fig. 7. Relationship between ionic strength of pore water and liquid limit

5. Conclusions

In this study, we considered cohesion and repulsion force of sediment based on DLVO theory. We measured the change in the liquid limits due to the ionic strength of pore water and adsorbed ions on sediments. The main conclusions obtained from the results are shown below.

(1) Liquid limits of sediments which the pore water were deionized water were almost the same when the amount of adsorbed Ca^{2+} ions increased by 18 cmol(+)/kg. From this result, we summarized that liquid limit do not change due to the adsorbed Ca^{2+} ion when the ionic strength of the pore water is equal.

(2) There was a strong correlation between the ionic strength of the pore water and the liquid limit. As the ionic strength of pore water becomes larger, the liquid limit became lower.

Acknowledgements

The authors gratefully acknowledge partial funding from Japan Society for the Promotion of Science: Grant-in-Aid for Science Research (Grant No. 15K18124, 15K00514, 15H05221, 16H04418, 16K14311). The constructive comments of anonymous reviewers are also appreciated.

References

1. K. Otsubo, K. Muraoka, Physical properties and critical shear stress of cohesive bottom sediments. Journal of Japan Society of Civil Engineers, Vol. 363, pp. 225-234 (1985).

2. C. Chassagne, F. Mietta, J.C. Winterwerp, Electrokinetic study of kaolinite suspensions, Journal of Colloid and Interface Science, Vol. 336, pp. 352–359 (2009).

3. Mahir Alkan, Özkan Demirbaş, Mehmet Doğan, Electrokinetic properties of kaolinite in mono- and multivalent electrolyte solutions, Microporous and Mesoporous Materials, Vol. 83, pp. 51-59 (2005).

4. H. Tanaka, H. Hirabayashi, T. Matsuoka, H. Kaneko, Use of fall cone test as measurement of shear strength for soft clay materials, Soils and Foundations, Vol. 52, pp. 590-599 (2012).

5. B. P. Warkentin, Interpretation of the Upper Plastic Limit of Clays, Nature, Vol. 190, pp. 287-288 (1961).

6. S. Nakashita, Y. Morimoto, N. Kinjo, T. Hibino, Effect of adsorbed ion on organic mud for liquid limit., Journal of Japan Society of Civil Engineers, Ser. B2 (Coastal Engineering), Vol. 72, pp. 1321-1326 (2016).

7. A. S. Michaels, J. C. Bolger, The Plastic Flow Behavior Flocculated Kaolin Suspensions, I&EC Fundamentals, Vol. 1 (3), pp. 153-162 (1962).

8. Y. Adachi, K. Nakaishi, M. Tamaki, Viscosity of a Dilute Suspension of Sodium Momtmorillonite in a Electrostatically Stable Condition, Journal of Colloid and Interface Science, Vol. 198, pp. 100-105 (1998).

Study on Characteristics of Wave Deformation and Transport of Coral Gravels Around Swash Zone

Shota Seto and Yoshimitsu Tajima

Dept. of Civil Eng., The University of Tokyo,
Bunkyo-ku, Tokyo,113-8654, Japan
E-mail: seto2@coastal.t.u-tokyo.ac.jp

This study investigates characteristics of wave deformation on a coral beach and behavior of coral gravels under such forward-leaning waves through two different laboratory experiments. All these experiments were captured by multiple video cameras mounted beside the glass wall of the flume and water surface fluctuations, topography change, near-bottom current velocities and gravel velocities were quantitatively extracted through the analysis of recorded images. Primary findings of the experiments are as follows: (i) deformation of nearshore waves and bed profiles dynamically interacted with each other and developed steeply sloping coral bed around the swash zone enhanced the formation of partial standing waves; (ii) initiation of motion of the gravels depends not only on the local bottom shear stress but also on the positions of each gravel relative to the others; (iii) after the initiation of motion, behavior of gravels were dominantly determined by the surrounding flow velocity but not by the bed conditions.

Keywords: Coral gravels; Forward-leaning waves; Surf zone; Porous bed.

1. Introduction

Understanding physical mechanisms of formation of coral cay is one of key tasks for designing and planning appropriate protection and conservation strategies of coral coasts. Ballast island, for example, is a small coral cay on the isolated coral reef located offshore of Iriomote-jima Island in Okinawa, Japan. This small island with the width of just several meters, the length of around thirty meters and the height of a couple of meters above the mean sea level, has been kept formed on the reef for more than 50 years although waves and currents on the reef have often changed the shape and locations of the island[1]. Mechanisms of such natural formation process of coral cays can be applied to sustainable coastal protection measures.

Authors carried out a field survey focusing on such natural processes of formation of Ballast Island and pointed out that wave-associated transport of coral gravels plays dominant role to form the coral cay through comparisons of numerical analysis and observed features[2]. This study, therefore, focuses on wave-associated transport of coral gravels and tries to capture the mechanism of forming coral gravel beach in the experiment.

In this study, two laboratory experiments were carried out; The first experiment, A, mainly focused on understanding the relationship between the morphology change and the deformations of waves on the coral gravel bed such as breaking, reflection and penetration. The second experiment, B, focused on the characteristics of each moving gravels on porous gravel beds.

2. Laboratory Experiments

Figures 1(a) and 2(a) respectively show the setups of laboratory experiments A and B. In both experiments, coral gravels with grain size of around 1cm, as shown in Figure 1(c), were placed on the plane fixed bottom of the flume with length of

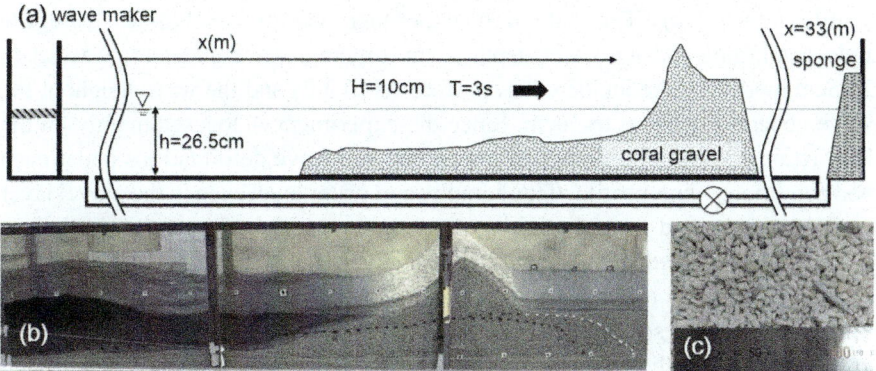

Fig. 1. (a): Setups of the Experiment A, (b): pictures of coral bed when the profile reached equilibrium state, (c): coral gravels used in the experiments.

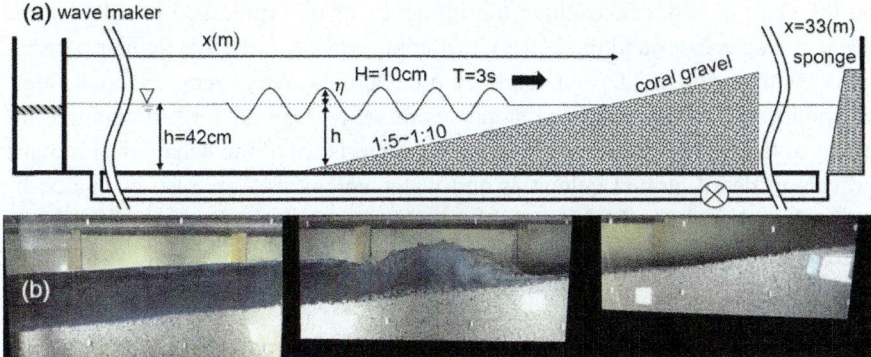

Fig. 2. (a): Setups of the Experiment B and (b): an example of rectified image.

33m and the width of 60cm. A piston-type wave generator was set to the left end of the flume and the stream function method[3] was applied for determination of time-varying displacement of the piston to generate linear and non-linear wave profiles. Two valves located in front of the wave generator and the right end of the flume were kept open to circulate the water so that the water levels on both side of the coral cay nearly equal to each other. The water in the flume was colored in blue and one side wall of the flume was colored in yellow so that fluctuating water surface, coral gravel movement and change of topography can be easily captured by three video cameras with flame rate of 60fps, mounted at the outside of the other side glass wall of the flume. Water surface fluctuation, behavior of coral gravels and topography change were extracted through the analysis of succeeding still images recorded by these video cameras.

In the experiment A, plane bed with uniform slope of 1/20 was prepared and then periodic regular waves were applied until the coral bed reaches equilibrium profile (Figure 1(b)). The scale of the experiment was assumed to be 1/10 of the field conditions and the water depth at the offshore wave generator was set to 26.5cm, period of the incident waves were set to 3.0s and the wave height at the wave generator was set to 10cm. Since the experiment A focused mainly on the interaction between the change of gravel bed and wave deformations under each bed conditions, time-profile of topography and water level even in the gravel layer were kept extracted from the initiation of wave generation to equilibrium state. The detail of procedure of image-based analysis will be explained in the next section. In addition to the image-based analysis, two wave gauges were installed to record the water surface fluctuations.

According to the results of Experiment A, formation of steep slope around run-up end is one of key elements to understand the formation of the coral beach. The experiment B, therefore, focused on the behavior of individual coral gravels on the sloping bed. To exclude the influence of deformed bed profiles on the wave characteristics and the behavior of coral gravels, only five regular periodic waves with period of 3.0s and height of 10cm at the wave generator were incident on uniformly sloping bed. Bed slopes were set to 1/5 and 1/10 to compare the effect of the bed slopes. To capture the characteristics of the behavior of moving gravels in detail, gravel velocities and water velocities were also estimated by image-based analysis, PIV, in addition to the water level. To apply PIV, tracer particles with specific density close to the water were put in the flume.

3. Image-based Analysis

This section outlines procedures of various image-based analysis for quantitative estimations of water surface fluctuations, current velocities and moving gravels.

First, recorded movies were converted to consecutive still images. Then, following Tajima et al. (2009)[4], recorded images were rectified based on the XY-coordinates on the side glass wall of the flume. Figure 2(b) shows one example of rectified still images recorded by three video cameras in the case of Experiment B with bed slope of 1/10. Since these rectified images are based on the plane rectangular coordinate system along the side glass wall, horizontal and vertical displacement or movement of fluid and gravels can be directly quantified once the target is specified in these rectified images. Thus, water surface fluctuation, movement of gravels and change of topography are quantitatively observed by these images.

In Experiment A, time profile of water surface and bed level were extracted by detecting the RGB-values of each pixel data. As shown in Figure 3, water level in the gravel layer was also observed by the following procedures: (i) convert the rectified images into gray-scale images based on the parameter, $Z = B - 1.2(R + G)$ with R, G and B representing RGB-values of each pixels; (ii) extract the pixel data along the vertical line at arbitrary horizontal position from succeeding images; (iii) place extracted vertical line data of succeeding images side by side in the order of the recorded time of each image. Figure 3(c) and (d), produced in this manner, show the time-varying water surface profiles in the gravel layer respectively at X=17.94m and X=18.02m. Horizontal axis of Figure 3(c) and (d)

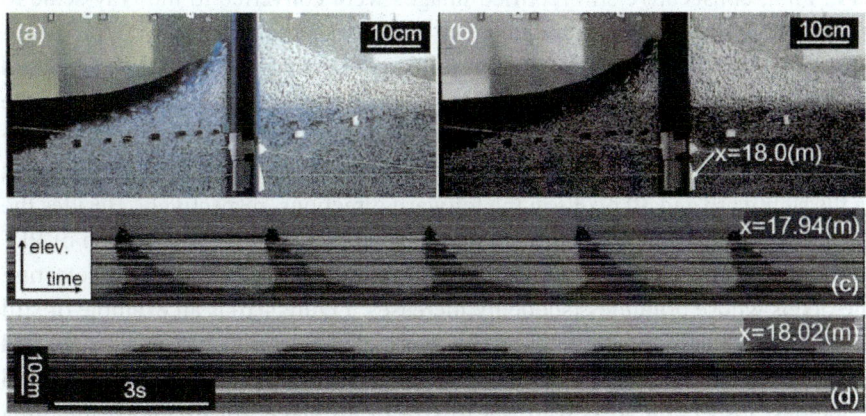

Fig. 3. An example of rectified image (a), gray-scale image (b) and time profiles of water level in the gravel layer at X=17.94 (c) and at X=18.02 (d).

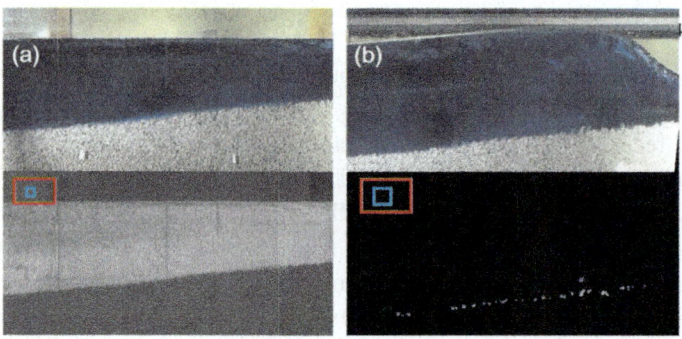

Fig. 4. An example of tracer particle-highlighted image (a) and extracted image of moving gravels (b).

therefore indicates the time while the vertical axis of these figures corresponds to the vertical distance. As seen in Figure 3(c), which shows the water surface profile in the layer of coral gravels at around the run-up end, waveform of the colored water has fluctuation of more than 10cm and shows forward-leaning or even overhanging profiles. Figure 3(d) shows the similar water surface fluctuations under the coral bed but the horizontal location is 8cm landside from the one of Figure 3(c). As seen in Figure 3(d), we can still observe certain water surface fluctuation while its amplitude largely decays. In this study, these procedures are applied to many horizontal positions in gravel layer and colored water elevation were observed at many locations.

In Experiment B, in addition to time-profile of water level as Experiment A, near-bottom horizontal velocities of fluid and gravel velocities were estimated by PIV. First, consecutive still rectified images were converted to the gray-scale to highlight the tracer particles based on the following parameter, $P2 = B/(R + G + B)$, as shown in Figure 4(a). In the figure, size of the range of inspection area used in PIV is shown in red rectangular with size of 20 pixels square and the size of the range of search area is also shown in blue rectangular which extends the inspection area up to 20 pixels in the upward and downward directions, 50 pixels in the rightward and 30 pixels in the leftward directions, respectively. Besides the current, velocity of moving gravels were also obtained by the application of PIV to the images of extracted gravels. Moving gravels were extracted by detecting the pixels whose brightness suddenly increases. Figure 4(b) is an example of still images where only the moving gravels were extracted with inspection area and search area shown in the red and blue rectangular. For estimation of gravel velocities, inspection areas were set to size of 40 pixels square and search area were the same as Figure 4(a).

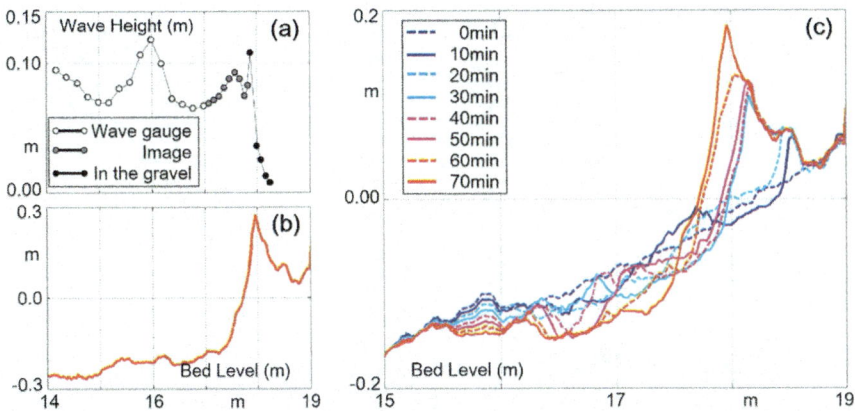

Fig. 5. Cross-shore distribution of wave heights (a) and the topography (b) at equilibrium profile
and time profile of bed level (c).

4. Interaction between Wave Deformation and Topography Change

This section outlines the observed relationship between the changing bed profiles
and wave deformations on the coral gravel bed.

Figure 5(a) shows the estimated wave heights, when bottom bed reached the
equilibrium profile shown in (b). Measured wave heights were determined by the
value of standard deviation of water surface fluctuation η multiplied by $2^{1.5}$.
White circles represent the estimated wave heights based on the values recorded
by wave gauges, gray circles represent those of image-based analysis and black
ones represent those in gravel layer estimated as explained in the former section.
White circles and gray circles are consistently connected at X=15m and it insures
the validity of the image-based analysis. As seen in the figure, partial standing
waves were developed in the surf zone. It is also interesting to note that sudden
rise of the wave heights were observed around the run-up end and they attenuated
in the gravel layer. This feature may be consistent with the overhanging wave
profiles shown in Figure 3(c), i.e., run-up waves partially and relatively slowly
permeate through the coral gravel bed layer in the downward direction.

To further investigate the relationship between wave and coral gravel bed,
time-profile of cross-shore distribution of wave heights and bed level were
analyzed from the initiation of wave generation (0min) to equilibrium profile
(70min). Figure 5(c) shows the time-varying bed profiles in every 10min. As seen
in the figure, steep bed was formed in less than 10min around the swash zone. It
can also be observed that developing speed of steep bed was not constant;
topography changed dramatically in the periods of 20~30min and 50~60min from

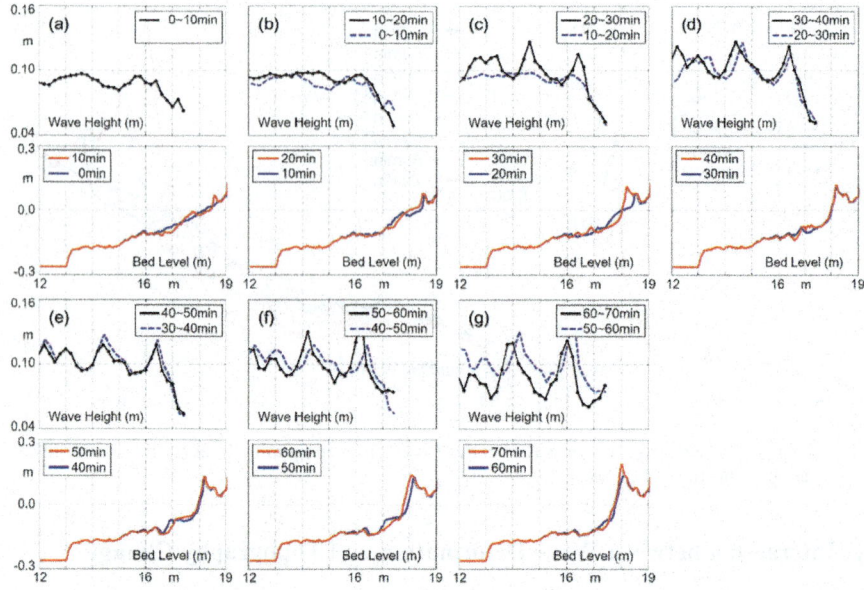

Fig. 6. Time-varying profile of cross-shore distribution of wave height (upper) and bed level at that time (lower).

the initiation of wave generation. Wave heights were calculated in every 10min based on all the waves within the each periods. Figure 8 shows the cross-shore distribution of measured wave heights (black solid line in upper panel) with two bed levels (lower), at the beginning of the period (blue) and the end of the period (red). Blue dot lines in the upper panels represent the measured wave heights 10min before that of black solid lines to compare the change of wave heights. At the beginning of the experiment, as seen in (a), steep bed around the surf zone was produced immediately and partial standing wave had started to develop. In the case of 20~30min (c) and 50~60min (f), partial standing waves were dominant and steep bed developed rapidly. On the other hand, when the characteristics of wave heights had relatively less changes, i.e., the case of 10~20min (b) and 30~40min (d), steep bed profiles around the swash zone also showed little changes. However, in those times, topography in relatively deeper water showed gradual changes. In (d) and (e), for example, a certain amount of gravels at X=13~15m moved to X=15~17m. From these features, it can be deduced that: (i) partial standing waves and steep bed around the swash zone grow rapidly in the beginning when gravels around the swash zone were dominantly transported to the shore; (ii) after the rapid development of steep bed around the swash zone, partial standing waves are developed and increased wave orbital velocity transports gravels at relatively deeper water to the onshore-ward direction. Based on these discussions, the modeling for the formation of coral gravel beach may require the following two essential features: (i) threshold condition for initiation of motion of coral gravels; and (ii) behavior of gravels on the developed steep bed where run-up waves were partially permeate.

5. Characteristics of Gravel Movement on Uniformly Sloping Bed

Following the discussions in the previous section, the second Experiment B was conducted to investigate the characteristics of each coral gravels on uniformly sloping bed under forward-leaning nearshore waves. In this experiment, behavior of each gravels were analyzed under the first five periodic waves with period of 3.0s propagating on uniformly sloping bed. Figure 7(a) and (b) shows the temporal and spatial distribution of water surface elevation with track of representative coral gravels around the swash zone. In the figure, magnitude of water level is indicated in the color map and vertical and horizontal axis indicate the time positive in downward direction and vertical coordinates of the bed level whose origin is set to mean water level respectively. Black areas mean no water was detected above the bed level. Blank circles represent the instantaneous horizontal locations of representative gravels. Therefore, the track of circles in the right down direction indicates the onshore movement of gravels. Figure 7(b) is

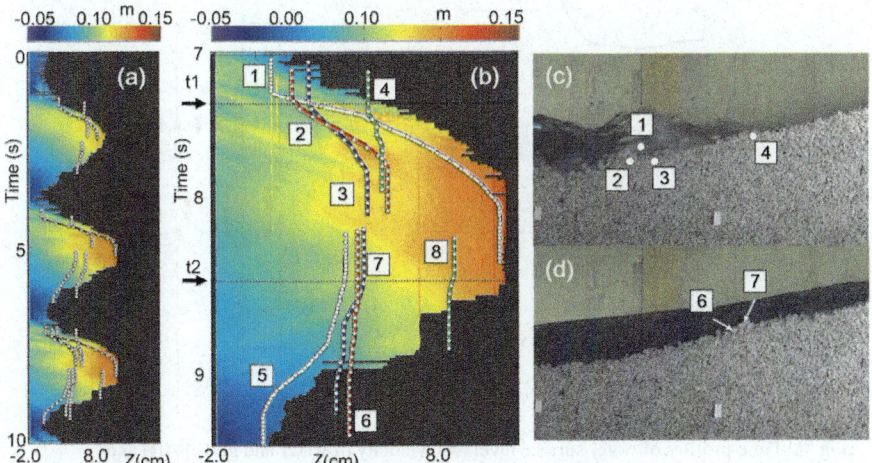

Fig. 7. Temporal and spatial distribution of water surface level with track of moving gravels for the first 3 waves (a) and enlarged view of the third wave (b). Snapshot images at time t1 (c) and t2 (d) indicated in (b). The identification number of gravels in (c) and (d) corresponds to the ones in (b).

418

the enlarged view of the third wave in (a). As seen in the figure, gravels tend to travel longer distance in the onshore-ward direction and shorter distance in the offshore-ward direction. This asymmetry around run-up end plays an important role in forming steep bed mentioned in the former section. Moreover, when waves are running up, the gravels located offshore side were transported for longer distance than the gravels at onshore side. The gravel with the identification number, 1, shown in (b), for example, overtakes the other representative gravels such as 2, 3 and 4. Figure 7(c) and (d) are the snapshots at the time of t1 and t2 respectively shown in (b). In Figure 7(c), gravel No.1 passes the others. Such different onshore-ward velocity of gravels, 1, 2, 3 and 4, may be explained by instantaneous vertical locations of each gravels. The gravel No.1 is located above the gravel bed and may have little influence of the bottom friction force. Also, these Figures 7(b) and (c) indicate that once a gravel is slightly floated, it moves with the surrounding fluid velocity. When the waves are running down, on the other hand, no gravels passes the others in the offshore-ward direction except for the gravel No.6 and 7. This exception may be due to positions of each gravel: while No.6 was partially buried under the gravel layer, No.7 was located on top of the gravel bed.

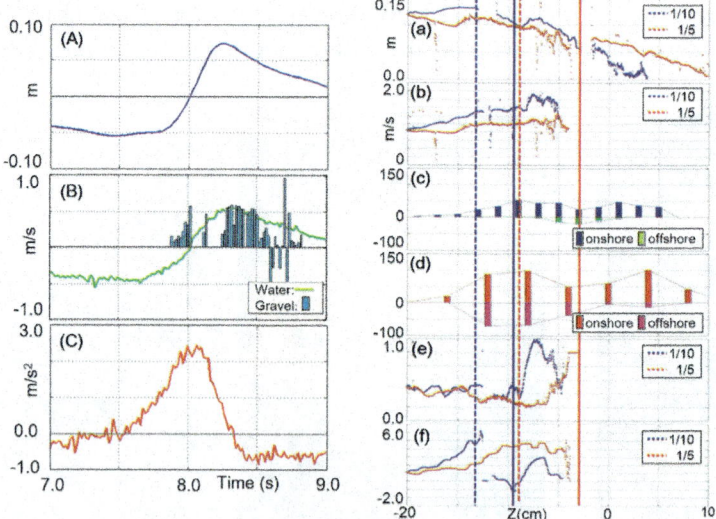

Fig. 8. Time-profiles of water surface level (A), velocity of water and gravels (B) and acceleration (C) in the left. The right figure shows the cross-shore distribution of wave height (a), estimated velocity (b), number of gravels which pass the each horizontal positions onshore-ward and offshore-ward per single wave in the case of 1/10 (c), in the case of 1/5 (d), skewness of water surface elevation (e) and skewness of time derivative of water surface elevation (f).

To further investigate this feature, gravel velocities crossing each horizontal locations are analyzed. Figure 8 shows the time-profile of water surface level (A), horizontal velocity of fluid (B), acceleration (C) and gravel velocities as blue bars in (B), positive in onshore-ward direction and negative in offshore-ward direction. Each values were observed at the depth of 17cm in the case of bed slope was 1/10 with period of 3.0s. The acceleration of the fluid was estimated based on the filtered time-series of the horizontal flow velocity. The low-pass filter based on FFT was applied and the frequency components higher than 50Hz was excluded. As seen in the figure, gravels start to move before currents turned to onshore-ward direction while the acceleration was at its peak. This feature indicates that return flow at the bottom and inertia force have an influence on the initiation of motion of gravels. However, after the initiation of motion, gravel velocities are nearly equal to the surrounding fluid velocity. These features indicate that bottom friction force has little impact on the behavior of gravels after its initiation of motion.

Finally, Figure 8(c) and (d) show the number of gravels which pass the each horizontal positions positive in the onshore-ward and negative in the offshore-ward directions under the single wave in the case of 1/10 (c) and 1/5 (d) bottom slopes, respectively. Horizontal axis of these figure is the vertical coordinates of the bed level. The number of gravels are analyzed with the cross-shore distribution of wave heights: H (a), estimated velocities based on the assumption of linear

$$S_{k,\xi} = \frac{\left\langle \left(\xi - \xi_{avg} \right)^3 \right\rangle}{\left\langle \left(\xi - \xi_{avg} \right)^2 \right\rangle^{\frac{3}{2}}}$$

shallow water equation, which is defined as $U = H\sqrt{g/h}$ (b), skewness of water surface elevation defined as the following equation (e) and skewness of the time derivative of water surface elevation (f).

Compared blue and red lines representing the case of 1/10 and 1/5 respectively in each panels, wave heights (a), estimated velocities (b) and skewness of water surface elevation (e) have the opposite trend with the number of moving gravels while only the skewness of time derivative of water surface elevation (f) has the same trend. This feature indicates that asymmetric acceleration, which has large influence of the bottom shear stress, affects the amount of gravel transport and other factors don't affect much. It can be said that while the velocities of gravels transported under the waves are decided mainly by the fluid condition, the amount of them is determined mainly by the bottom shear stress. Therefore, to estimate the topography change of coral bed, it is necessary to capture both the induced forces to the gravel bed and the current velocities.

6. Summary and Conclusion

Laboratory experiments and image-based analysis were carried out to investigate the important factors in formation of coral gravel beach. Based on the results of them, steeply sloping bed is formed by shoaling waves around run-up end and it develops partial standing waves. Increase of the wave heights due to the development of partial standing waves promotes the onshore-ward transport of gravels. In terms of each gravel movements, it was revealed that while coral gravels start to move mainly from bottom shear stress, after the initiation of motion, their behavior are determined mainly by the fluid condition around it. It also seems to be important that their horizontal and vertical positions relative to the others affect to the behavior of moving gravels.

Acknowledgments

Authors acknowledge that a part of this research was based on the research project funded by the Ministry of Land, Infrastructure, Transport and Tourism, Japan, and the University of Tokyo Ocean Alliance.

References

1. Kayanne, H., Aoki, K., Suzuki, T., et al., 2016. Eco-geomorphic processes that maintain a small coral reef island: Ballast Island in the Ryukyu Islands, Japan, *Geomorphology*, 271, pp. 84-93.
2. Takemori, R., Tajima, Y., Fujikawa, H. and Kayanne, 2015, Investigation on characteristics of local accumulation of coral gravels on an isolated reef, Journal of JSCE, B2 (Coast. Eng.), 71, No. 2, pp. I_721-I_726 (in Japanese).
3. Dean, R.G., 1965. Stream function representation of non-linear ocean waves, *Journal of Geophysical Research*, vol. 70(18), pp. 4561-4572.
4. Tajima, Y., Liu, H. and Sato, S., 2009. Dynamic changes of waves and currents over the collapsing sandbar of the Tenryu river mouth observed during Typhoon T0704, *Proc. Coastal Dynamics 2009*, World Scientific.

Hydraulic Model Experiment and Numerical Calculation on Scour

Kohei Suzuki

Civil Engineering, Chuo University,
Tokyo, Bunkyo ku, Kasuga 1-13-27, 112-8551, Japan
E-mail: a13.nddr@g.chuo-u.ac.jp
www.chuo-u.ac.jp

Katsumi Seki

Research and Development Initiative, Chuo-university,
Tokyo, Bunkyo ku, Kasuga 1-13-27, 112-8551, Japan
E-mail: seki-k.15e@g.chuo-u.ac.jp

Taro Arikawa

Faculty of Science and Engineering, Chuo-university,
Tokyo, Bunkyo ku, Kasuga 1-13-27, 112-8551, Japan
E-mail: taro.arikawa.38d@g.chuo-u.ac.jp

It is important to estimate the scouring depth for the design of coastal facilities against tsunamis. The past researches indicated that the maximum scouring depth was proportional to the overflow time. In this study, in order to clarify the relation between overflow time and scour depth behind the structure, experiments and numerical calculations were carried out and analysis of the past experiments was conducted and discussed. The results showed that the scour depth could be evaluated approximately using the total flow rate and the verification of the numerical simulations were confirmed

Keywords: Scour; Overflow; Tsunami; CADMAS-SURF.

1. Introduction

Huge tsunamis caused numerous damages to coastal structures in Tohoku area in the Great East Japan Earthquake of 2011. In breakwaters, Arikawa et al. (2013) indicated that many were washed away by the water level difference outside and inside the harbor and some were falling down due to backwashing due to overflow[1]. They also clarified that the scouring behind reduced the bearing capacity of the ground and lead to collapse of the caisson.

Therefore, it is important to estimate the scouring depth for the design of coastal facilities against tsunamis. Noguchi et al. (1997) conducted the scouring test in front of the quay walls by using receding flow due to solitary waves[2]. The results indicated that the maximum scouring depth was almost same as the size of the vortex formed by the falling water mass. The different scale experiments behind the caisson were performed by Arikawa et al. (2014)[3]. The results indicated that the scale effect could be applied to the sedimentation rate matching method and that the maximum scouring depth was proportional to the overflow time. So, the coefficient in the scouring depth formula submitted by Arikawa is almost 3 times as that in the Noguchi's formula. The reason is considered to depend on the overflow time. They also pointed out that the scouring depth will not be deeper any further when the falling height of the water mass is above a certain level. Arikawa et al. (2015) also showed this limitation by using the small scale experiments[4]. So, it is necessary to clarify the relationship between overflow time and scour depth.

In this study, in order to clarify the relation between overflow time and scour depth behind the structure, experiments and numerical calculations were carried out and analysis of the past experiments was conducted and discussed

2. Experiment

2.1. *Summary of experiment*

Experiments were conducted in a 0.3 m wide and 0.5 m deep steel flume in which a model of upright seawall was installed. The schematic diagram of the experiment is shown in Figure 1. The geomaterial (2.0 m long, 0.3 m wide, and 0.4 m height) used was Tohoku Keisa No. 6 sand (D_{50} = 0.34mm), and a pump were used to circulate in the flume to simulate long time overflow. The conditions of each of the experimental cases is summarized in Table 1.

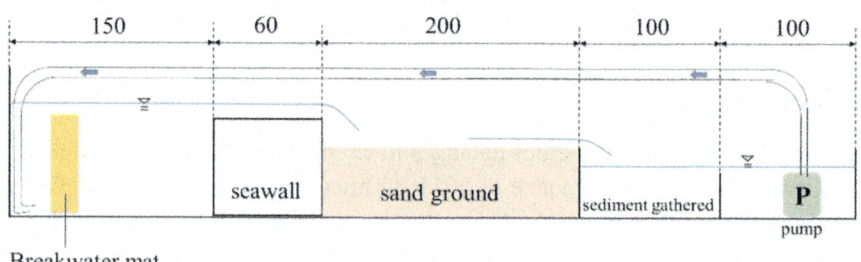

Fig. 1. Schematic diagram of scour experiment (unit : cm)

Table 1. Experimental Conditions

Case No.	$\dfrac{Zf}{[cm]}$	$q[m^2/s]$	Case No.	$\dfrac{Zf}{[cm]}$	$q[m^2/s]$	Case No.	$\dfrac{Zf}{[cm]}$	$q[m^2/s]$
1	5		6	15		9	15	
2	10		7	20	0.0024	10	20	0.0055
3	15	0.0043	8	25		11	25	
4	20							
5	25							

Where, Zf indicates the overflow height and q represents the per unit width flow rate. Five different overflow heights are considered (5 cm, 10 cm, 15 cm, 20 cm, and 25 cm), for an overflow rate of 0.0043 m^2/s.

Furthermore, for drop heights of 15 cm, 20 cm, and 25 cm, the overflow rate was changed (0.0024 m^2/s and 0.0055 m^2/s). The experimental simulations were recorded using a video camera (59.94 fps) to observe scouring.

2.2. Experiment results

Figure 2 shows the scour phenomenon for case 5 and Figure 3 shows the temporal change in scour depth.[b] The sand ground was scoured vigorously for up to approximately 60 seconds after the start of overflow, and thereafter, it was observed to scour slowly.

The scour depths when Zf are 15 cm, 20 cm, 25 cm respectively are similar. According to the past experiment conducted by Arikawa et al. (2015), the maximum scour depth increases as the overflow height increases to a particular height, however, the maximum scour depth decreases exceed a particular height because the diameter of the vortex decreases at such overflow heights.[4] For this reason, it is considered that the scour depth when Zf are 15 cm, 20 cm , 25 cm are similar because of the maximum scour depth height around 20cm in these experiments.

The vortex caused by the overflow water-mass was investigated in this experiment. Figure 4 shows the vortex for case 3. The diameter of the vortex was measured from the locus of the sand that was rolled up. Figure 5 shows the relationship between the vortex diameter and scour depth. The scour depth increased approximately in proportion to the vortex diameter till the diameter reached approximately 15 cm. After that, however, the diameter appeared to be approximately constant, while the scour depth increased.

---

[b] In figure 3, the past experimental results conducted by Arikawa et al. (2015) are included. (Zf = 45 cm)

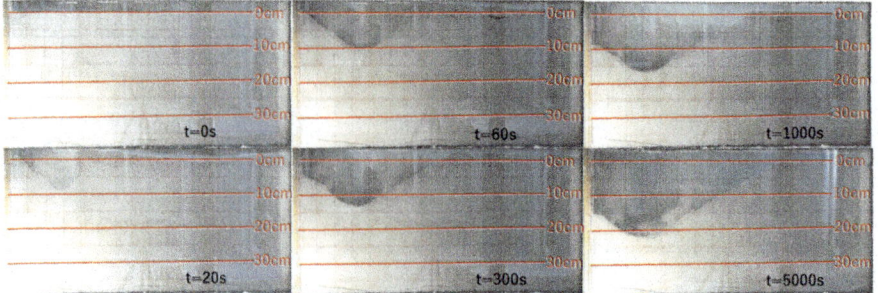

Fig. 2. Change of scouring(case 5)

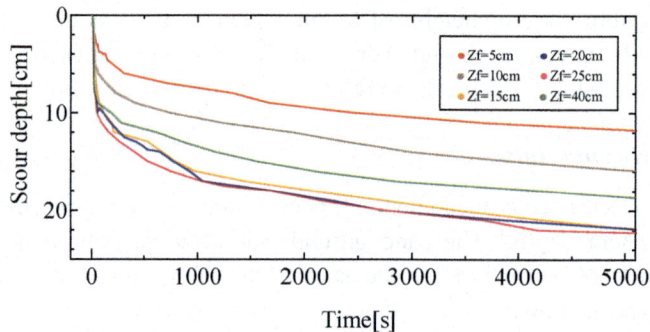

Fig. 3. Temporal change of scour depth

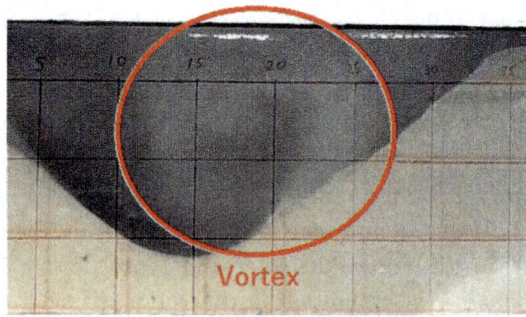

Fig. 4. The observed vortex

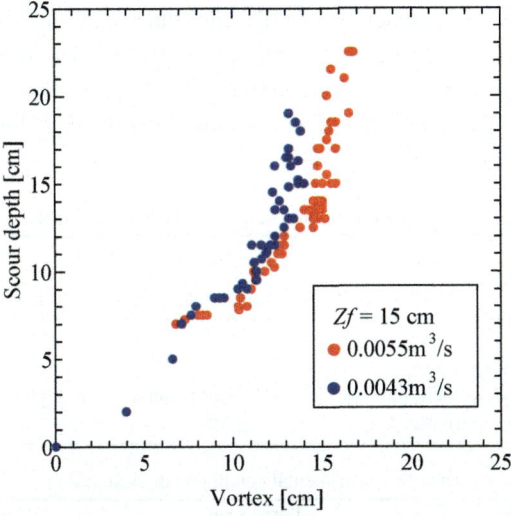

Fig. 5. The relationship between scour depth and the vortex

3. Law of Similarity

In order to confirm the difference in scale, the results of the past experiments conducted by Arikawa et al. (2014) were reviewed[3]. Since the grain size was different in these experiments, the similarity rule was considered.

3.1. Similar method

Based on the method of Yamano et al. (2013), a law of similarity using sedimentation velocity was applied as shown in Eq. (1)[5].

$$W_{0m}\Big/W_{0p} = \left(l_m\big/l_p\right)^{\frac{1}{2}}, \qquad (1)$$

where 'W_0' is the sedimentation velocity (m/s), 'l' is the representative length, 'm' denotes the model, and 'p' represents the prototype. The equation of the sedimentation velocity is calculated from Eq. (2) and Eq. (3).

$$W_0 = \sqrt{Sgd_N}\,(0.954 + \tfrac{5.12}{S_*})^{-1} \qquad (2)$$

$$S_* = \tfrac{d_n}{4v}\sqrt{Sgd_N}, \qquad (3)$$

where S is the specific gravity of water, g is the gravitational acceleration, d_N is the considered grain diameter ($=d/10$, d is the sediment diameter), and v is the coefficient of kinematic viscosity. Table 2 shows the sedimentation velocity in these experiments. Assume that the small scale experiment of Arikawa et al. (2014) is 1/42, the sedimentation velocity of local sand is 15~16 cm/s, and the others model scale that satisfies this sedimentation velocity is 1/7 (large scale), and 1/12 (this experiment). The experimental conditions equivalents in local scale for each of the experiments are shown in Table 3.

Table 2. Sedimentation velocity in each experiment

| | Arikawa et al. (2014) | | This experiment |
	Small scale	Large scale	
Grain diameter[cm]	0.021	0.43	0.034
Specific gravity in water		1.65	
Coefficient of kinematic velocity[cm²/s]		0.0011(pure water; 15degrees [Celsius])	
W_0(sedimentation velocity)[cm/s]	2.38	5.92	4.58

Table 3. Experimental conditions in local scale

| Small scale | | | Large scale | | | This experiment | | |
Case No.	Zf [m]	q[m²/s]	Case No.	Zf [m]	q[m²/s]	Case No.	Zf [m]	q[m²/s]
s-1	10.1	7.1	1-1	7	2.23	t-1	2.4	0.229
s-2	14.3	12.2	1-2	7	3.15	t-2	3.0	0.229
s-3	20.2	11.7	1-3	7	5.33	t-3	3.0	0.357

3.2. Results

Noguchi et al. (1997) proposed the evaluation formula of the scour depth as shown in Eq. (4) and (5)[2].

$$R = g^{-\frac{1}{4}}q^{\frac{1}{2}}Z_f^{\frac{1}{4}} \tag{4}$$

$$D = 2.1R , \tag{5}$$

where R is the vortex diameter evaluated from energy disposal, g is the gravitational acceleration, q is the per unit width flow rate, Zf is the drop height, and D is the scour depth. However, the relationship between the scour depth and time series is not clear.

The prediction formula considering the time series for scour depth was designed from these experimental results. Figure 6 shows the temporal change of scour depth considering the law of similarity in each experiment. The result of these experiments are verified using Eq. (4) and (5). Figure 7 shows the relationship between scour depth and R from Noguchi's formula. The vertical line shows scour depth, and the horizontal one depicts R. In addition, the scour depth plots four types, 100, 200, 300, and 400 seconds after the overflow. As an example, D_{100} represents the scour depth at the 100 second after the overflow. it is observed that the scour depth at a certain time is in proportion to R.

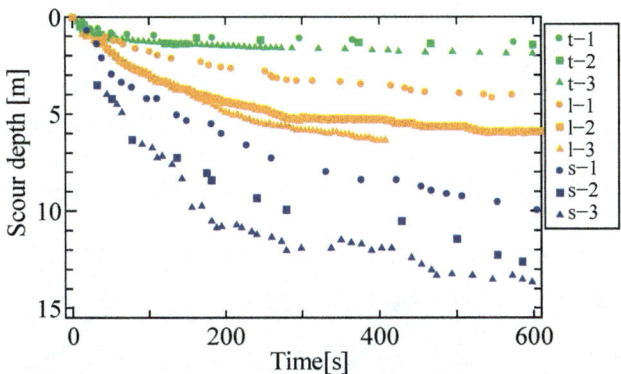

Fig. 6. Temporal change of scour depth in local scale

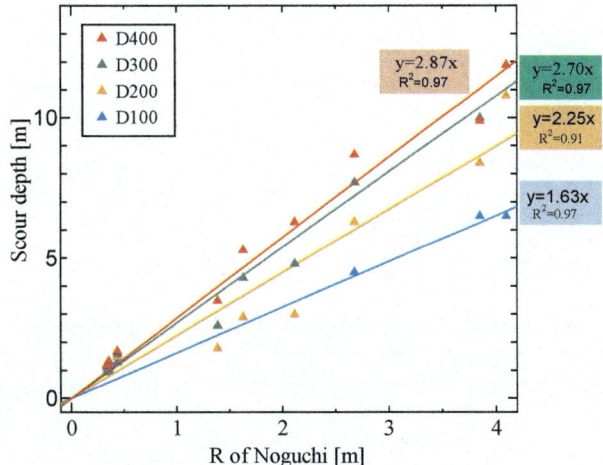

Fig. 7. Relationship between scour depth and R of Noguchi in local scale

Next, assume that the temporal change of scour depth is given Eq. (6) in term of the total flow rate.

$$D(t) = 0.2 \times \sqrt{qt} \tag{6}$$

428

The number 0.2 is a coefficient determined from these experimental results. The graph of this estimation equation in local scale is shown in Figure 8. The vertical line shows the scour depth and the horizontal line represents the time. Figure 8 shows that the scour depth in local scale can be substantially predicted using Eq. (6).

In addition, it was not found the relationship the change of initial scour depth and overflow height Zf. However, It seems that the initial scour depth does not greatly exceed Eq. (6) because the overflow height Zf in these experiments is the overflow height that takes the maximum scour depth at the flow rate. Although, the future task is to take into account the overflow height into Eq. (6).

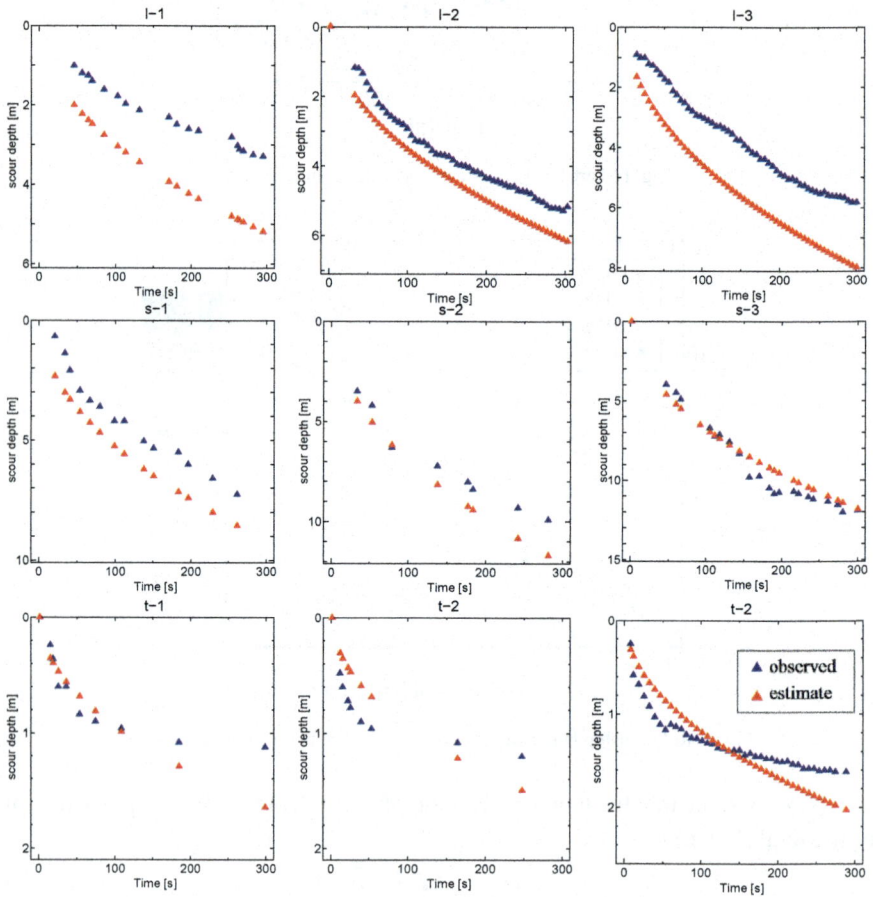

Fig. 8. Estimate and observed value on scour in local scale

4. Numerical Calculation

This study attempt to calculate the results of past experiment using Three-dimensional numerical calculation and sediment transport calculation model (CADMAS-STM). CADMAS-STM is consist of two solver. One is fluid analysis (CADMAS) and the other is sediment transport analysis (STM)(Takahashi et al., 1999, 2011)[6][7].[7] The fluid analysis is based on the CADMAS-SURF/3D (Arikawa et al. (2005)), which are continuous equation and 3D N-S equation with VOF[8]. The sediment transport model (STM) is incorporated in CADMAS-SURF/3D. CADMAS is a calculation of three dimensional model and STM is a two dimensional model. The CADMAS cell size in the axis of $X Y Z$ is $L \times M \times N$, the STM cell size in the axis of $X Y$ correspond to $L \times M$, and the date communications of both programs is $L \times M$.

4.1. *Fundamental equation*

The continuous equations of Sediment Transport Model (STM) are shown below.

$$\frac{\partial z}{\partial t} + \frac{1}{1-\lambda}\left\{\frac{\partial q_{Bx}}{\partial x} + \frac{\partial q_{By}}{\partial y} + w_{ex}\right\} = 0 \tag{7}$$

$$\frac{\partial (C_S h_S)}{\partial t} + \frac{\partial (M C_S)}{\partial x} + \frac{\partial (M C_S)}{\partial y} - w_{ex} = 0 \tag{8}$$

where, Z is the height from the base level, λ is the porosity, q_{Bx} and q_{By} are the bed load transport per unit width and unit time in x and y directions, w_{ex} is the sediment discharge in the vertical direction per unit area and unit time, C_S is the average suspended sediment layer concentration, hs is the suspended sediment layer thickness, and C_S is the flow flux. In addition, q_B and w_{ex} are formulated as shown below based on the results of experiments.

$$\Phi_B = 2.6\tau_*^{3/2} \tag{9}$$

$$\Psi_{ex} = 0.000016\tau_*^2 - \frac{w_0 C}{\sqrt{sgd}}, \tag{10}$$

where, Φ_B is the bed load transport, Ψ_{ex} is the pickup sand, τ_* is the shields number, s is the sand specific gravity in water, g is gravitational acceleration, d is the grain diameter, w_0 is the sedimentation velocity according to the Rubey equation, and C is the suspended sediment layer concentration. Furthermore, the bed load transport and the pickup sand are made dimensionless by following equation.

$$\Phi_B = \frac{q_B}{\sqrt{sgd^3}} \tag{11}$$

$$\Psi_{ex} = \frac{w_{ex}}{\sqrt{sgd}} \tag{12}$$

4.2. *Calculation conditions*

Next, calculation on the large scale scouring experiment conducted by Arikawa et al. (2014) is performed.[3] The calculation chart and conditions are shown in Figure 9. In this calculation, the water level h is set to 0.5m from the experimental result.

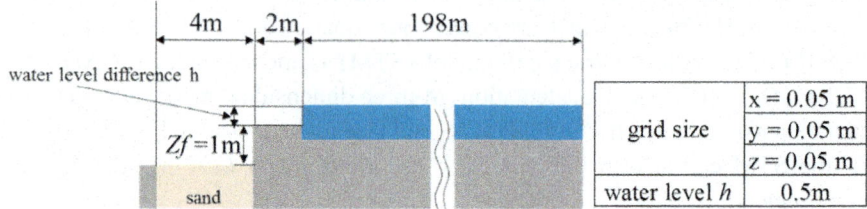

Fig. 9. Calculation chart and conditions

4.3. *Calculation results*

The water level date at overflow section in Figure 10. The calculated values correspond to the experimental data average of 0.148 m. The graph comparing the experimental and calculation results is shown in Figure 11. The scour depth was approximately 1.0 m in the experiments, whereas, the calculation was approximately 0.85 m. In addition, the calculation resulted in a higher rate of progress of scouring as shown in Figure 11.

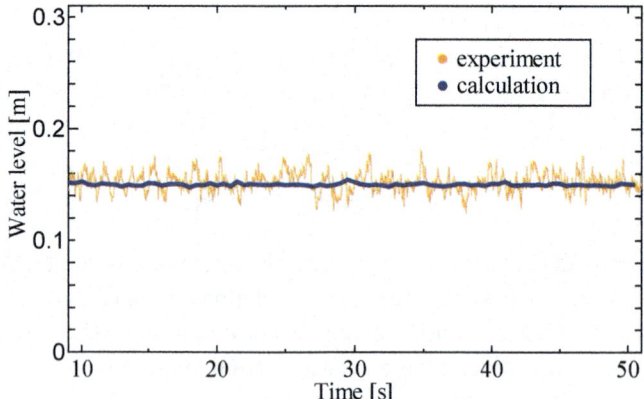

Fig. 10. Water level on calculation

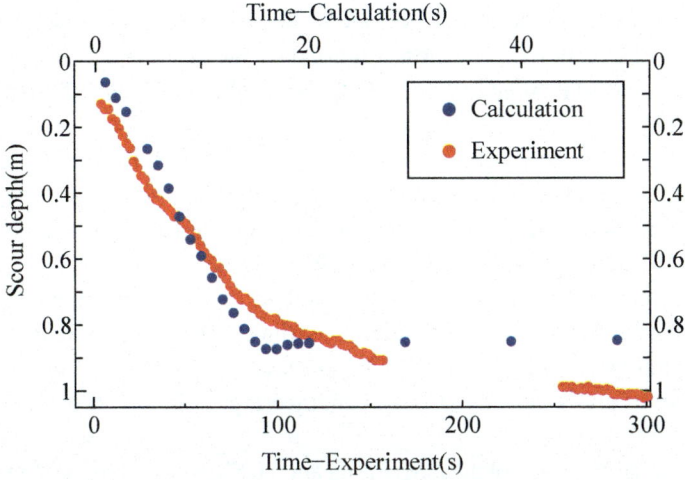

Fig. 11. Comparing experiment and calculation results

5. Conclusion

(1) It was found that the scour depth at a certain point of time is in proportion to the parameter R.

(2) The scour depth can be evaluated approximately using the total flow rate.

(3) Result from scouring calculation based on a lattice method using Sediment Transport Model showed that the maximum scouring depth was underestimated by approximately 15%. In addition, investigation of the mesh size and coefficients of the STM equations is future tasks.

References

1. T. Arikawa, M. Sato, K. Shimosako, T. Tomita, G. S. Yon and T. Niwa, Failure Mechanism and Resiliency of Breakwaters under Tsunami, Technical note of the Port and Airport Research Institute, No. 1269, 2013

2. K. Noguchi, S. Sato and S. Tanaka, Large scale model experiment on seawall overtopping and front scouring due to tsunami run up(in Japanese), Proceedings of coastal engineering, Vol. 44 (JSCE, Japan, 1997), pp. 296-300.

3. T. Arikawa, T. Ikeda and K. Kubota, Experimental sudy on scour behind seawall due to tsunami overflow(in Japanese), Journal of Japan society of civil engineers, Ser. B2, Coastal engineering, Vol. 70, No. 2 (JSCE, Japan, 2014) p. I_926-I_930.

4. T. Arikawa, S. Ueda, H. Igarashi and K. Seki, Effect on scour depth of falling height of overflow(in Japanese), Journal of Japan society of civil engineers, Ser. B2, Coastal engineering ,Vol. 72, No. 2 (JSCE, Japan, 2016), p. I_1087-I_1092.

5. T. Yamano, R. Fujiwara, K. Nomura and K. Shiraki, Field observation and reproduction experiment of local scour around the pile(in Japanese), Journal of Japan society of civil engineers, Ser. B3, Ocean Engineering, Vol. 69, No. 2 (JSCE, Japan, 2013) p. I_874-I_879.

6. T. Takahashi, N. Shuto, F. Imamura and D. Asai, Development of movable bed model for tsunamis considering sedimentation transportation between bed load layer and suspended sediment layer(in Japanese), Journal of Japan society of civil engineers, Coastal engineering, Vol. 42(JSCE, Japan, 1999), pp. 606-610.

7. T. Takahashi, T. Kurokawa, M. Fujita and H. Shimada, Hydraulic experiment on sediment transport due to tsunamis with various sand grain size(in Japanese, Journal of Japan society of civil engineers, Ser. B2, Coastal Engineering, Vol. 67, No. 2 (JSCE, Japan, 2011), p. I_231-I_235.

8. T. Arikawa, F. Yamada, M. Akiyama (2005): Study of Applicability of Tsunami Wave Force in a Three-dimensional Numerical Wave Flume, Collected Papers on Marine Engineering, Vol. 52, pp. 46-50.

Morphological Interactions between Navigation Channel and Estuary after the Improvement Works in Yangtze Estuary

Han Yufang[†] and Lu Chuanteng

Nanjing Hydraulic Research Institute,
Nanjing, 210024, China
[†]E-mail: yfhan@nhri.cn
www.nhri.cn

Yangtze Estuary deepwater channel improvement project started in 1998. The channel depth have been dredged from 7.0m(under Theoretical Lowest Water Level) to 8.5m, 10m and 12.5m. After the project, which formed the regulating line along the groins heads and adjusted the flow fields, deepened the mouth bar area and merged the ebb with flood channel, a faintly-curved channel was obtained in the North Passage. At the same time, in other passages such as in South Passage and North Channel, the mouth bar is still persisting after the project as long as mouth bar terrain has disappearing in north passage. In this study we will focus on the Yangtze Estuary, with major aims: (a) to reveal the evolution process through field data analysis before and after the project reveal, and compare the different evolution model of three main passages in Yangtze Estuary, (b) to understand and quantify the effect of measures like the navigation works on the natural morphological system, (c) to obtain more insight in the large-scale morphological development in estuaries and the underlying processes and mechanisms, with specific interest in the bifurcating channels and the development of the inter-tidal flats, (d) to predict the long-term impact of the human interventions to the large scale morphological development of the estuary, to understand and quantify the large-scale response of the morphological system to human interventions with emphasis to those related to navigation channels. It is aimed to improve our fundamental knowledge on small-scale processes around navigation channels and increase the understanding of the large-scale response of estuaries, to explain the complex phenomenon of the high siltation in the North Passage of Yangtze estuary. The results will be useful for a proper management of the estuary.

Keywords: The Deep Waterway Improvement Works; Yangtze Estuary; North Passage; Mouth bar; Fine sediment; Morphological interaction.

1. Background

Yangtze Estuary is located at the north of Shanghai, China, which of the longitudinal length is about 160 kilometers. The width at the entrance is about 90 km. Yangtze Estuary is a delta characterized by ample flow and sediment and obvious tidal influence, which create a basic regime of the estuary characterized as three-stage bifurcation, four-river mouth split, shoal developed, available navigation channel alternated, sandbar and submerged delta stretching.[1,2,3,5] Estuaries like the Yangtze Estuary in China serve as important waterways for navigation. However, waterways in natural estuaries often have shallow parts forming obstacles for navigation.In Yangtze Estuary, the ETM (Estuary Turbidity Maximum) is always accompanied by a board shallow area (basically around 6 m water depth) in the mouth zone, called mouth bars. Deepening of navigation channels through these shallow areas are major human interferences. They can have two major consequences: (1) The channel is silting up; and (2) The channel changes the natural morphological system, harming the integrity of the estuary. The first consequence leads to (costly) maintenance dredging. The second consequence may lead to a different system, like silting up of complete branches and causing serious environmental impact.

Before Yangtze Estuary deepwater channel improvement project in 1998 in North Passage, ebb channel is located at south of the river and flood channel is closed to north, between the ebb and flood channel, mouth bar (sill) formed due to the complex morph-dynamic conditions and sediment flocculation in the maximum turbidity area (same area with mouth bar). After more than 40 years study[1,2,4,5]. North Passage was selected as the deep-draft navigation channel and the regulation works started in 1998. The channel depth had been dredged phasing from 7.0m (under Theoretical Lowest Water Level) to 8.5m, 10m and 12.5m. Thus extensive engineering works, consisting two long training dikes and 19 groins, are implemented to achieve deeper water depth with the help of dredging activities in the North Passage of the Yangtze Estuary.

After the project, which formed the regulating line along the groin's heads and adjusted the flow fields, deepened the mouth bar area and merged the ebb channel with flood channel, a faintly-curved channel was obtained in the North Passage. At the same time, in other passages such as in South Passage and North Channel, the mouth bar is still persisting after the project as long as mouth bar terrain has disappearing in north passage.

2. The Main Projects in Yangtze estuary and the Impacts to Morphological System

With the rapid development of China's economy and society, the demand for cargo transport increases dramatically. It is necessary to have a deep navigation channel in Yangtze Estuary to meet the demand of large-sized ships to come into and out of Shanghai harbor and the Yangtze River Valley. After more than 40 years study, North Passage was selected as the deep-draft navigation channel and the regulation works started in 1998.

The Improvement Project was selected to locate in South Channel and North Passage, where there were the best river pattern and construction conditions. The project, combining regulating structures with channel dredging, constructs the diversion gap control works at the bifurcation area of North Passage and South Passage, two 50km long training dikes and 19 groins alongside the passage, see Fig. 1. The total length of the structures reaches 141.484km. The navigation channel would be dredged and maintained to the depths of 12.5m under theoretical lowest tide level with the bottom 350-400m wide. Thus the third/fourth generation container ships will be able to enter Yangtze Estuary under all weather conditions, while fifth generation container vessels and 100,000-ton bulk carriers can use a tidal window.

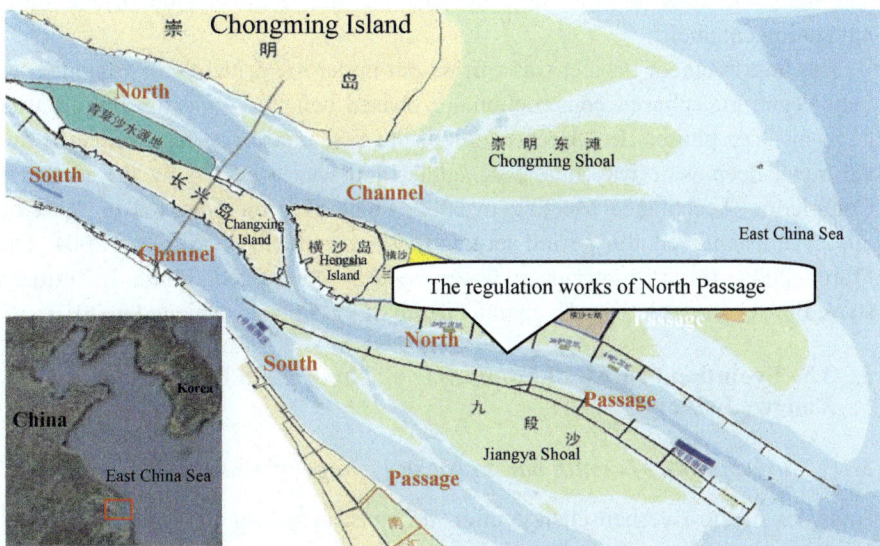

Fig. 1. General Plan of Yangtze Estuary Deepwater Channel Improvement Project

The regulation works of North Passage, combining regulating structures with channel dredging works, based on channel dredging as the guideline for the design, is set to construct training dikes and groins along with the side of passage, so as to give full play to the functions of water diversion, sand retention and siltation reduction by improving the flow field conditions of North Passage.

The regulating structures include two training dikes, one split point control works between North Passage and South Passage, and 19 groins along the two dikes.

The diversion gap control works are constructed to stabilize South Jianya Shoal and keep it from eroding to maintain the river regime of North Passage and South Passage, retain flow and sediment distribution rate between North Passage and South Passage.

South and North training dikes are constructed to further shape river behavior of North Passage and provide support for the groins to form the regulating lines, stop the influence of sediment stirred up by wind wave on the shoals beside the channel, block ebbing current diverted from Jiangya Gully to North Passage, and prevent sediment of North Channel from discharging to North Passage through Hengsha East Shoal Gully.

The Groins are constructed to form reasonable regulating lines of the channel and maintain flow field benefit the shaping and maintenance of the deep navigation channel.

The Improvement Project was carried out under the principle of "planning at very beginning, phased construction for phased benefits", and the Project was done in three phases. In March of 2000, the first phase project completed the 8.5-meter-deep waterway before schedule, and was accepted by the National Completion Acceptance Meeting in 2002. Phase II Project started in April of 2002, and achieved the period target of 9-meter-depth in May of 2004. On March 2005, the 10-meter-deep waterway was completed. Phase III Project started in April of 2009, and a channel depth of 12.5m was achieved in 2010.

3. The Evolution Process of Mouth Bar of Three Main Passages in Yangtze Estuary

3.1. *The evolution of mouth bar in North Passage of Yangtze estuary*

Analyses of the riverbed change after the construction of the first and second phase regulation works show that the regulation principle was presented which is compared with nature morphology change in the Yangtze Estuary.

The general effects of the regulation works in North Passage can be shown on Fig. 2 and Fig. 3.

The construction of split point control work between the North Passage and South Passage stabilized the central shoal and stopped the continuous erosion and recession of the shoal ridge, which provides a favorable boundary for the distribution of runoff and sediment transport.

The two training dikes give a stable boundary for North Passage, which block the water and sediment interchange between shoal and channel and give a effective barrier for sediment transport from shoal to channel during typhoon season or winter storm.

Before the project, ebb channel is located at south and flood channel is closed to north, between the ebb and flood channel, mouth bar (sill) formed due to the complex morph-dynamic conditions and sediment flocculation in the maximum turbidity area (same area with mouth bar).

After the project, which formed the regulating line along the groins heads and adjusted the flow fields, deepened the mouth bar area and merged the ebb with flood channel, a faintly-curved channel was obtained in the North Passage.

In the groin fields, sedimentation happened due to the velocity decreasing in these areas. The channel transferred from a wide-shallow river pattern to a narrow-deep pattern, which is favorable for channel maintenance.

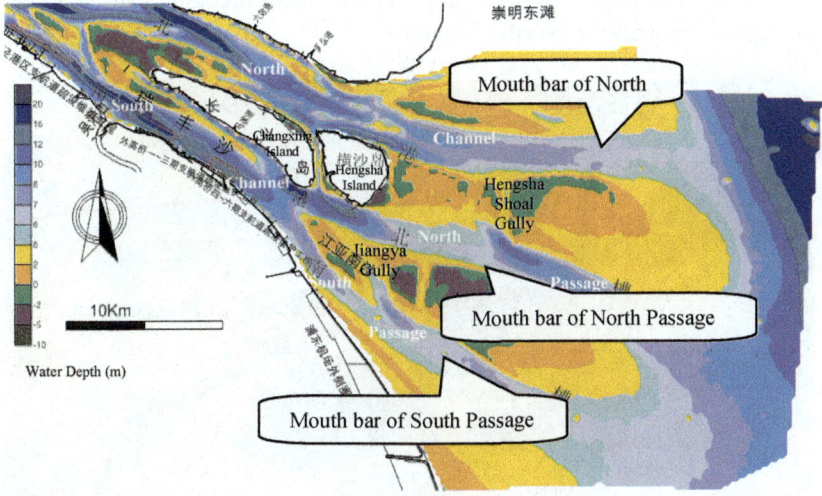

Fig. 2. Topography of the Yangtze Estuary (1998) with isobaths in meters.

At the entrance of North Passage, sedimentation happened and the cross-section area decreased due to the fact that runoff distribution from the South Channel declined caused by the river friction increasing in North Passage. And near the split point, south training dike blocked the gully used to discharge some runoff from North Passage to South Passage, which was also reduced the hydrodynamic condition near the entrance. But in this area, the water depth is still enough to maintain the development and equilibrium of North Passage.

In the middle bend area, there was a gully connected North Passage with North Channel. After the construction of north training dike, it blocked the water and sediment interchange between the two channels, thus in this area, velocity decreased and siltation of suspended sediments happened.

The lower reach of the North Passage, after the project, the flow field changed from rotating tidal current to alternating tidal current, and the hydrodynamics increased, thus erosion of sea bed happened.

Out of the North Passage, some sedimentation happened. With the morphological adjustment inside the North Passage, some sediment was eroded and transported outside of the regulating structure. Out of the regulating structures, the channel was getting wide and velocity decreasing, which can cause sedimentation. In this area, it is relatively deep and the strong rotating velocity field will help to take away the sediment.

In summary, the regulation works in North Passage is successful. The principle of regulating works combining with channel dredging can be adopted for the regulation works in North Channel.

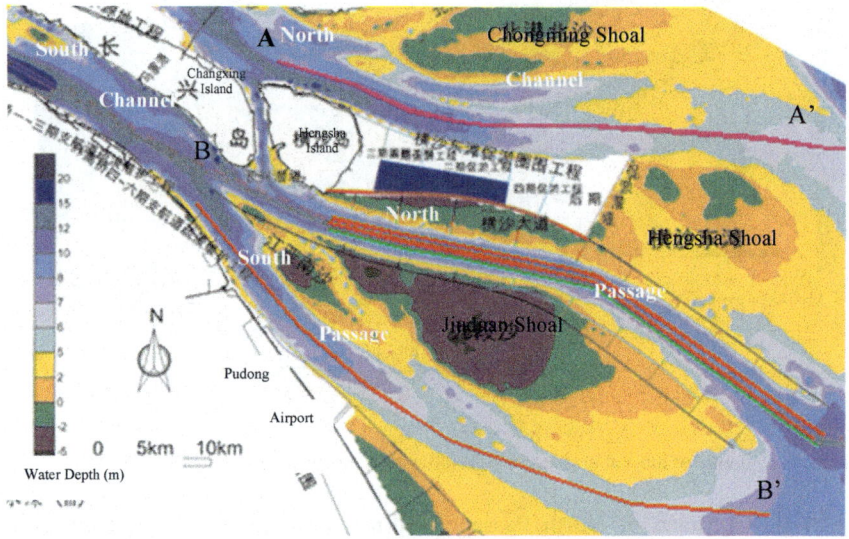

Fig. 3. Topography of the Yangtze Estuary (2015) with isobaths in meters.
The alignment A-A' is the thalweg of the North Channel and B-B' is the thalweg of the South Passage.

3.2. *The evolution of mouth bar in Another Two Passages of Yangtze Estuary*

After the project, which formed the regulating line along the groins heads and adjusted the flow fields, deepened the mouth bar area and merged the ebb with flood channel, a faintly-curved channel was obtained in the North Passage. At the same time, in other channel such as in the South Passage and the North

Channel, the mouth bars are all still persisting (Fig. 4, Fig. 5) after the project as long as mouth bar terrain has disappearing in north passage.

North Channel is total length of 90km, located between Chongming Island and Changxing Island, Hengsha Island. North Channel is more important convenient channel for ships from the northern coastal ports entering and leaving the mouth of the Yangtze River to the sea, and economic development and foreign exchange in Chongming, Changxing, Hengsha Islands coastal areas.

Now North Channel is a natural depth channel, water depth is up to 10m, sand bar segment water depth is insufficient 5.5m, which is now only navigable to some of the fishing boats and passenger ferries and small cargo ship. North channel further management needs to study the engineering measures, the outer section of the constraints are more need to be closely integrated with other projects.

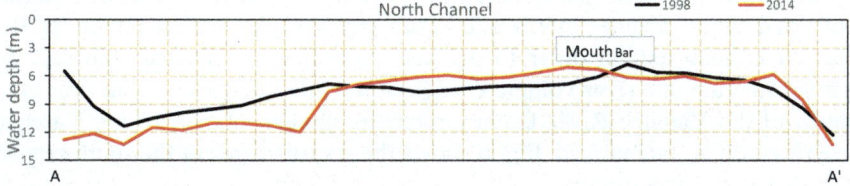

Fig. 4. water depth in North Channel（A-A'in Fig. 3, 2014）

Before 1998 (In Fig. 4), the shallowest water depth of mouth bar in North Channel is about 5m, which is almost same with the depth in 2014.

South Passage is total length of about 86km, the main diversion from the south Channel, which is the main way of small ships from the Yangtze River estuary into the sea. In 2013, the local dredging of the south channel is carried out. The scope of the project is 19.2km downstream from the Jiuduansha warning area. The effective width is 250m and the design depth of the channel is 5.5m. It can meet the capacity of 5,000 tons of bulk carriers with two- Class bulk carriers to reduce the need for two-way navigation.

The influence of the deep channel in the north passage has been washed out in recent years, but the change of the terrain of the barrier is small, and the next step is to adjust the function of the upstream port.

In Fig. 5, the shallowest water depth of mouth bar in South Passage is similar before and after the regulation works in North Passage.

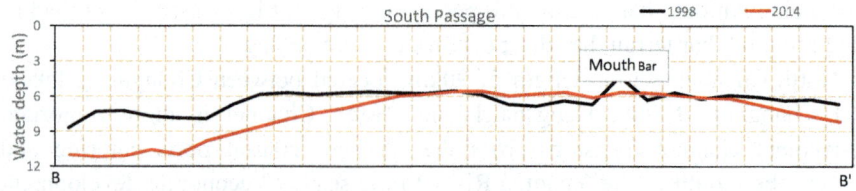

Fig. 5. water depth in South Passage (B-B'in Fig. 3, 2014)

4. The Response of the Morphological System to Human Interventions

The change of the river channel volume reflects the combined effect of the hydrodynamic adjustment and the river bed section adjustment. The total volume change reflects the sediment supply to the downstream of the river, without considering the supply of sediment transport in the upper reaches of Xuliujing. Based on the analysis of the volume change of the south branch, south and north channel ,north and south passage, the sedimentation of the riverbed is analyzed. In order to analyze the influence of terrain change on channel siltation, the river volume change of the following sections in the south branch of the Yangtze River Estuary after the implementation of the Yangtze River Estuary is summarized. Fig. 6 shows the statistics area of the south branch of the Yangtze River estuary.

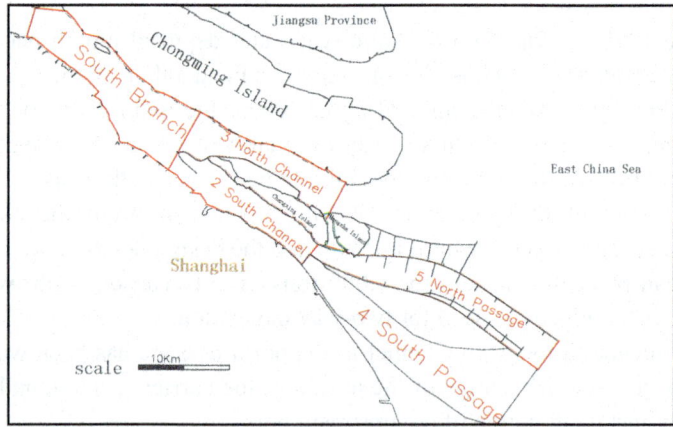

Fig. 6. Statistics area of the south branch of the Yangtze River estuary
The names of calculating areas are as follows: (1) the South Branch, (2) South Channel, (3) North Channel, (4) South Passage, (5) the North Passage.

The evolution of the South Branch seriously influences the morphological process of North Channel. According to the measured data, the current-rushing point of each shoals in the South Branch were continuous slowly moving downward. In recent years, although the thalweg in the South Branch moved with the recession of the shoals, the thalweg in the lower reach of South Branch keeps stable with little swaying. This gives a good condition that, if the split point between the South Channel and North Channel is stabilized with engineering structures.

During the period from 1998 to 2016, the volume of the south branch of the Yangtze River estuary was gradually increasing, and the trend of channel volume erosion was maintained for 12.5 m channel maintenance in the third stage (Fig. 7). Compared to the recent years, the rate of increase in the volume of the river channel increased slightly.

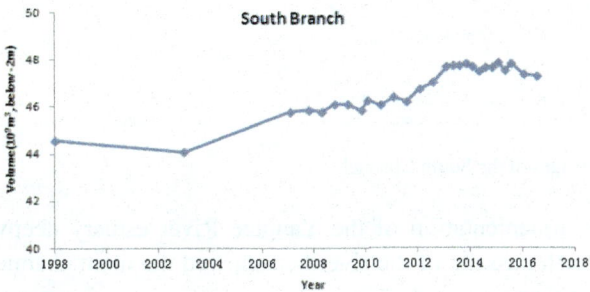

Fig. 7. Volum changes of the South Branch

After 1998, the volume change of the South channel was directly related to the evolution of Ruifeng Shoal in this river. The change in the volume of the river channel continued to increase after 2012. The erosion trend of the South Channel was beneficial to the maintenance of the river channel, but the erosion provided some part of the sand source to the downstream river.

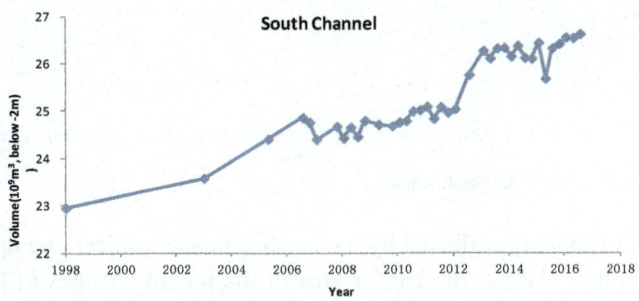

Fig. 8. Volum changes of the South Channel

From 1998 to 2012, the volume of the river in the North Channel is generally increasing (Fig. 9). However, due to the implementation of the Qingcaosha Reservoir, the area of the high beach decreased between 2007 and 2010. There was no obvious change in the diversion ratio between the north and the south Channel, and the Volume below 5m and 10m increased more. The riverbed reach to narrow channel.

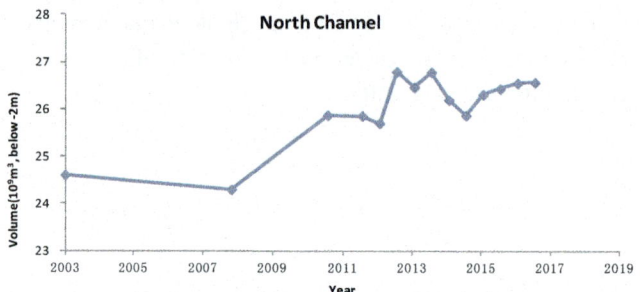

Fig. 9. Volum changes of the South Channel

Since the implementation of the Yangtze River estuary deepwater channel project, the north passage of the river bed showed the main channel erosion, the dam site deposition.

The total volume of north passage since 1998 to 2016 is generally reduced (Fig. 10), while the 5m isobath is increasing continuously.

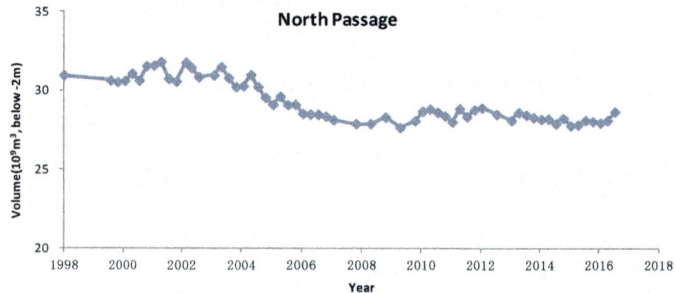

Fig. 10. Volum changes of the South Channel

The south channel is affected by the north passage project, the upper part of the erosion rate is larger, the barrier area of the mouth changes in the smaller. The total volume of the south passage increased before 1998 to 2003, then the change will be reduced from 2004 to 2013; during 2014 ~ 2016 channel volume increasing trend has accelerated.

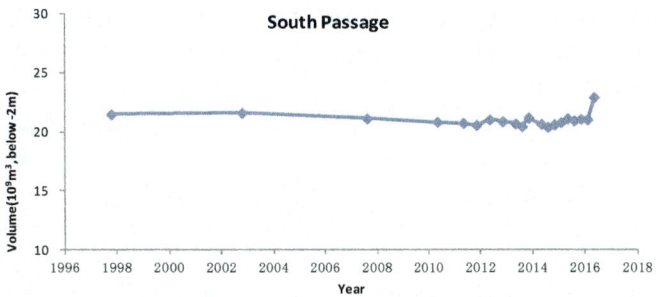

Fig. 11. Volum changes of the South Channel

5. Conclusion

The basic regime of three-stage bifurcation, four-river mouth split of Yangtze Estuary came into being under specific boundary conditions, and the general river regime will remain stabilized for a long time.

The regulating structures at the split area of South and North Channel and the reclamation works along the North Channel will provide a favorable boundary condition for the channel regulation works. As a result of the works, the present regime will be stabilized. It is favorable to maintain a split ratio of flow and sediment for main bifurcated channels. From controlling the river regime point of view, to regulate the North Channel, it is important to block the gullies connecting to North Branch, stabilize the boundary in mouth bar area, to minimize the water/sand interchanging between shoal and channel.

To compare the different evolution model of three main passages in Yangtze Estuary, the navigation works effects on the natural morphological system only within the north passage, not on the the large-scale morphological development in estuaries, not change the long-term development of the inter-tidal flats.

References

1. CHEN Zhi-chang, and GU Pei-yu, and ZHU Yuan-sheng, and ZHAO Xiao-dong, 1995. Research on deep water channel of Yangze estuary. Hydro-Science and Engineering. 1995(3), 210-220.
2. CHEN Zhi-chang, and LE, Jia-zuan, 2005. Regulation principle of Yangze River estuary deep channel. Hydro-Science and Engineering. 2005(1), 1-7.
3. Han Yufang, Chen Zhichang. Effects of spur dikes in adjustment of bed topography of wide-shallow river. Hydro-Science and Engineering, 2004 (2) : 23~28.

4. LE, Jia-zuan, and CHEN Zhi-chang, 2005. Selection and training principle of deep channel in the Yangze River estuary. Hydro-Science and Engineering. 2005(2), 1-8.

5. Wan, Y., D. Roelvink, W. Li, D. Qi, and F. Gu (2014), Observation and modeling of the storm-induced fluid mud dynamics in a muddy-estuarine navigational channel, Geomorphology, 217(0), 23-36.

6. Han Yu-fang, Chen Zhi-chang, Luo Xiao-feng, Physical Model Research on the Improvement Project of the Deep-Draft Channel of the Yangtze Estuary, Proceedings of the Seventh International Conference on Asian and Pacific Coasts(APAC), Sep. 2013 Bali Indonesia (APAC-2013), P209~213.

Lessons Learned from Deploying Low Crested Breakwaters in North Coast of Java

Dede M. Sulaiman, Dedi Junarsa, Huda Bachtiar

Experimental Station for Coastal Engineering, Research Center for Water Resources, Ministry of Public Works and Housing, Jalan Gilimanuk-Singaraja Km 122, Gerokgak, Buleleng, Bali, Indonesia
E-mail: dedems@ymail.com; dedi_junarsa@yahoo.com; huda.bachtiar@gmail.com

Fegi Nurhabni

Ministry of Marine and Fisheries Affair, Jalan Medan Merdeka Timur No.16 Jakarta, Indonesia
E-mail: f.nurhabni@gmail.com

Beyond the boundary and design criteria of Low Crested Breakwater (LCB), the success of LCB applications largely depends on the layout of the installation. The layout includes vertical and horizontal properties. Vertical layout related to the peak elevation position of the structure to the sea level, both to Mean Sea Level (MSL) and High Water level (HWL). Horizontal layout regarding LCB placements related to the optimal distance from the shoreline, the length of the structure, and the width of the gap between LCBs. These three parameters determine sediment volume accumulated behind LCB. This paper presents a summary of field experiences that is useful for the development of LCB as one of the methods of coastal protection. The objective of developing LCB structure is to make LCB concept as preferred coastal structure that is applicable to all types of materials commonly used for coastal protection structures.

Keywords: Low Crested Breakwater; Geotube; Beach erosion; Coastal protection.

1. Introduction

The North Coast of Java is one example of coastal ecosystems that experienced bad coastal degradation due to long time improper management of coastal areas. Conversion of mangrove into fishponds and settlement lead to various impacts such as silting, coastal erosion, pollution, tidal flooding, declining fisheries productivity and various social problems. Climate change that occurs globally also gives a tremendous impact on coastal areas. The increase of sea surface level causing coastal areas with sloping contours which can be inundated by sea water. One of efforts to countermeasure coastal erosion, the government has installed low-crested protection structures made of geotextile tube. LCB geotubes has been deployed along the North Coast of Java from Tuban in East

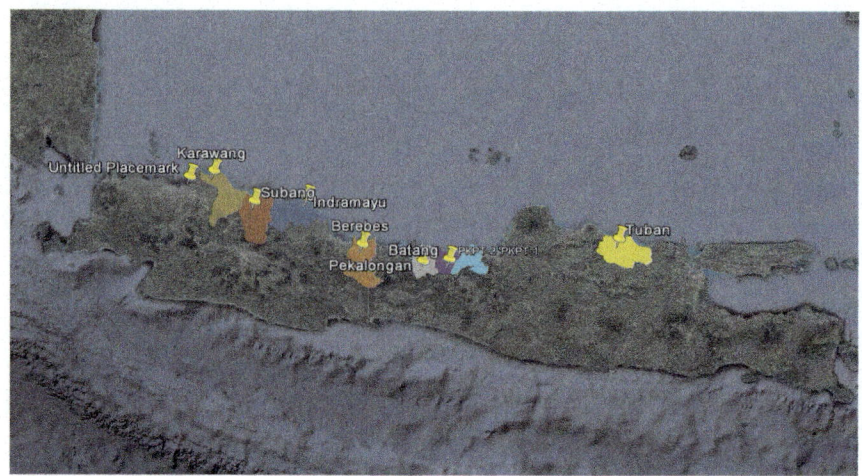

Fig. 1. Location of geotube LCBs along the North coast of Java (Google Earth, 2017)

Java to Karawang in West Java Province with the length approximately 25 Km of (Figure 1).

LCB's research and development activities have been started since 2009. A physical model testing in the laboratory was first carried out and it was followed by LCB prototype model test in the field. The purpose of this R & D activity is to get coastal protection technology that is effective, economically affordable, and friendly to the surrounding coastal environment. A total of ten LCB prototypes made of geotextile tube (geotube) material and have been applied in three different coastal types. White sands beach, Anyer, represents sand beach with high wave conditions and a rather steep coastal slope. Tanjung Kait Beach, Tangerang, Banten Province, represents sandy sloping beaches with relatively calm wave conditions. While the beach Pisangan, Karawang, used as a test for a muddy beach with medium wave height. The first LCB implementation by Local Government of Pekalongan City along 5.5 Km of Sari Beach, Slamaran beach, and Muara Rejeki beach in 2012 (Basyir Ahmad et al., 2015). Furthermore, the Marine and Fisheries Agency of Batang Regency, Central Java in 2014 and 2015 had installed LCB geotube to overcome erosion in Sigandu Beach (Sulaiman et al., 2015a). Until 2016, the Ministry of Marine Affairs and Fisheries, had installed 12.6 km Beach Belt long from geotextile sack, which is another name of LCB geotube along the North Coast of Java (MMAF, 2016).

This paper is a lessons learned from several applications and field experiences in an effort to obtain an effective, affordable, and environmentally friendly coastal protection method. The ultimate goal of the development of

LCB structures is to make LCB as a preferred coastal protection method that is capable of controlling and restoring coastal damaged into stable conditions with affordable cost.

2. Methodology

2.1 *Physical and Numerical Model Test*

Physical model test was conducted at JTSL Hydraulics Laboratory Gadjah Mada University, Yogyakarta. Tests were conducted to determine the effect of LCB elevation on wave transmission and coastal response behind the structure. The type of material used consists of geotextile tubes and rubble mounds. The test results show that the coastal response formed behind the structure is much influenced by the freeboard, the distance of the structure from the shore, and the wave height.

In numerical model test, the study of the existence of LCB structure to the current pattern, wave transmission, and coastal profile is performed by MIKE 21. The module used is MIKE 21 SW (Spectral Wave) for wave model, MIKE 21 HD FM (Bathimetry Meshing) for Current patterns, and MIKE 21 FM ST for sediment transport. The numerical modeling is also to determine the best formation of the LCB structure to be installed in relation to the emerging current pattern and the coastal response to be formed behind the LCB (Figure 2).

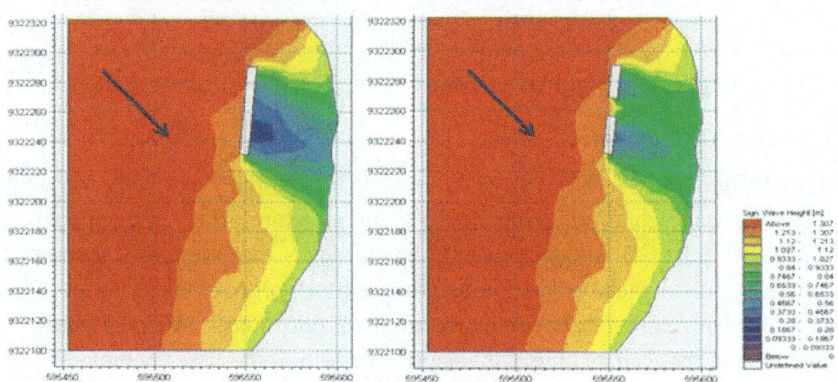

Fig. 2. Numerical Model for Current and Wave for Single and
Segmented LCB (Sulaiman et al., 2012)

2.2 *Deploying LCB Geotube along North Coast of Java*

LCB prototype has been implemented in three coastal locations, namely Pasir Putih, Anyer beach, Tanjung Kait beach, Tangerang, and Pisangan beach,

Karawang. The geotube LCB prototype on the Anyer coast is carried out in two phases, ie Phase 1 done in December 2010 with 3 pieces of geotube LCB installed with a gap at low water level (LWL) position. While the Phase 2 installation was carried out in November 2011, two LCB geotube prototypes were installed on average water level (MSL) 50 m from the shore.

The installation of LCB geotube prototypes at Tanjung Kait beach was conducted in May 2011 to represent sloping beaches with relatively quiet wave conditions, LCB geotube installed at a position about 20 cm above average water level. A total of 3 pieces of geotube LCB were installed with a gap of 10 m at a distance from the coast about 120 m. The geotube LCB condition after six months of installation is shown in Figure 3.

Fig. 3. LCB geotube in Tanjung Kait Beach (Sulaiman et al., 2012)

3. Field Experiences and Discussion

3.1 *LCB vertical layout*

The LCB function is to force a breaking wave and let the wave be forwarded so that a calm wave condition occurs behind the structure. The sediment transport capacity behind LCB is also reduced, which means that sand will accumulate. Runoff also shows the presence of water mass transport over the structure. There is a strong correlation between wave transmission, mass transport above the structure, and reduced sediment transport. In addition the wave parameters such as wave height, period, and steepness, LCB geometry, such as threshold distance, distance from shore, and length of structure also play a significant role in determining its effectiveness. Therefore, LCB design positioning determines the purpose and usefulness of LCB construction.

3.1.1 *As stabilizer and sand stopper*

When we use beach stabilizer and sand stopper, the LCB placement should be designed and placed in a low water level (LWL) position. The LCB position of the Pasir Putih beach, Anyer against the water level is shown in Figure 4. The submergency degree of the LCB is very influential both on wave transmission and on the coastal profile formed behind the structure. In the installation of LCB geotube Phase I, which is on December 2010, the placement of 3 LCB units was conducted at low sea level elevation. In such LWL positions, the breakwater building includes fully submerged about six months after installation, the three LCB have been buried in sand and formed a new beach profile (Sulaiman et al., 2012). The coastal conditions around the LCB show a sloping profile that indicates stable coastal conditions and it is different from the coastal profile before LCB is installed.

The results of monitoring LCB conditions two years after installation indicate that the presence of all three LCBs has been buried in sand sediments, but the coastal profile behind the structure appears stable and sloping. The sedimentation process and the formation of stable beaches behind the LCB Anyer can be analyzed that sediments derived from coastal shipping, when transported back to the offshore, the sediments are obstructed and settle behind the LCB structure. Within a period of 6 months until the last monitoring of March 2013, the sediment covered the entire LCB body. The coastal profile formed between the LCB's location toward the coast shows a stable beach. In other processes, the reflection wave that occurs in front of the LCB structure causes local scouring and slowly leads to instability and subsequent settlement and accumulated sand from onshore-offshore and longshore transport.

3.1.2 *Coastal erosion controller*

The effectiveness of LCB in returning eroded beaches is strongly influenced other than wave parameters, also by geometry of LCB structures, especially threshold distances, distance from shore, and length of structure. Hanson and Kraus (1991) show that the shoreline response to the presence of breakwaters is controlled by at least 14 variables, eight of which are highly variable (1) distance from the coast; (2) the length of the structure; (3) the height and the width of the light; (4) coastal slope; (5) wave height, (6) wave period; (7) the angle orientation of the structure; And (8) the direction of the dominant wave. A practical approach to producing an effective LCB for erosion control and beach shore is to place the LCB structure in a position above MSL. LCB prototypes of Tanjung Kait beach (Sulaiman et al., 2012) as presented at Figure 3, geotube

450

LCB of Sari beach, Pekalongan (Basyir Ahmad et al., 2015) as showed on
Figure 5, Sigandu beach (Sulaiman et al., 2015b) as presented on Figure 6, and
Pasir Putih beach Karawang (MMAF, 2016) as presented on Figure 7, are LCB
structures with crest level between the mean sea level and the highest water
level (Figure 4). The geotube LCB at the coastal site provides a positive coastal
response that formed salient or tombolos, which is a new beach formed by
LCBs.

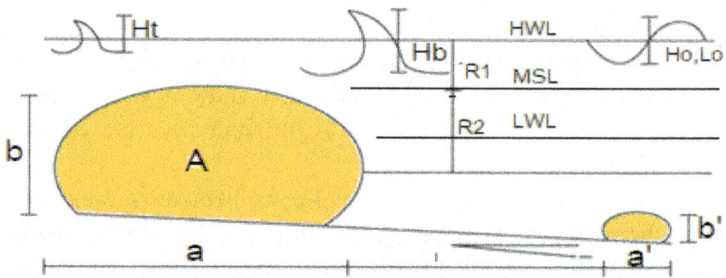

Fig. 4. Crest level of LCB At Mean Sea Level (Alvarez et al., 2006)

Fig. 5. LCB geotube and the Result in Sari Pekalongan Beach (Ahmad et al., 2015)

Fig. 6. Sigandu Beach, Before and After LCB Installation (Sulaiman et al., 2015)

Fig.7. Sedimentation Behind LCB at Pasir Putih Beach, Karawang (MMAF, 2016)

3.2 *Dimension of geotube*

The geotube used in LCB prototypes is a large sack made of non-woven geotextiles, which, once filled with sand, will be shaped like a bolster pillow. The geotube size used after the sand has a dimension of length, width and height of about 20 m, 2 m, and 1.2 m. In accordance with the height, the geotube type works well as an LCB structure for a maximum coastal depth of 2 m. For deeper water conditions, for a depth of about 4-5 m, two stacking geotubes may be used with the formation of mounting as shown in Figure 8. Alternatively, for deeper waters can be used sand filled megacontainer with a height dimension of 3 m and a length of about 20-60m.

Fig. 8. Two stacking geotubes at Pusong beach, Aceh (Sulaiman et al., 2012)

3.3 *Advantages use of LCBs*

In terms of elevation and dimensions, the use of LCB has some advantages, namely (1) aesthetically, LCB does not obstruct the view towards the sea, because it is installed at a depth below the high water level, but when the LCB recedes can appear to the surface; (2) the wave is not completely obstructed so that the coastal response behind the LCB is relatively uniform in the long shore direction, (3) The wave energy behind the LCB has been reduced so that the waters behind it are safe for swimming, and (4) the LCB-generated impact is less than Conventional breakwater, therefore LCB is more environmentally friendly (RCWR, 2016). With LCB elevation lower than conventional breakwater, the shoreline change process and Tombolo or salient formation will be slower than conventional breakwater.

4. Conclusion and Recommendation

4.1 *Conclusion*

1) The Successful in handling coastal erosion using geotube LCB is characterized by the formation of new coastlines behind the structure. This success is supported by proper technical planning taking into account the characteristics of the coast. Coastlines that were previously eroded, in less than a year, have formed new beaches with large, thick sedimentary deposits.

2) Planning of the LCB's layout, either vertically or horizontally is crucial and determines its success in protecting and rehabilitating the coast. Vertical layout is related to the elevation position of LCBs crest relative to mean sea level. While horizontal layout is related to the distance from the beach, the length of the structure, and the width of the gap.

3) In addition to its fuction as a wave absorber, geotube LCBs act as sediment catchers and retaining sediments. The sediment transported by waves and settles behind the structure, continuously settles and forms a new coastline called salient and at some points the sediment is connected or merged with the structure called as tombolos. Sediment source that accumulated behind LCBs, coming from offshore-onshore transport and a longshore transport.

4) Applying a space or a gap among LCB structures contribute to the sedimentation processes behind the structure. In addition to the supply of sediments through wave overtopping, even at low tide periode, the sediment supply continues through the gap.

5) In contrast, unsuccessful application of LCB structures, caused by inaccuracies in the LCB placement. It results from the position of LCBs is too close to the beach. Similarly, the crest level of LCBs is too low which results in wave dissipating so small that it does not produce a relatively quiet situation behind the structure that allows the sedimentation processes to occur.

6) LCBs made of geotubes are designed as a coastal protection. Coastal rehabilitation through LCB sructures will be more effective when combined with coastal vegetation. The combination of geotube LCB as a wave energy absorver and coastal vegetation behind it will drive the sedimentation processes and accelerate coastal changes towards the sea. In its development, coastal vegetation such as mangrove will serve as a natural protector against extreme conditions in coastal areas.

4.2 *Recommendation*

1) To avoid structural deformation and extend the life of structures, it is necessary to install bamboo piles and rafts; The need for proper LCB positioning from the coast; And the need for an armor layer on the outside of the geotube.

2) LCB structure will experience local scour on the front and rear legs of the structure, it is necessary to install foot protector both in front and behind the structure.

Acknowledgements

The authors are grateful to both individuals and institutions over the data, information, and materials so that this paper can be arranged. My thanks go to the Direktorat Jenderal Pengelolaan Ruang Laut, Direktorat Pendayagunaan Pesisir, Ministry of Marine and Fisheries Affair; Director of Research Center for Water Resources; Chief of Experimental Station for Coastal Engineering; for the opportunity and support to complete this paper.

References

1. E. Alvarez, R. Rubio, and H. Ricalde, Shoreline restored with geotextile tubes as submerged breakwaters, (Geosynthetics Magazine, 2006), Volume 24, Nimber 3, pp. 1-8.

2. B. Ahmad, M. Ismanto, S. Miftakhudin, and D.M. Sulaiman, Coastal Erosion and tidal flood countermeasuring at Pekalongan Beach, in Proc. Annual Meet of 32nd HATHI, (Malang, Indonesia, 2015).

3. Google Earth, http://www.googleearth.com/Java, (down load June 15, 2017).

4. H. Hanson and N.C. Kraus, Shoreline Response to a Single Transmissive Detached Breakwater, in Proc. 22nd Coastal Engineering Conf. ASCE. The Hague, (1990).

5. Ministry of Marine Affairs and Fisheries, Monitoring of Coastal Belt along the North coast of Java, (Jakarta, Indonesia, 2016).

6. Research Center for Water Resources, Draft Guidelines for Planning and Design of Low Crested Breakwaters, Concensus Meeting, (Bandung, Indonesia, 2016).

7. D.M. Sulaiman, Beach rehabilitation using geotube LCBs at Tanjung Kait Beach Tangerang, J.Keairan, Vol. 2. No. 2, (Dessember, 2012).

8. D.M. Sulaiman, "Piling-up and Current pattern behind segmented low crested breakwaters", Doctoral Dissertasion, Graduate Program Prahyangan Catholic University, (Bandung, Indonesia, 2014).

9. D.M. Sulaiman, H. Bachtiar, and A.Taufiq, Beach profile changes due to low crested breakwaters at Sigandu Beach North Coast of Central Java, in Proceedia Engineering, 8th International Conference on Asian and Pacific Coasts, (Elsevier, 2015) Volume 116, pages 510-519.

10. D.M. Sulaiman, S.S. Effendi, R.M. Azhar, and Suprapto, Coastal restoration using segmented LCBs: Case study Sigandu beach, Batang, Central Java Province, in Proc. of 33rd Indonesian Hydraulics Engineering Conference, (Semarang, Indonesia, 2016).

Shoreline Variation at Tip of Cuspate Foreland in Response to Change in Wave Direction

Shiho Miyahara[a], Takaaki Uda[b] and Masumi Serizawa[a]

[a]Coastal Engineering Laboratory, Co., Ltd.
1-22-301 Wakaba, Shinjuku, Tokyo 160-0011, Japan
E-mail: qqqu4vdd@aroma.ocn.ne.jp

[b]Head, Shore Protection Research, Public Works Research Center,
1-6-4 Taito, Taito, Tokyo 110-0016, Japan
E-mail: uda@pwrc.or.jp

The morphological features of the cuspate foreland extending at the east end of Mikawa-oshima Island in Mikawa Bay were investigated by field observation together with the examination of aerial photographs taken between 1977 and 2014. The change in the shoreline configuration of the Konose-bana cuspate foreland on the south shore of Sado Island was also studied while referring to the results of the previous study. On the basis of these field data, the BG model was used to predict the cyclic change in shoreline around the tip of the cuspate foreland in response to the change in wave direction.

Keywords: Cuspate foreland; Numerical simulation; BG model; Konose-bana; Mikawa-oshima Island; Shoreline changes.

1. Introduction

When waves are incident from two opposite directions, a land-tied island or a cuspate foreland may develop. Miyahara et al. (2014)[1] predicted the development of the land-tied island of Oyoshima Island offshore of Shodo Island using the BG model (a model for predicting three-dimensional beach changes based on Bagnold's concept)[2]. In their study, the wave field was evaluated by the angular spreading method for irregular waves, and the sand transport equation using the wave energy flux at the breaking point was employed in calculating the wave-sheltering effect owing to the sand bar itself. Then, Serizawa et al. (2015)[3] developed an improved BG model for predicting the development of a cuspate foreland under the condition that waves are incident from two opposite directions, taking the Futtsu cuspate foreland and a cuspate foreland located at the northeast end of Graham Island in British Colombia, Canada, as examples. In their model, the wave field was calculated using the energy balance equation given by Mase (2001)[4], in which not only the wave-sheltering effect but also wave refraction and wave decay by wave breaking were included. As a result, the formation mechanism of the cuspate foreland was successfully explained by these numerical simulations. On the other hand, as another example, shoreline variation with a cyclic mode in response to the change in wave direction can be observed at the tip of the cuspate foreland under the condition that waves are incident from two opposite directions. Such a phenomenon, however, has not yet been investigated, and examples are seen at the east end of Mikawa-oshima Island in eastern Mikawa Bay and at the Konose-bana cuspate foreland. In this study, the topographic features around the cuspate foreland on Mikawa-oshima Island were investigated by field observation, and the variation in the shoreline configuration of the cuspate foreland was investigated using aerial photographs and satellite images taken between 1977 and 2014. Regarding the shoreline changes in the Konose-bana cuspate foreland, we referred to the study results obtained by Uda and Yamamoto (1990)[5]. Finally, the mechanism of variation of the shoreline at the tip of the cuspate foreland was explained using the BG model given the wave field in which the wave direction changes in a cyclic mode.

2. Cuspate foreland extending at east end of Mikawa-oshima Island

2.1. *General conditions*

Mikawa-oshima Island with an area of 21.6 ha and a shoreline length of 4 km is located 3 km southwest of Mitani fishing port in Mikawa Bay (Fig. 1). Figure 2 shows an enlarged satellite image of this island. Because the island is composed of granite, the beach material is composed of decomposed granite sand. A rock protrudes at the northeast end of the island, and a sandy beach (Nishihama Beach) of 320 m length extends west of the rock because of the blockage of longshore sand transport by this rock. Similarly, Higashihama Beach, which has a shoreline approximately parallel to the shoreline of Nishihama Beach, develops on the lee of the island on the eastern side, and a cuspate foreland with a sharp tip develops. The locations of St. 1–St. 14, where the photographs were taken, are shown in Fig. 2.

From the wind rose in 2013 measured at AMEDAS Gamagori 6.4 km north of Mikawa-oshima Island, the probability of W or WNW winds is extremely high with a lesser probability in other directions with fetch distances of 3.95 and 3.75 km, respectively. Owing to these wave incidences, a cuspate foreland was formed at the east end of Mikawa-oshima Island, and the elongation of the

Fig. 1. Location of Mikawa-oshima and Kojima Islands in eastern Mikawa Bay.

Fig. 2. Satellite image of Mikawa-oshima Island and location number of site photographs.

cuspate foreland ceased up to the present length because of the limited source of sand on the island.

2.2. *Field observation*

Field observation was carried out on May 29, 2014. We landed on the island from the west jetty, and coastal conditions were observed from the southwest

458

Fig. 3. View of tip of cuspate foreland from vicinity of east jetty of Mikawa-oshima Island.

end of Nishihama to Higashihama Beach along the shoreline, while turning around the rock located at the northeast end of the island. Figure 3(St. 9) shows the tip of the cuspate foreland, taken in the immediate vicinity of the east jetty, and a sharp triangular sand bar extending eastward. Figures 3(St. 10) and 3(St. 11) show the tip and base of the cuspate foreland, respectively, taken from Sts. 10 and 11 near the tip of the cuspate foreland. The width of the cuspate foreland was approximately 4 m. On the other hand, the foreshore slope at St. 12 on Higashihama Beach southwest of the cuspate foreland was 1/10 with a berm height of 1.75 m above MSL (Fig. 3(St. 12)).

2.3. Long-term changes in cuspate foreland on Mikawa-oshima Island

Figure 4 shows the variation in the shape of the cuspate foreland extending at the east end of Mikawa-oshima Island between 1977 and 2014. In 1977, the shoreline at the base of the cuspate foreland markedly protruded northward with a slender cuspate foreland, whereas its tip slightly protruded southward. By 1982, the shoreline at the base of the cuspate foreland extended straight, and the same condition continued by 1987. By 1991, the shoreline at the base of the cuspate foreland had retreated, while forming a concave shoreline together with an advance in the shoreline on the other side. This implies that the cross-shore sand transport across the base of the cuspate foreland from the north to the south side occurred. However, in 2003, the cuspate foreland extended straight, and the base of the cuspate foreland protruded northward as a whole in 2006, and the same condition had continued to 2010. The shape of the cuspate foreland in this period showed opposite features in terms of the curvature of the shoreline relative to that in 1977. By 2014, sand had been deposited north of the base of the cuspate foreland forming a straight shoreline on the north side of the cuspate foreland. Although the shoreline at the base varied in the north-south direction, the cuspate foreland had been stable over the long term. Because the slender cuspate foreland has long been stable, it is considered that the wave energy from both directions balanced each other at the tip of the cuspate foreland in the long term.

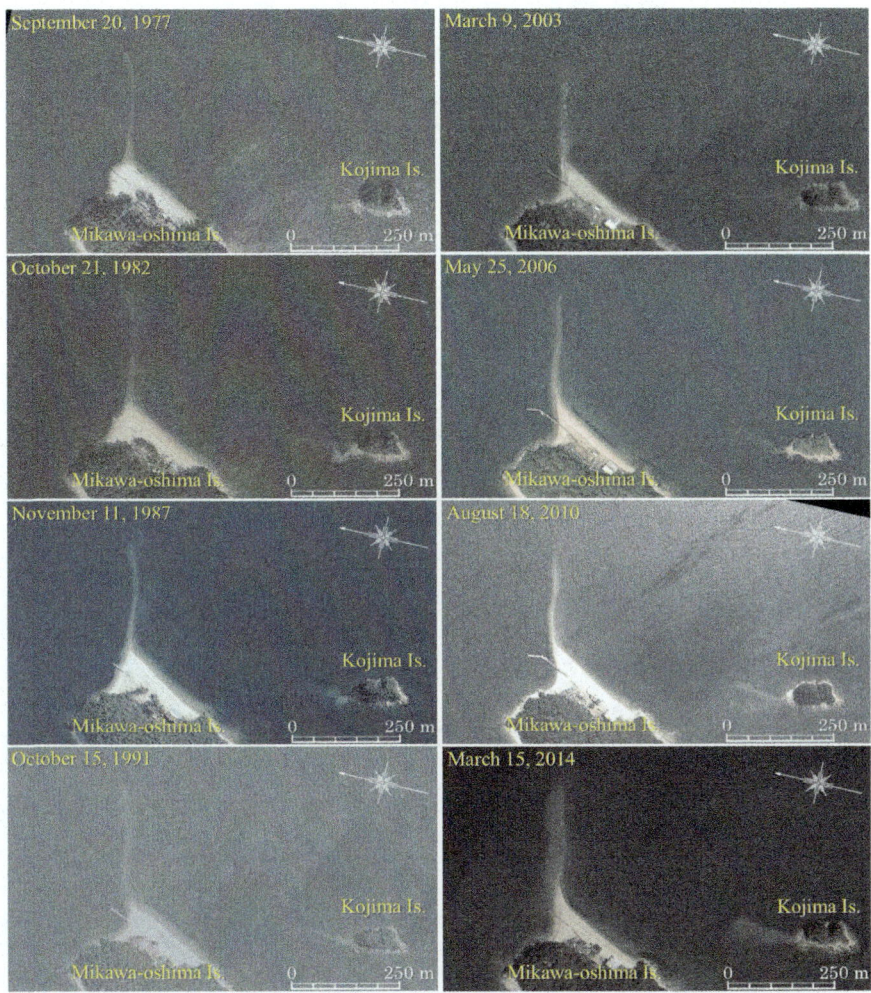

Fig. 4. Time changes in configuration of cuspate foreland between 1977 and 2014.

3. Konose-bana cuspate foreland extending at central part of Matsugasaki coast

The Matsugasaki coast is located southeast of Kosado and has a 2.5 km length separated by the Tokura tunnel and Benten-zaki Point, and has the Konose-bana cuspate foreland in the center, as shown in Fig. 5. The coastline of the Matsugasaki coast significantly changes its direction from SW-NE to SSW-NNE on the south and north sides of Konose-bana. Uda and Yamamoto (1990)[5] investigated the shoreline changes around Konose-bana on the basis of aerial

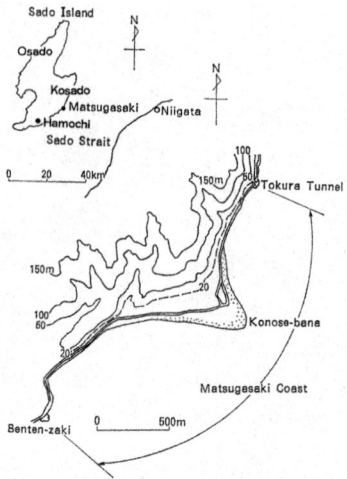

Fig. 5. Location of Matsugasaki coast and Konose-bana cuspate foreland on Sado Island.

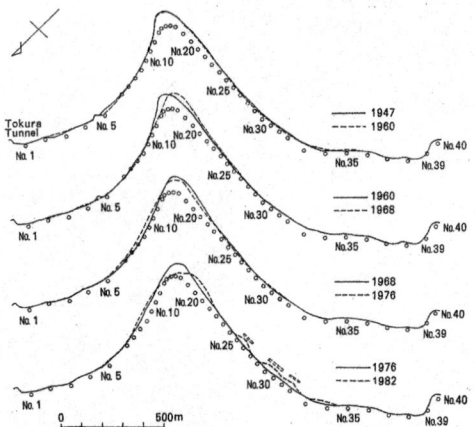

Fig. 6. Shoreline variation at Konose-bana cuspate foreland.

photographs taken between 1947 and 1985. Figure 6 shows the results. Transects No. 1–No. 40 are set between the Tokura tunnel and Bentenzaki Point, and transect No. 15 coincides with the tip of the cuspate foreland in 1947. In Fig. 6, the shoreline positions in 1947 and 1960 superimposed each other. When comparing the shoreline configurations in 1960 and 1968, the shoreline retreated between No. 11 and No. 16 near the tip of the cuspate foreland, whereas the shoreline advanced between No. 16 and No. 23 on the south side, implying that southward longshore sand transport occurred while turning around the tip of the cuspate foreland. Between 1968 and 1976, shoreline change with a reversed mode compared with that between 1960 and 1968 took place, such that the shoreline advanced between No. 5 and No. 14, and it receded between No. 15 and No. 21. This implies that northward longshore sand transport occurred around the tip of the cuspate foreland. Then, the shoreline advanced further between No. 6 and No. 12 in 1982 compared with that in 1976, whereas the shoreline significantly retreated between No. 12 and No. 18. In contrast, the shoreline advanced between No. 18 and No. 22 north of the cuspate foreland. Thus, it is clear that the shoreline near the tip of the cuspate foreland fluctuated in each observation period. Uda and Yamamoto (1990)[5] carried out wave forcasting using the wind rose between 1981 and 1987 at Hamo approximately 18 km southwest of the Matsugasaki coast, and concluded that the shoreline fluctuation is due to the fact that waves are incident from two opposite directions: 18% of the entire energy flux of waves incident to the Matsugasaki coast is from NNE, whereas 77% of all are from SSW and SW.

4. Numerical simulation of shoreline variation of cuspate foreland using BG model

4.1. *Calculation conditions*

The BG model employed in the numerical simulation of the variation in the shoreline configuration of a cuspate foreland by Serizawa et al. (2015)[3] was used in this study. Since the fundamental equations and the procedure of the numerical simulation are the same as those in the previous study, their descriptions are omitted in this paper and only the calculation conditions are shown. First, Cartesian coordinates (x, y) are selected, as shown in Fig. 7. A rectangular calculation domain of 1.6 km length in the x and y directions was selected. The calculation domain was assumed to have a constant depth of 4 m, and a sand mound of 1 m height was provided as the initial sand source in the area between x = -400 and 400 m, and between y = 0 and 300 m. The initial slope of the mound was assumed to be 1/20, and sand necessary for the formation of a cuspate foreland was supplied from this mound.

Serizawa et al. (2015)[3] investigated the development of a cuspate foreland under the condition that waves of 1 m height, which propagate upward and downward with a probability of occurrence of 0.5, were incident to the initial topography, similar to that in the present study. Under this condition, longshore sand transport toward the y-axis does not cease forever, because waves are continuously incident obliquely to the shoreline on both sides of the cuspate foreland. The cuspate foreland, therefore, continued to develop with time.

However, both the cuspate forelands extending at the east end of Mikawa-oshima Island and the Konose-bana cuspate foreland have remained stable, although the shoreline configuration of the cuspate foreland changes with time. This implies that after the cuspate foreland extended into a slender triangular form, it became stable, and the shoreline orientation on both sides of the cuspate foreland is normal to the direction of incident waves. For this reason, the wave directions were assumed to be $-10°$ and $190°$ with respect to the x-axis in the calculation, so that the final shoreline will be stabilized, because the shorelines are normal to the wave directions, although waves are incident from almost opposite directions. The time interval was set to be $\Delta t = 0.5$ h, and the wave direction was determined by a random number every 10 steps in the calculation of beach changes. The total number of steps is 5×10^4. After the formation of a cuspate foreland of a stable form, waves were incident from each side with a cyclic mode. Finally, a cuspate foreland under a dynamical equilibrium condition was obtained.

Table 1. Calculation conditions.

Wave conditions	Incident waves: $H_I = 1$ m, $T = 4$ s, wave direction $\theta_I = -10°$ and $190°$ with respect to $-x$ axis
Berm height	$h_R = 1$ m
Depth of closure	$h_c = 4$ m
Equilibrium slope	$\tan\beta_c = 1/20$
Coefficients of sand transport	Coefficient of longshore sand transport $K_s = 0.2$ Coefficient of Ozasa and Brampton (1980)[6] term $K_2 = 1.62K_s$ Coefficient of cross-shore sand transport $K_n = K_s$
Mesh size	$\Delta x = \Delta y = 20$ m
Time intervals	$\Delta t = 0.5$ h
Duration of calculation	2.5×10^4 h (5×10^4 steps)
Boundary conditions	Shoreward and landward ends: $q_x = 0$, right and left boundaries: $q_y = 0$
Calculation of wave field	Energy balance equation (Mase, 2001)[4] • Term of wave dissipation due to wave breaking: Dally et al. (1980)[7] model • Wave spectrum of incident waves: directional wave spectrum density obtained by Goda (1985)[8] • Total number of frequency components $N_F = 1$ and number of directional subdivisions $N_\theta = 8$ • Directional spreading parameter $S_{max} = 25$ • Coefficient of wave breaking $K = 0.17$ and $\Gamma = 0.3$ • Imaginary depth between minimum depth h_0 (0.5 m) and berm height h_R • Wave energy $= 0$ where $Z \geqq h_R$ • Lower limit of h in terms of wave decay due to breaking: 0.5 m
Wave conditions	Incident waves: $H_I = 1$ m, $T = 4$ s, wave direction $\theta_I = -10°$ and $190°$ with respect to $-x$ axis

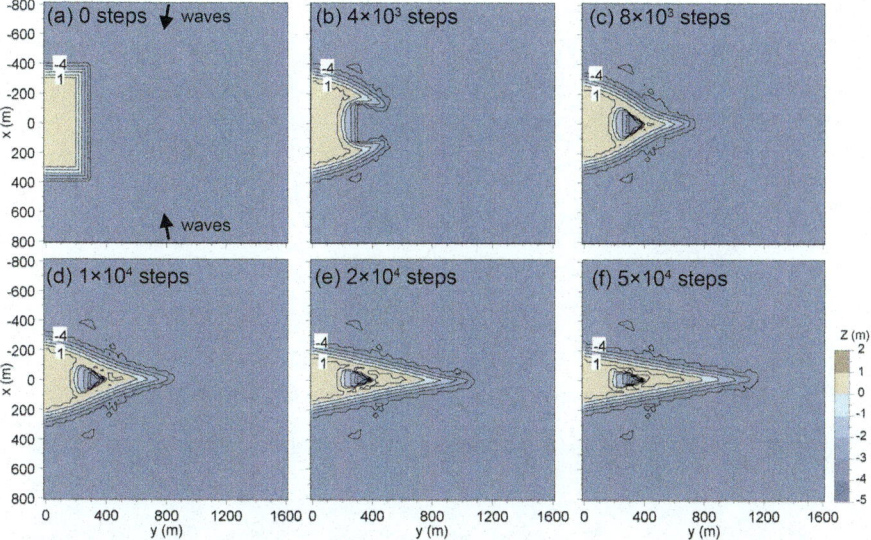

Fig. 7. Calculation results of elongation of cuspate foreland under wave incidence from two directions (Case 1).

4.2. Calculation results of elongation of cuspate foreland

Figure 7 shows the calculation results for a cuspate foreland simply extending, as in Case 1, owing to waves incident from two directions. Because waves were incident in the upward and downward directions randomly with the same probability, a sand spit elongated from the right end of the sand mound up to 4×10^3 steps, and the two tips approached each other owing to the wave-sheltering effect. After 8×10^3 steps, two sand spits that elongated from the upper and lower corners turned toward each other and connected, leaving a triangular closed water body in the central part and forming a concave shoreline. After 1×10^4 steps, the connected sand bars elongated rightward to a length of 800 m. Then, the cuspate foreland further elongated up to 2×10^4 steps, but after 5×10^4 steps it reached a completely stable form with no topographic changes even if waves were further incident to the cuspate foreland. A triangular water body was enclosed near the point of connection of the sand mound with the cuspate foreland, which was already predicted by Serizawa et al. (2015)[3]. Figure 8 shows the aerial photograph of the Konose-bana cuspate foreland taken on September 8, 1976. At the base of the cuspate foreland, a pond of 250 m length and 70 m width was enclosed. The formation of a closed water body is in good agreement with the results shown in Fig. 7. Thus, the formation of a slender cuspate foreland and the fact that it had been stably maintained for a long

464

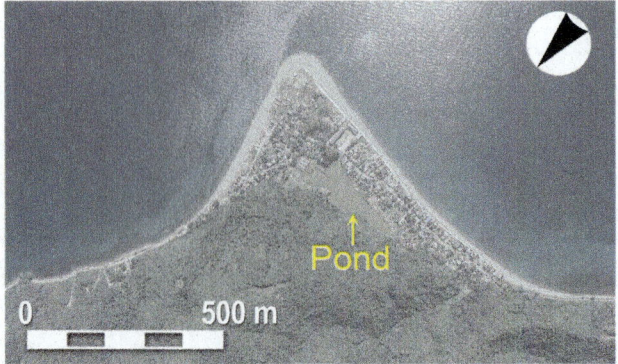

Fig. 8. Aerial photograph of Matsugasaki coast taken on September 8, 1976.

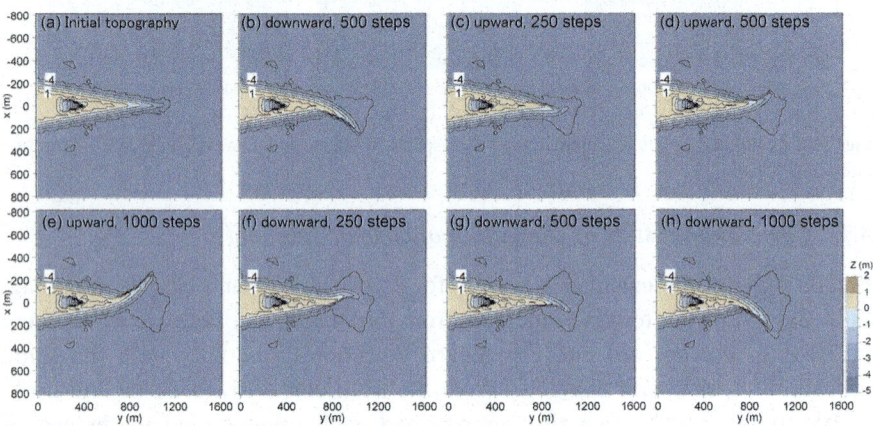

Fig. 9. Predicted variation of tip of cuspate foreland (Case 2).

time while the shoreline varied were successfully explained by the numerical simulation of the BG model.

4.3. Calculation results of topographic changes at tip of cuspate foreland

After the formation of a stable cuspate foreland in the calculation up to 5×10^4 steps in Case 1, waves were incident alternately from the upward or downward directions, and the topographic changes of the tip of the cuspate foreland were predicted (Fig. 9). Given the cuspate foreland (Fig. 9(a)) determined from the calculation up to 5×10^4 steps in Case 1 as the initial topography, waves were incident in the downward direction for 500 steps (Fig. 9(b)). The tip of the cuspate foreland turned downward by wave incidence, and a steep slope was

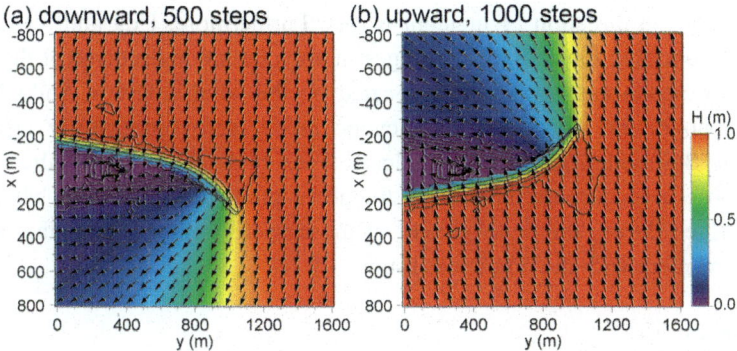

Fig. 10. Wave field after downward (upward) wave incidence for 500 (1,000) steps.

formed along the down side of the cuspate foreland owing to the falling of sand. Figure 10(a) shows the wave field corresponding to Fig. 9(b). Because waves propagate downward, a wave-shelter zone was formed on the down side of the cuspate foreland, whereas on the upside of the cuspate foreland, waves were incident from the left relative to the direction normal to the shoreline, inducing longshore sand transport toward the tip of the cuspate foreland.

Then, waves were incident in the upward direction for 1,000 steps given the initial topography shown in Fig. 9(b). Figure 9(c) shows the results after 250 steps. The tip of the cuspate foreland returned to the central position owing to the upward wave action. Figures 9(d) and 9(e) show the results after continuous wave action up to 500 and 1,000 steps, respectively, whereby an upward curved foreland was formed. The shape of the cuspate foreland after 1,000 steps is symmetric with that after downward wave incidence for 500 steps shown in Fig. 9(b). Figure 10(b) shows the wave field at 1,000 steps, and it is an upside-down inversion of the wave field shown in Fig. 10(a). Then, waves were incident to the cuspate foreland in the downward direction for 1,000 steps, given the initial topography shown in Fig. 9(e). Figures 9(f), 9(g), and 9(h) show the calculation results after 250, 500, and 1,000 steps, respectively. It is clear that the wave incidence with upward and downward cyclic modes causes cyclic change in the shoreline at the tip of the cuspate foreland.

5. Conclusions

The analysis of aerial photographs and satellite images of Mikawa-oshima Island taken between 1977 and 2014 revealed that although the cuspate foreland that formed at the east end of Mikawa-oshima Island maintains its stable form in the long term, the shape of the cuspate foreland changes in response to the variation in the direction of incident waves. Furthermore, Uda and Yamamoto (1990)[5] found that the shoreline changes of a cyclic mode occurred at the tip of the Konose-bana cuspate foreland located on the south coast of Sado Island. These topographic changes with a cyclic mode were successfully predicted using the BG model. In addition, a closed water body was found at the base of the Konose-bana cuspate foreland in the aerial photograph in 1976, which had been formed during the development of the cuspate foreland. This fact was in good agreement with the predicted results.

References

1. S. Miyahara, T. Uda, M. Serizawa, Prediction of formation of land-tied islands, *Proc. 34th ICCE*, (2014), pp. 1-14.
2. M. Serizawa, T. Uda, T. San-nami, K. Furuike, Three-dimensional model for predicting beach changes based on Bagnold's concept, *Proc. 30th ICCE*, (2006), pp. 3155-3167.
3. M. Serizawa, T. Uda, S. Miyahara, Model for predicting formation of a cuspate foreland, *Coastal Sediments '15*, CD-ROM, No. 65, (2015), pp. 1-14.
4. H. Mase, Multidirectional random wave transformation model based on energy balance equation, *Coastal Eng. J., JSCE*, 43(4), (2001), pp. 317-337.
5. T. Uda, K. Yamamoto, Beach changes around Konose-bana Spit on Sado Island, *Tran. Japan. Geomorph. Union*, Vol. 11, (1990), pp. 13-28. (in Japanese)
6. H. Ozasa, A. H. Brampton, Model for predicting the shoreline evolution of beaches backed by seawalls, *Coastal Eng.*, Vol. 4, (1980), pp. 47-64.
7. W. R. Dally, R. G. Dean, R. A. Dalrymple, A model for breaker decay on beaches, *Proc. 19th Inter. Conf. on Coastal Eng.*, (1984), pp. 82-97.
8. Y. Goda, *Random Seas and Design of Maritime Structures*, (University of Tokyo Press, Tokyo, 1985), 323 p.

Seasonal Movement of Sand Spits on a Coral Cay
of Embudu Village Island in the Maldives

Takaaki Uda[a], Masumi Serizawa[b] and Shiho Miyahara[b]

[a]Head, Shore Protection Research, Public Works Research Center,
1-6-4 Taito, Taito, Tokyo 110-0016, Japan
E-mail: uda@pwrc.or.jp

[b]Coastal Engineering Laboratory, Co., Ltd.
1-22-301 Wakaba, Shinjuku, Tokyo 160-0011, Japan
E-mail: coastseri@nifty.com

On Embudu Village Island in the Maldives, the seasonal movement of a pair of sand spits can be observed, which are generated by the seasonal changes in wave direction associated with the tropical monsoon. Field observation of the beach changes on this coral cay was carried out in May 1992. Satellite images of the cay taken between February 2005 and November 2013 were compared to investigate the planar changes of the cay. The BG model (a model for predicting three-dimensional beach changes based on Bagnold's concept) was used to predict beach changes of the coral cay. The calculated beach changes were in good agreement with those measured. It was confirmed that the wave-sheltering effect of the island itself played a decisive role in the beach changes.

Keywords: Maldives; Embudu Village Island; Coral cay; Sandy beach; Sand spit; Shoreline change; BG model.

1. Introduction

Coral reefs generally develop in tropical countries, and a cay surrounded by a coral reef may develop with the shoreward transport of sand produced from the vicinity of the reef edge, which is composed of foraminifera remains and coral debris[1]. Many cays have been used as beach resorts, which are constantly visited by many tourists. The Maldives in the Indian Ocean is a nation composed of approximately 1,200 islands. Male Island, which includes the capital of the nation, is located at the south end of the northern Male Atoll, as shown in Fig. 1. Embudu Village Island (4°5'N, 73°30'E) is located near the north end of the South Male Atoll. Because the Maldives is located in the tropical monsoon area, a northeast wind blows during the dry season from November to April with calm wave conditions, whereas in the rainy season from May to October, a southwest

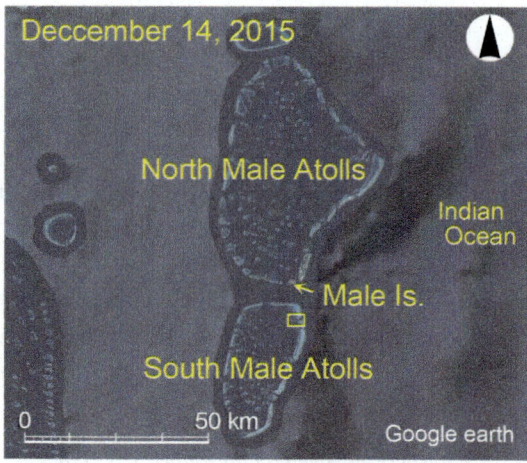

Fig. 1. Location of North and South Male Atolls, and Male Island in Maldives.

wind is predominant with a high probability of strong wind and rough waves. Thus, in the Maldives, the wind direction seasonally changes. However, because Embudu Village Island is located immediately west of a gap in the reef of the Atoll, waves are incident both from the east and southwest, resulting in seasonal variation of the shoreline around the coral cay[2]. Uda (1993)[2] carried out a field observation on Embudu Village Island at the end of May 1992 and produced a schematic diagram of the deformation mechanism of the sandy beach of a coral cay. In this study, the planar changes in the shoreline of this island were investigated using satellite images on the basis of the previous study at the end of May 1992. Then, the mechanism of beach changes on a coral cay was investigated using the BG model employed in Uda et al. (2012)[3], San-nami et al. (2013)[4], and Miyahara et al. (2015)[5]. Regarding the seasonal variation of a coral cay, Suzuki et al. (2013)[6] and Iwatsuka et al. (2015)[7] carried out a field investigation around a ballast island composed of only coral debris on an isolated reef flat north of Iriomote Island in Okinawa Prefecture, Japan. They found that the location of this ballast island moved in the east-west direction in a two-dimensional manner by over 10 m in response to the seasonal variation of waves. The aim of the present study was to investigate the three-dimensional changes of the sandy beach in response to the changes in the wave direction.

2. Shoreline changes of Embudu Village Island

Figure 2 shows an enlarged image of the rectangular area in Fig. 1. The Embudu Village Island is located 1.2 km from the west end of a gap with 1.1 km width in

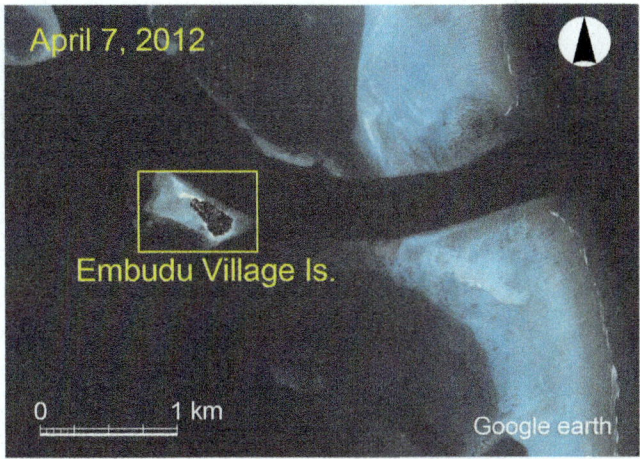

Fig. 2. Location of Embudu Village Island.

a reef extending in the north-south direction. Because of its proximity to the reef gap, wind waves generated in the atoll are incident from the southwest of the island during the southwest monsoon, whereas waves propagating from the Indian Ocean into the atoll via the reef gap are incident from the east of the island during the northeast monsoon. Thus, waves are alternately incident to the island from two directions. Figure 3(a) shows enlarged satellite images of the rectangular area including Embudu Village Island, as shown in Fig. 2. This island is located near the east end of an isolated coral reef, as shown in the image in Fig. 3(a) taken on February 2, 2005, and a shallow reef with approximately 1 m depth extends over a width of 200 m on the west side. The distance across the center of the island is 310 m, and the distance between the east and west jetties is 185 m. Because the wave direction seasonally varies around this island, the west and east jetties are used during the northeast (November – April) and southwest monsoons (May – October), respectively, (Uda, 1993)[2]. The average elevation of the island is 1.5 m, and the flat surface is densely covered with tropical vegetation. Furthermore, the maximum sea water level during high tide is 0.7 m above mean sea level[9].

Regarding the development of sand spits, sand spit A extended approximately westward on February 2, 2005, together with the ongoing development of sand spit B at the southwest end of the island, as shown in Fig. 3(a). Moreover, cuspate foreland C had formed near the foot of the west jetty. This satellite image was taken on February 2 during the northeast monsoon, and therefore, waves passing through the reef gap were incident to the island from the east, resulting in the formation of the pair of sand spits A and B, at the

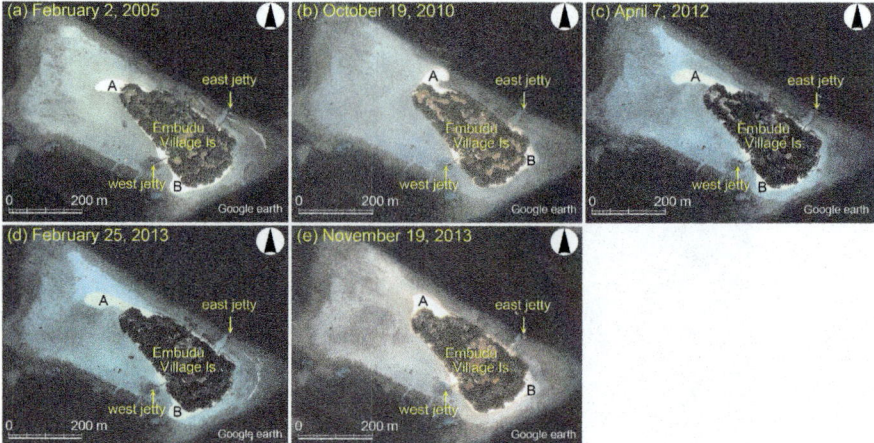

Fig. 3. Satellite image of Embudu Village Island.

north and south ends of the island, respectively. By October 19, 2010, sand spits A and B had moved to the northeast and east ends of the island, respectively (Fig. 3(b)). Because the satellite image on October 19 was taken during the southwest monsoon season, it is concluded that the location of the sand spit changed from the west side to the east side of the island owing to the wave incidence from the southwest. By April 7, 2012, a topography similar to that on February 2, 2005, was observed and sand spits A and B extended on the west side of the island (Fig. 3(c)). Although the direction of the extension of the tip of each sand spit was similar, the sand spit A extended northward with a more slender form than that observed on February 2, 2005. On the other hand, it is inferred that oblique wave incidence, by which sand is transported northward along the west shoreline, also occurred at the same time from the fact that sand spit B was smaller than that on February 2, 2005, and that the size of cuspate foreland C was also reduced. On February 25, 2013, a sand spit similar to that on April 7, 2012, could be observed (Fig. 3(d)). In this figure, the refraction pattern of waves incident from the east can be clearly seen on the reef edge, which had a convex shape near the east end of the island. By November 19, 2013, the location of sand spits A and B had again moved to the east side of the island owing to the wave action during the southwest monsoon (Fig. 3(e)). As mentioned above, waves are alternately incident from the southwest and east sides on Embudu Village Island, respectively, resulting in the development of sand spits on the east and west sides of the island in response to the changes in wave direction.

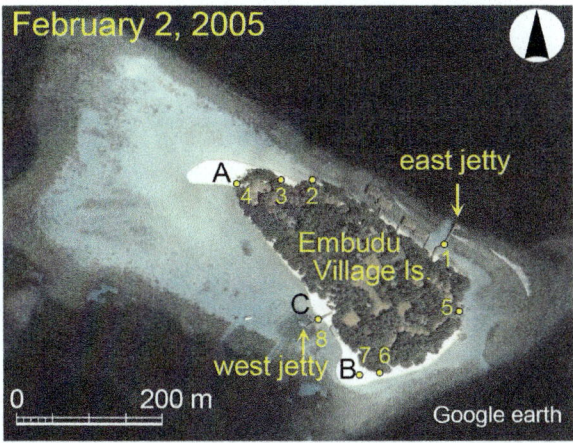

Fig. 4. Locations where site photographs were taken.

3. Field observation of Embudu Village Island

At the end of May 1992, after the southwest monsoon season had begun, the beach conditions of Embudu Village Island were observed. Figure 4 shows the satellite image of the island taken on February 2, 2005, together with the location numbers of the site photographs. Because waves during the northeast monsoon season had already had a marked effect on the beach topography, this satellite image was used, even though the field observation was carried out at the end of May. In the field observation, the shoreline conditions were observed counterclockwise along the shoreline from St. 1 on the east jetty to St. 4 at sand spit A located at the north end of the island. Then, the beach conditions were observed along the shoreline on the southeast side of the island between St. 5 and St. 8. Figure 5 shows the results of the field observation.

Figure 5(St. 1) shows the coastal conditions of the northeast side of the island viewed from Fig. 5(St. 1) on the east jetty. Because the southwest wind blew during the observation, a calm wave zone had formed along the east coast of the island with high transparency of the sea water, and tropical vegetation grew very close to the shoreline. At Fig. 5(St. 2), located 200 m northwest of St. 1, there was a beach rock, and a hooked shoreline had formed on the lee of this beach rock (Fig. 5(St. 2)), As shown in Fig. 5(St. 3), the roots of a coconut tree were exposed to waves, implying that this area with a hooked shoreline had recently eroded. The formation of a hooked shoreline downcoast of the beach rock corresponds to the fact that westward longshore sand transport prevailed immediately before the field observation, which was generated by the wave incidence from the east. Figure 5(St. 4) shows the foreshore of sand spit A

Fig. 5. Beach conditions around the island in the end of May, 1992.

located at the northwest end of the island with a view of the vegetation on the right side. A sandy beach with a flat foreshore extended at the west end of the island.

At St. 5, located on the southeast end of the island, the beach was severely eroded and scarps had formed with the exposure of vegetation roots (Fig. 5(St. 5)). Many roots were exposed around this area and they appeared to help the beach resist wave abrasion. Furthermore, the roots of a coconut tree in the east of the area had been scoured and the tree was about to fall down. The erosion in this area was considered to be due to the action of waves incident from the east, taking into consideration the fact that the relative height of the scarp decreased westward with a maximum value near the roots of the coconut tree.

Figure 5(St. 6) shows sand spit B formed at the southwest end of the island, where a large amount of fine sand had been deposited. Moreover, it can be seen that the coral sand had been extensively deposited offshore of the sandy beach. Figure 5(St. 7) shows the beach in the north part of sand spit B and the west jetty. A semicircular ridge extended and a lowland existed landward of this ridge. From these facts, it is realized that the size of sand spit B had been increased by the successive deposition of sand transported along the shoreline of the spit by longshore sand transport. Finally, Fig. 5(St. 8) shows sand spit B.

Although sand spit B significantly protruded westward, erosion of this sand spit was under way because waves were incident from the southwest during the field observation. Thus, the field observation at the end of May 1992 was carried out during the transition period from the general northeast monsoon period between November and April to the southwest monsoon period between May and October. Sand spit A extended westward at the northwest end of the island and sand spit B extended southwestward, similarly to in Fig. 4, and the field observation was carried out during the time when beach changes due to waves incident from the southwest were about to begin.

4. Numerical simulation of movement of sand spits

On Embudu Village Island, the sand spits develop on the east and west sides of the coral cay because of the alternate action of waves incident from the southwest and east, respectively. Similar beach changes on an extremely shallow reef can often be observed in the deformation of a sand bar on a tidal flat, and such changes can be predicted using the BG model, as reported in Miyahara et al. (2015)[5], in which the deformation, movement, and mergence of a sand bar on a tidal flat were analyzed. Here, we use the same model. In the calculation, the x- and y-axes were taken alongshore and normal to the shore, respectively, the calculation domain of a square area with a length of 600 m in the x- and y-directions was selected, and a model island was set in this domain. Taking the shallowness of the reef into account, the water depth around the island was assumed to be 2 m and the reef was given by a solid bed. The shape of Embudu Village Island was modeled as a trapezoid with a height of 320 m, and 60 and 160 m lengths along the top and bottom, respectively, as shown in Fig. 6.

First, the wave field was calculated using the energy balance equation for irregular waves[8]. Neglecting the wave refraction owing to the changes in the marginal topography of the coral reef, the wave field was calculated under the condition that waves were directly incident to a sandy island on the reef flat, and it was used for the numerical simulation of beach changes. Because of the proximity of Embudu Village Island to the reef gap to the east, the direction of incident waves was assumed to be both from the east and southwest, and the wave field for the initial topography, as shown in Fig. 6, was calculated. As the wave characteristics, a significant wave height of 0.5 m was first selected for typical waves, which can approach to near the shoreline without breaking on the reef flat with a depth ranging between 1 and 2 m, and a wave period of 4 s was reversely calculated, given the wave steepness of 0.02 of the normal waves as an intermediate type between wind waves and swells. In the calculation of beach

Table 1. Calculation conditions.

Wave conditions	Incident waves: $H_I = 0.5$ m, $T = 4$ s, and wave direction: E and SW
Berm height	$h_R = 1$ m
Depth of closure	$h_c = 4H$ (H: wave height at a local point)
Equilibrium slope	$\tan\beta_c = 1/10$
Angle of repose slope	$\tan\beta_g = 1/2$
Coefficients of sand transport	Coefficient of longshore sand transport $K_s = 0.05$ Coefficient of Ozasa & Brampton (1980)[10] term $K_2 = 1.62\ K_s$ Coefficient of cross-shore sand transport $K_n = K_s$
Mesh size	$\Delta x = \Delta y = 10$ m
Time intervals	$\Delta t = 0.02$ h
Duration of calculation	80 h (4×10^3 steps)
Boundary conditions	Shoreward and landward ends: $q_x = 0$, left and right boundaries: $q_y = 0$
Calculation of wave field	Energy balance equation (Mase, 2001)[8] • Term of wave dissipation due to wave breaking: Dally et al. (1984)[11] model • Wave spectrum of incident waves: directional wave spectrum density obtained by Goda (1985)[12] • Total number of frequency components $N_F = 1$ and number of directional subdivisions $N_\theta = 8$ • Directional spreading parameter $S_{max} = 25$ • Coefficient of wave breaking $K = 0.17$ and $\Gamma = 0.3$ • Imaginary depth between minimum depth h_0 and berm height h_R : $h_0 = 1$ m • Wave energy = 0 where $Z \geq h_R$ • Lower limit of h_1 in terms of wave decay due to breaking : 0.5 m

changes, we assumed that the trapezoidal island was surrounded by a seawall of 1.5 m height, as shown in Fig. 6, because (1) the central part of the island is covered with tropical trees, the roots of which can increase resistence to wave abrasion similarly to a seawall, and (2) the sand that can be transported by waves is only distributed along the marginal area of the island. The berm height was approximated by $h_R = 1$ m, by referring to the beach conditions shown in Fig. 6. Furthermore, the depth of closure was assumed to be $h_c = 2$ m, the same as the depth of the reef flat. Given these conditions, we set $\Delta x = \Delta y = 10$ m and $\Delta t = 0.02$ h, and the beach changes up to 4,000 steps (80 h) were predicted. Table 1 summarizes the calculation conditions.

Figure 6 shows the calculation results of the changes in beach topography up to 4,000 steps (80 h) under the condition that waves are incident via the channel on the east side of the island when the northeast wind blows in the dry season from November to April. Waves incident from the east propagate toward the east end of the island E, and longshore sand transport is generated along sides EF and EH. The sand spits start to form near the corners F and H, and their size increases with time. Although the sand spits continue to develop, no beach changes occur along side GH because the area is included in the wave-shelter zone of the island against waves incident from the east. In addition, the sand

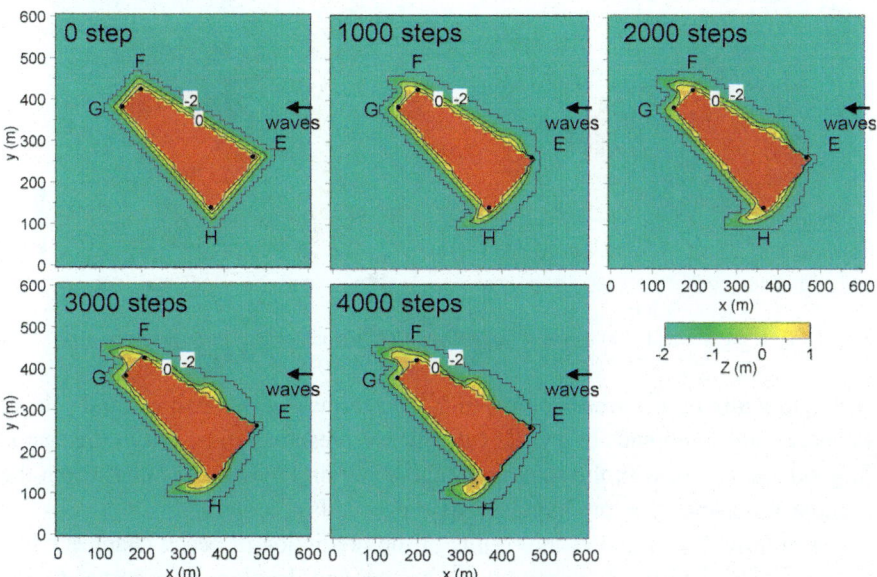

Fig. 6. Topographic changes on coral cay under condition of wave incidence from east.

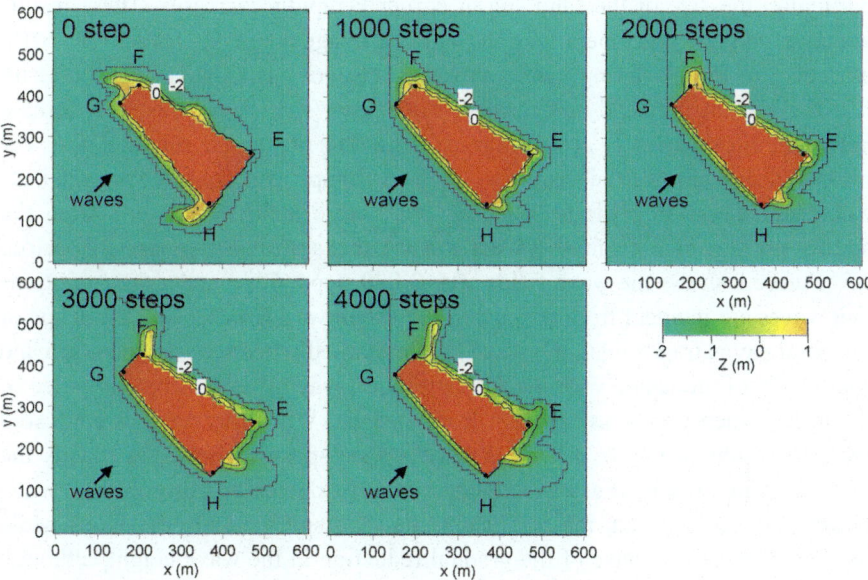

Fig. 7. Topographic changes on coral cay under condition of wave incidence from southwest.

476

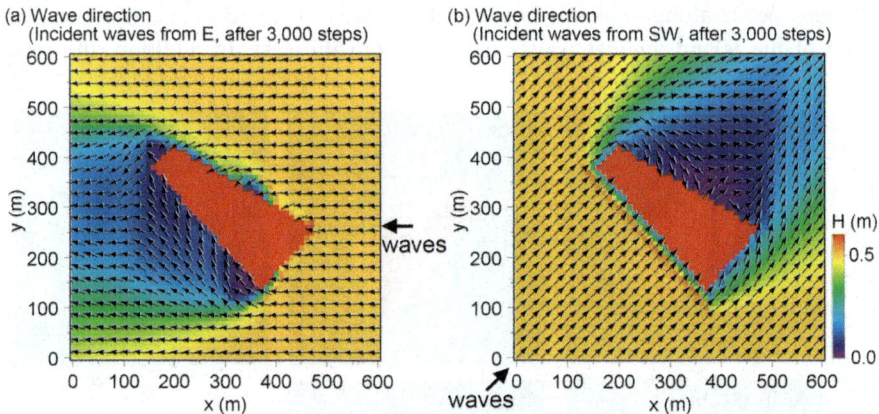

Fig. 8. Wave fields ((a) incident waves from east, (b) incident waves from southwest)

body gradually moves from point E to F. The formation of a pair of sand spits at the north and south ends of a coral cay when waves are incident from the east is in good agreement with the satellite image taken on February 2, 2005, when the northeast monsoon was predominant, as shown in Fig. 3.

Similarly, Fig. 7 shows the results of the prediction of beach changes when waves are incident from the southwest. A sand spit develops near corner F, and another sand spit develops with gradual movement at corners H and E. In particular, the size of the sand spit at corner F rapidly increases. These results are again in good agreement with the satellite image taken on October 19, 2010, as shown in Fig. 3, during the southwest monsoon. Also, the calculation result that the size of the sand spit formed near corner F is larger than that formed between the corners E and H agrees with the measured result. Here, the measured direction of the extension of the sand spit is more eastward than the calculated direction. This is because waves were incident not only from the southwest but also from the west during the southwest monsoon period.

Figure 8 shows the wave fields after 3,000 steps (60 h) under the conditions that waves are incident from the east (a) and southwest (b). Because the shape of the coral cay is trapezoidal, the wave height markedly decreases at the north and south ends of the island with the formation of a wave-shelter zone on the lee of the island when waves are incident from the east. When this result is compared with the results in Fig. 6, it is clear that the sand spit was formed at exactly the same location where the wave height abruptly decreased. In contrast, when waves are incident from the southwest, the sand spit gradually develops while moving eastward because of the gradual reduction in the wave height, although the result that a sand spit is formed at the north end of the cay where rapid reduction in the wave height occurs is the same as that when waves are incident from the east, as shown in Fig. 7. Thus, the wave-sheltering effect of the coral cay itself is a key factor in the variation of a sandy beach on a coral cay.

5. Conclusions

Seasonal variations of a sandy beach on Embudu Village Island were investigated using satellite images taken between 2005 and 2013 along with a field observation carried out at the end of May 1992. Then, the beach changes around a coral cay were predicted using the BG model. It was found that sand spits were formed on the west side of the cay during the dry season because of the predominance of the northeast wind, whereas during the rainy season between May and November, sand spits were formed on the east side because of the predominance of the southwest wind. In this study, we assumed that a sandy beach continuously distributed along the entire shoreline of a trapezoidal coral cay as the initial beach conditions for calculating the seasonal variation. Therefore, the recovery mechanism from the condition that a pair of sand spits have developed at the corners of the coral cay to the initial topography given in this study has not yet been studied, and further study is required. On the other hand, when artificial alterations, such as the construction of jetties, groins, and detached breakwaters, are carried out, there may be an imbalance in longshore sand transport along the shoreline of the coral cay, resulting in local erosion and accretion. Since a coral cay has a closed system of sand transport, as mentioned in San-nami et al. (2013)[4], such an alteration will impact the entire shoreline of the cay. On Embudu Village Island, the direction of longshore sand transport periodically varies in response to changes in the wave direction, and therefore, a large amount of longshore sand transport occurs on the southeast shore of the island and at the central part of the northwest shore. Accordingly, if an artificial structure such as an impermeable jetty or a detached breakwater is constructed on the southeast or northwest shore of the island, the construction will have an impact on both sides of the structure because of the blockage of longshore sand transport. Sufficient attention should be paid for such points to protect the coral cay. Once a cay is subjected to anthropogenic impact, the balance of sand transport around the cay is severely damaged.

References

1. E. Bird, *Coastal Geomorphology, An Introduction* (John Wiley and Sons, New York, 2000) p. 322.

2. T. Uda, Some notes on coastal features of Embudu Village Island in Maldives, *Civil Eng. J.*, Vol. 35, No. 11, (1993), pp. 69-73, (in Japanese)

3. T. Uda, M. Serizawa, S. Miyahara, BG model based on Bagnold's concept and its application to analysis of elongation of sand spit and shore - normal sand bar (Chap. 16), in *'Numerical Simulation - From Theory to Industry'* (Andriychuk, M. ed., INTECH, 2012), pp. 339-374, http://www.intechopen.com/books/numerical-simulation-from-theory-to-industry/bg-model-basedon-bagnold-s-concept-and-its-application-to-analysis-of-elongation-of-sand-spit-and-s

4. T. San-nami, T. Uda, M. Serizawa, S. Miyahara, Prediction of devastation of natural coral cay by human activity, *Proc. Coastal Dynamics 2013*, No. 138, (2013), pp. 1427-1438.

5. S. Miyahara, T. Uda, M. Serizawa, T. San-nami, Elongation of sand spit and profile changes on sloping shallow seabed, *8th Int. Conf. on Asian and Pacific Coasts* (APAC 2015), Procedia Engineering, Vol. 116, (2015), pp. 245-253, http://www.sciencedirect.com/science/article/pii/S1877705815019426

6. T. Suzuki, H. Kayanne, Y. Iwatsuka, H. Katayama, T. Sekimoto, M. Isobe, Studies on the topographic change mechanism of coral cays, *J. JSCE*, Vol. 69, No.2, (2013) , pp. I_838-I_843, (in Japanese)

7. Y. Iwatsuka, T. Kotoura, H. Katayama, R. Takemori, Y. Tajima, H. Kayanne, Field survey of formation and maintenance mechanism on ballast island, *J. JSCE*, Vol. 71, No. 2, (2015), pp. I_455-I_460. (in Japanese)

8. H. Mase, Multidirectional random wave transformation model based on energy balance equation, *Coastal Eng. J., JSCE*, Vol. 43, No. 4, (2001), pp. 317-337.

9. H. Kan, Degradation danger of a nation (Maldives) composed of atolls - a chain reaction between disaster and development, in Chap. 3, *'Natural disasters in the age of global warming: six field reports from around the world'*, ed. by Commission of Disaster Responses of the Association of Japanese Geographers (AJG), Y. Hirai, T. Aoki, (Kokon Shoin Publishers, Tokyo, 2009), (in Japanese)

10. H. Ozasa, A. H. Brampton, Model for predicting the shoreline evolution of beaches backed by seawalls, *Coastal Eng.*, Vol. 4, (1980), pp. 47-64.

11. W. R. Dally, R. G. Dean, R. A. Dalrymple, A model for breaker decay on beaches, *Proc. 19th Inter. Conf. on Coastal Eng.*, (1984), pp. 82-97.

12. Y. Goda, *Random Seas and Design of Maritime Structures*, (University of Tokyo Press, Tokyo, 1985), 323 p.

Rapid Formation of a Sand Spit at River Mouth and its Prediction Using BG Model

Masumi Serizawa[a], Takaaki Uda[b] and Shiho Miyahara[a]

[a]Coastal Engineering Laboratory, Co., Ltd.
1-22-301 Wakaba, Shinjuku, Tokyo 160-0011, Japan
E-mail: coastseri@nifty.com

[b]Head, Shore Protection Research, Public Works Research Center,
1-6-4 Taito, Taito, Tokyo 110-0016, Japan
E-mail: uda@pwrc.or.jp

Several years ago, the river mouth bar disappeared at the Shimanto River mouth in Kochi Prefecture, Japan. After the disappearance of the river mouth bar, wave invasion into the river was enhanced and a sand bar in front of the left bank of the river, which had been stable for a long time, was rapidly eroded and sand was transported upstream, resulting in the formation of a sand spit. The formation of the sand spit was investigated by shoreline analysis using satellite images together with a field observation. Then, the topographic changes of the sand spit were numerically reproduced using the BG model.

Keywords: River mouth; River mouth bar; Sand spit; Shoreline changes; Satellite image; BG model.

1. Introduction

On a coast where waves are incident to the coastline with a large angle greater than 45°, a sand spit may be formed. Because the wave incidence angle is usally less than 45° on the ordinary coasts, the formation of a sand spit is rare. In a bay or the river mouth with an embayed shoreline, however, a sand spit can be formed because the orientation of the mean shoreline may extend approximately parallel to the direction of incident waves. Therefore, if there is a sufficient volume of sand available for the formation of a sand spit, a sand spit can be formed at the bay mouth or river mouth. Such a sand spit acts as a natural levee against the invasion of waves into the bay mouth or river mouth. Moreover, the brackish water upstream of the sand spit becomes an important habitat for many kinds of animals and young fishes with important environment qualities. In this study, we investigated the development of a sand spit, taking the Shimanto River mouth in Kochi Prefecture, Japan, as an example, where a sand spit was formed on the left bank of the river owing to longshore sand transport induced by breaking waves along the left river bank. In this river, the mouth bar disappeared owing to the excavation of the sand deposited in the navigation channel behind the river mouth bar, and after the disappearance of the river mouth bar, the intensity of wave invasion into the upstream area of the river mouth increased, resulting in the generation of longshore sand transport along the left bank. The shoreline changes associated with the development of the sand spit were first investigated by field observation and using satellite images, and then the topographic changes were predicted using the BG model (a three-dimensional model for predicting beach changes based on Bagnold's concept) (San-nami et al., 2014)[1].

2. Overall changes around river mouth

Figure 1 shows an aerial photograph of the study area around the Shimanto River mouth taken on May 24, 2002, and a satellite image of the same area taken on February 10, 2013. At this river mouth, a river mouth bar of 300 m length extended between the left bank and the training jetty of the Shimanto River in 2002. This sand spit acted as a breakwater, resulting in wave attenuation upstream of the river mouth. At that time, there was a stable sand bar at points A along the left bank and B immediately downstream of the sand in the river because of the wave-sheltering effect of the river mouth bar (Fig. 1(a)). During storm wave conditions, a large amount of sand was transported into the navigation channel immediately upstream of the river mouth bar shown in Fig. 1(a), resulting in heavy sedimentation of the navigation channel. As a measure against sand deposition, dredging was repeatedly carried out along the navigation channel, resulting in a decrease in the volume of sand of the river mouth bar (Furuike et al., 2009)[2]. The river mouth bar had disappeared by February 10, 2013 (Fig. 1(b)). After the disappearance of the river mouth bar, sand was transported upstream to form a sand spit at A' together with the deformation of the sand bar from B to B', as shown in Fig. 1(b), because of the increased wave action in the river mouth area.

3. Field observation of a sand spit

On December 8, 2015, a field observation was carried out around this sand spit. Figure 2 shows a satellite image of the Shimanto River mouth and Shimoda Port taken in 2014. The location numbers of the site photographs taken during the observation are shown in Fig. 2, where St. 1 is located on the hill top on the right bank and Sts. 2-10 are located around the sand spit.

(a) May 24, 2002

(b) February 10, 2013

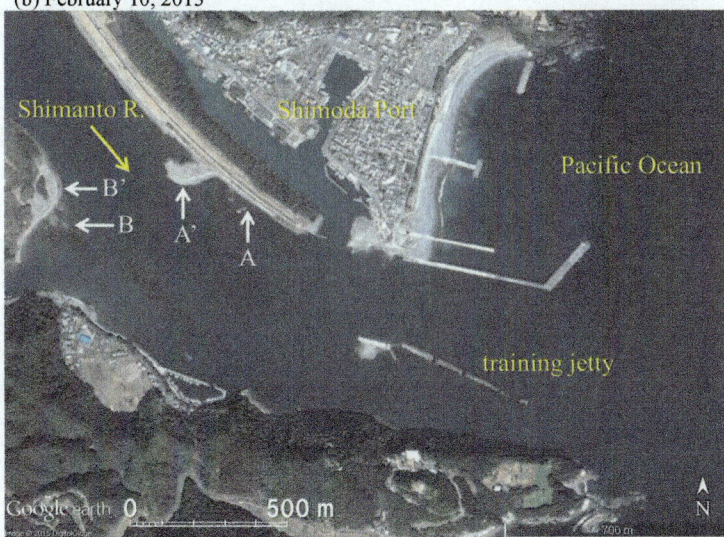

Fig. 1. Aerial photograph and satellite image of area around Shimanto River mouth taken on May 24, 2002 and February 10, 2013.

Figure 3(St. 1) shows the view of the river and the sand spit from St. 1 on top of the hill south of the river (Fig. 2). The sand spit extends straight toward the center of the river and is approximately normal to the left river bank.

Fig. 2. Satellite image of Shimanto River mouth and Shimoda Port taken in 2014 and location numbers of site photographs.

Fig. 3. Results of field observation at Sts. 1-9.

The elongation of the sand spit is closely related to not only the wave intrusion further upstream along the left bank but also the local scouring at the toe of the left bank, which requires foot protection. Figure 3(St. 2) shows the photograph of the sand spit taken from the top of the left bank while looking southwestward.

A ridge extends along the seaward shoreline of the sand spit, whereas there is a gentle slope on the back of the berm with a large amount of debris at the toe of the gentle slope. This implies that wave overtopping has taken place on the sand spit during storm wave conditions. Furthermore, a slender sand spit extends further upstream at the tip of the main body of the sand spit.

Figure 3(St. 3) shows the downstream side of the left bank protected by a revetment and the foot of the sand spit. Because sand has been transported upstream by longshore sand transport caused by waves propagating upstream from the river mouth, the concrete revetment installed for foot protection of the river bank has been exposed to waves with scattering of some of the foundation stones. Figure 3(St. 4) shows the downstream side of the sand spit. A large amount of driftwoods was found on the berm, implying that high storm waves were incident to the sand spit. Figure 3(St. 5) shows a photograph of the berm where it was well developed, taken from the shoreline. The berm height was +2 m above mean sea level (MSL) and steep slope of over 1/3 on the beach face. Taking into account the fact that the berm height is approximately equal to the wave height at this point during high tide, it is concluded that the berm was formed by waves with approximately 2 m height. Figure 3(St. 6) shows the back slope of the berm, which is less steep than the shore-facing slope of the berm. Some traces owing to wave overtopping remained on the surface of the back slope. A closed water body had formed upstream of the sand spit owing to the extension of a recurved sand spit. Figure 3(St. 7) shows the condition of the tip of the sand spit covered with gravel. The area upstream of the sand spit was completely sheltered from incident waves from the river mouth by the sand spit itself. Figure 3(St. 8) shows the sand deposited upstream of the recurved sand spit, where fine sand was deposited because of a lack of strong wave action, in contrast to the area downstream of the sand spit shown in Fig. 3(St. 7). Finally, Fig. 3(St. 9) shows the shallow lagoon remaining behind the tip of the sand spit. This closed water body was formed because the tip of the sand spit extended upstream while rotating clockwise. Thus, the main geomorphic agent for the development of the sand spit is considered to be the breaking of waves incident from the river mouth that propagated along the left river bank, which in turn induced longshore sand transport in the upstream direction.

4. Investigation of sand spit development using satellite images

On the basis of five sets of satellite images taken from August 2006 to December 2014, the rectangular area shown in Fig. 2 was selected and enlarged

484

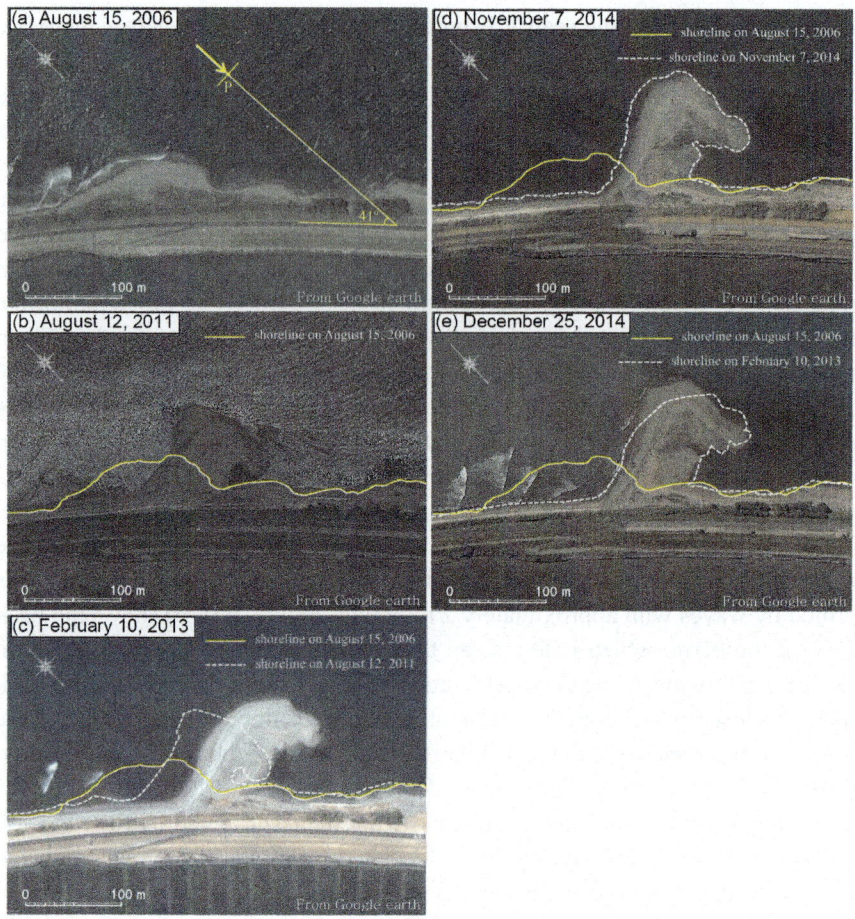

Fig. 4. Satellite images of sand spit.

satellite images were obtained to investigate the development of the sand spit near the left river bank. Figures 4 shows the selected satellite images, where the shoreline in August 2006 and that in previous occasion when the image was taken are superimposed in Fig. 4.

In Fig. 4(a), taken on August 15, 2006, a slender sand bar with 160 m length and 60 m width and with a neck in the upstream direction existed in front of the left river bank. On the south side of this sand bar, waves were obliquely incident to the river bank, and wave breaking along the shoreline can be seen. In addition, wave crest lines propagating obliquely to the river bank can be seen at point P, approximately 80 m offshore of the sand bar. The angle of the wave crest line relative to the river bank is as large as 41°, as shown in Fig. 4(a). These high-angle waves, which propagate obliquely to the shoreline with a large angle, cause longshore sand transport in the upstream direction, resulting in the deformation of the sand bar. By August 12, 2011, the sand making up this slender sand bar had been transported upstream, where a sand spit started to form (Fig. 4(b)). The downstream half of the sand bar observed in 2006 had eroded, whereas a sand deposition zone extended upstream obliquely from the upstream end of the sand bar. As a result, the sand spit protruded by 100 m from the river bank while traversing the river. By February 10, 2013, the shoreline of the sand spit on the downstream side had moved by 35 m, resulting in the upstream migration of the sand spit (Fig. 4(c)), and the longest distance of the sand spit from the river bank reached 106 m. Moreover, recurved spits were formed owing to the successive sand transport from upcoast. The river bank composed of sand observed on August 15, 2006, had eroded, resulting in the exposure of the revetment to waves.

By November 7, 2014, the southwest end of the sand spit had extended in such a way that the angle between the seaward shoreline of the sand spit and the river bank line had increased, as shown in Fig. 4(d), whereas the shoreline had receded slightly at the base of the sand spit. As a result, the longest distance of the sand spit from the river bank reached 117 m. On December 25, 2014, the shape of the sand spit was similar to that on November 7, 2014, with overall recession of the shoreline seaward of the sand spit by 6-9 m (Fig. 4(e)). During the period between November 7 and December 25, 2014, the plane area of the sand spit was reduced, and this may have been caused by the landward transport of sand, raising the berm of the sand spit. As mentioned above, a large sand spit elongated in the direction normal to the river bank line on December 25, 2014. The sand source for the formation of the sand spit was the sand bar located at the toe of the river bank, and the revetment of the river bank was exposed to waves, because sand had been transported away upstream by longshore sand transport by waves propagating in the upstream direction.

Figure 5 shows the bathymetry around the river mouth in 2011. Since the bathymetric survey was carried out in 2011, the satellite image shown in Fig. 4(b) corresponds to this bathymetry. A flat wave-cut terrace with 1.5 m depth remained in the area downstream of the sand spit, and the sand spit was formed in a zone with elevation between -1 and +1.5 m.

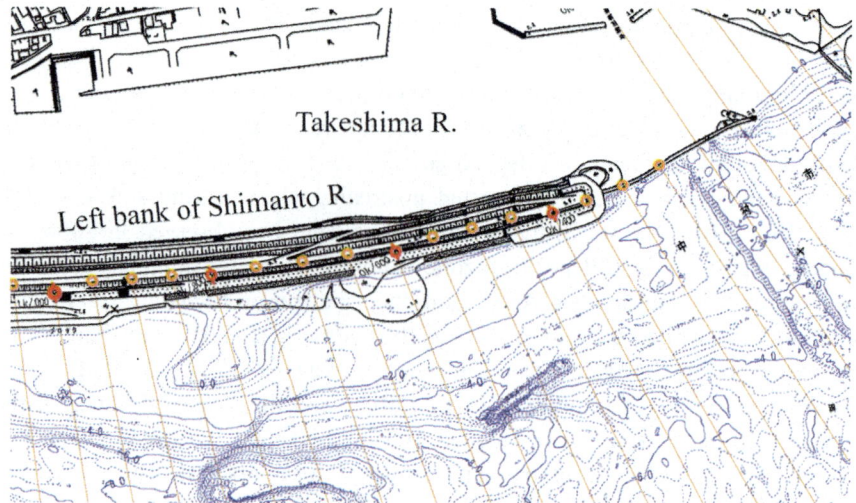

Fig. 5. Bathymetry around river mouth in 2011.

5. Numerical simulation of development of sand spit using BG model

5.1. *Calculation conditions*

In the numerical simulation of the development of a sand spit, the BG model was employed, similarly to in the numerical simulations of the elongation of a sand spit on a seabed with different water depths and slopes, and of the deformation of an isolated offshore sand bar on a tidal flat (San-nami et al., 2014; 2015)[1,3]. In this model, sand transport flux due to waves is calculated using an expression involving the third power of the amplitude of the bottom wave orbital velocity for the sand transport equation. The x- and y-axes are taken alongshore and normal to the shore, respectively. A rectangular domain with 600 m width and length and with a flat solid bed of 2 m depth was selected. The initial sand bar was placed on the flat solid bed. Although the berm height was + 2 m above MSL, as shown in Fig. 5, this height includes the effect of the change in the tide level of approximately 1 m. Accordingly, in the calculation of the development of the sand spit with the reference to MSL, the berm height h_R was assumed to be 1 m. Furthermore, the water depth in front of the sand spit becomes approximately 1 m during low tide, because topographic changes around the sand spit occur between -2 m (h_c) and 1 m (h_R) along with the tidal range of 1 m. Under this condition, the upper limit of the height of waves which can propagate to the sand bar without breaking is approximately given by 0.5 m. The incident wave height, therefore, was assumed to be 0.5 m. In addition, the

Table 1. Calculation conditions.

Wave conditions	Incident waves: $H_I = 0.5$ m, $T = 4$ s, and wave direction: $\theta_I = -10°$ with respect to x axis
Berm height	$h_R = 1$ m
Depth of closure	$h_c = 4H$ (H: wave height at a local point)
Equilibrium slope	$\tan\beta_c = 1/10$
Angle of repose slope	$\tan\beta_g = 1/2$
Coefficients of sand transport	Coefficient of longshore sand transport $K_s = 0.05$ Coefficient of Ozasa & Brampton (1980)[4] term $K_2 = 1.62\ K_s$ Coefficient of cross-shore sand transport $K_n = K_s$
Mesh size	$\Delta x = \Delta y = 10$ m
Time intervals	$\Delta t = 0.05$ h
Duration of calculation	250 h (5×10^3 steps)
Boundary conditions	Shoreward and landward ends: $q_x = 0$, left and right boundaries: $q_y = 0$
Calculation of wave field	Energy balance equation (Mase 2001)[5] • Term of wave dissipation due to wave breaking: Dally et al. (1984)[6] model • Wave spectrum of incident waves: directional wave spectrum density obtained by Goda (1985)[7] • Total number of frequency components $N_F = 1$ and number of directional subdivisions $N_q = 8$ • Directional spreading parameter $S_{max} = 25$ • Coefficient of wave breaking $K = 0.17$ and $G = 0.3$ • Imaginary depth between minimum depth h_0 and berm height h_R : $h_0 = 1$ m • Wave energy $= 0$ where $Z \geq h_R$ • Lower limit of h_1 in terms of wave decay due to breaking F: 0.5 m

wave period was given by 4 s, assuming that the wave steepness of the normal waves is 0.02. The depth of closure h_C was assumed to be 2 m on the basis of the bathymetry, as shown in Fig. 5. The calculation domain was discretized by a mesh of $\Delta x = \Delta y = 10$ m, Δt was selected as 0.05 h, and the calculation was carried out for 5,000 steps (250 h). First, a trial and error calculation was carried out for wave directions of 0, 10, and 20° clockwise relative to the x-axis, and the best-fit result was obtained when the wave direction was 10° clockwise relative to the x-axis. Table 1 shows the calculation conditions.

5.2. Results of numerical simulation

Figure 6 shows the results of the numerical simulation of the development of a sand spit. Although the shape of the initial sand bar had an approximately symmetrical form in the upstream and downstream directions, the downstream

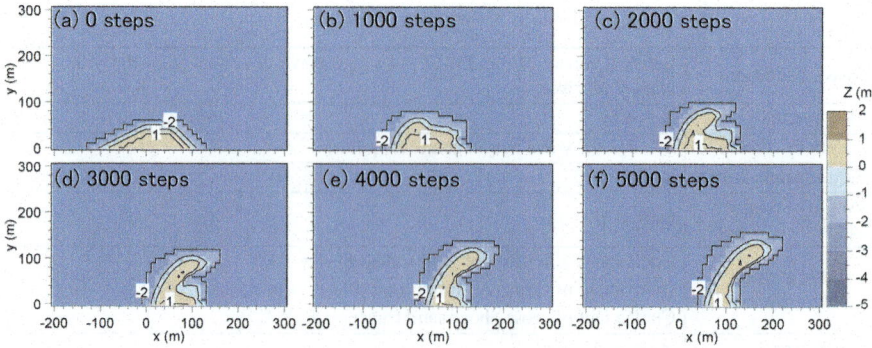

Fig. 6. Results of numerical simulation of development of a sand spit.

part of the sand bar had eroded by 1,000 steps and the eroded sand was transported upstream and deposited near a location with $x = 80$ m while forming a small protusion owing to the oblique wave incidence with a counterclockwise angle of $10°$ relative to the x-axis. The disapperance of the sandy beach along the downstream bank corresponds to the exposure of the revetment of the river bank, as shown in Fig. 4(b). After 2,000 steps, the downstream side of the sand bar had further eroded and a sand spit had started to form in the upstream direction. The shoreline immediately upstream of the tip of the sand spit became significantly concave and the longshore sand transport along the concave shoreline decreased because of the wave-sheltering effect, by which the elongation of the sand spit was further enhanced. After 3,000 steps, the sand spit was elongated in the direction with an angle of $45°$ relative to the river bank and had a constant width of 30 m. Figure 7 shows the wave field after 3,000 steps. Waves with an oblique angle of $10°$ relative to the x-axis were incident to the tip of the sand spit with a large angle and a large wave-shelter zone existed on the upstream side of the sand spit.

After 5,000 steps, the entire sand spit had moved upstream and the sand spit had become slender (Fig. 6(f)). Figure 8 shows the overall shoreline changes until 5,000 steps. The calculation result, i.e., the elongation of the sand spit in the river with increasing distance from the river bank, was in good agreement with the measured result. The development of the sand spit is very similar to the deformation of solitary waves due to the breaking on a sloping bed. In the measured deformation of the sand spit, as shown in Fig. 4(e), the sand spit was elongated approximately normal to the left bank and then the tip turned around to form a recurved spit. In contrast, in the calculation, the tip of the sand spit elongated in the upstream direction obliquely as a simple sand spit. This is the main difference between the calculated and measured results. In the wave field

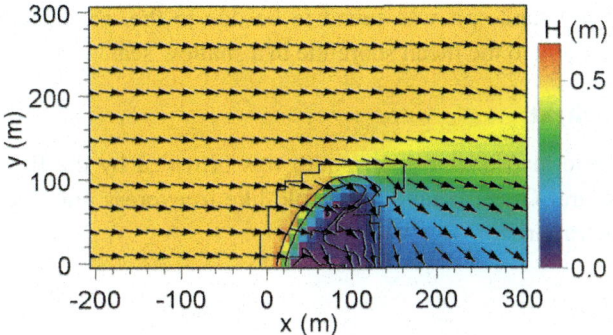

Fig. 7. Wave field after 3,000 steps.

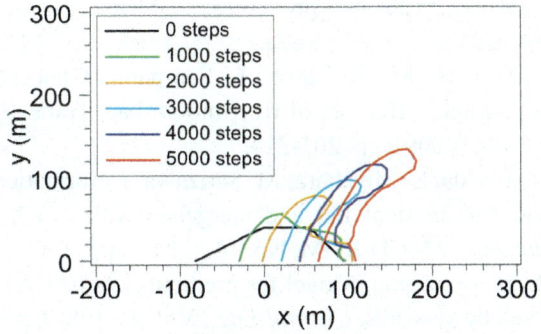

Fig. 8. Shoreline changes of a sand spit.

employed in the numerical simulation, waves were incident from a direction approximately parallel to the *x*-axis, as shown in Fig. 7. Under this condition, the upstream side of the sand spit was completely sheltered by the sand spit itself, resulting in less sand movement. For the development of the recurved sand spit in Fig. 4(e), wave incidence from the west side is necessary. This must have been due to the wind waves generated in the river. In this study, waves from the river mouth were considered and this caused the discrepancy between the calculated and measured results. Future study is needed in this regard.

6. Conclusions

The sand bar in front of the left river bank at the Shimanto River mouth in Kochi Prefecture, Japan, was eroded owing to the waves incident from the river mouth, and a slender sand spit was formed by the successive movement of sand. The mechanism of the development of the sand spit was investigated by field observation, and topographic changes were reproduced using the BG model. The formation of the sand spit was found to be due to the wave incidence with a large angle relative to the mean shoreline in a shallow sea.

Acknowledgement

The bathymetric map of the river mouth was provided by the Port Division of Kochi Prefectural Government. We would like to express our sincere gratitude for their cooperation.

References

1. T. San-nami, T. Uda, M. Serizawa, S. Miyahara, Numerical simulation of elongation of sand spit on seabed with different water depths and slopes, *Proc. 34th ICCE*, (2014), pp. 1-13, `https://journals.tdl.org/icce/index.php/icce/article/view/7112/pdf_415`
2. K. Furuike, T. Uda, M. Serizawa, T. San-nami, Conceptual model for predicting topographic changes of river-mouth bar, Trans. *Jpn. Geomorph. Union*, Vol. 30-3, (2009), pp. 201-217.
3. T. San-nami, T. Uda, S. Miyahara, M. Serizawa, Deformation of an isolated offshore sand bar on tidal flat and mergence with beach due to waves, *Coastal Sediments '15*, CD-ROM, No. 14, (2015), pp. 1-14.
4. H. Ozasa, A. H. Brampton, Model for predicting the shoreline evolution of beaches backed by seawalls, *Coastal Eng.*, Vol. 4, (1980), pp. 47-64.
5. H. Mase, Multidirectional random wave transformation model based on energy balance equation, *Coastal Eng. J., JSCE*, 43(4), (2001), pp. 317-337.
6. W. R. Dally, R. G. Dean, R. A. Dalrymple, A model for breaker decay on beaches, *Proc. 19th Int. Conf. on Coastal Eng.*, (1984), pp. 82-97.
7. Y. Goda, *Random Seas and Design of Maritime Structures*, (University of Tokyo Press, Tokyo, 1985), 323 p.

Simulation of Nearshore Coastal Processes for Development of Fishery Harbour

D.P.C. Laknath[†] and K. Raveenthiran

Lanka Hydraulic Institute Ltd,
Katubedda, Moratuwa,10400, Sri Lanka
[†]E-mail: chanaka.laknath@gmail.com
www.lhi.lk

K.K.P.P. Ranaweera

Ministry of Fisheries and Aquatic Resources Development
New Secretariat, Maligawatta, Colombo,00900, Sri Lanka
E-mail: prabathr26@gmail.com

This study was carried out for a proposed fishery harbour in Wennappuwa, in North-western coastal stretch of Sri Lanka. Simulation of nearshore coastal processes such as hydrodynamic and sediment transport processes to ensure safe navigation and continues harbour operations are the main objective of this study. MIKE 21 modelling system was used to simulate hydrodynamics and sediment transport processes. With the hindcast nearshore wave climate, coastal processes were simulated for representative nearshore wave climate conditions. According to the MIKE 21 HD (Hydrodynamics) model results, it was identified that proposed port configuration is favored for its hydrodynamic performance. MIKE 21 ST (Sediment Transport) model was used to understand sediment transport patterns, especially at harbour entrance and approach channel to investigate the siltation possibilities. According to the simulated results which are based on average wave conditions, it was found that net sediment transport is directed from south to north direction. However, sediment transport movement has not supported to move sediment significantly in to the harbour through its entrance. Hence, sediment bypass can be expected through the harbour entrance with the proposed configuration. Thus, suitability of the harbour location and layout were justified with the numerically simulated results.

Keywords: MIKE 21; Wave climate; Hydrodynamic; Sediment transport; Harbour entrance.

1. Introduction

Before implementing the Coastal Resource Management Project (CRMP) in Sri Lanka between 2000 - 2006 periods, North-western coast line of Sri Lanka was identified as a critical coastal stretch, mainly due to the coastal erosion. Hence, implementation of development projects such as fishery harbours was discouraged in the same coastline. On the other hand, due to the construction of some fishery harbour with lack of understanding on prevailing coastal processes, fishery community has been adversely affected [1,2]. Therefore, detailed and accurate assessment about the coastal processes in the proposed area is essential to understand the natural coastal processes. The correct knowledge about the behaviour of water bodies subjected to a variety of natural forcing such as wave, tide and wind is essential to decide the configuration of harbour layout, ensure safe navigations and sedimentation problems. Accordingly, the main objective of this study is to simulate the nearshore coastal processes such as hydrodynamic and sediment transport processes to ensure safe navigation and harbour operations of the proposed harbour throughout the year under different climatic conditions.

2. Study Area

This numerical model study was carried out for a proposed Fishery Harbour in Wennappuwa. The selected location for proposed harbour is shown in Figure 1. Generally, the wave climate in west coast of Sri Lanka is mainly governed by the southwest monsoon. Comparatively, northeast monsoon is very much milder. The tides occurring in Sri Lanka is semi-diurnal type with an average tidal range of less than 1m. Due to that, magnitudes of tidal currents along the coast are not significant. In case of wave induced currents, the long shore coastal

Fig. 1. Study area and selected harbour location (Source: *Google earth*)

currents are strongest during the respective months. Thus, during the southwest monsoon period, wave induced currents are in a northerly direction while currents in a southerly direction could be experienced during the northeast monsoon time. In terms of sediment process, this study area was recognized as a highly sensitive area for coastal erosion before implementing the CRMP in Sri Lanka between 2000 - 2006 periods. Though this area has been stabilized with beach nourishment and hard structures under the CRMP, still this coast can be considered as a highly sensitive and critical coastal stretch for sediment processes.

3. Related Previous Numerical Model Studies

3.1. *Proposed harbour*

The proposed harbour layout is consisted of two breakwaters, quay wall and onshore facilities (Figure 2). Thus, main and northern breakwaters of proposed layout are connected to the existing offshore breakwaters (OB1 and OB2). The dredging depth below datum in the harbour basin was taken as 4.0 m MSL. At the entrance of harbour, including the effect of waves, dredging depth is further increased up to 5.0 m MSL below the datum.

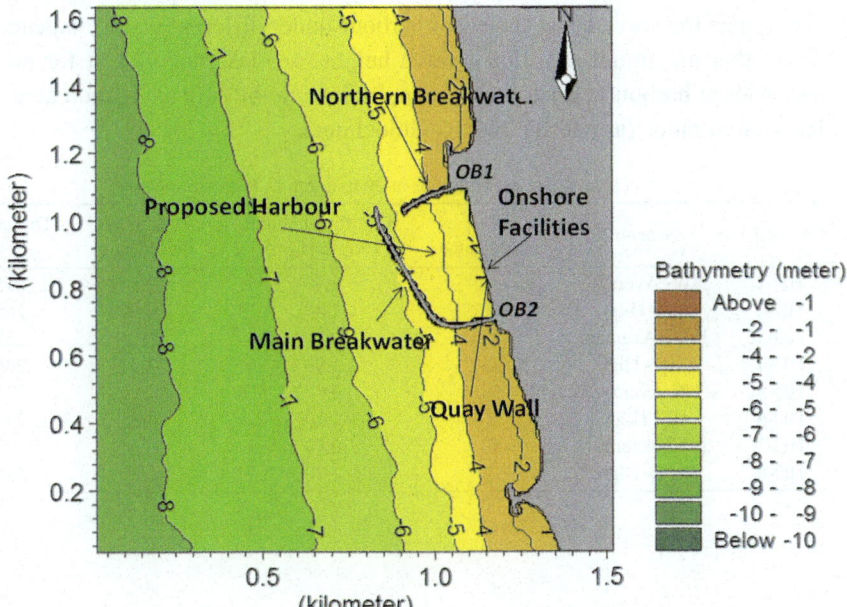

Fig. 2. Proposed fishery harbour and marine structures

3.2. *Previous studies*

Wave Transformation Matrix and MIKE 21 Spectral Waves (SW) models have been used by a recent study [3] to hindcast the wave climate in the study area. In this study, directional wave data, recorded at 15m ~ 16m depths in Colombo Port during 1998 – 2015 periods was transformed to the offshore boundary of the SW model and thereafter transformed back to the desired destination locations of North-western coast. Wave data was analyzed for Southwest Monsoon (SW), Inter-Monsoon 1(IM-1), Northeast Monsoon (NE) and Inter-Monsoon 2 (IM-2) of each year for the period of June 1998 – September 2015. Thus, main wave parameters (i.e. significant wave height, peak wave period, and directions) were identified. With further analysis of wave data, wave characteristics for each season under average (50% exceedance) and maximum (2 % exceedance) categories have been identified. Accordingly, representative nearshore wave climate was established at 20 m water depth in the study area. From that analysis, highest percentage of annual waves at 20 m depth in study area has been identified between $250^0N - 260^0N$ directional range. Seasonal wise, southwest monsoon season was recognized as dominant season. Wave parameters of 4 main seasons and Wave Rose Diagrams at 20 m depth for study area are illustrated in Table 1 and Figure 3 respectively. Further, MIKE 21 BW (Boussinesq Wave) model [4] has been used in a seperate study [5] to investigate the wave disturbances of harbour under different wave incidence and found that maximum significant wave heights are less than 0.4 m for most of scenarios at harbour entrance and basin, indicating the safe navigation across the harbour entrance (as per the PIANC Guidelines).

Table 1. Wave conditions at 20 m depth in Wennappuwa

Case 1	Season	% Exceedance	Significant Wave Height H_s(m)	Peak Wave Period T_p(s)	Direction (^0N)
HD-1	SW- Average	50	1.30	9.5	255
HD-2	SW- High	2	2.30	8.3	265
HD-3	IM-1-Average	50	0.70	11.3	235
HD-4	IM-1- High	2	1.90	7.4	245
HD-5	NE-Average	50	0.50	12.3	235
HD-6	NE- High	2	1.10	10.0	245
HD-7	IM-2-Average	50	0.70	11.9	225
HD-8	IM-2- High	2	1.30	6.2	255

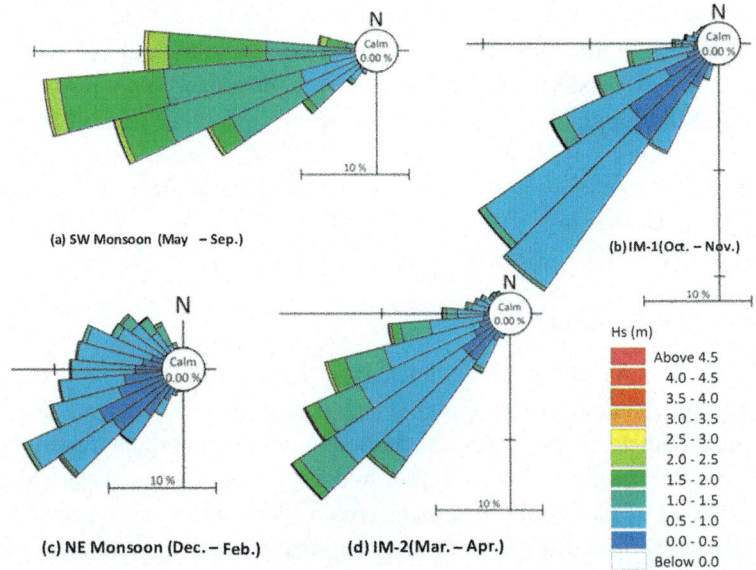

(a) SW Monsoon (May – Sep.)

(b) IM-1(Oct. – Nov.)

(c) NE Monsoon (Dec. – Feb.)

(d) IM-2(Mar. – Apr.)

Hs (m)

	Above 4.5
	4.0 - 4.5
	3.5 - 4.0
	3.0 - 3.5
	2.5 - 3.0
	2.0 - 2.5
	1.5 - 2.0
	1.0 - 1.5
	0.5 - 1.0
	0.0 - 0.5
	Below 0.0

Fig. 3. Wave rose diagram on seasonal basis for Wennappuwa – 20 m depth (Source: [3])

4. Methodology

4.1. *Modelling framework*

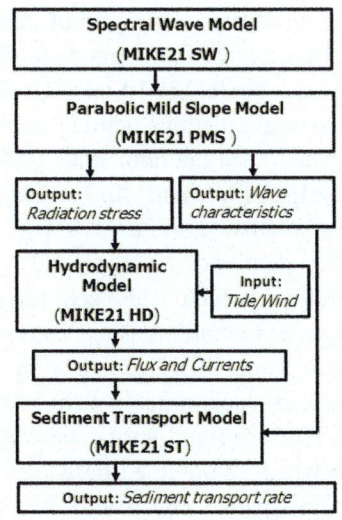

Fig. 4. Flow chart for modeling sequence

Mathematical model study was carried by using MIKE21 2D modelling system. As illustrated in Figure 4, predicted nearshore wave climate from MIKE 21 SW (Spectral Waves Model) [6] was used as the boundary of MIKE 21 PMS (Parabolic Mild Slope Model) [7] wave model. This approach assists to convert unstructured mesh results to rectangular grid results. To assess the hydrodynamic conditions in and around the harbour area, MIKE 21 HD (Hydrodynamics Model) [8] was used. To understand sediment transport patterns, MIKE 21 ST (Sediment Transport Model) [9] was used.

4.2. *Hydrodynamic study*

Current circulation and harbour sedimentation are some of the key topics relying on a clear understanding of hydrodynamics. Hence, proper understanding about the hydrodynamics in the proposed harbour area is important. In this study, MIKE 21 HD module assessed the hydrodynamic of the basin and the neighbourhood of the proposed harbour area. Computations were performed on a nested grid set-up starting from a larger regional model and gradually reducing to smaller models while moving towards the area of interest. The "Regional Model" was set-up with a grids resolution of 1000 m × 1000 m with the grid size of 89 km × 293 km. The depth at off shore of the Regional Model bathymetry is around 100 m. The boundary conditions for the larger "Regional Model" were based on the tidal constituents from Colombo and Kalpitiya. Two small and more refined models ("Intermediate" and "Local") were nested within this larger model (see Figure 5). The resolution of the grid of the "Intermediate Model" is 100 m × 100 m. The size of the grid of "Intermediate Model" is 50 km × 98 km. The grid resolution of "Local Model" is 10 m × 10 m and extent of the local model is 10 km × 10 km. The simulations within a sub grid model are based on boundary conditions extracted from the immediately higher model. Simulations were performed for the forcing of tide and wind conditions. Regional model and intermediate models were simulated for a period of 2 weeks covering both spring and neap periods for identified "average" and "high" wave conditions of each season. Wave conditions which had been used in "Local model" are presented in Table 1. These wave conditions are corresponded to the wave data extracted at 20 m depth from MIKE 21 SW model simulation results. Considering the time constrain, local models were simulated only for one tidal circle (i.e. 7 days), including neap and spring tide. HD model scenarios (i.e. Case No. HD-1 – HD-8) which are defined based on the wave conditions are presented in Table 1. In this paper, HD results for most dominant scenarios during the southwest monsoon (Case HD -1 & HD -2) are presented.

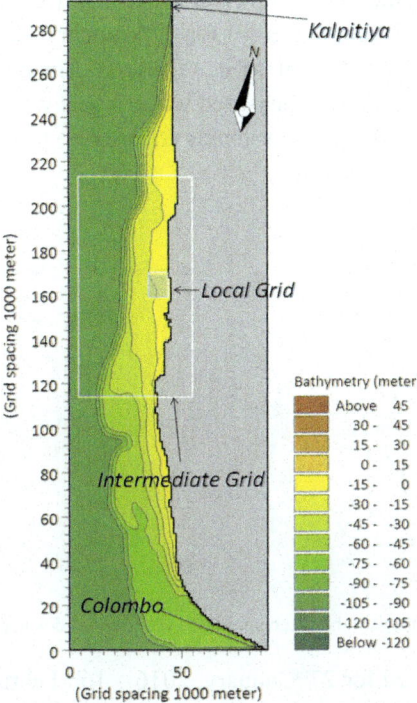

Fig. 5. Grids – Regional, Intermediate and Local models

The established tidal constituents for the northern and southern boundaries of the Regional Model are listed in Table 2.

Table 2. Derived tidal constituents for the northern and southern boundaries

Regional Model Boundaries	Derived Amplitudes of Tidal Constituents (m)				Derived Phase of Tidal Constituents (deg.)			
	M2	S2	K1	O1	M2	S2	K1	O1
North - Kalpitiya	0.18	0.12	0.09	0.03	53	118	65	55
South - Galle	0.16	0.11	0.05	0.01	56	99	21	73

Spatial wind boundary conditions under "average" and "high" wave conditions for each monsoon are presented in Table 3.

Table 3. Spatial wind boundary conditions for HD model

Season		SW	IM - 1	NE	IM - 2
Wind Speed (m/s)	Average	5	4	4	4
	High Wave	10	8	8	8
Wind Direction (⁰ N)		SW	W	NW	W

To verify HD model results, water levels and currents were measured at project site (i.e. 94,459.8 m E, 233,646.0 m N) for 27th January, 2016 – 16th February, 2016 period at a depth of 10.2 m MSL. These measurements were used for model verification. Accordingly, measured water levels were compared with the MIKE 21 HD model simulated water levels (*see* Figure 6). As seen in Figure 6, both values are in good agreement.

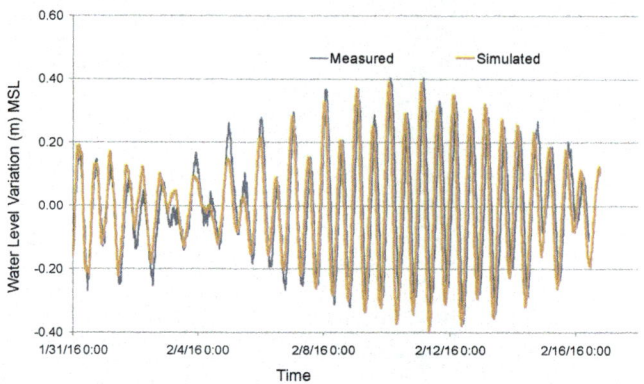

Fig. 6. Comparison between simulated and measured water levels

The average current speed for 27th January, 2016 – 16th February, 2016 period is 0.07 m/s. According to the simulated current speed for the same period, average speed was identified as 0.13 m/s. Since Inter Ocean S4DW Electromagnetic Current Meter (S4DW ECM) has measured current speed at 9.3 m MSL depth and numerical model results are depth average value, numerically simulated current speeds must be higher than the measured values. Thus, HD model can be validated.

4.3. *Sediment transport study*

MIKE 21 ST calculates the sediment transport rates on a rectangular grid covering the area of interest on the basis of the hydrodynamic data obtained from a simulation with MIKE 21 HD and the wave parameters calculated by MIKE 21 PMS, together with information about the characteristics of the bed material. The computation is based on wave and current theory with Bijker's method [10]. Sediment properties (D_{50}, $\sqrt{D_{84}/D_{16}}$, porosity(n), specific gravity), bed roughness values (M) and bed load transport coefficient (B) were used to calibrate the model. Information on the size of seabed sediment is obtained by the field surveys carried out by Lanka Hydraulic Institute at 21 locations on 27th January 2016 in the study area. Based on the analysis data, a mean sediment size (D_{50}) of 0.55 mm and geometric spreading factor ($\sqrt{D_{84}/D_{16}}$) of 1.8 are good representation and were used for model assuming uniform spatial distribution. Further, porosity and specific gravity of the sand is specified as a constant for the whole area and taken as 0.4 and 2.65 respectively. A value of B (bed load transport coefficient) was taken as values between 1 and 5. For the transport rates calculation outside the surf zone, it was taken as 1 and whereas a value of B was taken between 1-4 values within the surf zone. In the case of bottom friction constant, Manning number was taken as 32. Table 4 presents the selected combinations of hydrodynamic and wave conditions for sediment transport modeling.

Table 4. Boundary conditions at 20 m depth in Wennappuwa

Case No	Season& Condition	Wind Speed (m/s)	Tide	Wave Parameters		
				H_s(m)	T_p(s)	Dir. (^{0}N)
ST-1	SW - Average	5	Spring/neap	1.30	9.5	255
ST-2	IM-1- Average	4	Spring/neap	0.70	11.3	235
ST-3	NE-2- Average	4	Spring/neap	0.70	8.9	285
ST-4	IM-2- Average	4	Spring/neap	0.70	11.9	225

Accordingly, representative "average" wave conditions of each monsoon and related hydrodynamic conditions were used for the simulations and subsequent sediment transport rate calculations. Simulated results were validated qualitatively considering the order of sediment transport rates identified in previous studies.

5. Results and Discussion

5.1. *Hydrodynamic model results*

2D currents plot for most dominant scenario (i.e. Case No. HD-1(SW-Average)) is illustrated in Figure 7. To represent high tidal current impact, results at spring tidal conditions are presented. After comparing all 2D current plots (i.e. all cases in Table 1), it has identified that current movement direction is generally directed from south to north direction. For SW monsoon average condition, current movements at the harbour entrance do not strongly support the possible sand transport in to the harbour. In case of IM -1 and IM – 2 seasons, current effects are not very strong at the harbour entrance for average wave conditions. Similarly, current actions for average wave conditions of NE monsoon seasons are also not directed to the harbour through the entrance. Detailed analysis of simulated data was carried out for further understanding about the current speed and directions quantitatively.

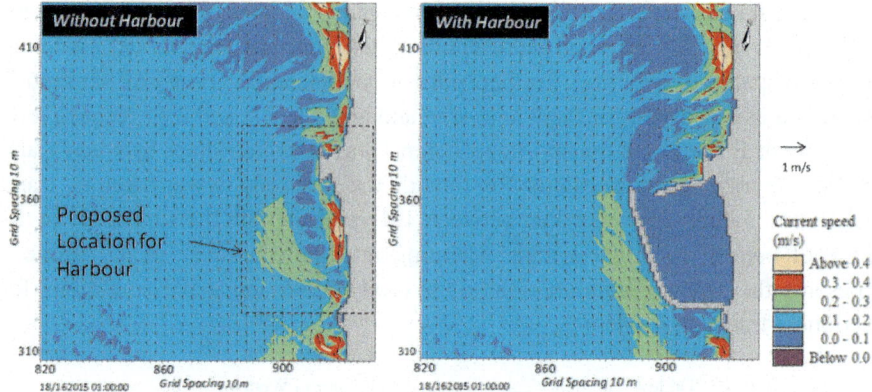

Fig. 7. 2D currents plot of Case No. HD-1(SW- Average) for spring tide condition

Accordingly, representative 5 locations were selected just inside and outside the harbour entrance. At each location, current speed was extracted for the simulation period of all scenarios (i.e. Case No. HD-1 – Case No. HD-8). Current speed at each location for (a) without harbour (existing condition) and (b) with harbour conditions were compared. Before construction of harbour, current speed varies between 0 m/s – 0.65 m/s at considered representative location in the harbour area. After the proposed harbour, this range has varied for 0 m/s – 0.46 m/s range. It was observed that there is a considerable reduction of current speed at harbour entrance points for Case HD-1 (SW – Average). Since SW monsoon prevails long period of time and its impact is dominant for the proposed harbour in terms of current and sediment movement, reduction of current speed in entrance area is essential for the easy navigation. It is noted that most of the current velocities, generated due to wave and tide remain less than 0.5 m/s in the selected location with the proposed harbour. In general, proposed port configuration is favorable for hydrodynamic performance.

5.2. *Sediment transport model results*

To identify possible sediment transport patterns for the selected dominant case (i.e. Case No. ST-1(SW- Average) in the vicinity of harbour, model output is presented in two-dimensional sediment transport rate plots (Figure 8). Similar to Figure 8, after comparing sediment transport pattern for "without" and "with" harbour conditions for all seasons, it was recognized that proposed harbour configuration does not support significantly to increase the sediment transport rate toward harbour through the harbour entrance. Further, it was identified that sediment transport rate during the SW monsoon is significant than other seasons.

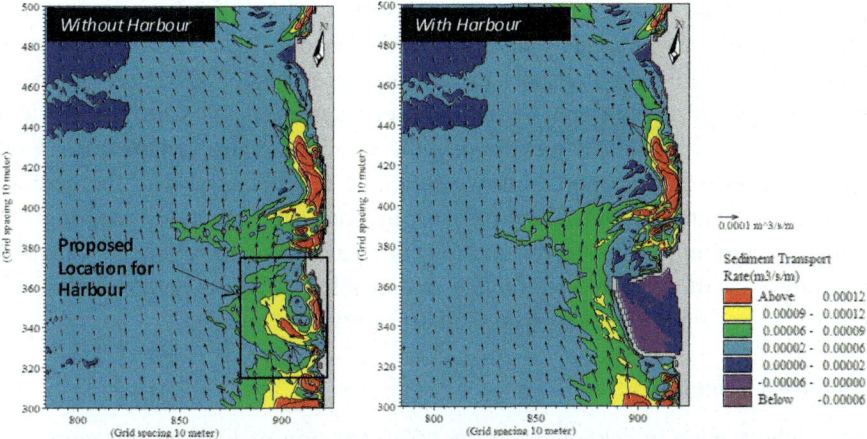

Fig. 8. 2D Sediment transport rate plots (left) without harbour and (right) with harbour:
(Case No. ST-1(SW- Average))

It is noted that average current velocity is approximately 0.05 m/s at the entrance point of the proposed harbour for southwest monsoon. The main direction of sediment transport is directed from south to north during the SW monsoon. However, sediment transport movement has not supported to move sediment significantly in to the harbour through its entrance. Hence, sediment bypass can be expected through the harbour entrance with the proposed configuration. Thus, as a result of reduced siltation at harbour entrance, safe navigation could be achieved.

6. Conclusion and Recommendations

In this study, MIKE 21 modelling system was used to simulate hydrodynamics and sediment processes. For this purpose, already established representative nearshore wave climate in the study area was used. According to the MIKE 21 HD (Hydrodynamics) model results, it was identified that current movement direction is generally directed from south to north direction. It is noted that average current velocity is approximately 0.05 m/s at the entrance point of the proposed harbour for southwest monsoon. In general, proposed harbour configuration is favoured for its hydrodynamic performance. MIKE 21 ST (Sediment Transport Model) was used to understand sediment transport patterns, specially at harbour entrance and approach channel to investigate the siltation possibilities. The net sediment transport direction is directed from south to north direction. Considering the order of sediment transport rates identified in previous studies, simulated results was validated qualitatively. It was recognized that sediment transport rate during the SW monsoon is significant than other seasons. The sediment transport pattern at the entrance was observed to be smoothly passed the entrance. Hence, the orientation of the structure provides smooth by pass of the sediment across the entrance. Thus, sediment transport movement does not support to move sediment significantly in to the harbour through its entrance with the proposed harbour configuration, ensuring the required minimum depth (i.e. for draft) for safe navigation due to naturally controlled siltation. However, for the sustainable use of navigational channel, moderate dredging operations would be necessary.

Acknowledgments

The authors wish to acknowledge the *"Construction of Fishery Harbours and Anchorages Project"* of MFARD (Ministry of Fisheries and Aquatic Resources Development) in Sri Lanka for various supports for the success of this research.

References

1. D Premasiri, Policy Review and Suggestions for Improvements to Coastal Structures in Sri Lanka: a Case Study in Kirinda and Hikkaduwa Areas, Master's thesis, Asian Institute of Technology (2004).
2. D P C Laknath and J Sasaki, Assessment of the tsunami rehabilitated fishery harbours in Sri Lanka. *Journal of Coastal Research*, SI 64 (*Proceedings of the 11th International Coastal Symposium*), 1245 – 1249, (Szczecin, Poland, 2011).
3. D P C Laknath, H P G M Caldera, and D P L Ranasinghe, "Wave Hindcasting and Extreme Value Analysis for North-western Coast of Sri Lanka", *Proceedings of the 27th International Offshore and Polar Engineering Conference*, (ISOPE-2017), (San Francisco, USA, 2017).
4. Danish Hydraulic Institute, MIKE 21 BW User Guide, (Denmark, 2009).
5. D P C Laknath, D E N Senarathne, and K K P P Ranaweera. "Wave Tranquillity Study for a Fishery Harbour using Boussinesq Type Wave Model" *Proceedings of the 111th Annual Sessions, The Institution of Engineers*, (Colombo, Sri Lanka, 2017).
6. Danish Hydraulic Institute, MIKE 21 SW User Guide, (Denmark, 2009).
7. Danish Hydraulic Institute, MIKE 21 PMS User Guide, (Denmark, 2009).
8. Danish Hydraulic Institute, MIKE 21 HD User Guide, (Denmark, 2009).
9. Danish Hydraulic Institute, MIKE 21 ST User Guide, (Denmark, 2009).
10. E W Bijker, Littoral Drift as a Function of Waves and Current. Publication No. 58, Delft Hydraulics Laboratory, Delft, (The Netherlands, 1969).

A Study on Coastal Erosion and Deposition Processes in Subang, Indonesia

S. Kikuyama

Graduate school of Urban Innovation, Yokohama National University
79-5 Tokiwadai, Hodogaya, Yokohama 240-8501, Japan
E-mail: kikuyama-seiko-mw@ynu.jp

T. Suzuki

Faculty of Urban Innovation, Civil Eng. Dept., Yokohama National University
79-5 Tokiwadai, Hodogaya, Yokohama 240-8501, Japan
E-mail: suzuki-t@ynu.ac.jp

J. Sasaki

Graduate School of Frontier Sciences, the University of Tokyo
5-1-5 Kashiwanoha, Kashiwa, Chiba 277-8563, Japan
E-mail: jsasaki@k.u-tokyo.ac.jp

H. Achiari

Ocean Engineering Study Program, Institut Teknologi Bandung
Jl. Ganesha No.10, Lb. Siliwangi, Coblong, Kota Bandung, Jawa Barat 40132, Indonesia
E-mail: achiarihendra@gmail.com

S.A. Soendjoyo

School of Civil Engineering, Institut Teknologi Bandung
E-mail: omstephanus@gmail.com

H. Higa

Faculty of Urban Innovation, Civil Eng. Dept., Yokohama National University
E-mail: higa-h@ynu.ac.jp

A. Wiyono

Ocean Engineering Study Program, Institut Teknologi Bandung
E-mail: agungwhs59@gmail.com

A coastal land area of approximately five km2 around the Pondok Bali beach in Subang, Indonesia, has been lost to the sea for the past decade. Before this land loss, there were several fish ponds with mangrove trees. The local people have suffered from the relative sea level rise, as they have had to abandon their houses because of the frequent inundations. To understand the status and causes of this coastal problem, the perception and adaptation strategies of the people, and potential countermeasures, we performed interview surveys in the associated villages and local governments. By using a sea chart of 1965 and satellite images, we calculated the cumulative volume change of deposition and erosion around the Cipunagara River mouth areas, where Pondok Bali is situated. Further, we conducted field measurements of turbidity and velocity, as well as bathymetry in the Cipunagara River to estimate the sediment supply. From the interviews, we found that a decrease in sediment discharge is a major cause of the erosion around Pondok Bali, which is due to the change of the river channel in 1965, when a floodway was constructed. At the present river mouth, a huge river delta has been formed owing to the sediment supply from the river. We also found that the local villages and governments have been considering countermeasures for reducing land erosion and promoting sustainable land use in the Pondok Bali areas, such as a plan for creating a new branch off from the main channel of the river to the Pondok Bali areas. However, further discussion on the new branch route is required among the associated villages because the effect of the new branch on Pondok Bali is unknown. From the field survey in the river, we calculated the sediment flux and estimated the total sediment supply to the river mouth, and this information is used for assessing the feasibility of the plan.

Keywords: Sediment transport; River delta; Coastal erosion; Change of river channel; Mangrove.

1. Introduction

Mangrove forests are distributed throughout the coastal areas of semi-tropical and tropical climate zones. However, during the past several decades, the mangrove forest area has been reduced, mainly owing to human activity[1,2]. Mangrove forests have been studied from not only the forestry aspect but also from an engineering viewpoint. Blasco[3] proposed that mangroves could be used as indicators of coastal change or sea level rise. Alongi[4] reported their resilience, protection from tsunamis, and response to global climate change. Because the mangroves were also distributed in mud coasts, Winterwerp[5] proposed a sustainable use of mangrove mud coastal systems.

Indonesia contains the largest area of mangrove forests in the world[6]. However, the decrease of mangrove forests started before the 1980s[7]. The research on mangrove forests was reviewed. Armitage[8] showed the method of mangrove forest conservation and Brown[9] applied the resilience concepts to the mangroves of Indonesia. Van Oudenhoven[10] analyzed the effects of different management regimes on the ecosystem services in the mangroves of Java. Thus, although a lot of countermeasures and suggestions have been published, we need to adjust them to the various target areas.

The Subang Regency of Indonesia is located around 100 km to the east of central Jakarta. At the Pondok Bali beach of Legon Kulon in the Subang Regency, severe coastal erosion has occurred[11]. It is reported that 56.8 m/year of shoreline retreat has affected the surrounding land use or development since 2002. It was also suggested that land transformation from mangrove forests to fish and shrimp ponds might affect the shoreline retreat. However, the major cause of this erosion is unknown. The Cipunagara River had been flowing about 2.5 km east from the Pondok Bali area until 1965. In 1965, to reduce the damage from floods, the direction of the river channel was changed to the present channel, i.e., farther to the east of the Pondok Bali area. Some local events in the area have been considered as causes of coastal erosion, such as mangrove cutting, subsidence owing to groundwater use, and reduction of the sediment supply from the river. The construction of a new flood channel toward the Pondok Bali area is planned in the coming years. However, the effects of the new flood channel on the reduction of beach erosion and/or the accumulation effect on the area are unknown.

Thus, the objective of this study is to understand the sediment transport of the Cipunagara River and confirm the feasibility of the new channel plan.

2. Outline of the Field Site

In Mayangan of Legon Kulon, in the Subang Regency, the Cipunagara River had been flowing toward the Pondok Bali area. In 1965, to reduce floods, the river channel was changed from the Pondok Bali area to the present channel, as mentioned in the previous chapter. Since 2007, more than 500 ha of land around the Pondok Bali beach has been lost to the sea owing to coastal erosion. Although the government and villages in Legon Kulon succeeded in suppressing the erosion by installing some simple revetments and geotubes, the future status of erosion is not clear. As of 2016, 1,700 ha of river deltas have formed in the current river mouth. Local people have constructed dikes to stabilize the mangrove forest protecting the land, and developed rice-fields and fish-ponds supporting their daily life. For the purpose of reclamation in Mayangan, the government and Legon Kulon are discussing the construction of a new discharge channel towards Pondok Bali. In addition, the construction of a new commercial port adjacent to the east side of the Cipunagara River mouth, the Patimban Port, is planned by the Indonesian government, in collaboration with the Japanese government. The large amount of sediment supply from the river should be a major concern for the design of the new port.

506

3. Methodology

In this study, interview surveys were performed at governmental offices and related villages to collect information about the history of coastal erosion, land use, and opinions about the new discharge channel. In addition, the annual sediment flux of the river was estimated based on the time variation of the land shape in the deposited area by overlaying satellite images on a sea chart from 1965. Furthermore, we compared this estimation with the annual sediment flux of the river derived from its turbidity, the cross-sectional area of the river channel, and the current velocity measured at Stations 1 and 2 in Figure 1. Thus, based on the above comparison, the correlation between the sediment supply and amount of deposited sediment was clarified.

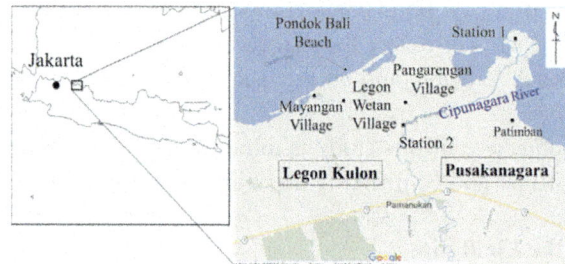

Fig. 1. Legon Kulon and Pusakanagara in Subang Regency, Indonesia (Google map).

3.1. *Interview survey*

To understand the perceptions and opinions of all the villages regarding the plan of the new discharge channel, status of coastal erosion, history of the land, and land use, among other issues, interview surveys were performed in the villages, governmental offices, and forestry offices. We carried out interviews with officers of the three villages of Mayangan, Legon Wetan, and Pangarengan, which are related to the coastal erosion and new discharge channel plan, and officers of the sub-district consisting of these three villages. Moreover, we interviewed officers at the Bandung Ministry of Public Works (BBWS), who have historical data around Pondok Bali, at the Public Works Subang District, who have land information, and at the Forestry Office, who manage the forests and mangroves. All the interviews were conducted for 30 – 60 min, face-to-face.

3.2. *Estimation of eroded/deposited area*

The route of the Cipunagara River was changed in 1965, and the sediment deposition started to occur at the new river mouth (Figure 2(a)). At the Pondok

Bali beach area, severe land erosion has occurred since around 2007 (Figure 2(a)). In this study, both areas are chosen as the target areas, and the variations of land area in both cases were calculated to determine the annual change in area for both deposition and erosion. To calculate the areas of both eroded and accumulated regimes, satellite images from Google Earth were used. To calculate the designated area, polygons including the coastline were created in Google Earth. In the polygon function, the boundary between land and seawater needs to be ascertained from the Google Earth images. At the deposition area, the boundary was ascertained where the place could be recognized as a green color (mangroves), whereas the erosion area was judged where the land was covered with sea water. The amount of variation in the area was determined as the difference between the image of the target year and that of the previous year.

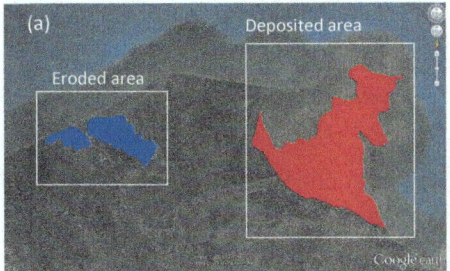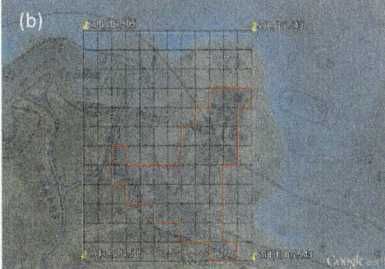

Fig. 2. (a) Deposited and eroded areas in Subang Regency, (b) Mesh system for volume calculations of deposited area (Google Earth, Image Landsat / Copernicus).

3.3. Estimation of accumulated sediment volume in deposited area

For the calculation of the deposited area, the four corners of latitude and longitude were determined at the target area (Figure 2(b)). The selected rectangular area was divided into 12×12 meshes (1 mesh = 660 m×610 m). Using a sea chart, the bathymetry data were extracted and linearly interpolated on each of the meshes. The coastline in 1965 was set as a baseline and each mesh in each year was classified into land or sea. If 50 % of a mesh is covered with green color (mangroves), it is considered as a land mesh. The total deposited area was calculated by counting the number of land meshes that had changed from sea meshes. The volume of deposited area was calculated as the summation of the product of the area and depth for all the meshes converted to land. From the field survey, the elevation of the accumulated land from the surface water level was approximately 0.5 m, and thus, the volume for this land area with 0.5 m thickness was added to the previous total volume.

3.4. *Turbidity and current velocity survey at the Cipunagara River*

Turbidity, current velocity, and water depth were measured at Stations 1 and 2 (Figure 1) in the Cipunagara River on October 3 and 4, 2016, using two turbidity meters (INFINITY Turb ATU75W2-USB, JFE Advantech; TB-31, TOA-DKK), a current meter (VR-301, KENEK), and a water depth meter (PS-7, HONDEX). The locations of Stations 1 and 2 are approximately 150 m and 11 km upstream from the river mouth, respectively. The measurements were performed at the depths of 0.0 m, 0.25 m, 0.5 m, 0.75 m, and 1.0 m at Station 1, and 0 m, 0.5 m, 1.0 m, 1.5 m, and 2.0 m at Station 2. In addition, a cross sectional bathymetry of the river was measured at Station 2.

4. Results and Discussion

4.1. *Interview survey*

4.1.1. *Major river basin board office of ministry of public works in Bandung (BBWS Citarum)*

The interview was conducted at the Ministry of Public Works in Bandung on April 27, 2016, to understand and consider the countermeasures for the beach erosion at the Pondok Bali area. The Ministry of Public Works has installed jetties, breakwaters, dikes, and sand-tubes to mitigate the beach erosion. The ground level of the land and roads was increased as a countermeasure against floods caused by the relative sea level rise

4.1.2. *Public works, Subang regency*

At the Ministry of Public Works of the Subang Regency, which mainly focuses on the irrigation sector, the management of paddy fields was interviewed on April 28, 2016. The paddy fields of the Subang Regency are divided into four classes, depending on their size. The Subang Public Works manages paddy fields with areas of less than 1,000 ha. The paddy field areas larger than 1,000 ha and less than 3,000 ha are managed by the Province office in Bandung, and the fields which areas larger than 3,000 ha are managed by the Ministry of Public Works in Bandung.

4.1.3. *Forestry office in Bandung*

The interview was conducted at the Forestry Office in the Subang Regency on April 28, 2016. They are focused on the conservation and management of mangroves. The mangrove forest areas in Subang are managed by the Forestry Office and Perhutani, a public corporation. Some of the local people said they were encouraged by Perhutani to construct dikes to develop fish and/or shrimp ponds in the newly extended Cipunagara River mouth area. Shrimp ponds, however, require complex technology for their management. The local people have been supported by large companies and by the Ministry of Fishery. Thus, there are conflicts between two agencies, forestry and fishery, regarding the policies preserving mangrove forest or developing farming ponds.

4.1.4. *Mayangan village*

At the Pondok Bali beach in the Mayangan village, we conducted an interview with the village leader on April 28, 2016. The Pondok Bali beach is located at the ocean side of the Mayangan and Legon Wetan villages. Most of the people in Mayangan are engaged in local or offshore fishery. Owing to the coastal erosion and salt-water intrusion, some people have changed their jobs from farming to fish culturing, which has led to a change from paddy fields to fish ponds. Once the field changes from paddies to fish ponds, salt pollution around the ponds becomes more serious, which results in a conflict between farmers and fishers. In the interview in 2014, local people mentioned that 30 cm of increment in the sea level suddenly occurred after the 2004 Indian Ocean earthquake. The village leader, however, denied the rise in sea level due to the earthquake. The sea level rise is considered to occur gradually, possibly caused by ground subsidence. As a remedy, the central government has planned to create a discharge channel from the Cipunagara River to the Pondok Bali area, which may contribute to increase the sediment discharge to that area, and restore the lost land. The leader agrees with this construction plan.

4.1.5. *Legon Wetan village*

The Legon Wetan village is located in the east side of Mayangan, and it suffers from the most serious coastal erosion. The interview was conducted on October 4, 2016. The leader of the fishing union considers the principal cause of erosion to be the 2004 Indian Ocean tsunami and the increase in the tidal range. According to the leader, the tidal level has gradually become higher after the tsunami, and recorded its highest point in 2007. After 2012, however, the rate of sea level rise has lowered. In front of the Pondok Bali beach, a 900-m long breakwater was installed, and a 700-m-long breakwater was added at the east side. Further, a 600 m long geotube was constructed by the central government in 2015. However, the breakwaters and geotube seem to be ineffective in suppressing the beach erosion. Even under the threat of a sea level rise and an increasing risk of inundation, many of the locals continue to live in their original houses adjacent to the channel, because the cost of relocation is not covered by the government and their way of life as fishers has not changed. It even seems that family incomes have increased because people do not need to go offshore to catch fish anymore, and thus, they can reduce the fuel fee. The village leader and many of the residents of Legon Wetan agree with the plan to construct the new discharge channel from the Cipunagara River to the Pondok Bali side.

4.1.6. *Pangarengan Village*

The Pangarengan village is located in the east side of Legon Wetan. We conducted an interview with the head of the village on October 4, 2016. The village is not affected by the coastal erosion. Most of the people in the village are farmers, except for some fishers around the coastal area. According to the interview, the Pangarengan village has never been involved in the planning of the new discharge channel. The village disagrees with the plan because the effectiveness of the new discharge channel in the suppression of coastal erosion is not very clear and the planned route of the channel is going to divide the village into two separated areas. The village prefers to help the other villages suffering from erosion with the construction of breakwaters or geotubes.

4.1.7. *Legon Kulon Sub-district*

The interview survey was conducted at the sub-district covering Mayangan, Legon Wetan, and Pangarengan. In these villages, the population growth rates are now around 5 % per year. In this sub-district, a small river existed and floods had occurred owing to heavy rains. In order to protect land and residents from floods, a discharge channel was constructed towards the east in 1965, which is the current channel of the Cipunagara River. Regarding a new discharge channel towards Pondok Bali, the Pangarengan village expresses its disagreement, despite the severe coastal erosion in the Pondok Bali area. The reason is that the planned discharge channel is going to divide their villages, and they do not understand whether the planned discharge channel could effectively suppress the coastal erosion.

4.2. *Time series of eroded/deposited area*

The temporal variations of the deposited and eroded areas (Figure 2(a)) were estimated using Google Earth satellite images from 1985 to 2016. Figure 3(a) shows the temporal variations of the cumulatively deposited and eroded areas in Legon Kulon and Pusakanagara, respectively. Although the estimation of the deposited area should start from 1965, when the flood channel (the current Cipunagara River channel) was constructed, and only the results after 1985 are shown in Figure 3(a) because of the unavailability of image data. The average rates of deposition and erosion areas are approximately 0.19 km^2/year and 0.49

km²/year, respectively. Figure 3(b) shows the change in volume of deposited area from 1985 to 2015, which reveals that the rate of increase in the volume is approximately $2.26×10^6$ m³/year.

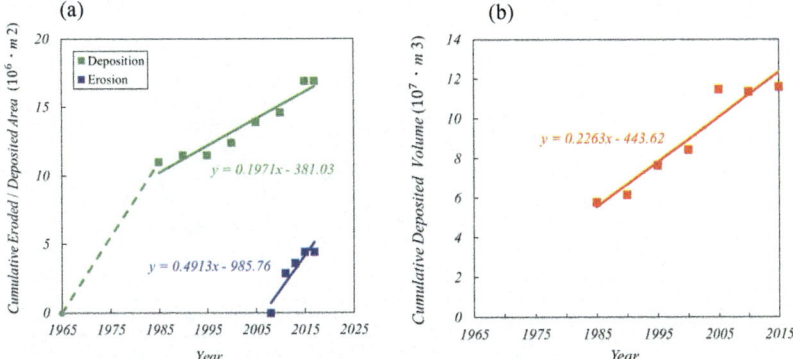

Fig. 3. (a) Change in cumulatively deposited and eroded areas in Legon Kulon and Pusakanagara, (b) Cumulative volume at the Cipunagara River mouth delta.

Table 1. Turbidity and current velocity with water depth at Stations 1 and 2.

Station 1 (h = 1.47 m)			Station 2 (h = 7.86 m)	
Water depth [m]	Turbidity [mg/L]	Current velocity [cm/s]	Water depth [m]	Current velocity [cm/s]
Surface	385.4	79.4	Surface	54.0
0.25	379.4	83.4	0.5	55.8
0.5	382.5	93.8	1.0	56.3
0.75	378.1	88.4	1.5	48.5
1.0	379.3	76.0	2.0	43.1
Average	380.4	84.2	Average	51.5

4.3. Turbidity, current velocity, and cross-sectional profile of the Cipunagara River

The measured turbidity and current velocity were measured at the side of the each section and the values at Stations 1 and 2 are shown in Table 1. The water depths at Stations 1 and 2 were 1.47 m and 7.86 m, respectively. Because the sensors were fixed on the tip of a level staff of 2 m length, the maximum possible measuring depth was limited to 2 m. The vertical profiles were almost uniform, probably because the water mixing was very strong due to high current velocity. The mean vertical current velocity and turbidity at Station 1 were 0.84 m/s and 380.9 mg/L, respectively, whereas the mean current velocity at Station 2 was 0.52 m/s. The turbidity at Station 2 could not be measured because it exceeded the maximum range of the sensor. The cross sectional bathymetry of Station 2 was measured at 7 points from the pier shown in Figure 4. . Because of the limitation

512

of the measuring method, including the cable length of the echo sounder, the measurement was only performed in the vicinity of the river bank. Thus, the remaining part of the bathymetry was estimated using a Google earth image, assuming symmetry of the channel section and a uniform bottom around the center, as shown in Figure 4.

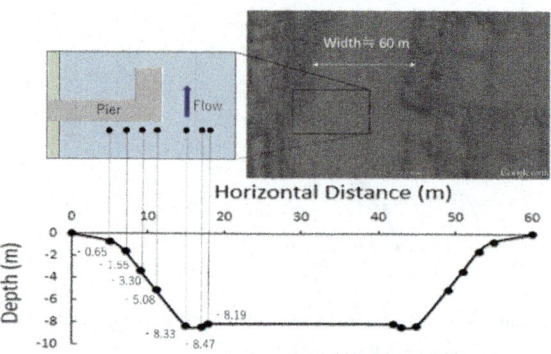

Fig. 4. Cross sectional bathymetry of Cipunagara River at Station 2 (Google Earth, Image Landsat / Copernicus).

4.4. *Estimation of sediment flux from Cipunagara River*

The cross-sectional area of the river was calculated as 334.7 m² by using the measured bathymetry and the assumptions explained in Subsection 4.3. Using the measured velocity and the cross-sectional area at Station 1, the discharge of the Cipunagara River was estimated as 227.3 m³/s. Combining this discharge and the measured mean turbidity, the annual sediment flux from the river was calculated as 2.73×10^6 t/year. This result was then compared with the accumulated sediment volume at the river mouth. From the satellite image analysis described in Subsection 4.2, the rate of increase in the volume of the deposited area was 2.26×10^6 m³/year (see Figure 3(b)). Supposing that the dry density of soil is 1.4 g/cm³ [12], the rate of change in volume can be derived as 3.16×10^6 t/year, which is consistent with the estimated annual sediment flux of 2.73×10^6 t/year. Although the field measurements were conducted at a specified location, the limited period of a year, and used the simplified estimation method, this approach of flux estimation is considered useful for assessing the feasibility of the new discharge channel plan.

In the deposited area of the river mouth, because the water depth increase with increasing distance from the mouth, it is reasonable to affirm that the deposition rate decreased after 2005, as shown in Figure 3(b). Thus, it can be considered that the rate of increase in area of land formation will decrease.

Moreover, the front of the accumulation area will be more exposed to wave and current forces, which may result in larger sediment transport to the offshore and adjacent coastal areas. This possible increase in sediment flux should be taken into account when the newly planned Patimban Port is designed.

5. Conclusions

An area of coastal land around the Pondok Bali beach in Subang, Indonesia, has been lost to the sea for the past decade. To understand the status and causes of this coastal erosion problem, perception of the people, and adaptation strategies, interview surveys were conducted at the associated villages and local governments. Field measurements and satellite image analyses were performed to estimate the eroded and deposited areas in Pondok Bali, and the surrounding areas of the Cipunagara River mouth, respectively, and the sediment flux of the river.

From the interview surveys, a decrease in sediment discharge is considered as a major cause of erosion around Pondok Bali, which is due to the change of the Cipunagara River channel in 1965, when a floodway was constructed to suppress inundations in the area during floods. As a result, a huge river delta has been formed because of the large amount of sediment supply from the river.

The local villages and governments have been considering countermeasures for suppressing erosion and promoting sustainable land use in the Pondok Bali area, which have led to the plan of creating a new branch channel to Pondok Bali from the main channel of the river. However, more discussion on the route of the new branch channel is required among stakeholders, not only because the currently planned route separates one of the villages (and thus the village leader is against the plan), but also because the effectiveness of the plan is still unclear. The sediment flux of the river based on satellite image analyses and field measurements was estimated at 3.16×10^6 t/year and 2.73×10^6 t/year, respectively, which are consistent with each other. This information will be useful for assessing the feasibility of the plan of constructing a branch channel of the river for supplying sediment to the Pondok Bali area to suppress erosion. Furthermore, it may be useful for considering the effect of sediment transport on a newly planned commercial port in Patimban, in an area adjacent to the South-East side of the Cipunagara River.

Acknowledgments

This study was partially funded by JSPS KAKENHI (Grant No. 25303016) and the Strategic Research Foundation Grant-aided Project for Private Universities from MEXT (Tomoya Shibayama, Waseda University, No. S1311028).

References

1. E. T. Choong, R. S. Wirakusumah and S. S. Achmadi: Mangrove Forest Resources in Indonesia, Forest Ecology and Management, 33/34, 45-57, (1990).
2. V. P. Upadhyay, R. Ranjan and J. S. Singh: Human-mangrove conflicts: the way out, Current Science, 83, 11, 1328-1336, (2002).
3. F. Blasco, P. Saenger and E. Janodet: Mangroves as Indicators of Coastal Change, Catena, 27, 167-178, (1996).
4. D.M. Alongi: Mangrove Forests: Resilience, Protection from Tsunamis, and Responses to Global Climate Change, Estuarine, Coastal and Shelf Science, 76, 1-13, (2008).
5. J. C. Winterwerp, P. L. A. Erftemeijer, N. Suryadiputra, P. van Eijk and L. Zhang: Defining Eco-Morphodynamic Requirements for Rehabilitating Eroding Mangrove-Mud Coasts, Wetlands, 33, 515-526, DOI 10.1007/s13157-013-0409-x, (2013).
6. Food and Agricultural Organization: The world's mangroves 1980-2005, FAO Forestry Paper 153, (2007).
7. P. R. Burbridge: Management of Mangrove Exploitation in Indonesia, Applied Geography, 2, 39-54, (1982).
8. D. Armitage: Socio-Institutional Dynamics and the Political Ecology of Mangrove Forest Conservation in Central Sulawesi, Indonesia, Global Environmental Change, 12, 203-217, (2002).
9. B. Brown: Resilience Thinking Applied to the Mangroves of Indonesia, IUCN & Mangrove Action Project, Yogyakarta, Indonesia, (2007).
10. A. P. E. Van Oudenhoven, A. J. Siahainenia, I. Sualia, F. H. Tonneijck, S. Van der Ploeg, R. S. de Groot, R. Alkemade and R. Leemans: Effects of Different Management Regimes on Mangrove Ecosystem Services in Java, Indonesia, Ocean & Coastal Management, 116, 353-367, (2015).
11. H. Achiari, N. Wulandari, Y. M. Yustiani and D. Harlan: Study Erosion and Coastal Destruction at Pondok-Bali, North Coast-West Java of Indonesia, Proceedings of 34th the IIER International Conference, Singapore, ISBN: 978-93-85465-79-6, (2015).
12. C. P. K. Gallage and T. Uchida: Effects of Dry Density and Grain Size Distribution on Soil-Water Characteristics Curve of Sandy Soils, Soils and Foundations, 50(1), 162-172, (2010).

Investigation of Beach Erosion Near New Washington and Deformation of Sand Spit on Panay Island in the Philippines

Takaaki Uda[a], Shingo Ichikawa[b], Susumu Onaka[b], Shubun Endo[c] and Masatoshi Izumi[c]

[a]*Public Works Research Center*
1-6-4 Taito, Taito, Tokyo 110-0016, Japan
E-mail: uda@pwrc.or.jp

[b]*Nippon Koei, Co., Ltd.*
1-14-6 Kudan-Kita, Chiyoda, Tokyo 102-8539, Japan
E-mail: ichikawa-sn@n-koei.jp

[c]*Futaba Inc., Co., Ltd.*
2-76 Minami, Koriyama, Fukushima 963-0115, Japan

As a study on beach erosion in a developing country, beach erosion in Kalibo City, located in the northern part of Panay Island in the Philippines, was investigated together with the deformation of a sand spit using satellite images. The study area is the shoreline of 7.7 km length between New Washington, south of Kalibo City, and the sand spit located southwest of the city. Field observation was carried out on January 16, 2014, two months after Typhoon Haiyan hit the area. In this area, the shoreline has severely receded owing to the decrease in sand supply from the Aklan River. Local optimization of shoreline protection using seawalls has been employed, resulting in no fundamental solution to the erosion.

Keywords: Philippines; beach erosion; coastal damage; sand spit; Panay Island; local optimization.

1. Introduction

Japan experienced rapid economic growth in the 1970s and 1980s, and many facilities such as the breakwaters of fishing ports and commercial ports were constructed nationwide in the coastal zone. Also, many large dams were constructed in the upstream basins of rivers, and rivers were subjected to riverbed mining to obtain construction materials. After these developments, severe beach erosion occurred, mainly due to the blockage of continuous longshore sand transport and the effect of the formation of wave-shelter zones behind the structures, resulting in the disappearance of natural sandy beaches (Uda, 2017)[1]. In developing countries, similar development is now being carried out without sufficient environmental protection. The importance of shore protection in developing countries is therefore increasing. Countermeasures against beach erosion should be taken by considering the comprehensive management of sand in the entire area. However, local optimization methods have been taken against beach erosion or wave overtopping over seawalls. Thus, a vicious cycle often occurs, with the solution of one problem causing another problem. The concept of integrated coastal zone management (ICZM) was proposed in Rio Earth Summit held in 1992. According to this concept, comprehensive coastal zone management is required to sustainably maintain coasts. Here, we considered a case in the Philippines as an example of a coast in a developing country. The present method of shore protection in the Philippines is not in accordance with the concept of ICZM, obstructing sound social development. In this study, we selected a coast in the Philippines as an example and carried out a field observation on January 16, 2014 in the suburb of Kalibo City, located in the northern part of Panay Island, and investigated the beach erosion in this area in terms of long-term stabilization along with the deformation of a sand spit in southeastern part of the study area.

2. General geomorphology of study area

Figure 1 shows the location of Panay Island in the middle of the Philippines. The study area is near Kalibo City, located in the northern part of Panay Island, as shown in an enlarged satellite image (Fig. 2) of the rectangular area in Fig. 1. The coast in Kalibo faces the Sibuyan Sea, and runs in the WNW-ESE direction. Figure 3 shows an enlarged satellite image of the rectangular area in Fig. 2. The Aklan River flows into the sea at Kalibo, and the shoreline around the Aklan River delta significantly protrudes offshore. The coastline of this river delta smoothly extends in the northwestern part, whereas the curvature of the coastline is large in the eastern part, forming an asymmetric river delta shoreline, and the tributary of the Aklan River significantly meanders southward. The formation of the asymmetric delta indicates that the predominant waves are incident from the north,

Fig. 1. Location of Panay Island in middle of Philippine.

resulting in the predominance of southeastward longshore sand transport near the Aklan River mouth. Because of the protrusion of the river delta, the incident wave angle is large on the southeast side of the delta and a hooked shoreline has formed. New Washington is located 10.5 km southeast of Kalibo, as shown in Fig. 3. A barrier island with a wide lagoon has developed from the vicinity of New Washington, and a sand spit extends at the tip of this barrier island. Also, another smaller sand spit has formed on the opposite shore of the tidal inlet. This again indicates that southeastward longshore sand transport prevails around New Washington.

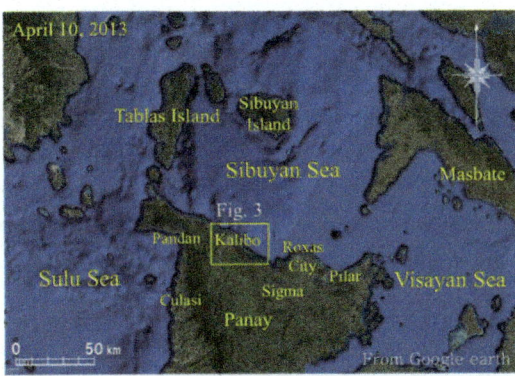

Fig. 2. Location of Kalibo in northern part of Panay Island.

Fig. 3. Satellite image of coastline in Kalibo and New Washington.

The coast in the vicinity of New Washington has been formed by sand supplied from the Aklan River and transported by southeastward longshore sand transport. However, riverbed mining has been extensively carried out in the Aklan River to obtain construction materials, resulting in a decrease in sediment yield from the river. On the other hand, a coastal road extends along the shoreline from New Washington up to the southeast end of the barrier island, and many houses have been built on both sides of this road. To protect the village and the road from storm surges and erosion, seawalls have been constructed locally. As a result of the significant reduction in the sand supply from the Aklan River, these seawalls have been destroyed at many locations due to erosion, and since the destroyed structures have been left in the sea while protruding offshore, they have functioned as groins, and the downcoast of the seawalls has been severely eroded owing to the blockage of continuous longshore sand transport.

Figure 4 shows an enlarged satellite image of the rectangular area in Fig. 3 together with the station numbers of the site observation on January 16, 2014 in the area between New Washington and the tip of the barrier island. In a suburb of New Washington, beach erosion was investigated at Sts. 1-10, and all site photographs were taken facing north, which is opposite to the direction of predominant longshore sand transport. Then, site observation was carried out at St. 11, which is located midway between New Washington and the sand spit, and finally, beach erosion around the tip of the sand spit was observed at Sts. 12-18.

3. Site observation

3.1. *Coast in New Washington City*

In New Washington, many houses have been built on both sides of the coastal road extending along the shoreline. To protect these houses and the road from waves, concrete seawalls have been built (Photo 1). There is no foreshore in front of the seawall, and almost all the concrete blocks placed along the seawall as foot protection have subsided, resulting in a decrease in their effectiveness in preventing wave overtopping. At St. 2, wave overtopping over the top of the seawall has been so severe that a wide pond has formed behind the seawall (Photo 2). Taking into account the fact that wave overtopping was severe even

Fig. 4. Satellite image of barrier island extending southeastward from New Washington.

though normal waves were incident during the field observation, this road should be closed to traffic when extraordinary high waves are incident during a typhoon.

Because a vertical seawall has been constructed upcoast, as shown in Photo 2, resulting in a discontinuity in longshore sand transport, beach erosion has been severe downcoast of the area protected by the seawall (Photo 3). At St. 4, the ground elevation in front of a newly built house had decreased by 1.5 m, and the house was about to be destroyed (Photo 4). Furthermore, near St. 5, the decrease in ground elevation reached 1.5 m, and the roots of tall trees grown near the coastline were exposed. Countermeasures against beach erosion using stones and timber have been carried out in this area, but they have been inadequate as measures preventing erosion (Photo 5). Downcoast of St. 5, a concrete seawall has been built to protect the houses, as shown in Photo 6. Concrete debris is attached to the seawall at the previous ground level over 1.4 m above the present ground level, implying that the ground elevation of the beach has decreased by 1.4 m owing to erosion. Furthermore, at St. 7 the seawall has been completely destroyed and the hinterland has been severely eroded (Photo 7). At St. 8, not only has the seawall fallen down but also a scarp of 1–2 m height has formed in the hinterland (Photo 8). The white house shown in this figure was under construction, and the foundations of the house were protected using solid concrete. However, downcoast of the structure has been severely

| Photo 1. Concrete seawall to protect houses and road from waves. | Photo 2. Wave overtopping over top of seawall. | Photo 3. Severe beach erosion downcoast of the area protected by seawall. |

| Photo 4. Destroyed houses. | Photo 5. Countermeasures against beach erosion using stones and timber. | Photo 6. Concrete seawall to protect house. |

eroded. At St. 9, located further southeast, a seawall with a parapet and 1.9 m height extended straight with a foreshore remaining in front of the seawall (Photo 9). This seawall was built to protect public property. The gradual shoreline recession in this area can be seen by comparing the satellite images taken on June 27, 2008 and November 19, 2013 (Fig. 5). This seawall slightly protruded toward the nearby shoreline, but a foreshore with 17 m width remained in 2008. By November 19, 2013, the seawall was exposed to waves and a hooked shoreline had formed downcoast owing to beach erosion. Photo 10 shows a beach scarp with 1.6 m height formed on the natural beach downcoast of the seawall, where many roots of trees have been exposed.

3.2. *Beach conditions at midpoint between New Washington and tip of sand spit*

The continuity of longshore sand transport was markedly affected by the seawall constructed to protect the shoreline in New Washington, resulting in the breakdown of the seawall itself and severe downcoast erosion. However, the natural sandy beach remained unchanged at St. 11 midway between New Washington and the tip of the sand spit. The shoreline in this vicinity ran straight in the NW-SE direction. Photo 11 shows the shoreline condition north

| Photo 7. Completely destroyed seawall and hinterland severely eroded. | Photo 8. Damaged seawall and formation of a scarp of 1–2 m height. | Photo 9. Seawall with a parapet and 1.9 m height in front of public property. |

Fig. 5. Shoreline changes in front of public property between 2008 and 2013.

of St. 11. A natural sandy beach of 20–30 m width and 1/10 slope extended alongshore without any artificial structures, and many coconut trees were thriving in the hinterland.

3.3. *Tip of the sand spit*

An enlarged satellite image of the rectangular area including a sand spit at the southeast end of the barrier island is shown in Fig. 6. The shoreline had a difference in the position by 38 m at point C because of the existence of hard rocks on the shoreline, which give a control point for longshore sand transport. The coastal road extending from New Washington along the coastline turns right near point C, and then runs inland. On the east side of this road, there is a wetland with many lagoons. The road extending to the south runs along the

Photo 10. Beach scarp with 1.6 m height formed on natural beach downcoast of seawall.   Photo 11. Shoreline condition north of St. 11.

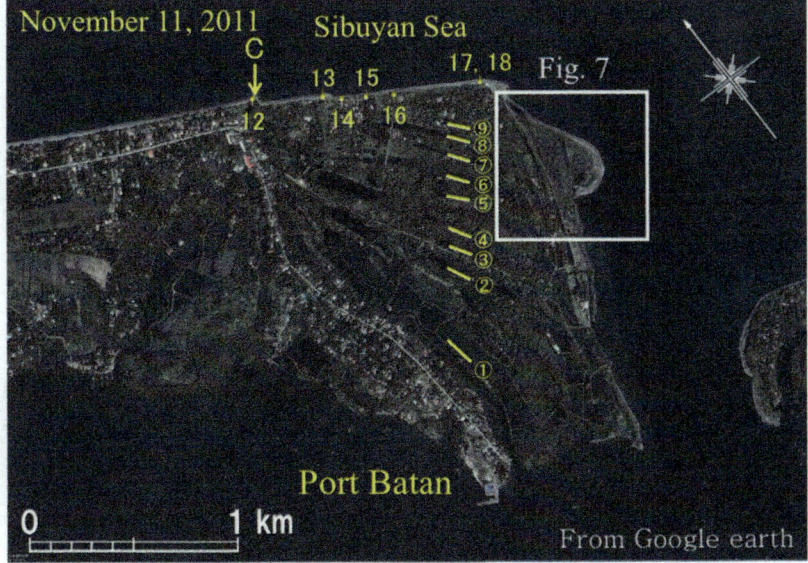

Fig. 6. Southeast end of the barrier island and sand spit.

ridge with an elevation slightly higher than the surrounding lowland, and a lowland extends on the east side of this road. Many lagoons have developed in this lowland, and ridges have developed between the areas of lowland. The total number of slender lagoons is nine, as numbered Nos. 1–9 from the south, and the angle between the direction of the lagoons and the shoreline extending southeastward from point C gradually decreases from lagoon No. 1 to No. 9. This is because sand spits have intermittently developed from the vicinity of point C, which is the control point for longshore sand transport, while separating from the previous sand spit.

In the vicinity of point C, field observation was carried out at St. 12. On the northwest side of point C, a wide foreshore has been formed by the successive deposition of sand transported from upcoast, as shown in Photo 12. In contrast, the beach was severely eroded downcoast (southeast) of point C (Photo 13). Along the narrow pavement between the backshore and the hinterland, a continuous scarp of 0.8 m height has formed, and the local residents have carried out a measure against beach erosion involving the use of sand bags and timber piles (Photo 13). However, the effect of the measure has been inadequate, as can be seen from the subsidence of sand bags. Furthermore, although local residents placed concrete blocks downcoast, as shown in Photo 14, many blocks have been scattered because of their insufficient weight. Photo 15 shows the damage to a house built 510 m southwest of point C. The foreshore width has

Photo 12. Wide foreshore formed on northwest side of point C.

Photo 13. Severely eroded beach downcoast (southeast) of point C.

Photo 14. Many scattered blocks.

Photo 15. Damage to a house built 510 m southwest of point C.

Photo 16. Fallen pole for electricity transmission.

Photo 17. Southeast end of straight shoreline protected by seawall.

been narrowed by beach erosion, and many roots of coconut trees were scattered on the shore. This damage was due to storm waves associated with Typhoon Haiyan, which made landfall on Leyte Island on November 7, 2013, causing damage to many houses near the coastline (Kawai et al., 2014)[2]. Photo 16 shows a fallen pole for electricity transmission immediately east of the coastal village. The roots of vegetation that had been transported by storm surges and become attached to the pole at a height of 1.8 m above the surface of the ground can be seen. From this, it was concluded that inundation with a height of approximately 2 m occurred in this area during Typhoon Haiyan.

At the southeast end of the barrier island, a photograph of the coastal condition along the strait was taken from the vicinity of the lighthouse, as shown in Photo 17. According to interviews with local residents, the lighthouse had been moved landward three times in the past owing to the preceding shoreline recession. The seawall along the shoreline had been entirely destroyed except for the part of the foundation. Photo 18 shows the rotating steps of the fallen observatory, as shown by an arrow in Photo 17.

4. Development of a new sand spit at southeast end

The southeast end of the straight shoreline has been protected by a seawall, as shown in Photo 17, causing the shoreline downcoast of the seawall to retreat and a hooked shoreline to form downcoast because of the discontinuity in longshore sand transport (Fig. 6). Further downcoast, the gradual movement of a sand spit can be seen. An enlarged satellite image of the rectangular area in Fig. 6 including the developing sand spit is shown in Fig. 7, where the shoreline of the sand spit measured on June 27, 2008, November 11, 2011, and November 19, 2013 is shown, and the shape of the sand spit on June 27, 2008 is superimposed on the satellite images of November 11, 2011 and November 19, 2013. On June 27, 2008, there was a sand spit south of the destroyed seawall, forming

Photo 18. Rotating steps of fallen observatory.

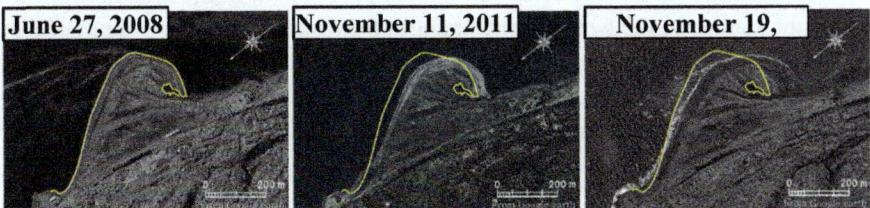

Fig. 7. Successive change in the form of a sand spit.

a hooked shoreline, and a semicircular shoreline with a diameter of 280 m protruded southward, leaving a narrow channel between the tip and the opposite shore. By November 11, 2011, the sand spit had moved southward by approximately 40 m in parallel to the mean coastline while the protrusion to the east had become smaller. Southward movement of the sand spit continued up to November 19, 2013. When a sufficient volume of sand is supplied from upcoast to the sand spit by longshore sand transport, successive erosion upcoast of the tip of the sand spit will not occur, as shown in Fig. 7. Therefore, this change in the shoreline is inferred to correspond to the decrease in the sand supply from upcoast. Similar shoreline changes have been observed at the Shimanto River mouth in Shikoku, Japan (Serizawa et al., 2017)[3]. On the other hand, lowland can be seen inland of the shoreline of the sand spit in the three different years, and this lowland is composed of a combination of a concave topography and beach ridges. Furthermore, the angle between the direction of the shoreline of the ridge around the tip of the sand spit and the present shoreline is large and the ridge is truncated by the present shoreline. This is because the upcoast side of the sand spit was eroded while the entire sand spit moved southward, as shown in Fig. 7, and the eroded sand was transported downcoast, resulting in a new sand spit. The same phenomena have been observed at Notsukezaki sand spit in eastern Hokkaido, Japan (Uda and Yamamoto, 1992)[4].

5. Offshore topography of study area

Figure 8 shows the bathymetry of the study area with the classification of the bed materials in the offshore zone (National Mapping & Resource Information Authority, Philippines, 2004)[5]. Immediately east of Floripon Point, a tidal inlet with a deep channel having a depth up to 8 m separates the opposite shore, and an ebb tidal delta formed by ebb tidal currents has developed in the offshore area. West of the deep channel offshore of the tidal inlet, a shallow area of 3 m depth covered with sand protrudes. Sand transported by southeastward longshore sand transport can be trapped in the channel after turning around the tip of the sand spit, resulting in its offshore deposition by the ebb tidal currents while forming a protruding sand bar. On the other hand, in the area north of the tip of the sand spit, the bed material changes from sand to mud at a depth of 5 m, and the 6 m contour line extends almost parallel to the shoreline, implying that the beach changes have occurred from the foreshore to 6 m depth, which is equal to the depth of closure h_c. The reason that h_c is smaller than the value of approximately 10 m measured on the coast exposed to the ocean is because the Sibuyan Sea is an inland sea surrounded by many islands and the distance to the

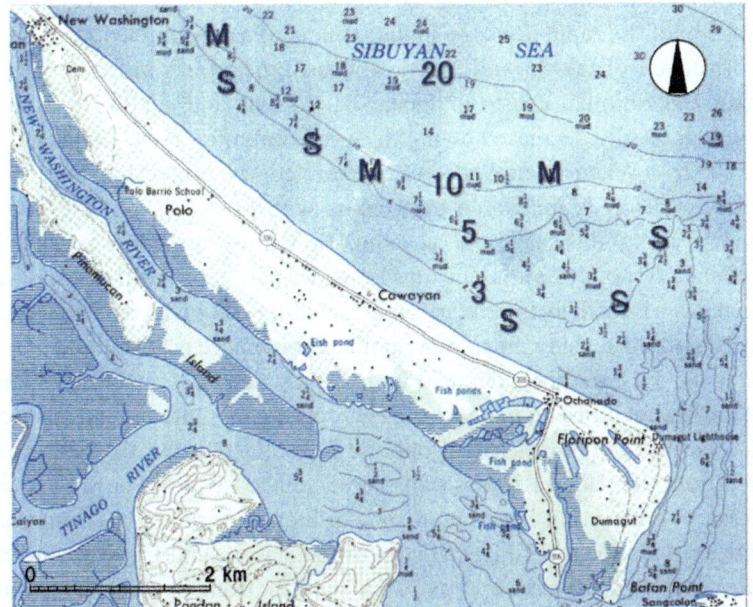

Fig. 8. Bathymetry and grain size distribution of bed materials offshore of barrier island.

opposite shore in the vicinity of Kalibo is too short (160 km) for the development of strong wind waves.

6. Discussion

In the study area, sand supplied from the Aklan River has been transported by southeastward longshore sand transport, and such sand has been deposited to form a barrier island and a sand spit at the southeast end of the barrier island. In recent years, extensive riverbed mining has been carried out in the Aklan River, resulting in a decrease in sediment yield. On the other hand, many houses have been built along both sides of the coastal road extending along the coastline, and a seawall was constructed to protect the road and houses against waves at locations where the coastal road is located close to the coastline. Owing to the beach erosion, the seawall was damaged and debris from the seawall has been left in the sea. As a result, the southeastward longshore sand transport has been blocked by these structures, resulting in further downcoast erosion. Moreover, the coast was further damaged and eroded by storm waves during Typhoon Haiyan. In this area, the comprehensive management of sand in the entire 21 km stretch between the Aklan River and the sand spit has not been considered, and only the local optimization of the coast involving the construction of seawalls has been carried out, preventing coastal erosion from being resolved. The combination of the rapid decrease in the volume of sand supplied from the river, the construction of the seawall and downcoast erosion is similar to that on the Shizuoka and Shimizu coasts in Japan, where the sand supply from the Abe River was markedly decreased owing to riverbed mining in the 1970s and 1980s (Uda, 2017)[1]. When a seawall is constructed to protect a coast locally, downcoast erosion is inevitable on a coast with predominant longshore sand transport. Taking this fact into account, protective measures against beach erosion should be determined while considering the overall comprehensive management of sand in the entire area including the river, and the coastline should be stabilized using the minimum number of shore protection facilities. As measures for the coastline examined in this study, land use should be controlled so that coastal facilities do not excessively protrude offshore, taking into account the fact that the sandy beach will only be maintained if the sand continuously supplied from the Aklan River is transported southeastward. For this purpose, it is necessary to carry out beach nourishment including sand bypassing after the construction of coastal facilities to control longshore sand transport, instead of measures, such as the construction of a seawall from upcoast to downcoast. Also, it is necessary to enhance local awareness of beach erosion and the effectiveness of shore protection measures over a wider area.

References

1. Uda, T., 2017. *Japan's Beach Erosion - Reality and Future Measures*, 2nd ed., p. 530 (World Scientific).
2. Kawai, H., Seki, K., Fujiki, T., 2014. *Storm surge and wave characteristics of Typhoon 1330 in central Philippines*, Proc. JSCE, B2 (Coastal Engineering), Vol. 70, No. 2, pp. I_1411-I_1415. (in Japanese)
3. Serizawa, M., Uda, T., Miyahara, S., 2017. *Rapid formation of a sand spit at river mouth and its prediction using BG model*, 9th Int. Conf. on Asian and Pacific Coasts (APAC 2015).
4. Uda, T., Yamamoto, K., 1992. *Formative process of Notsukezaki compound spit in Hokkaido*, Trans. J. Geomorph. Union, Vol. 13, pp. 19-33. (in Japanese)
5. National Mapping & Resource Information Authority, 2004. Philippines.

Numerical Simulation of 2D Topography Change in Harbor Due to Tsunami Using High Order WENO Scheme

Y. Kajikawa[†] and M. Kuroiwa[††]

Graduate school of Engineering, Tottori University,
Minami 4-101, Koyama-cho, Tottori City, 680-8552, Japan
[†]E-mail: kajikawa@cv.tottori-u.ac.jp
[††]E-mail: kuroiwa@cv.tottori-u.ac.jp

A two-dimensional numerical model using the fifth-order weighted essentially non-oscillatory (WENO) scheme is presented in order to estimate topography change due to tsunami with high accuracy. In the model, the Cartesian coordinate system is adopted, and the fractional area/volume obstacle representation (FAVOR) method is introduced into the governing equations in consideration of applying the estimation to such as harbor shape with complex topography. In order to verify the validity and applicability of the model, it is applied to small-scale laboratory experiments and to large-scale actual topography change. Consequently, although the model cannot reproduce local scouring around breakwater where three-dimensional flow is developed, it is clarified that the model can reproduce the topography change well by contracted flow around a harbor.

Keywords: Tsunami; Topography change; WENO scheme; FAVOR method.

1. Introduction

Topography change caused by tsunami affects serious damage to harbor functions, such as the damage to the harbor facilities owing to scouring and the use restriction of ships owing to sediment deposition. Therefore, it is extremely important for disaster prevention to predict the topography change due to tsunami quantitatively. Numerous studies on the prediction of topography change due to tsunami have been conducted so far. In particular, there are many studies using a two-dimensional (2D) nonlinear long-wave model for the calculation of tsunami propagation e.g., [1] [2] [3]. Recently, three-dimensional (3D) models for the calculation of the tsunami propagation have been also proposed e.g., [4] [5], and the application of these 3D models to large-scale actual topography change are advancing.

Incidentally, in terms of the calculation of tsunami propagation using a 2D nonlinear long-wave model, TUNAMI-N2 (Tohoku University's Numerical Analysis Model for Investigation of Near-field tsunamis) [6], COMCOT

(Cornell Multi-grid Coupled Tsunami Model) [7], and so on has been used widely. In these models, first-order upwind scheme is applied to discretization of the convection terms of the momentum equations. Although the first-order upwind scheme can calculate tsunami propagation stable, it has a possibility that the prediction accuracy decrease by numerical diffusion if calculation mesh is coarse. In order to avoid the decrease of prediction accuracy, the application of high-order schemes is effective [8]. In the present study, the authors propose a numerical model in which the fifth-order weighted essentially non-oscillatory (WENO) scheme is used for the calculation of tsunami propagation and topography change. The WENO scheme, an improved ENO scheme [9], is a technique for achieving high order accuracy and is able to calculate discontinuous flows such as shock waves stably [8]. Therefore, a model applied the WENO scheme has the possibility of being able to predict the tsunami propagation and topography change with stable and high accuracy even if calculation mesh is coarse.

According to the above, a 2D numerical model of topography change due to tsunami using the fifth-order WENO scheme is presented in this paper. In order to verify the validity and applicability of the model, it is applied to laboratory experiments conducted by Fujii et al. [10] and to an actual topography change in Yuriage harbor, which is located at Miyagi prefecture in Japan, caused by the 2011 off the Pacific coast of Tohoku Earthquake Tsunami.

2. Numerical Model

2.1. *Governing equations*

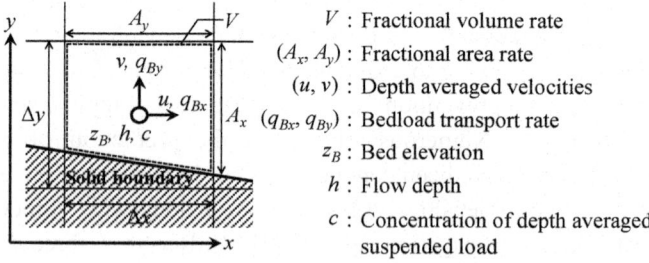

V : Fractional volume rate
(A_x, A_y) : Fractional area rate
(u, v) : Depth averaged velocities
(q_{Bx}, q_{By}) : Bedload transport rate
z_B : Bed elevation
h : Flow depth
c : Concentration of depth averaged suspended load

Fig. 1. Definition of calculating points.

In the presented numerical model, the 2D shallow-water equations based on the nonlinear long-wave theory are used for the calculation of tsunami propagation. The Cartesian coordinate system is adopted and the fractional area/volume obstacle representation (FAVOR) method [11] with its ability to impose the boundary conditions smoothly at complex boundaries such as harbor shape is introduced into the governing equations. In the FAVOR method, it is assumed that both fluid and solid boundary exist in an arbitrary grid shown in Fig. 1, the fractional volume rate V and the fractional area rate (A_x, A_y) occupied with fluid at the grid are introduced into the governing equations. The detail of the 2D shallow-water equations into which the FAVOR method is introduced is exhibited in reference [12].

In consideration of the development of the Ekman layer owing to vortex flow in a harbor, near-bed flow velocities which are important for the estimation of moving direction of bedload are calculated the following equations [13]:

$$u_b = u_{bs} \cos \alpha_s - v_{bs} \sin \alpha_s \tag{1}$$

$$v_b = u_{bs} \sin \alpha_s + v_{bs} \cos \alpha_s \tag{2}$$

$$u_{bs} = 8.5 u_* \tag{3}$$

$$v_{bs} = -N_* \frac{h}{r} u_{bs} \tag{4}$$

$$\frac{1}{r} = -\frac{u\left(u\dfrac{\partial v}{\partial x} - v\dfrac{\partial u}{\partial x}\right) + v\left(u\dfrac{\partial v}{\partial y} - v\dfrac{\partial u}{\partial y}\right)}{\left(u^2 + v^2\right)^{3/2}} \tag{5}$$

where (u_b, v_b) = near-bed flow velocity components in the x and y directions, respectively; α_s = arctan(v/u); (u, v) = depth-averaged velocity components in the x and y directions, respectively; u_* = shear velocity; h = flow depth; N_* = coefficient of the strength of the secondary flow; and r = radius of curvature of the streamline. The shear velocity u_* is calculated from the Manning formula, and the coefficient $N_* = 7.0$ [14].

Concerning the governing equations of topography change, the sediment continuity equation and the suspended sediment transport equation into which the FAVOR method was introduced are used:

$$\frac{\partial z_B}{\partial t} + \frac{1}{V(1-\lambda)}\left[\frac{\partial (A_x q_{Bx})}{\partial x} + \frac{\partial (A_y q_{By})}{\partial y} + V\left(q_{su} - w_f c_b\right)\right] = 0 \tag{6}$$

$$\frac{\partial ch}{\partial t} + \frac{1}{V}\left\{\frac{\partial (A_x cuh)}{\partial x} + \frac{\partial (A_y cvh)}{\partial y}\right\} = \frac{1}{V}\left[\frac{\partial}{\partial x}\left(A_x K_h \frac{\partial ch}{\partial x}\right) + \frac{\partial}{\partial y}\left(A_y K_h \frac{\partial ch}{\partial y}\right)\right] + q_{su} - w_f c_b \tag{7}$$

where V = fractional volume rate; (A_x, A_y) = fractional area rate in the x and y directions, respectively; z_B = bed elevation; λ = porosity of bed material; (q_{Bx}, q_{By}) = bedload transport rate in the x and y directions, respectively; q_{su} = suspended load transport rate per unit area; w_f = settling velocity; c_b = near-the-bed suspended load concentration; c = depth-averaged suspended load concentration; and K_h = horizontal diffusion coefficient.

The bedload transport rate q_B ($=\sqrt{q_{Bx}^2 + q_{By}^2}$) is calculated using the Ashida and Michiue's formula [15]:

$$q_B = \frac{17\rho u_{*e}^3}{(\rho_s - \rho)g}\left(1 - \sqrt{K_c}\frac{u_{*c}}{u_*}\right)\left(1 - K_c\frac{u_{*c}^2}{u_*^2}\right)$$ (8)

where ρ = density of water; ρ_s = density of sediment; g = gravitational acceleration; u_{*e} = effective shear velocity; u_{*c} = critical shear velocity estimated by Iwagaki's formula [16]; and K_c = correction factor attributed to the influence of bed inclination on sediment motion [17].

The suspended load transport rate per unit area q_{su} is evaluated by the following Itakura and Kishi's formula [18] which has high performance in a river flow:

$$q_{su} = K\left(\alpha_* \frac{\rho_s - \rho}{\rho_s}\frac{gd}{u_*}\Omega - w_f\right)$$ (9)

$$\Omega = \frac{\tau_*}{B_*}\frac{\int_{a'}^{\infty}\xi\frac{1}{\sqrt{\pi}}\exp(-\xi^2)d\xi}{\int_{a'}^{\infty}\frac{1}{\sqrt{\pi}}\exp(-\xi^2)d\xi} + \frac{\tau_*}{B_*\eta_0} - 1$$ (10)

where $\alpha' = B_*/\tau_* - 1/\eta_0$; $\eta_0 = 0.5$; $\alpha_* = 0.14$; $K = 0.008$; and $B_* = 0.143$.

The exponential distribution is assumed for the vertical distribution of suspended load concentration. Hence the near-the-bed suspended sediment concentration c_b is evaluated by the following equations [10]:

$$c_b = \frac{\beta}{1 - \exp(-\beta)}c, \quad \beta = \frac{w_f h}{K_z}$$ (11)

where K_z = diffusion coefficient in the vertical direction (= $\kappa u_* h/6$); and κ = the Von Karman constant (= 0.41).

2.2. Numerical methods

In the numerical model, regular grid is adopted for the calculation grid shown in Fig. 1, and the fifth-order WENO scheme [8] [12] is applied to discretization of the convection terms of the governing equations. Since calculated velocities are not defined at the gird boundaries in the regular grid, the Lax-Friedrich flux splitting scheme [8] is applied at the center of each grid when the WENO scheme is used. Moreover, the sixth-order central difference scheme and the forth-order central difference scheme are applied to the pressure term and the viscosity term of the momentum equations, respectively. The third-order total variation diminishing Runge-Kutta scheme is applied to the time integration [8].

3. Application to Laboratory Experiments

3.1. *Experimental and calculation conditions*

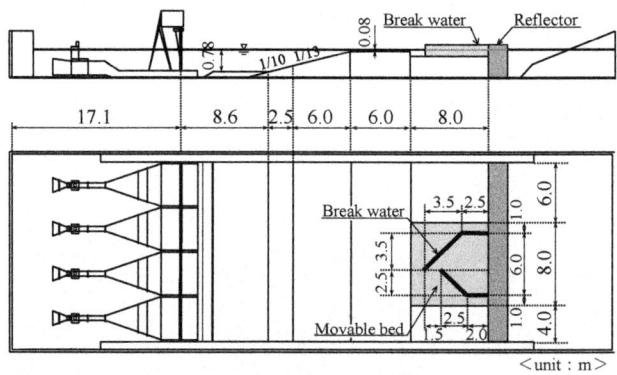

Fig. 2. Outline of experimental wave tank (Fujii et al., [10]).

In order to verify the validity of the presented model, numerical simulations were performed using experimental data by Fujii et al. [10] of both fixed- and movable-bed conditions in a large-scale wave tank at first. Figure 2 shows the outline of the experimental wave tank. In the experiments, a tsunami with half of 60 seconds period and 0.06 m half-amplitude was generated under the fixed- and the movable-bed conditions, and the height of breakwater was the height which the generated tsunami did not overflow. Only flow were measured in the fixed-bed condition, and the topography change was measured in the movable-bed condition using uniform sand with $d = 0.08$ mm.

In the calculation conditions, the grid size Δx and Δy were set to 0.10 m, the calculation time step Δt was set to 0.02 seconds, the Manning's roughness coefficient n was set to 0.012 and the horizontal diffusion coefficient K_h was set to 0.0001 m^2/s. Moreover, comparisons between the presented model and first-order upwind scheme model were conducted in order to confirm the effectiveness of the WENO scheme.

3.2. *Results and discussion*

As the results of fixed-bed condition, Figure 3 shows the horizontal velocity vectors by (a) experiment, (b) the presented WENO scheme model and (c) the first-order upwind scheme model. The upper figures of Fig. 3 show the situation at the time of the leading wave of tsunami, and the lower figures show the situation at the time of the backwash. From these figures, both models can

532

reproduce the flow situations of experiments on the whole. However, by comparisons between the WENO scheme and the first-order upwind scheme, the vortex flow velocity in the harbor is reproduced larger in the WENO scheme than the first-order upwind scheme. Moreover, regarding high speed outflow from the harbor entrance at the time of the backwash, although the flow calculated by the first-order upwind scheme is spreading, the flow spreading is reduced in the WENO scheme as with the experiments. Namely, it can be seen that the numerical diffusion is reduced in the WENO scheme because the WENO scheme is high order scheme.

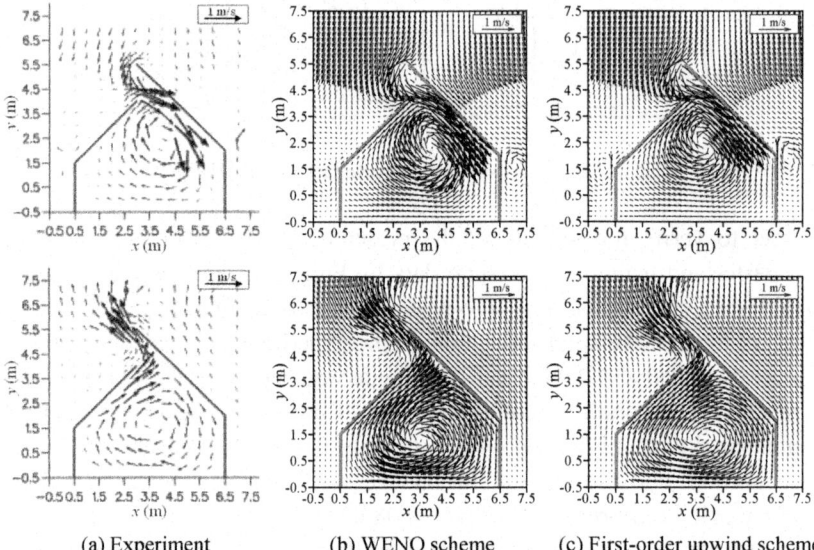

(a) Experiment   (b) WENO scheme   (c) First-order upwind scheme

Fig. 3. Comparisons between measured and calculated flow velocities on the fixed-bed.
(Upper: Leading wave of tsunami, Lower: Backwash of tsunami)

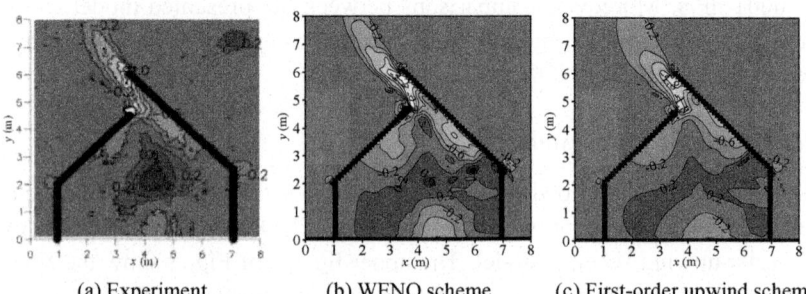

(a) Experiment   (b) WENO scheme   (c) First-order upwind scheme

Fig. 4. Comparisons between measured and calculated bed level change on the movable-bed.

Figure 4 show the comparisons of bed level change contour after tsunami attack in the movable-bed condition. Figure 4 (a) is the experiment, (b) is the presented WENO scheme model and (c) is the first-order upwind scheme model. Both models are not able to reproduce the sedimentation at the center of harbor as seen in the experiment. However, the sedimentation height is higher in the WENO scheme than the first-order upwind scheme, and the erosion around the harbor entrance calculated by the WENO scheme is good agreement with the experimental data. Conversely, the erosion calculated by the first-order upwind scheme around the harbor entrance is small compared with the experiment, and the erosion area tends to spread.

Given these facts, the validity of the presented model using the WENO scheme was confirmed for the phenomena in laboratory scale.

4. Application to Large-scale Topography Change

4.1. *Outline of Yuriage harbor*

Yuriage harbor, which is one of fishing ports, is located at Miyagi prefecture in the northeastern part of Japan. The location of Yuriage harbor is shown in Fig. 5, and the shape of the harbor is shown in Fig. 6.

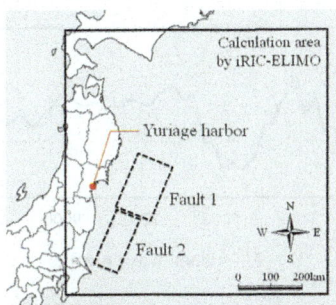

Fig. 5. Location of Yuriage harbor.

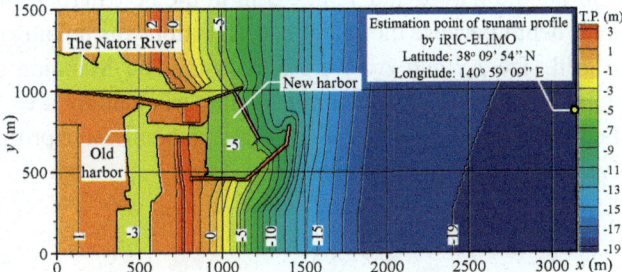

Fig. 6. Calculation area including Yuriage harbor.

534

Yuriage harbor consists of an old harbor and a new one (Fig. 6). This harbor was serious damaged by the 2011 off the Pacific coast of Tohoku Earthquake Tsunami. After the disaster, bathymetric survey of topography change in the harbor was conducted by Matsubara et al. [19]. In the present study, the authors examined the applicability of the presented model for the topography change in this harbor.

4.2. Calculation conditions

The calculation area for the presented model is shown in Fig. 6. Since the calculation area shown in Fig. 6 is only around Yuriage harbor, tsunami wave profile at the offshore boundary is needed. The tsunami wave profile was generated at the estimation point indicated in Fig. 6 by the calculation of tsunami propagation using iRIC-ELIMO (International River Interface Cooperative Easy-performable Long-wave Inundation MOdel) [20] from tsunami source indicated in Fig. 5. For the calculation of tsunami propagation from the source, 500 m mesh depth-sounding data which is available from Japan Oceanographic Data Center [21] and the fault models which are reported from Geospatial Information Authority of Japan [22] were used. The generated tsunami wave profile is shown in Fig. 7.

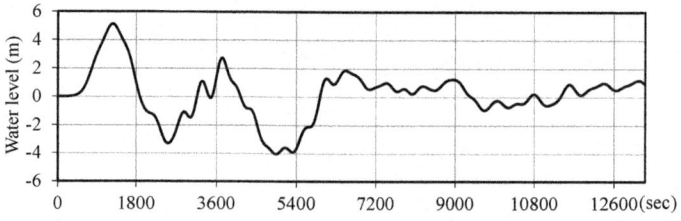

Fig. 7. Tsunami wave profile.

For the initial bed elevation in Yuriage harbor shown in Fig. 6, Tokyo Peil (T. P.) – 3 m in the old harbor and T. P. – 5 m in the new harbor were set from the planed water depth because the initial bed elevation was unknown before the disaster. The grid size Δx and Δy were 5.0 m and the Manning's roughness coefficient n was set to 0.03 uniformly in the calculation area. The mean diameter of bed material d was 0.55 mm, and the tsunami wave profile shown in Fig. 7 was given uniformly at the offshore boundary.

4.3. *Results and discussion*

Figure 8 shows the flow situation in Yuriage harbor at tsunami attack [23]. It can be seen that the separation flow at the tip of breakwater and the circulating flow in the harbor are generated. Figure 9 shows the calculated flow velocity vectors at tsunami attack. Figure 9(a) is the situation at the leading wave of tsunami and (b) is the situation at the backwash. By comparison between Fig. 8 and Fig. 9(a), the calculated result can reproduce the flow situation in the harbor at the leading wave of tsunami as with actual phenomenon. Moreover, from Fig. 9(b), the strong contracted flow is generated in the channel connecting the old harbor with the new one.

Fig. 8. Flow situation in Yuriage harbor at tsunami attack (Japan Coast Guard, [23]).

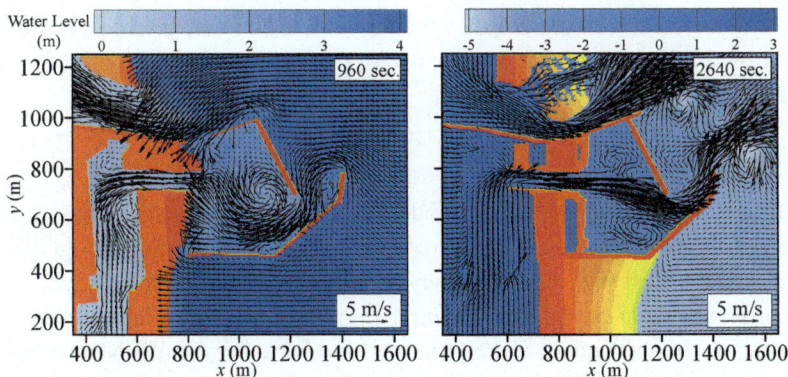

Fig. 9. Calculated flow velocity vectors. Left is at the time of the leading wave. Right is at the time of the backwash of tsunami.

536

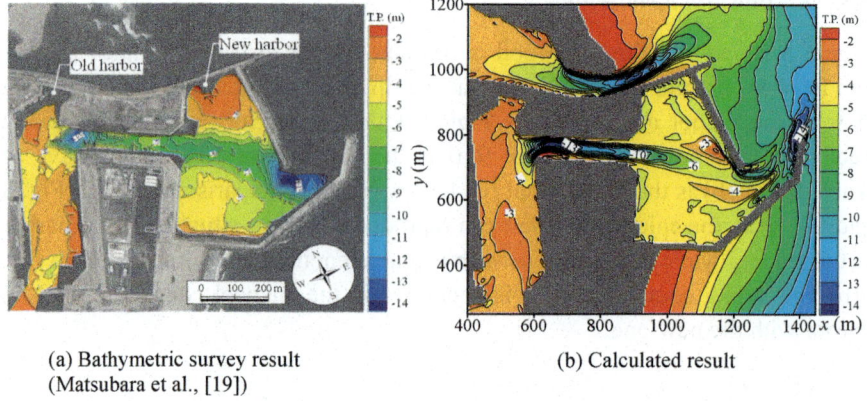

(a) Bathymetric survey result
(Matsubara et al., [19])

(b) Calculated result

Fig. 10. Comparison of bed elevation after tsunami attack.

As the comparison of topography change after tsunami attack, Figure 10(a) and (b) show the bathymetric survey result [19] and the calculated result, respectively. Although the local scouring at the tip of breakwater is not reproduced in the calculated result, the erosion in the channel between the old harbor and the new one is reproduced well. Furthermore, the calculated deepest bed elevation in the channel shows agreement with the survey data. The reason why the model was able to reproduce the erosion in the channel is because the model could reproduce the contracted flow in the channel shown in Fig. 9. In contrast, the cause which the model could not reproduce the local scouring at the tip of breakwater is seemingly because the model was not able to reproduce the complex flow in front of the breakwater where three-dimensional flow might be generated.

5. Conclusions

In the present study, a 2D numerical model using the WENO scheme based on the nonlinear long-wave theory was proposed in order to predict topography change due to tsunami with high accuracy; the model was applied to laboratory experiments and to an actual phenomenon regarding the topography changed due to tsunami. In consequence, the presented model using the WENO scheme was able to reproduce the phenomena more accurately than the low-order scheme model. Moreover, although the present model was not able to reproduce the local scouring at the tip of breakwater where three-dimensional flow was developed, it was clarified that the model could reproduce the topography change well owing to the contracted flow in the large-scale actual phenomenon.

References

1. Takahashi, T., Shuto, N., Imamura, F., Asai, D., Modeling sediment transport due to tsunami with exchange rate between bedload layer and suspended load layer, *Proc. Int. Conf. Coastal Eng. 2000*, 1508-1519 (2000).

2. Li, L., Qiu, Q., Huang, Z., Numerical modeling of the morphological change in Lhok Nga, west Banda Aceh, during the 2004 Indian Ocean tsunami: understanding tsunami deposits using forward modeling method, *Natural Hazards*, 64, 1549-1574 (2012).

3. Sugawara, D., Goto, K., Imamura, F., Sediment transport due to the 2011 Tohoku-oki tsunami at Sendai: Results from numerical modeling, *Marine Geology*, 358, 18-37 (2014).

4. Gelfenbaum, G., Vatvani, D., Jaffe, B., Dekker, F., Tsunami inundation and sediment transport in vicinity of coastal mangrove forest, *Coastal Sediments 07*, 2, 1117-1128 (2007).

5. Apotsos, A., Buckley, M., Gelfenbaum, G., Jaffe, B., Vatvani, D., Nearshore tsunami inundation model validation: toward sediment transport applications, *Pure and Applied Geophysics*, 168, 2097-2119 (2011).

6. Goto, C., Ogawa, Y., Shuto, N., Imamura, F., Numerical method of tsunami simulation with the leap-frog scheme (IUGG/IOC Time Project), IOC Manual, UNESCO, 35 (1997).

7. Liu, P. L. F., Cho, Y., Briggs, M. J., Kanoglu, U., Synolakis, C. E., Runup of solitary waves on a circular island, *J. Fluid Mech.*, 302, 259-285 (1995).

8. Shu, C.-W., High order finite difference and finite volume WENO schemes and discontinuous Galerkin methods for CFD, NASA/CR-2001-210865, ICASE Report No. 2001-11 (2001).

9. Harten, A., Engquist, B., Osher, S. and Chakravarthy, S., Uniformly high order accurate essentially non- oscillatory schemes III, *J. Comp. Phys.*, 71, 231-303 (1987).

10. Fujii, N., Ikeno, M., Sakakiyama, T., Matsuyama, M., Takao, M., Mukohara, T., Hydraulic experiment on flow and topography change in harbor due to tsunami and its numerical simulation. *J. Japan Society of Civil Eng.*, B2-65 (1), 291-295 (in Japanese with English abstract) (2009).

11. Hirt, C. W. and Sicilian, J. M., A porosity technique for the definition obstacle in rectangular cell meshes, *Proc. 4th Int. Conf. Numerical Ship Hydrodynamics*, Washington, D.C., 1-19 (1985).

12. Kajikawa, Y. and Hinokidani, O., Numerical simulation of 2-D bed deformation in a slit sabo dam, *Proc. of 35th IAHR World congress*, USB, Theme C, A11452, (2013).

13. Jang, C. and Shimizu, Y., Numerical simulation of relatively wide, shallow channels with erodible banks. *J. Hyd. Eng.*, ASCE, 131 (7), 565-575 (2005).

14. Engelund, F., Flow and bed topography in channel beds. *Journal of Hydraulic Division*, ASCE, 100 (HY11), 1631–1648 (1974).

15. Ashida, K., Michiue, M., Study on hydraulic resistance and bed-load transport rate in alluvial streams, *Transactions of the Japan Society of Civil Eng.*, 206, 59-69 (in Japanese) (1972).

16. Iwagaki, Y., Fundamental study on critical tractive force, (I) Hydrodynamical study on critical tractive force, *Transactions of the Japan Society of Civil Eng.*, 41, 1-21 (in Japanese with English abstract) (1956).

17. Takebayashi, H. and Okabe, T., Geometric characteristics of braided channel on non-uniform sediment bed, The Int. Cong. INTERPRAVENT 2002 in the Pacific Rim, Congress publication, 1, 79-89 (2002).

18. Itakura, T., Kishi, T., Open channel flow with suspended sediments, *Journal of Hydraulic Division*, ASCE, 106 (HY8), 1325-1343 (1980).

19. Matsubara, Y., Kuroiwa, M., Shibutani, Y., Ichimura, Y., Yonemura, S., Damage investigation of the Pacific coast of Tohoku Earthquake Tsunami in the Yuriage port using small side scan sonar. 68th Annual Conference of Japan Society of Civil Engineers, II-185, 369-370 (in Japanese) (2013).

20. iRIC Project, (2010). http://i-ric.org/ (accessed 2017-05-15).

21. Japan Oceanographic Data Center, (2005). http://www.jodc.go.jp/index.html (accessed 2017-05-15).

22. Geospatial Information Authority of Japan, The 2011 off the Pacific coast of Tohoku Earthquake: Crustal Deformation and Fault Model, (2011). http://www.gsi.go.jp/cais/topic110422-index-e.html (accessed 2017-05-15).

23. Japan Coast Guard, (2011). http://www.agreenring.com/cgi/ciel/topics.cgi (accessed 2017-05-15).

Morphodynamics of Sand Spit at the Tenjin River Mouth in Tottori, Japan

M. Kuroiwa[†], Y. Kajikawa[††], S. Toda, R. Anan and S. Yamamoto

Graduate school of Engineering, Tottori University,
Minami 4-101, Koyama-cho, Tottori City, Tottori, 680-8552, Japan
[†]E-mail: kuroiwa@cv.tottori-u.ac.jp
[††]E-mail: kajikawa@cv.tottori-u.ac.jp

T. Yamamoto

Pneta-Ocean Construction Co. Ltd.
Koraku 2-2-8, Bunkyo-ku, Tokyo, 112-8576, Japan

H. Kurashige[†††] and T. Katayama

Shinwa Engineering Consultant Co. Ltd.
Dosho-cho 4-12-24, Yonago City, Tottori, 683-0064, Japan
[†††]E-mail: h-kurashige@shinwa-giken.co.jp

A field investigation for deformation of a river-mouth sand bar in the Tenjin River, which is located at Tottori prefecture in Japan, was carried out. The topographic surveys using RTK-GPS, UAV and RC boat were conducted from July, 2015 to March, 2017. Moreover, 4 cameras were set at the both sides of the Tenjin River embankment in order to monitor the deformation of the river-mouth bar. From these measurements, the characteristics of the deformation of the sand bar tip in the opening of the river mouth were analyzed and the validity of an alternative survey technique using RTK-GPS, UAV and RC boat was confirmed.

Keywords: Morphodynamics; Sand spit; River mouth; RTK-GPS; UAV; RC boat.

1. Introduction

River-mouth bars, such as sand spits, are formed in river mouths connecting to sea areas with low tidal range mainly owing to wave action. The river mouth will be closed by the bars when the river discharge is small. The prevention of flooding due to rising water level by the river-mouth closing and the maintenance of the functions of river mouths require the prediction of morphodynamics around river mouths. However, the flow field around a river mouth is extremely complex owing to river discharge, waves, wave-induced currents, and tidal currents of different densities.

Topographic features around river mouths are roughly classified into three types from the following work conducted by Sawaragi [1]; In a river-flow-dominating type, deltas and terraces are formed in front of the river mouth. In a wave-dominating type, a river-mouth bar and asymmetric bars are formed under normal wave conditions, and a longshore bar is formed under an extreme wave condition. In a tidal-current-dominating type, sand bars are formed in the river under high tide and high wave conditions.

The Tenjin River mouth facing with the Sea of Japan has a sand spit, and the river mouth is often closed by sedimentation owing to waves in winter season. The river-mouth closing influences flood due to river flow and sand supply for sandy beaches around the river mouth. Therefore, the mechanism of the development and deformation of the river-mouth spit is needed to clarify in order to take some countermeasures for maintaining of the opening channel at the river mouth. However, the morphodynamics of sand spit at the river mouth is so complex as mentioned above [1] that the mechanism of the sand spit deformation is not cleared.

In the present study, authors performed an investigation of topographical changes in order to clarify the morphodynamics of sand spit at the Tenjin River mouth. In the field investigation, the alternative survey techniques, namely RTK-GPS (Real Time Kinematic Global Positioning System), UAV (Unmanned Aerial Vehicle) and RC (Radio Control) boat with sounding survey system, were used. Although studies on topographic survey using RTK-GPS and/or UAV have been conducted conventionally e.g. [2] [3], there are few studies which have targeted the detailed topographical change in river mouths using these techniques. Authors investigated the detailed topographical changes, namely seasonal changes, short-term topographic changes under stormy and flooding conditions between July, 2015 and March, 2017. From the measured data, the characteristics of the deformation of the sand bar tip in the opening of the river mouth were analyzed.

2. Field Investigation

2.1. *Field site and characteristics of the Tenjin River*

The Tenjin River is located at the middle part of Tottori Prefecture in Japan as shown in Fig. 1. The length of the river from the upstream to the river mouth is 32 km long, and the river mouth is facing the Japan Sea. The width of the river mouth is approximately 300 m, and the averaged flow discharge per year is 20 m³/s. As shown in Fig. 1, a large sand bar was formed and the left bank side was opened due to the river flow in July 2, 2015. The opening is frequently changed into right bank side. The field investigation for the sand bar deformation was carried out over the period from July 2, 2015 to January 7, 2017.

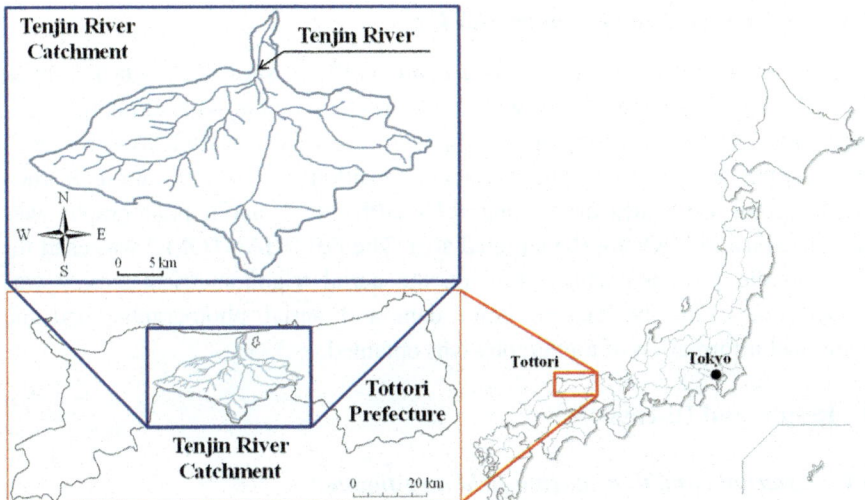

Fig. 1. Field site and river mouth bar at the Tenjin River.

2.2. Topographic survey with RTK-GPS

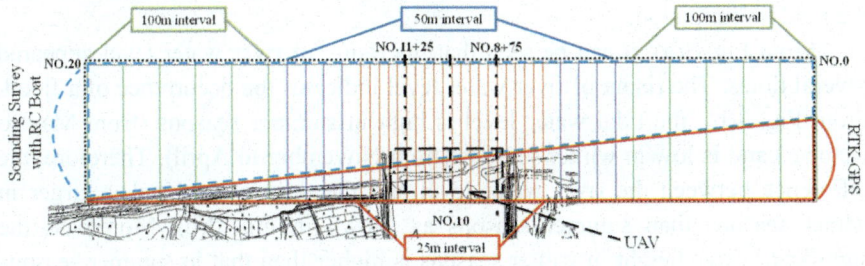

Fig. 2. Survey area using RTK-GPS, UAV and RC boat.

Figure 2 shows the whole topographic survey area in the presented study, and the red triangular frame indicates the survey area with RTK-GPS. The survey lines in the transversal direction on the sand bar were set at the interval of 25 m, and the survey lines of 50 m and 100 m interval were set in the out of the sand bar area. The data of three-dimensional position was measured every two seconds. The survey of topography changes of the river-mouth spit using RTK-

GPS were conducted 14 times between July 2, 2015 and January 7, 2017. Furthermore, the shoreline positions both of the left and right banks at the river mouth were measured by using RTK-GPS, and the sea bottom level around the river mouth were measured by RC sounding boat 7 times during the field investigation.

2.3. *Aerial photographic survey with UAV*

The aerial photographic survey area using UAV is shown the black broken frame in Fig. 2. In the survey with UAV, picrure points with airphoto signal which are reference for the horizontal position and the elevation were set up in the targeted area at first. The horizontal position and the elevation of each picrure point were measured using RTK-GPS. Next, aerial photography was conducted using UAV for the targeted area. The DJI PHANTOM 2 was used for UAV in the presented study. Finally, the aerial trigonometrical survey was carried out using the picrure point data and aerial photographs, and the numerical information of topography was obtained.

3. Results and Discussions

3.1. *External conditions during field investigation*

Figure 3 shows time variations of (a) river water level at the river mouth, (b) tidal water level, (c) difference between the river water level and the tidal water level, (d) significant wave height and wave period and (e) wave direction, during the field investigation.

From Fig. 3(a), it can be seen that the rising of river water level appeared several times. The rising of river water level indicates the occurrence of a flood. From Fig. 3(b), the tidal water level is high in summer seasons (from May to October) and is low in winter seasons (from November to April). Therefore, the difference between the river water level and the tidal water level is larger in winter seasons than summer seasons as shown in Fig. 3(c). Moreover, the significant wave height in winter seasons is higher than that in summer seasons as shown in Fig. 3(d), and the high waves which is more than 3.0 m height appeared several times in winter seasons. The river mouth in this river was often closed in winter seasons. Hence there is a possibility that the higher river-mouth bar is formed by the high waves when the difference between the river water level and the tidal water level is small in winter seasons. From Fig. 3(e), since the wave direction is almost north throughout the year, the waves always surge in from the sea to the river.

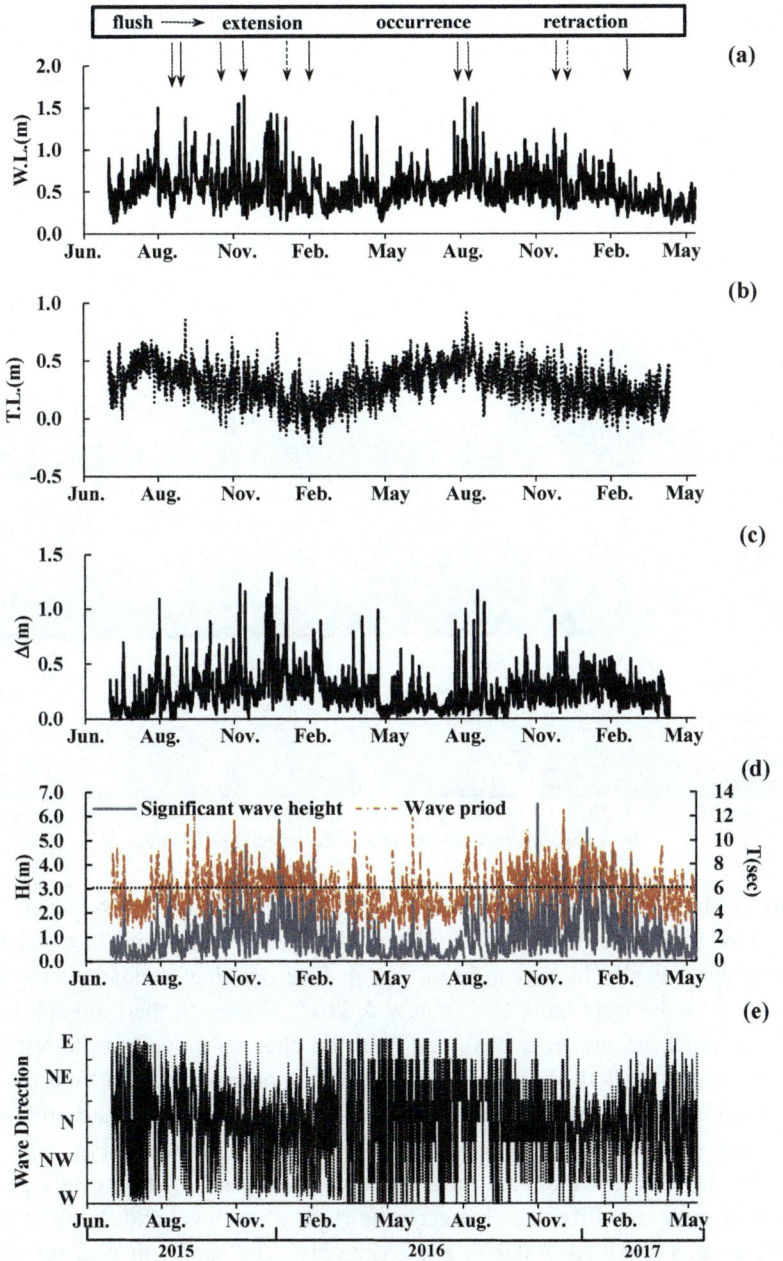

Fig. 3. External conditions during field investigation; (a) River water level at the River mouth, (b) Tidal water level, (c) Difference between river water level and tidal water level, (d) Significant wave height and wave period and (e) wave direction.

3.2. *Morphodynamics of sand spit under stormy and flooding conditions*

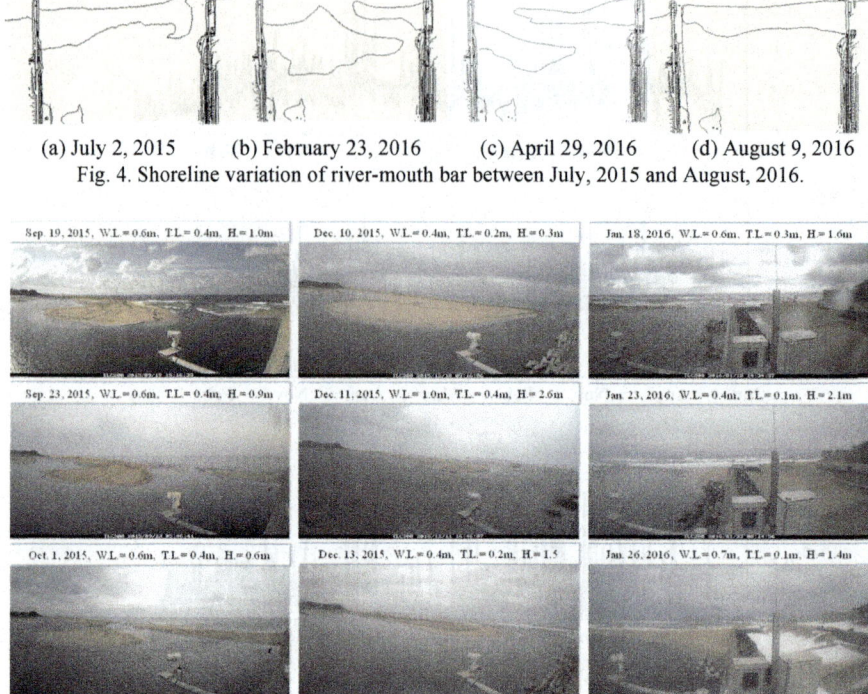

(a) July 2, 2015   (b) February 23, 2016   (c) April 29, 2016   (d) August 9, 2016

Fig. 4. Shoreline variation of river-mouth bar between July, 2015 and August, 2016.

(a) Sep. 19 – Oct. 1, 2015   (b) Dec. 10 – Dec. 13, 2015   (c) Jan. 18 – Jan. 26, 2016

Fig. 5. Circumstances at the time of river-mouth bar change.

Figure 4 shows the shoreline variation of the river-mouth bar between July, 2015 and August, 2016. It is found that the river-mouth sand bar had been varying dynamically in a year from Fig. 4. The opening channel of the river mouth was at the right bank side in July 2, 2015. However, the sand spit began to be formed from the right bank end, and the channel moved to the left bank side in August 9, 2016. Figure 5 shows the circumstances at the time of river mouth bar changing taken by the right bank side camera. The sand spit at the right bank was formed in a brief period of 10 days as shown in Fig. 5(a). The wave height was less than 1.0 m and comparatively small during this period. However, since the difference between the river water level and the tidal water level was also small ($\Delta = 0.2$ m approximately), the sand spit was seemingly formed by the small waves. Although the sand spit formed at right bank was flushed owing to a flood caused in December 11, 2015 (Fig. 5(b)), the sand spit was formed again from the right bank end in 10 days in January, 2016 (Fig. 5(c)). The wave height was more than approximately 1.5 m in January, 2016, and there was considered to be the condition which the sand spit was formed easily.

3.3. *Seasonal changes of sand spit*

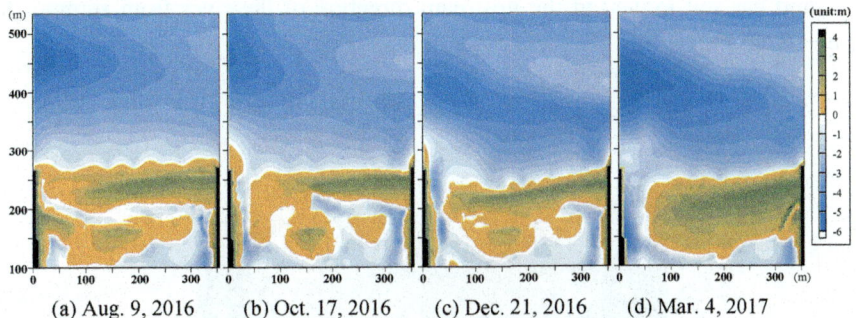

(a) Aug. 9, 2016   (b) Oct. 17, 2016   (c) Dec. 21, 2016   (d) Mar. 4, 2017

Fig. 6. Topography variation in the river mouth between August, 2016 and March, 2017.

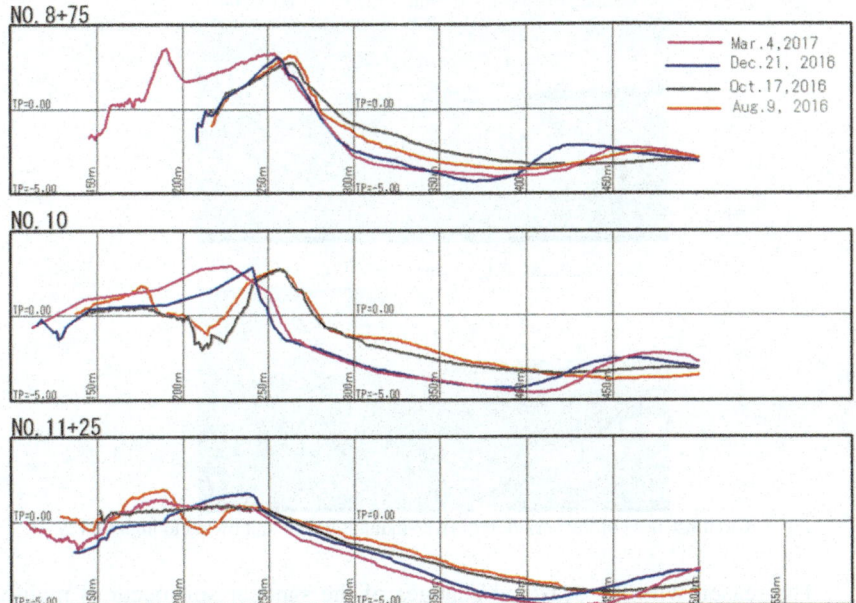

Fig. 7. Variation of river-mouth sand bar spit in survey lines. Top, middle and bottom figures are in the line number NO.8+75, NO.10 and NO.11+25 shown in Fig. 2, respectively.

On the seasonal changes of the river-mouth sand spit, the topography variation in the Tenjin River mouth between August, 2016 and March, 2017 is shown in

Fig. 6. The river-mouth sand bar was moved in the upstream direction of the river between August (summer) and December (winter), 2016. In the sea area, the area where the sea bottom elevation is less than - 5.0 m was also moved to the shore direction during this period. Figure 7 shows the variation of river-mouth sand bar profile in each survey line shown in Fig. 2. It is also found that the sand bar spit moved in the upstream direction of the river from summer to winter from Fig. 7.

Fig. 8. Circumstances of the variation of river-mouth sandbar spit owing to high waves.

The reason why the seasonal changes of the sandbar spit occur is mainly the effect of high waves in winter. The occurrence of the high waves more than 3.0 m in winter is indicated in Fig. 3(d). Moreover, Figure 8 shows the circumstances of the variation of sandbar spit owing to the high waves in December, 2016. When the wave height was more than 3m, the waves was run-up over the sandbar, and the sandbar spit moved to the inner of the river mouth in a few days. The sandbar spit moved to the inner of the river mouth in winter is considered to move in the offshore direction owing to floods in summer indicated in Fig. 3(a). In the Tenjin River mouth, these phenomena are considered to be occurring throughout the year.

4. Conclusions

In the presented study, the detailed topography survey at the Tenjin River mouth using RTK-GPS, UAV and RC boat were conducted in order to clarify the morphodynamics of the sandbar spit in the river mouth. In the river mouth, it is revealed that the sandbar spit is formed rapidly owing to waves with 1.0 m wave height approximately when the difference between the river water level and the tidal water level is less than 0.2 m. Moreover, the authors identified that the sandbar spit is moving to the inner of the river mouth due to the high waves with more than 3.0 m height in winter. The sandbar spits at the Tenjin River mouth have been varying dynamically throughout the year.

Acknowledgments

This research was conducted in the cooperation of Kurayoshi Office of River and National Highway, Chugoku Regional Development Bureau, Ministry of Land, Infrastructure, Transport and Tourism, Japan.

References

1. Sawaragi, T., Coastal Engineering −Waves, Beaches Wave-Structure Interaction, *Development in Geotechnical Engineering*, 78, Elsevier, 304-308, (1995).
2. Harley, M., Turner, I., Short, A., Ranasinghe, R., Assessment and integration of conventional, RTK-GPS and image-derived beach survey methods for daily to decadal coastal monitoring, *Coastal Engineering*, 58, 194-205 (2011).
3. Turner, I., Harley, M., Drummond, C., UAVs for coastal surveying, *Coastal Engineering*, 114, 19-24 (2016).

Erosion of Cai River Mouth in Nha Trang, Vietnam

Akio Kobayashi[a], Takaaki Uda[b] and Yasuhito Noshi[a]

*[a]Department of Oceanic Architecture & Engineering,
College of Science & Technology, Nihon University,
7-24-1 Narashinodai, Funabashi, Chiba 274-8501, Japan
E-mail: kobayashi.akio@nihon-u.ac.jp*

*[b]Head, Shore Protection Research, Public Works Research Center,
1-6-4 Taito, Taito, Tokyo 110-0016, Japan
E-mail: uda@pwrc.or.jp*

The beaches at the mouth of the Cai River, which flows into Nha Trang Bay, Vietnam, have been eroded, resulting in the disappearance of sandy beaches on both sides of the river mouth. Satellite images taken between 2002 and 2015, and the effect of a storm that occurred in 2008 were analyzed. The most important cause of the beach erosion was sand excavation from the sand bar on the right bank to obtain materials for land reclamation on the left bank. Another cause was sand movement due to waves in the upstream direction along the left bank of the river. To recover the sandy beach, the construction of training jetties is expected to be effective, similarly to the case of the bay mouth jetty of Dam Nai Bay in Phan Rang, Vietnam.

Keywords: Nha Trang; Cai River; River mouth bar; Beach erosion; Satellite image; River mouth jetty.

1. Introduction

A sandy beach with 4.3 km length, which faces Nha Trang Bay, extends in the central part of Nha Trang in Vietnam, as shown in Fig. 1. This beach has a gradually curved shoreline that is subject to the wave-sheltering effect of Vinh Nguyen Island, located offshore of the coast, and the sandy beach composed of fine granite sand is a well-known spot for oceanic recreation, attracting many tourists. The Cai River flows into the bay at the north end of the coast, and the Cai River mouth has been eroded in recent years, and sandy beaches located on both sides of the river mouth have disappeared. These beaches were used for bathing before erosion, similarly to Nha Trang Beach. The local government plans to restore the sandy beaches. However, the exact cause of the beach erosion is unknown. Thanh et al. (2015)[1] carried out field observation of the shoreline changes at the north end of Nha Trang Beach using a video camera, and showed

Fig. 1. Study area near Nha Trang City.

that the shoreline varied with the seasonal changes in the wave direction, resulting in the seasonal reversal of the direction of longshore sand transport at the north end of Nha Trang Beach. Although the shoreline seasonally changes near the north end of the beach, the long-term monotonic retreat of the shoreline has not been measured. Thus, the observation results of the shoreline recession cannot be explained by those of Thanh et al.[1] In this study, field observation was carried out at the Cai River mouth in 2015 and 2016, and the shoreline changes were investigated using satellite images taken between 2002 and 2015 together with the analysis of the erosion at the river mouth due to storm waves during a typhoon in 2008. Also, a meeting was held with members of the regional government, which is responsible for the management of the Cai River, regarding beach erosion at the Cai River mouth.

2. Beach erosion at Cai River mouth

2.1. *Field conditions*

A field observation around the Cai River mouth was carried out on October 27, 2016. Photographs were taken at the six locations shown in Fig. 2. First, the beach condition at St. 1 at the north end of Nha Trang Beach is shown in Photo 1. The berm height in this vicinity was approximately +1.4 m above MSL. Since the observation was carried out during the monsoon period with southwesterly waves, the beach at the north end of Nha Trang Beach was accretive with a flat sandy beach. Although the north end of the beach was protected by a seawall, the beach width in front of the seawall gradually decreased northward with the seawall becoming exposed to waves (Photo 2). Furthermore, there was no sandy beach and the foot of the seawall was exposed to waves at the Cai River mouth (Photo 3). On the other side of the bridge shown in Photo 3, an isolated house remained near the north end of the bridge (Photo 4). By measuring the elevation of the floor of the house using a measuring stick of 2 m length beneath the house, the elevation of the floor was found to be 3.3 m above MSL. In contrast, north of this isolated house, a coastal park has been constructed (Photo 5). Photo 6 shows the stairway attached to the east side of the park. The ground elevation along this stairway was measured to be 3.7 m above MSL.

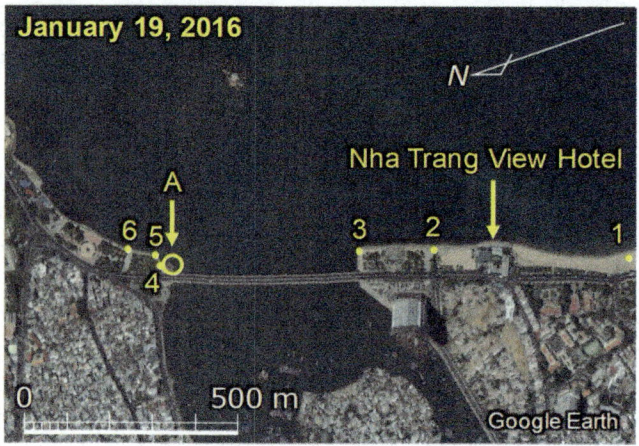

Fig. 2. Satellite image of Cai River mouth and location numbers of site photographs.

Photo 1. Sandy beach at north end of Nha Trang Beach.

Photo 2. Seawall on right bank of the river.

Photo 3. View of Cai River mouth from right bank.

Photo 4. Isolated house A offshore of left bank.

Photo 5. Seawall immediately north of isolated house.

Photo 6. Park constructed on left bank of Cai River mouth.

2.2. *Analysis of satellite images*

Beach changes were investigated in a rectangular area around the Cai River mouth, as shown in Fig. 1. Although there was no river mouth bar in the satellite image shown in Fig. 2, taken in January 2016, the sequential changes in the river mouth were investigated using satellite images taken between 2002 and 2012. In the following figures, the initial shoreline determined from the image of March 24, 2002 is shown, and the location of the isolated house shown in Photo 4 is designated by A.

On March 24, 2002, the construction of a bridge connecting both sides of the Cai River was underway at the river mouth (Fig. 3), and sand bars extended on both sides of the river mouth so as to almost close the river mouth. The sand bar on the right bank protruded further offshore than that on the left bank, and a smoothly curved shoreline extended on both sides. The river mouth was markedly narrowed by the river mouth bars and had an opening width of 95 m, and the house

552

was located 150 m north of the tip of the river mouth bar on the left bank. By August 13, 2003, the construction of the bridge had been completed and both sides of the river were connected by a road (Fig. 4). The outer shoreline of the right river mouth bar had significantly retreated by this time, although the location of the intersection between the left river mouth bar and the bridge was the same as that in March 2002, and the river mouth width had increased to 121 m. By enlarging this image, it can be confirmed that the channel at the tip of the right river mouth bar was a dredging hole.

On October 27, 2016, we had a meeting with researchers from Nha Trang Oceanographical Institute and members of the regional government, which is responsible for the management of the Cai River mouth. In this meeting, we were told that the right river mouth bar of the Cai River had been excavated in the past, and that most of the sand had been used for land reclamation on the left bank. This reclaimed land is presently used as a coastal park, as shown in Photos 5 and 6. For this reason, the hole at the tip of the right river mouth bar, as shown in Fig. 4, is considered to be the hole formed by the excavation.

Up to August 26, 2006, the shoreline shape of the sand bar on the south side of the river mouth changed from straight to curved, and the tip of the right bank

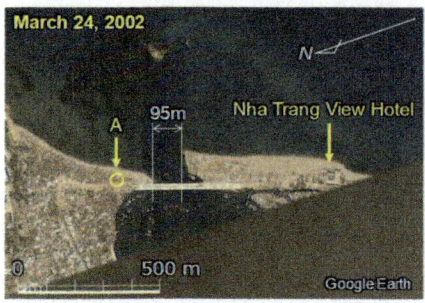

Fig. 3. Satellite image on March 24, 2002.

Fig. 4. Satellite image on August 13, 2003.

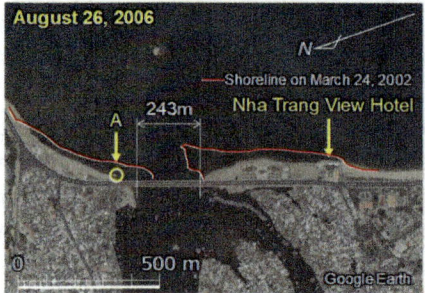

Fig. 5. Satellite image on August 26, 2006.

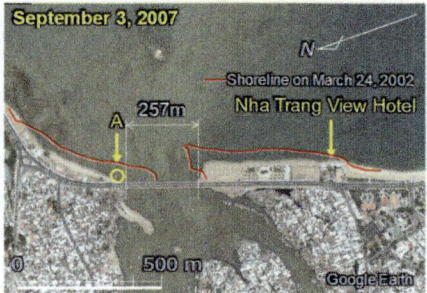

Fig. 6. Satellite image on September 3, 2007.

bar retreated by 68 m (Fig. 5). Furthermore, house A remained behind the shoreline. The increase in curvature of the shoreline along both river banks is assumed to have accelerated the longshore sand transport from the outside to the inside of the river mouth because of the increase in the oblique angle of waves relative to the direction normal to the shoreline. The river mouth width had increased to 243 m by 2006.

By September 3, 2007, large beach changes had occurred on the left bank of the river (Fig. 6). Although a river mouth bar was located immediately upstream of the bridge until August 2006, the sand composing this sand bar had eroded and been transported upstream. As a result, a slender sand spit with 156 m length protruded obliquely to the river. On the other hand, although the location of the tip of the right river mouth bar was the same as that in August 2006, the maximum width of the river mouth bar had increased to 108 m. Because the sand bar moved in the upstream direction of the river on the left bank, wave action causing the upstream movement of sand was predominant during this period. Under these conditions, it is difficult to explain by the physical reasons for the formation of the shoreline extending northward in a straight line from the Nha Trang View Hotel, located 470 m south of the river mouth, before turning by a right angle at the tip, implying that land reclamation was also carried out in the vicinity of the hotel to widen the sandy beach. The opening width increased to 257 m owing to the changes in the river mouth bars.

The sandy beach on the right bank, which expanded up to September 2007, had also severely eroded by January 22, 2012 (Fig. 7), and the right bank had retreated southward by 155 m, exposing the seawall to waves. Moreover, a slender sand spit attached to the river bank, losing its protruding shape, extended on the left bank. As a result, the river mouth width increased to 439 m, which permitted the direct entry of ocean waves into the river mouth, reducing the width of the mooring area of fishing boats, which had been as far as immediately upstream of

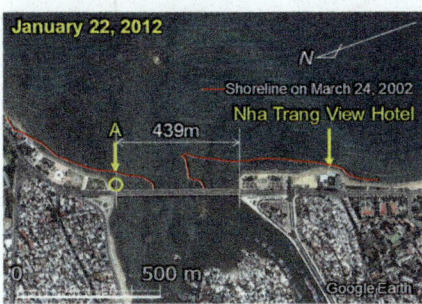

Fig. 7. Satellite image on January 22, 2012.   Fig. 8. Satellite image on January 13, 2015.

the bridge. Similarly to the present situation, house A remained behind at the left end of the bridge. In contrast, a coastal park had been constructed in the reclaimed land north of house A (Fig. 7). This park area according to the satellite image in April 2016 is drawn in Fig. 4 and the park has an area of 3.0×10^4 m². According to the satellite image before the construction of a park in August 2003, this area was a sandy beach. Here, we assume that a sandy beach in almost the same condition as the beach shown in Photo 1 existed there, and that the mean elevation of the beach was approximately 1.4 m on the basis of Photo 1. The ground elevation after the land reclamation was 3.7 m above MSL, and the difference of 2.3 m is assumed to be due to the addition of landfill. By multiplying the planar area by this height, the total volume of sand used for reclamation was 6.9×10^4 m³. On the right river mouth, the sand volume is considered to have decreased by this amount. Finally, no further changes were observed in the river mouth up to January 2015, as shown in Fig. 8, but the sand bar on the left river bank had moved further upstream.

Because the area of sandy beach rapidly decreased after 2002 on both sides of the river mouth, the area of the sandy beach on the right and left banks was calculated relative to that on March 24, 2002 (Fig. 9).

It was found that the area on the right and left banks decreased by 3.3×10^4 and 1.1×10^4 m² by 2015, respectively. In total, a foreshore area of 4.4×10^4 m² disappeared on both sides of the river mouth. Assuming that the berm height is equivalent to 1.4 m, as in Photo 1, the total amount of sand that disappeared was 6.2×10^4 m³. This is similar to the volume of sand of 6.9×10^4 m³ that was considered to have been used for land reclamation on the left bank. Furthermore,

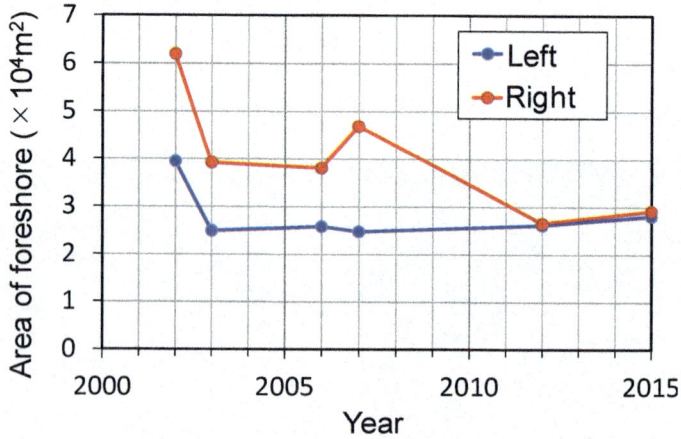

Fig. 9. Change in foreshore area on both sides of Cai River mouth.

the sand volume has monotonically decreased, except for a temporary increase in the sand volume in 2007 on the right bank. As reported by Thanh et al. (2015)[1], the direction of longshore sand transport seasonally changes at the north end of Nha Trang Beach in response to changes in the wave direction. However, the sand volume has monotonically decreased over time. Accordingly, another factor other than the seasonal variation in longshore sand transport must be considered as the cause of the decrease in sand volume around the river mouth.

2.3. *Changes in river mouth bars caused by storm waves in 2008*

The river mouth bars on both sides of the river were severely eroded between 2007 and 2012, as shown in Figs. 6 and 7. One of the causes of these changes is storm waves hitting the coast during the period. On August 6, 2008, a long sand bar extended on the seaward side of the bridge, as shown in Photo 7. Typhoon Noul made landfall at Nha Trang Beach on November 17, 2008 and storm waves hit the coast. The seawall on the right river bank was destroyed and the right river mouth bar had moved upstream to the location of the bridge pier by November 30, 2008, soon after the typhoon, as shown in Photos 8 and 9. The right river mouth bar, which extended seaward of the bridge on August 6, 2008, as shown in Photo 7, was moved upstream and was located below the bridge on November 30, 2008. However, the entire sand bar disappeared by January 22, 2012, as shown in Fig. 7. It is difficult to explain the cause of the disappearance of such a large amount of sand that had accumulated at the river mouth without considering human activities such as sand mining.

Photo 7. Sand bar on right bank of Cai River before typhoon waves (August 6, 2008).

Photo 8. Seawall damaged by storm waves and river mouth bar moved upstream to the location of bridge (November 30, 2008).

Photo 9. River mouth bar moved upstream up to the location of bridge (November 30, 2008).

3. Discussion

The river mouth bars of the Cai River were severely eroded between 2002 and 2012. Comparing the shoreline configurations in March 2002 and January 2012, as shown in Fig. 8, not only did the river mouth bars retreat simultaneously on both sides of the river but also the shoreline changes increased with increasing proximity to the river mouth. The shoreline variation, however, decreased in the vicinity of Nha Trang View Hotel. This implies that the cause of beach erosion may have originated from the river. From these findings, the main cause of shoreline recession is considered to be the excavation of the right river mouth bar to obtain the materials for the land reclamation to build the coastal park on the left bank of the river, which was also confirmed by one of the engineers of the regional government, and that part of the sand was transported upstream along the left bank by waves incident to the river mouth. The agreement of the sand volumes estimated on both banks is further evidence of the shoreline recession on the right bank. The sand bar that formed on the inner side of the left river bank first developed obliquely to the river, while forming a sand spit, and then the tip of the sand bar gradually approached and connected to the seawall of the river bank with successive upstream movement. A sand spit can elongate in the direction normal to that of wave propagation when the water depth where the sand spit elongates is sufficiently small for the wave energy to be greatly reduced; the characteristics of the sand spit are lost with increasing water depth, and sand is transported in the direction of wave propagation (Miyahara et al., 2015)[2]. Accordingly, the development of a sand spit inside the river mouth implicitly suggests that the water depth of the Cai River mouth has increased in recent years.

4. Concrete measures

As a measure to stabilize the river mouth, the extension of a jetty at the Dam Nai Bay entrance in Phan Rang, 77 km south of Nha Trang, where the bay mouth shoreline was stabilized after the construction of the jetty, is considered for reference. Figure 10 shows the shoreline changes around the jetty, which was constructed at the Dam Nai Bay entrance to block sand transport to the bay entrance, between 2003 and 2012. The east jetty of 574 m length blocked sand transport into the entrance, resulting in the deposition of sand east of the jetty. In this case, the amount of blocked sand reached 4.1×10^4 m³/yr (Noshi et al., 2015)[3]. In the Cai River mouth, the extension of the training jetties on both sides of the mouth is necessary to prevent sand from moving inside the Cai River mouth, because sand can be transported from both sides of the river mouth. If beach nourishment is carried out on the far side of the training jetties after the construction of the jetties, stable sandy beaches could be restored. To determine the size of the jetties and the method of beach nourishment, a numerical simulation, such as the one based on BG model (a model for predicting three dimensional beach changes based on Bagnold's concept) (Uda, 2017)[4] is required.

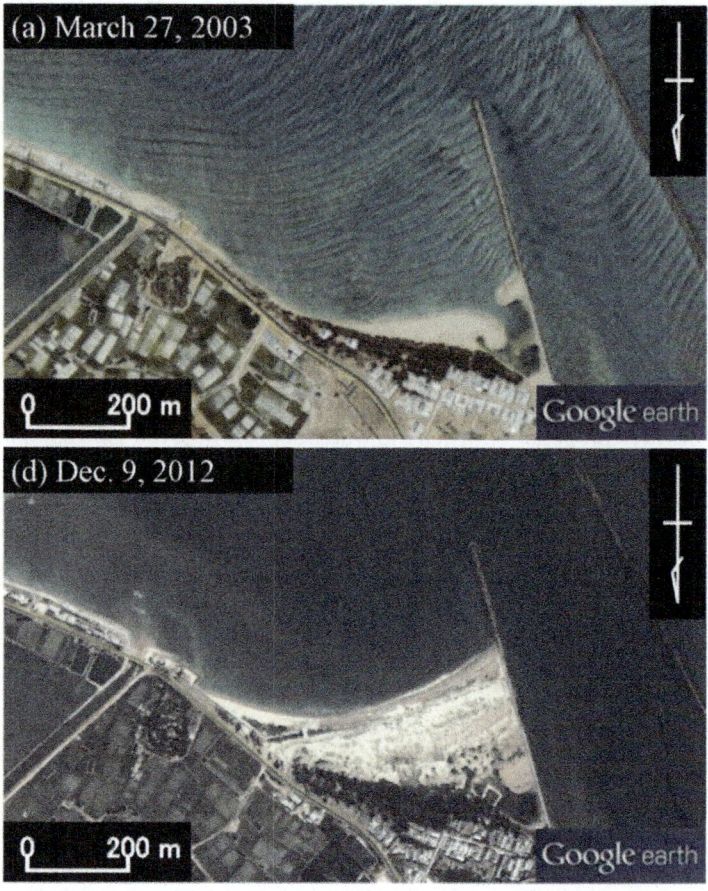

Fig. 10. Shoreline changes around jetty constructed at Dam Nai Bay entrance.

5. Conclusions

It is true that the beach erosion around the Cai River was triggered by wave action, but the anthropogenic factor, such as the dredging around the river mouth to acquire the sand for land reclamation, is also related to the beach erosion. In the development of the river mouth area, the impact of such an activity should be carefully considered to minimize the effect. Also, comprehensive management of sand is required to maintain the river mouth topography because many factors are related to the topographic changes.

References

1. Thanh, T. M., Tanaka, H., Viet, N. T., Mitobe, Y., Hoang, V. C., Evaluation of longshore sediment transport on Nha Trang coast considering influence of northeast monsoon waves, *J. JSCE, Ser. B2 (Coastal Engineering), Vol. 71, No. 2, pp. I_1681-I_1686,* (in Japanese, 2015).
2. Miyahara, S., Uda, T., Serizawa, M., San-nami, T., Elongation of sand spit and profile changes on sloping shallow seabed, *8th Int. Conf. on Asian and Pacific Coasts (APAC 2015), Procedia Engineering, Vol. 116, pp. 245-253,* (2015).
3. Noshi, Y., Uda, T., Kobayashi, A., Miyahara, S., Serizawa, M., Beach changes observed in Phan Rang City in southeast Vietnam, *8th Int. Conf. on Asian and Pacific Coasts (APAC 2015), Procedia Engineering, Vol. 116, pp. 163-170,* (2015).
4. Uda, T., *Japan's Beach Erosion - Reality and Future Measures,* 2nd ed., World Scientific, p. 530, (2017).

Topographic Changes Around Co May River Mouth Located in Vung Tau, Vietnam

Yasuhito Noshi[a], Takaaki Uda[b] and Akio Kobayashi[a]

[a]*Department of Oceanic Architecture & Engineering,*
College of Science & Technology, Nihon University,
7-24-1 Narashinodai, Funabashi, Chiba 274-8501, Japan
E-mail: noshi.yasuhito@nihon-u.ac.jp
[b]*Head, Shore Protection Research, Public Works Research Center,*
1-6-4 Taito, Taito, Tokyo 110-0016, Japan
E-mail: uda@pwrc.or.jp

Topographic changes around the Co May River mouth located 9 km east of Vung Tau, Vietnam were investigated. An asymmetric shoal develops in the east–west direction around the river mouth, suggesting the predominance of westward longshore sand transport in this area. A large amount of sand was transported into the mangrove forest in the hinterland by storm waves, resulting in the widening of the sand bar, and then this sand bar was eroded by longshore sand transport along the river bank, resulting in the development of a sand spit at the upstream end. Moreover, the exposure of a mud layer composed of cohesive material showed that the right river bank was subjected to sand deposition under storm wave conditions and wave abrasion owing to longshore sand transport.

Keywords: Vietnam; Vung Tau; Co May River; River mouth sand bar; Mangrove; Longshore sand transport.

1. Introduction

In developing countries, beach erosion has been occurring with rapid economic development, similarly to the past situation in Japan, and should the situation be left as it is, it may become a factor causing significant external diseconomies. To prevent this situation from occurring, the understanding of sand transport by waves and resultant beach changes on a coast is important, and the effect of various anthropogenic factors should be evaluated beforehand in order for effective measures to be taken. However, there are many cases in which environmental protection is considered with development taking a higher priority. The authors are interested in shore protection in developing countries experiencing rapid economic growth, and field observations have been carried

Fig. 1. Location of Co May River and study area

out in several countries[1,2,3]. Furthermore, Noshi et al. (2015)[4] have recently reported the beach changes on coasts in Phan Rang City, 270 km east of Ho Chi Minh City, as carried out in cooperation with Ho Chi Minh City University of Vietnam. In their paper, the effectiveness of the combined method of satellite image analysis and site observations is shown, even if sufficient data necessary for the analysis was difficult to obtain, such as in the case of developing countries. In this study, the coasts in Vung Tau 67 km southeast of Ho Chi Minh City were selected as the study site, as the second stage in cooperation with Ho Chi Minh City University of Vietnam, and field observations were carried out on October 25 and 26, 2014. Here, the river mouth changes around the Co May River located 9 km east of Vung Tau, as shown in Fig. 1, were investigated.

2. General conditions of study area

The satellite images taken on February 14, 2010 and December 1, 2012 of the

Fig. 2. Satellite image of Co May River mouth (February 14, 2010).

Fig. 3. Satellite image of Co May River mouth (December 1, 2012).

rectangular area shown in Fig. 1 are shown in Figs. 2 and 3, respectively. At the Co May River mouth, a river mouth sand bar of 1.7 km length extends westward and obstructs the river mouth. The channel, therefore, extends westward. In addition, the width of the opening immediately west of the tip of the left river mouth bar was 525 m in February 14, 2010, but it increased by 69 m to a total width of 594 m by December 1, 2012. A narrow stream separated by the offshore shoal extends southwestward along the right river bank, as shown in Fig. 2. The narrowest width of this stream was 115 m on February 14, 2010, but it decreased to 99 m on December 1, 2012, i.e., a change in a reversed mode was observed between the width of this stream and that of the opening immediately west of the left river mouth bar. This is assumed to be explained by the entire opening width of the river being determined by the discharge of the ebb and flood tidal currents, whereby the increase in the opening width of the main stream then caused the decrease in the width of the secondary stream.

Because a shallow offshore shoal develops around the river mouth, the marginal line of the shoal can be drawn by the broken line, as shown in Figs. 2 and 3. This line has an asymmetric form in the east–west direction. East of the sand spit on the left bank, it smoothly approaches the shoreline and then extends parallel to the shoreline, whereas offshore of the river mouth, it markedly protrudes and then it approaches the west shoreline at a large angle near the west end of the shoal. The development of a sand spit on the left bank and the asymmetric shoal clearly demonstrates the predominance of westward longshore sand transport at this river mouth, and the shoal is considered to be dynamically maintained by continuous longshore sand transport.

In Figs. 2 and 3, study areas I, II and III, where the shoreline change was investigated, are shown. Of the three areas, it was found in area I that a mud layer, which was originally formed in a lagoon, was buried under the sandy beach by storm waves, and again exposed by erosion. In areas II and III, the development and reduction of a cuspate foreland and successive elongation of a sand spit were observed, respectively. In particular, the accretion and erosion of a sand bar in area I were studied by the combined analysis of satellite images and field observation.

Meteorological observation at Vung Tau showed that the wind direction alternates between NE and SW in the study area, because the coast is located in a tropical monsoon region. In February and March during the NE monsoon, the mean wind velocity ranges from 5.2 to 5.7 m/s, but the wind velocity is weak as low as 3 m/s in August during the SW monsoon. In this period, however, wind as strong as 30 m/s may occur during the passage of a tropical depression. The mean significant wave heights are 0.8 m during the NE monsoon and 0.7 m during the SW monsoon. As for the tide level, in reference to the observational data of tide level in 2012 at Vung Tau, the monthly mean tidel level was -70 cm above the datum level (DL) and the high water level (HWL) was at +116 cm and low water level (LWL) was at -256 cm.

3. Results of investigation using satellite images and site observation

3.1. *Re-exposure of mud layer once buried under sandy beach*

Figure 4 shows the five satellite images of study area I taken between January 30, 2008 and March 15, 2014, together with the location of Sts.1-10, where the photographs were taken on October 26, 2014. Moreover, the locations of A, B

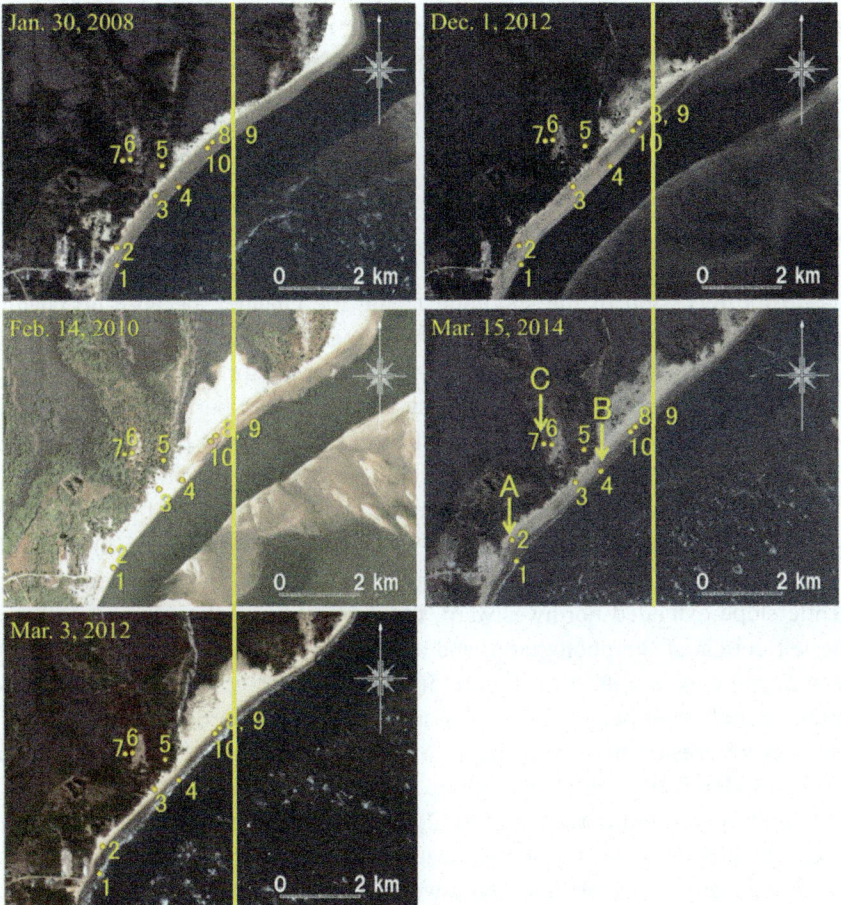

Fig. 4. Deformation of sand bar in study area I.

and C, where characteristic features can be observed as described later, are also shown in the image taken on March 15, 2014, and a straight line was drawn through a point 2.94 km east of St. 1 in each image for facilitating the comparison of the changes in the sand bar width. When referring to this reference line, the dry beach width along this line was 330 m in January 30, 2008, but it markedly increased to 1,720 m by February 14, 2010. This change in sand bar width is assumed to be caused by the landward transport of a large amount of sand under storm wave conditions. If sand was transported deep into the hinterland from the previous shoreline, erosion near the shoreline and sand deposition in the hinterland should occur simultaneously. In fact, however, a marked deposition of sand in the hinterland occurred without beach changes near the shoreline. It is difficult to assume that a large amount of sand was transported from the offshore shoal to the right bank by shoreward sand transport, because the offshore shoal and the right river bank are separated by a deep channel. Even if sand were transported shoreward across the offshore shoal, such sand should be deposited on the seaward steep slope of the channel. On the other hand, because sand was also deposited in an area south of the marked sand deposition zone, sand was assumed to be mainly transported from the southwest side along the channel.

By March 3, 2012, the dry beach width was narrowed to 1440 m. This is due to the recession of the shoreline facing the channel, because no changes were observed at the landward end of the expanded sand bar. By December 1, 2012, coastal vegetation started to grow on the newly formed sand bar, whereas a previous mud layer covered by cohesive material and buried under the sandy beach was exposed in the area north of St. 4, as shown in the image taken on December 1, 2012. Almost the same condition continued until March 15, 2014.

Figure 5 shows the photographs taken at the location of St. 1 - St. 10 in Fig. 4, which corresponds to (a) - (j) in Fig. 5, respectively. First, Fig. 5(a) shows the coastal condition at St. 1, looking northwest. Although a sandy beach with a gentle slope extended northwestward, the debris of buildings were found near the left corner of the photograph, and a scarp formed along the sand dune was seen at point A in Fig. 5(a). Figure 5(b) is the close-up view of the scarp of approximately 4 m height. Thus, this area was severely eroded in the past, and such severe erosion may correspond to the erosion occurred between February 2010 and March 2012, as shown in Fig. 4.

At St. 3, 1660 m northeast of St. 2, an unconsolidated mud layer composed of cohesive material was observed near the present shoreline, as shown in Fig. 5(c), and many roots of dead trees were left in the mud layer. The detailed condition in the vicinity of point B is shown in Fig. 5(d). The existence of a unconsolidated mud layer and roots of many dead trees implies that this area was originally formed as a tidal flat with vegetation such as mangroves grown in the intertidal zone, and this mud flat was exposed owing to erosion. Because a mud flat composed of cohesive material is difficult to form on beaches directly exposed to ocean waves, this tidal mud flat was assumed to be formed in a lagoon surrounded by a sandy beach in the past, and then the sandy beach surrounding the lagoon was eroded, resulting in the exposure of the mud layer composed of cohesive material. The same phonomena were observed in Banzu tidal flat developing around the Obitsu River mouth in Tokyo Bay, although Phragmites australis dominated instead of mangroves in the study area (Mikami

Fig. 5. Beach conditions along right bank of Co May River (a to f).
(a)-(f) correspond to the photographs taken at Sts. 1-6 shown in Fig. 4. Arrows A and B show the locations where scarp was formed along sand dune, and many roots of dead trees were found in mud layer, respectively. C is the location where mangrove forest grows.

Fig. 5. Beach conditions along right bank of Co May River (g to j).

et al., 2016)[5].

When crossing the sandy beach landward of the area with the exposure of the mud layer and a small sand dune, there was a flat field covered with grass (Fig. 5(e)). Although few trees grew in this field, trees were observed beyond the grass field, as denoted by arrow C, as shown in Fig. 5(e). At St. 6, another inland sand dune can be seen, and beyond this sand dune, a mangrove forest has developed (Fig. 5(f)). Figure 5(g) shows the condition of the tidal flat behind the sand dune, and it was confirmed that the mangrove forests indeed grew on this tidal flat composed of cohesive material behind the sand dune. It was confirmed from this finding that the mud layer composed of cohesive material, as shown in Fig. 5(d), really connected to this mangrove forest in the hinterland.

After returning to the shoreline, the roots of the trees remaining in the mud layer were investigated, as shown in Fig. 5(h). Because the attachment of corn barnacles to dead trees terminated at 1 m height above the ground, where the roots of the dead trees were exposed, the water depth above the mud layer was assumed to be approximately 1 m. Figure 5(i) shows the location of the beach with the same elevation as the high tide level, in which the foot of the measuring stick has the same elevation as that 1 m above the ground, as shown in Fig. 5(g). Accordingly, the thickness of the sand layer at the location of the measuring stick was approximately 1 m, and a mud layer was assumed to be buried under this sandy beach. Furthermore, as one of the lines of evidence that a mud layer composed of cohesive material remained under the sandy beach, many crab balls composed of cohesive material, which is considered to be deposited underneath the sandy beach, were found around the nest hole of a Stimpson's ghost crab (*Ocypode stimpsoni*) living in a hole on the sandy beach, as shown in Fig. 5(j), implying that the Stimpson's ghost crab dug a hole through the sand layer. It is concluded that the black band along the shoreline, as shown in Fig. 5(d), appeared following the erosion of the sandy beach covering the mud layer.

3.2. *Development and disappearance of cuspate foreland in the middle of the sand bar*

The shoreline of the sand bar on the right bank protrudes at its middle, while forming a cuspate foreland, as shown in Fig. 2, and the shoreline orientation of the right river bank changes counterclockwise by 20° at the cuspate foreland. This change in shoreline orientation is assumed to be due to the fact that the shoreline north of the protrusion was affected by the waves incident from the opening of the main stream of the Co May River, as shown in Fig. 2, resulting in a local increase in northward longshore sand transport north of this cuspate foreland.

Figure 6 shows the satellite images around this cuspate foreland between January 2008 and March 2014. Because the boundary between the white dry beach and the wet beach can be clearly identified in Fig. 6, this line is defined as the shoreline, and the shoreline in the previous satellite image was also shown in each image. Here, the shoreline change due to the variance of the tide level was neglected because of the absense of the beach slope.

On January 30, 2008, a symmetrical sand bar of 260 m length in the south–north direction and a maximum width of 60 m in the east–west direction developed. Offshore of the cuspate foreland, a longshore bar extended in the ENE–SW direction beyond the channel of the 160-m-wide south stream, and an exposed sand bar above the sea surface was observed at point E. In the following, the locations of points D, E and F at the south end of the vegetation zone on January 30, 2008 are shown.

By February 14, 2010, the sand bar markedly developed northward over point D and the beach width increased by 86 m. Simultaneously, the shoreline receded by 30 m over a distance of 490 m southwest of point D, and the boundary of the vegetation zone receded near point F by 34 m. On the other hand, during the same period, the landward boundary of the offshore sand bar moved by 20 m to the right bank at the middle of points D and F. This suggests that although shoreward sand movement occurred on the offshore sand bar, such

568

sand was difficult to transport to the shoreline on the right bank across the channel, and the change in the sand bar on the right bank was mainly triggered by the updriftward sand transport along the channel. On the surface of the offshore sand bar, many slender sand bars were observed. The elongation of such slender sand bars in the extremely shallow sea denotes that waves were incident from SSE, and these waves were obliquely incident to the shoreline on the right bank, causing northeastward longshore sand transport, suggesting that the deposition and erosion of point D were triggered by this northeast longshore sand transport under waves incident from the southeast. By March 3, 2012, this sand bar was significantly eroded, and point D was left in the river. The beach was further eroded near the shoreline protrusion, resulting in the marked

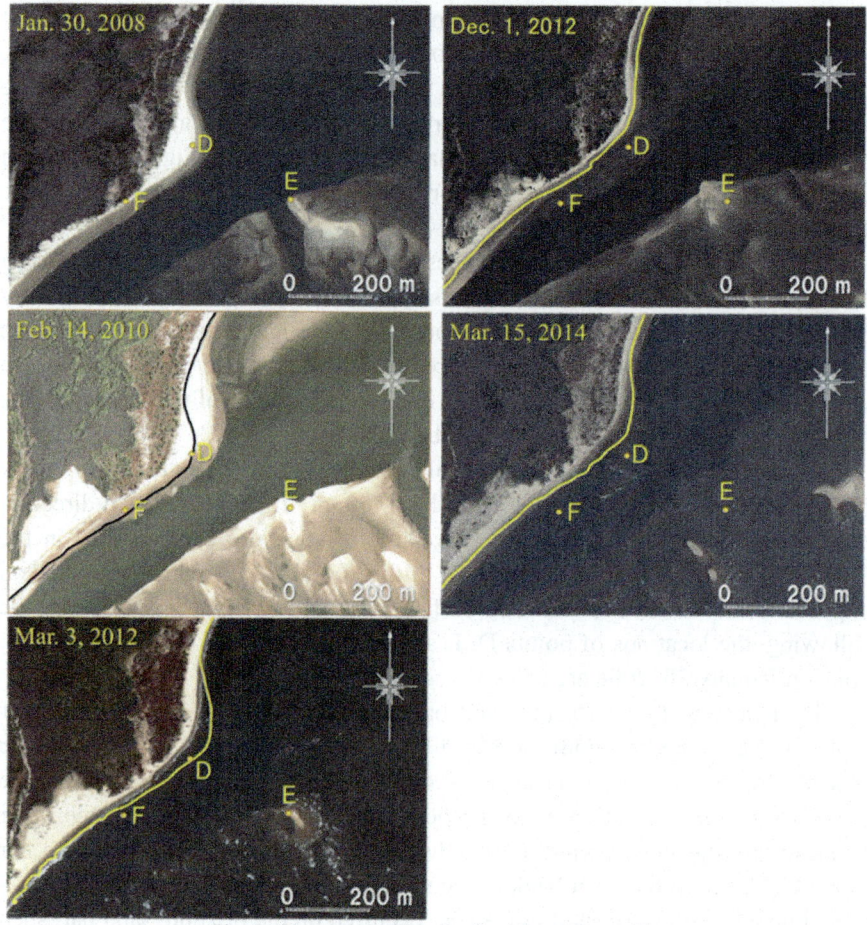

Fig. 6. Change in cuspate foreland in study area II.

decrease in beach width at point D. The erosion further continued up to March 15, 2014 with a large shoreline recession.

As another point, although an isolated, exposed sand bar was formed between 2008 and 2012 at point E on the offshore shoal, this sand bar disappeared in 2014 owing to the effect of a flood that might have occurred between 2012 and 2014. In study area I, a large amount of sand was transported landward between January 30, 2008 and February 14, 2010, and the sand bar markedly developed in study area II during the same period, indicating that waves incident from the southwest were predominant in this period.

3.3. *Development of sand spit at northeast end of right river bank*

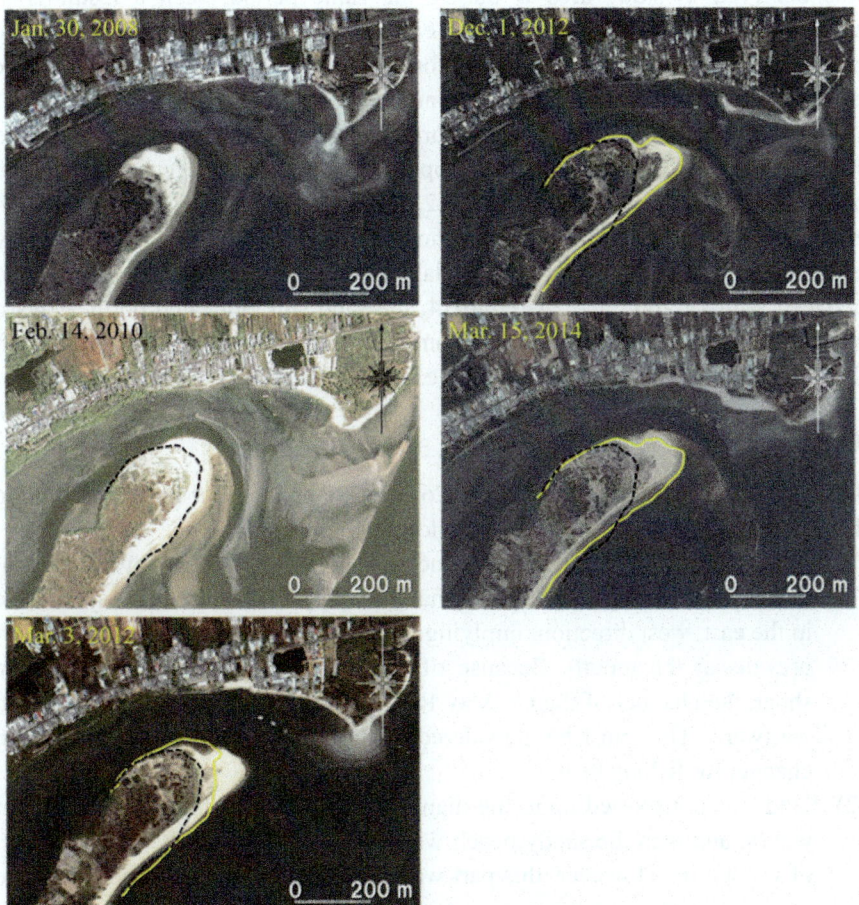

Fig. 7. Extension of sand spit in study area III.

At the northeast end of the right river bank, a tributary that extends in the northeast direction joins with the main stream, and a sand spit is formed at the corner. Figure 7 shows satellite images of the area including the sand spit on five occasions between 2008 and 2014, as well as the initial shoreline on January 30, 2008 and the shoreline in the previous satellite image. On January 30, 2008, the sand spit had a semicircular shape of 210 m diameter, and a deep channel surrounded the shoreline of the sand spit from the tributary to the main stream. Moreover, a cuspate foreland extended toward the offshore shoal on the opposite shore. By February 14, 2010, the sand bar extended northeastward by 59 m, and simultaneously, the tip of the cuspate foreland on the opposite shore disappeared. Although the development of northward longshore sand transport was observed in study area II during the same period, such a result is in accordance with the northeastward elongation of the sand spit and the disappearance of the tip of the cuspate foreland in study area III as a result of the action of waves propagating northeastward along the stream.

By March 3, 2012, the sand spit further elongated northeastward, and sand started to be deposited again on the opposite shore where a cuspate foreland once existed. Up to December 1, 2012, a slender sand spit of 210 m length elongated westward on the opposite shore. Furthermore, the sand spit slightly extended northwestward, creating a lagoon. Finally, in the six-year period between January 2008 and March 2014, the sand spit elongated northeastward by 140 m owing to the sand supply from the south and the cuspate foreland on the opposite shore disappeared over time.

4. Conclusions

The topographic changes around the Co May River mouth 9 km east of Vung Tau, Vietnam can be summarized as follows.
(1) West of the left river mouth sand bar of the Co May River, a shoal develops. The shape of the marginal boundary of this shoal is asymmetric in the east–west direction, implying that westward longshore sand transport prevails at the mouth. Because of the southwestward elongation of the shoal, the channel of the Co May River is considered to be forced to bend westward. This must be considered in the maintenance of the navigation channel for fishing boats.
(2) Sand was transported up to the mangrove forest in the hinterland by storm waves, and then the sandy beach was stabilized temporarily by the growth of vegetation. However, this part was eroded again owing to the longshore sand transport toward the upcoast, resulting in the formation of a sand spit

at the upstream end in study area III. Large shoreline changes were observed along the right bank of this river and their changes mainly depended on the occurrence of storm waves incident from the southwest.

(3) An unconsolidated mud layer composed of cohesive material appeared near the present shoreline in study area I, and dead trees were left on this mud layer. This demonstrates that although sand was transported deep into the hinterland owing to high waves, the beach was gradually eroded.

(4) The southwestward development of a shoal may cause the westward movement of the channel, and the right bank of the channel could be eroded by the currents during floods.

Acknowledgements

This study was carried out in cooperation with Ho Chi Minh City University for Natural Resources and Environment (HCMUNRE) of Vietnam. The authors would like to thank Associate Professor Bay Nguyen of the Department of Fluid Mechanics, HCMUT, for arranging the workshop meeting and the field trips to the coasts of Phan Rang City.

References

1. Onaka, S., Endo, S., Uda, T., 2013. Bali beach conservation project and issues in beach maintenance after completion of project, Asian and Pacific Coasts 2013, Proc. 7th International Conf., pp. 198–203.

2. San-nami, T., Uda, T., Onaka, S., 2013. Long-term shoreline recession on eastern Bali coast caused by riverbed mining, Asian and Pacific Coasts 2013, Proc. 7th International Conf., pp. 275–282.

3. Uda, T., Onaka, S., Serizawa, M., 2015. Beach erosion downcoast of Pengambengan fishing port in western part of Bali Island, Proc. 8th International Conf. on Asian and Pacific Coasts (APAC 2015), Procedia Engineering, Vol. 116, pp. 494-501.

4. Noshi, Y., Uda, T., Kobayashi, A., Miyahara, S., Serizawa, M., 2015. Beach changes observed in Phan Rang City in Southeast Vietnam, Proc. 8th Int. Conf. on Asian and Pacific Coasts (APAC 2015), Procedia Engineering, Vol. 116, pp. 163-170.

5. Mikami, Y., Kobayashi, A., Uda, T., Noshi, Y., 2016. Long-term shoreline changes on marginal coast of tidal flat in Tokyo Bay and rapid deformation of sand bars owing to tsunami, Proc. 35th, ICCE, management.1, pp. 1-13.

Model for Predicting Formation of Blowout on Coastal Sand Dune Using Cellular Automaton Method

Takuya Yokota[a], Akio Kobayashi[a], Takaaki Uda[b], Masumi Serizawa[c], Atsunari Katsuki[d] and Yasuhito Noshi[a]

[a]*Department of Oceanic Architecture & Engineering, College of Science & Technology, Nihon University, 7-24-1 Narashinodai, Funabashi, Chiba 274-8501, Japan
E-mail: csta17030@g.nihon-u.ac.jp*

[b]*Head, Shore Protection Research, Public Works Research Center, 1-6-4 Taito, Taito, Tokyo 110-0016, Japan*

[c]*Coastal Engineering Laboratory, Co., Ltd., 1-22-301 Wakaba, Shinjuku, Tokyo 160-0011, Japan*

[d]*Nihon University, 7-24-1 Narashinodai, Funabashi, Chiba 274-8501, Japan*

On a sand dune, a blowout is often formed owing to wind effect. The formation of a blowout was observed on November 25, 2016 at the Node coast facing the Pacific Ocean, where a blowout has been formed to leave a concave topography. Then, a model for predicting the formation of a blowout was developed using a cellular automaton method, in which two important factors of saltation and avalanche were taken into account. The results of the numerical simulation were compared with the measured results on the Node coast, and the predicted and measured shapes of the blowout were in good agreement.

Keywords: Sand dune; Blowout; Cellular automaton method; Node coast; Ishikari-hama beach; Field observation.

1. Introduction

On a coast composed of fine sand, windblown sand could be deposited to form a coastal sand dune along the shoreline. Such a sand dune is not only an effective barrier against the inundation of sea water into the land in occasions of storm waves or tsunamis but also very important area in the protection of coastal environment, because the seaward marginal line of the sand dune covered with vegetation is often selected as a hatching area for the loggerhead turtle *Caretta caretta*. In such a coastal dune, part of sand dune may be eroded by the effect of wind as a natural agent, resulting in the formation of a blowout. In other case, when a beach access across the sand dune from the land to the shoreline is produced, the same phenomena that occur at the blowout naturally formed by the action of wind can be observed. Once a blowout is formed, the function of the sand dune as a coastal dike will be lost, because the elevation of the sand dune locally decreases around the blowout. It is important, therefore, to investigate the formative process of a blowout and to develop a model for predicting the formation of a blowout on the sand dune from the engineering point of view. However, the studies on the formative mechanism of a blowout and the predictive model are rare, although there are many studies on windblown sand itself in the previous studies. In this study, the formation of a blowout on the coastal sand dune was investigated by field observations, and a model for predicting the formation of a blowout was developed using a cellular automaton method on the basis of field observations at the Node coast facing the Pacific Ocean and Ishikari-hama Beach in Hokkaido.

2. Field observation of blowout on sand dune at Node coast

2.1. General conditions of study area

The formation of a blowout was investigated on a sand dune at the Node coast facing the Pacific Ocean in Chiba Prefecture, as shown in Fig. 1. On this coast, artificial headland No. 9 has been constructed together with headland No. 10 at a location 1,200 m south of headland No. 9 as a measure against beach erosion, and a sand dune area extends alongshore between the shoreline and the coastal residential area. In area A of the rectangular area in Fig. 1, a blowout has been formed, and the sand dune is truncated by this blowout, as shown in Fig. 2. Sand was transported inland across the sand dune, and was deposited at the landward end of this blowout while forming a steep slope with the angle of repose of sand,

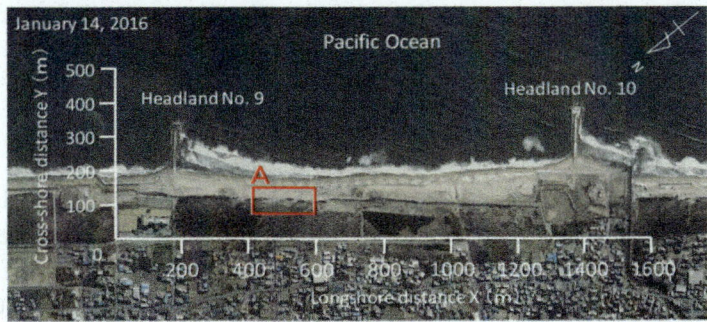

Fig. 1. Aerial photograph of Node coast and study area of blowout on sand dune.

Fig. 2. Blowout formation on sand dune on Node coast.

Fig. 3. Steep slope of angle of repose of sand at landward end of blowout.

Fig. 4. Sand dune vegetation *Carex kobomugi* covering seaward slope of sand dune.

Fig. 5. Vegetation densely covering landward slope of sand dune.

as shown in Fig. 3. On the other hand, coastal vegetation such as *Carex kobomugi* covers the seaward slope of the sand dune, including the top of the sand dune, as shown in Fig. 4. This coverage of the sand dune by coastal vegetation is effective in reducing the windblown sand. Similarly, coastal vegetation densely covers the landward slope of the sand dune except the blowouts, because there is few impact owing to salinity and windblown sand, as shown in Fig. 5, and it contrasts well with the exposed sand surface in the blowout area. In this study, topographic survey around this blowout was carried out on November 25, 2016, and the changes in the seaward marginal line of the sand dune and the shoreline of this coast were investigated using aerial photographs taken between 2012 and 2016. Furthermore, wind rose at the Yokoshiba-hikari observatory of Meteorological Agency was referred to investigate the occurrence of wind in this area.

2.2. Results of field observation

Figure 6(a) shows the changes in the seaward marginal line of the sand dune and the shoreline on the Node coast, and Fig. 6(b) is an enlarged figure of the rectangular area in Fig. 6(a). It is found from Figs. 6(a) and 6(b) that several blowouts have gradually developed on the sand dune over time. In particular, the development of a blowout is significant between $X = 300$ and 700 m, and the

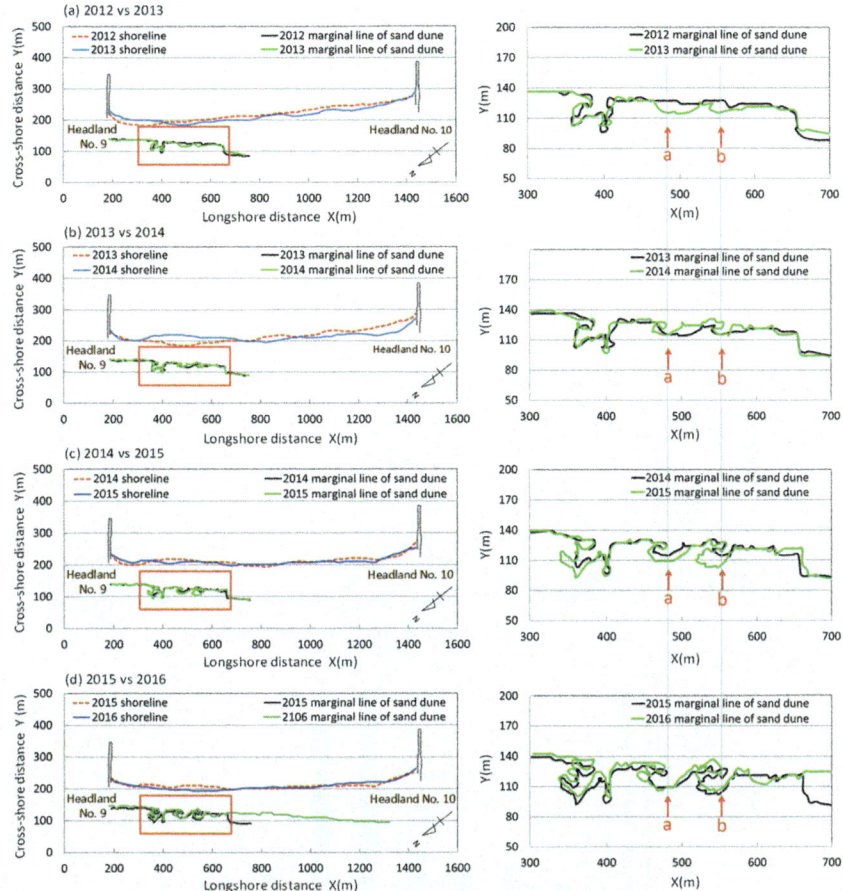

Fig. 6. Changes in shoreline position and marginal line of sand dune in study area.

seaward marginal line of the sand dune have become irregular, while the shoreline maintained almost the same position with variation over time. Figure 7 also shows the overall changes in the shoreline and the seaward marginal line of the sand dune in the entire period between 2012 and 2016. The irregularity of the seaward marginal line of sand dune increased with time because of the formation of a blowout, whereas a swing motion in the shoreline position occurred.

To investigate the development of a blowout on the sand dune, the data set of wave rose measured at the Yokoshiba-hikari observatory since 2012 were examined. Since the direction normal to the shoreline in the study area is approximately equal to the SE (N135°E), as shown in Fig. 1, and the shoreline

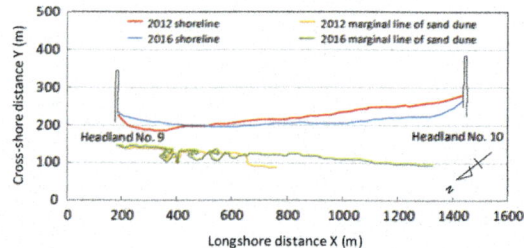

Fig. 7. Changes in shoreline position and marginal line of sand dune between 2012 and 2016.

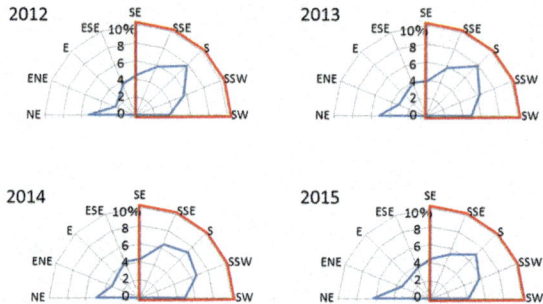

Fig. 8. Probability distribution in wind direction.

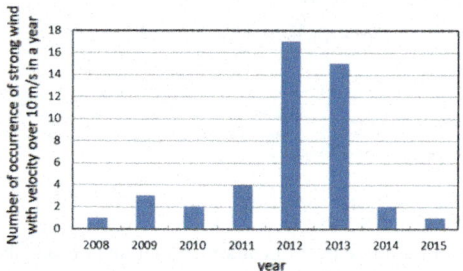

Fig. 9. Number of occurrence of strong wind with wind velocity over 10 m/s.

runs in the direction of the NE-SW, the northerly wind does not contribute to the development of sand dune via landward transport of windblown sand from the shoreline. Therefore, wind rose between 2012 and 2015 was drawn as in Fig. 8, except the northerly wind out of the measured data. In this area, the prevailing wind blows from the S in all seasons with an oblique incidence angle of 45° relative to the direction normal to the mean shoreline.

The probability of occurrence of strong wind with a velocity over 10 m/s from the S is shown in Fig. 9. The number of occurrence of strong wind with a velocity over 10 m/s markedly increased in 2012 compared with that between

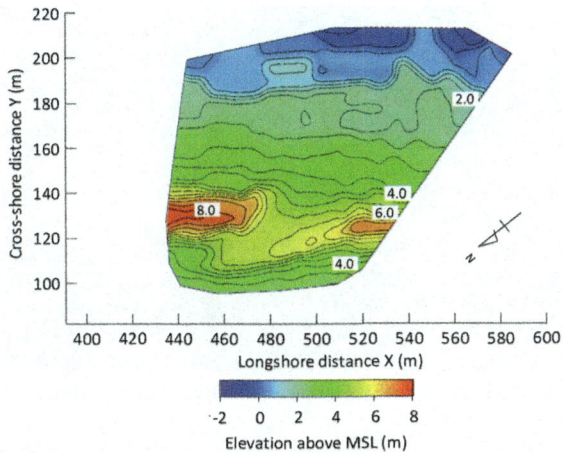

Fig. 10. Topography around blowout measured on sand dune on Node coast.

2008 and 2011, and this corresponds well to the fact that the blowout significantly developed since 2012.

Thus, the rapid development of a blowout, as shown by arrows **a** and **b** in Fig. 6, is considered to be due to the marked increase in the probability of occurrence of strong wind since 2012. Figure 10 shows the measured topography of the blowout formed at the location shown by arrow **a** in Fig. 6 on the sand dune. The concave contours were formed on the seaward slope of the blowout, whereas a steep slope was formed at the landward end of the blowout owing to windblown sand. Furthermore, a mound was formed on the south slope of the blowout, whereas the undercut steep slope was formed on the north slope.

3. Field observation of sand dune on Ishikari-hama Beach

On the Node coast, a blowout has been formed naturally on the sand dune by wind, whereas on Ishikari-hama Beach in Ishikari City in Hokkaido, topographic changes observed around a natural blowout on sand dune can be seen in a beach access connecting the shoreline with the land. This situation was investigated on November 3, 2016. Figure 11 shows the satellite image of the bathing area of Ishikari-hama Beach. This bathing area has a longshore length of 550 m and 40 m width, and a sand dune of 5 m height and 50 m width extends alongshore behind the beach, and a parking lot is located inland of the sand dune. The beach access from the parking lot to the shoreline extends at seven locations at 90 m intervals alongshore. At St. 1 located in the central beach access a large amount of windblown sand was transported from the beach to the parking lot, as shown

Fig. 11. Satellite image of Ishikari-hama Beach facing Ishikari Bay.

Fig. 12. Sand deposition at landward entrance to coast (November 3, 2016).

Fig. 13. Sand deposition around exit to parking lot (November 3, 2016).

in Fig. 12. In this case, the ground elevation on the right (north) side of the beach access was lower than that on the left (south) side, and the roots of the coastal vegetation were exposed as well, whereas a small mound was formed on the left side because of deposition of windblown sand. In contrast, a large amount of sand was deposited on the right (north) side of the entrance of the beach access to the parking lot at St. 2, as shown in Fig. 13. From these facts, it is concluded that the slope on the right (north) side of the beach access was severely eroded by wind as the undercut slope which is subject to direct action by wind blowing from the direction counterclockwise with respect to the direction of the beach access, whereas on the left (south) slope of the beach access sand was deposited because of weaker wind velocity. Furthermore, at the entrance of the beach access, a large amount of sand was assumed to deposit on the right (north) side. These observation results are in good agreement with that observed at the blowout on the Node coast, as shown in Fig. 10.

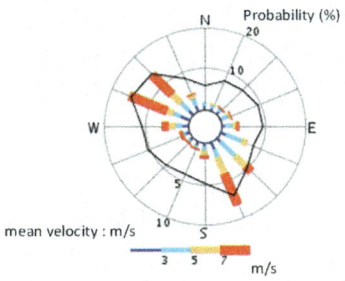

Fig. 14. Wind rose at Ishikari-hama Beach (NEDO).

The direction normal to the mean shoreline in this area is N54°W, as shown in Fig. 11, and the beach access extends parallel to this direction. Asymmetric topographic changes along the beach access were assumed to occur, because the predominant wind blows obliquely with respect to the direction of the beach access. Figure 14 shows the wind rose in this area, referring to the local wind rose map produced by NEDO[1]. The predominant wind direction in this area is WNW (N67.5°W), and NW (N45°W) follows this. Most predominant direction of strong wind with the velocity over 7 m/s is the WNW. Since the direction of the beach access is perpendicular to the shoreline, i.e., N54°W, wind from the WNW makes at 13.5° counterclockwise relative to the direction of the beach access. Because of the predominance of the WNW wind in this area, the marked effect by wind from the WNW is considered to be left behind in the beach access, and thus the topographic features around the beach access as mentioned above can be explained. The relationship between the direction of a beach access crossing the sand dune and the predominant wind direction is a key factor to determine the impact of a blowout to the sand dune.

4. Model for predicting blowout formation using cellular automaton method

4.1. Predictive model

A model for predicting the formation of a blowout on the sand dune was developed referring a numerical model for predicting the formation of a sand dune by Katsuki and Kikuchi (2006)[2] on the basis of the field observation on the Node coast. First, the two-dimensional meshes were taken on the Cartesian coordinates (x, y), and the elevation at mesh point is set $h(x, y, t)$. Assume that the mesh size is sufficiently large compared with the size of the sand particle.

In this study, two most important processes of the saltation and avalanche were taken into account in the formation of a blowout on the sand dune together with the reduction effect of wind-blown sand owing to the coverage of the sand dune by vegetation. The saltation is a process that sand particle is transported by the action of wind, and the saltation distance L_s was assumed to be defined by Eq. (1).

$$L_S = a + bh(x, y, t) - ch^2(x, y, t) \qquad (1)$$

Here, we set a = 1.0, b = 2.0, and c = 0.01 in this study. Eq. (1) shows that the higher the elevation where sand is deposited, the longer distance the sand is transported, but the saltation distance has a limit, as illustrated in Fig. 15. Eq. (1) is the simplest polynominal expression which can evaluate the observed results of the sand flux on a sand dune including multiphase flow (Andreotti et al., 2002)[3], and the sand flux after the maximum value is regarded as a constant, and the value of Eq. (1) was evaluated only within the domain of increasing function, because Eq. (1) has a quadratic form in which a maximum value appears at an elevation, and L_s decreases in the elevation higher than the elevation where a maximum L_s occurs. Furthermore, taking into account of the observation facts that saltation does not occur because a vortex is formed owing to the separation of the flow downwind slope of the sand dune (Pye and Tsoar, 1990)[4], saltation is assumed to only occur on the upwind slope of the sand dune. Originally, the sand flux is given by the product of the moving mass and the saltation distance, and therefore the sand flux can be expressed by Eq. (1) when the wind velocity is a constant, assuming that the moving mass is a constant. When the wind velocity changes, the coefficient of Eq. (1) can be changed in response to the wind velocity (Katsuki et al., 2011)[5]. As a result, the relationship between the wind velocity and the formation of a sand dune can be well explained by Eq. (1). The moving mass q was set to 0.01. In the blowout, since sand movement is considered to become active with the increase in wind velocity, the moving mass should be locally increased. However, because there

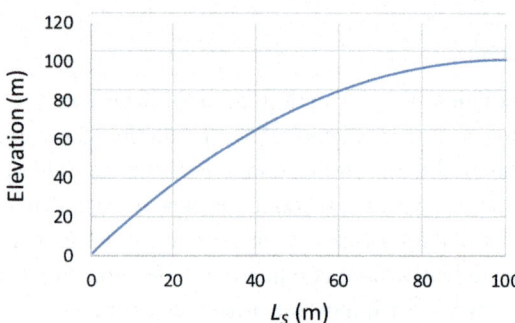

Fig. 15. Relationship between saltation distance L_s and the elevation.

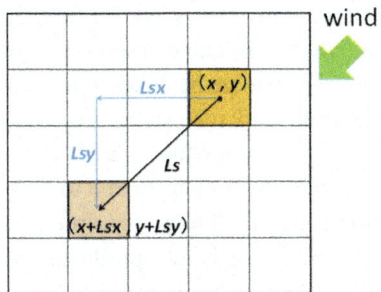

Fig. 16. Schematic diagram of saltation process.

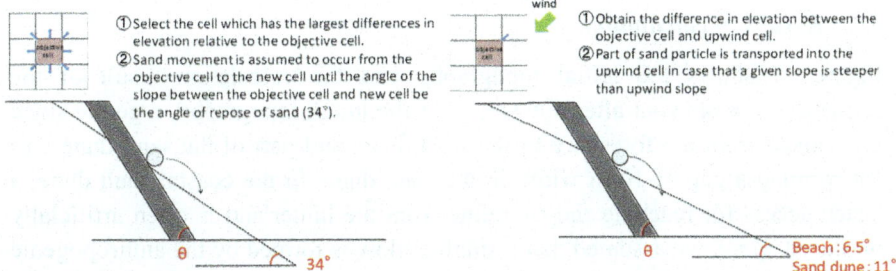

Fig. 17. Schematic diagram of avalanche process.

Fig. 18. Schematic diagram showing formation of foreshore slope.

is no general expression to evaluate the increase in the wind velocity in the blowout, the moving mass in the blowout was assumed to be $10q$ in this study. In the cell model, the space has a nondimensional form, and the elevation h can be normalyzed using the mean grain size of sand.

Consider that the sand particle moves from a point of (x, y) to another point of $(x+L_{sx}, y+L_{sy})$ at each time step, as shown in Fig. 16. Another process of sand movement of sand particle is the avalanche, which is a process that sand particle moves down the most steep slope when the slope is larger than that of the slope of angle of repose until the slope becomes smaller than the angle of repose slope (Fig. 17). In this study, the slope of angle of repose of sand particle was assumed to be 34°. Regarding the sand movement at a mesh point on the front surface of the sand dune, the slope is calculated from the difference in the elevation between at a mesh point and another mesh point immediately upwind of the point, and sand particle is moved to the upwind direction, so that the slope is smaller than the assumed slope (Fig. 18). Here, the front slope of the sand dune was assumed to be 11° and the backshore slope was 6.5° on the basis of the measured slope on the Node coast.

582

In the numerical simulation, the sand particle is first moved by the saltation, and the sand particle is moved corresponding to the slope in front, and finally sand movement by the avalanche takes place. Sand movement by the difference in the slope in front and real slope, and that by the avalanche are recurrently carried out until the stable topography is obtained. After a stable topography is obtained, the sand movement by the saltation is assumed to occur. In the calculation, saltation is assumed to be possible in the direction of wind at the small gap with low elevation artificially produced on a part of the slope of the sand dune, and saltation is incapable because of the coverage by vegetation in the other area including the sand dune and the hinterland.

4.2. Results of calculation

Figure 19 shows the initial topography and the calculation result of the formation of a blowout after 600 steps. As the initial topography, a gentle slope is assumed from the foreshore to the sand dune, and part of the sand dune was cut forming a gap of 10 m width in the sand dune. In the coastal sand dune, a beach access for reaching the shoreline from the hinterland is often artificially produced. Here, we assumed that a small hollow is formed by the anthropogenic factors, such as walking on the sand dune, resulting that coastal vegetation is withered by walking on the sand dune. The results of the calculation is shown in Fig. 19(b). Because wind from the south blows to the hinterland through a narrow gap of the sand dune, a blowout was formed. Furthermore, the north slope of the hollow which is the undecut slope against south wind was severely eroded, whereas sand was deposited on the south slope. In addition, it is clear that sand transported inland by the windblown sand was deposited on the

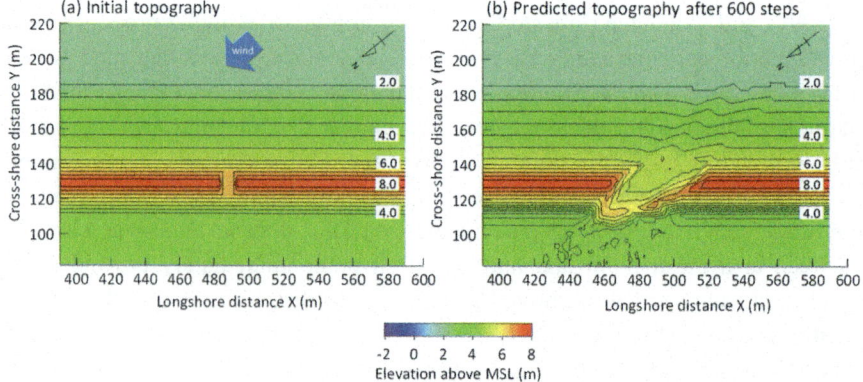

Fig. 19. Initial topography and predicted topography after 600 steps.

backslope of the sand dune to form the slope of angle of repose of sand. These characteristics well explain the results shown in Fig. 2 and the observation results at Ishikari-hama Beach.

5. Conclusions

The morphological features of a blowout formed on the sand dune on the Node coast were investigated in detail, and the topographic changes owing to windblown sand in the beach access extending from the land to the shoreline across the sand dune at the bathing area of Ishikari-hama Beach facing Ishikari Bay were observed. Furthermore, on the Node coast, time changes in the topography of the blowout and in shoreline position were investigated using aerial photographs. Owing to the analysis of the wind records in this area, the development of a blowout on the sand dune has begun since 2012, because the probability of strong wind with a velocity over 10 m/s markedly increased since 2012. This is the cause of the rapid development of a blowout. Finally, a model for predicting the formation of a blowout was developed using a cellular automaton method and numerical simulation was carried out given a small gap on the sand dune. The results of the numerical simulation were in good agreement with those measured on the Node coast.

References

1. NEDO: http://app8.infoc.nedo.go.jp/nedo/
2. Katsuki, A., Kikuchi, M., 2006. Simulation of barchan dynamics with inter-dune sand stream, RIMS Kôkyûroku, Vol. 1472, pp. 67-70. (in Japanese)
3. Andreotti, B., Claudin, P., Douady, S., 2002. Selection of dune shapes and velocities Part 1: Dynamics of sand, wind and barchans, Eur. Phys. J., B 28, pp. 321-339.
4. Pye, K., Tsoar, H., 1990. Aeolian Sand and Sand Dunes, Unwin Hyman, London, pp. 42-43.
5. Katsuki, A., Nishimori, H., Kikuchi, M., Endo, N., Taniguchi, K., 2011. Cellular model for sand dunes with saltation, avalanche and strong erosion: collisional simulation of barchans, Earth Surface Processes and Landforms, Vol. 36, pp. 372-382.

584

Study on Sedimentation for Waterway Regulation of the Daliao River Estuary[*]

Gao Xiang-Yu[†], Jiao Jian, Dou Xi-Ping, Ding Lei

Nanjing Hydraulic Research Institute, State Key Lab of Hydrology-water Resources and Hydraulic Engineering, Nanjing 210029, China
[†]E-mail: xygao@nhri.cn
www.nhri.cn

Aiming at the siltation problem at the waterway in the east branch of the Daliao River Estuary, the two-dimensional tidal current sediment numerical model has been established based on the analysis of hydrodynamic and sediment characteristics. In combination with the experience of channel governing of the Daliao River Estuary, three regulation schemes have been compared by using numerical model. The results show that there is a small influence on hydrodynamic force in the Daliao River after training works and dredging is not sufficient to maintain the channel flow. After building eastern and western jetties, the maximum thickness of sediment siltation in channels is only 0.62 m/a and the regulation effects are relatively good.

Keywords: Branch channel; Sediment; Numerical model.

1. Introduction

Yingkou Old Harbor is Located on concave bank of the Daliao River Estuary in the north of Liaodong Bay with good water depth conditions. The natural water depth at the front of the coastline for 3 km can meet the draft requirement of a ship for 5000 t. The mouth of the Daliao River has a sandbar. The east branch and west branch are separated by the west shoal. The main channel is the east branch (Figure 1). The minimum water depth of channels is only about 1.6 m, the 3000 t cargo ship with deloading for 1/3 can enter and go out with a spring tide. The development scale and economic benefits of Yingkou old Harbor are seriously affected. The waterway regulation of the Daliao River Estuary has

---

[*] This work is supported by the National Natural Science Foundation of China (51479122), "333" Science and Technology Support Project of Jiangsu Province (BRA2015459), Central water resources fee project (126153-0210283), the central public fund of Nanjing Hydraulic Research Institute (Y216001, Y216012, Y216019, Y216020).

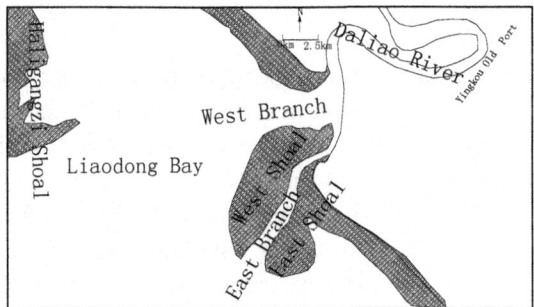

Fig. 1. Pattern of the Daliao River Estuary.

always been focused by researchers. The research of sediment siltation in channel is of great importance to decision-making of waterway regulation.

2. Characteristics of hydrodynamic force and sediment

2.1. Tide

The tide in the Daliao River Estuary is irregular semi-diurnal tide with a large tidal range and remarkable diurnal inequality. According to the tidal level data statistics of Sidaogou station, the historically highest tidal level is 5.06 m, the lowest tidal level is -0.43 m and the extreme tidal range is 4.42 m; the average high water level is 3.25 m and the average low water level is 0.58 m; the flood tide duration is about 5 hours and the ebb tide duration is about 7 hours. The ebb tide duration is longer than that of the flood tide.

2.2. Tidal current

According to hydrologic data actually measured during April to May, 2007, the main directions of tidal current in the open sea of the Daliao River Estuary are northerly and southerly, where the N direction is the direction of flood tides and the S direction is the direction of ebb tides; The direction of tidal current in the Daliao River are basically corresponding with bank line; The tidal current is basically reciprocating (Figure 2). The maximum of tidal current occurs in the Daliao River, the maximum current velocity is about 1.70 m/s. The mean current velocity of each ebb tide is between 0.19 m/s and 0.82 m/s, the mean current velocity of flood tide is between 0.25 m/s and 0.91 m/s, the current velocity of the flood tide is greater than that of the ebb tide; The coefficient to distinguish tidal stencils is $(W_{O1}+W_{K1})/W_{M2} \leq 0.5$, the tidal currents are regular semi-diurnal current.

586

2.3. *Wave*

In open sea out of the Daliao River, there are mainly stormy waves for waves and few surges. The frequent waves in sea areas are in the direction of SW with the frequency to occur of 22.2% and the less frequent waves are in the directions of SSW and WSW with the frequencies to occur of respectively 16.6% and 16.4%, The largest height of the strongest waves in the direction of SSW is 3.2 m with $H_{1/3}$ of 2.22 m; the less strongest waves are respectively in the direction of WSW, W and SW, the largest waves are respectively as high as 2.7 m, 2.6 m and 2.5 m and their $H_{1/3}$ are respectively 1.7 m, 1.8 m and 1.8 m. The average height of waves in each month is between 0.2 and 0.6 m. Among waves throughout the whole year, the frequency of waves with $H_{1/3}$ less than 0.7 m occupies about 66.6%, the frequency of waves with $H_{1/3}$ less than 1.0 m occupies 89.1%, and the frequency of waves with $H_{1/3}$ more than 1.0 m occupies 10.9%. Among waves more than 1.0 m, the frequency of occurrence in the direction of SSW to WSW occupies 8.9% of that during the whole year.

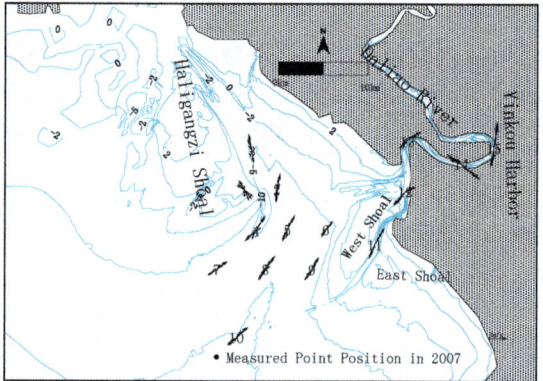

Fig. 2. Tidal current vectors of spring tide in the Daliao River Estuary.

2.4. *Sediment characteristics and sediment concentration distribution*

It is shown from measured data that the median size of suspended load is 0.0060~0.0347 mm. The clay content is between 20% and 29.9% and the average value is 26.6%. The suspended sediment is mainly clay silt. The median size of bed load is 0.0071~0.081 mm. The types of surface deposit sediments are clay silt, silt sand and sand.

The sediment concentration in the Daliao River and East waterway is more than that of the western sea area on the western beach. The average sediment concentration in the Daliao River and East waterway is between 0.165 and 0.418

kg/m³, and the average sediment concentration of the western sea area is 0.020~0.248 kg/m³. The change of sediment concentration has some relations with the change of flow velocity of flood and ebb tides. In general, the sediment concentration of spring tides is larger than that of middle tides and neap tides; the average sediment concentration of spring tides is 0.199 kg/m³, the average sediment concentration of middle tides (windy) is 0.21 kg/m³ and the average sediment concentration of neap tides is 0.098 kg/m³. When it is windy, stormy waves will lift sediments of Gelizigang Shoal and Western Shoal and the sediment concentration increases. After a windy day (wind speed of 14 m/s), the average sediment concentration of the bottom in East Water Channel reaches 1.76 kg/m³.

3. Evolutions of the Daliao River Estuary

The Daliao River Estuary is located on Liaodong bay head. In tectonic, it belongs to the second giant zone of subsidence in Neocathaysian. The tidal current conditions are complex near the Daliao River Estuary while the bar is complex and varied at the entrance. The western shoal at the estuary in the 1960s formed an entire bottomland. The West shoal divides the Daliao River Estuary into two channels of West and East. Until 1990s, the western shoal was washed as several isolated bottomlands while the sedimentation area of the eastern shoal was further increased. The interpreting results of remote sensing pictures show that the eastern shoal at the estuary of the Daliao River was slightly silty during 1958 to 1977 with the average rate of change of 0.2-0.3 m/a, there was no big change in the western shoal. During 1978 to 1991, the change of scour and silting in various sandbanks is small, and except the larger extension on the east shoal, there is a light change of scour and silting on the western shoal. Shoals on both sides are slightly rushed back. During 1991 to 2000, Siltation speed is fast on various shoals and sedimentation mainly develops in the northwest and west direction on the west shoal. For the west water channel in the north of the western shoal, the width over water with the depth more than 0 m is 2044~3563 m in 1980s. Since 1990s, for the development of river island, the currently over-water width has greatly been shortened with the only width of about 770 m. The area of the western shoal with the water depth less than 0 m has grown from 18.4 km² in 1991 to 30.4 km² in 2000 and grown by 65%. The east shoal has extended to the sea with the average velocity of 2.14 m every year (by Wu et al. (1997) and Liu et al. (2005)).

4. Establishment and verification numerical model

4.1. *Governing equations*

With the hydrostatic assumption, 2D unsteady flow equations under the orthogonal curvilinear coordination system are described as:

$$\frac{\partial \zeta}{\partial t} + \frac{1}{g_\xi g_\eta} \frac{\partial}{\partial \xi}\left(Hu g_\eta\right) + \frac{1}{g_\xi g_\eta} \frac{\partial}{\partial \eta}\left(Hv g_\xi\right) = 0 \tag{1}$$

$$\frac{\partial u}{\partial t} + \frac{u}{g_\xi}\frac{\partial u}{\partial \xi} + \frac{v}{g_\eta}\frac{\partial u}{\partial \eta} + \frac{uv}{g_\xi g_\eta}\frac{\partial g_\xi}{\partial \eta} - \frac{v^2}{g_\xi g_\eta}\frac{\partial g_\eta}{\partial \xi} + g\frac{u\sqrt{u^2+v^2}}{C^2 H} - fv + \frac{g}{g_\xi}\frac{\partial \zeta}{\partial \xi} = E\left(\frac{1}{g_\xi}\frac{\partial A}{\partial \xi} - \frac{1}{g_\eta}\frac{\partial B}{\partial \eta}\right) \tag{2}$$

$$\frac{\partial v}{\partial t} + \frac{u}{g_\xi}\frac{\partial v}{\partial \xi} + \frac{v}{g_\eta}\frac{\partial v}{\partial \eta} + \frac{uv}{g_\xi g_\eta}\frac{\partial g_\eta}{\partial \xi} - \frac{u^2}{g_\xi g_\eta}\frac{\partial g_\eta}{\partial \xi} + g\frac{v\sqrt{u^2+v^2}}{C^2 H} + fu + \frac{g}{g_\eta}\frac{\partial \zeta}{\partial \eta} = E\left(\frac{1}{g_\xi}\frac{\partial B}{\partial \xi} + \frac{1}{g_\eta}\frac{\partial A}{\partial \eta}\right) \tag{3}$$

$$A = \left[\frac{\partial\left(ug_\eta\right)}{\partial \xi} + \frac{\partial\left(vg_\xi\right)}{\partial \eta}\right] / g_\xi g_\eta \tag{4}$$

$$B = \left[\frac{\partial\left(vg_\eta\right)}{\partial \xi} + \frac{\partial\left(ug_\xi\right)}{\partial \eta}\right] / g_\xi g_\eta \tag{5}$$

$$g_\xi = \sqrt{x_\xi^2 + y_\xi^2} \tag{6}$$

$$g_\eta = \sqrt{x_\eta^2 + y_\eta^2} \tag{7}$$

Under the orthogonal curvilinear coordinate system, the basic equation of 2-D suspended load transport is described as:

$$\frac{\partial(HS)}{\partial t} + \frac{1}{g_\xi g_\eta}\left[\frac{\partial}{\partial \xi}(HuSg_\eta) + \frac{\partial}{\partial \eta}(HvSg_\xi)\right] + a\omega(S - S_*) = 0 \tag{8}$$

Where: g_ξ and g_η are Lami coefficients, u and v are respectively velocity components (m/s) on the directions of ξ and η; ζ is the water level (m) and H is the total depth of water (m); C is Chezy coefficient, $C=H^{1/6}/n$, n is roughness coefficient; f is Coriolis Force Coefficient; E is the flow turbulent viscosity coefficient (m²/s); S is the average sand content of vertical lines (kg/m³); η_s is the erosion and deposition thickness caused by suspended load (m); a is the sedimentation probability and can be determined through verification calculation; ω is the settling velocity of sediment (m/s); S_* is the sediment transport capacity of tidal current and wind waves, Dou et al.'s formula (1995b) is described as:

$$S_* = \alpha_0 \frac{\gamma \gamma_S}{\gamma_S - \gamma} \left[\frac{(u^2 + v^2)^{3/2}}{C^2 H \omega} + \beta_0 \frac{H_w^2}{HT\omega} \right] \qquad (9)$$

Where: γ and γ_S are respectively the volume weight of water and sediment. H_w and T are respectively the mean wave height and the mean wave period. The α_0 and β_0 are undetermined coefficients, models are adopted with $a_0 = 0.023$ and $\beta_0 = 0.00004$.

Models shall be dispersed in the finite difference method, the tidal numerical model shall be calculated in the ADI method, and the sediment model shall be solved in display.

4.2. Calculating condition

Models shall be operated in a cold starting mode. The initial velocity of velocity shall be regarded as zero in calculation. The initial tidal level shall be a constant, and the initial sediment concentration shall be a given constant. The effects of initial conditions will gradually disappear after a period of calculation.

The open boundary in open sea of large models shall be controlled by tidal level and the upstream of Shuangtaizi River and the Daliao River shall be controlled by measured discharge and water level respectively. Boundaries of open seas in small models shall be provided by large models and the upstream boundaries shall be the same with those of the large models. The south boundaries of sediment models shall be valued as 0.01 kg/m³ and the upstream and the western boundary shall be gained by interpolation according to measured data.

4.3. Calculating area and grid

Numerical models shall be calculated with the nested method of large models and small models. The large model has the range in the north of Taizi Mountain in Liaodong Gulf. The south border of the small model is the Taiping Cape. West border is from 15 km away from the Shuangtaizi Estuary. North border is 13 km away from Shuangtaizi Estuary. The upstream boundary of the Daliao River is more than about 140 km away from the estuary. The orthogonal curve gridding shall be adopted for models. The minimum grid scale of large model is 200 m while the minimum grid scale of small model is 30 m. Figure 3 shows the generalized terrain and the area of small model.

590

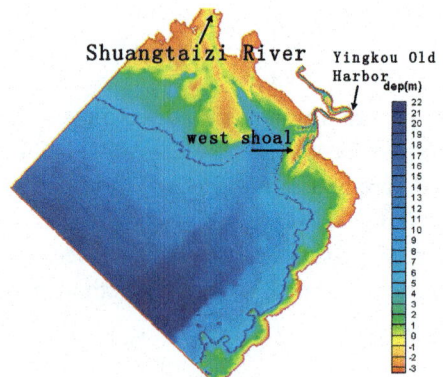

Fig. 3. The generalized terrain of the area of small model.

4.4. *Verification of models*

The data measured during April to May in 2007 is used to verify the model. The data includes tidal levels, velocities, current directions and sediment concentrations. The positions of measuring points are in Figure 2.

Using linear Interpolation method, roughness coefficients from the sea to the up stream of The Daliao River and Shuangtaizihe River are set as $n = 0.012 + 0.01 / H \sim 0.025 + 0.01 / H$. Turbulent viscosity coefficient is 15 m^2/s. The time steps in the model of tidal currents and sediment transport are 1.5 s and 120 s respectively. the unit volume-weight of sediment particles is 2650 kg/m^3. The sediment deposition probability is 0.1.

Figure 4 and Figure 5 are the comparison between calculated and measured tidal levels, velocities, current direction and sediment concentrations respectively. It shows that the established models can well demonstrate hydrodynamic and sediment transport here.

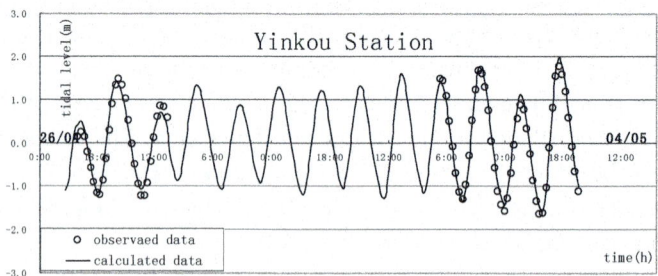

Fig. 4. Verification of tidal level.

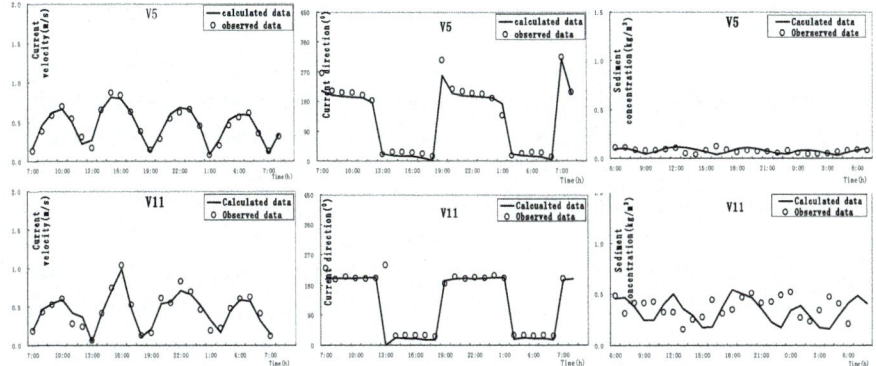

Fig. 5. Verification of current velocity, its direction and sediment concentration (spring tide).

5. Analysis of influence and effects of the regulation project

Based on the analysis of sediment data for many years, the sediment transport of the Daliao River and the nearby rivers has less effect on siltation of external channel. Siltation of external channel is mainly caused by the lifted sediment on the eastern and western shoals under the action of wind waves. To comprehend the effect of the flow field after dredging channel on the riffle of the Daliao River Estuary, three different schemes are studied (Figure 6).

The main regulation project includes dredging and building jetty. The first scheme only dredge channel, the depth of channel is 3.5 m. The second scheme is to expand the east jetty, based on the first scheme. The third is to repair the west jetty on the basis of the second scheme. The length of the east and west jetty is about 14.4 km and 2.8 km respectively. The elevation of the east jetty top is from 0.5 m to 2 m and that of the west jetty top is 2 m.

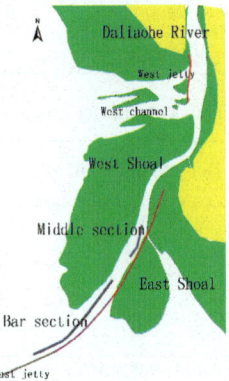

Fig. 6. The map of the regulation project at the Daliaohe Eustuary.

5.1. *Impact of the project on tidal current*

Table 1 and Table 2 are respectively statistical tables for changes in tidal level and current velocity before and after the projects. It is observed from the table: after regulation Scheme I, the low tidal level has lowered 0.35%~0.13% and the high tidal level has grown for 0.11%~0.17% from Panjin Old Harbor District to Yingkou Old Harbor District in the Daliao River. The maximum current velocity and the average current velocity have slightly decreased for 4.40%~8.66%; after regulation Scheme II, the high and low tidal levels have slightly elevated for 1.31%~2.15% from Panjin Old Harbor District to Yingkou Old Harbor District in the Daliao River. The maximum current velocity has increased slightly for 0.96%~1.12% and the average current velocity has decreased slightly for 2.22%~4.20 %; after regulation Scheme III, the high tidal level from Panjin Old Harbor District to Yingkou Old Harbor District in the Daliao River has been lowered for 3.81%~5.25% and the low tidal level has been elevated for 18.23%~22.43%. The maximum current velocity and the average current velocity have decreased 7.18% to 15.07%. After building guiding jetty and dredging channels, the boundaries of estuaries have been adjusted in some degree with slight influence for hydrodynamic force fields in the Daliao River, with small influence for purely dredging channels, and with slightly big influence for building and extending guiding jetty and dredging channels (by Gao (2007)).

Table 1. Tidal level change table before and after project.

Station	Tidal level	Before project (m)	Scheme I	Scheme II	Scheme III
Yingkou Old	High	3.86	0.11%	1.31%	-3.81%
Harbor District	Low	0.56	-0.13%	2.15%	18.23%
Panjin Old	High	3.79	0.17%	1.26%	-5.25%
Harbor District	Low	0.51	-0.35%	1.76%	22.43%

Table 2. Tidal level change table before and after project.

Station	Current velocity	Before project (m/s)	Scheme I	Scheme II	Scheme III
Yingkou Old	Maximum	1.04	-5.75%	0.96%	-7.18%
Harbor District	Average	0.57	-8.66%	-2.22%	-16.11%
Panjin Old	Maximum	0.98	-4.40%	1.12%	-8.43%
Harbor District	Average	0.55	-7.59%	-4.20%	-15.07%

Note: changing value in the table=(value after project-value before project)/value before project.

5.2. *The calculating siltation of the channel*

Siltation thickness of suspended sediment P can be calculated by using the formulas for navigation channels after dredging in estuaries (by Luo (1987)).

$$P = \frac{\alpha \omega_f ST}{\gamma_c} \left[1 - \left(\frac{V_2}{V_1} \right)^2 \left(\frac{H_2}{H_1} \right) \right] (\cos n\theta)^{-1} \tag{10}$$

Where a is the probability of sediment settling, ω_f is the sediment flocculation settling velocity, T is the siltation period, H_1, V_1, H_2, V_2 are the depths and current velocities before and after regulation project respectively, θ is the intersection angle between current direction and the axis of navigation channel, n is the transient coefficient.

The average current velocity, current direction and sediment concentration before and after the regulation project are calculated by the tidal sediment numerical model. Annual siltation thickness of each regulation scheme is calculated by using the formula (6). Table 3 is the statistical table of calculating siltation results. It can be seen from the siltation condition: the purely dredging channel has a large siltation strength and the annual siltation thickness can nearly reach dredging thickness. The maximum value of the siltation thickness is in the middle section and the siltation thickness is 0.83 m/a, which is very difficult to meet water depth of 5,000 t channel. After repairing the eastern guiding jetty, the eastern guiding jetty has some function on guide and retaining sediment. But the siltation thickness of bar section is still large with the maximum siltation thickness of 0.71 m/a. The current velocity of ebb tide is not enough to maintain channels. The scheme of repairing the western guiding jetty and eastern guiding jetty have slightly good effects for reducing siltation with the maximum siltation thickness of 0.67 m/a.

Table 3. Annual siltation thickness table after regulation schemes.

Channel position	Value	Dredging thickness(m)	Siltation thickness (m/a)		
			Scheme I	Scheme II	Scheme III
	Minimum	0.32	0.20	0.05	
Middle section	Maximum	0.91	0.83	0.27	-
	Average	0.63	0.54	0.15	
	Minimum	0.48	0.20	0.07	0.02
Bar section	Maximum	2.00	0.96	0.71	0.62
	Average	1.54	0.68	0.46	0.35

Note: "-" shows no siltation in the table.

6. Conclusion

The water depth condition of the Daliao River channel is well and the riverbed is stable, which meets the requirement of extending 5,000 t waterway. However, since there are sandbars in channels off the estuary, the water depth only maintains for about 1.7 m, which has limited navigation capacity of waterway. The 3,000 t cargo ship needs to take the tide to enter and exit. The three regulation schemes of the east channel are researched by the verified numerical models in the sea area in the Daliao River. The results show that the influence for hydrodynamic force within the Daliao River is small after the regulation project with pure dredging. The channel is difficult to maintain. The siltation in channels is relatively small if the eastern and western guiding jetty are built and extended.

Acknowledgements

This work is supported by the National Natural Science Foundation of China (51479122), "333" Science and Technology Support Project of Jiangsu Province (BRA2015459), Central water resources fee project (126153-0210283), the central public fund of Nanjing Hydraulic Research Institute (Y216001, Y216012, Y216019, Y216020).

References

1. M.Y. Wu, H.B. LIU, S.S. Yang. Treatment Engineering of Liaohe Estuary. [J] China Port Engineering. 7-11, Feb., 1997.
2. Y.P. Liu, J. Zhang. Evolution Characteristics of Water and sediment in Liaohe Estuary. water resources & hydropower of northeast china [J]. Vol. 23, N. 253:38-41. Aug. 2005.
3. Dou Guoren, Dong Fengwu and Dou Xiping. Sediment Transport Capacity of Tidal Currents and Waves [J]. Chinese Science Bulletin, 1995b, 40(13): 1096~1101(in Chinese).
4. Luo Zhaosen. Computations of Siltation in Dredged Channel in Estuaries [J]. Sediment Research. 1987. (2) (in Chinese)
5. X.Y. Gao. Analysis on Sedimentation and Calculation of Numerical Model for 5000T Channel Expansion Project of east Waterway of Yingkou Port [R]. Nanjing Hydraulic Research Institute. Dec. 2007.

Study of Tide Data Processing and Utilization by NAMRIA – Application to Astronomic Tide Synthesis of South Harbor Port Manila

R. E. Dabu

*Institute of Civil Engineering, University of the Philippines-Diliman,
Quezon City, 1011, Philippines
E-mail: redabu1@up.edu.ph
www.up.edu.ph*

Water level determines flooding and considerably effects wave climate. Hence, it is important to consider it in both disaster mitigation and coastal structure design. Unlike other water level components such as storm surge and waves, astronomic tide is periodic and predictable. The oceans and bays are forced-oscillating system, allowing astronomic tide to oscillate with the same frequencies as the tide-producing forces. While astronomy determines the tide constituents' frequencies, it is the basin hydrodynamics that controls their amplitudes and phase lags. Tide time series record is needed in harmonic analysis to predict astronomic tides. The predictability of tide allows the analysis of storm tide level through the extraction of storm surge component in the water level time series. This method to synthesize the storm surge was validated in this study as an application to disaster mitigation. To de-trend the 1969-2015 hourly water level from National Mapping and Resource Information Authority (NAMRIA), sea level rise was examined. Least-square linear solution to the monthly mean values (1901-2015) was utilized. Results showed that the slopes of the trends of sea level are rapidly increasing over time because a varying trends for 1901-1968 (1.76 mm/year) and 1969-2015 (13.6 mm/year) was obtained. The de-trended 1969-2015 hourly water level (through 13.6 mm/year trend) was used to hindcast the astronomic tide. The results are the storm surge values of 47 historical typhoons that tracked within Manila Bay (study area). Finally, to apply the NAMRIA tide data in coastal structure design, tidal datum values were compared to the computed values (through zero-crossing method). Results showed that the selection of wave period on which an individual wave will be defined causes large variation on the compared tidal datum values (NAMRIA-specified versus computed). NAMRIA uses the tidal day (24.84 hours) as the wave period while results in zero crossing method showed that the length of periods of individual waves being analyzed is ranging from 12 to 275 hours.

Keywords: Storm surge; Sea level rise; Manila Bay; NAMRIA; Tidal datums.

1. Introduction

Water level is an important factor in storm surge analysis and wave climate studies. National Mapping and Resources Information Authority (NAMRIA) is the government agency that collects, process and synthesize water level data along its 18 tide stations in the coastal areas of the Philippines.[1]

NAMRIA tide data is composed of tide time series and tidal datums (tide normals and tide extremes). Tide time series is used to analyze a historical storm by separation of the non-tidal component to water level time series. The remaining tidal component is the astronomic tide which can be predicted by harmonic analysis. Long term trends in the tide time series, which corresponds to the sea level rise of the local study area, affects the results of harmonic analysis so a thorough study of the trends should be done prior to predicting the astronomic tide. Tidal datums are composed of tide extremes and tide normals. Tide extremes (Highest and lowest tide) are used to determine design periods of coastal structures while tide normals (MLLW, MLW, MSL, MHW, MHHW) are used as input parameter in wave climate simulations (i.e. through REF/DIF software) to obtain the wave height distribution and wave direction in the study area. Figure 1 shows the framework of this study which is basically the uses of NAMRIA tide data. Although they can be classified through their uses, tide data are interrelated and each contributes to other's applications. Further elaboration is discussed.

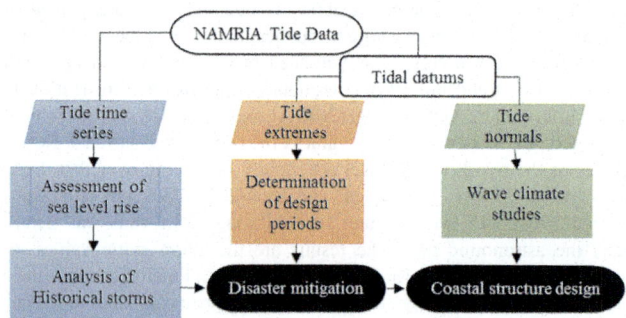

Figure 1: Framework of the study

1.1 Tide time series

According to National Oceanic and Atmospheric Administration (NOAA), storm tide is the combination of astronomic tide and storm surge. Astronomic tide is the long-period wave that oscillates through the ocean due to the gravitational pull of the moon and sun to the earth. Conversely, storm surge is the unusual rise of water level over the predicted tide level, generated mainly by strong winds during typhoons.[2] Storm surge or the non-tidal water level component is meteorological and stochastic but astronomic tide is a cyclic phenomenon. The oceans and bays are forced-oscillating system. Thus tides oscillate with the same frequencies of the astronomic tide-producing forces, which are determined by the relative movement of the sun, moon and earth. The complex motions of the earth's orbit around the sun, and moon's orbit around the earth lead to many different tidal frequencies but astronomers have precisely determined all of the required astronomical frequencies for tide prediction.[3]

Because the tide harmonics (amplitude and phase lags) are controlled by the basin hydrodynamics, tide time series record is needed to predict astronomic tides for a specific location. The length of tide data needed is 19 years which is also the national tidal datum epoch (NTDE).[4] (see section 1.2)

1.2 Tidal datums

Local water level measurements are reckoned from a standard elevation called tidal datum. Tidal datum are defined by a certain phase of tide. Hence, their values vary with different locations, and cannot be extended into areas with different oceanographic characteristics. A specific 19-year period designated as the National Tidal Datum Epoch (NTDE) is selected because it is the closest full year to the 18.6-year nodal cycle, the time required for the regression of the moon's node to a complete revolution.[5]

Currently, the NTDE for South Harbor Tide Station is 1989-2007. NTDE allows one to separate all the astronomic frequencies needed to define the astronomic tide constituents for tidal prediction. It is also long enough to average out all the local meteorological effects on water level.[4]

The hydrodynamics controls the behavior of tides in oceans and connected bays. It introduces different types of tides (semi-diurnal, diurnal, and mixed) that can be present in a given area. The type of tides present at a coast defines the number of high and low tides encountered in a day. The Philippines has a semi-diurnal tides, hence it experiences 2 high tides (MHW and MHHW) and 2 low tides (MLW and MLLW) in a tidal day (24.84 hours). The 2 high tides and 2 low tides must be approximately equal in height. When there is a significant difference, the type of tide is mixed. The diurnal-type of tide has one low tide and one high tide in a tidal day. If a coast has a diurnal type of tide, then there is no MLLW or MHHW in its list of tidal datums.[3]

Tide data varies with different locations, hence this study is limited to analyzing tide data from a specific tide station. Since the main concern for studying tide data utilization is because of its importance in both disaster mitigation (particularly storm surge analysis) and coastal structure design (wave climate studies), this study chose a specific site where the hazard of storm surge is considerably high and mitigating solutions such as coastal structures are needed. Hence, this research specified South Harbor tide station as the study area because of its location in Manila Bay.

The general objective of this research is to study how NAMRIA collects, process and utilize tide data. Specifically, assess the tide data suitability to storm surge analysis and wave climate studies. Methodological framework for the specific objectives is shown in Figure 2.

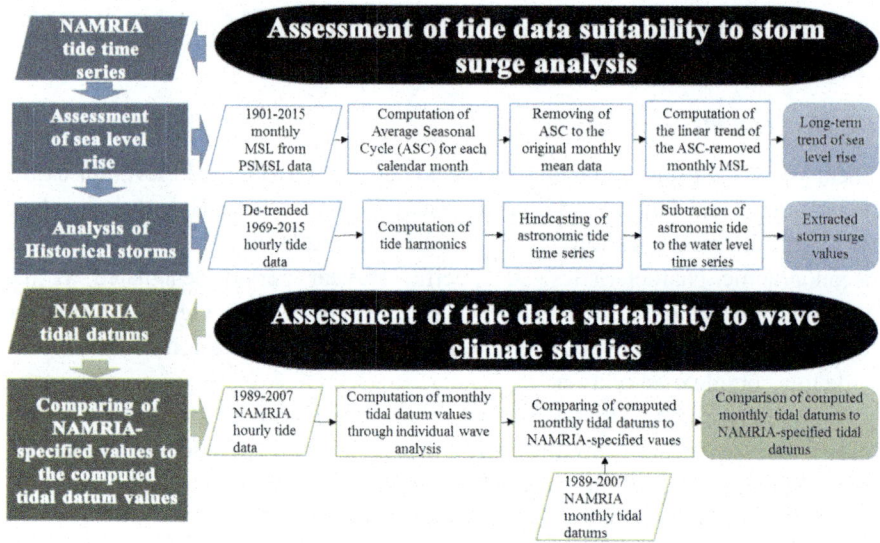

Figure 2: Methodological framework

2. Results and Discussion

2.1 Assessment of tide data suitability to storm surge analysis

Least-square linear equation (equation 1) was used to analyze the 1969-2015 monthly mean sea level trend.

$$y_i = bt_i + m_j \tag{1}$$

where y_i is the monthly mean sea level , b is the linear trend, t_i is the time in fractional years and m_j is the average seasonal cycle for calendar months.[6]

The monthly data was grouped based on calendar months. The results are 12 plots of monthly mean sea level (with spacing equals one year) corresponding to the calendar months. Getting the y-intercepts of each plot, the results are the average seasonal cycle for each calendar month.

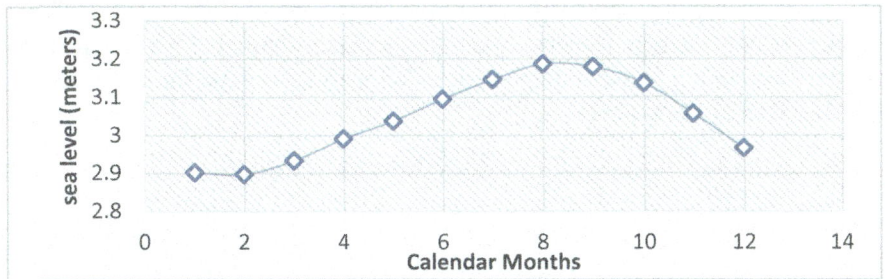

Figure 3: Average seasonal cycle for each calendar month

Figure 3 shows the average seasonal cycle (ASC) for 12 calendar months for the period of 1969- 2015. By definition, average seasonal cycle in coastal water levels are combination of the effects of air pressure, wind, water temperature, salinity, ocean currents and river discharge.[6] Hence, the increasing trend from February to May (summer months) may be due to the water level rise because of thermal expansion. The continuous increasing trend for June-August (rainy seasons) is due to the river discharge from Pasig river to Manila Bay (study area) because of the intense and frequent rainfall for these months. A decreasing trend from September to December (cold season) is due to water level drop because of contraction of water. To calculate only the linear trend of sea level rise without the meteorological effects, the ASC values for each calendar month was subtracted for each monthly mean as shown in Figure 4. The observed increasing trend is still clearly be seen. The computed slope of the resulting linear regression line which is 13.6 mm/year is the local sea level rise for Manila Bay for 1969- 2015. This increasing linear trend is relative with respect to the absolute global sea level rise and the vertical stationarity of the local study area.

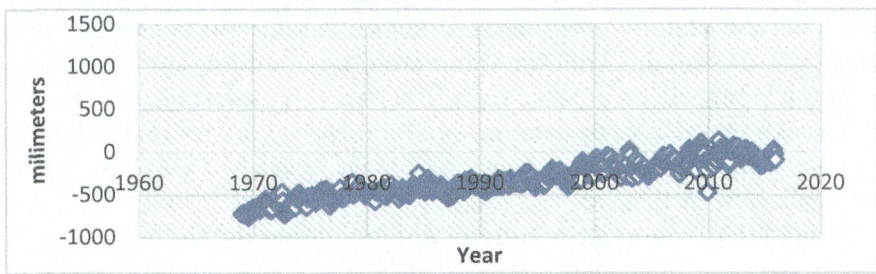

Figure 4: Removed average seasonal cycle on the monthly MSL with linear trend 13.6 mm/year

To further quantify the sea level rise, the 1901-2014 monthly mean sea level from Permanent Service for Mean Sea Level (PSMSL) data was used for the same procedure. Figure 5 clearly shows a different trend for 1901-1968 monthly mean sea level and 1969-2014 monthly data. The trend for the former is milder (1.75 mm/year) compared to the 1969-2014 (14 mm/year) monthly mean sea level. To see the effect of average seasonal cycle for the large variation of the linear trends for this two periods, a separate procedure for analysis is done for these periods (1901-1968 and 1969-2014).

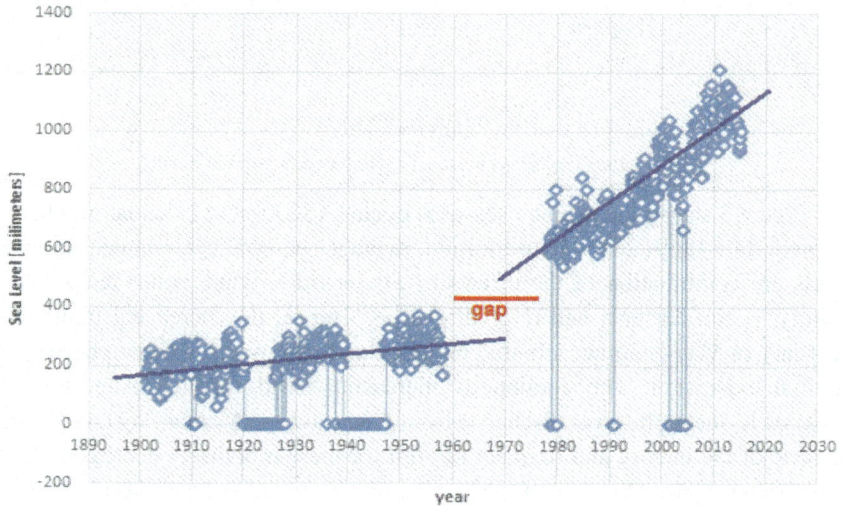

Figure 5: Monthly MSL for 1901-2014 and 1901-1968 from PSMSL data

There are two different sets of average seasonal cycle for periods 1901-1968 and 1969-2014 that are subtracted to their corresponding monthly mean sea level. The results for both periods have no significant change in sea level trend. For 1901-1968, the linear trend is from 1.75 mm/year to 1.76 mm/year (ASC-removed linear trend) and for 1969-2014, the sea level trend is from 14mm/year to 13.77 mm/year (ASC-removed linear trend). Therefore, the large variation for the linear trends of 1901-1968 and 1969-2014 is not due to climate-driven average seasonal cycle but mostly effects of relative sea level rise. This method of analyzing sea level trends are validated by doing the same procedure to analyze sea level rise in Legaspi, Davao, and Jolo and comparing it to the results by NOAA. All results are summarized in Table 1.

The sea level rise which is 13.6 mm/year was used to de-trend the hourly tide time series prior to harmonic analysis. Harmonic analysis is done piecewise annually, defining the tidal constituents of each year. The averages of the amplitude of the principal tide constituents (O1, K1, M2 and S2) for 1969-2015 tide data are shown in Table 2. The type of tide present in the study area can be determined based on the amplitudes of the principal tide constituents. Table 3 shows the classification criteria from NOAA.

Table 1: Summary of results for sea level rise

Data Source	Data Period	Station	Sea level Rise (mm/yr)	
			Computed	NOAA[7]
NAMRIA	1969-2015	South harbor	13.6	-
PSMSL	1901-2014	South harbor	7.96	-
PSMSL	1901-1968	South harbor	1.76	-
PSMSL	1969-2015	South harbor	13.73	-
PSMSL	1948-2014	Legaspi	5.45	5.38
PSMSL	1948 – 2015	Davao	5.29	5.32
PSMSL	1947-1996	Jolo	0.017	0.19

Table 2: Averages of O1, K1, M2, S2 amplitude for 1969-2015

	M2	K1	O1	S2
Amplitudes (mm)	296.9	275.7	187.25	67.15

Table 3: Classification of tides based on amplitude of K1, O1, M2, S2[3]

NOAA CO-OPS Tides and Glossary	Semi diurnal	Mixed mainly semi diurnal	Mixed mainly diurnal	Diurnal
r=(K1+O1)/(M2+S2)	r<0.25	0.25<r<1.5	1.5<r<3	r>3

Computing the value of r, r = 1.27166598, the type of tide is mixed mainly-semi diurnal. Manila Bay experiences two high tides and two low tides in a tidal day, with significant variation on the heights of tide. The computed tide constituents are used to hindcast the astronomic tide. The residual time series with significant values, after the astronomic tide has been removed in the water level time series (STL), are selected. The height of the residual is the storm surge height during a corresponding typhoon. Table 4 summarizes the storm surge values with their corresponding astronomic tide level, STL rank and astro rank. Storm tide analysis involves separation of storm surge values to the water level time series. The risk posed by storm tide depends on its two factors, astronomic tide and storm surge. For instance, both typhoon Pedring and typhoon Faye, which is rank 1 and rank 2 in storm tide level occurs at a high tide with their surges also relatively higher. Most of the top 10 in STL values occurs at a high tide. However, typhoon Frankie/Edeng, is rank 8 based on storm tide with its surge occurring at a low tide compared to the other top 10 in STL. The lowest based on storm tide level, typhoon Nanmadol, is also lowest on astronomic tide level.

Table 4: Summary of extracted storm surge from water level time series

Typhoon	Date of Occurrence of	STL (mm)	Astro (mm)	Surge (mm)	STL rank	Astro Rank
Georgia	11-Sep-70 : 07	985.8	645.3	340.5	7	2
Ora	25-Jun-72 : 11	1061.3	638.5	422.7	5	4
Ruth/Narsing	28-Nov-74 : 13	528	56.1	471.9	19	29
Elaine/Wening	25-Aug-75 : 04	217.8	44.4	173.5	39	32
Irma/Bidang	20-Oct-75 : 17	355.7	159.6	196.1	32	20
Dinah/Openg	27-Oct-78 : 02	884.3	374.5	509.7	9	10
Rita/Kading	16-Aug-79 : 18	633.2	74	559.2	14	27
Herbert/Huaning	13-Jun-81 : 03	368.2	-16.6	384.7	31	37
Joe/Nitang	18-Jul-81 : 05	526.9	207.2	319.7	20	17
Betty/Aring	24-Nov-81 : 15	392	54.5	337.5	29	30
Irma/Anding	04-Dec-82 : 04	497.9	63.9	434	22	28
Winona/Emang	05-Jun-83 : 20	271	7.3	263.7	36	35
Faye/Norming	15-Jul-83 : 15	1149.5	641.1	508.4	2	3
Vera/Bebeng	03-Oct-84 : 10	212.8	-56	268.8	41	41
Ike/Nitang	03-Sep-85 : 15	760.2	364.2	396	11	11
Tess/Milling	29-Apr-86 : 18	211.2	-101.5	312.7	42	43
Peggy/Gading	24-Oct-87 : 02	330.8	-28.7	359.5	34	40
Lynn/Pepeng	28-Jul-88 : 12	680.3	460.6	219.8	13	8
Ruby/Unsang	19-Oct-89 : 14	163.5	-129.4	292.9	43	44
Elsie	04-Oct-90 : 21	450.3	246.7	203.6	24	15
Ruth/Sendang	26-Oct-92 : 08	152.1	-28.3	180.3	44	39
Ted/Maring	05-Oct-93 : 17	499.1	49.4	449.7	21	31
Koryn/Goring	10-Jul-94 : 15	398.7	34.6	364	28	33
Flo/Kadiang	07-Aug-94 : 06	607.6	389.3	218.3	15	9
Ira/Husing	21-Oct-94 : 13	424.8	98.2	326.6	26	26
Tim/Iliang	30-Jul-95 : 06	214.2	3.1	211.1	40	36
Teresa/Katring	03-Nov-95 : 10	550.6	144.4	406.2	18	23
Sibyl/Mameng	15-Sep-96 : 20	328.7	156.3	172.3	35	21
Angela/Rosing	10-Nov-96 : 17	346.6	164.2	182.4	33	19
Frankie/Edeng	26-May-97 : 15	889.2	326.7	562.4	8	13
Levi/Bening	24-May-98 : 05	225.5	8.9	216.6	38	34
Bilis/Isang	22-Aug-00 : 18	444.6	142.5	302.2	25	24
Nanmadol/Yoyong	02-Dec-04 : 21	685.9	332.3	353.6	12	12
Milenyo	28-Sep-06 : 16	570.9	104.9	465.9	16	25
Halong	17-May-08 : 17	258.4	-78.9	337.3	37	42
Fengshen/Frank	22-Jun-08 : 07	817.1	487.5	329.6	10	7
Megi Juan	19-Oct-10 : 19	495.2	214.4	280.8	23	16
Bebeng	09-May-11 : 00	-62.4	-420.8	358.4	46	46
Falcon	24-Jun-11 : 10	555.9	184.2	371.6	17	18
Nanmadol	27-Aug-11 : 16	-156.5	-427.1	270.6	47	47
Pedring	27-Sep-11 : 11	1352.3	539.3	813	1	6
Saola	30-Jul-12 : 00	990.8	290.5	700.2	6	14
Kai-Tak/Helen	14-Aug-12 : 22	400.2	149.1	251.1	27	22
Nari	12-Oct-13 : 00	1074.3	712.7	361.6	4	1
Glenda	16-Jul-14 : 11	1123.8	606	517.8	3	5
Mario	19-Sep-14 : 17	51.4	-301.6	353	45	45
Lando	18-Oct-15 : 05	386.6	-20.8	407.4	30	38

*Measurements are reckoned from MSL, negative values for astro means tide occurs at low tide.

2.2 Assessment of tide data suitability to wave climate studies

By definition, mean higher high water (MHHW) and mean high water (MHW) are the averages of all the higher high waters and high waters respectively in a given period of time. The higher high water is the higher peak and the high water is the lower peak in an individual wave. The same is true for lower low water (lower trough) and low water (higher trough). NAMRIA uses a uniform wave period equivalent to a tidal day (24.84 hours) to form the individual waves for tidal datums computations. For instance, if there are 2484 hours of tide record to be analyzed, there will be 100 individual waves to be considered.

Zero-crossing method uses the same procedure to compute for tidal datums except that the wave periods used are not uniform but have varying lengths of time. The tide time series are divided based on zero-crossing points (up or down) to form the individual waves. Hence, the only difference between the two methods is the selection of wave periods on which the tide time series record will be divided to form the individual waves. To see its effect on tidal datum values, the 1989-2007 NAMRIA monthly tidal datums (MHHW, MHW, MLW, and MLLW) were compared to the computed values (through zero-crossing). Figure 6 shows the variability of the lengths of wave periods when using zero-crossing method. Figure 7 shows sets of graphs representing the comparison of the NAMRIA-specified values versus computed tidal datum values.

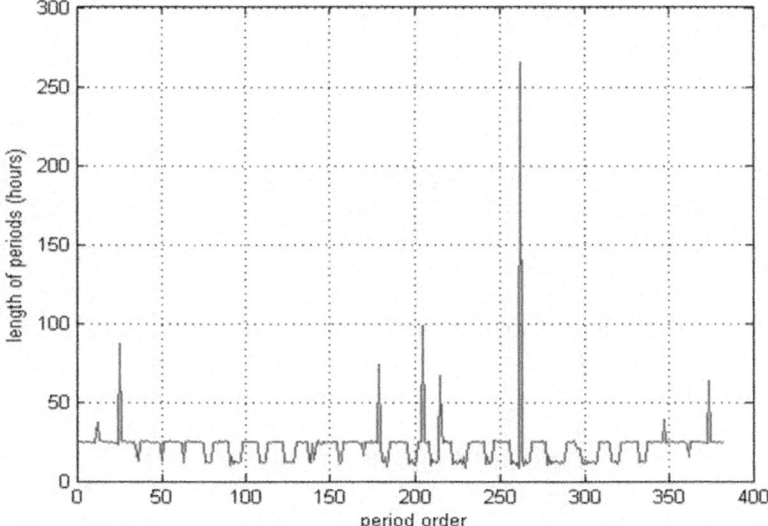

Figure 6: Variation of lengths of wave periods of individual waves using zero-crossing method

604

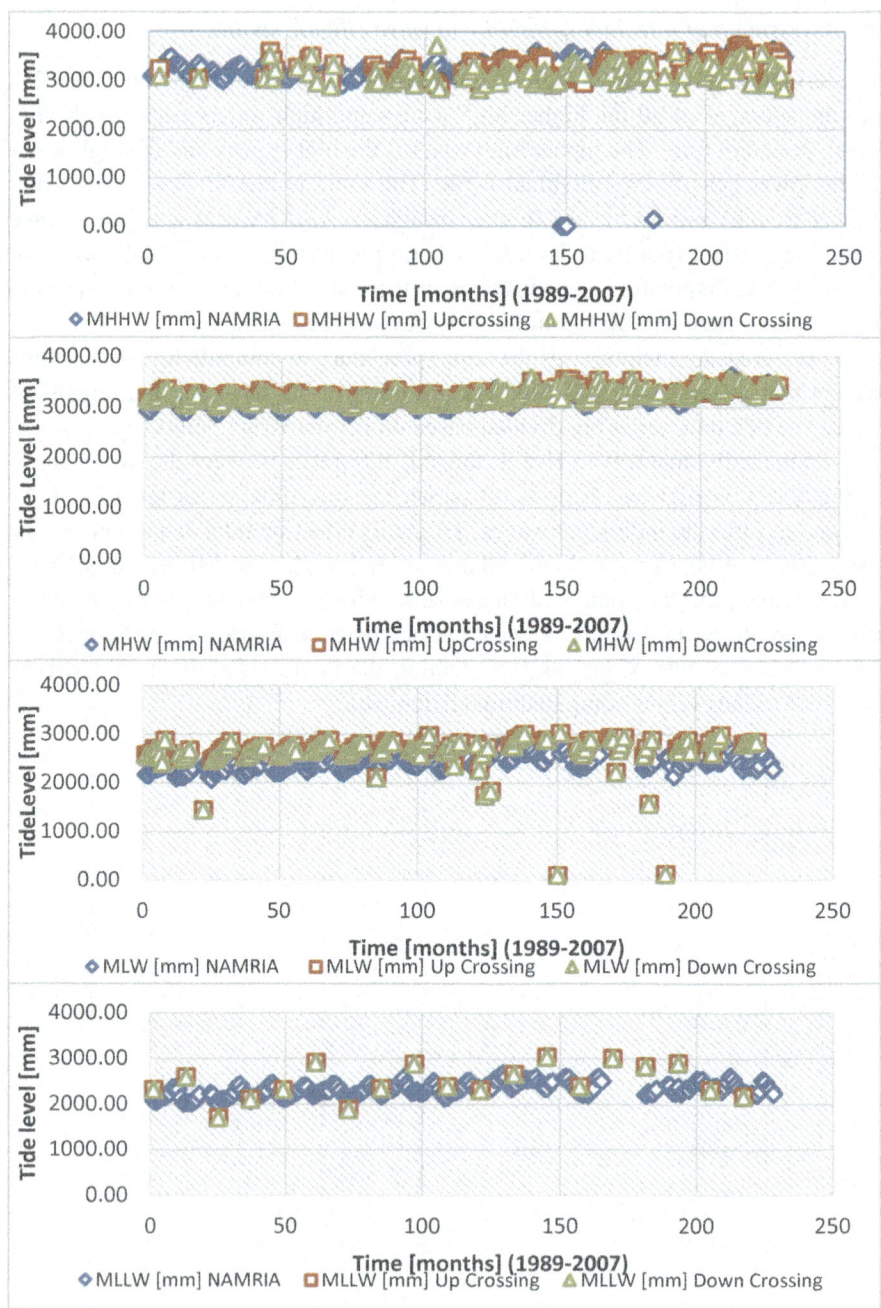

Figure 7: NAMRIA versus computed monthly tidal datums (MHHW, MHW, MLW, MLLW)

Zero values in the graphs indicate gaps in the hourly tide record. For both MHHW and MHW, monthly values from zero-crossing method do not match with the NAMRIA-specified values. Although, their averages for the whole 19-year period (1989-2007) do not vary significantly. (see Table 5)

Moreover, computed monthly values for MLW and MLLW also deviate considerably to the NAMRIA-specified monthly MLW and MLLW. As shown in Figure 7, there are less number of computed MLLW (zero-crossing values). This is because more number of the individual waves formed, by dividing the tide time series record based on zero-crossing points (up or down), have only one trough (that is considered to be a low water not a lower low water). This variation also effects the 19-year averages of LLW and LW. (see Table 5).

The mean sea level (MSL), highest tide and lowest tide were also compared to the specified values by NAMRIA. MSL, by definition, is the average of all the 1969-2015 hourly tide data. The same is true for highest tide and lowest tide. Although the same definitions are used to compute for the MSL, highest tide and lowest tide, there is still variations in the compared values. All results are summarized in Table 5.

Table 5: Summary of comparison for 1989-2007 monthly tidal datums

Tide Normals	NAMRIA	Zero Up Crossing	Zero Down Crossing	Abs error (up-crossing)	% error
MHW	3160.74	3222.69	3231.54	61.95	1.96
MHHW	3231.55	3236.55	3140.1	5	0.15
MSL	2776.55	2754.05	2754.05	22.5	0.81
MLW	2401.57	2620.99	2621.27	219.42	9.14
MLLW	2305.88	2459.91	2460.31	154.03	6.68
Tide Extremes					
Highest Tide	4210	4280		70	1.66
Lowest Tide	2480	2670		190	7.66

3. Conclusion

To assess the tide data suitability to storm surge analysis, assessing the sea level rise prior to harmonic analysis was done. Based on the results on sea level rise assessment, least-square linear solution method is not suitable for assessing the sea level rise of longer periods of observation. Using this method for longer periods of observation will underestimate the relative sea level rise of the study area. In the analysis of historical storms, it was proven that the extraction of storm surge values in the water level time series is possible by predicting the astronomic tide through harmonic analysis.

For the second specific objective, comparing the computed (through zero-crossing method) tidal datum values to the NAMRIA-specified tidal datums was done. Results showed that the selection of wave periods on which an individual wave is to be analyzed causes large variations in the compared values MLW and MLLW. NAMRIA uses a uniform wave period (24.84 hours) while zero-crossing has a varying values for wave period (range from 12 to 275 hours with 25 hours as the mode of values). Manila bay experiences two high tides and two low tides in a tidal day, with significant differences in height. To fully represent this variation in tidal datum values, the researcher recommends the use of uniform wave periods for individual wave analysis which is also equal to a tidal day (24.84 hours) for tidal datum values computation.

References

1. N. Baloran, *National Report of the Philippines* (Oceanography Division Coast and Geodetic Survey Department National Mapping and Resource Information Authority, 2006).
2. National Hurricane Center, National Oceanic and Atmospheric Administration (n.d.), `http://www.nhc.noaa.gov/surge/#TIDE`
3. S. Gill. and J. Schultz, *Tidal datums and their applications* (U.S. Department of Commerce, National Oceanic and Atmospheric Administration National Ocean Service Center for Operational Oceanographic Products and Services, Special Publication NOS CO-OPS 3, 2007).
4. B. Parker, *Tidal analysis and prediction* (U.S. Department of Commerce, National Oceanic and Atmospheric Administration National Ocean Service Center for Operational Oceanographic Products and Services, Special Publication NOS CO-OPS 3, 2007).
5. D. Evans, C. Lautenbacher, R. Spinrad and M. Szabados, *Computational technique for tidal datums handbook* (U.S. Department of Commerce, National Oceanic and Atmospheric Administration National Ocean Service Center for Operational Oceanographic Products and Services, Special Publication NOS CO-OPS 3, 2003).
6. C. Zervas, *Sea Level Variations in the United States* (U.S. Department of Commerce, National Oceanic and Atmospheric Administration National Ocean Service Center for Operational Oceanographic Products and Services, Technical Report NOS CO-OPS 053, 2009)
7. National Oceanic and Atmospheric Administration (n.d.), `https://tidesandcurrents.noaa.gov/sltrends/sltrends_global_country.htm?gid=1271`

Investigation of Alongshore Sediment Transport and Shoreline Deterioration Around Submerged Discharge Pipelines in Fuji Coast, Shizuoka, Japan

Kevin Bobiles* and Shinji Sato[†]

*Department of Civil Engineering, The University of Tokyo,
7-3-1, Hongo, Bunkyo-ku, Tokyo, 113-8656, Japan*
*E-mail: kevinbobiles@coastal.t.u-tokyo.ac.jp
[†] E-mail: sato@coastal.t.u-tokyo.ac.jp

Recent floodings in urban areas have made local government unit to discharge excess runoff in the open sea. Discharge pipelines are then installed normal to the shoreline to help mitigate inundation in lowland areas. However, severe erosion in the downdrift side and significant change in nearshore topography are observed in the vicinity of the discharge pipeline. Field survey, with the use of UAV, is conducted in Fuji Coast to investigate nearshore topography around the structure. Field data analysis reveals that updrift side of the structure exhibits a convex cross-shore profile whereas downdrift side exhibits a concave cross-shore profile. A photogrammetric technique known as Structure from Motion (SfM) is then used to develop the three-dimensional model of the topography around the structure. It can be shown that estimation of nearshore topography is possible through UAV-based measurements. Finally, preliminary experiments are carried out with the aim of reproducing the erosion-accretion process observed in the field. It is shown in this paper that the process can be reproduced in a laboratory scale model. Further experiments are conducted to reproduce the process in a larger scale and quantitative measurements will follow so that more pertinent information can be obtain to elucidate the erosion-accretion mechanism occurring around these discharge pipelines protruding across a shoreline.

Keywords: Fuji Coast, alongshore sediment transport, discharge pipeline, UAV-based measurements, Structure from Motion, nearshore topography.

1. Introduction

Fuji Coast is located between the cities of Fuji and Numazu in Shizuoka Prefecture of Japan. Shizuoka is one of the Prefecture of Japan in the Chubu region and is found in the Pacific side of Honshu Island. The coast is part of the Suruga Bay which is bounded by the Honshu Island in the southwest and west and the Izu Peninsula in the east. The bay is facing the Pacific Ocean in the south and is somewhat protected from high waves due to the presence of Izu Peninsula (refer to Figure 1). Fuji Coast is found north of Suruga Bay and near the foot of Mt. Fuji.

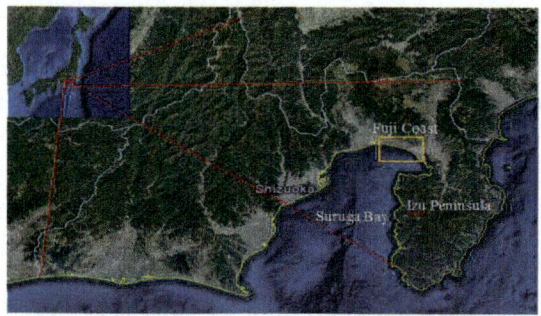

Fig. 1. Google Earth view of Shizuoka Prefecture together with Suruga Bay and Izu Peninsula. Orange borders shows the location of Fuji Coast.

Along the stretch of Fuji Coast, channels and pipelines connected to Numagawa River were installed to prevent inundation of low-lying areas in Numazu and Fuji by discharging excess runoff into the Suruga Bay. Such structure which is also the study area of this paper is shown in Figure 2.

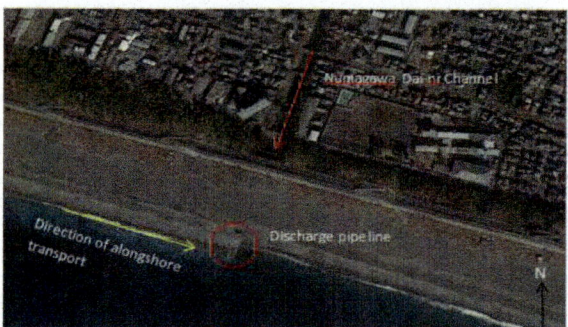

Fig. 2. Google Earth view of Fuji Coast with its discharge channel and pipeline shown.

However, with the presence of the discharge pipeline protruding across the shoreline, interruption of alongshore sediment transport causes severe erosion in the downdrift side of the channel and accretion in the updrift side of the channel. Coupled with this shoreline deterioration, significant topography change can be observed around the vicinity of the channel leading to a more convex cross-shore profile in the updrift side and a concave cross-shore profile in the downdrift side.

In this study, attempts to elucidate the mechanism for such shoreline deterioration and corresponding topographical change are carried out through field survey and laboratory experiments. Specifically, UAV-based field measurements are utilized to analyze the topography change as well as the shoreline deterioration that is occurring in the vicinity of the discharge pipeline. A photogrammetric technique known as Structure from Motion (SfM) is then used to estimate the nearshore topography around the pipe. Furthermore, preliminary small-scale laboratory experiments using a two-dimensional wave basin are also carried out to reproduce the coastal erosion found in the field.

2. Field survey analysis

2.1. *Field survey set-up and preliminary observations*

Field survey along Fuji Coast in Numazu, Shizuoka Prefecture was conducted last April 17, 2017. The survey is composed of two main steps (refer to Figure 3); first is the collection of overlapping images using UAV within the 800m x 100m stretch of Fuji Coast with the discharge pipeline serving as the centerline of the study area for nearshore topography estimation and second is the measurement of elevation of ground control points (GCPs) depicted by the blue markers using Real Time Kinematic GPS (RTK-GPS). For measurements using DJI Phantom UAV which has an image resolution of 4000 x 3000, overlapping images are collected through 2 secs interval snapshots at a height of 30 m with an angle of 90 degrees from horizontal. Overlapping images are also collected but this time at a height of 20 m with an angle of 30-45 degrees from horizontal facing the shoreline.

Preliminary observations of the study area show indication of severe erosion in the downdrift side and accretion in the updrift side. Figure 4 shows some images near the vicinity of the pipe taken during the field survey. In the downdrift side, it can be seen that the shoreline has significantly retreated into landward side therefore exposing some portion of the pipe in the action of waves and sediments. Significant decrease in elevation is also observed in the downdrift side which makes the swash zone steeper. On the other hand, large sediments have piled up near the shoreline in the updrift side. The profile of the uprift side seems to bulge out due to significant accretion of sediments in the vicinity of the pipe.

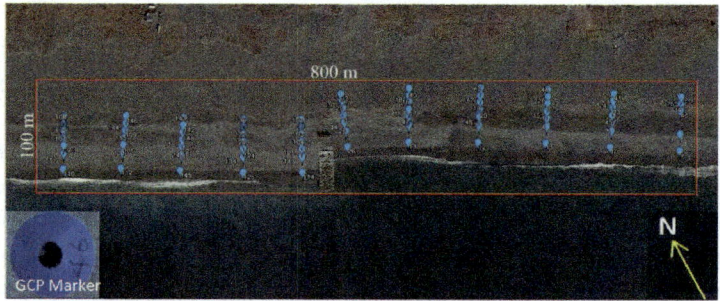

Fig. 3.   Field survey setup with ground control points shown.

Fig. 4.   Top images: Downdrift side at the immediate vicinity of the pipe.   Bottom images: Updrift side at the immediate vicinity of the pipe.

2.2. *Cross-shore profiles revealed by RTK-GPS data*

The elevation data of the GCPs measured by RTK-GPS is then used to plot several cross-shore profiles both in the updrift and downdrift side.

Figure 5 shows the GCPs together with the cross-shore profile designations. RTK-GPS data are marked on Google Earth for reference. Red line is designated as reference shoreline (i.e. 0 m for horizontal axis) for plotting the cross-shore profiles. Covered cross-shore distance, measured from the red line is about 80 m. Figure 6 shows the average cross-shore profile by taking three profiles for both updrift and downdrift side. As seen from the plot. The profile in the updrift side resembles a convex shape (blue plot) whereas in the downdrift side a concave profile (green plot) is observed.

It can be seen also from Figure 6 that the cross-shore profile in the

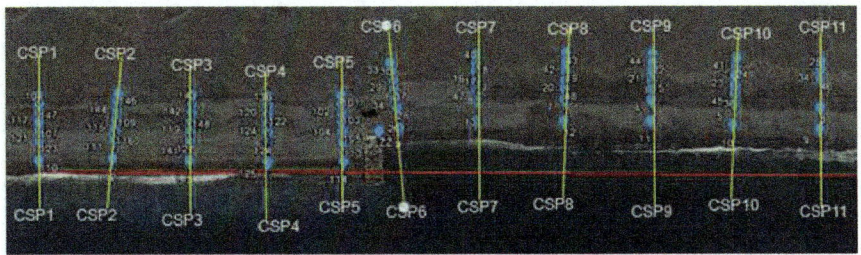

Fig. 5.   Transect profiles in updrift and downdrift side.

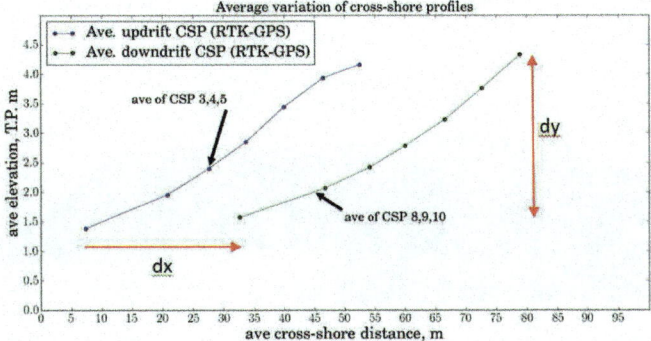

Fig. 6.   Average cross-shore profile in updrift and downdrift side. Elevations are measured in T.P.

downdrift side has significantly retreated landward by an amount of dx. Because of this retreat, a loss of sand volume in the downdrift side is expected by an amount of dx*dy*L where L is some distance in the downdrift side. Furthermore, more sand volume loss is expected due to the concave profile in the downdrift side. Therefore, we have to include the change in profile in estimating the sand volume loss.

2.3. *Nearshore topography estimation using SfM model*

Overlapping images collected by the UAV during the field survey are then used as input data for the development of Structure from Motion (SfM) model of the study area. Structure from Motion (SfM) is a photogrammetric technique for estimating three-dimensional structure from series of overlapping two-dimensional images. The model is then georeferenced by assigning ground control points in the images. Figure 3 shows the setup for the estimation of nearshore topography though UAV-based measurements.

612

Blue markers labelled in Figure 3 are used as ground control points (GCPs) since its actual elevations are measured using RTK-GPS. These GCPs will serve as reference points in georeferencing the SfM model. With the GCPs now set in the field, overlapping images with 2 secs interval are collected by the UAV at different angles at a height of 20-30 m. Images are then input as data in the model and camera internal parameters such as focal length and lens distortion are precalibrated. The model will build dense cloud based from aligned and estimated positions of the images. The model is then georeferenced by assigning the GCPs its real coordinates and elevations obtained from RTK-GPS. With these processes, mesh is built from which reconstructed surface and textured digital elevation model can be obtained.

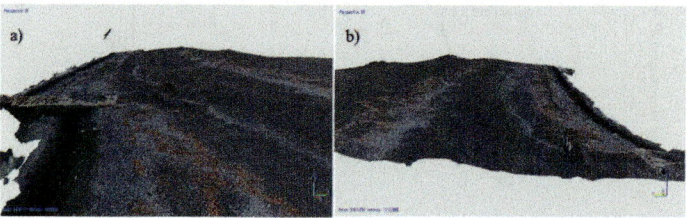

Fig. 7.   Reconstructed 3D SfM model. a.) Updrift side; b.) Downdrift side

Figure 7 shows the reconstructed three-dimensional SfM model for Fuji Coast. The model was able to capture the convex and concave profile for the updrift and downdrift side, respectively.

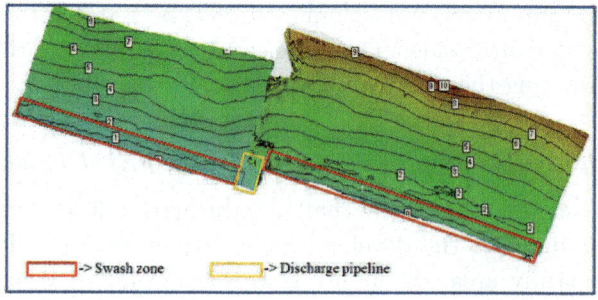

Fig. 8.   Digital Elevation Model (DEM) for Fuji Coast

Finally, a digital elevation model of the study area is developed by combining the model for the updrift and downdrift side. Contour lines are also shown for visualization purposes. Note that contour lines in the downdrift side have retreated landward relative to its corresponding contour lines in the updrift side. This is consistent with the preliminary observation in the field and with the cross-shore profile revealed by RTK-GPS data. However, accuracy of SfM model within the swash zone is still subject to some verification.

To get an insight on the extent of swash zone in the study area, tidal variations near the site are obtained to estimate the run-up height and finally run-up elevation.

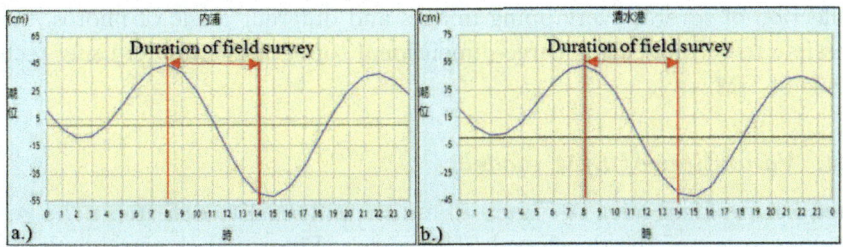

Fig. 9. 04/17/2017 Tide Chart at a.) Uchiura Station and b.) Shimizuminato Station (source: Japan Meteorological Agency)

Figure 9 shows the tidal chart for the two stations nearest to Fuji Coast namely, Uchiura Station and Shimizuminato Station. Tide level are measured from zero datum which is T.P. This tide level is then added to the run-up height to estimate run-up elevation in the study area.

Run-up height is estimated by the deepwater wave height and surf similarity parameter as defined by Mase (1989). The significant wave run-up height for irregular runup on plane beach with slopes 1:5 to 1:30 is given by

$$\frac{R_{1/3}}{H_o} = 1.38\xi_o^{0.7}, \tag{1}$$

where $R_{1/3}$ is the significant wave run-up, H_o is the deepwater wave height and ξ_o is the surf similarity parameter defined by $\frac{m}{\sqrt{H_o/L_o}}$.

Knowing the wave period, deepwater wave height and wavelength together with the beach slope, run-up height can be estimated. At the time of field survey, wave period and deepwater wave height are estimated to be 10 secs and 1 m at most, respectively. Beach slope is assumed to be 1:10.

The estimated significant wave run-up is found to be 1.613 m. The tide levels at Uchiura Station and Shimizuminato at 10am during the field survey are 0.25 m and 0.34 m from T.P., respectively. Therefore, the run-up elevations at Uchiura Station and Shimizuminato at time of measurement are 1.863 m and 1.953 m from T.P., respectively. From these run-up elevations, it can be inferred that swash zone extends 2 m at most from T.P. Therefore, it can be said that the area 1 m-2 m above T.P. is constantly dry and wet by the action of wave run-up. Further validation is needed in considering the accuracy of the SfM model within the swash zone (refer to Figure 8).

It can be said also that measurements can be done by UAV. Through collection of several overlapping images and different angle of photos, it is possible to reconstruct a three-dimensional topographic model using technique of SfM.

2.4. *Validation of SfM model*

Transect profiles (refer to Figure 5) are then plotted to determine the vertical accuracy of the developed SfM model. Figure 10 shows the plot of updrift cross-shore profiles based from RTK-GPS (i.e. blue plots) and SfM model. It can be noticed that profiles from SfM (i.e. red plots) especially for CSP 1 and CSP 2 yield significant deviation from that of RTK-GPS. However, measurements from RTK-GPS and SfM agrees quite well as it moves from CSP 3 to CSP 5.

To improve more the vertical accuracy of the SfM model, study area (refer to Figure 5) are then divided into segments by classifying the images based from its longitude. Overlapping the boundaries between two segments is preferred so that accuracy is ensured at both ends. Green plots in Figure 10 show the improved SfM cross-shore profiles after the classification of images is done. Vertical accuracy is now even better compared with the initial model. It can be seen in the plots that after the improvement, SfM model estimates the topography almost the same with that of RTK-GPS.

Therefore it can be said that by confining the SfM model into smaller areas or segments, an improved elevation accuracy is obtained and thus estimation of nearshore topography becomes better also.

3. Preliminary laboratory experiments

Preliminary laboratory experiments are carried out to reproduce the process that is occurring in the study area (i.e. Fuji Coast). The goal of

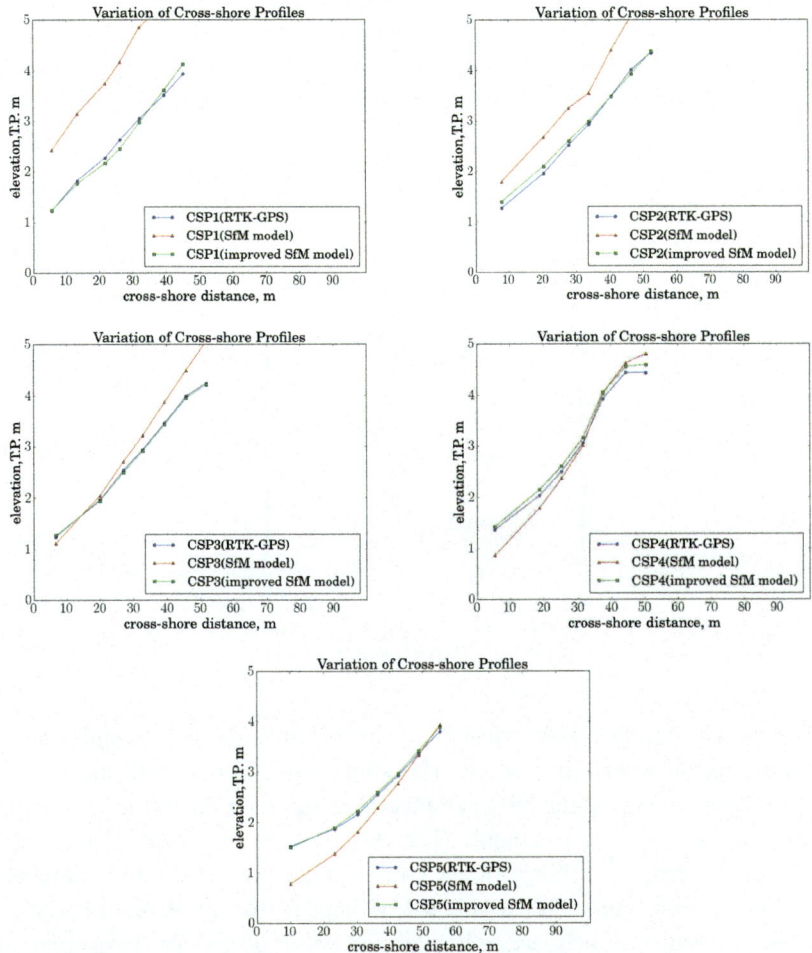

Fig. 10. Comparison of updrift cross-shore profiles based from RTK-GPS, SfM and improved SfM models

the experiments is to quantitavely measure the parameters that cause the erosion around the pipeline structure.

3.1. *Laboratory set-up*

Laboratory experiments are conducted in a two-dimensional wave basin with monochromatic wave propagating in normal direction (see Figure 11). Wave absorber composed of sponge layer is put behind the wave generator

616

to eliminate reflection. Guide walls on both ends of the set up are placed to prevent wave disturbances in the study area. The guide walls extend up to the wave generator to ensure monochromatic sinusoidal waves inside the study area. The width of the study area is about 2.55 m from both ends of guide walls.

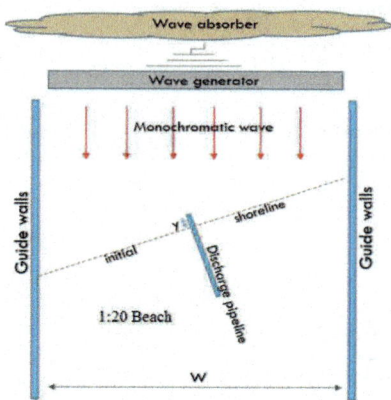

Fig. 11.   Experimental setup

The beach configuration consists of an initially oblique straight shoreline and a pipe protruding across the shoreline. The initial shoreline is made oblique with respect to wave direction in order to induce a natural alongshore transport of sediment. The beach slope is initially set to 1:20. Protrusion length, y, is initially set to about 20 cm. Sediment diameter, D_{50}, is about 0.15 mm. With this setup, initial test cases are conducted with varying water depth, wave height and wave period to determine the appropriate wave condition and beach configuration that will reproduce the process observed in the field.

3.2. Methodology and test cases

Initial trial experiments are conducted with the following parameters shown in Table 1. Piston stroke refers to the displacement of the wave generator with respect to its mean position.

In each trial experiment, wave period, offshore water depth and piston stroke are changed in order to find the optimum values that will reproduce the process found in actual condition. Duration of each trial experiment is

Table 1. Experimental conditions.

Trial Expt No.	Wave Period (T), secs	Offshore water depth (h_o), cm	Piston stroke, mm
1	3.0	15.5	60
2	1.0	18.0	30
3	1.0	17.0	15

about 50-60 minutes so that the system is allowed to stabilize and reach
equilibrium. Piston stroke is also important since wave height depends
on how much the piston is displaced. In trial 1, large disturbance was
observed inside the study area primarily due to wave period and piston
stroke. Wave period is quite long making reflection and diffraction around
the shore become significant therefore disturbing the incident waves through
wave breaking far from the shore. In trial 2, wave period was decrease to
1 sec so that unnecessary wave reflection and diffraction is prevented inside
the study area. However, wave height is quite large with respect to the
beach configuration which produces wave overtopping and inundation in
the beach setup. It can now be inferred that a lower piston stroke is needed
to reproduce the actual condition in the field.

3.3. *Preliminary results*

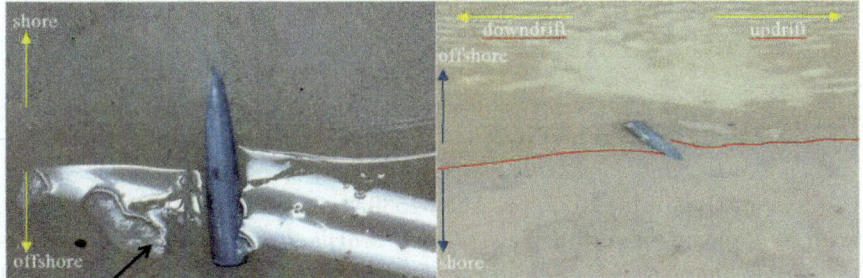

Fig. 12. Trial 3 experimental result.

Piston stroke is decreased from 30 mm to 15 mm for trial 3 of the
experiment. Wave period and offshore water depth is set to 1.0 sec and
17.0 cm, respectively. Experiment is allowed to run for about 50 minutes
to let the system stabilize and reach equilibrium.

Figure 12 shows the configuration of the beach setup after the test. It is
worth noting that in this trial experiment, the setup and methodology was
able to reproduce the expected shoreline response like the one observed

in the field survey. Accretion is observed in the updrift side of the pipe while erosion is observed in the downdrift side. However, such observation is confined only on a small scale. Succeeding experiments are now aim at reproducing the process on a large scale so that measurement of shoreline response is possible. From this scenario, quantitative measurements will subsequently follow to obtain more pertinent information about the erosional mechanism around these discharge pipeline structures.

4. Conclusion

This paper presents field survey analysis and preliminary laboratory experiments to study the erosion mechanism occurring around discharge pipelines protruding across the shoreline of Fuji Coast in Shizuoka Prefecture, Japan. Resulting topography change in the vicinity of the structure is investigated through field survey particularly by the use of UAV to conduct measurements. It is shown that estimation of nearshore topography is possible with UAV-based measurements.

Preliminary experiments are also conducted with the aim of reproducing the process observed in the field in a laboratory scale model. It is shown that coastal erosion observed in the field can be simulated in a laboratory scale model. Further tests are following to reproduce the process in a large scale so that further quantitative measurements can be done.

References

1. Y. Matsuba, S. Sato and K. Hadano, *Rapid Change in Coastal Morphology due to Sand–bypassing captured by UAV–based monitoring system.* Coastal Dynamics 2017, pp. 1529-1539.
2. R.S. Ranasinghe and S. Sato, *Beach Morphology behind Single Impermeable Submerged Breakwater under Obliquely Incident Waves.* Coastal Engineering Journal, Vol. 49, No. 1, pp. 1-24, 2007.
3. USACE Coastal Engineering Manual Part II. 2002. *Surf Zone Hydrodynamics.* EM 1110 - 2 - 1100.
4. Japan Meteorological Agency. https://www.data.jma.go.jp/kaiyou/db/tide/suisan/suisan.php, accessed last July 10, 2017.
5. Shizuoka Prefecture. https://en.wikipedia.org/wiki/Shizuoka_Prefecture, accessed last July 9, 2017.

Development of a UAV-based System
for High Resolution Beach Sediment Mapping

Konomi Goto and Shinji Sato

*Department of Civil Engineering, The University of Tokyo,
7-3-1, Hongo, Bunkyo-ku, Tokyo 113-8656, Japan
E-mail: konomi@coastal.t.u-tokyo.ac.jp*

A high resolution beach sediment mapping technique was developed on the basis of serial pictures taken from a UAV (Unmanned Aerial Vehicle) mobilized alongshore with a walking speed at an altitude of 10 m above ground. Two dimensional wavelet transform was introduced to determine the size of sediments in the range from 10^0 cm to 10^2 cm. The system was applied to the 16 km long Fuji Coast, composed of sand and gravel originated from the Fuji River. Distributions of sediment size in the foreshore zone were successfully estimated both in the cross-shore and the alongshore directions.

Keywords: Wavelet analysis; UAV; Grain size; Monitoring; Beach morphology.

1. Introduction

Beach topography and sediment size are essential parameters for understanding nearshore processes. However, it generally needs much labor and costs to obtain them in high resolution been appropriate for beach topography highly variable on time and space. In monitoring of sea bottom topography, various methods have been developed; a method using a sonar on a fishing vessel (Okabe et al.[1]) and methods for estimating the water depth from wave velocity which is obtained by using X band radar (Takewaka et al.[2]) or images taken by UAV (Matsuba et al.[3]), which enable high frequency and flexible topographic survey. On the other hand, analysis of sediment characteristics have been indicated to be useful for understanding the sediment movement in long-term and wide areas (Sato et al.[4]), though both sampling and analysis need abundant labor and costs. Therefore, technological innovation is needed.

On gravel sampling on the beach, it is known that sediment size varies from the swash zone to backshore especially on sand gravel mixed beaches. Hence, representativeness of the sampling location becomes indispensable. Although surface sampling by line grid method and area grid method and direct sampling by using heavy equipment (Yamamoto[5]) have been suggested, the representativeness and large amount of labor remains an annoying issue.

On sediment size analysis, in addition to the standard sieve analysis, the image analysis taken by digital camera has been proposed as the correlation method (Tsujimoto et al.[6]) and the one-dimensional wavelet analysis method (Buscombe[7]). However, because these studies focused close-up images taken in ideal condition, the applicability for a field survey remains unverified.

Based on these backgrounds, the purpose of this study was to develop a method to estimate the sediment size distribution and to achieve a sediment mapping by analyzing the images taken by UAV in a sediment transport system scale. The proposed method is intended to reveal continuous sediment size distribution with low costs and high mobility.

2. Field survey

2.1. *Target beach*

Target beach is the Fuji Coast, which is 16 km long gravel beach in Shizuoka Prefecture, Japan. Field surveys were conducted from the Kano River mouth to the Shouwa floodway in October, 2016, and from the Tagonoura Port to the Fuji River mouth in December, 2016. Detailed survey was conducted around the Numagawa Dai-ni floodway in April, 2017. The sediment transport system in the Fuji Coast is as shown in Figure 1. The three floodways for draining water from inland low marsh area interrupt the littoral drift from west to east. As a result, coastal erosion has been developed on the east side of the Shouwa floodway and the Numagawa Dai-ni floodway. Gravel beach nourishment has been conducted on the east side of the Shouwa floodway, where coastal erosion is severe. The gravel image shown in Figure 1 is the comparison between the natural beach sediment (on the right side) and the gravels for beach nourishment (on the left

Fig. 1. Sediment transport system in Fuji coast

side) at the location where beach nourishment is conducted. It could be seen that the gravels for beach nourishment are larger than the natural beach sediment.

2.2. *Survey*

In the surveys in October and December, 2016, UAV was operated along the shore above the berm keeping 10m altitude from the berm top according to displayed altitude on the receiver. The pictures of whole foreshore beach were taken vertically downward by UAV. The latitude and longitude where pictures were taken was recorded by the built-in GPS of UAV (Phantom4, DJI) Sample image is shown in Figure 2. The partial section between the west side of Tagonoura port and the west side of the Shouwa floodway where many large wave-dissipating concrete blocks were installed near the shoreline was excluded. In addition to the alongshore investigation, cross-shore investigation was conducted in October, 2016, on the point indicated in Figure 1. Since the topography varies more largely in the cross-shore direction than in the alongshore direction, markers were set as shown in Figure 3. All the pictures were taken from the 10m altitude above the berm top and each picture size was 3000 by 4000.

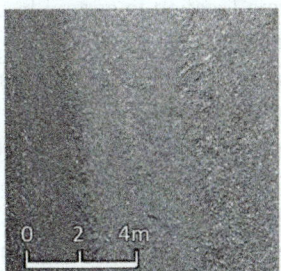

Fig. 2. An example of the image taken by UAV from 10 m height and flying UAV

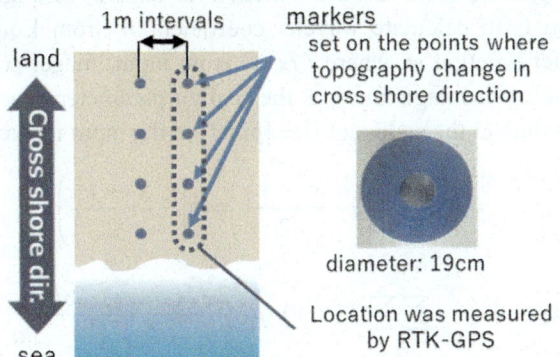

Fig. 3. The schematic diagram of cross-shore investigation in Oct., 2016

In the detailed survey in April, 2017, the target area was 500 m in the alongshore direction with a central focus on the Numagawa Dai-ni floodway by 50 m in cross-shore direction from shoreline. Markers were situated in the target area with 50m intervals in the alongshore direction and at 14 m, 7 m, 6 m, 6 m, 6 m and 6 m intervals started from the berm top in the cross-shore direction. The location of all the markers were measured by a VRS type RTK-GPS. The continuous pictures covering the target area were taken by UAV from 10 m altitude for sediment size analysis and from 30 m altitude for topography analysis. The UAV height is defined as the height from the elevation of take-off points. Take-off points were near the markers and they were recorded at the time of the change of take-off positions for every battery exchange. The size of each picture was 3000 by 4000.

3. Method for calculating sediment size

3.1. *Wavelet transformation*

Wavelet transformation is one of the frequency analysis methods. In this method, the wavelet coefficient of a signal at certain location in certain frequency domain is obtained by multiplying the signal by the scaled and shifted basic function called "mother wavelet function". Compared to the Fourier transformation, which is another representative frequency analysis technique, wavelet transformation has advantages that it can keep the location information and can correspond to the wide-range frequency.

Among various functions suggested as the mother wavelet function, two dimensional Mexican-hat wavelet, which is introduced by Freeman et al.[8], was adopted for this study. The Mexican-hat wavelet function is expressed as Equation (1) and the general form of the function is shown in Figure 4. Wavelet transformation is to calculate wavelet coefficient W from Equation (2) using mother wavelet function ψ, where $f(x, y)$ is an input image, (x_0, y_0) is a target location of the input image and a is the scaling parameter of ψ. The scale a is adjusted according to the sediment size [pixel] in the input image.

$$\psi(x,y) = \left(1 - \frac{x^2 + y^2}{2}\right) \exp\left(-\frac{x^2 + y^2}{2}\right). \tag{1}$$

$$W(x_0, y_0) = \frac{8}{a^2} \sum_x \sum_y f(x,y) \cdot \psi\left(\frac{2\sqrt{2}(x - x_0)}{a}, \frac{2\sqrt{2}(y - y_0)}{a}\right). \tag{2}$$

3.2. *Wavelet transformation in this study*

The distortion of photographed images generated by lens distortion were corrected according to Brown's distortion model[9]. Then, the images were converted to gray scale images. Two dimensional wavelet transformation were performed with the scale a of 2 to the power of 0, 0.2, 0.4 ..., 7 for the each of the gray-scaled images. D is the mean value of the scale weighted by the value of squared wavelet coefficient, which is expressed as Equation (3) and calculated for each pixel. a_n is a certain scale and W_n is the wavelet coefficient correspond to the scale. A sample image and the distribution of the D calculated for the image is shown in Figure 5.

$$D = \frac{\sum_n a_n \cdot W_n^2}{\sum_n W_n^2}. \tag{3}$$

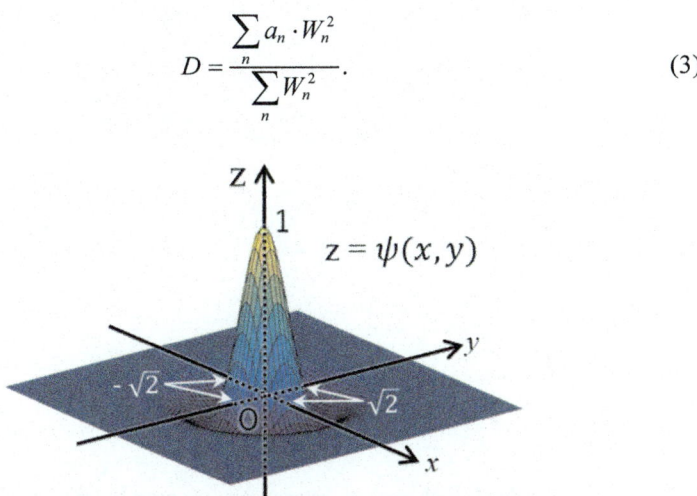

Fig. 4. The outline of mother two dimensional Mexican-hat wavelet function

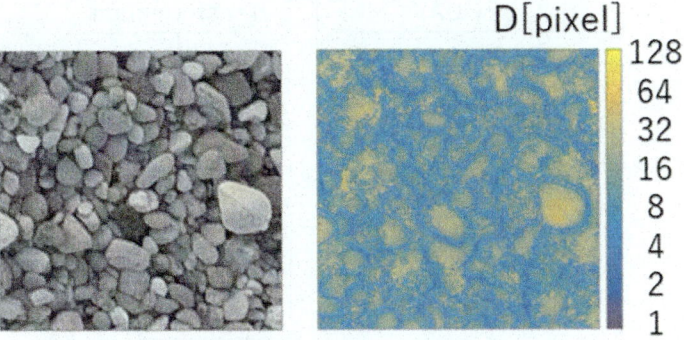

Fig. 5. The distribution of the weighted scale average D

624

3.3. *Calculation of mean sediment size*

There are a sample image on the upper row and a graph of the D value in the red line on the sample image on the middle row in Figure 6 shows a sample image (top) and the D value along the red line (middle). It is confirmed that the values of D are biggest at the center of each gravel. The bottom panel in Figure 6 shows the D value after noise removal by an 8-neighbor median filter. It could be seen that the sediments size was successfully estimated. Finally, the mean value of the peak values of D was multiplied by the real length per pixel in the image to estimate the mean sediment size. A whole process is shown in Figure 7.

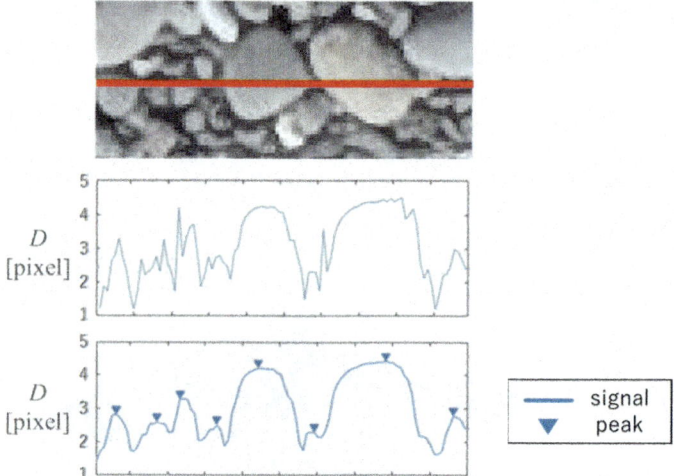

Fig. 6. High-accuracy estimation of the mean sediment size

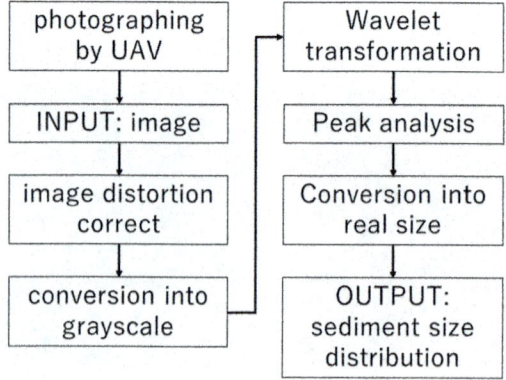

Fig. 7. The whole process of estimating the sediment size distribution

3.4. *Validation*

The whole process was applied to nine sample images which had been taken in the field survey. Figure 8 is a scatter diagram of the mean sediment size as compared with the real mean sediment size. The estimation by using one dimensional wavelet transformation by Buscombe[7] is also shown in Figure 8. A high-accuracy sediment size estimation method was established since mean absolute error of Buscombe's method was 48% and the one of this method was 4.7%. As shown in Figure 5, it is one of the advantages that the location information of the sediment is not lost.

4. Sediment size mapping for Fuji coast

4.1. *Sediment size mapping in the along shore direction*

The method was applied to the images which were selected from pictures taken in October and December in 2016 at about 200 m intervals. The mean sediment size distribution in the alongshore direction is shown in Figure 9. Real length per pixel was determined by the linear relationship with the constant height from surface ground, 10 m. Sediment size distribution in 1996[10] is also shown in Figure 9. It seems from Figure 9 that the mean sediment size around the Kano River mouth and the Fuji River mouth does not largely change for 20 years from 1996 to 2016. On the other hand, the mean sediment size from the east side of the Shouwa floodway to the Shin-Nakagawa floodway in 2016 seems to be from 10 mm to 20 mm larger than that in 1996. It is considered that this is due to beach nourishment with relatively larger materials than the native beach gravel.

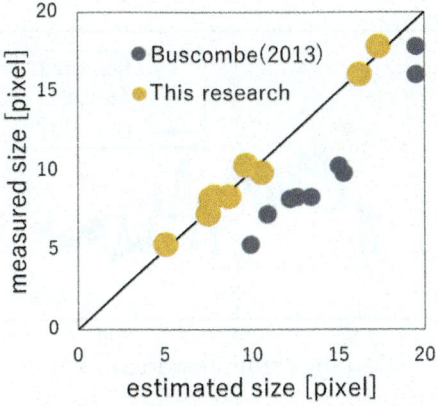

Fig. 8. Accuracy verification of the sediment size estimation method

The comparison of the sediment size distribution in October, 2016 and April, 2017 is shown in Figure 10. The sediment size distribution appears unchanged around the floodway. On the other hand, the sediment sizes about 100 m away from the floodway in October, 2016 seem to be larger than in April, 2017. The relationship between seasonal wave change and sediment size change could be discussed by conducting more frequent field surveys.

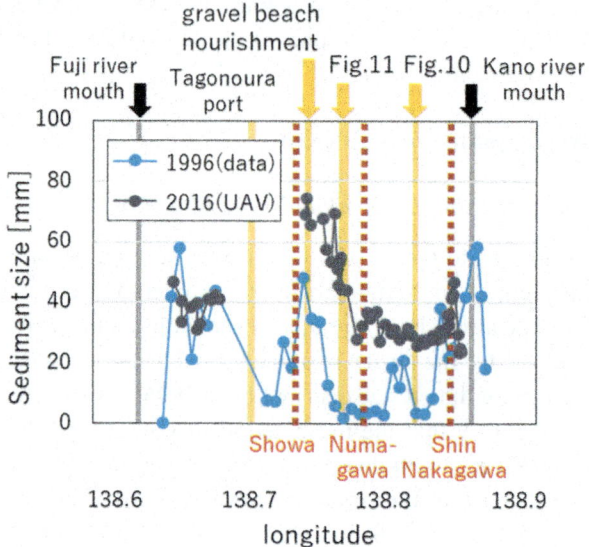

Measured data: from a material by Numazu work office (1998)

Fig. 9. Along-shore mean sediment size distribution near the shoreline in Fuji coast

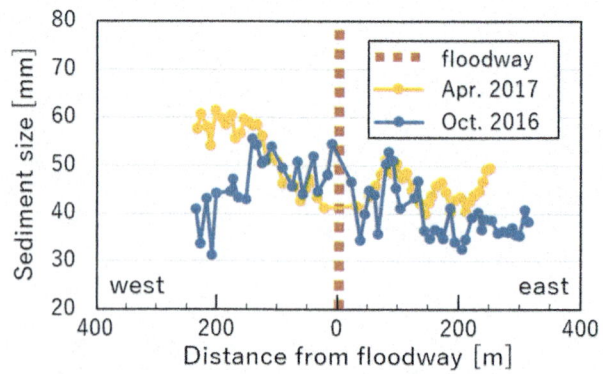

Fig. 10. The comparison of the mean sediment size near the Numagawa Dai-ni floodway between Apr., 2017 and Oct., 2016

4.2. *Sediment size mapping in the cross-shore direction*

The sediment size distribution in the cross-shore direction in October, 2016 plotted by the latitude is shown in Figure 11. The target area is indicated in Figure 1 and Figure 9. Unlike the alongshore direction, the real length per pixel in each image was determined by the number of pixels of the 1 m marker intervals, because topography change is larger in the cross-shore direction than in the alongshore direction. The cross-shore profile measured by RTK-GPS and that calculated by the real length per pixel in each image are also shown in Figure 11. It is suggested that the survey by UAV is capable of sediment size investigation considering topography change because estimated profile agrees well with the RTK measurements. Judging from Figure 11, the sediment size changes from 10 mm to 50 mm in 80 m in the cross-shore direction. The sediment size is from 40 mm to 50 mm, which is largest size at the target point, on the berm, is stably from 30 mm to 40 mm in the inland area, and sharply become smaller from the point where topography become more steep. Considering the above-mentioned observation, it is considered that sediment sizes on the Fuji Coast significantly vary in the cross-shore direction and conventional sampling method where sampling points are decided by the certain distance from shoreline can develop a large fluctuation in sediment size.

4.3. *Two dimensional sediment size mapping*

The wavelet transformation was applied for the images around Numagawa daini floodway in the field survey in April, 2017, and the D values and peak locations were calculated for each image. Then, the mean sediment size was calculated for every 1 m square mesh. Real length per pixel in each of the images were determined by the liner relationship with the various height from surface ground which was calculated using the elevation of every of the take-off points measured

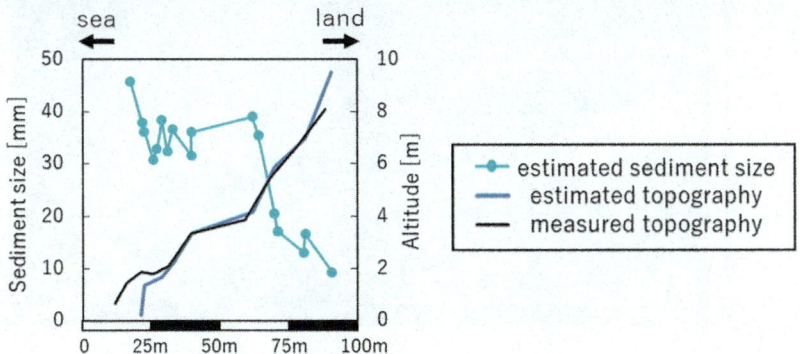

Fig. 11. The cross-shore sediment size distribution

by RTK-GPS, and the elevation of each photographed point was obtained from three dimensional topography data constructed by the SfM (Structure from Motion) method. The SfM method can reconstruct three dimensional form from multi-angle images, using images taken from 30 m height and RTK-GPS information of markers. A contour map from 1 m to 6 m created from the three dimensional topography data is shown in Figure 12. The mean sediment size in each 1 m square mesh mapped on two dimensional coordinate based on latitude and longitude by built-in GPS of UAV is shown in Figure 13. Comparing the map and the original images shown in Figure 13, it is confirmed that various sediment size and the boundary of sediment size change are well expressed. In addition to that, it is found that the elevation changes 6 m in 50 m of the cross-shore distance. Considering what mentioned above, it is shown that this method can be applied to gravel beaches where the bed slope is quite large. Furthermore, it could be confirmed that sediment size largely change in the cross-shore directions which is 15 m long. Thus, it is suggested that the two dimensional mapping is useful for topography monitoring around a floodway.

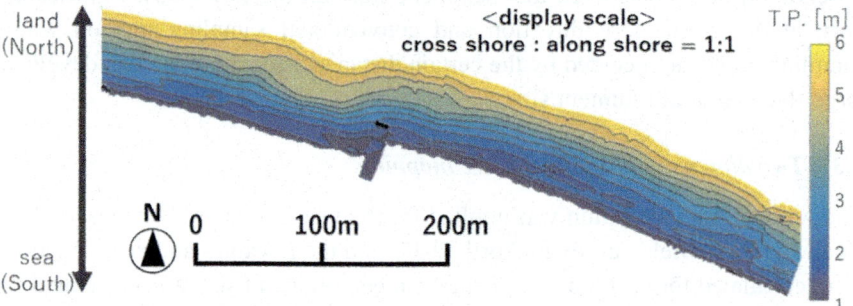

Fig. 12. Contour map near the Numagawa Dai-ni floodway with altitude of from 1 m to 6 m created from the three dimensional topography data by SfM

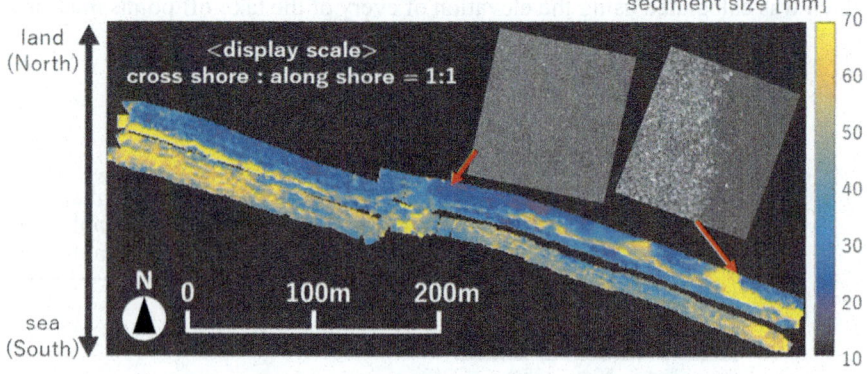

Fig. 13. Two dimensional sediment size distribution map near the Numagawa Dai-ni floodway

4.4. *Error evaluation*

Although the method developed in this study is a powerful method which can obtain the sediment size information for wide area with low costs, there are some points which should be reminded in order to secure sufficient accuracy. First, tree branches and debris drifted by waves have significant influence on the estimation. The estimated sediment size tended to be overestimated.

Second, the contrast of brightness in the picture also has influence on estimation results. To observe the influence of the contrast change, seven sample images were prepared and their lightness was changed to 5 levels (-100%, -50%, 0%, +50%, +100%). The errors of the estimated mean sediment size caused by the difference of lightness were as small as 2%. Therefore, the developed method is found robust against the brightness level in the whole image. However, in case that many small shades caused by human footprints were present in a part of images, the shade area was recognized as one large sediment. In addition to that, in case that a relatively large shade which was larger than the largest wavelet function caused by wave-dissipating concrete blocks were present in a part of images, the wavelet coefficients at relatively large scales tended to be large at the boundary between light area and shade area. The shade area must be influenced by the weather, sunny or cloudy, though it is not investigated yet.

Finally, overlapping between images is one of the factors which needs to be carefully adjusted. There were some data gaps in Figure 13. This is because some rims of each image was eliminated by two dimensional wavelet transformation because of low estimation accuracy. Thus, overlapping between images should be carefully determined. Another factor is the flying height of UAV. Although keeping the flying height of UAV is important for this method, the height would be unstable when strong wind occurred. This instability is proportional to the error size of the sediment size estimation because UAV altitude is proportional to the real length per pixel in the image. Thus, some countermeasures like photographing standardization scale appropriately in the images are needed to ensure the accuracy.

5. Conclusions

A method for mapping beach topography and sediment size distribution with high accuracy, high resolution and mobility by applying two dimensional wavelet transformation to continuous images taken by UAV from constant height was developed. In addition to that, applicability for fields was confirmed by applying the method to Fuji coast, which is one of the gravel admixture beach in Japan, and the impacts of gravel beach nourishment material and structures on the beach

630

process were discussed. Although there are still some issues such as influences caused by rubbish or structures' shade, this method innovated the sediment size investigation with high costs and labor and enable to conduct field survey for wide range involving the sediment transport system.

Acknowledgments

This study is partially supported by the KAKENHI (H, No.17K18898) provided by the Japan Society of Promotion of Science.

References

1. T. Okabe, S. Aoki, T. Uda, M. Serizawa and S. Kato, Analysis of offshore bathymetric changes using GPS and fishfinder data, *Journal of JSCE*, Ser. B2 (Coastal Engineering), Vol. 66, pp. 696-700 (2010) (in Japanese).
2. S. Takewaka, I. Gotoh and H. Nishimura, Survey on foreshore topography using X-band nautical radar, *Proceedings of Coastal Engineering, JSCE*, Vol. 50, pp. 546-550 (2003) (in Japanese).
3. Y. Matsuba, S. Sato and K. Hadano, Application of UAV based nearshore topography monitoring to a sand-bypassed beach on Fukude-Asaba coast in Shizuoka prefecture, *Journal of JSCE*, Ser. B2 (Coastal Engineering), Vol. 72, No. 2, pp. I_851-I_858 (2016) (in Japanese).
4. S. Sato, Sediment movement mechanisms in the fluvial system inferred from analysis of coastal sediments, *JSCE Summer Lectures on Hydraulics*, B-3, pp. 1-22 (2001) (in Japanese).
5. K. Yamamoto, Sampling of gravel river bed and statistical analysis, Doboku Gijutsu Siryo, Vol. 13, No. 7, pp. 40-44 (1971) (in Japanese).
6. G. Tsujimoto, F. Yamada and T. Kakinoki, A study on measuring sand particle size using digital images of sediment by Rubin's method, *J. JSCE*, Ser. B3 (Ocean Engineering), Vol. 24, pp. 1207-1212 (2008) (in Japanese).
7. D. Buscombe, Transferable wavelet method for grain-size distribution from images of sediment surfaces and thin sections, and other natural granular patterns, *Sedimentology*, Vol. 60, No. 7, pp. 1709-1732 (2013).
8. P. E. Freeman, Kashyap V., Rosner, R. and Lamb, D. Q., A wavelet-based algorithm for the spatial analysis of Poisson data, *Astrophysical. J. Suppl. Series*, Vol. 138, pp. 185-218 (2002).
9. D. C. Brown, Decentering distortion of lenses, *Photogrammetric Engineering*, Vol. 32, No. 3, pp. 444-462 (1966).
10. Fundamental Plan of Erosion Control of the Fuji Coast, Numazu Work Office, Ministry of Construction (1998) (in Japanese).

Study on Applicability of Synthetic Aperture Radar for Shoreline Monitoring

Sorayuki Akamatsu, Yoshimitsu Tajima, Takenori Shimozono and Shinji Sato

Coastal Eng. Laboratory, The University of Tokyo,
Hongo 7-3-1, Bunkyo, Tokyo 113-8656, Japan
E-mail: yoshitaji@coastal.t.u-tokyo.ac.jp

Monitoring shoreline change is one of essential tasks for appropriate design of coastal management and protection strategies. As one of remote sensing technologies, synthetic aperture radar, SAR can be a promising option because SAR is affected neither by cloud coverage nor by sun illumination. In this study, various kind of parameters, such as the grain size of bed materials, incident angles and wave height, were analyzed and it was found that the grain size is one of the most important factors for the shoreline monitoring skills using SAR.

Keywords: Synthetic Aperture Radar; ALOS-2; Coastal erosion; Monitoring shoreline.

1. Introduction

For better planning of coastal protection and conservation strategies, development of frequent monitoring of shoreline change is one of important tasks in the field of coastal engineering. Although use of aerial photographs or optical satellite imagery is often preferred as "cost-effective" monitoring techniques, such image-based monitoring technique still has disadvantages especially in frequency of obtained shoreline locations. Monitoring frequency may be crucial for understanding of long-term trend of shoreline change. The instantaneous shoreline location may be affected by tide and waves and also affected by seasonal or shorter term morphology change.

Use of synthetic aperture radar, SAR, can be one of good options to replace the conventional monitoring techniques. Since SAR is one of active sensors, which generate microwaves and observe their reflections (Figure 1), and the radar is not sensitive to clouds, the monitoring is affected neither by cloud coverage nor by sun illumination. SAR-based monitoring thus has an advantage in monitoring frequency. SAR has been used for many studies in the field of coastal engineering such as the estimation of swell spatial behavior[1] or the estimation of tsunami-induced building damage.[2] Some studies about the estimation of shoreline location using SAR are also reported by Tajima et al.[3] or Yukikawa et al.[4] However, those studies focused only on one target area and at most ten SAR scenes, and the study which inspects the parameters affecting the back scattering observed by SAR is not known. Therefore, the goal of this study is to comprehensively inspect these parameters such as off nadir angles, incident angles of microwaves (Figure 1), grain size of bed materials, beach slopes and wave heights and to disclose the condition under which SAR can be applied for the shoreline monitoring.

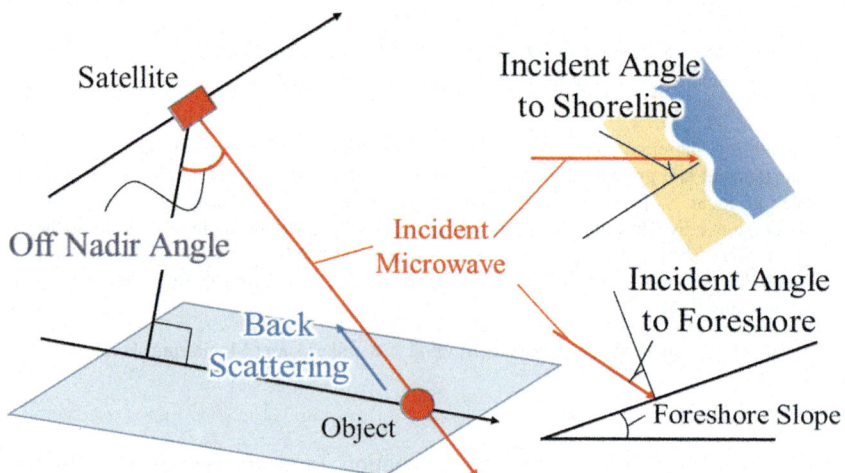

Fig. 1. SAR overview and incident angle of microwave.

2. Target Areas and Used SAR Data

To inspect the influence of various parameters on SAR-based observations in a comprehensive manner, it is necessary to select the several target areas whose characteristics are different from each other and also to analyze many SAR scenes. Figure 2 shows the selected six target areas with their characteristics. In each area, a field survey was conducted to measure the actual shoreline locations which will be compared with SAR data. Shoreline locations were measured by the handy GPS (GPSmap 60CSx, GRAMIN). The total length of the measured shoreline in the six target areas was approximately 58.3km. The error of the handy GPS is around 3.0m. These field surveys were conducted within several days from the data acquisition date of the SAR observation and when the tide level was nearly equal to the one at the time of SAR observation.

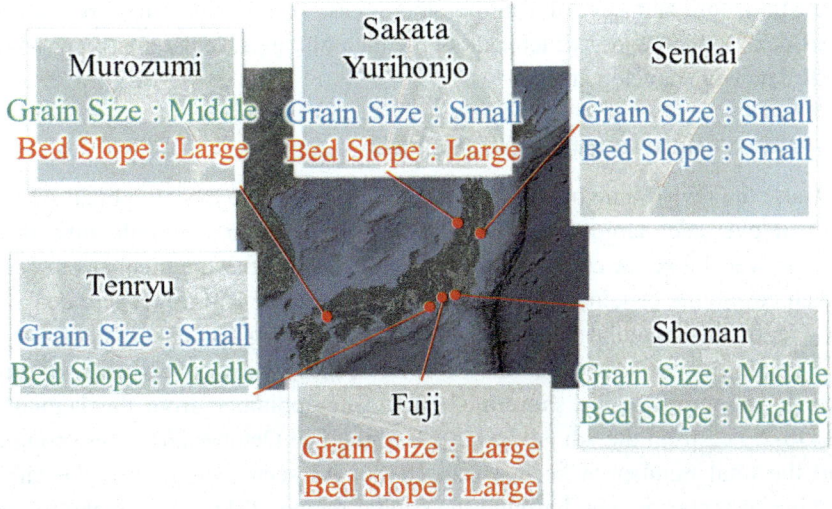

Fig. 2. Target areas and their characteristics.

In this study, SAR data which were obtained by ALOS-2, a satellite launched by Japan Aerospace Exploration Agency (JAXA) on the 24th May 2014, were analyzed. ALOS-2 has a L-band synthetic aperture radar, PALSER-2, and its spatial resolution, as fine as 3m, was improved compared with the conventional ALOS and revisit time is 14 days. Therefore, ALOS-2 is expected to achieve the development of SAR-based shoreline monitoring. This study used all the SAR scenes which were observed from the initiation of ALOS-2 operation, August 2014, to the field survey to analyze many SAR scenes. Table 1 summarizes the number of obtained SAR scenes at each target site and the date of the field survey. This study used the PALSAR-2 scenes with the observation mode of Stripmap and the spatial resolution of 3m in the case of single or dual polarization and 6m in the case of full polarization. SAR was obtained in the CEOS format with the processing level of Level 1.5 in Sendai

Table 1. The number of obtained SAR scenes and the date of field survey.

Target Area	SAR Scenes	Dual Pol. Scenes	Full Pol. Scenes	Date of Field Survey
Sendai	21	0	1	15 Aug. 2016
Murozumi	27	0	2	26 Aug. 2016
Tenryu	33	0	2	23 Nov. 2016
Shonan	35	5	2	30 Nov. 2016
Fuji	40	0	2	3 Dec. 2016
Sakata, Yurihonjo	64	0	4	15 and 16 Dec. 2016

and Murozumi, and Level 1.1 in the other areas. To efficiently analyze multiple SAR data, Python codes which extract the data of the particular area from whole SAR data were developed.

3. Observation Frequency of ALOS-2

Because the revisit time of ALOS-2 is 14 days, one can expect to obtain ALOS-2 scene over each target area every 14 days. To confirm the availability of the scenes over the entire coast of Japan, Figure 3 compared the number of scenes which covers the certain part of the shoreline along the entire Japanese coast. In the Figure, the shorelines were separated into 1km-long section and the color on this image represents the number of SAR scenes which covers each of these 1km-long section of the shoreline. The scenes obtained from the launch of ALOS-2, i.e., from the 4th August 2014, to the 2nd October 2016 were counted and the total number of scenes was 10580. As seen in Figure 3, the larger number of scenes is found around the capital region, Tokyo, in Tohoku region,

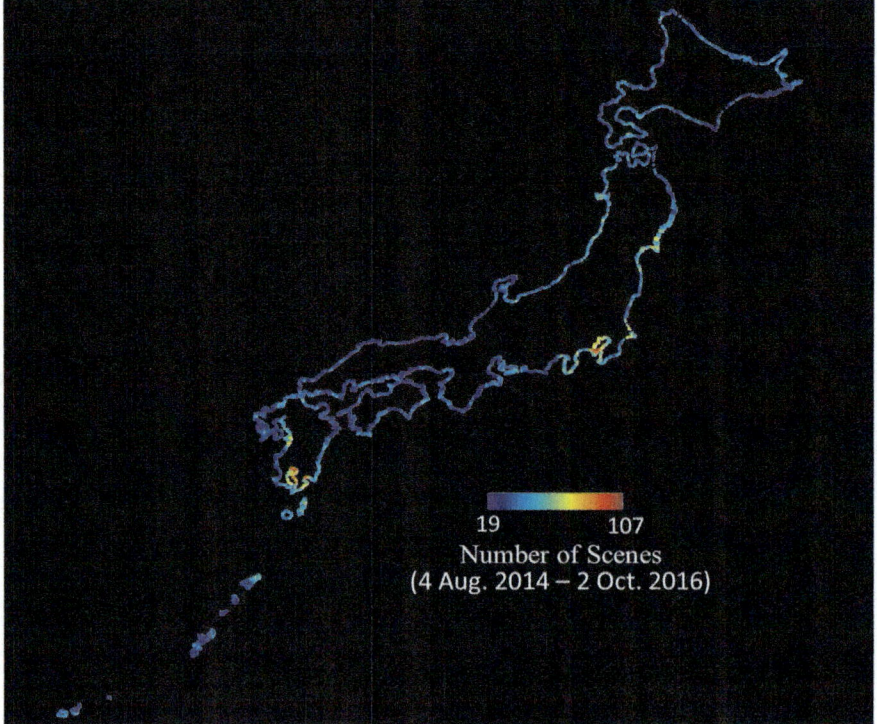

Fig. 3. The observation Frequency by ALOS-2 in Japan.

the north-eastern part of the main island, where suffered the 2011 Tohoku earth quake tsunami disaster, and around Kumamoto and Kagoshima, the south part of Kyushu island where volcanic eruption occurred. Although the number of scenes are not equally distributed, on the other hand, the minimum number of the counted scenes was 19 over the two years, i.e., 9 scenes in each year. Therefore, it can be concluded that the enough number of scenes are obtained to estimate the shoreline change.

4. Analysis of the Suitable Conditions for the Monitoring Shoreline by Using SAR data

4.1. *Quantification of the possibility of extracting shorelines*

To analyze the suitable conditions for the monitoring shoreline by using SAR data, it is necessary to evaluate the ability of SAR images for detection of the shoreline in a quantitative manner. In this study, rectangular areas on the foreshore and on the sea near the shoreline were respectively selected as shown in Figure 4, and the mean back scattering intensity of these two areas was calculated, and the ratio of the mean intensity in foreshore area to the one in sea area was computed as the indicator of the reliability of SAR-based shoreline detection. In the other words, shoreline detection can be easier and more reliable if the the value of this indicator becomes larger. Based on the visual observations of SAR images, the shoreline detection appeared to be easy if this indicator is larger than 1.5. The spatial scale of the selected rectangular domains

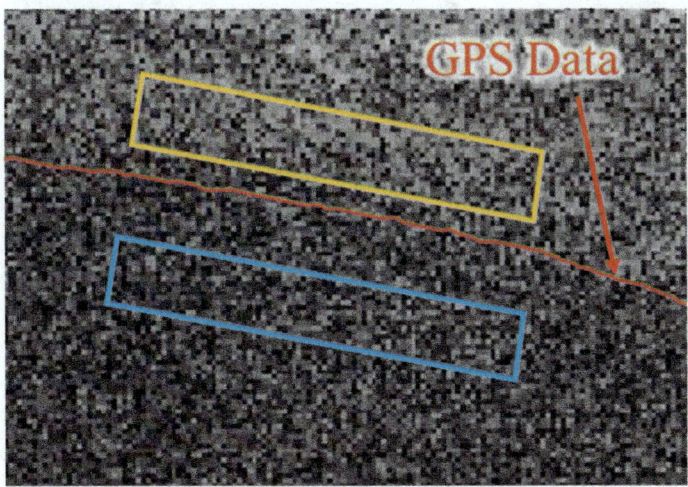

Fig. 4. An example of cutting the foreshore and sea area.

was basically about 30m in the offshore direction, and 100m in the shoreline direction while the size of these domains was modified accordingly on the relatively narrow beach. The analysis was conducted by calculating this ratio of all obtained SAR data. The following section investigates the influence of various observation parameters on the value of proposing indicator.

4.2. *Grain size of bed materials*

Grain size of bed materials, which is considered as the roughness of the beach, may be of the most important parameter in this study. Figure 5 shows the proposed indicators, i.e., the ratio of the back scattering intensity on the sea and on the beach obtained at Fuji and Sendai. The horizontal axis represents the off nadir angle and each circle indicates the value of indicator obtained from each of the SAR scenes. The grain size in Fuji is the largest in all the target sites of this study while the one in Sendai is the smallest. In the case of Fuji coast, as seen in Figure 5, most of the computed ratios of each SAR scene are larger than 1.5 and this feature indicates that the shorelines can be estimated in almost all the scenes. Therefore, monitoring shorelines by SAR can be conducted at the coast which has the same or larger grain size of Fuji. In the case of Sendai, however, most of the computed ratio were below 1.5 and it may be difficult to detect the shoreline in most of the scenes.

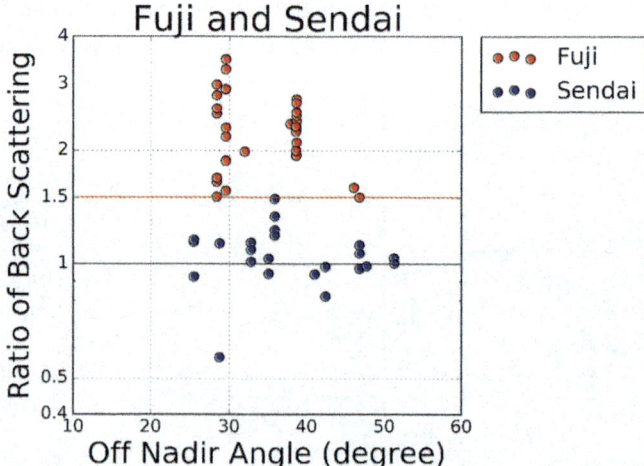

Fig. 5. The ratio of the mean back scattering intensity of the foreshore and sea in Fuji and Sendai.

4.3. *Incident angle of microwaves*

The most significant effect of the incident angle of microwaves was observed in the case of Murozumi coast because the Murozumi coast has the steeper foreshore slope and the shoreline stretching in the north-south direction, which is almost perpendicular to the incident direction of the microwaves radiated from ALOS-2. Figure 6(a) shows the ratio of the back scattering, and "Right" indicates the incident direction of microwaves is from the left (west) to the right (east) and thus from the offshore to the onshore in the case of Murozumi coast and vice versa in "Left". When the incident direction was right, in other words, the incident angle of the microwave to the foreshore becomes larger and thus the computed ratio tends to be larger. As seen in Figure 6(a), the computed ratio was bigger than 1.5 in most of the scenes and thus one can expect high reliability in SAR-based shoreline detection.

Figure 6(b) shows the wider and the narrower views of the SAR images. The locations of the narrower close images are indicated green and yellow squares in the wider image. As seen in the figure, clear land-sea boundary, i.e., the shoreline, is observed in the domain of yellow square while the land-sea boundary is not very clear in the domain of green square. This feature indicates that the incident angle of the microwave toward the shoreline may be one of important factors for reliable SAR-based shoreline detections.

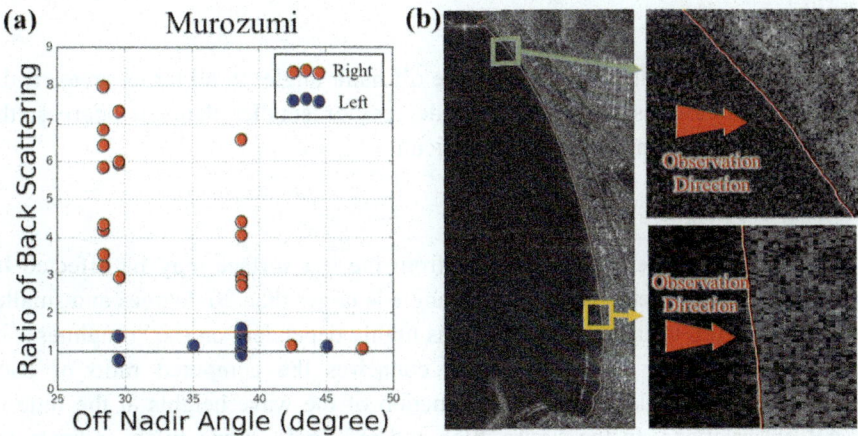

Fig. 6. (a) The ratio of the mean back scattering intensity of the foreshore and sea in Murozumi. (b) SAR image of Murozumi.

638

Figure 7 shows the plot of computed ratio of back scattering at the coast around Shonan, Sakata and Tenryu, respectively. In this figure, the values of Sakata were dispersed compared with the other two areas. This feature is because the shoreline stretches in the north-south direction but the foreshore slope is not stable due to high wave in Sakata. On the other hand, the foreshore slope seems not to be important in Shonan or Tenryu, where the shorelines are in east-west direction, parallel to the incident microwaves.

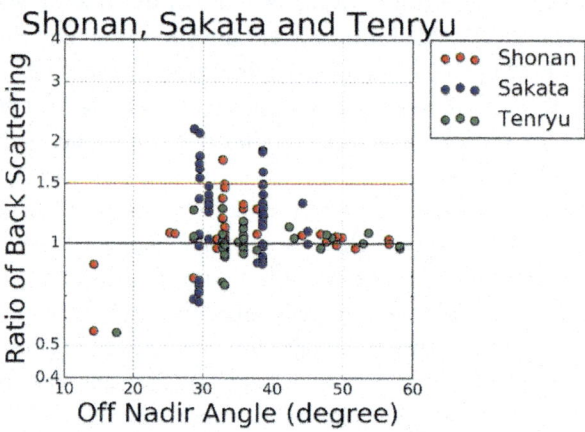

Fig. 7. The ratio of the mean back scattering intensity of the foreshore and sea in Shonan, Sakata and Tenryu.

The incident angle depends on the off nadir angle. In all target areas, if the off nadir angle was larger than 40 degrees or smaller than 20 degrees, the shoreline detection appeared to be difficult.

4.4. Wave height

The intensity of the back scattering from the sea surface may be affected by wave conditions even if the off nadir angle is larger than 20 degrees and smaller than 40 degrees because the intensity is highly dependent on the "roughness" of the surface of the target. Figure 8 compares the computed ratio of back scattering at the Sakata coast as a function of the wave heights at the time of SAR observations. In the figure, blue and red circles respectively indicate the computed ratios of each scene when the off nadir angle was larger (blue) and smaller (red) than 30 degrees. As seen in the figure, the computed ratio was clearly decreased under the waves higher than 2m in the case red circles, i.e., the case with off nadir angle of less than 30 degrees while no clear dependency on the wave height was observed in the case of blue circles, i.e., the case with the off nadir angle of over 30 degrees. This feature indicates that the relatively higher back scattering intensity from the rough sea surface is expected if the off nadir angle is less than 30 degrees and shoreline detection may be difficult under such conditions. In the area where the wave height is high, the off nadir angle should be large to estimate shorelines.

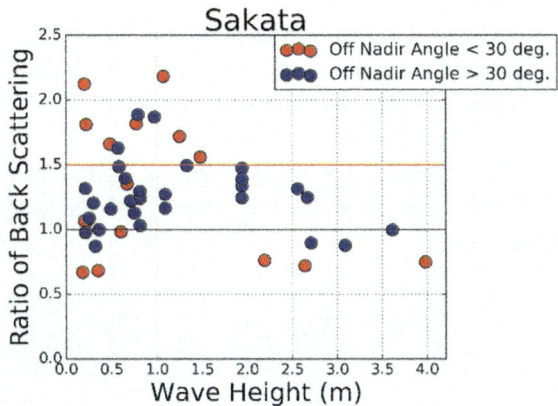

Fig. 8. The relationship between the wave height and the ratio of the mean back scattering intensity of the foreshore and sea in Sakata.

5. Conclusion

This study investigated the optimum SAR-based observation conditions for detection of the shoreline location. The suitable conditions of the SAR-based monitoring of the shorelines are summarized below: (1) if the grain size of bed materials was bigger than 20mm, the shoreline location can be easily detected regardless of the other conditions; (2) in the case of smaller grain size, the shoreline can be detected in some scenes if the wave height is smaller than 0.8m and the incident angle to the shoreline is smaller than 30 degrees, or the one to foreshore is less than 40 degrees; (3) the best off nadir angle for detection of the shoreline ranges from 30 to 40 degrees, which minimizes the influence of the sea wave.

Acknowledgments

Authors acknowledge that this study was supported by the research project funded by the Ministry of Land, Infrastructure, Transport and Tourism, Japan.

640

References

1. Nishimura, Y. and Matsuura, T., 2016. Estimation of swell spatial behavior from ALOS-2 / PALSAR-2 satellite image, Journal of JSCE, B2 (Coast. Eng.), 72, No. 2, pp. 169-174.
2. Gokon, H. and Koshimura, S., 2015. Estimation of tsunami-induced building damage using L-band synthetic aperture radar data, Journal of JSCE, B2 (Coast. Eng.), 71, No. 2, pp. 1723-1728.
3. Tajima, Y., Mochizuki, S., Funatake, S., et al., Investigation of nearshore sedimentary characteristics along the west coast of Sri Lanka based on the analysis of satellite images and feldspar thermoluminescence, Journal of JSCE, B2 (Coast. Eng.), 67, No. 2, pp. 631-635.
4. Yukikawa, S., Matsumoto, A. and Takewaka, S., 2016. Shoreline positions and bottom profiles retrieved from synthetic aperture radar, Journal of JSCE, B2 (Coast. Eng.), 72, No. 2, pp. 1735-1740.

Experimental Study on Shoreface Nourishment from Offshore Bar for Quick Recovery of Shoreline After a Storm

S. Aoki

Osaka University, Department of Civil Engineering
Suita, Osaka 565-0871, Japan
E-mail: aoki@civil.eng.osaka-u.ac.jp

G. Himori

Sekisui Chemical Co., LTD.
Iwamizawa, Hokkaido 068-0014, Japan
E-mail: himori001@sekisui.com

D. T. Chu

Penta-Ocean Construction Institute of Technology
1534-1 Nasushiobara, Tochigi, 329-2746, Japan
E-mail: ducthang.chu@mail.penta-ocean.co.jp

In this paper, a new method of beach nourishment, in which sediment dredged from an offshore bar is deposited at the shoreface, is proposed as one of the measures for beaches that suffer from irreversible beach erosion due to a single storm event. A series of laboratory experiments were carried out to investigate the effectiveness of the method. Some parameters that may influence the effectiveness were discussed based on the experimental results by regular waves. Effect of irregularity of the waves is also shown.

Keywords: Shoreface nourishment; Offshore bar; Shoreline recovery.

1. Introduction

There are some sandy beaches suffering from irreversible erosion after a great erosion caused by high storm waves, where the erosion does not recover even by the ordinary waves after the storm and the erosion becomes worse in the succeeding storms.[1] Although such an irreversible beach deformation may be related to both on-offshore and alongshore sediment transport, we focus on the former that allows us to study it as a two dimensional problem, where the irreversible erosion occurs because of the formation of an immobilized offshore bar by storm waves.

Figure 1 shows an example of beach profile change on the Enshu-Nada coast in Japan. Six major typhoon events influenced the coast from 2011 to 2014 in which waves higher than 4m were observed four times. In the years after 2012, the bar-trough profile has not changed very much even in the calm season of winter. For the coasts with such irreversible erosion, protecting the beach by armor units or any structures such as detached breakwaters may deteriorate

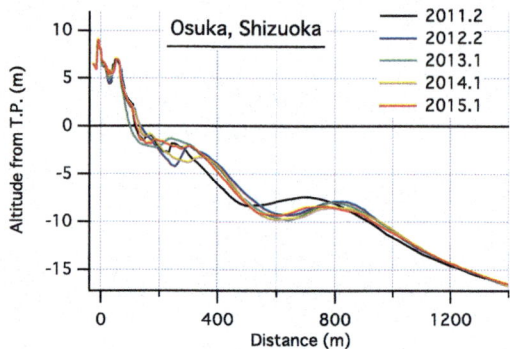

Fig. 1. An example of annual change of beach profile, showing an immobilized offshore bar (Enshu-Nada Coast, Japan).

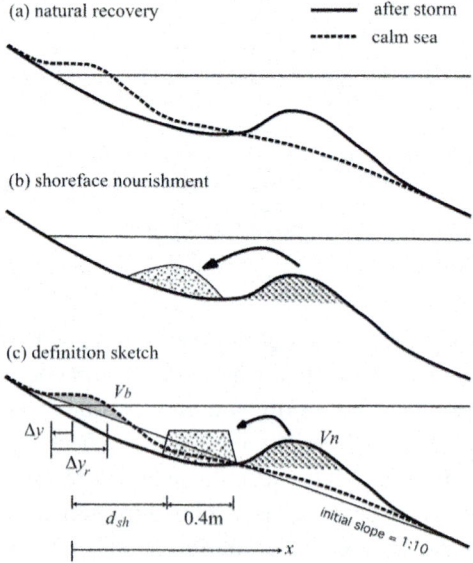

Fig. 2. Concept of the shoreface nourishment from offshore bar and definition sketch.

Table 1. Experimental cases. (E21S and E23S are irregular wave cases)

Case	Waves		Dredge area	Deposition point	Sand volume	Water depth
No.	H (m)	T (s)	x (m)	d_{sh} (m)	V_n (m³/m)	(m)
E01S	0.04	1.5	-	-	-	0.4
E02S	0.08	1.5	$2.2 < x < 2.75$	1.6	0.014	0.38
E05S	0.08	1.5	$2.2 < x < 2.75$	1.6	0.014	0.37
E07S	0.04	1.5	$2.2 < x < 2.85$	1.6	0.02	0.4
E08S	0.04	2.0	$2.2 < x < 2.85$	1.6	0.02	0.4
E09S	0.06	1.5	$2.2 < x < 2.85$	1.6	0.02	0.4
E10S	0.08	1.5	$2.7 < x < 3.05$	1.6	0.008	0.4
E11S	0.08	1.5	$2.2 < x < 2.75$	1.6	0.014	0.4
E12S	0.08	1.5	$2.4 < x < 3.1$	1.6	0.03	0.4
E13S	0.08	1.5	-	-	-	0.4
E16S	0.10	1.5	$2.2 < x < 2.85$	1.6	0.02	0.4
E21S	**0.057**	**1.5**	**$2.2 < x < 2.85$**	**1.6**	**0.02**	**0.4**
E23S	**0.113**	**1.5**	**$2.2 < x < 2.85$**	**1.6**	**0.02**	**0.4**
E25S	0.08	1.5	$2.2 < x < 2.85$	1.3	0.02	0.4
E26S	0.08	1.5	$2.2 < x < 2.85$	1.4	0.02	0.4
E27S	0.08	1.5	$2.2 < x < 2.85$	1.5	0.02	0.4
E28S	0.08	1.5	$2.2 < x < 2.85$	1.6	0.02	0.4
E29S	0.08	1.5	$2.2 < x < 2.85$	1.7	0.02	0.4

beach environment and self-recovery function of the natural beach. Although the beach nourishment by additional volume of sand is one of the practical measures, it is not always possible because of availability of sand and operating cost.

In this study, we propose a new method of beach nourishment without any additional volume of sand, in which a certain volume of sand is dredged from an offshore bar and transported to the foreshore in order to accelerate the beach recovery after a storm (Fig. 2). As a relevant study, the authors investigated the influence of sand dredging from the offshore bar on the shoreline retreat.[2,3]

The feasibility of the proposed method was studied by a series of hydraulic experiments in a wave flume of 40cm water depth using a sloped beach of 1/10 with sand of which median grain diameter d_{50} = 0.22mm. In the experiments, high regular waves with wave height H = 0.14m and wave period T = 1s were first generated for 60 minutes resulting in shoreline retreat and formation of bar and trough, and then with this equilibrium profile as the initial profile, small waves were generated for 120 minutes to see how the eroded beach recovers by the small waves. Before the generation of small waves, a volume of sand, V_n, was dredged and placed at a distance of d_{sh} as indicated in Fig. 2(c). The beach profiles were measured by an optical profiler at 0, 30, 60 and 120 minutes after

644

the generation of the small waves. All the experimental cases are shown in Table 1, in which the dredge area is expressed as a range of x in Fig. 2(c). Among the experiments, in the cases of E21S and E23S, irregular waves having the indicated significant wave height and period were used. The spectrum of the incident waves of these random wave cases were determined based on the Bretschneider-Mitsuyasu spectrum. In the cases of E02S and E05S, water depths were slightly decreased than other cases to see the effect of water level.

2. Results and Discussion

2.1. *Effectiveness of the shoreface nourishment from an offshore bar*

To investigate the effectiveness of the proposed method, we compare the results in a pair of cases with and without nourishment. The comparison was made

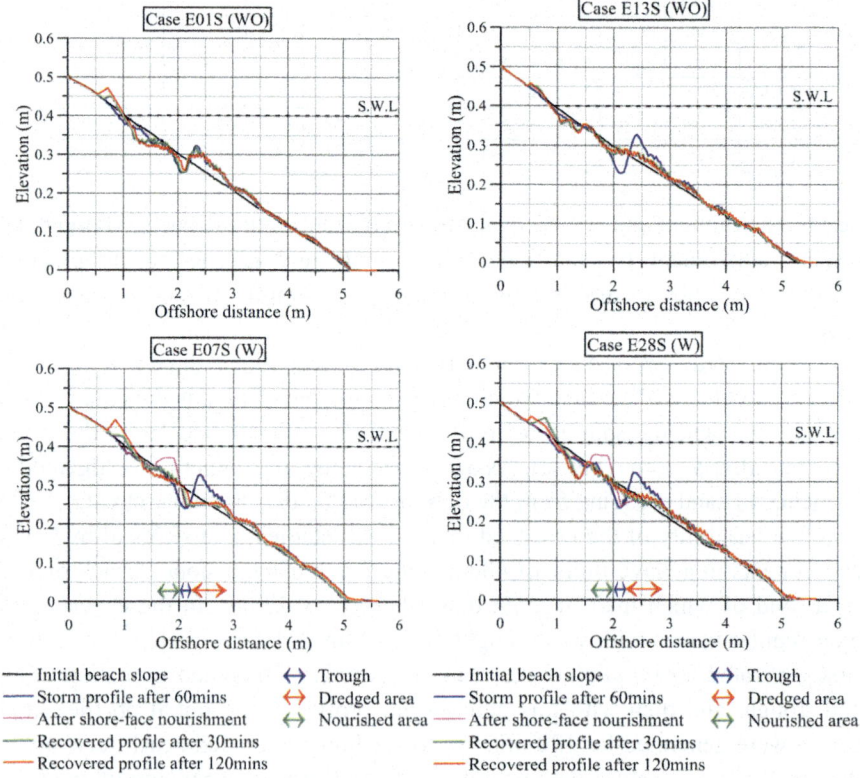

Fig. 3. Comparison of profile change
w/ and w/o nourishment (E01S, E07S) .

Fig. 4. Comparison of profile change
w/ and w/o nourishment (E13S, E28S).

between the cases E01S and E07S, and also between E13S and E28S. The latter pair is different form the former pair only in the wave height in the process of beach recovery. The results are shown in Figures 3 and 4. In the cases without nourishment, smaller waves (E01S) yielded smaller deformation of the bar and larger berm on the shore. In contrast, in the case E13S, the bar disappears quickly. For the cases with nourishment (E07S and E28S), the shoreline advanced and berm developed more quickly than the cases without nourishment.

Figure 5 shows time variation of the shoreline position, where $\Delta y_r/\Delta y$ denotes recovery ratio of the shoreline to the initial retreated distance of the shoreline, Δy. The 100% recovery means that the shoreline recovers to the initial position. Figure 5 shows that the shoreline recovered quickly in the case with nourishment. In the case E28S, the shoreline showed retreat after the quick recovery. This is because the berm moved more onshore. Figure 6 shows the time variation of the berm volume, V_b. The volume is defined as the area deviated from the initial profile as shown in Fig. 2(c). The large berms were formed in the cases with shoreface nourishment and its effectiveness was confirmed.

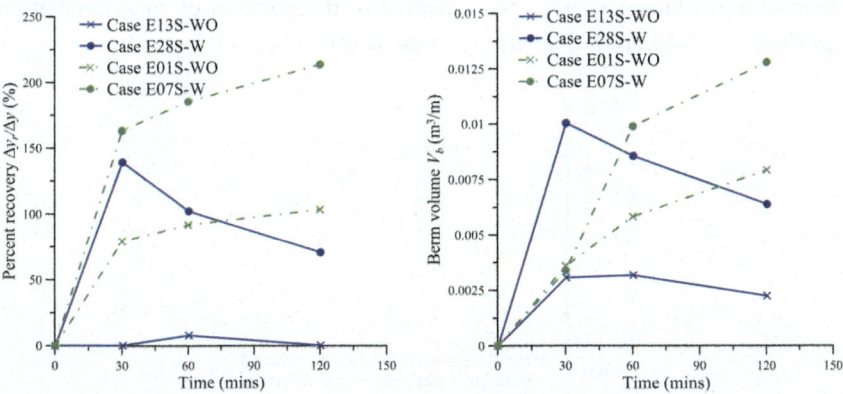

Fig. 5. Time variation of recovery ratio (E01S, E07S, E13S, E28S).

Fig. 6. Time variation of berm volume (E01S, E07S, E13S, E28S).

Figure 7 shows the distribution of the sediment transport rate estimated from the profiles of 0 and 30 minutes for the cases E13S and E28S. In the case of the shoreface nourishment (E28S), the negative or onshore sediment transport rate became greater than the case without nourishment (E13S), although the positive or offshore sediment transport were seen in the trough.

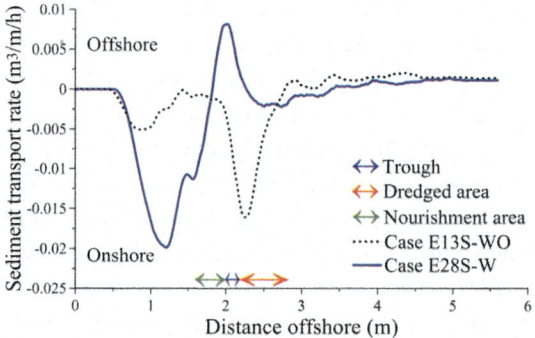

Fig. 7. Comparison of sediment transport rate(0-30 min.: E13S, E28S).

2.2. Influence of nourishment position

Comparison among the cases E25S, E26S, E27S, E28S and E29S shows influence of the position of depositing the dredged sand. Figure 8 shows the berm volume V_b as a function of the distance, d_{sh}. The effectiveness of the shoreface nourishment is not very sensitive to the position of sand deposition. Depositing the sand around the trough may be effective.

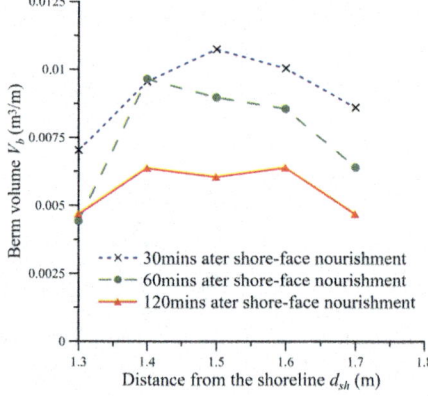

Fig. 8. Influence of nourishment position on the berm volume (E25S-E29S).

2.3. Influence of wave condition in the process of recovery

Comparison among the cases of E07S, E08S, E09S, E16S and E28S indicates influence of the wave condition in the recovering process. Figure 9 shows the time variation of recovery ratio, which shows the waves having small wave steepness show high efficiency in shoreline recovery. Figure 10 shows the berm volume as function of wave steepness. For large wave steepness, the berm volume tends to be decreased.

2.4. *Influence of nourishment volume*

Figure 11 shows the effects of nourishment volume on the shoreline recovery ratio. As the volume increases, the shoreline recovery becomes significant. Figure 12 also shows the effect of the nourishment volume.

2.5. *Influence of water level*

In the cases E02S and E05S, the experiments were carried out with slightly smaller water depth in the recovery process considering that the mean water level falls a little than that in the storm. Figure 13 shows that the recovery ratio is very sensitive to the small decrease in the water level.

2.6. *Comparison between regular and irregular waves*

The results and discussion shown above are for regular waves. To see if the proposed method is effective in irregular waves, the results of regular (E07S and E28S) and irregular (E21S and E23S) waves were compared. The significant wave heights of E21S and E23S were chosen so that the wave energies of these cases were the same as the cases of regular waves, E07S and E28S, respectively. Figure 14 shows the profile change in the case E21S. Comparing with the case

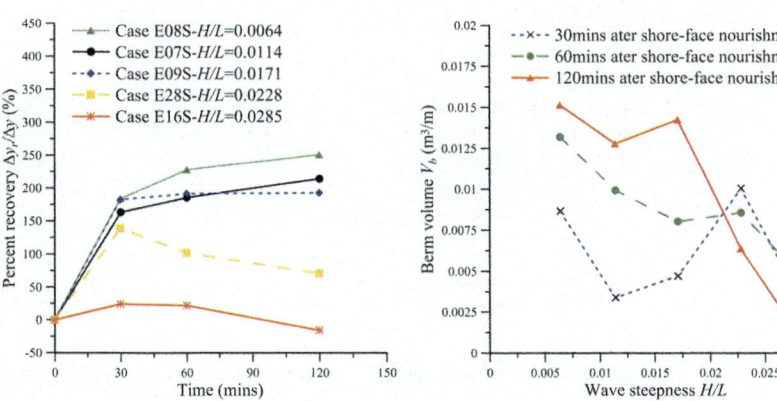

Fig. 9. Influence of wave steepness on recovery ratio.

Fig. 10. Influence of wave steepness on berm volume.

648

E07S in Fig. 3, the final profile in irregular waves became flatter and the berm developed in higher position of the beach. Figure 15 shows the comparison of the berm volume between regular and irregular waves. Although the volume becomes smaller, the quick recovery was still observed in irregular waves.

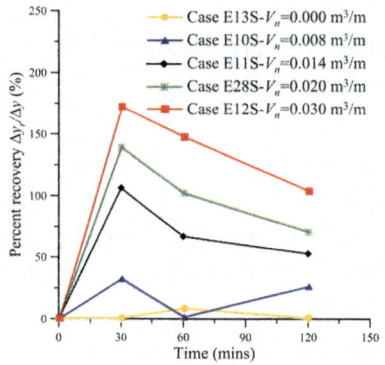

Fig. 11. Time variation of shoreline recovery recovery ratio (E10S~E13S, E28S).

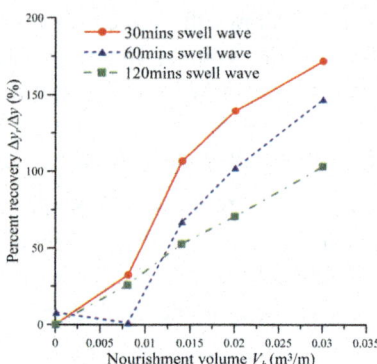

Fig. 12. Influence of nourishment volume on shoreline recovery ratio.

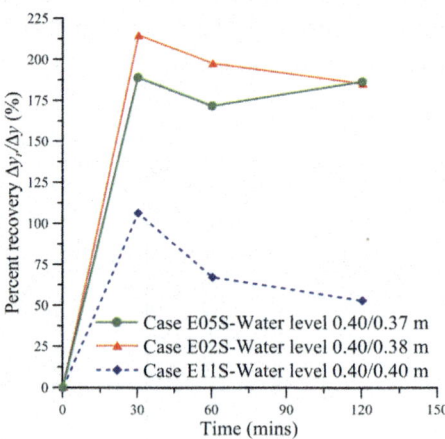

Fig. 13. Influence of water level in the process of recovery (E02S, E05S, E11S).

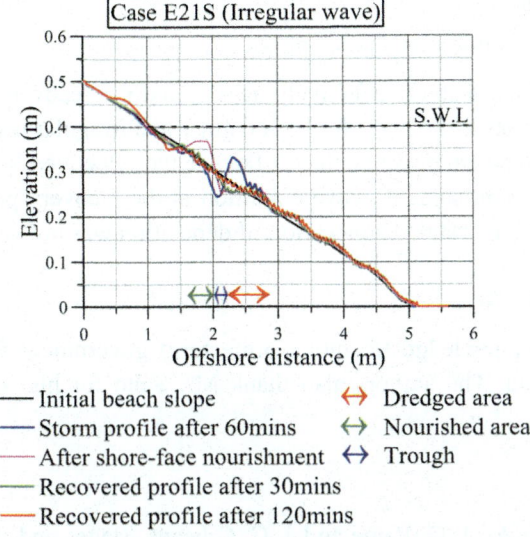

Fig. 14. Profile change in irregular waves.

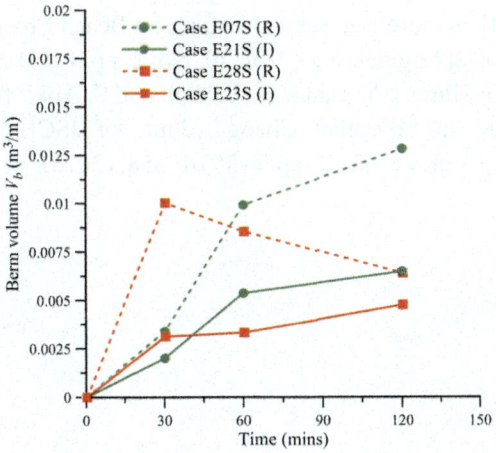

Fig. 15. Change in berm volume by regular (R) and irregular (I) waves
(E07S, E21S, E28S, E23S).

3. Conclusions

Major findings of this study are as follows:

(1) The proposed method effectively accelerated the recovery of the shoreline and formation of the berm for both regular and irregular waves.
(2) Among the parameters that may influence the effectiveness of the method, a volume of nourishment, wave condition in the recovery process, and water level were important. The location of deposition was not very important.

Acknowledgement

The authors acknowledge Shizuoka prefectural government for providing the bathymetric data. The authors also thank Mr. Saito for his cooperation in the experiments.

References

1. Morton, R. A., J. G. Paine and J. C. Gibeaut, Stages and durations of post-storm beach recovery, Southeastern Tex-as Coast, U.S.A., J. Coastal Research, Vol. 10, No. 4, pp. 884-908, (1994).
2. Chu D. T., G. Himori, B. T. Vinh and S. Aoki, An Experimental Study of the Effect of Offshore Bar Sand Dredging on Beach Erosion, Jour. of JSCE, Ser. B2 (Coastal Engineering), Vol. 70, No. 2, pp. I_531-I_535, (2014).
3. Chu D. T., G. Himori, Y. Saito, B. T. Vinh and S. Aoki: Impact of Dredging at Sand Bar on Shoreline Change, Jour. of JSCE, Ser. B2 (Coastal Engineering), Vol. 71, No. 2, pp. I_553-I_558, (2015).

Effectiveness of Gravel Beach Nourishment on Pacific Island

Susumu Onaka[a,†], Shingo Ichikawa[a], Masatoshi Izumi[b], Takaaki Uda[c], Junichi Hirano[d]
and Hideki Sawada[d]

[a]Nippon-Koei Co., Ltd., 1-14-6 Kudankita, Chiyoda, Tokyo 102-8539, Japan
[†]E-mail: onaka-ss@n-koei.jp, ichikawa-sn@n-koei.jp

[b]Futaba Co., Ltd., 2-76 Minami, Koriyama, Fukushima 963-0115, Japan
E-mail: izumi@futasoku.co.jp

[c]Public Works Research Center, 1-6-4 Taito, Taito, Tokyo 110-0016, Japan
E-mail: uda@pwrc.or.jp

*[d]Japan International Cooperation Agency (JICA), 5-25 Nibancho, Chiyoda,
Tokyo 102-8012, Japan*
E-mail: Hirano.Junichi@jica.go.jp, Sawada.Hideki@jica.go.jp

The South Pacific Island nation of Tuvalu, which is composed of coral gravel and sand, is vulnerable to storm waves and sea level rise, resulting in beach erosion. Beach nourishment with self-produced coral gravel and sand was implemented in Tuvalu as the first trial of a user- and eco-friendly type of coastal conservation measure in Pacific Island countries. In order to examine the applicability of this type of coastal conservation measure, continuous monitoring has been carried out for one year to check the change in shoreline and beach profile. Beach monitoring for large-scale reclamation project, which was executed at the neighboring coast in almost the same period, was also conducted to compare the change of beach in the two different projects. The results show that the executed gravel beach nourishment can maintain stability under seasonal and extreme condition of wave actions.

Keywords: Gravel beach; Nourishment; Wave overtopping; Coral reef; Climate change; Adaptation; Monitoring; Tuvalu; JICA; ODA.

1. Introduction

The South Pacific Island nation of Tuvalu, which is located 1,000 km north from Fiji, consists of four small islands and five atolls. The total area is about 26 km^2 and this is the fourth smallest country in the world. Fongafale Island, which is located in Funafuti Atoll with a 2.4 km^2 area (Fig. 1), is the capital island in Tuvalu and more than half of the Tuvaluan population of about 6,000 people are

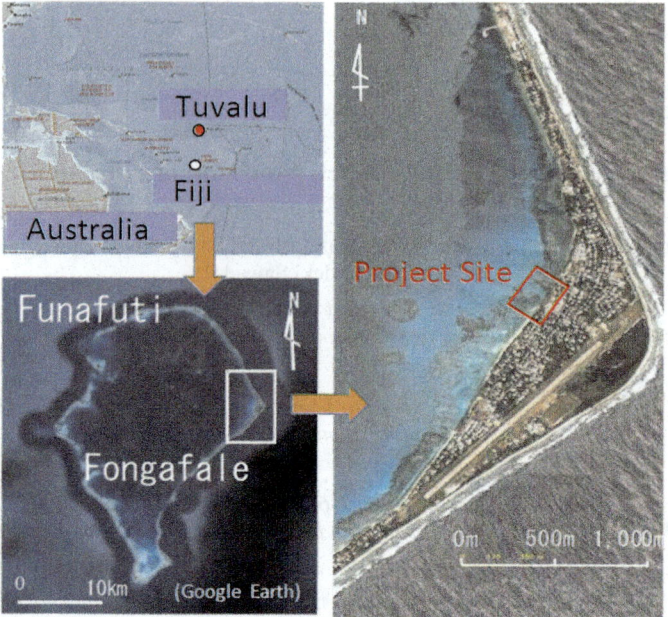

Fig. 1. Satellite Image of the Project Site in Tuvalu.

concentrated in this small island. The land was formed by the accumulated coral sand and gravel; the elevation of the land is very low and the maximum height is only about +4.5 m from the CDL. Due to the rapid concentration of the population to Fongafale Island in the last 20 years, the coastal area which was an undeveloped area before was highly utilized as a residential area, and coastal issues such as coastal erosion, wave overtopping, flooding, deterioration of water quality, and degradation of corals have been a serious problem in Tuvalu.

As one of the remedies, a pilot scale project of beach nourishment using coral gravel and sand was implemented (Ichikawa et al., 2016, Uda et al., 2013). The construction was completed in December 2015 and the beach monitoring has been carried out continuously since the completion of the construction. Apart from this project, another big-scale reclamation project was implemented at the neighboring area by the Tuvaluan government in the same period, and the sand filling was undertaken on the existing beach. This study aims to show the change in beach behavior due to gravel beach nourishment based on the monitoring results observed for one year. Also, the change in beach behavior for the two different projects, namely, gravel beach nourishment and reclamation project, was compared.

2. Project Outline

2.1. *Outline of the gravel beach nourishment project*

The Project was executed as a Japanese technical cooperation project in order to enhance the protection function against the risk of coastal disaster including the impact of climate change, and to examine the effectiveness and applicability of user- and eco-friendly type of coastal conservation measure of beach nourishment method in the small Pacific Island. The Project is divided into three phases. The first phase was the "planning and design phase", which was executed from March 2012 to March 2013. The second phase was the "construction phase", which was undertaken from January to December 2015. The third phase was the "monitoring and adaptive management phase", and this phase started in January 2016 and will continue until December 2017. This study is based on the monitoring results taken for one year which was from January 2016 to December 2016.

2.2. *Selection of coastal conservation method*

The reasons why the gravel beach nourishment method was selected are as follows:

- Each island in Tuvalu was originally formed by the accumulated coral gravel and sand due to the effect of wave action (Uda et al., 2015). The natural beach consists of coral gravel and sand which existed on the lagoon side of the Funafuti Atoll in the past. Basically, it is desirable to learn the natural process for the formation of the beach and land for the selection of the costal conservation measures.
- The coastal area at the lagoon side is now highly utilized as a residential area. The community strongly requested to consider both protection function and usage of the beach in the selection of the coastal conservation measures.
- The land of Fongafale Island in Funafuti Atoll consists of only coral gravel and sand, and the land area is very limited. Common construction materials, such as armor rock and concrete, were very hard to obtain in the country of Tuvalu. In addition, it is also desirable to use self-produced materials in Tuvalu and to select an easy protection method taking into account the sustainable beach maintenance work which will be carried out by the Tuvaluan side with their strong initiative after the construction under their own social and economic condition.

These are the reasons why beach nourishment using coral gravel and sand, which can only be procured in Tuvalu, was selected as an adequate coastal conservation measure in Tuvalu to recover the same image of the previous natural beach.

2.3. *Beach condition before and after the project*

Due to the existence of the main community hole and church in Fongafale Island behind the target project area, a lot of residents are commonly gathering in this area since it is a main public area. However, the condition of the beach in front was seriously deteriorated. The sandy beach that existed in the past completely disappeared and the existing concrete block type seawall which was constructed as a remedy against coastal erosion also collapsed. Wave run up and overtopping in the hinter residential area frequently occurred as shown in Fig. 2. Due to such deterioration of beach condition, people could not use the beach space. Such change in the condition of the beach was causing the reduction of people's awareness on the beach environment, and the beach area became one of the dumping sites for garbage (Fig. 3).

After the implementation of the project, the beach drastically changed as shown in Fig. 4. About 3,300 m³ of coral gravel and 4,500 m³ of coral sand were filled into the project area (Fig. 5), with a longshore distance of 180 m. Coral gravel was taken from the surrounding island which is located in the same atoll. These coral gravels were washed ashore during the Cyclone Bebe which attacked the Funafuti Atoll in 1972 (Maragos et al., 1973). The coral sands on the other hand were provided by the Tuvaluan government. This sand was dredged from the seabed at the lagoon side of Funafuti Atoll and stocked on land under the other implemented project. Average width of the nourished beach is about 20 m, in which gravel was filled at the backshore side with 6 m width and sand was filled at the foreshore side with an approximately 15 m width. Rock armor type groins (armor rock was imported from Fiji) were constructed at both ends of the project area in order to minimize the unexpected future loss of gravel

Fig. 2. Wave Overtopping at Hinterland. (Before the Project)

Fig. 3. Deposition of Garbage on the Beach. (Before the Project)

and sand due to wave action. Construction work was commenced in August 2015 and completed in December 2015.

2.4. *Outline of the reclamation project by the Tuvaluan government*

There were several sunken areas on the land in Fongafale Island. These were leaving holes (called "borrow pit"), which were made by the excavation activity of the US military during the second world war to obtain construction material for the runway of the warplane. To rehabilitate this, a new project to backfill into the borrow pits was undertaken through the New Zealand grant project in 2015. In this project, a large amount of coral sand was dredged from the se abed at the lagoon of Funafuti Atoll by using a pump dredger, and backfilled into the borrow pits.

Fig. 4. Beach Condition Before and After the Project.

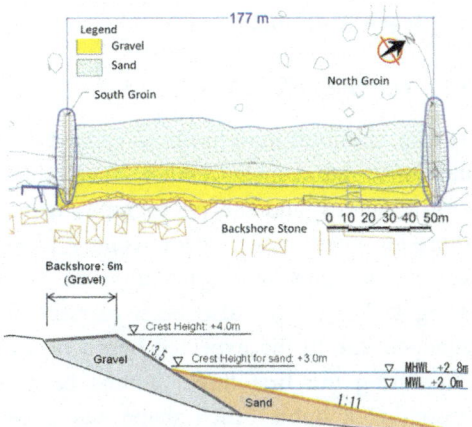

Fig. 5. Layout and Typical Cross Section of Beach Nourishment.

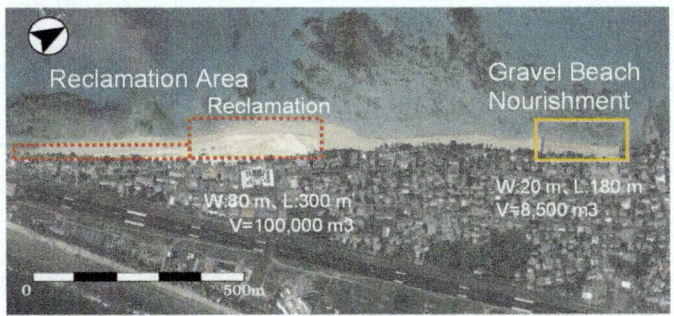

Fig. 6. Location of the Gravel Beach Nourishment and Reclamation Project.

After the completion of the project, the big-scale reclamation project was also undertaken by the Tuvaluan government at the neighboring beach in November 2015 using the same dredger. As shown in Fig. 6, the location of reclamation area is 0.7 km far from the gravel beach nourishment site to the north. About 100,000 m³ of dredged sand was directly filled into the coastal area with a 300 m distance and 80 m width without any enclosure. This reclamation activity was completed in the latter part of December 2015. However, additional sand filling and construction of two groins were conducted due to significant sand loss which was caused by the attack of Cyclone Ula which approached to Funafuti Atoll from the end of December 2015 to the beginning of January 2016. Finally, it was completed in June 2016.

3. Beach Monitoring

3.1. *Outline of beach monitoring*

The beach monitoring has been carried out continuously since the completion of the construction in December 2015. Main items for monitoring are the beach profile survey and the photographs taken from the fixed points. Further, the shoreline positioning survey using a handy type GPS to cover a wide coastal stretch including the reclamation project area and taking oblique photograph using a drone have also been carried out. Time interval for monitoring was changed due to the expected beach behavior. Monthly monitoring was carried out during the first three months because significant profile change toward becoming a stable beach shape was expected. After that, the monitoring was carried out every three or six months. Six times of monitoring work were undertaken up to December 2016. Fig. 7 shows the position of the lines for the beach profile survey. The survey lines were set every 20 m at both the inner and

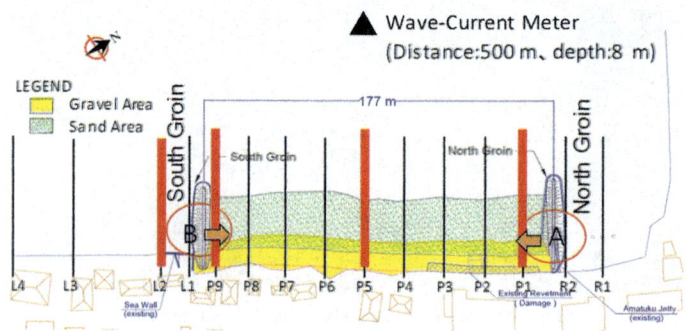

Fig. 7. Monitoring Line.

outer part of the project area. Wave observation has been conducted using a bottom mount-type self-recorded wave-current meter (Wave-Hunter) since March 2015 before the commencement of the construction work. The wave-current meter was installed offshore in front of the project site at the lagoon area with 8 m depth, and wave, current and water temperature were measured every two hours.

3.2. *Wave observation during monitoring period*

Two seasons exist in Tuvalu, i.e., winter (dry) season from April to October and summer (rainy) season from November to March. During the winter season, main wind direction is from the southeast (means wind blows from the landside at the lagoon side coast). On the other hand, during the summer season, main direction is from northwest (means wind blows from offshore side at the lagoon side coast). Average wind seed is mainly less than 4 m/s. Due to such seasonal change of wind direction, wave commonly becomes calm during the winter season. Cyclone commonly approaches Tuvalu from November to March in the summer season. During the cyclone approach, strong wind from the west of more than 10 m/s blows at the atoll, and waves at the lagoon side become rough due to this strong wind. If this happens during spring tide, high wave run-up, overtopping, and intrusion into the land side occur significantly.

Figure 8 shows the wave observation result for one year from December 2015 (completion of the construction). In this figure, the wave forecasting result (orange colored line) estimated from wind speed is also shown together with the observed one. Wave height ($H_{1/3}$) at the lagoon side was commonly less than 0.1 m; however, wave of more than 0.5 m height was observed during the storm. During this observation period, two cyclones approached the atoll; one is "Cyclone Ula", which approached from late December 2015 to early January 2016, and the other is "Cyclone Winston" which approached in the middle of February 2016. During Cyclone Ula, a 1.3 m wave height ($H_{1/3}$), which is the same level of the predicted wave height with 10-year return period, was observed.

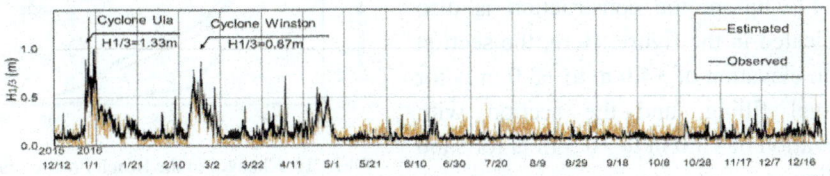

Fig. 8. Wave Height at the Lagoon Side During the Monitoring Period.

658

Fig. 9. Beach Change Based on Photos Taken From Fixed Point (Upper: Point A (North Side),
Lower: Point B (South Side)).

3.3. *Beach change for gravel beach nourishment*

The photos taken from the fixed points at both end points (Point A and Point B in Fig. 7) for one year (just after the construction, 6 months later and 1 year later) are shown in Fig. 9. At Point A (north side of the project area), sand at the foreshore part seems to decrease and gravel is exposed. On the other hand, at Point B (south side of the project area), sand seems to be accumulated.

Figure 10 shows the change in beach profile at the four representative monitoring lines (three lines for inside of the project area and one line (L2) for outside of the project area). The beach profile before the construction is also indicated in the figure. Here, the section with elevation of +4.0 m to +3.0 m is for gravel filling, and the section with elevation of +3.0 m to +0.5 m is for sand filling.

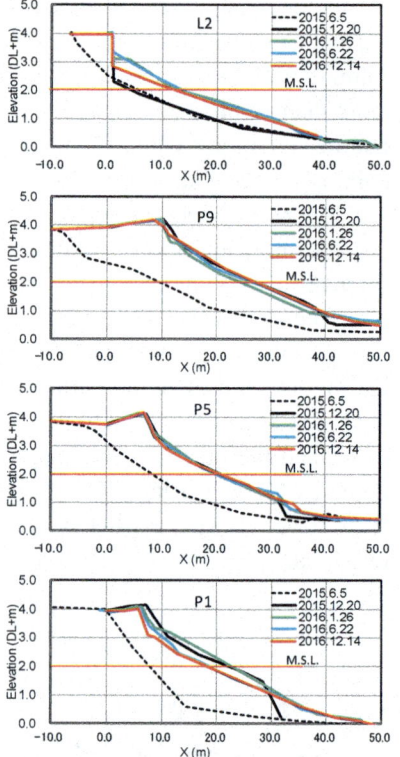

Fig. 10. Changes in the Beach Profiles as
Represented by the Four Lines.

To make clear the change of the position for both sections of gravel and sand, differences of the distance from the initial position just after completion of construction at these two elevations (+3.7 m is the representative elevation for gravel section and +2.0 m is for sand section) are also shown in Fig. 11. Form these figures, the following tendency was observed:

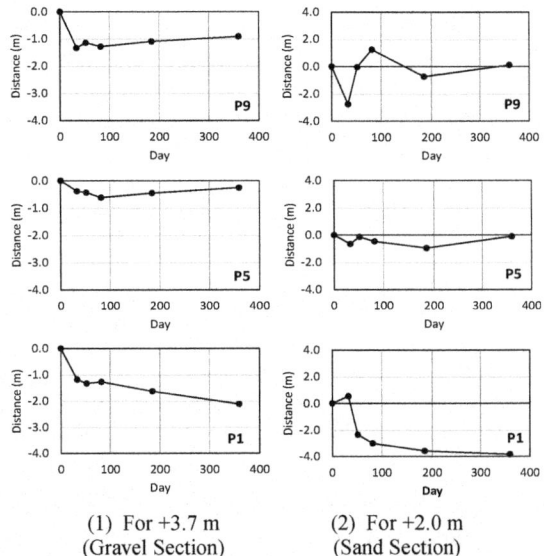

(1) For +3.7 m
(Gravel Section)

(2) For +2.0 m
(Sand Section)

Fig. 11. Change in On-offshore Position at +3.7 m and +2.0 m.

- The difference of beach profiles, especially the beach slope for each section was not really significant, and almost the same slope was secured for one year. This means that the initially designed beach slope, which was determined based on the actual observed beach slope at a nearby beach, could be appropriate.

- Due to the strong waves during Cyclone Ula from late December 2015 to early January 2016, significant sand accumulation occurred at the southern outer part of the project area (L2 in Fig. 10). On the other hand, inside the project area, decrease of sand section was observed at the south side (P9 in Fig. 10) and accumulation was observed at the north side (P1 in Fig. 10). Also, the steep slope at the toe part of the sand section at P1, which existed just after the completion, became mild. Wave direction during this period was observed from the west (this means wave was incident from left side obliquely to the shoreline). From this, northward littoral drift might be significant during this period. Due to this northward littoral drift, some quantity of sand at the outer west coast flowed into the project area.

- As shown in Fig. 11, about 1 m retreat of the gravel section (backshore area) was observed at the south side (P9) due to strong wave action during Cyclone Ula. However, after that, no significant change of gravel section was observed. Retreat of the sand section of about 3 m was temporarily observed at P9 during Cyclone Ula. However, after 52 days (10 February 2016), the sand section was recovered at an almost the same level as the initial condition and can keep the stable condition.

- At the north side (P1), the gravel section retreated about 1 m during Cyclone Ula. After that, degree of retreat became small but it still decreased gradually. Total retreat for one year was about 2 m at P1. The sand section at P1 was temporarily accumulated during Cyclone Ula. However, after 52 days (10 February 2016), sand section retreated about 2 m and still decreased gradually.
- Significant change in beach profile at the center point (P5) for both sand and gravel sections was not observed during the one year monitoring period.

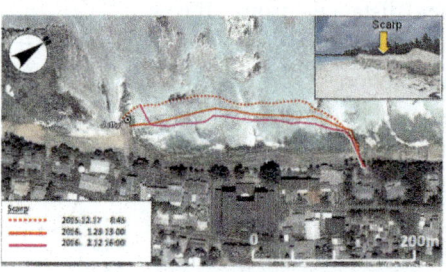

Fig. 12. Change in Shoreline at the Reclamation Area.

3.4. *Shoreline change at the reclamation project area*

The change in position of the shoreline at the reclamation project area was measured by using a handy type GPS. During the construction work, dredged sand was directly pumped to the beach area through the sand discharge pipe and no leveling work at the foreshore area was undertaken; therefore, it was difficult to identify the position of the shoreline visually. However, significant scarp was formed after the pumping due to the wave action at the foreshore part (shown in the upper photo in Fig. 12). Thus, the position of the shoreline was used to measure the position of the scarp, and the adjustment due to change in position for tidal condition was not required. Figure 12 shows the

Fig. 13. Change in Beach Condition
(Upper: Before Reclamation,
Middle: Before Groin Construction,
Lower: After Groin Construction).

change in the position of the scarp forming for three months from December 2015 just after the completion of the sand pumping work. During this period, sand was filled into the existing beach with 300 m distance and 80 m width, and no coastal structure, such as groin, was constructed. After the approach of Cyclone Ula on 23 January 2016, significant retreat of the beach with average width of about 20 m and maximum width of 25 to 30 m was observed. Most of the disappeared sand was moved to the north side for a distance of 170 m by wave action. Such retreat continued until 12 February 2016 and roughly 10 m retreat was further observed for 20 days, even though wave was calm in this period. After that, two groins were constructed at both sides of the reclamation area to minimize further sand loss and additional filling of sand was conducted. Finally, this reclamation project was completed in June 2016.

Figure 13 shows the change in beach condition in the three stages, i.e., before the reclamation, completion of sand filling (but without groins), and the completion of the project (completion of groin construction and additional sand filling). After June 2016, which was the completion of the project, no storm waves came as shown in Fig. 8, and significant sand loss which occurred before the construction of groins was not seen even though some amount of sand near each groin flowed out to the outer area.

3.5. Change in volume of gravel and sand

Figure 14 shows the changes in volume of gravel, sand, and both, which were calculated from the result of the beach profile survey as shown in Fig. 10. Vertical axis means the change in the ratio of the volume to the initial one. Three lines are indicated in each figure, which means the volume rate for the total area (black line), that for 60 m interval at the south side (yellow line), and that for 60 m interval at the north side (blue line).

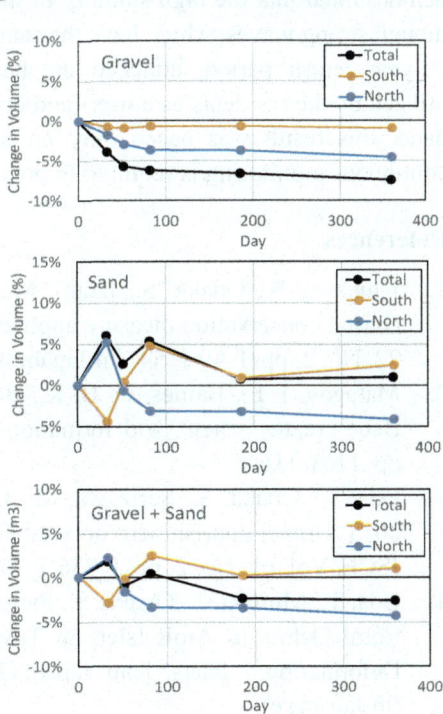

Fig. 14. Change in Volume of Gravel and Sand.

As shown in the upper figure in Fig. 14, about 7% of gravel was moved from the original backshore section to the foreshore sandy section; it was especially significant at the north side. The volume of sand was temporarily increased by the approaching two cyclones. After that, some decrease of sand volume was observed and almost the same volume of sand as in the initial condition was finally kept (middle figure in Fig. 14). The total volume of combined gravel and sand decreased a little compared to the initial condition by about 3% for one year as shown in the lower figure in Fig. 14. On the other hand, the sand loss at the reclamation area from January to February 2016 was estimated to be about 34% (about 34,000 m^3 of sand) based on the analysis for area change. This significant sand loss was mainly caused by the lack of consideration for construction of supplementary coastal structure even though the sand was filled with wide width of 80 m.

4. Conclusion

As a result of beach monitoring of gravel beach nourishment for one year, it was demonstrated that the high stability of the nourished beach can be secured even though strong waves, which have the same level as the predicted wave height for 10-year return period, attacked the beach. Nourished beach is now highly utilized by the residents as a user- and environment-friendly beach. On the other hand, this result was based only on one year monitoring data, and further continuous monitoring is required to prove the long-term stability of the beach.

References

1. Ichikawa, S., Onaka, S., Izumi, M., Endo, S., Uda, T., 2015. Sustainable coastal conservation measure applied in island nation, Tuvalu, J. JSCE, Vol. 72, No. 2, pp. I_49-I_54. (in Japanese)
2. Maragos, J. E., Baines, G. B. K., Beveridge, P. J., 1973. Tropical cyclone Bebe creates a new land formation on Funafuti Atoll, Science, Vol. 181, pp. 1161-1164.
3. Uda, T., Onaka, S., Serizawa, M., Izumi, M., San-nami, T., Miyahara, S., 2013. Gravel nourishment on west coast on Fongafale Island in Tuvalu, J. JSCE, Vol. 69, No. 2, pp. I_736-I_740. (in Japanese)
4. Uda, T., Mimaki, J., Onaka, S., San-nami, T., 2015. Rapid Accumulation of Coral Debris to Atoll Islets in Tuvalu by Storm Waves and Subsequent Deformation of Islets, Tran. Japan. Geomorph. Union, Vol. 36, pp. 285-308. (in Japanese)

Approaches to Establish a Community-based Beach Management in the Pacific Island Country

Shingo Ichikawa[†], and Susumu Onaka

Nippon Koei Co., Ltd., 1-14-6 Kudan-Kita, Chiyoda-ku, Tokyo 102-8539, Japan
[†]E-mail:ichikawa-sn@n-koei.jp
http://www.n-koei.co.jp/english

Takaaki Uda

Head, Shore Protection Research, Public Works Research Center, 1-6-4 Taito,
Taito-ku Tokyo 110-0016, Japan
E-mail: uda@pwrc.or.jp

Junichi Hirano[†] and Hideki Sawada

Japan International Cooperation Agency (JICA), Nibancho Center Building 5-25,
Niban-cho, Chiyoda-ku, Tokyo 102-8012, Japan
[†]E-mail: Hirano.Junichi@jica.go.jp

As a part of the Pilot Gravel Beach Nourishment Project in Tuvalu by Japan International Cooperation Agency (JICA), approaches to establish a community-based beach management had been implemented as the first trial among the Pacific Island countries. A community-based beach management requires an active public participation; however, public awareness on beach seemed to be quite low in the beginning of the Project because of the terrible beach condition with dumped rubbish scattered on the beach. Therefore, raising public awareness on beach and making people understand and experience the benefits from the beach were considered necessary to establish the community-based beach management. In the Project, several public relations and educational activities were implemented for these purposes and one notable effective activity was the beach sports festival that was firstly held in Tuvalu. Public awareness on beach, especially on beach use and environment, was much improved through these activities so that community and residents voluntarily started beach management activities such as periodic beach cleaning and public notification of illegal activities on the beach. These changes contributed to maintain the good condition of the Project beach for almost one and half year after the construction.

Keywords: Community-based beach management; Public relations; Environmental education; Gravel beach nourishment; Pacific Island country.

1. Background and objective

Gravel beach nourishment was implemented on Fongafale Island in Tuvalu[1,2] in 2015, as the first application of beach nourishment in Pacific Island countries, under a Japanese official development assistance (ODA) Project to protect low-elevated coastal area and to restore functions of beach use and environment (Fig. 1 and Fig. 2). The gravel beach nourishment consists of two layers of gravel and sand, respectively, which was originally designed based on a healthy beach in Tuvalu. The gravel layer was adopted to improve its stability against waves and sand layer was done to improve beach use and environment. To maintain a beach with good condition for a long term requires continuous beach management by the community in which the beach is located. Since a beach is a part of the local community life, their active participation is essential for beach management. People in the small pacific islands, however, have little knowledge and experience in beach management and, therefore, it was not easy to obtain their cooperation directly. Thus, changing public awareness on beach through public relations and educational activities was considered as the first approach to realize community-based beach management.

In this paper, the methodologies applied to establish a community-based beach management through the Project were described, and their effects were evaluated in terms of beach management.

2. Policy on activities to establish a community-based beach management

Beach condition before the Project implementation directly showed the people's lack of interest on the beach; concrete blocks used for temporary protection were scattered and rubbish dumped by residents was accumulated and smelled bad on the beach as shown in Fig. 3. Furthermore, it is anticipated that even if beach environment has been improved by the Project, it would go back to the same bad situation sooner or later if nothing changes regarding public awareness on beach.

Thus, the following step-by-step approaches were considered to be necessary to realize community-based beach management: 1) people become interested on the beach, 2) people experience and understand the benefits from the beach, and 3) people begin to take concrete actions on beach management. To accomplish these objectives, approaches through several public relations and educational activities were applied during the entire Project period, which were for planning, design, construction, and post construction phases.

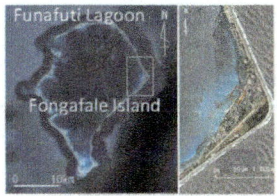
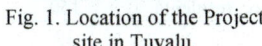

Fig. 1. Location of the Project site in Tuvalu.

Fig. 2. Before and after the Project (gravel beach nourishment).

Fig. 3. Beach condition before the Project with scattered concrete blocks (left) and accumulated rubbish on the beach (right).

3. Activities implemented to establish a community-based beach management

The following sections show activities implemented to establish a community-based management mainly with two purposes, i.e., to improve public interest on beach as a first step and to make people recognize and experience the benefits from beach as a second step.

3.1. *Activities to improve public interest on beach*

The activities had been implemented mainly from the beginning of the Project to the completion of the construction, which took about one year. Expecting a widespread effect to men and women of all ages, various types of public relations activities were implemented as shown in Table 1 and Fig. 4.

For example, a radio announcement, which was the only public media in Tuvalu, could widely broadcast the Project information to the public not only in Fongafale Island where the Project was implemented but also in other islands. On the other hand, beach cleaning event gathered nearby stakeholders such as local community members and residents who live near the Project site to make them more interested in the beach condition and a beach tour during construction targeted primary school students to raise the interest on beach of younger generations.

Table 1. Activities implemented to improve public interest on beach.

Type of activity	Phase	Target	Involved numbers
Radio announcement	All	Public	N/A
Stakeholder meeting on the Project	Planning, design and construction	Community, residents	150
Beach cleaning event	Planning, design	Community, residents	150
Singing competition	Planning, design	Public	30
Drawing competition	Planning, design	School students	30
Beach tour during construction period	Construction	School students, community leaders from all islands	100
Opening ceremony of beach	Construction	Government, community, residents	70

Fig. 4. Activities to improve public interests on beach; left) beach cleaning event before construction, center) beach song competition, right) beach tour during construction.

3.2. *Activities to make people recognize and experience the benefits from the beach*

Since gravel beach nourishment is a measure with multiple functions, i.e., protection, beach use, and environment, activities were implemented to make people recognize and experience benefits in each aspect as follows.

3.2.1. *Activities to make people realize the benefits in protection*

A cyclone hit Tuvalu in December 2015 just after the construction was completed and residents living near the Project area firstly experienced high waves after the construction. A brief interview survey was conducted with the residents to see whether they recognize any Project effects or not. Results showed that more than 80% answered that there was a significant effect in protection against high wave during the cyclone compared with the situation before the construction and with adjustment beaches. This survey results were also utilized as materials to publicize the protection function of the Project beach for public relations and educational activities afterwards.

3.2.2. *Activities to make people recognize the benefits in beach use and environment*

Typical beach uses in Tuvalu were mainly for bathing in the morning and early evening and for boat parking and landing for local fishery industries. To make people recognize other ways of beach use, the beach sports festival involving primary school students was planned and implemented through the cooperation between the government and the local community. It was the first trial in Tuvalu to hold a sports festival on the beach area and more than 800 people including the students' family participated in the festival. It should be noted that this beach sports festival also aimed to create opportunities for participants to consider their roles in using the beach area safely and pleasantly.

Firstly, to create an opportunity for school students to think about the importance of beach, classes on beach environment were held in cooperation with the teachers in the school. Secondly, to make participants understand and experience their roles of using the beach area pleasantly, a beach cleaning event was held a few days before the beach sports festival involving school students, teachers, community, and the government. Finally, the beach sports festival was held to make participants experience the pleasantness of playing on the beach. Figure 5 summarizes this stepwise sequence of activities, i.e., environmental education, beach cleaning event, and beach sports festival. In the beach sports festival, two types of program were prepared, namely, one that uses sandy beach area and another that uses shallow water area. These programs were organized depending on the beach area changes due to tidal fluctuations during the festival so that participants could experience different types of beach use as shown in Fig. 6.

Fig. 5. Sequence of public relations and education activities related to beach sports festival.

Fig. 6. Programs that use sandy beach during low tide (Left) and that use shallow water area during high tide (right).

4. Outcome of the activities

Outcomes of the activities in terms of public awareness and actual activities related to beach management were described as follows:

4.1. *Changes in public awareness on beach*

Interview survey to residents was conducted in January 2017 to grasp the effects of the aforementioned public relations and educational activities quantitatively. The sample number was 62 and it was conducted not only near the Project site but also outside of the site as shown in Fig. 7 and was done irrespective of age or sex to grasp the extent of the effect. At first, the survey results were quite positive because 92% of the respondents replied that they already know the Project beach. The following four sections show changes in public awareness based on the interview survey results.

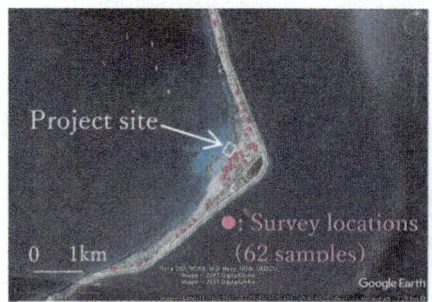

Fig. 7. Locations of interview survey (January 2017) and photo during survey.

4.1.1. *Shift from protection-oriented awareness to that of beach use- and environment-conscious*

Figure 8 shows the interview survey results on both "project purpose" and "project effect". More than 90% of the respondents recognized that the Project was implemented to protect the vulnerable coastal area. This seems not to be a particular trend as protection is one of the most important functions but it was also indicated that they did not put weight on the other benefits based on the Project purpose. On the other hand, in terms of Project effect, they highly evaluated environment (85%) and beach use (56%) compared with protection (29%). Therefore, the shift from protection-oriented awareness to that of beach use- and environment-conscious was concluded from these results.

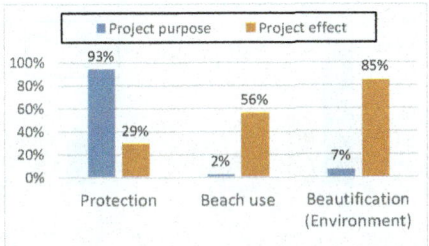

Fig. 8. Interview survey results on project purpose and project effect.

4.1.2. *Awareness on beach use and environment was developed for a wider range of people especially through the beach sports festival*

Interview survey results on "opportunities that people become familiar with the Project beach" showed that 70% of the respondents become familiar with the beach through the events related to the beach sports festival (Table 2). On the other hand, the radio announcement scored the lowest among these opportunities though it is the only public media in Tuvalu. Therefore, experience-based activities are considered more effective to improve or change public awareness.

Table 2. Interview survey results on opportunities that people become familiar with the Project beach.

Opportunity that people become familiar with the Project beach	Ratio
Stay nearby	23 %
Visiting beach area	35 %
Construction work	37 %
Radio announcement	12 %
Sequence of events related to beach sports festival (Sec. 3.2.2)	70 %
Other PR activities (singing competition, drawing contest, opening ceremony)	47 %

4.1.3. Importance of beach maintenance was recognized due to improvements of public awareness on beach use

Figure 9 shows the interview survey results on "concerns on the Project beach in the future" for both surveys conducted in January 2016 and January 2017. In January 2016, which was just after the completion of the construction in December 2015, the gravel collapse had the highest proportion while in January 2017, beach maintenance had the highest proportion of 29%, which was 0% in 2016. Most of these respondents replied that maintenance is required to remove some scattered gravel on the beach so that people can use sandy beach area pleasantly. Therefore, changes in public awareness on beach maintenance were concluded from these results.

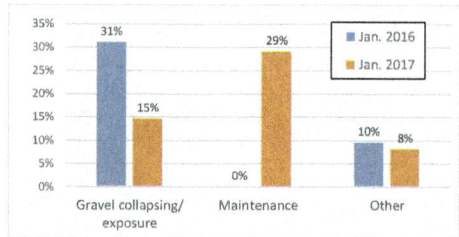

Fig. 9. Interview survey results on concerns on the Project beach in the future (if any).

4.1.4. People firstly realized the effects of gravel beach nourishment and desired to extend the same measure to other coastal areas in Tuvalu

As aforementioned, typical coastal protection measure applied in Tuvalu is mainly concrete block wall and this Project was the first trial to apply beach nourishment in Tuvalu. Interview survey results on "desirable coastal conservation measure to be applied to other areas in Tuvalu" showed that beach nourishment became well recognized and 73% of the respondents replied that they prefer beach nourishment to concrete block wall as shown in Table 3. It should be noted that most of them did not even know beach nourishment before the Project started, about one and half year ago.

Table 3. Interview survey results on desirable coastal conservation measure to other areas in Tuvalu.

Type of Measure	Beach Nourishment	Seawall	Both	Not Sure
Ratio	73%	3%	19%	6%

Finally, changes in people's awareness related to the beach through the Project can be summarized in Fig. 10. Awareness on protection remains at a high level at all times because it is directly related to their safety and their property. On the other hand, awareness on beach use and environment was initially very low as these concepts had never been developed in Tuvalu before. However, it was much improved just after the construction as people confirmed its effects by themselves and the awareness was continuously improved after the construction through public relations and educational activities, especially through the sequence of events related to the beach sports festivals. Awareness on beach maintenance was newly developed after public awareness on beach use had increased.

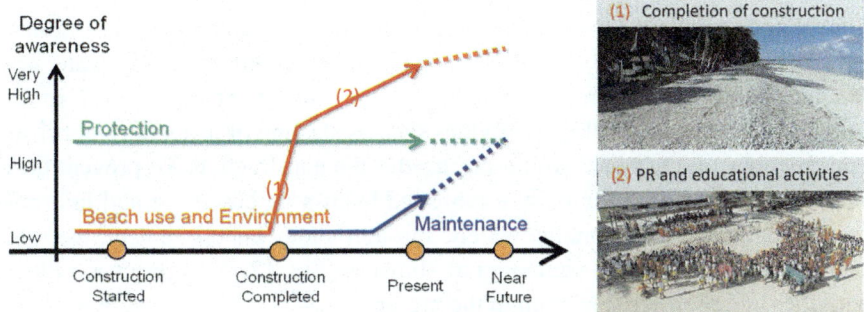

Fig. 10. Changes in peoples' awareness on beach through the Project period.

4.2. *Changes in activities related to beach management and beach use*

The following three sections show changes in actual activities related to beach management and beach use:

4.2.1. *Community started periodic beach cleaning and beach environment was kept at a good condition*

Community started beach cleaning voluntarily involving residents just after the completion of the construction and continued it every two weeks until the present for one and half year (Fig. 11). At the same time, dumping of garbage from residents was reduced to almost nothing because their manner was also improved through public relations and educational activities. As a result, the environment of the Project beach has been maintained well.

Fig. 11. Periodic beach cleaning by community.   Fig. 12. Signboards installed on the beach.   Fig. 13. A new beach use: fishing at groin.

4.2.2. *Illegal activities on the beach had been prevented through multiple measures implemented by the community*

Gravel and sand used for the beach nourishment are common materials utilized for housing construction in Tuvalu. There also existed several private boat landing slopes at the Project area before the implementation. Thus, it was initially anticipated that illegal activities such as stealing of materials and private constructions on the beach would occur after the construction. To prevent these illegal activities, the community established bylaws on beach use and informed the regulations to the public widely through radio announcement, periodic patrol, and installation of signboard as shown in Fig. 12. As a result, no illegal activities had been confirmed up to the present.

4.2.3. *A new way of beach use became widespread among the public*

Through the beach sports festival, a new way of beach use became widespread among primary school students and the local community. They enjoyed some sports on the beach referring to programs that had been implemented during the beach sports festival. In addition, fishing from groins, which are located at both ends of the Project site to prevent sand outflow, were getting more popular among local residents as groins functioned as a good fish bed (Fig. 13). Furthermore, since primary school, community, and the government plan to make beach sports festival an annual event, continuous and active beach use is highly expected in the following years.

5. Preparation for maintenance works initiated by the community and the government

Beach nourishment in general requires periodic maintenance works as adaptive management such as re-nourishment and reprofiling after implementation.

Details of maintenance work and its frequency differ based on purpose, natural conditions, and nourishment materials.

In this Project, as public awareness on beach use and environment was rapidly increased, they are required to maintain the beach at a level where it can be used pleasantly like a playground. To respond to such requests, the community and the government decided to implement maintenance works and would continue periodic re-nourishment and reprofiling on a yearly basis. The first maintenance work would be implemented under the supervision of the JICA Expert Team to learn basic procedure and methods. In addition, to make the maintenance works sustainable in the future, a win-win relationship among schools, community, and the government was proposed as shown in Fig. 14. One example of this relationship is that the school will hold a beach sports event involving the community and the community will do beach maintenance in return so that the beach area can be used for the festival. It might not be so easy to establish such kind of sustainable maintenance structure in practice, however, considering the drastic changes in people's awareness and activities through this Project, it is possible that this can be realized.

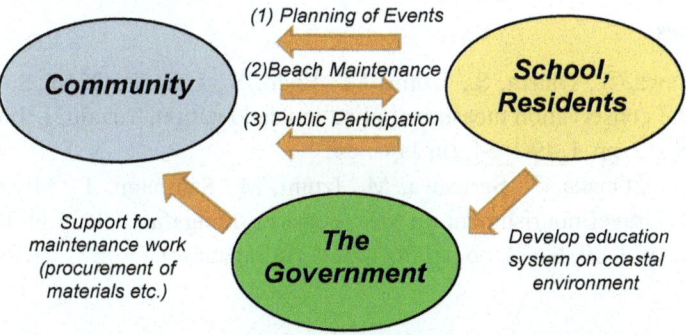

Fig. 14. Idea on sustainable beach management structure.

6. Conclusion

Public awareness on beach was much improved through public relations and educational activities under the Project. People experienced benefits from the beach through these activities and realized their roles in using the beach area pleasantly. The community started voluntary beach cleaning involving residents and prepared bylaws on beach use and disseminated these rules to the public. As a result of these activities, beach environment had been kept at a good condition and no illegal activities such as dumping of rubbish, stealing of materials, and private construction on the beach had occurred for one and half year after the completion of the construction. In addition, with an increased desire for beach maintenance, the community and the government decided to implement periodic maintenance works such as reprofiling and re-nourishment. It is, therefore, concluded that a community-based beach management was established and functioned successfully during the Project period.

However, it also should be noted that continuous public relations and educational activities will be needed to maintain public awareness on beach at a high level so that they would voluntarily keep participating in the beach management in the future.

Acknowledgement

This paper described the Pilot Gravel Beach Nourishment Project in Tuvalu which was executed by JICA as one of the Japanese ODA projects. The authors would like to express their deep appreciation to the Nauti Primary School, Funafuti communities, and the Tuvaluan government for their active support for the Project.

References

1. Ichikawa, S., Onaka, S., Izumi, M., Endo, S., Uda, T., 2016. Sustainable coastal conservation measure applied in island nation, Tuvalu, J. JSCE, Vol. 72, No. 2, pp. I_49-I_54. (in Japanese)
2. Uda, T., Onaka, S., Serizawa, M., Izumi, M., San-nami, T., Miyahara, S., 2013. Gravel nourishment on west coast on Fongafale Island in Tuvalu, J. JSCE, Vol. 69, No. 2, pp. I_736-I_740. (in Japanese)

Consensus Building Among Conservation of Surfing Ground, Shore Protection and Use of Coast — An Example on Ichinomiya Coast, Japan

Satoquo Seino[a], Takaaki Uda[b], Takeo Kondo[c], Ryoji Yoshida[d] and Takahiro Todoroki[d]

*[a]Graduate School of Engineering, Kyushu University,
Motooka, Nishi, Fukuoka, Fukuoka 819-0395, Japan
E-mail: seino@civil.kyushu-u.ac.jp*

*[b]Head, Shore Protection Research, Public Works Research Center,
1-6-4 Taito, Taito, Tokyo 110-0016, Japan
E-mail: uda@pwrc.or.jp*

[c]Nihon University, 7-24-1 Narashinodai, Funabashi, Chiba 274-8501, Japan

[d]Chiba Prefectural Government, 1-1 Ichiba-machi, Chuo, Chiba 260-8667, Japan

On the Ichinomiya coast located on south Kujukuri Beach, a world famous surfing ground was threatened. A conflict regarding the construction of artificial headlands as a measure against beach erosion took place among the administration of coast, surfers, fishermen and users of the coast in 2010. Roundtable discussions to reach a consensus regarding conservation and use of the coast were held to solve the conflict since June 27, 2010. Meetings were held five times between June 27, 2010 and September 3, 2011, and shore protection measures of the coast and the impact of the construction of the artificial headlands on the beach and surfing grounds were discussed. Finally, a consensus was reached among the administration of coast, surfers and users of the beach. This beach will be the place of surfing Olympic games to be held in 2020.

Keywords: Consensus building; Ichinomiya coast; public meeting; shore protection; use of coast; surfing ground.

1. Introduction

Movements against shore protection measures often occurred in the 1990s in Chiba Prefecture, Japan, and the procedure of the protection works and coastal engineering were questioned. For this purpose, stakeholders, such as the engineers in the coastal administration office, coastal engineers and local people, gathered and discussed the issue via a trial and error method at the Shirasuka coast. In this meeting, a new methodology was adopted, that is, no coastal plan was explained to the local people after the decision making within the administration, but the local people participated in the meeting from the beginning of the planning stage on the basis of the basic flow chart of consensus building among the municipalities, NPO, local people, related organizations, fishermen, and users of the coast, as shown in Fig. 1. Although this method, in which various stakeholders participated and open discussions were held, is socially accepted as one of the methods of consensus building in Japan at present, the policy of Chiba Prefectural Government, in which the procedure of the participation and management method were determined, was established via trial and error in the 1990s. In this study, on the basis of discussions in public meetings at the Ichinomiya coast, the process of consensus building based on scientific data, and the role of engineers were studied. On this coast, the same conflicts took place in 2010, and public meetings for consensus building were held seven times between 2010 and 2013, and by February 23, 2011, a consensus was reached. In this study, various conditions necessary for the process of consensus building among stakeholders were studied: the participation of various stakeholders and researchers as facilitators, the provision of information to the public, and the activity of the administration (Seino et al., 2013)[1].

2. Long-term changes on Ichinomiya coast where conflict arose

Figure 2 shows an aerial photograph of the Ichinomiya coast in 2010, where the

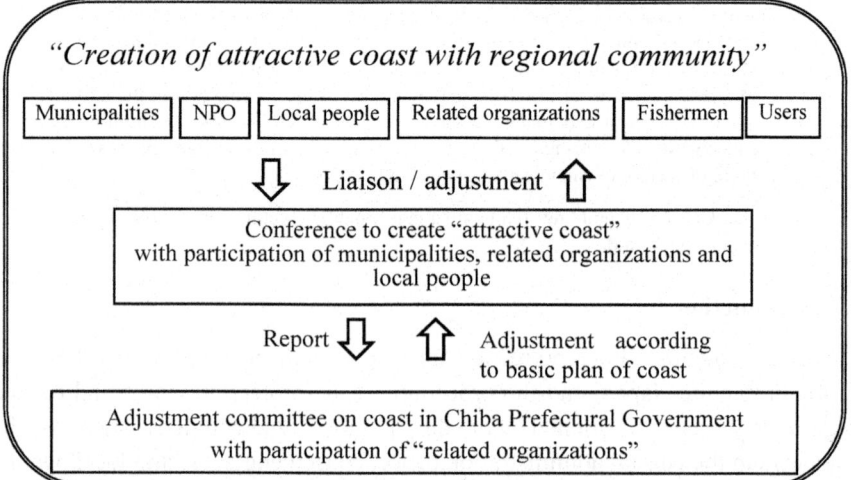

Fig. 1. Basic flow of consensus building among municipalities, NPO, local people, related organizations, fishermen, and users of a coast.

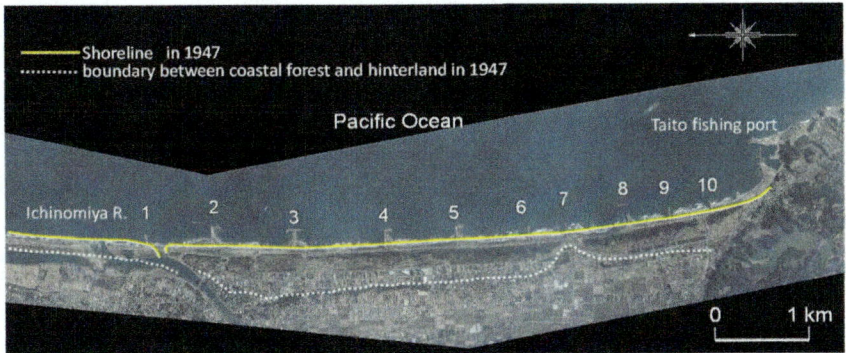

Fig. 2. Aerial photograph of Ichinomiya coast in 2010 and shoreline in 1947.

boundary between the natural sand dune and the hinterland, and the shoreline in 1947 are denoted by broken and solid lines, respectively. Although a natural sand dune with a maximum width of 600 m expanded near the shoreline in 1947, the coastal forest and residential area markedly advanced seaward until 2010 together with the construction of ten artificial headlands as numbered. As typically seen in the area between headland Nos. 2 and 6, half of the natural sand dune in 1947 altered the residential area of the town until 2010 with the formation of a coastal forest zone seaward, resulting in the narrowing of the natural sandy beach. Although the shoreline locally advanced in the vicinity of Taito fishing port, the shoreline receded in almost the entire study area. Thus, in the study area, the seaward expansion of the residential area associated with the artificial alteration of land was much greater than the shoreline recession, even though people believed that beach erosion had occurred, which resulted in the narrowing of the sandy beach in the past. At first, many local people misunderstood this point, and they considered that the sandy beach disappeared as a result of severe erosion.

Figure 3 shows the shoreline changes of the Ichinomiya coast between 1980 and 2010 with reference to the shoreline in 1947. The shoreline markedly advanced in the vicinity of Taito fishing port until 1980, whereas the beach was eroded between $X = 3.5$ and 7 km north of the sand accumulation zone near the fishing port. From these facts, the cause of the beach changes was considered to be the sharp decrease in northward longshore sand supply from Taito sea cliffs, extending south of the fishing port, and the erosion zone successively expanded northward from the vicinity of the fishing port. On the basis of this, the construction of ten artificial headlands was planned in the 1980s. However, simultaneously, the Taito fishing port breakwater was extended until 1991, and then the shoreline advanced in 2000 in the area near the fishing port, where the

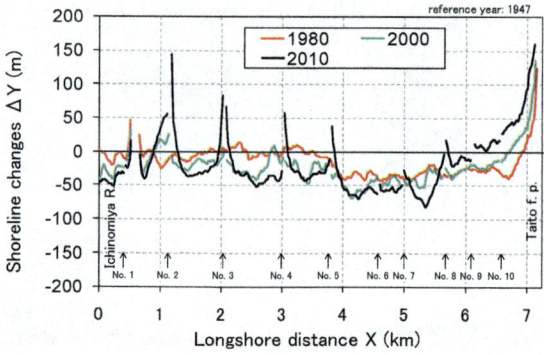

Fig. 3. Shoreline changes up to 1980, 2000, and 2010 with reference to the shoreline in 1947.

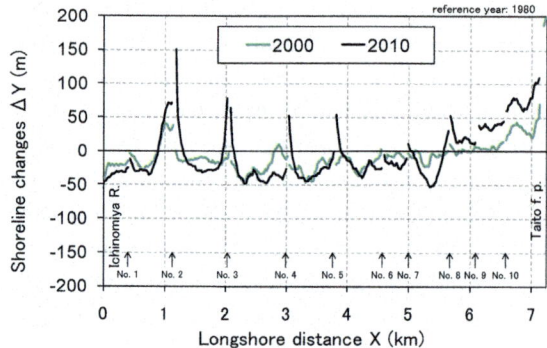

Fig. 4. Shoreline changes up to 2000 and 2010 with reference to the shoreline in 1980.

shoreline significantly receded until 1980, owing to the wave-sheltering effect of the fishing port breakwater. On the other hand, the shoreline receded in the area between $X = 2$ and 5 km. With the successive extension of the artificial headlands, the shoreline behind the headlands markedly advanced as a spike, whereas the shoreline continued to retreat between the headlands. In contrast to the shoreline recession in the area of $X < 4$ km, the shoreline markedly advanced immediately north of Taito fishing port.

Figure 4 shows the shoreline changes until 2010 with reference to the shoreline in 1980. The shoreline advanced over time south of $X = 5.5$ km including headland Nos. 8, 9, and 10 with a particularly large shoreline advance in the vicinity of the Taito fishing port. The shoreline advance near the fishing port was considered to be due to the southward longshore sand transport locally induced in the wave-shelter zone of the fishing port breakwater. This result totally differed from the previous explanation that was made to the local people at the start of the construction of the headlands: the shoreline would be stabilized, while forming a triangular foreshore south of each headland owing to the prevailing northward longshore sand transport, and in contrast, the shoreline advanced only in the vicinity of the headlands with the shoreline recession between the headlands. Thus, the discrepancy between the initial explanation to the people at the start of the construction of the headlands and the observed shoreline changes around the headlands resulted in fundamental questions concerning the effectiveness of the headlands to the local people. This also showed the inadequacy of the previous design method of the artificial headland system in the 1990s, and thus the explanations had to be revised on the basis of monitoring data and numerical simulations using more advanced technology.

3. Conflicts on conservation and use of coast, and consensus building

Conflicts have arisen between the people who agreed with and were against the shore protection works, such as the surfers, and the administration of coast, on the Ichinomiya coast, as to whether the construction of the artificial headlands should be continued. A campaign to collect signatures against the construction of the artificial headlands was carried out in 2010. Given this situation, public meetings were held by Ichinomiya Town, and a wide range of people, i.e., the administration of coast, local residents, representatives of the users of the beach, and fishermen, together with the coastal engineers as facilitators, participated. These meetings were held seven times between June 27, 2010 and February 23, 2013, and consensus building was carried out to solve the conflict. The main processes of these public meetings were managed by Ichinomiya Town, instead of the prefectural government, which was responsible for the shore protection works, and various investigations on coastal engineering were supported by the prefectural government. In this paper, the results of the discussions made during five meetings between June 27, 2010 and September 3, 2011 are summerized.

4. Major points of discussions

Consensus building was enhanced via discussions among the stakeholders on the following points, as follows.
(1) Meeting on June 27, 2010
•   to confirm the fact that a public meeting was initiated as a result of the campaign to collect signatures by surfers and local people against the construction of the artificial headlands

- to confirm the necessity of the open-to-the-public discussion regarding the aim of the construction of the headlands as shore protection works on the basis of field data
- to confirm the mechanism of the headlands in controlling longshore sand transport
- to reconfirm the importance of beach nourishment to maintain the shellfish ground

(2) Meeting on September 19, 2010

- to resolve the misunderstanding associated with the construction of the headlands among stakeholders through the explanation of the aim and the present situation of the headlands
- to clarify the decrease in the number of sea turtles landing on the Ichinomiya coast mainly owing to the destruction of their spawning ground by constructing gabions along the shoreline
- to understand the importance of preserving the vegetation zone on the backshore as the spawning ground of the sea turtles
- to understand that the surfing ground on the Ichinomiya coast was activated after the extension of the longitudinal part of the headland, because the rip currents along the headlands were effectively used by surfers to approach the offshore surfing grounds

(3) Meeting on December 23, 2010

- to let the stakeholders know the mechanism of the occurrence of beach erosion near Taito fishing port (Many people considered at first that sand turning around the tip of the fishing port was deposited immediately north of the fishing port, but they realized that the direction of longshore sand transport was changed by the formation of the wave-shelter zone associated with the extension of the fishing port breakwater.)
- to explain to the surfers the formation mechanism of the surfing ground due to the wave refraction around the natural submerged reefs offshore of the Ichinomiya coast

(4) Meeting on March 6, 2011

- to explain the effect of the extension of artificial headland No. 6 and the impacts on the surrounding area
- to clarify that the extension of the offshore breakwater of the headland causes the shoreline recession between the headlands owing to their wave-sheltering effect

(5) Meeting on September 3, 2011

- to explain the plan of the beach nourishment between headland Nos. 2 and 3, and headland Nos. 7 and 8 to recover the sandy beach of 40 m width
- to agree on the beach nourishment plan
- to commonly acknowledge among stakeholders that the final target is not the construction of the headlands but the recovery of the sandy beach

5. Importance of scientific explanation to local people

Most important was the open-to-the-public discussions on information regarding the real situation of the coast, including the shore protection plan of the coast. To enhance the understanding of the shore protection works for stakeholders, such as the local residents, surfers, and fishermen, and to clearly answer their questions, every scientific data had to be discussed in a manner that every stakeholder can easily understand. Particularly, major points to be discussed were the reversal of the direction of longshore sand transport after the extension of the fishing port breakwater, the continuation of beach erosion, even though the headlands have been constructed, and the explanation of the mechanism of the formation of the surfing ground on the Ichinomiya coast to the surfers.

5.1. *Seabed topography of Ichinomiya coast*

Figure 5 shows the bathymetry offshore of the Ichinomiya coast in 2010. The bottom contours locally protrude offshore of headland Nos. 2-5. When the headlands were extended on a coast under a dynamically equilibrium condition, sand is transported offshore turning around the tip of the headlands, because longshore sand transport is partially blocked by the headlands, and such sand is transported shoreward downcoast of the headlands, resulting in the meandering of offshore contours (Uda et al., 2008)[2]. This shows that part of the longshore sand transport discharges offshore of the tip of the headlands, and thus a statically stable beach cannot be formed on a coast composed of fine sand, such as the Ichinomiya coast. This was the evidence that a statically stable beach, which was planned in the beginning, was impossible to form on this coast. Thus, the discrepancy between the initial plan and the real situation of the Ichinomiya coast had to be explained to the local people, implying the necessity of the revision of the initial plan concerning the headlands.

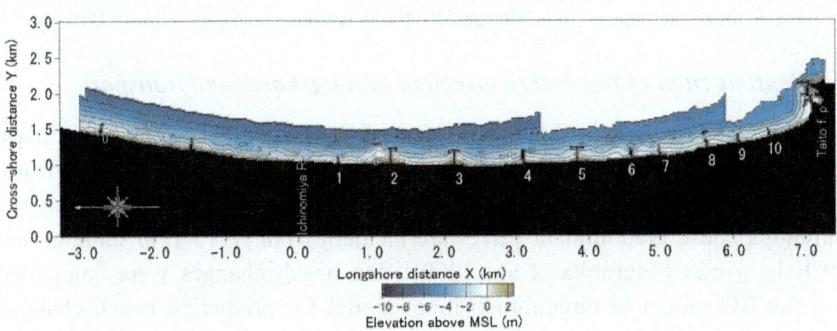

Fig. 5. Shoreline changes up to 2000 and 2010 with reference to the shoreline in 1980.

5.2. Shoreline changes between artificial headlands

Seino et al. (2015)[3] showed the shoreline changes between artificial headland Nos. 2 and 3 measured after the construction of these headlands. In their study, they argued that although the shoreline between the headlands extended straight in 1990 and 2000 without the T-shaped part of the breakwater, the shoreline on the lee of the headland markedly advanced after 2005 when the T-shaped part began to be extended. Because the T-shaped part of the headlands was extended according to the initial plan after the recession of the shoreline, the wave-sheltering effect of the offshore breakwater appeared, resulting in the marked shoreline recession between the headlands and excess sand deposition on the lee of the headlands. This situation was detrimental to the use of the beach for bathing in summer, and significantly differed from the initial explanation concerning the recovery of the beaches, in which the local people believed that a wide sandy beach will be recovered between the headlands. From these discussions, an adaptive management method based on monitoring surveys was introduced, instead of pursuing the previous plan.

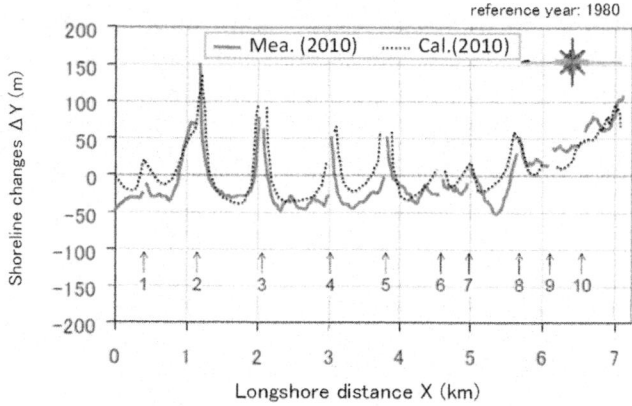

Fig. 6. Shoreline changes up to 2000 and 2010 with reference to the shoreline in 1980.

5.3. Reproduction of reversal in direction of longshore sand transport

During the same period of the construction of the headlands on the Ichinomiya coast, the south breakwater of Taito fishing port was extended near Point Taito, and the wave-shelter zone was formed near the breakwater. Because on the Ichinimoya coast, predominant waves are incident from N113°E in summer and N79°E in winter (Tsuruoka et al., 2009)[4], the beach changes were calculated using the BG model (a three-dimensional model for predicting beach changes based on Bagnold's concept) proposed by Serizawa et al. (2006)[5], given the superimposed wave field taking these seasonal changes in wave direction into account.

Figure 6 shows the results of the calculated and measured shoreline changes in 2010 with reference to the shoreline in 1980. Owing to the construction of the headlands, the shoreline between the headlands markedly receded, whereas the shoreline near the Taito fishing port advanced to a large extent. The shoreline advance and recession between headland Nos. 1 and 8 and the marked shoreline advance near the fishing port were well reproduced by the numerical simulation using the BG model. These results were explained to the local people, and they understood that the direction of the predominant longshore sand transport, which was originally northward in the past, was altered to become southward as a result of the extension of the south breakwater of Taito fishing port.

5.4. *Effect of beach nourishment*

The effect of the beach nourishment offshore of headland Nos. 2 and 3 was investigated using the contour-line-change model considering the change in grain size (Uda et al., 2004)[6]. They showed that when sand back pass was carried out using fine sand (grain size of 0.15 mm) dredged from inside Katagai fishing port located 17 km north of the Ichinomiya coast, beach nourishment was effective only for temporarily increasing the offshore seabed elevation because of northward longshore sand transport. In contrast, if beach nourishment was carried out using coarse sand of 0.25 mm grain size deposited immediately south of Taito fishing port, the beach was successfully restored. After the explanation of the results of these numerical simulations, the effectiveness of beach nourishment using coarser material to recover the sandy beach was realized by all stakeholders. Since this consensus, beach nourishment using coaser sand deposited immediately south of Taito fishing port has been successfully carried out up to the present.

6. Present conditions of Ichinomiya coast

After the consensus building on the Ichinomiya coast, beach nourishment has been carried out continuously, and the recovered beach has been extensively used in various usages. Photos 1 and 2 show the scene of the dragnet on the gentle slope, which had been famous at Kujyukuri Beach for a long time, and many people gathered to catch fishes as a recreation on a beach. Photo 3 shows a horse trotting on the shoreline during low tide. On the other hand, Photo 4 shows a hatching area for the loggerhead turtle *Caretta caretta*, which has also been very famous on Kujyukuri Beach. Finally, Photos 5 and 6 show the beach use for bathing, and many people gathered to the beach in summer. Thus, the beach became an important natural resource in this coastal area, and it is useful in the development of this area.

7. Concluding remarks

The most difficult factors for local people to understand in consensus building were the fundamental causes of the beach erosion of the Ichinomiya coast and the effect of the measures against beach erosion. Particularly, the reversal of the direction of longshore sand transport from northward to southward associated with the extension of the fishing port breakwater was difficult at first even for the engineers of the administration. The reversal of the direction of longshore sand transport was found on the basis of monitoring data collected over a long period, although the construction of the headlands took over 20 years, and an adaptive management method was required owing to this fact. On the other hand, the discrepancies in opinion between the administration and the fishermen regarding the devastation of the shellfish ground were minimal, and the effect of the beach nourishment was easily realized by the stakeholders.

Photo 1. Dragnet on gentle slope.   Photo 2. Caught fishes in dragnet on beach.

Photo 3. Horse trotting on shoreline during low Photo 4. Hatching area for the loggerhead turtle
tide.   *Caretta caretta.*

Photo 5. Beach use for bathing.   Photo 6. Suntan on coast.

In this consensus building, not only the administrative officials in Chiba Prefecture and Ichinomiya Town but also stakeholders, such as local people, surfriders, and fishermen, participated in the roundtable discussion. Most of them had no interest in coastal engineering, and lengthy explanation on coastal engineering hampered their understanding, resulting in loss of their interest on the phenomenon, even though such discussion was important in understanding the beach changes. In these cases, it was necessary for the engineers that the results should be explained in plain language.

Many local people including surfers eager to know how to construct artificial headlands in the surf zone, where they use as surfing ground. They, however, are not interested in the continuous activities to maintain the beach as a coastal management, and they did not have incentive to participate and to carry out their activities on such movement. In particular, although consensus meeting was held mainly on weekend holiday, it was frequent for surfers to play on the same day, and they selected surfing instead of attending the roundtable discussions when wave conditions are appropreate for surfing. For this reason, useful discussion was often delayed in consensus meeting.

In general, the loggerhead turtle *Caretta caretta* hatches near the boundary between the backshore and the vagetation zone in front of sand dune of the coast, and the people related to the protection of sea turtle are eager to observe hatching on the beach. They, however, are lacking to understand the variation in sandy beach owing to wave action. Furthermore, they did not understand the beach changes associated with artificial alteration on coasts, and only argued no alteration on the beach, even though mitigation against beach erosion will be carried out.

In the public meetings for consensus building on the Ichinomiya coast, related information was open to the public, while permitting the participation of groups and individuals with different opinions on shore protection. However, since the administration naturally considered a construction plan of shore protection facilities, the discrepancies in opinion between the local people who disliked or rather wanted to demolish the facilities and the administration became more pronounced. In such a case, the engineers had to understand the background of the opinions made by the local people, and alternatives that were applicable to the coast had to be proposed on the basis of numerical simulations. Furthermore, it was useful in consensus building among stakeholders that the coastal engineers clearly explained the reality of the coast to the local people on the basis of scientific views. In the final consensus building, it was important not to exclude any discussion among stakeholders and to continue the discussion, even if they were difficult for the administrators. Finally, it was concluded that the consensus building among stakeholders with different background was important in the accomplishment of the necessary public works under complicated conditions at present.

Acknowledgement

This study was supported by various participants and organizations. Ichinomiya Town provided consensus building platform. Environment Research and Technology Development Fund S-13 supported part of the coastal consensus building studies. We express our sincere gratitude for their cooperation.

References

1. S. Seino, T. Uda, T. Kondo, H. Mizugaki, K. Uno, Conflict and consensus building on conservation and use of the Ichinomiya coast, Chiba Prefecture, J. Coastal Zone Studies, Vol. 26, No. 3, (2013) , pp. 79-91.
2. T. Uda, S. Watanabe, K. Furuike, Y. Hoshigami, H. Nagayama, Changes in bathymetry and longshore sand transport around artificial headland with various shapes and scales, Annual J. Coastal Eng., JSCE, Vol. 55, (2008), pp. 566-570.
3. S. Seino, T. Uda, Y. Ohtani, Y. Ohki, Essential aspects of beach erosion - Lessons from devastation of Ichinomiya coast, Japan, 8th Int. Conf. on Asian and Pacific Coasts (APAC 2015), Procedia Engineering, Vol. 116, (2015), pp. 446-453.
4. H. Tsuruoka, T. Uda, M. Serizawa, K. Furuike, T. Fukumoto, Y. Hoshigami, S. Miyahara, Numerical simulation of diffusion of nourishment sand offshore of south Kujukuri Beach and effect of continuous beach nourishment, J. Coastal Eng., JSCE, Vol. 56-1, (2009), pp. 711-715.
5. M. Serizawa, T. Uda, T. San-nami, K. Furuike, Three-dimensional model for predicting beach changes based on Bagnold's concept, Proc. 30th ICCE, (2006), pp. 3155-3167.
6. T. Uda, T. Kumada, M. Serizawa, Predictive model of change in longitudinal profile in beach nourishment using sand of mixed grain size, Proc. 29th ICCE, (2004), pp. 3378-3390.

Power Take Off (PTO) Performance of Wave Energy Converter Based on Water Mass Gravity Force Under Container Volume Variation [*]

Masjono Muchtar[†], Salama Manjang, Dadang A Suriamihardja, M Arsyad Thaha
and Muslimin

Politeknik ATI Makassar,
Makassar, Sulawesi Selatan, Indonesia
[†]*E-mail: masjono@kemenperin.go.id*
www.atim.ac.id

A.N. Authors

Hydraulic Laboratory
Department of Civil Engineering Hasanuddin University,
Makassar, Sulawesi Selatan, Indonesia

To date there were few studies of wave energy converter that utilize the gravity force of sea water as a prime mover of wave energy converter to generate electricity. In this work, physical model of wave energy converter based on water mass gravity force to generate electricity were investigated. Experiment was conducted at Hydraulic Laboratory, Hasanuddin University, Makassar Indonesia. The physical model of wave energy converter be made up of connecting chain, gravity weight container (Mg), counter weight (mc), rotating shaft, gear box and flywheels. The amount of extracted power was calculated based on converter shaft radial speed (RPM) and torque. Experiment result showed that gravity weight volume is strongly determined the amount of power generated by the converter. The maximum harvested power was 1.66 Watt at wave height of 7 cm, wave period 1.2 second and water container volume is 4.5 liters. In this experiment five gravity weight pairs were employed that successfully extracted 85 % of theoretical available wave power. This experiment result showed that the proposed wave energy converter that utilizes the gravity force of water trapped in a container, gives significant contribution to the existing wave energy converter technology in harvesting the ocean wave power as a promising source of renewable energy.

Keywords: Renewable Energy; Wave Energy Converter; Electricity.

---

[*] This work is supported by Center of Industrial Education and Training, Ministry of Industry Republic of Indonesia.
[†] Lecturer at Politeknik ATI Makassar, Ministry of Industry Republic of Indonesia.

1. Introduction

Energy is one of major interest to most countries around the globe as a foundation of their future existence [1]. Primary energy consumption from 2005 – 2015 indicated that fossil based fuel remain dominant. In 2015 oil contribute 23.85% followed by coal 29.20% and gas 32.94 %. These are blamed as the major CO_2 and Green-House-Gases (GHG) emission sources that causes environmental problem [2]. An attempt to reduce these environment impact is carried out by increasing the renewable energy share on energy consumption. One source of renewable energy that has not been much utilized is marine energy since ocean covering 71% of Earth surface. Wave power is one of clean energy that can be extracted as a substitution of fuel based energy [3].

Wave energy utilization as an energy source has long been known since the introduction of The Wave Motors of Southern California in 1890-1910, however, it did not develop due to several failures experienced by that time [4]. Until the mid-20th century waves of energy has not been attract a lot of attention. Due to the fuel crisis in 1973-1974, researcher and politician began to turn their attention on alternative energy sources. The idea to maximize the renewable energy source was conveyed by the President of the United States Richard Nixon's during the energy crisis [5]. Research on the utilization of the waves has caught attention of some researchers, such as utilizing ocean waves as a source of renewable energy by harnessing the waves up and down movement to drive a hydraulic pump which is connected to the turbine to rotate electric generators [6]. To cope with the variations of the electrical energy produced can be achieved by combining the floating wave power stations with wind power plant [7]. The subsequent conversion of potential energy of buoy vertical translation up and down to drive linear permanent magnet generators was introduced by Eriksson [8]. To address discontinuity of produced electric energy caused by the non linearity properties of ocean waves was anticipated by means of mechanical smoothing techniques [9]. The dimension of the wave energy converters buoy affect the efficiency of converter generated power. To obtain optimum power, then the converters and ocean wave resonances need to be set at the same frequency [10]. Wave energy converter should be able to survive in extreme ocean waves conditions. Therefore, these circumstances could anticipate by utilizing floating wave energy converter [11]. Numerical simulation based on the theoretical equations model on the domain of time, both for regular or non regular waves is conducted to predict the quantity of energy that can be captured by wave energy converter [12]. The efficiency of power generated of a wave energy converter can be predicted based on hydrostatic transmission topology model using generic components. The model is used to optimize the dimensions of component to achieve the resulting energy efficiency [13]. Base on these pervious research it can be identified a common similar principle in wave energy conversion technique. Most of them are utilizing the water buoyancy force as the prime mover. Hence, the harvested power or power take off is the difference between gravity and buoyancy forces of wave energy converter buoy. Moreover, these converter PTOs are strongly determined by significant wave height and period. This paper propose a new approach to increase the efficiency of wave energy converter by using the wave gravity force instead of buoyancy force. The converter physical model consists of water container connected to one-way gear. The amount of extracted energy that related to container volume is specifically presented in this paper. Experiment result showed that the variation of the container volume is proportionally related to the amount of proposed converter PTO.

2. Method and Materials

Physical model experiment to investigate the influence of container volume variation of proposed wave energy converter PTO was carried out at a wave tank or flume in hydraulic laboratory Department of Civil Engineering Hasanuddin University Indonesia. Lab scale physical model. In this experiment there were four rectangular shape plastic container was employed. The fabrication of these model was carried out at Politeknik ATI Makassar Indonesia.

The proposed wave energy PTO performance in fact is determined by several independent variables, however, to investigate the influence of water volume then the other variable were need to be set at a constant value. Therefore, wave length was set at λ=160 cm, gear distance (G_s) = 40 cm, water depth (d) =25 cm, wave height (H) = 7 cm, number of gravity weight = 5 and wave period T = 1.2 second.

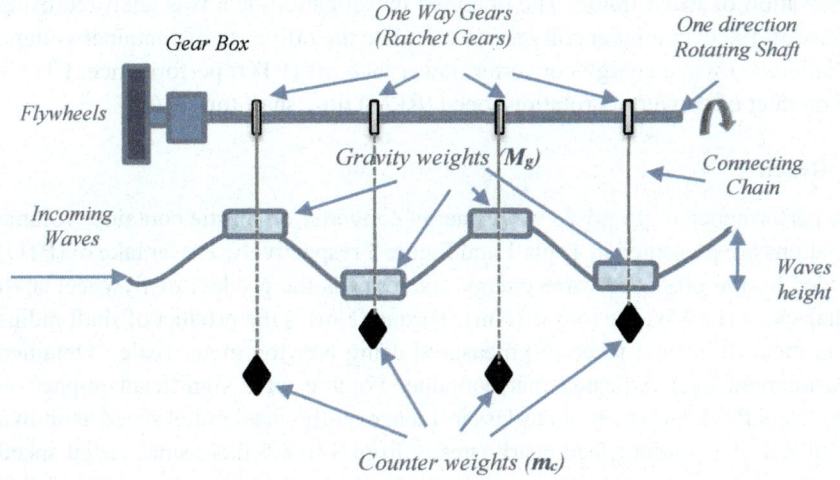

Figure 1. Illustration of proposed wave energy converter

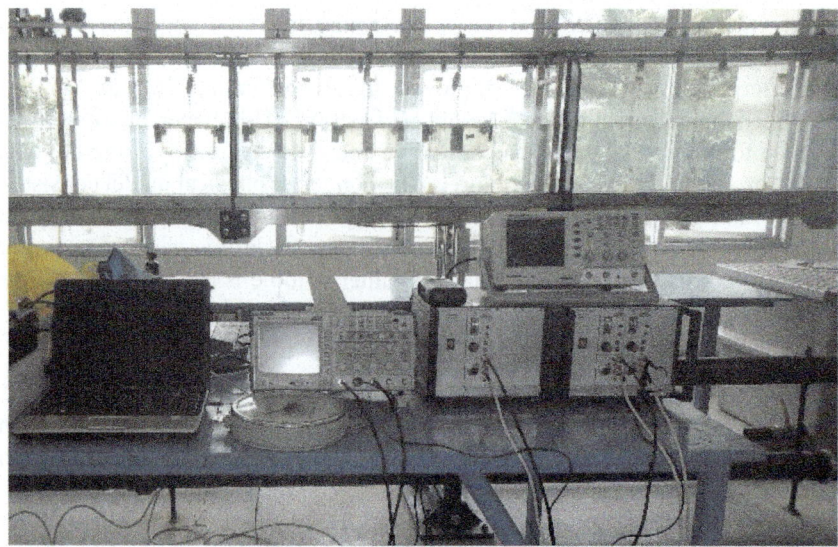

Figure 2. Physical model experiment set-up

Volume of water filled into the plastic container (gravity weight) was varied from 1 to 4.5 liters. Investigation was carried out by means of several instrument consists of wave monitors, wave height sensors, four channel digital oscilloscope, digital tachometer, newton meter scale and supporting computer software. Investigation data was acquired based on oscilloscope screen display and visual observation of wave flume. The obtained investigation data was analyzed using Microsoft Excel computer software to calculate the influence of container volume variations on wave energy converter power take off (PTO) performance. PTO is the product of converters rotation speed (RPM) time shaft torque (τ).

3. Result

The performance of proposed wave energy converter on plastic container volume variations are presented in Table 1 and Figure 3 respectively. Power take of (PTO) yielded by the proposed wave energy converter is the product of flywheel shaft radial speed (RPM) time torque (Nm). Torque (Nm) is the product of shaft radius (r) in meter unit time force (N) measured using Newton meter scale. Obtained measurement data indicated that container volume gave significant impact on converters PTO. However, it has less influence on flywheel radial speed as shown in Table 1. The radial speed nearly steady from 3 to 4.5 liters since radial speed depends on wave propagation speed which out of the scope presented in this paper.

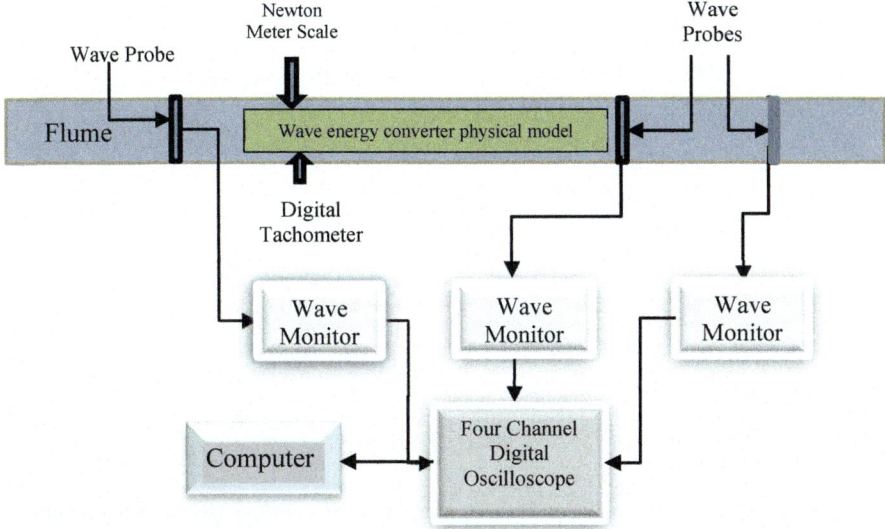

Figure 3. Lab scale experiment set-up of wave energy converter based on water mass gravity force

Graphical relationship obtained using statistical tools in Microsoft excel computer software shown in Figure 3. Result shows very significant influence of container volume on converter power take off (PTO) which is nearly straight line in Figure 4.

Table 1 Experiment result of container's volume variation

Gravity weight Volume (liter)	Number of gravity weight	Mass of Counter Weight (kg)	Flywheel radial speed (RPM)	Torque (τ) (Nm)	Power Take Off (Watt)
1.0	5	0.328	147	0.0225	0.35
1.5	5	0.328	162.1	0.0375	0.64
2.0	5	0.328	162.2	0.045	0.76
2.5	5	0.328	182.1	0.0525	1.00
3.0	5	0.328	196.7	0.0555	1.14
3.5	5	0.328	193.5	0.06	1.22
4.0	5	0.328	204.4	0.0675	1.44
4.5	5	0.325	191.8	0.0825	1.66

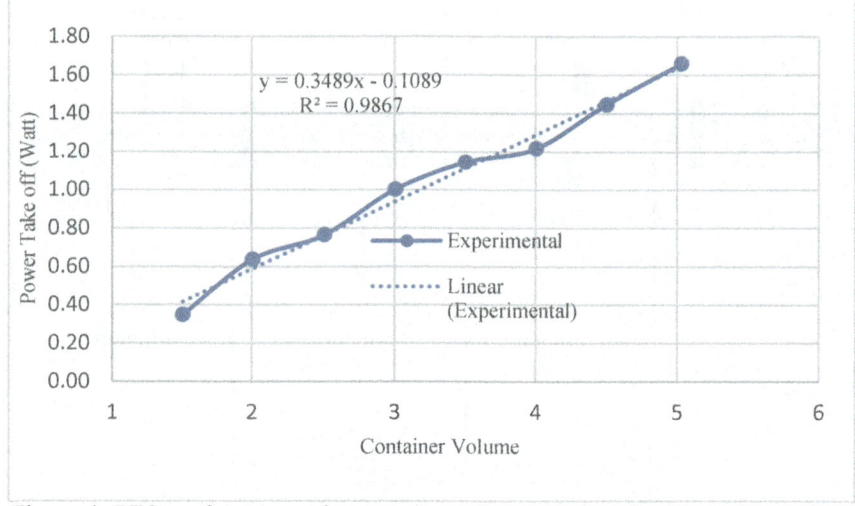

Figure 4. PTO performance of wave energy converter based on container volume
variation of wave energy converter based on water mass gravity force.

Due to the limited size of the container and the wave flume used during the experiment, then the optimum point is not yet achieve in this experiment. However, the results obtained provide preliminary promising information that the volume of water in the gravity weight container give significant contributing on harvested wave energy converter power.

4. Discussion

The experiment results showed that the volume of water filled into plastic container gives significant influence on extracted wave energy. This container is used as an interface between converter and waves. Its up and down movement is set in resonance with wave frequency to gain maximum extracted power. Since the converter utilizes one-way gear at converter shaft that connect the gravity weight with counter weight, then only container down movement potential energy is captured. Therefore, the extracted power not only determined by wave height and period but the volume of water filled into container plays an important role in wave energy conversion mechanism.

The advantages of using the method proposed in this research is the capability to extract different amount of energy at constant wave height and period. The result obtained is valid for regular waves since the experiments were conducted on two-dimensional flume that can only able to simulate regular waves. Therefore, experiment on real ocean wave environment is needed to validate the preliminary result gained from this experiment.

The advantages of this model when compared to the buoyancy based converter is the stability of the converter's interface since the center of gravity of container filled with water is under water surface. When compared with the model of such as the popular oscillating water column (OWC) model, this proposed model advantages rely on It's capability to adapt with changes in water level due to high and low tides period. Moreover, this model's component that come into contact with sea water is the water plastic container. Hence, installation and maintenance are easy to be carried out. Furthermore, this proposed converter do not require high wave to be able to create significant power take off (PTO) due to the contribution of the amount of water filled into to gravity weight container.

5. Conclusion

Physical model investigation of the influence of water container variation of power take off (PTO) performance of wave energy converter based on water mass gravity force is presented in this paper. Experiment result indicated that there is a positive correlation between the container's volume and the amount of extracted wave energy. It was found that the proposed wave energy converter presented in this paper does not depend entirely on the wave height and period. The amount of energy produced can be adjusted according to our needs by just changing the volume water in the converter's gravity weight container. However, the results obtained are still in the laboratory test under regular wave's environment. Therefore, further full scale test at the real sea condition is required. Hopefully, energy policy in Indonesia in particular will take into account of ocean wave energy as one of the promising renewable energy source.

Acknowledgments

This work was supported by The Centre of Industrial Education and Training and Politeknik ATI Makassar, Ministry of Industry Republic of Indonesia.

References

1. G. Lavidas, V. Venugopal and D. Friedrich, "Wave energy extraction in Scotland through an improved nearshore wave atlas," *International Journal of Marine Energy,* vol. 17, p. 64–83, 2017.
2. W. E. Council, "World Energy Resources 2016," World Energy Council, London, UK, 2016.
3. Masjono, S. Manjang, D. A. Suriamohardja and A. Thaha, "MODELLING OF ONE WAY GEARS WAVE ENERGY CONVERTER FOR

IRREGULAR OCEAN WAVES TO GENERATE ELECTRICITY," *Jurnal Teknologi,* vol. 78 , no. 5-7, p. 37–41 , 2016.

4. C. Miller, "A Brief History of Wave and Tidal Energy Experiments in San Francisco and Santa Cruz," 2004. [Online]. Available: http://www.outsidelands.org/wave-tidal3.php. [Accessed 12 May 2016].

5. US Public Affair, "US Department of State Office of the Historian," 31 October 2013. [Online]. Available: http://history.state.gov/milestones/1969-1976/oil-embargo. [Accessed 5 June 2017].

6. R. H. Hansen, M. M. Kramer and E. Vidal, "Discrete Displacement Hydraulic Power Take-Off System for the Wavestar Wave Energy Converter," *Energies,* vol. 6, no. 8, pp. 4001- 4044, 2014.

7. T. Kelly, T. Dooley, J. Campbell and J. V. Ringwood, "Comparison of the Experimental and Numerical Results of Modelling a 32-OscillatingWater Column (OWC), V-Shaped Floating Wave Energy Converter," *Energies,* vol. 6, no. 8, pp. 4045-4077, 2013.

8. M. Eriksson, J. Isberg and M. Leijon, "Hydrodynamic modelling of a direct drive wave energy converter," *International Journal of Engineering Science,* vol. 43, no. 17-18, p. 1377–1387, 2005.

9. S. Sahlin, M. Sidenmark and T. Andersson, "EWTEC 2013 - Evaluation of a Mechanical Power Smoothing system for Wave Energy Converters," 2013 . [Online]. Available: http://www.oceanharvesting.com/resource/EWTEC2011.pdf. [Accessed 18 September 2014].

10. S. Bozzi, A. M. Miquel, A. Antonini, G. Passoni and R. Archetti, "Modeling of a Point Absorber for Energy Conversion in Italian Seas," *Energies,* vol. 6, no. 6, pp. 3033-3051, 2013.

11. A. Pecher, J. P. Kofoed and T. Larsen, "Design Specifications for the Hanstholm WEPTOS Wave Energy Converter," *Energies,* vol. 5, no. 4, pp. 1001-1017, 2012.

12. A. Wacher and K. Neilsen, "Mathematical and Numerical Modeling of the AquaBuOY Wave Energy Converter," *Mathematics-in-Industry Case Studies Journal,* vol. 2, pp. 16-33, 2010.

13. R. H. Hansen, T. O. Andersen and H. C. Pedersen, "MODEL BASED DESIGN OF EFFICIENT POWER TAKE-OFF SYSTEMS FOR WAVE ENERGY CONVERTERS," Tampere, Finland, 2011.

Water Wave Interaction by Dual Cylindrical Caisson Breakwater with Partial Porous Area[*]

Min-Su Park and Youn-Ju Jeong

Structural Engineering Research Institute, Korea Institute of Civil Engineering and Building Technology, 283 Goyangdaero Ilsanseogu Goyangsi, 10223, Republic of Korea E-mail: mspark@kict.re.kr and yjjeong@kict.re.kr

In order to design a reliable breakwater that will withstand severe environmental conditions, accurate prediction of hydrodynamic interactions with multi-bodied structures must be considered. However, the interaction processes among dual cylindrical caisson breakwaters, which consist of a partial porous outer cylinder and an impermeable inner cylinder, are very complex and numerous. In the present study, the 3D numerical analysis method is developed with eigenfunction expansion method to evaluate the water wave interaction. The wave forces on the array of dual cylindrical caisson breakwaters with partial porous area are presented for various conditions.

Keywords: Dual Cylindrical Caisson Breakwater; Partial Porous Area; Eigenfunction Expansion Method; Water Wave Interaction; Multi-bodied Structures.

1. Introduction

Breakwaters are structures designed to protect the area on its leeside against the agitating wave climate either to provide tranquil conditions for the safe maneuvering of vessels or to provide relief against the coastal erosion on its leeside. In order to design a reliable breakwater that will withstand severe environmental conditions, accurate prediction of hydrodynamic interactions with multi-bodied structures must be considered. Since a circular cylinder has the advantage of a hydrodynamic effect without directivity, an array of vertical circular cylinders is commonly used for coastal structures. However, impermeable vertical structures result in considerable wave reflection leading to an increase in wave agitation on the seaside of the breakwater as well as the local scour. Permeable caissons offer a solution to such problems of wave reflection by dissipating partially the wave energy due to turbulence generated,

---

[*] This work is supported by Korea Institute of Marine Science & Technology Promotion and Ministry of Land, Infrastructure and Transport.

as the waves intrude into the wave chamber through the perforated wall. Perforated structures are considered to be economical and are easy to construct at locations where, certain degree of wave disturbance can be tolerated [1].

In case of porous cylinders, the wave motions in the exterior and all interior fluid regions are expressed by Williams and Li [2, 3]. Wang and Ren [4] were the earliest to study the wave interaction with a concentric surface piercing porous outer cylinder protecting an impermeable inner cylinder. It was reported that the outer porous cylinder is significantly effective to reduce the hydrodynamic force and wave run-up on the inner cylinder compared to be exposed to direct wave impact. Wang and his research group (Wang et al. [5], Wang and Jiang [6], Jiang and Wang [7], Wan and Ren [8, 9]) carried out a systematic numerical analysis of solitary and cnoidal waves interacting with a vertical cylinder or cylinder arrays. Sankarbabu et al [1, 10] extended this study to an array of concentric, porous, two-surfaced cylinders and calculated the wave forces, free surface elevations and overturning moments analytically.

However, the interaction processes among dual cylindrical caisson breakwaters, which consist of a partial porous outer cylinder and an impermeable inner cylinder, are very complex and numerous since scattered waves from each dual cylinder affect the others in the array of dual cylinders. In the present study, the 3D numerical analysis method is developed with eigenfunction expansion method, which is expressed by Williams and Li [3], and Wang and Len [4], to evaluate the water wave interaction by an array of dual cylindrical caisson breakwaters with partial porous area. Darcy's law is also applied to the porous body boundary condition. Firstly, the results obtained from the developed numerical method are compared with Wang and Ren [4] to verify the developed numerical analysis method. The wave forces on the array of dual cylindrical caisson breakwaters with partial porous area are presented for various porosity depths of the outer cylinder and distances between the inner cylinder and the outer cylinder. From these results, the present method is very useful to evaluate the wave force acting on the array of dual cylindrical caisson breakwaters with partial porous area. The dual cylindrical caisson breakwater with partial porous area is remarkably effective to reduce the wave forces the effects of water wave interaction.

2. Mathematical Formulation

It is assumed that the computational fluid domain is inviscid and incompressible, and its motion is irrotational. An arbitrary array of N dual cylindrical caisson breakwater with partial porous area is situated in water of uniform depth d and the draughts of each permeable and impermeable area of dual cylindrical caisson breakwater are h and c, respectively. The outer and the inner radius of the jth dual cylindrical caisson are a_j and b_j, respectively. Also, the global cartesian coordinate system (x, y, z) is defined with an origin located on the sea bed with the z-axis directed vertically upwards. The center of each dual cylindrical caisson at (x_j, y_j) is taken as the origin of a local polar coordinate system (r_j, θ_j), where θ_j is measured counterclockwise from the positive x-axis. The center of the lth dual cylindrical caisson has a polar coordinate (R_{jl}, α_{jl}) relative to the jth dual cylindrical caisson. The coordinate relationship between the jth and lth dual cylindrical caisson is shown in Fig. 1. Moreover, the fluid domain is divided into two regions: region 1 which is interior to the dual cylindrical caisson ($b_j\leq r_j\leq a_j$, $d-h\leq z\leq d$) and region 2 which is exterior to the dual cylindrical caisson and extends to infinity in the horizontal plane ($r_j\geq a_j$, $0\leq z\leq d$).

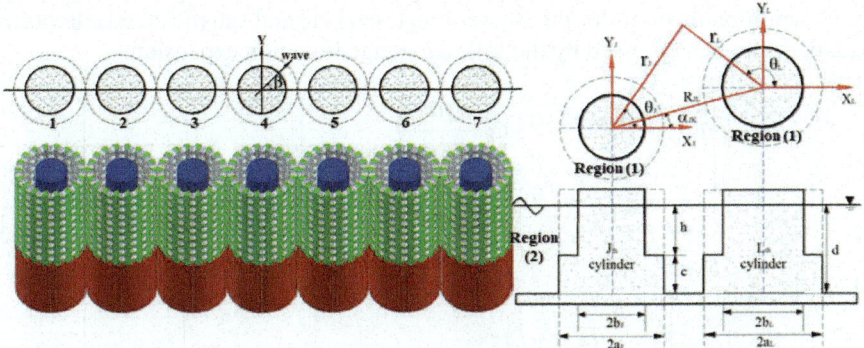

Fig. 1. Coordinate system for an array of dual cylindrical caissons with partial porous area.

The wave potential of the jth cylindrical caisson in the interior region (region 1), which satisfies the linearized free surface and body boundary conditions, can be expressed by the following eigenfunction expansions:

$$\phi_1^j = \sum_{n=-\infty}^{\infty}\left[\begin{array}{l} A_n^j\left\{J_n(k_0 r_j)-\dfrac{J_n'(k_0 r_j)}{Y_n'(k_0 a_j)}Y_n(k_0 a_j)\right\}\dfrac{\cosh\{k_0(z-d+h)\}}{\cosh(k_0 h)} \\ +\sum_{q=1}^{\infty}A_{nq}^j\left\{\begin{array}{l}I_n(k_{0q}r_j)\\ -\dfrac{I_n'(k_{0q}r_j)}{K_n'(k_{0q}a_j)}K_n(k_{0q}a_j)\end{array}\right\}\dfrac{\cos\{k_{0q}(z-d+h)\}}{\cos(k_{0q}h)}\end{array}\right]e^{in\theta_j} \quad \text{on } d-h\leq z\leq d \tag{1}$$

in which J_n and Y_n denotes the Bessel function of the first and the second kind of order n, and J_n' and Y_n' are the first derivatives of the Bessel function, respectively. I_n and K_n denotes the modified Bessel function of the first and the second kind of order n, and $I_n{}'$ and $K_n{}'$ are the first derivatives of the modified Bessel function. A new wave number k_0 is introduced from the dispersion relation $\omega^2 = gk_0 tanhk_0 h$. $A_n{}^j$ and $A_{nq}{}^j$ are unknown complex coefficients, and h is the vertical length of the porous-surfaced body. k_{0q} (positive real roots) is related to the evanescent waves, which satisfy the dispersion relation $\omega^2 = -gk_{0q} tan(k_{0q}h)$.

The incident wave potential in the jth local polar coordinate system can be expressed using Jacobi-Anger expansion of Bessel function as follows,

$$\phi_{in}^j = \frac{\cosh kz}{\cosh kd} T_j \sum_{n \to -\infty}^{\infty} J_n\left(kr_j\right) e^{in\left(\pi/2 + \theta_j - \beta\right)} \tag{2}$$

where $T_j = e^{ik\left(x_j \cos\beta + y_j \sin\beta\right)}$ is a phase factor associated with the cylinder j from the global origin.

The wave potential in the exterior region (2), which is expressed by using Graf's addition theorem for the Bessel functions [11] and satisfies the Helmholtz equation, can be expressed by the following eigenfunction expansion.

$$\phi_2^j\left(r_j, \theta_j\right) = \sum_{n=-\infty}^{\infty} \left[\begin{cases} T_j J_n\left(kr_j\right) e^{in(\pi/2-\beta)} + C_n^j \dfrac{H_n\left(kr_j\right)}{H_n'\left(ka_j\right)} \\ + \displaystyle\sum_{\substack{l=1 \\ l \neq j}}^{N} \sum_{m=-\infty}^{\infty} C_m^l \dfrac{J_n\left(kr_j\right)}{H_n'\left(ka_j\right)} H_{m-n}\left(kR_{lj}\right) e^{i(m-n)\alpha_{lj}} \end{cases} \times \dfrac{\cosh kz}{\cosh kd} \\ + \displaystyle\sum_{q=1}^{\infty} \begin{cases} C_{nq}^j \dfrac{K_n\left(k_q r_j\right)}{K_n'\left(k_q a_j\right)} \\ + \displaystyle\sum_{\substack{l=1 \\ l \neq j}}^{N} \sum_{m=-\infty}^{\infty} C_{mq}^l \dfrac{I_n\left(k_q r_j\right)}{K_n'\left(k_q a_j\right)} K_{m-n}\left(k_q R_{lj}\right) e^{i(m-n)\alpha_{lj}} e^{-im\pi} \end{cases} \times \dfrac{\cos k_q z}{\cos k_q d} \right] \times e^{in\theta_j} \tag{3}$$

Where k is the incident wave number related to the wave frequency through the dispersion relation $\omega^2 = gktanh(kd)$, and d is the water depth. $C_n{}^j$ and $C_{nq}{}^j$ are unknown complex potential coefficients. H_n is a Hankel function of the first kind of order n. $H_n{}'$ is the first derivatives of the Hankel, respectively. The scattered wave potential in Eq. (3) is composed of the propagating wave (first term) and the local (evanescent) wave mode (second term). The evanescent wave mode represents a standing wave system near the body, and exists only near the body. The wave numbers (k_q, positive real roots) associated with such local waves satisfy the dispersion relation $\omega^2 = -gk_q tan(k_q d)$.

The fluid flow passing through the porous surface of dual cylindrical caisson is assumed to obey Darcy's law. Therefore it can be written as follows [3],

$$\frac{\partial \phi_1^j}{\partial r} = \frac{\gamma}{\mu} \rho i \omega \left[\phi_1^j - \phi_2^j \right] \text{ on } r_j = a_j, \, j = 1, 2, 3, ..., N \tag{4}$$

where μ is the coefficient of dynamic viscosity, γ is a material constant having the dimension of length and ρ is the fluid density, respectively. Subsequently, the porosity of the dual cylinder will be characterized by the dimensionless parameter, G. The body boundary condition on the porous surface of dual cylinder can be expressed with the G.

$$\frac{\partial \phi_1^j}{\partial r} = ik_0 G \left[\phi_1^j - \phi_2^j \right], \quad Here \quad G = \frac{\rho \omega \gamma}{\mu k_0} \tag{5}$$

In addition to applying the body boundary conditions associated with the free surface conditions, the present boundary value problem must satisfy the matching conditions at the interface between the regions, which are given by

$$\phi_1^j = \phi_2^j \quad on \quad r_j = a_j, \, d - h \le z \le d$$

$$\frac{\partial \phi_1^j}{\partial r} = ik_0 G \left[\phi_1^j - \phi_2^j \right] \quad on \quad r_j = a_j, \, d - h \le z \le d \tag{6}$$

After solving the velocity potentials, the wave excitation forces on each dual cylindrical caisson are obtained using the integration of pressure on the wetted surface of cylinder. Wave forces in x-direction (F_x) and in y-direction (F_y) are calculated as follows,

$$F_x^j = -i\rho\omega \int_{d-h}^d \int_0^{2\pi} \frac{-igH}{2\omega} \left\{ \phi_2^j - \phi_1^j \right\} a_j \cos\theta d\theta dz$$

$$F_y^j = -i\rho\omega \int_{d-h}^d \int_0^{2\pi} \frac{-igH}{2\omega} \left\{ \phi_2^j - \phi_1^j \right\} a_j \sin\theta d\theta dz \quad on \quad d - h \le z \le d \tag{7}$$

$$F_x^j = -i\rho\omega \int_{d-h}^d \int_0^{2\pi} \frac{-igH}{2\omega} \left\{ \phi_1^j \right\} b_j \cos\theta d\theta dz$$

$$F_y^j = -i\rho\omega \int_{d-h}^d \int_0^{2\pi} \frac{-igH}{2\omega} \left\{ \phi_1^j \right\} b_j \sin\theta d\theta dz \quad on \quad d - h \le z \le d \tag{8}$$

$$F_x^j = -i\rho\omega \int_0^{d-h} \int_0^{2\pi} \frac{-igH}{2\omega} \left\{ \phi_2^j \right\} a_j \cos\theta d\theta dz$$

$$F_y^j = -i\rho\omega \int_0^{d-h} \int_0^{2\pi} \frac{-igH}{2\omega} \left\{ \phi_2^j \right\} a_j \sin\theta d\theta dz \quad on \quad 0 \le z \le d - h \tag{9}$$

where H is the incident wave height. Eq. (7), (8) and (9) present the wave forces acting on the porous outer cylinder, the impermeable inner cylinder and the impermeable outer cylinder, respectively.

3. Results and Discussion

In order to verify the accuracy of the calculated wave forces on a dual cylinder with full porous area, which the outer cylinder is porous and the inner cylinder is impermeable, the present numerical results are compared with the numerical results of Wang and Ren [4]. It is indicated that the hydrodynamic force on the inner cylinder can be effectively reduced by the existence of outer porous cylinder. Fig. 2 shows the comparison of wave forces on a dual cylinder. The various parameters are $b/a=0.5$, $G=1, 2, 5$ and $\beta=0^o$, respectively. The abscissa denotes the dimensionless frequency, $g/\omega^2 d$, and the vertical axis is the dimensionless wave force normalized with $\rho gbdH$ and $\rho gadH$, respectively. The wave force on the inner cylinder is gradually increased and the wave force on outer cylinder is inversely decreased as the porosity parameter G increase. It means that the wave force on inner cylinder is increased by the wave infiltration which percolates in the inside as the porosity of outer cylinder increase. The wave force of inner cylinder has the peak value at wave frequency 0.5; on the other hand, the wave force of outer cylinder has the lowest value. The calculated wave forces are in good agreement with the results from Wang and Ren. Therefore, the present method is very available to evaluate the wave forces on the dual cylindrical structures.

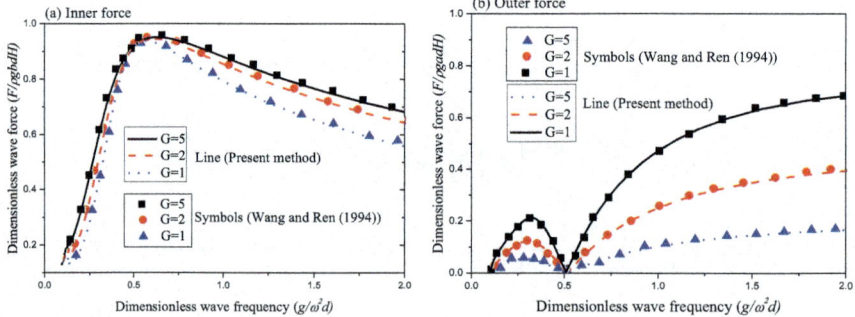

Fig. 2. Comparison of dimensionless wave forces on a dual cylinder for $b/a=1/2$ and $\beta=0^o$: (a) Inner Force, (b) Outer Force.

Fig. 3 shows the wave forces on the array of dual cylindrical breakwaters as shown in Fig. 1 for various porosity depths (h/d) of outer cylinder. The calculation conditions are $a=5.0$m, $b/a=0.5$, $d/a=4$, $G=2.0$ and $\beta=90.0^o$. The dual cylindrical caisson breakwaters are numbered from 1 to 7, and situated at (-30.0, 0.0)m, (-20.0, 0.0)m, (-10.0, 0.0)m, (0.0, 0.0)m, (10.0, 0.0)m, (20.0, 0.0)m, and (30.0, 0.0)m, respectively. The calculated total forces, which are sum of outer forces and inner forces acting on the dual cylindrical caisson, are normalized by $\rho g(H/2)a^2$. Since the amount of incident wave passing through porous area is increased as the porosity depth increases, the total wave forces are increased. The pattern of wave forces is strongly influenced by the porosity depth and the peak wave force is appeared between ka=0.1 and ka=0.2. The difference of total wave forces between the porosity depth 0.5 and 1.0 is very small in the short wave region ($ka\geq1.0$). It means that the dual cylindrical caisson breakwater with porosity depth 0.5 is very efficient to reduce the effect of wave force in the short wave region like the case of full-body porous cylindrical caisson breakwater.

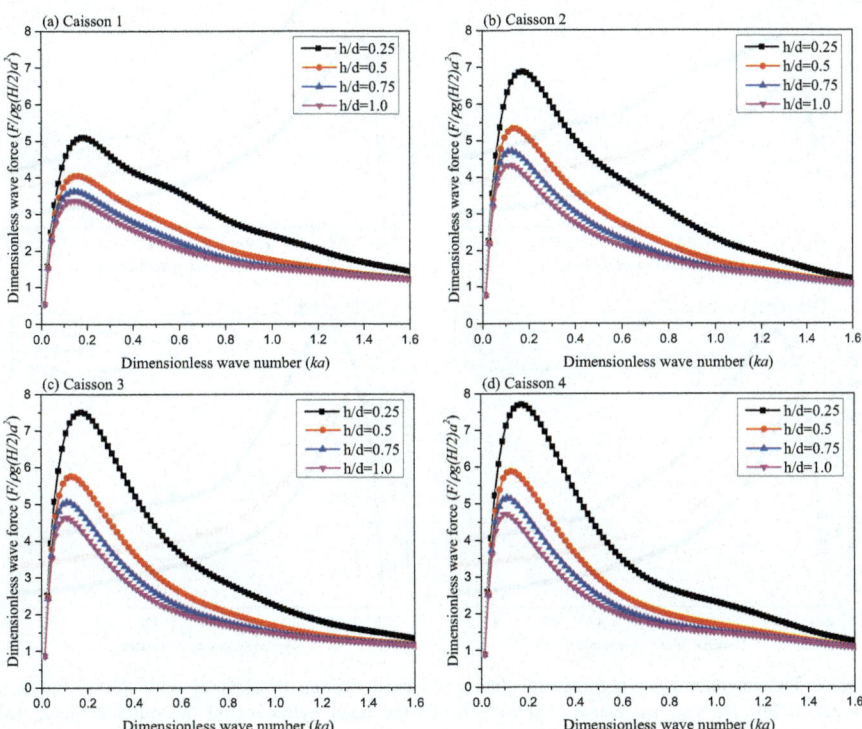

Fig. 3. Dimensionless wave forces on the dual cylindrical caisson breakwater with a=5.0m, b/a=0.5, d/a=4, G=2.0 and β=90^0 for various porosity depths (h/d) of outer cylinder: (a) Caisson 1, (b) Caisson 2, (c) Caisson 3, (d) Caisson 4.

Fig. 4 shows the comparison of wave forces on an array of seven dual cylindrical caisson breakwaters to examine the effect of the distance between the inner cylinder and the outer cylinder. The calculation conditions are a=5.0m, h/d=0.5, d/a=4, G=2.0 and β=90.0^0. The total wave forces are increased as the distance ratio increases but the pattern of wave forces between the distance ratio of 0.5 and 0.25 is very similar in a given range of wave number. This means that the dual cylindrical caisson breakwater with the distance ratio (b/a) 0.5 is significantly efficient to reduce the wave force in the short waver region. The peak wave forces acting on the caisson 4 have the largest values compared to the other caissons and these are decreased as the location of caisson breakwater is far from the middle point.

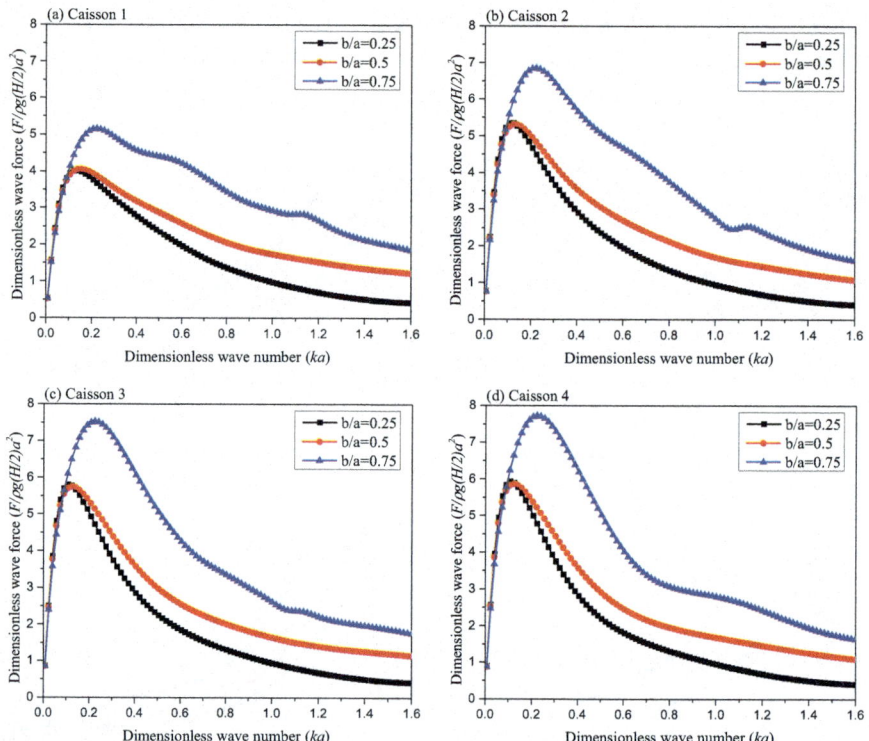

Fig. 3. Dimensionless wave forces on dual cylindrical caisson breakwater with d/a=4, h/d=0.5, G=2.0, β=90° for various distances (b/a) between the inner cylinder and the outer cylinder: (a) Caisson 1, (b) Caisson 2, (c) Caisson 3, (d) Caisson 4.

4. Conclusion

Under the assumption of potential flow and linear wave theory, the 3D numerical method for the array of dual cylindrical caisson breakwater with partial porous area, which consist of an impermeable inner cylinder and a partial porous outer cylinder, is developed using the Eigenfunction expansion method. In order to realize the effect of porosity on the cylinder surface, Darcy's law is applied to the porous body boundary condition. For verification of this new method, the calculated numerical results for a dual cylinder are compared with Wang and Ren [4]. It is found that the numerical results give the good agreement to the results of them. Therefore it is suggested that the developed numerical method is very useful to evaluate the wave force acting on the array of dual cylindrical caisson breakwaters with partial porous area. The water wave interaction due to the array of dual cylindrical caisson breakwaters is demonstrated to examine the effects of porosity of outer cylinders. The wave force caused by the interaction effects between wave and cylinders is significantly diminished as the porosity of the outer cylinder increases. Therefore the porosity of structure is very effective on the reduction of the wave force acting on the caisson breakwater. It is also found that the dual cylindrical caisson breakwater with porosity depth (h/d) 0.5 is remarkably efficient to reduce the wave force like the full-body porous cylindrical caisson breakwater.

Acknowledgements

This research was supported by Korea Institute of Marine Science & Technology Promotion (KIMST) through the research project "Development of design and construction technologies of modular revetment structure (20160281)" and by the Ministry of Land, Infrastructure and Transport through the research project "Development of Life-Cycle Engineering Technique and Construction Method for Global Competitiveness Upgrade of Cable Bridges (16SCIP-B119960).

References

1. K. Sankarbabu, S. A. Sannasiraj, and V. Sundar, Hydrodynamic performance of a dual cylindrical caisson breakwater, *Coastal Engineering*, Vol. 55, pp. 431-446 (2008)
2. A. N. Williams, and W. Li, Approximate hydrodynamic analysis of multi-column ocean structures, *Ocean Engineering*, Vol. 21, pp. 519-573 (1998)
3. A. N. Williams, and W. Li, Water wave interaction with an array of bottom-mounted surface-piercing porous cylinder, *Ocean Engineering*, Vol. 27, pp 841-866 (2000)
4. K. H. Wang, and X. Ren, Interactions of cnoidal waves with cylinder arrays, *Ocean Engineering*, Vol. 26, pp. 1-20 (1999)

704

5. K. H. Wang, T. Y. Wu, and G. T Yates, Three-dimensional scattering of solitary waves by vertical cylinder, *Journal of Waterway, Port, Coastal and Ocean Engineering*, Vol. 118, pp. 551-566 (1992)

6. K. H. Wang, and L. Jiang, Interactions of solitary waves with cylinder arrays, *Proceeding of the International Conference on Offshore Mechanics and Arctic Engineering-OMAE*, Vol. 1, pp. 99-107 (1994)

7. L. Jiang, and K. H. Wang, Hydrodynamic interactions of cnoidal waves with a vertical cylinder, *Applied Ocean Research*, Vol. 17, pp. 277-289 (1995)

8. K. H. Wang, and X. Ren, Water waves on a flexible and porous breakwater, *Journal of Engineering Mechanics*, Vol. 119. pp 1025-1048 (1993)

9. K. H. Wang, and X. Ren, Interactions of cnoidal waves with cylinder arrays, *Ocean Engineering*, Vol. 26, pp. 1-20 (1999)

10. K. Sankarbabu, S. A. Sannasiraj, and V. Sundar, Interaction of regular waves with a group of dual porous circular cylinders, *Applied Ocean Research*, Vol. 29, pp. 180-190 (2007)

11. M. Abramowitz, and I. A. Stegun, *Handbook of mathematical function* (New York, 1972)

Numerical Analysis of Interlocking Caisson Breakwater using Modular System[*]

Sung-Hoon Song[†], Min-Su Park, Young-Jun You, Youn-Ju Jeong and Yoon-Koog Hwang

Structural Engineering Research Institute, KICT,
Goyang-Si, Gyeonggi-Do 10223, Korea
†E-mail: songsunghoon@kict.re.kr
www.kict.re.kr

Breakwaters have been severely damaged quite often since they are directly exposed to large waves. And damage level of breakwaters has been scaled up because of abnormal global climate changes. So, recently, an interlocking concept has been much attention to enhance the structural stability of conventional caisson structures designed individually to resist waves. However, interlocking methods proposed so far have not been actively applied to the design of actual breakwaters because of structural stability, construction and economic issues. In the present study, a modular caisson was proposed that can maximize the existing caissons and improve structural stability and workability by using the modular interlocking system like LEGO ; and behavior and interlocking effect of the modular caisson breakwater are evaluated by numerical analysis method. From the results of the study, because of clearance between caisson units, the modular caisson units are interlocked by two interlocking behaviors : (a) indirect interlocking behavior by frictional force at contact surfaces between interlocking parts, (b) direct interlocking behavior by shear resistance between interlocking parts ; the direct interlocking effect by shear resistance is larger than the indirect interlocking effect by frictional force ; interlocking effect of the modular caisson breakwater is similar to the all-bonded long caisson breakwater.

Keywords: Caisson breakwater, Interlocking, Modular caisson, Wave force reduction

1. Introduction

Breakwaters are very important to keep harbors calm. However, they have been severely damaged quite often since they are directly exposed to large waves. Because of abnormal global climate changes, damage level of coastal structures has also been scaled up according to increase of wave height and duration of storms. In particular, according to the report of caisson breakwater damages in Japan(Takayama et al, 2002[1]), it was found that damages caused by caisson

---

[*] This work is supported by the Korea Institute of Marine Science & Technology Promotion.
[†] Researcher, Korea Institute of Civil Engineering and Building Technology(KICT)

sliding, see Figure 1, account for more than 70% of breakwater failure modes. So, the design criteria for new breakwaters is being intensified to cope with abnormal global climate changes and unusual waves.

Recently, an interlocking concept has been much attention to enhance the structural stability of conventional caisson structures designed individually to resist waves[2,3]. Since the total wave force acting on long breakwaters[4~6] is reduced, interlocked caisson breakwaters may be survival even if unusual high waves occur. As shown in Figure 2, several interlocking methods have been proposed until now due to the advantage of wave force reduction in long caisson breakwaters. However, the interlocking methods proposed so far have not been actively applied to the design of actual caisson breakwaters because of structural stability, construction and economic issues. Therefore, in the present study a modular caisson was proposed that can maximize the existing caisson and improve structural stability and workability by using the modular interlocking system like LEGO. And behavior and interlocking effects of the modular caisson breakwater were evaluated by numerical analysis method.

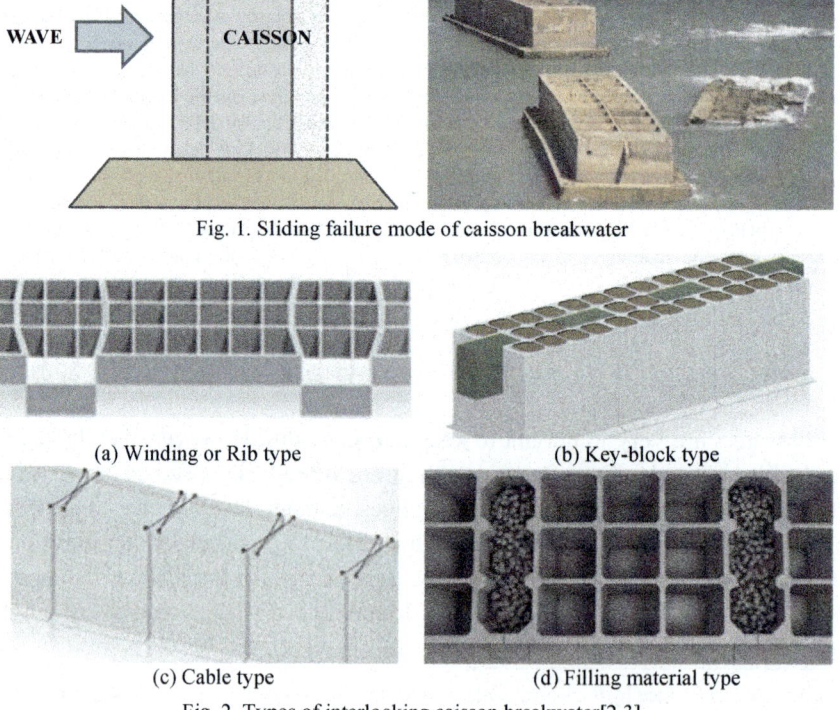

Fig. 1. Sliding failure mode of caisson breakwater

(a) Winding or Rib type   (b) Key-block type

(c) Cable type   (d) Filling material type

Fig. 2. Types of interlocking caisson breakwater[2,3]

2. Shape and features of modular caisson

Figure 3 shows names and dimensions of major parts of the modular caisson unit. As shown in Figure 3, the interlocking part of the modular caisson consists of the insertion part and the bottom plate with shear walls. In general, gap is always created between caisson units owing to manufacturing tolerance and construction errors. Therefore, the modular caisson units are interlocked by two interlocking behaviors. The first is the indirect interlocking behavior by frictional force at the contact surface between the lower surface of the insertion part and the upper surface of the bottom plate, see Figure 4(a). At this time, there is a gap between the insertion part and the shear wall of a bottom plate. The second is the direct interlocking behavior by shear resistance between the insertion part and the shear wall of the bottom plate without clearance, see Figure 4(b). As a result, the modular caisson breakwater becomes a long caisson breakwater by the two interlocking behaviors. It is then possible to reduce wave forces acting on caisson units and may be survival even if unusual high wave occurs. Moreover, the modular caisson is also advantageous to the structural safety during construction by the two interlocking behaviors between the insertion part and the bottom plate.

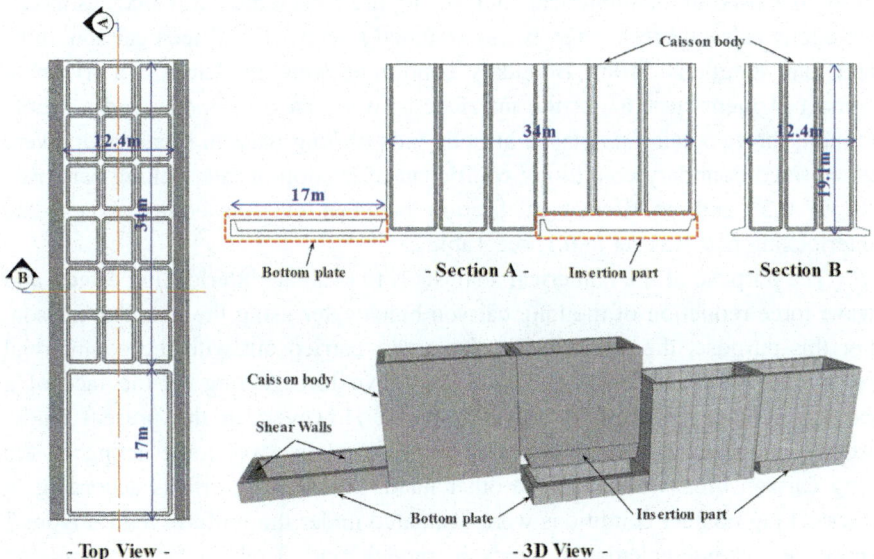

Fig. 3. Shape and dimensions of Modular caisson unit

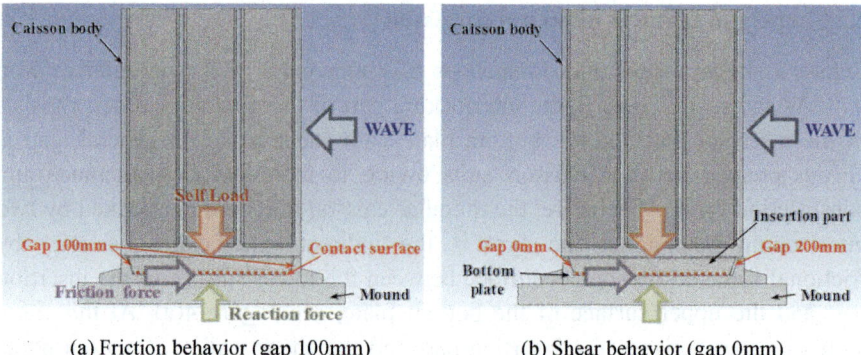

(a) Friction behavior (gap 100mm)   (b) Shear behavior (gap 0mm)

Fig. 4. Interlocking behavior of modular caisson breakwater

3. Numerical analysis

The numerical analysis model of the modular caisson breakwater was determined by referring to the caisson breakwater applied to the Busan New Port in Korea. As shown in Figure 3 and 5, the modular caisson unit presented width of 12.4m, height of 19.1m, total length of 51m ; length of the interlocking part was designed to be 50% of length of the caisson body. Table 1 shows material properties used in the numerical analysis. Figure 6 presents a full analysis model for a long caisson breakwater. It was assumed to consist of fifteen caisson units, and total length is 510m. Boundary conditions were modeled as a frictional contact element (μ=0.6[7]) at the interface between caisson bodies and a rubble mound; the two caissons placed at both ends of long caisson breakwaters were given fixed boundary condition ; coefficient of friction of interlocking parts was set to 0.5[7] and coefficient of friction between caisson bodies except the interlocking part was set to 0.1, see Table 2.

The purpose of the numerical analysis is to evaluate interlocking effects and wave force reduction of the long caisson breakwater using the modular caisson. For this purpose, the numerical analysis was carried out with the commercial program ANSY and AQWA. Generally, wave force acting on the face of a caisson can be obtained by Goda formula. However, in the present study, AQWA program was used to evaluate and simulate wave force acting on the long caisson breakwater. On the other hand, interlocking effects according to interlocking contact conditions were evaluated under the uniform load. Figure 7 shows the wave calculation analysis model, and Table 4 presents marine environmental conditions applied to the Busan New Port in Korea. Wave force reduction of the long caisson breakwater was evaluated according to the incidence angle of 0° and 60°.

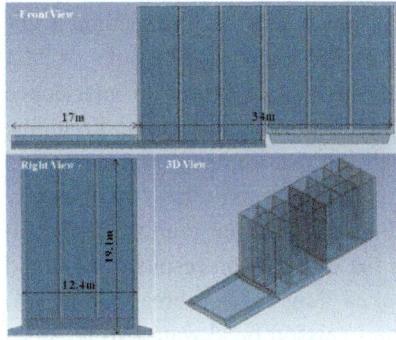

Fig. 5. Numerical analysis model

Table 1. Material properties

Material property	Caisson	Mound
Mass (kg/m³)	2300	2000
Young's modulus (GPa)	30	24
Poisson's ratio	0.18	0.3

Table 2. Coefficient of Friction

Part	μ
Interlocking part	0.5
Con'c - Con'c except Interlocking part	0.1
Caisson bodies - Mound	0.6

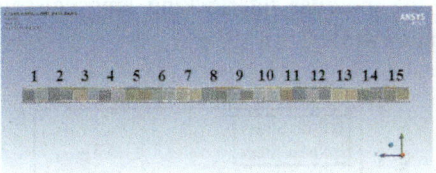
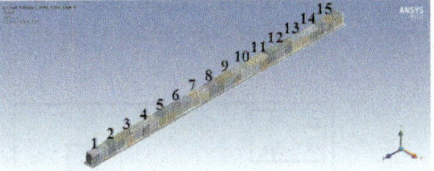

Fig. 6. Full analysis model(15 units) of a long caisson breakwater

Table 3. Contact Conditions of model

Model	Caisson – Mound contact condition	Interlocking contact condition
M-1	Frictionless	Bonded
M-2	Frictional (μ=0.6)	Bonded
M-3	Frictional (μ=0.6)	Frictional (μ=0.5) & Gap 0mm
M-4	Frictional (μ=0.6)	Frictional (μ=0.5) & Gap 100mm

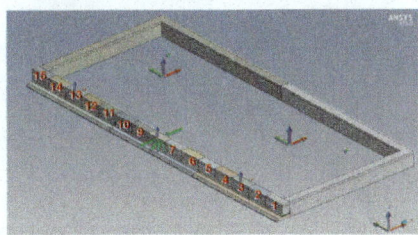

Fig. 7. Wave calculation analysis model

Table 4. Marine environmental conditions

Condition	Value
Water level	21.5 m
Wave height	8 m
Period	12 sec

Note: Refer to the Busan New Port

4. Results

4.1. *Interlocking effects according to contact conditions*

Figure 8 and 9 present the maximum displacement and stress results according to the interlocking contact conditions ; the displacement of X-direction is

710

deformation by sliding of a caisson ; the displacement of Y-direction is the deformation by overturning of a caisson. From the comparison results of M-1 and M-2, it was found that most of external wave force action on interlocked long caisson breakwaters was transmitted in the mound direction. From the comparison results of M-2 and M-3, although the maximum displacement and stress of M-3 were larger than M-2, the displacement and stress of the fully shear-interlocked long caisson breakwater(M-3), were similar to those of the all-bonded long caisson breakwater(M-2). From the comparison results of M-3 and M-4, the stress of the partially shear-interlocked long caisson breakwater(M-4) was larger than the fully shear-interlocked long caisson breakwater(M-3). Since clearance between the insertion part and the shear wall of the bottom plate is always present, the actual behavior of the modular interlocking long caisson breakwater will be similar to M-4, see Figure 4(a).

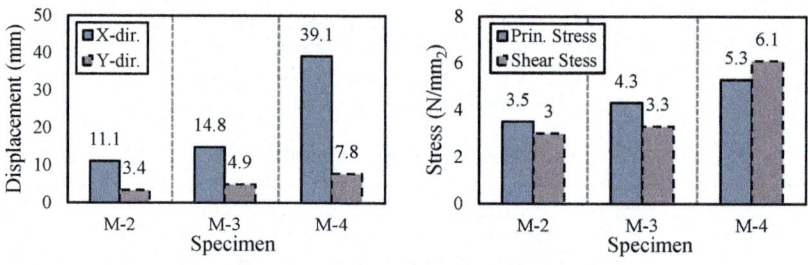

Fig. 8. Displacement and stress results

Fig. 9. Stress results according to the interlocking contact condition

4.2. *Wave force reduction of long caisson breakwater*

Figure 10 presents calculation results of wave force acting on each caisson unit of long caisson breakwaters according to the incidence angle of 0° and 60°. In the graph of Figure 10, square shows the total wave force acting on each caisson unit ; broken line shows the average of total wave force acting on a caisson breakwater ; straight line is the wave force calculated by the Goda formula. Compared with the incidence angle of 0°, the maximum wave force at the incidence angle of 60° was about 10% larger, and the minimum wave force was about 1.2% smaller. As shown in Figure 10(c) and (d), the average of wave force was smaller than one calculated by Goda formula, the wave force reduction, as one of the advantage of characteristics of a long caisson breakwater, was confirmed.

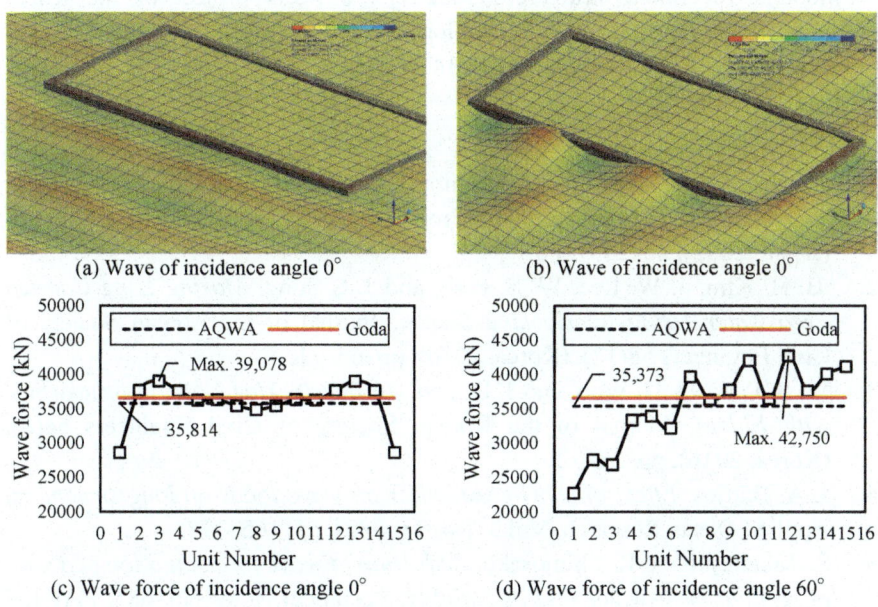

(a) Wave of incidence angle 0°   (b) Wave of incidence angle 0°

(c) Wave force of incidence angle 0°   (d) Wave force of incidence angle 60°

Fig. 10. Wave force calculation results

5. Conclusions

In the present study, the modular caisson was proposed that can maximize the existing caisson and improve the structural stability and workability by using the modular interlocking system like LEGO. And the behavior and interlocking effect of the modular caisson breakwater were investigated by numerical analysis method. The following knowledge was obtained through this study.

1. Clearance between caisson units is created because of manufacturing tolerance and construction errors. So, the modular caisson units are interlocked by two interlocking behaviors: (a) the indirect interlocking behavior by frictional force at the contact surface between the interlocking parts, (b) the direct interlocking behavior by shear resistance between the interlocking parts.

2. The direct interlocking effect by shear resistance is larger than the indirect interlocking effect by frictional force.

3. The interlocking effect of the modular caisson breakwater is similar to the all-bonded long caisson breakwater.

Acknowledgments

This work is part of the "Development of design and construction technologies of modular revetment structure(20160281)", a R&D project of the Korea Institute of Marine Science & Technology Promotion (KIMST) funded by the Korea Ministry of Maritime Affairs and Fisheries in 2016.

References

1. T. Takayama and K. Higashira, *Statistical analysis on damage characteristics of breakwaters*, Proc. of Ocean Development Conf., 18 (Japan, 2002), pp. 263–268.
2. B. H. Kim, J. W. Lee, W. S. Park, and J. S. Jung, *Making Long Caisson Breakwater Using Interlocking System*, Journal of the Korean Society of Civil Engineers 58(12), (Korea, 2010), pp. 65–71.
3. W. S. Park, D. H. Won, and J. H. Seo, *An Interlocking Caisson Breakwater with Fillers*, Journal of the Korean Society of Civil Engineers 64(8), (Korea, 2016), pp. 28–32.
4. J. A. Battjes, *Effect of short-crestedness on wave loads on long structures*, Applied Ocean Research, Vol 5, No. 3 (1982), pp. 165–172.
5. S. Takahashi and K. Shimosako, *Reduction of wave force on a long caisson of vertical breakwater and its stability*, Technical Notes No. 685, Port and Harbour Research Institute (Yokosuka, Japan, 1990).
6. H. F. Burcharth and Z. Liu, *Force reduction of short-crested non-breaking waves on caissons*, Section 4.3, Part 4, Class II Report of MAST II Project: PROVERBS, Technical University of Braunschweig, (Germany, 1998)
7. *Port and Harbor Design Standards*, Korea Ministry of Oceans and Fisheries, (Korea, 2014).

Calculation of Damage Progression and Total Cost of Wave Dissipating Works Considering Repair Process

H. Kawamura

Nishi-Nihon branch, Sanshosuiko Co. Ltd.
Fukuoka, 812-0013, Japan
†E-mail: h-kawamura@sanshosuiko.co.jp

T. Ota and Y. Matsumi

Graduate School of Engineering, Tottori University
Tottori, 680-8552, Japan
E-mail: ohta@sse.tottori-u.ac.jp

T. Hirayama

Higashi-Nihon branch, Sanshosuiko Co. Ltd.
Tokyo, 150-0045, Japan
E-mail: hirayama@sanshosuiko.co.jp

Based on the experimental results for profile change of wave dissipating blocks covering a caisson breakwater, the model profiles for the damaged block layer are made and the relation between the damage progression and change of performance is investigated. The results of the numerical experiment using the model profiles show that the overtopping rate increases by a factor of 1.5 compared to that of the initial profile with damage progression. Furthermore a statistical model based on the Markov chain is applied to the damage progression process and the total cost for the maintenance of wave dissipating blocks is estimated by using an assumed repair cost and loss. This method can be applied to find a policy of repair that minimizes the total (life cycle) cost, however, it is necessary to evaluate the practical amount of repair and loss reasonably.

Keywords: Wave dissipating works; Damage progression; Markov chain model; Repair.

1. Introduction

The infrastructures are required to keep a certain level of performance during the duration of service. Because the performance of the infrastructures including harbor and coastal structures deteriorate due to aging and damage that is caused by the action of external forces, it is necessary to perform appropriate maintenance. It is also needed to understand the deterioration process of the structure and make plans for the maintenance. Several studies dealt with the estimation of life cycle cost and method of the maintenance for wave dissipating works and so on by using Monte Carlo approach (e.g. Miyata et al.[1], Tsujio and Yasuda[2]). With a goal of finding a repair policy for wave dissipating blocks covering a caisson breakwater to minimize the total (life cycle) cost, the followings are investigated in this study. (1) Based on the experimental results for profile change of wave dissipating blocks, the model profiles for the damaged block layer are made and the relation between the damage progression and change of performance is investigated by the numerical experiment. (2) Using the data of the normalized eroded area S, the ranks that indicate the degree of damage are determined and the frequency ratio for each rank is calculated. The Markov chain model is applied to the transition process of the frequency ratio. (3) Combining the damage progression process based on the Markov chain model with the repair process, the total cost for the maintenance of wave dissipating blocks is estimated by using an assumed repair cost and loss.

2. Overview of Experiment to Measure Damaged Profile

The data of damage progression of the wave dissipating works were obtained from a series of model experiments. The experiments were conducted in two wave flumes (8.6 m long, 0.6 m wide and 0.6 m high) that were set up in a multi-directional wave basin using the irregular waves in the same way as Ota et al.[3] The multi-directional wavemaker was the serpent-type, and the wave flumes were made by dividing the wavemaker between the paddles by plywood. The plywood was placed through the clearance between the paddles to divide completely including the moving range of paddle. A conventional wave dissipating work that consisted of wave dissipating blocks was constructed on a mound made by plywood in each wave flume. The crest height h_c was 10 cm. The experimental setup in a flume is shown in Figure 1. The mean mass and

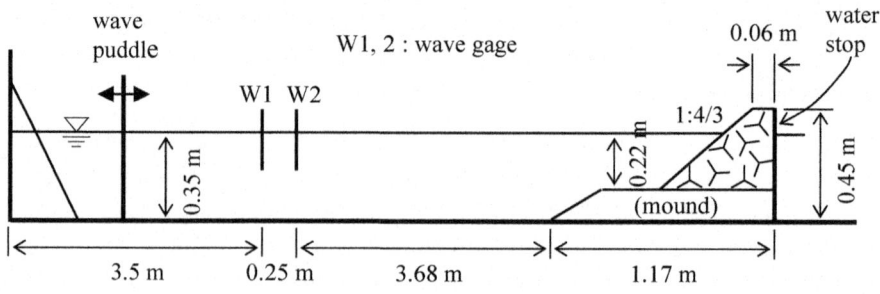

Fig. 1. Experimental setup

nominal diameter of the wave dissipating block were $M = 37.6$ g and $D_n = 2.56$ cm respectively.

The JONSWAP spectrum with the shape parameter $\gamma = 3.3$ and significant wave period $T_{1/3} = 1.2$ s was used as the target spectrum of the incident irregular wave. The length of the input signal for wavemaker was 20 min and two wave paddles were driven by the signal in the same phase. The number of generated waves was approximately 1100 and the significant wave height $H_{1/3}$ was about 0.11 m.

The profiles of each revetment were measured every 20 min along three cross-shore lines (the distance between the lines was 0.15 m) using a laser displacement sensor at intervals of 1 cm horizontally. The normalized eroded area (damage level) $S = A_e/D_n^2$ with A_e = eroded area was used as a parameter which showed the deformation quantity. The eroded area A_e was calculated using the damaged and initial (undamaged) profiles on each measurement line of the profile. Moreover, another normalized eroded area S_* was obtained using the profile that was given as the average of three profiles in each flume.

The waves were generated in bursts of 20 min repeatedly and the total time of wave generation was 200 min (10 times) in a test. The number of tests was 10 and the experiments were conducted under the same conditions of the incident wave and the initial profile of the wave dissipating work.

3. Modeling of Damage Progression

In this study, a set of model profiles of the damaged block layer are made for the purpose of understanding of the relation between the damage progression and the change of performance. The model profiles are made using the intersections between the initial profile and damaged profiles, the decrements of crest height η at two points of the crest and the horizontal displacement of the toe x_t, which are shown in Figure 2 schematically. Because the positions of the intersections

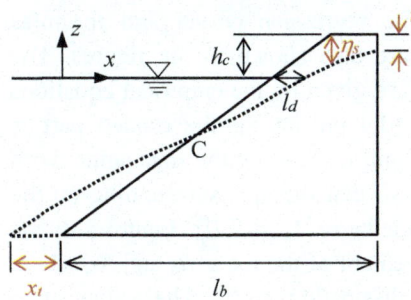

Fig. 2. Quantities used for model profile

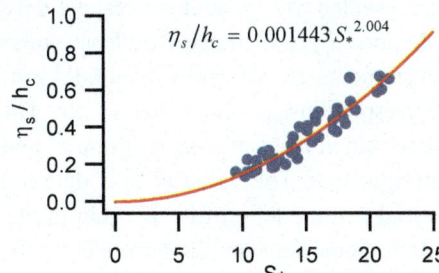

Fig. 3. Relation between S_* and η_s

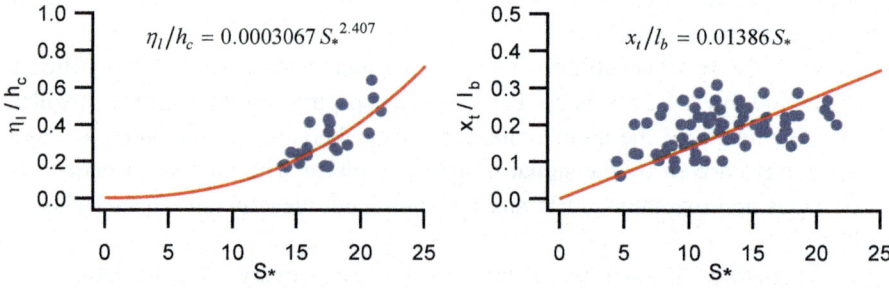

Fig. 4. Relation between S_* and η_l Fig. 5. Relation between S_* and x_t

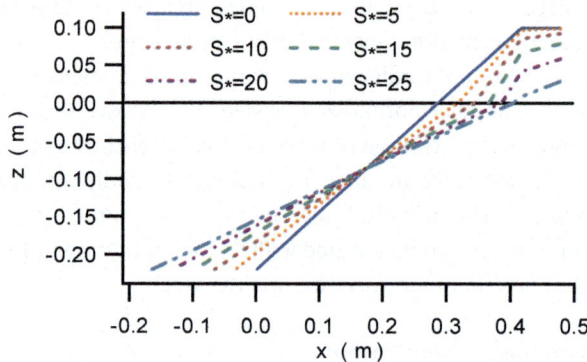

Fig. 6. Model profiles of damaged block layer

don't change much in spite of the damage progression, the coordinate of Point C is given by the average of obtained x-coordinates and the corresponding z-coordinate on the initial profile. The values of η and x_t are given by the empirical equations based on the experimental data. Figures 3 and 4 show the relation between S_* and η at the seaward and landward edge of the crest that are represented by η_s and η_l respectively. The regression curves and formulas obtained by the method of least squares are also shown in the figures. The decrements η_s and η_l for the model profile are given by the empirical equations corresponding to the value of S_*. The model profile for the eroded part is determined by using η_s, η_l, Point C and l_d, and connecting the four points with straight line. The value of l_d is determined so that the area surrounding by the initial and model profiles is equal to the eroded area $A_e = S_* D_n^2$. Figure 5 shows the relation between the normalized displacement of the toe x_t / l_b and S_*, where l_b = initial length of the base of wave dissipating work. The value of x_t corresponding to S_* is given by the regression formula shown in Figure 5. The model profile for the accumulated part is determined by connection Point C and the toe of wave dissipating work with a straight line for simplicity. The model profiles for the case of S_* =0 (initial profile), 5, 10, 15, 20 and 25 are shown in Figure 6.

4. Numerical Experiment Using Model Profiles

Numerical experiment is conducted to evaluate the performance change of wave dissipating work. The overtopping rate is taken as the performance index. A two-dimensional numerical wave flume "CADMAS-SURF" Ver. 5.1 [4] is used to predict the overtopping rate. The governing equations are the continuity equation and Navier-Stokes equation for incompressible and viscous fluid. The water surface variation is computed using the volume of fluid (VOF) method. The computational domain between the wave paddle and water stop is set so as to correspond to the experimental setup shown in Figure 1. A measuring box is set behind the water stop and the volume of overtopped water is computed from the increment of the VOF function F in the box. The horizontal grid interval is 1 cm around the wave dissipating work and is 2 cm in other regions. The vertical grid interval is 1 cm in all regions. The porosity, inertia and drag coefficient for the region of the wave dissipating work are 0.5, 1.2 and 1.0 respectively. The modified Bretschneider-Mitsuyasu spectrum with $T_{1/3}$ = 1.6 s and $H_{1/3}$ = 0.11 m is used as the given target spectrum of the incident irregular wave. The analysis time of the computation is 300 s.

The model profiles corresponding to S_* = 0, 5, 8, 10, 12, 15, 20 and 25 are used in the computation. Table 1 shows the relation between S_*, η_l/h_c and q_o'. The value of q_o' means the overtopping rate normalized by that of the initial profile. The overtopping rate shows a tendency to increase with increasing S_* and the maximum q_o' is about 1.5 at S_* = 20. Moreover, it is conceivable that the

Table 1. Relation between damage progression and overtopping rate

S_*	η_l/h_c	q_o'
0	0	1.00
5	0.0148	1.06
8	0.0458	1.17
10	0.0783	1.22
12	0.121	1.30
15	0.208	1.38
20	0.415	1.47
25	0.710	1.30

exposed water stop and gentle slope of the damaged wave dissipating work influence the decrease of $q_o{}'$ at $S_* = 25$ as Kajima et al.[5] and Ota et al.[3] mentioned.

Hirayama and Naganuma[6] proposed a relational expression between the wave transmission coefficient K_T and overtopping rate q.

$$K_T = -0.259q^{-0.204} + 0.582 \tag{1}$$

The value of K_T becomes up to 1.4 times when q becomes 1.5 times, and it might influence the operating rate of cargo handling in a port. The result shown in Table 1 cannot give a definite criterion for the repair of wave dissipating work. However, it is necessary to repair in the case of the damage level corresponding $S_* = 20$ at least and to consider the loss caused by the decrease in function of a port.

5. Modeling of Damage Progression

A simple analysis was also made for the variation of damage level S that was mentioned in chapter 2. The values of S were rounded off to the integers and the ranks of damage level were determined as follows[7]; Rank a: $S = 0 - 3$, Rank b: $S = 4 - 7$, Rank c: $S = 8 - 11$, Rank d: $S = 12$ or more. The relative frequency for each rank of damage level at each N_r (hereafter, "ratio of damage level") is obtained, where N_r = number of iteration of irregular wave action ($N_r = 0$-10). Figure 7 shows the transition of the ratio of damage level. Because the damage levels of all wave dissipating works are zero at $N_r = 0$ that means the initial state, the ratio of Rank a is one. As N_r increases, the ratio of Rank a decreases rapidly and that of Rank d increases gradually.

A few models using Markov chain with stationary transition probability are applied to the damage progression process in this study. The following three models[7] are applied:

Model 1; the damage level changes one rank at a time and all transition probabilities are same

Model 2; the damage level changes one rank at a time and the transition probability differs from each other

Model 3; the damage level changes more than one rank at a time and the transition probability differs from each other.

The state transition diagrams of these models are shown in Figures 8-10 and the character 'p' in the figures denotes the transition probability. The damage level does not decrease, Rank d is the final phase of damage and the probability to stay in this rank is one when the repair of wave dissipating work is not

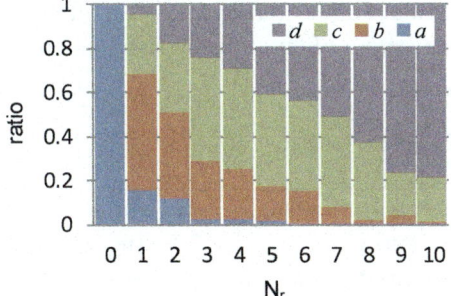

Fig. 7. Ratio of damage level

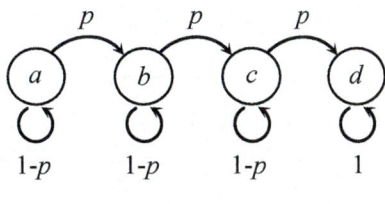

Fig. 8. State transition diagram for Model 1

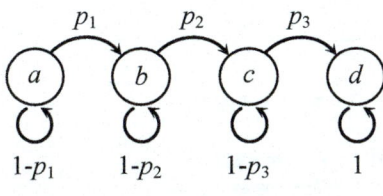

Fig. 9. State transition diagram for Model 2

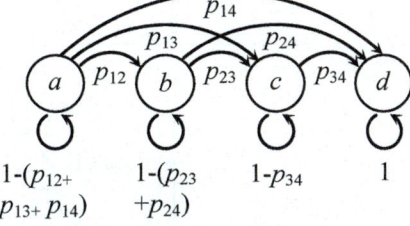

Fig. 10. State transition diagram for Model 3

considered. The ratio of damage level at N_r is given by Eq. (2) for Model 3 as an example.

$$\begin{pmatrix} a \\ b \\ c \\ d \end{pmatrix} = \begin{pmatrix} 1-(p_{12}+p_{13}+p_{14}) & 0 & 0 & 0 \\ p_{12} & 1-(p_{23}+p_{24}) & 0 & 0 \\ p_{13} & p_{23} & 1-p_{34} & 0 \\ p_{14} & p_{24} & p_{34} & 1 \end{pmatrix}^{N_r} \begin{pmatrix} 1 \\ 0 \\ 0 \\ 0 \end{pmatrix} \quad (2)$$

The transition probabilities are estimated by using Microsoft Excel 2007 (hereafter, "Excel"). The procedure is as follows[8].

1) The initial values of the transition probabilities are arbitrarily given in the range of (0, 1) for each model.
2) The ratio of damage level is calculated for each N_r.
3) The error sum of squares between the calculated and experimental values of the ratio of damage level at $N_r = 5$, which is the maximum number whose ratio of Rank a is not zero, is computed.
4) The values of the transition probabilities that minimize the error sum of squares are calculated by using the Solver in Excel. The constraint condition

in the parameter setting is that the all values of the transition probabilities are less than one. The addition item in the setting of option is the assumption of nonnegative number.

Table 2. Estimated transition probabilities

Model 1	value	Model 2	value	Model 3	value
p	0.4563	p_1	0.5897	p_{12}	0.4410
		p_2	0.4507	p_{13}	0.1006
		p_3	0.3526	p_{14}	0.04809
				p_{23}	0.2963
				p_{24}	0.09096
				p_{34}	0.1687

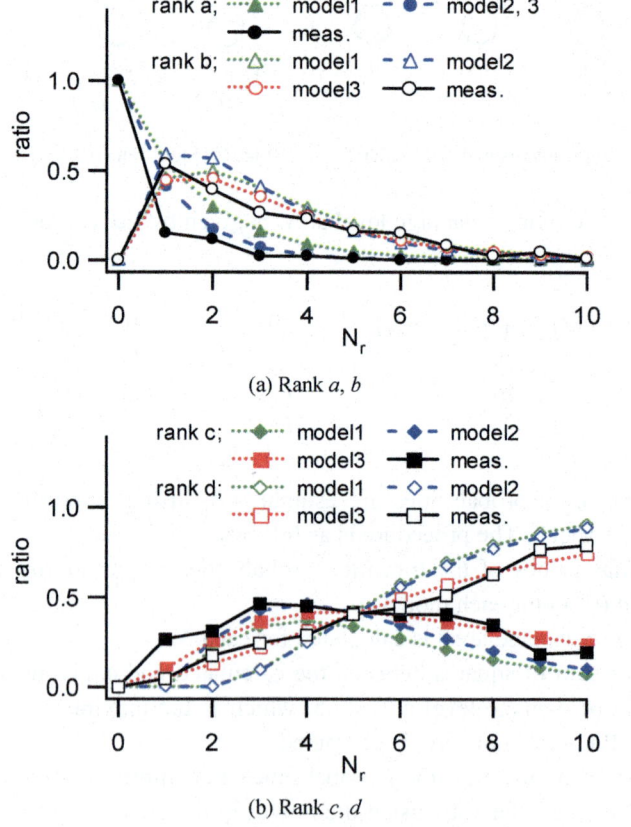

(a) Rank *a, b*

(b) Rank *c, d*

Fig. 11. Variations of ratios by Rank

The estimated values of the stationary transition probabilities are shown in Table 2.

The ratios of damage level up to $N_r = 10$ are calculated using the estimated transition probabilities. Figure 11 shows the change of the calculated ratios together with that of the experiment by Rank. Among the Markov chain models, the result of Model 3 shows the similar change of the ratio to the experimental results. It is conceivable that the model including the transition to more than two ranks can predict the transition process of the damage level.

6. Prediction of Damage Progression Considering Repair and Estimation of Total Cost

In the calculation model, the process of damage progression is given by the Markov chain model and the repair is done at the ratio of r_{21}, r_{31}, r_{41} from the Rank b, c, d as shown in Figure 12 because the damage of wave dissipating works is caused by the action of high waves and then the repair is performed. It is assumed that the repair is to be restored to original form and the state transitions Rank a. It is also supposed that the wave dissipating work is subjected to the next action of high waves after the repair is completed. Figure 13 shows the ratio of damage level for the case of repair of Rank d in Model 3. This figure shows the state right after the action of high waves for each N_r, and that the ratio becomes constant in $N_r > 7$.

The average of S in each Rank that is an object of repair is used to determine the eroded area and the number of wave dissipating block per unit length of wave dissipating work. The cost of repair is given by the ratio of damage level in each Rank, the ratio of repair, the number of wave dissipating block and the unit repair price of block. The cost C_R that is needed in a repair is expressed as Eq. (3).

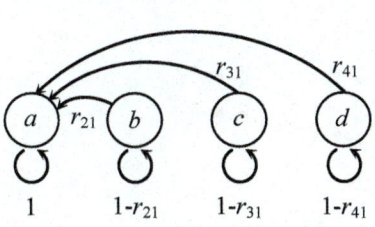

Fig. 12. State transition diagram for repair

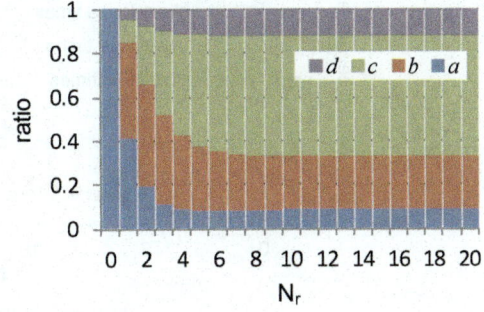

Fig. 13. Ratio of damage level (Model3, with repair)

722

$$C_R = P_u \left(N_b \cdot r_{21} \cdot R_b + N_c \cdot r_{31} \cdot R_c + N_d \cdot r_{41} \cdot R_d \right) \tag{3}$$

where P_u = unit repair price of wave dissipating block, N_b, N_c, N_d = required number of block per unit length in Rank b, c, d, R_b, R_c, R_d = ratio of Rank b, c, d. The required number of block is given by the average of S in each Rank (for Rank b; 5.5, Rank c; 9.5) and is expressed as Eq. (4) for the case of Rank c for example.

$$N_c = 9.5 \left(1 - n_p \right)/D_n \tag{4}$$

where n_p = porosity of block layer. The average of S for Rank d is 15.2 based on the experimental result.

As the policy of repair, it is considered that the repair is done every time of action of high waves that cause the damage or every two times or every three times etc., and the total cost is estimated including the loss due to the absence of

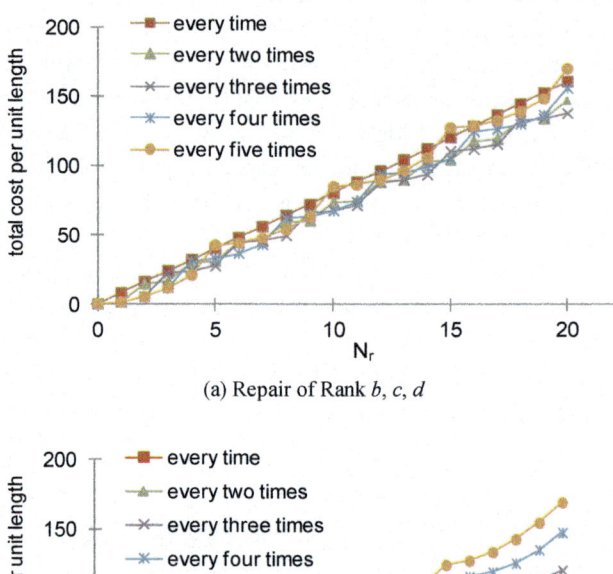

(a) Repair of Rank b, c, d

(b) Repair of Rank d

Fig. 14. Variation of total cost

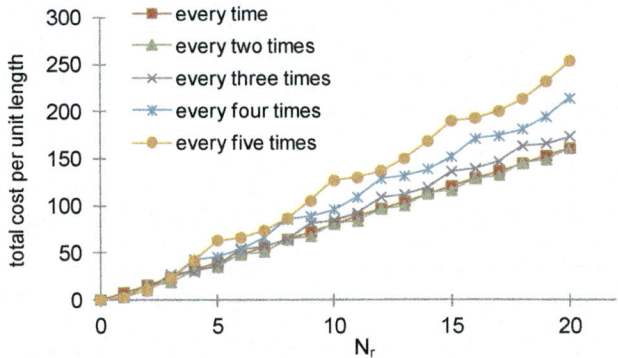

Fig. 15. Variation of total cost (repair of Rank b, c, d, the loss is twice as much)

repair except for the case that the repair is performed every time of the high waves. Figure 14 shows the computational examples of the total cost per unit length of the wave dissipating work for the cases that the repair is done from every time to every five times of the high waves. Figure 14(a) shows the case of repair of Rank b, c, d with $r_{21} = r_{31} = r_{41} = 1.0$, and Figure 14(b) shows the case of repair of Rank d only with $r_{21} = r_{31} = 0$, $r_{41} = 1.0$. The unit repair price $P_u = 10$ for the calculation of cost and it is assumed that the loss is equal to the repair cost of Rank d. Although the difference of the total cost due to the policy of repair in Figure 14(a) is not large, the cost for the case that the repair is done every three times is almost the least. The cost for the case of every time is the least as shown in Figure 14(b). If the loss becomes twice in the case of $r_{21} = r_{31} = r_{41} = 1.0$, the total costs for the cases of every time and every two times are almost same and the least as shown in Figure15. And if the half price of the repair cost of Rank c is added as the loss due to the absence of repair of Rank c, the total costs for the cases of every two times is the least. The repair policy that minimizes the total cost is dependent on the loss and to determine the practical repair price and loss are the future issues.

7. Conclusions

In this study, the model profiles for the damaged wave dissipating block layer covering a caisson breakwater were made on the basis of the experimental results for profile change of the block layer and the relation between the damage progression and change of performance is investigated. Furthermore, using the data of the normalized eroded area S, the ranks that indicate the degree of damage are determined and the models based on the Markov chain are applied to the damage progression process. Combining the Markov chain model with the repair process, the total cost for the maintenance of wave dissipating blocks is estimated by using an assumed repair cost and loss. This method can be applied to find a policy of repair that minimizes the total cost, however, it is necessary to evaluate the practical amount of repair and loss reasonably.

Acknowledgments

This study was supported by JSPS KAKENHI Grant Number 25420524. The authors express gratitude to Mr. Kengo Nishimura, Mr. Katsumasa Kido and Mr. Masaya Gonmori for their contribution to the experiments and data analysis.

References

1. M. Miyata, K. Kumagai, D. Tsujio and Y. Okubo, A study on maintenance methods considering expected maintenance costs of caisson breakwaters covered with wave-dissipating concrete blocks, *J. of Coastal Eng.*, JSCE, Vol. 56, 911-915. (2009) (in Japanese)
2. D. Tsujio and T. Yasuda, Optimum design for breakwaters covered with wave dissipating blocks considering life cycle cost, *J. of Coastal Eng.*, JSCE, Vol. 56, 916-920. (2009) (in Japanese)
3. T. Ota, Y. Matsumi, N. Kato and K. Ohno, Modeling of damage progression of rubble mound revetment and application to performance evaluation, *Proc. of 33rd ICCE*, Paper #: structures.50. https://icce-ojs-tamu.tdl.org/icce/index.php/icce (2012)
4. Coastal Development Institute of technology, *CADMAS-SURF*; *Examples of practical calculation*, 306p. (2008) (in Japanese)
5. R. Kajima, T. Sakakiyama, M. Matsuyama, T. Sekimoto and O. Kyoya, On deformation and change of protection function of wave dissipating works caused by extremely large wave, *Proc. of Coastal Eng.*, JSCE, Vol. 39, 671-675. (1992) (in Japanese)
6. K. Hirayama and J. Naganuma, Effects of transmitted waves through breakwaters due to wave overtopping for harbor tranquility, *J. JSCE, Ser.B2*, Vol. 70, No.2, I_761-I_765. (2014) (in Japanese)
7. T. Ota, Y. Matsumi, A. Hatono and T. Satow, Modeling of damage progression for rubble mound revetment, *Proc.of APAC 2013*, 332-337. (2013)
8. Port and Airport Research Institute, *Manual of maintenance technology for harbor facilities*, 229p. (2007) (in Japanese)

Effect of Vertical Slots on the Vertical Low Reflected Wave Dissipating Structures

Nyein Zin Latt

*Department of Port and Harbour Engineering, Myanmar Maritime University,
Thilawa, Yangon, Myanmar
E-mail: nzlatt@gmail.com*

Takayuki NAKAMURA

*Visiting Professor, Department of Port and Harbour Engineering
Myanmar Maritime University (former Prof. of Ehime University)
E-mail: nakamura@cee.ehime-u.ac.jp*

In the previous studies on low reflective double curtain seawalls comprised of curtain walls and a rear-side vertical wall, it was presumed that an upper side of the water chamber of the seawall was open to the atmosphere. Therefore, there is no restriction on the behavior of piston mode wave motions in the water chambers and consequently significant wave energy dissipation due to such wave motions is active. However, in the real site of wharves and piers, the upper part of low reflective seawalls is generally capped by ceiling slabs to be able to use the space for loading and unloading cargos to boats and ships. Under such general situations, it may be necessary to keep air ventilations above the water chamber to be able to activate piston mode wave motions and consequently to dissipate reflected waves from the pier. In this study, we have proposed a method to set vertical slots on curtain walls for the ventilation. Effectiveness of such vertical slots on the curtain walls has been examined experimentally and theoretically. Additional examinations have also been carried out to study the size of vertical slot on the wave dissipation effect

Keywords: Low reflective vertical seawalls; wave dissipating; curtain wall, slots.

1. Introduction

The effectiveness of the low reflective curtain walls with water chamber and the characteristics of wave forces on the structure has been verified in the previous study [1], [2]. Moreover, it was verified that the energy dissipation mechanism was achieved by the piston mode of water column inside the chamber made up of front curtain walls and the back wall of the structure. In those studies, it was presumed that an upper side of the water chamber of the seawall was open so

726

that there is no restriction on the behavior of piston mode wave motions in the water chambers. However, in the real site situation of wharves and piers, the upper part of the water chamber is generally capped by ceiling slabs and loading and unloading operations are carried out on the slabs. Under such general situations, it may be necessary to keep air ventilations above the water chamber not to disturb the piston mode wave motions and consequently to reduce the reflected waves from the pier.

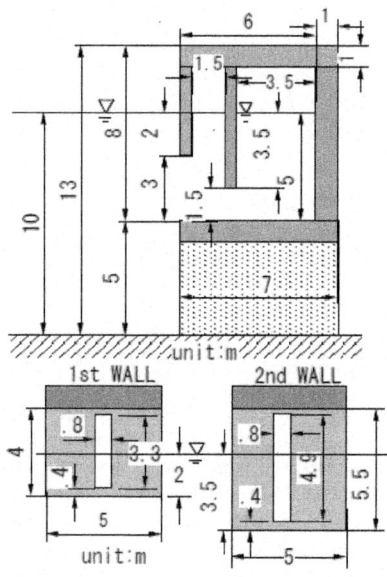

Fig. 1. Double curtain wall typed low reflective structure (opening ratio of slots is 15%).

In this study, based on the double curtain wall type structure (Figure 1) which is very effective in reducing the wave reflection for the wide range of wave periods, we proposed the vertical slots on the curtain walls of the structure as a way to let out the trapped air inside the upper parts of the water chamber freely. The effect of vertical slots was examined numerically and experimentally.

2. Numerical Analysis

To predict the effect of vertical slots on the curtain walls as shown in the Figure 1 and to determine the effective dimensions of the structure, damping wave theory[3][5][6] which was proposed by one of the authors was adopted. In this method, energy dissipation due to the vortex formation around the sharp edges or corners of the structure are approximately treated by considering that the damping fluid forces induced by the vortices. These damping fluid forces are linearized in the calculation for the simplicity.

In the analytical model, the matching boundary of separating the two different regions are considered to estimate the energy dissipation of the fluid inside the two water chambers of doubled curtain walls type structure and its proximity. Based on the previous research[4], the matching boundary line is set *L/60* distance away from the structure at the offshore side (*L*: wave period at deep water). The nearshore side from this line is treated as damping region and the offshore side is non-wave damping region.

This analytical model can be applied to the structures in two dimensions only so that the vertical slots as shown in the Figure 1 is impossible to include in analyzing the model. To compensate this, the equivalent horizontal slots were used as shown in the Figure 2 in the numerical analysis of this study. The opening ratio of the horizontal slots in the numerical analysis is equal to those of

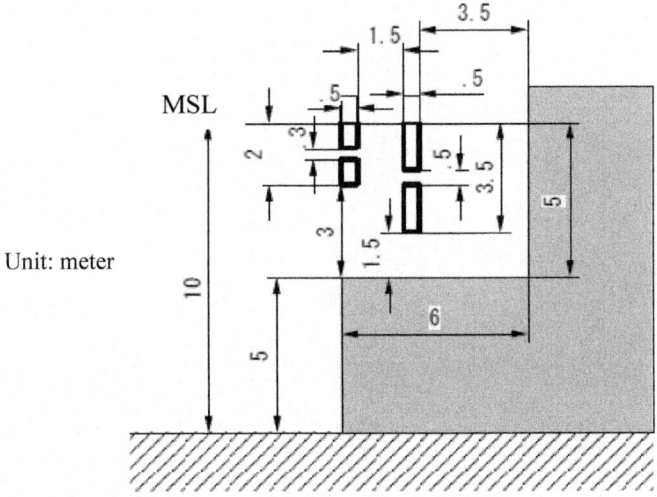

Fig. 2. Cross section of the double curtain wall type structure with vertical slots used as an analytical model by damping wave theory. (15% opening ratio)

the vertical slots of the real structure and the position of the horizontal slots were set at the middle of the underwater portions of the curtain walls.

728

3. Experimental Set up and Method

3.1. *Equipment*

The experiments were carried out at the hydrodynamic center of Myanmar Maritime University. The wave flume of (*10m* long, *2.5m* wide and *0.7m* high) was used for the experiments. There is a flap type wave maker driven by servo motor at the one end of the wave flume and a wave dissipating rubble mound at the other side. The wave flume was divided by a partition wall along its length. The narrower side (*1m* wide) of the flume was used to place the models and the wider side was used to monitor the incident waves.

The model scale of *1:30* for the structure shown in Figure 1 was used in the experiments. The Figure 3 and Figure 4 show the detailed dimensions and photo of the slotted doubled curtain walls structure with opening ratio of *15 %* which is the representative structure in this study. In the figure, the positions of the mini pressure sensors can be seen. The pressure sensors at the ceiling walls (PU1 and PU2) measure the air pressure inside the first and second water

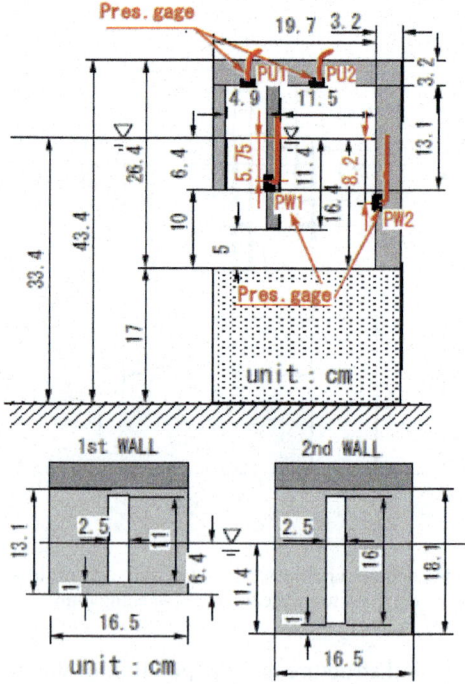

Fig. 3. Detailed dimensions of the double curtains wall typed structure with vertical slots used in experiments (15% opening ratio, 1/30 model scale).

chambers respectively. The pressure sensors (PW1 and PW2) set at the middle of the second curtain wall and back wall of the water chamber measure the change in water level inside the two water chambers.

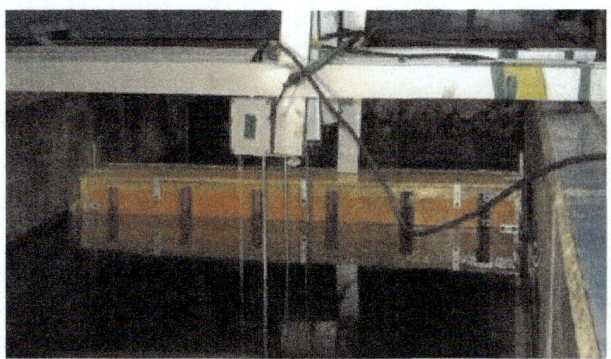

Fig. 4. Photo of the model installed inside the narrower side of the wave flume.

3.2. *Method of experiments and wave conditions*

In order to investigate the effects of vertical slots in reducing the air pressure inside the water chambers, four different type of cases as shown in the Table.1 were used.

Table.1 Experimental conditions.

CASE	Characteristics of the models
CASE1	Curtain walls with slots of 15% opening ratio
CASE2	Curtain walls with no slot (double curtain wall-type structure)
CASE3	Curtain walls with slots of 7.5% opening ratio
CASE4	slotted caisson-type with 15% opening ratio

The 15 different regular waves with model wave periods, T_m= 0.6 ~ 2.0 s (in prototype, wave periods, T= 3~10s) and model wave heights, H_m=4~ 6.5 cm (in prototype, wave height, H= 1.2 ~ 2m) were used in the experiments. The nominal wave condition of the proposed site is expected as the wave period is not more than 8s and wave height is less than 2m. The objective of the research is to reduce the wave reflections less than 50 percent against this nominal wave conditions.

Five capacitance type wave gauges were used in the experiments. Of these five wave gauges, in the model side of the wave flume, two wave gauges were

730

set for the resolution of the incident and reflected waves. One wave gauges was used to measure the wave just in front of the structures. In the other side of the wave flume, a wave gauge was used to measure the incident waves and the last wave gauge was set at the location of the model. In the experiments, the measurement of five wave gauges and the pressure sensors were done at the same time.

4. Effectiveness of the Vertical Slots for Ventilation in Reducing the Reflected Waves

4.1. *Effectiveness of the vertical slots for ventilation*

In Figure 5, the relations between the wave period T (prototype) and reflection coefficient, Cr, for the structure with vertical slots (CASE1) is shown as the experimental and numerical results. For the numerical analysis, three different

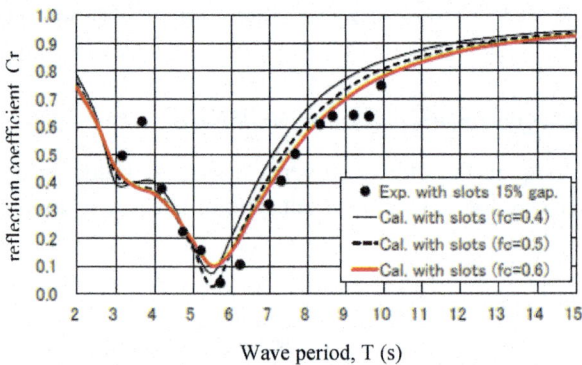

Fig. 5. Comparison of reflected coefficients between numerical results and experimental results (CASE1, opening ratio 15%,).

calculation results with different wave damping coefficient (fc=0.4,0.5,0.6) are shown. According to the numerical results, there is no significant difference with the wave damping coefficients for the shorter wave periods. But for the longer wave periods, the larger the value of fc, the lower the reflection coefficient, Cr. And there are also good agreements between the experimental results and numerical ones. It can be said that the numerical results with fc=0.6 shows best agreement with the experimental results. From this case, it can be seen that the reflection coefficients are less than 50 percent for the wide range of wave period from 3s to 8s, the objective of this study is fulfilled.

The comparison between the double curtain type structures with slots(CASE1) and without slot(CASE2) is shown in Figure 6. Without slots, the air inside the upper parts of the water chambers are entrapped. In this figure, it can be seen that the reflection coeffective becomes higher if there is no vertical slot on the curtain walls. The effective range of wave periods also shifts from shorter wave periods to longer wave periods.

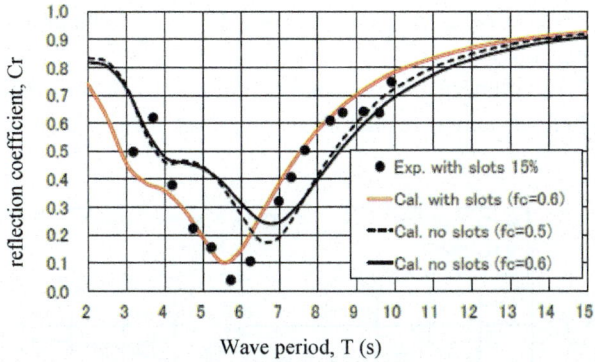

Fig. 6. Experimental and numerical results of the structure with vertical slots of 15% opening ratio and numerical results of the structure without slots with different damping coefficients.

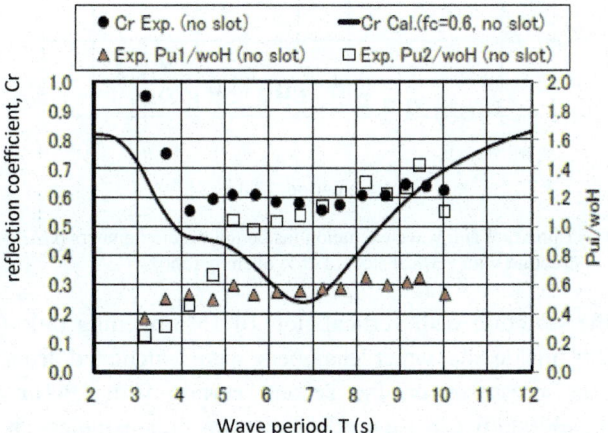

Fig. 7. Comparison between the numerical and experimental results of the structure with no vertical slot (CASE 2).

732

The experimental results for the double curtain walls type structure without slots (CASE2) can be seen in Figure 7. In this figure, reflection coefficients, Cr and the air pressure inside the water chambers by waves, Pu1 and Pu2 are shown. The pressure by waves Pu1 and Pu2 are made dimensionless by dividing with the water pressure of the incident waves, woH. According to the results, reflection coefficient is considerably increased at the degree of 0.6 if there is no vertical slot. By checking the pressure inside the water chambers, it can be said that the high pressure of entrapped air inside the water chambers disturb the piston mode water movements.

In Figure 8, the numerical and experimental results of dimensionless wave height ratio and experimental results of air pressures inside the water chambers

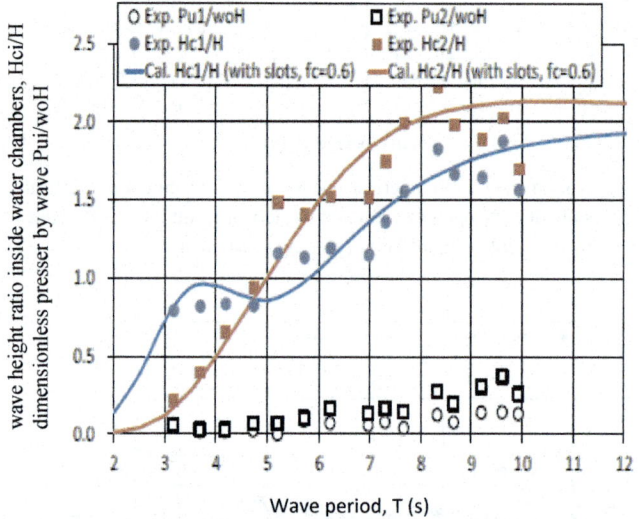

Fig. 8. Air pressure and wave excitation inside the water chambers (CASE 1, structure with vertical slots of 15% opening ratio).

are shown for the structure with vertical slots of 15% opening ratio (CASE1). The wave heights inside the water chambers were calculated from the data obtained from the pressures on the second curtain wall (PU1) and back wall(PU2) as already shown in Figure 3 and made dimensionless by dividing with the incident wave heights. The value of air pressure in CASE1 is less than 0.4 and this value is very small compared to the CASE2. It is very clear that the vertical slots can release the air pressure inside the water chambers.

4.2. *Effects of the opening ratio of the slots*

Here, how the size of the slots affects the performance of the structure will be examined. In Figure 9, the reflection coefficients, Cr, and dimensionless air

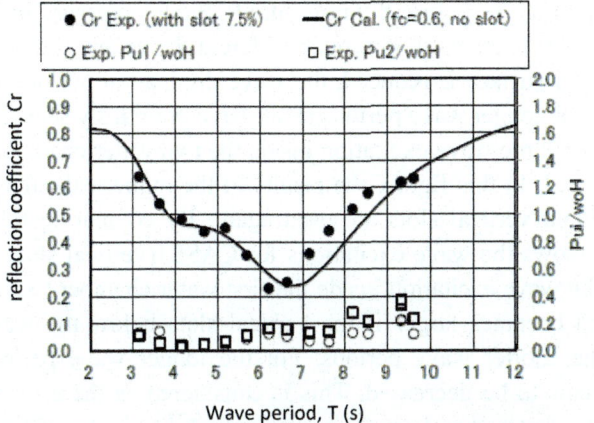

Fig. 9. Comparison between the experimental results of the structure with vertical slots of 7.5% opening ratio (CASE 2) and numerical results of the structure with no vertical slot.

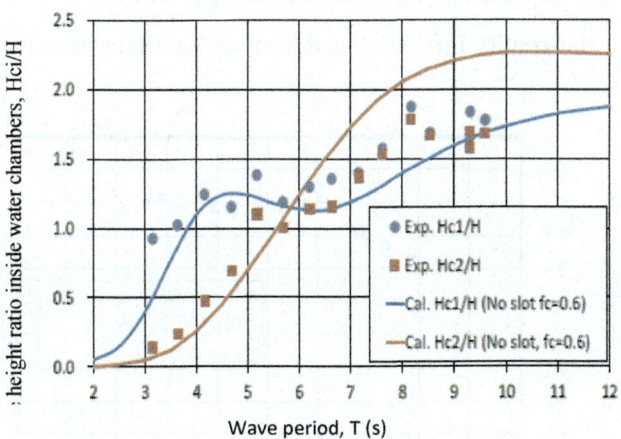

Fig.10. Experimental results of wave excitation inside the water chambers for the structure with vertical slots of 7.5% opening ratio (CASE 2) and numerical results of the structure with no vertical slot.

pressure inside the water chambers of the double curtain walls type structure with half opened vertical slots of opening ratio 7.5% (CASE3) are shown. In

734

CASE3, the lower part of the vertical slots of curtain walls of the structure of CASE1 are closed. Since it is impossible to include the vertical slots above the water level in the numerical calculations, the numerical results shown in this figure is used for the structure without vertical slots.

By comparing the Figure 8 and Figure 9, the air pressure inside the upper part of the water chambers in not much different. If the amount of opening ratio below the water surface is reduced, the wave dissipation performance becomes improved for the longer wave periods as we have previously seen in Figure 6.

In Figure 10, the wave excitation inside the two water chambers for CASE3 can be examined. In this figure, the results of the numerical calculation for the structure without vertical slots are used again. By comparing this figure with Fig.8 which shows the wave excitations for CASE1(vertical slots with opening ratio 15%), the wave excitation inside the first water chamber (water chamber at offshore side) becomes larger if the vertical slots below the water surface is closed for the shorter wave periods. For the longer wave periods, the wave excitation seems to be decreased. This is considered as the effects of the fluid flow under the water chambers. The other possibility is the effect of the water flow through the slots over the water surface and further study needs to be done to examine the mechanism.

4.3. Results for the case of slotted vertical wall type structure

In Figure 11, the results for the slotted vertical wall type structure (CASE4)

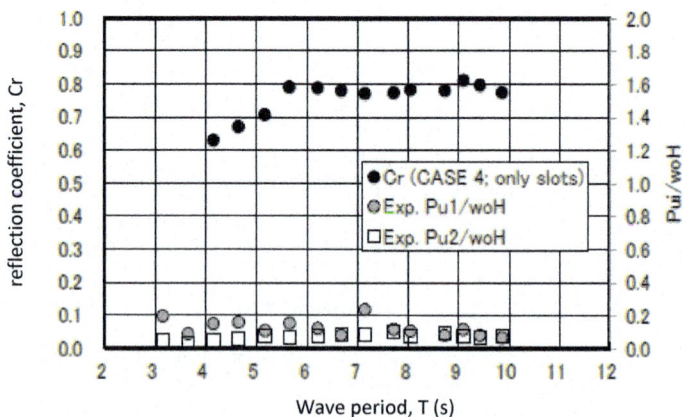

Fig. 11. Experimental results for the structure of CASE 4.

were shown. In this case, the opening below the 1st curtain wall of CASE1 is closed but keeping the vertical slots open fully. The air pressure inside the water chamber can also be seen in this figure. Since the area of the openings is very small (about 5%), the wave dissipation is failed in this case. The reflection coefficient is as high as 0.8. The air pressure inside the water chamber is very low (less than 0.2) and it means that the wave is almost not moving inside the water chamber since the openings are too small.

5. Conclusions

If the ceiling slab is closed on the double curtain wall type structure, the entrapped air inside the water chambers interrupt the piston mode wave resonance. The wave dissipating performance is not satisfactory since the reflection coefficients becomes significantly high.

If the vertical slots are applied above the water surface of the curtain walls, the air ventilation of the water chamber is possible and low reflection coefficient can be seen under the target wave conditions. The opening ratio of about 8% is sufficient and we can expect the same results as the structure without ceiling slab.

If the vertical slots on the curtain walls are extended below the water surface, the effective range of wave periods in reducing the reflection coefficients shifts to the shorter wave periods.

References

1. Nakamura, T., Kimikawa, H., About the curtain wall type breakwater for reducing the transmitted and reflected waves, 46th Japanese Conference on Coastal Eng., JSCE, pp.786-790, 1999. (in Japanese)
2. Nakamura, T., Onoduka, T. Kato, K., Jinno, M., Resonance phenomenon and governing parameters inside the water chambers of the curtain wall typed breakwaters, 21st Proceedings of civil engineering in the ocean, pp 535-540, 2005. (in Japanese)
3. Nakamura, T., Ide, Y., Analysis on Wave Transformations and Wave Forces about an Angular Body Considering Wave Energy Dissipations, Proceedings of civil engineering in the ocean, JSCE,Vol.13, pp.177-182, 1997.(in Japanese)
4. Husain, F., Inouchi, K., and Nakamura, T.: Development of a wave power extraction seawall and its effectiveness for wave dissipation, J. of Japan Soc. of Civil Eng., Ser. B2(Coastal Eng.), Vol. 70, No.2, pp.I_1311-I_1315, 2014.

736

5. Nakamura, T., MODELING AND ANALYSIS OF LOW REFLECTIVE SEAWALLS BY DAMPING WAVE THEORY AND ITS APPLICABILITY, Proceedings of the Japanese Conference on coastal engineering, 71(2), I_829-I_834, 2015 (in Japanese).

6. Fujiwara, R., Yamano, T., Three dimensional calculation for wave absorption of permeable breakwater due to the difference in the slit-type, Proceedings of the Japanese Conference on coastal engineering, 71(2), I_805-I_810, 2015 (in Japanese).

Assessment of the Ability of Mangrove Structures for Attenuation of Coastal Wave Energy: A Case Study in Bac Lieu Province, Vietnam

Ly Trung Nguyen[1]*, Van Pham Dang Tri[1], Le Tan Loi[1], Jun Sasaki[2], Hisamichi Nobuoka[3]
*[1]College of Environment and Natural Resources, Can Tho University,
Can Tho City, Vietnam*

[2]Department of Socio-Cultural Environmental Studies, the University of Tokyo, Japan
[3]Department of Urban and Civil Engineering, Ibaraki University, Japan
*E-mail: ltnguyen@ctu.edu.vn
www.ctu.edu.vn

Mangrove forests play an important role in coastal disaster mitigation as a natural measure by dissipating wave energy, which performance would change according to mangrove structures, including mangrove density, tree height, root height and tree diameter. This study was aimed at assessing the wave energy dissipating function in terms of mangrove structures and determining the necessary thickness of the mangrove forest to reduce the impacts of waves on the Bac Lieu coastal zone, Vietnam. Standard plots with the unit distance of 20 m mangrove were set for each transect, ranging from the seaside edge of mangrove to 100 m distance inland. Properties of mangrove structures were measured at three transects to ensure the representation of the study area. Waves were measured along the transects using wave gauges and characteristics of wave attenuation due to the drag force of one existing mangrove species, Avicennia sp., were analyzed. The density, height and diameter of mangrove trees were found to be distributed relatively evenly, with a slight increase of the number of trees per standard plot from 183 at the seaside edge to 218 at the 100 m distance inland. Regarding mangrove root, while its density was relatively uniform along transects, its height varied significantly. The amplitude of wave in the Bac Lieu coastal zone changed substantially with the tidal phase ranging from 0.09 to 0.22 m. The wave reduction between the seaside edge and 20 m and 100 m distances inland from the edge, respectively, increased from 32 % to 91 %. A number of mangroves characteristics influenced the rate of reduction of wave height per unit distance, most notably the physical structures of mangrove trees, especially their densities. The wave reduction coefficient was positively correlated with the thickness of mangroves ($r = 0.96$). When the distance of wave energy transmission through mangroves was greater or equal to approximately 76 m, the wave energy reduction reached 80 % or above, which assures the wave energy would not influence the coastal zone. Therefore, in Bac Lieu with dense mangrove trees, a thin band of mangroves could provide an adequate defense to protect the coastal area in this tidal range under normal wave conditions. However, for the case of storms with high waves and surges, more thickness of mangroves would be required to guarantee the protection function, which should be analyzed in future studies.

Keywords: Mangrove tree and root structures, field measurements, wave energy dissipation, coastal disaster mitigation.

1. Introduction

Given the expansion of the tidal-driven inundation in the coastal plains of the Vietnamese Mekong Delta (VMD) [1], waves propagating further inland is projected to become more severe [2;3], leading to adverse effects on socio-economic and environmental settings of the coastal areas. It is projected that the VMD will become one of the most heavily impacted areas of which coastal plains were considered to be highly vulnerable to sea level rise and changes of the hydrological regimes [4;5]. One of the solutions to mitigate impacts of waves on seashore is to enhance ecosystem services of local mangrove [6]. Mangrove forests have a function to help dissipate wave energy in the coastal zone [7;8;9], which function depends on a number of factors, including forest density, the size of trunks and roots and the water depth [10].

Bac Lieu, a coastal province of the VMD with a coastline of about 56 km, is considered to be strongly affected by sea level rise and changes in tidal regime [11], and the its coastal area has been invested in restoration of mangrove forests to mitigate impacts of waves [11]. Considering climate change and resultant increase in impacts of waves, it is of great significance to understand the relationship between characteristics of mangrove forests, including their types, and their functions of reducing wave energy, and to achieve successful protection and sustainable development in the mangrove coastal zones [12]. As a first step towards this goal, we perform an assessment of the ability of mangrove structures for attenuation of coastal wave energy under normal wave condition and propose required widths of mangrove forests for effective protection the coastal zone.

2. Materials and Methods

2.1. *Field survey on structural properties of mangrove forests*

Field surveys were conducted from 21 July 2015 to 13 June 2016 (the period of high waves in a year) in Bac Lieu as shown in Figure 1. Three transects were chosen, which were characterized by different mangrove structures. Three standard-plots (STP) with the area of 400 m² (20 m × 20 m) were set along each transect, ranging from 0 m to 20 m, 40 m to 60 m and 80 m to 100 m from the seaside edge of mangrove forests (Figure 1). Structural parameters of mangrove forests, including trunk diameter (D_T), root height (H_R) and tree height (H_T) were measured by using a tape and ruler. These properties are used to consider characteristics of wave reduction in mangrove forests along with the properties of incident waves and sea levels in coastal zone.

2.2. Field survey on structural properties of mangrove forests

Wave measurements were performed on three periods (including: from 21th to 27th July 2015; from 9th to 11th Jan 2016 and from 13th to 16th June 2016) using wave gauges (Infinity AWH-USB, JFE Advantech) installed at 6 stations starting from the seaside edge of each of the three transects with a 20 m interval to capture properties of waves propagating in the different types of mangrove forests (Figure 1). Individual wave heights and periods were determined and then significant wave heights (H_s) and periods (T_s) were obtained. Spatial distribution of significant wave heights along each of transects were used to estimate the wave reduction coefficient $r_{\Delta Hs}$ given by Equation (1) [9]:

$$r_{\Delta H_s} = (H_{Si} - H_{Sp})/H_{Si} \times 100 \ (\%) \tag{1}$$

where H_{Si} and H_{Sp} are the significant wave height of incident waves and transmitted wave heights at each of the points, respectively.

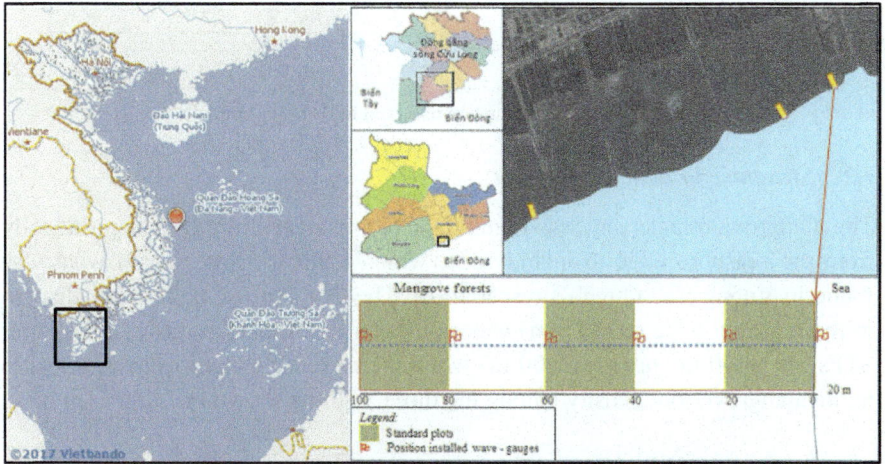

Figure 1: Study site and arrangement of three transects with three standard-plots
(Source: Edited from vietbando.com and Google Earth Pro)

3. Results and Discussion

3.1. The amplitude of wave

The result showed that the amplitude of wave ranging from 0.09 to 0.22 m (Figure 2). The wave has the highest tidal amplitude at the 2nd measurement with the amplitude: Hs = 0.22 m (Hmax = 0.37 m and Hmin = 0.001 m), followed by the 1st measurement with the amplitude value of the wave: Hs = 0.21 m (Hmax = 0.35 m and Hmin = 0.001 m, respectively) and the lowest tidal amplitude the 3rd measurement with: Hs = 0.09 m (Hmax = 0.2 m and Hmin = 0.0006 m). The results of this surveyed were consistent with the findings of the Institute of Marine Engineering (2016) showed that the tidal amplitude in the coastal area of Bac Lieu province ranged from 0.25 to 0.30 m. Thereby, it showed that the amplitude of the wave in Bac Lieu coastal zone were not high and relatively equal between the measuring periods. Due to the time of measurements in the dry season, the wave were not high, the fluctuation of the wave amplitude were not large, it agreed with what was found in the research result of Marine Institute of Technology (2015).

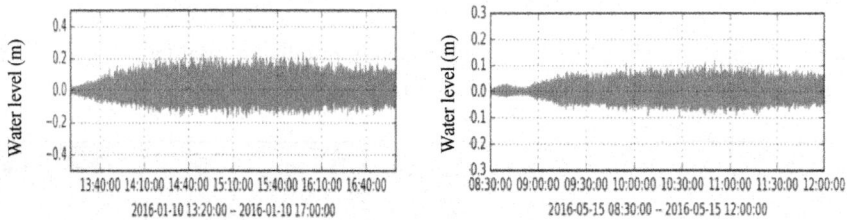

Figure 2: Water level fluctuation observed in Bac Lieu coastal zone

3.2. Structure in mangrove forest

The mangrove tree density was found to have a tendency of increasing gradually from the seashore edge to inland zone, which tendency agrees with what was found in Vuong Van Quynh's research [6]. The tree density in the STP adjacent to the seashore edge (0 to 20 m) was significantly lower than those in the other STPs (40 to 60 m and 80 to 100 m) while there were no significant differences in the mangrove root density among the three STPs as shown in Table 1.

Table 1. Mangrove tree density and root density in three STPs

STPs	Tree density (number of trees/ha)	Root density (number of roots/ha)
0 - 20 m	1,691[b]	1,542,200[a]
40 - 60 m	6,500[a]	1,558,000[a]
80 - 100 m	6,808[a]	1,280,000[b]

[a, b] *Different letter are significantly different was recognized at 5% (P ≤ 0.05) according to the Duncan test.*

Along each of the three transects, tree density, tree hight and tree diameter tend to increase as the ground elevation increases, which means that high trees with large diameters were dominant at the high elevation. Figure 3 presents difference in structures of mangrove among three STPs along the three transects.

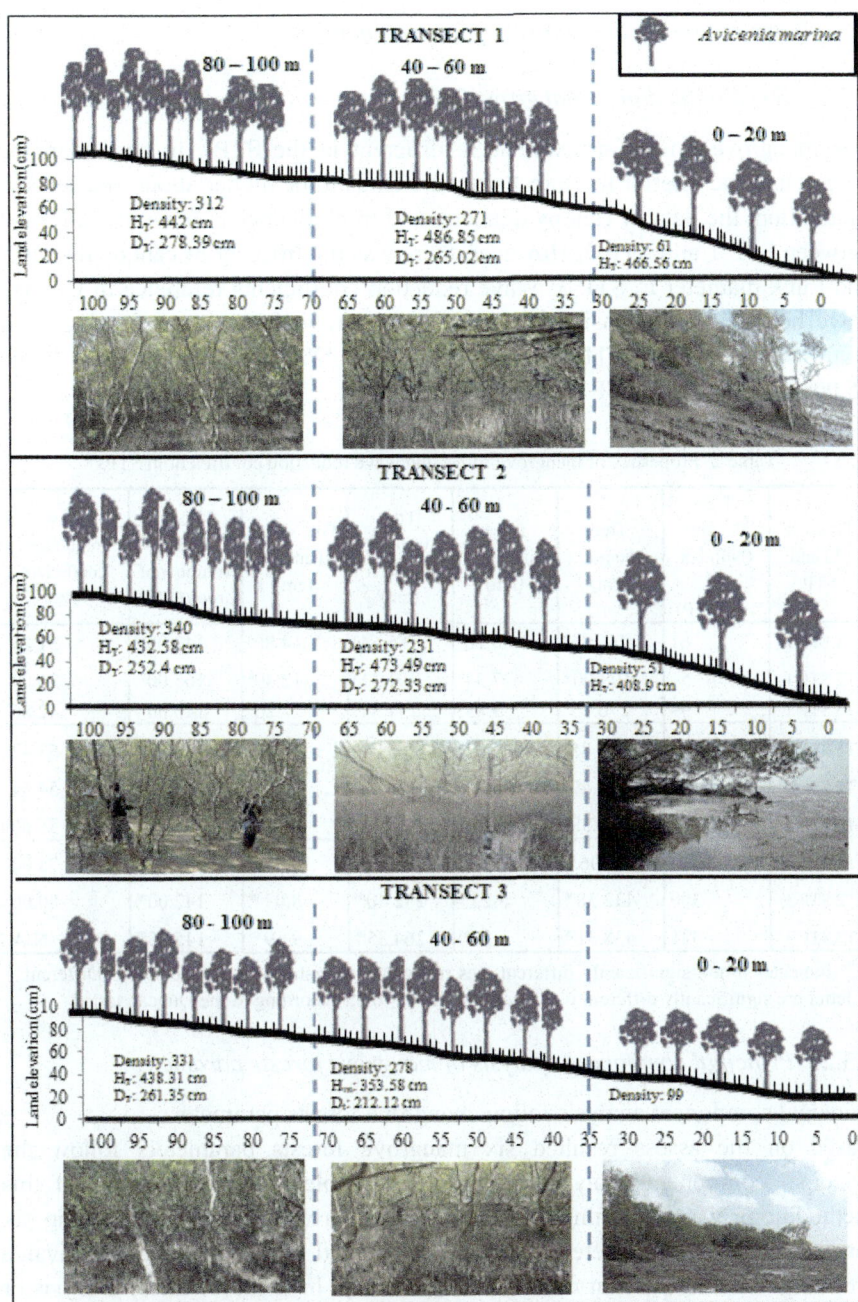

Figure 3: Structures of mangrove forests in three STPs along three transects

3.3. Wave reduction in mangrove forests structure

3.3.1. Wave reduction in mangrove forests

The mangrove forests structure were different in the STPs. In particular, the deeper into the interior the forest structure tends to be higher about tree density. In addition, the data of canopy diameter and trunk diameter shown the different between STP 1 and STP 2. However, there was no different of canopy diameter and trunk diameter in STP 3. Wave reduction coefficients reference to incident wave heights at the seaside edge of the mangrove forest were ranging from 26.09% to 42.86% in STP 1; It mean, from 53.42% to 55.00% in STP 2, and 75.00% to 80.00% in STP 3 as listed in Table 2.

Table 2. Properties of mangrove forest and wave reduction coefficient in STPs

Transect (T) and STP	Tree density (Number of tree/STP)	Tree height (cm)	Trunk diameter (cm)	Tree canopy diameter (cm)	Tree stump diameter (cm)	Root density (number of root/STP)	Wave reduction coefficient (%)
T1 STP 1	61	466.56[a]	6.26[a]	385.64[a]	14.80[a]	142.00[ns]	31.25
T2 STP 1	51	408.90[a]	7.34[a]	401.31[a]	12.49[a]	165.00[ns]	26.09
T3 STP 1	99	276.73[b]	3.83[b]	234.34[b]	7.20[b]	140.00[ns]	42.86
T1 STP 2	271	486.85[a]	5.07[a]	265.02[a]	8.47[a]	169.33[ns]	53.42
T2 STP 2	231	473.49[a]	6.23[a]	272.33[a]	9.54[a]	160.67[ns]	54.54
T3 STP 2	278	353.58[b]	3.37[b]	212.12[b]	6.62[b]	107.00[ns]	55.60
T1 STP 3	312	442.06[ns]	5.2[ns]	278.39[ns]	10.35[ns]	151.33[ns]	75.00
T2 STP 3	340	432.58[ns]	5.23[ns]	252.40[ns]	8.03[ns]	142.00[ns]	80.00
T3 STP 3	331	438.31[ns]	5.28[ns]	264.35[ns]	9.19[ns]	145.25[ns]	N/A

[ns] Indicates a non-significantly different was recognized statistically at $P \leq 0.05$, [a, b] Different letter are significantly different was recognized at $P \leq 0.05$ according to the Duncan test

3.3.2. Principal component analysis of mangrove forests parameter

Principal component analysis follow mangrove forests parameter

Based on the assess resulted six mangrove forests parameters follow the principal component analysis methodology (Table 3). The objective of this methodology was determined quantity need to present the data (including six mangrove forests parameters). Scree plot showed the decrease of eigenvalue amplitude and the percent cumulative variability. In terms of factor analysis or principal component analysis, Scree plot helps analysts visualize the relative importance of components. This components must describe at least 80% of the cumulative percentages of variance (Shi et al., 2002). In this case, the components 1 and 2 have an eigenvalue greater than 1 and account for 87.03% of cumulative variance (Table 3).

Table 3. Analysis of the main composition based on six parameters of mangrove structure

Main parameter	F1	F2	F3	F4	F5	F6
Eigenvalue	**3.67**	**1.45**	0.56	0.13	0.06	0.02
Variance (%)	**62.24**	**24.29**	9.4	2.18	0.97	0.4
Cumulative variability (%)	**62.67**	**87.03**	98.62	98.6	99.6	100

The component 3 and 4 have a very small interaction with the variables, which can be easily seen through the cumulative percentile curve of the variance (unchanged from F3 onwards) and the sharp decrease in magnitude of the individual values of F3 and F4 (Resano et al., 2010). The magnitude of the components from 3 to 6 (F3 to F6) was very small compared to components 1 and 2, so there was no need to used components from F3 to F6 to represent the number of metric sets (Figure 4).

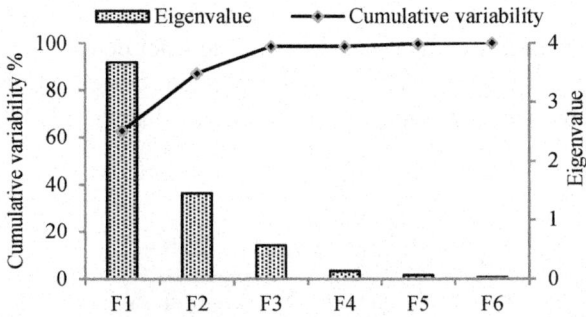

Figure 4: Eigenvalue and cumulative variability (%) presented by Scree plot

The resulted showed in Table 4, the first component and the second component were builted based on interaction with the attributes of the mangrove structure and shown in equation (1) and equation (2):

$$F1 = -0.465x - 0.056y - 0.217z + 0.131\alpha + 0.306\beta + 0.265\varepsilon \quad (1)$$

$$F2 = 0.305x + 0.339y + 0.486z + 0.234\alpha + 0.031\beta + 0.068\varepsilon \quad (2)$$

Table 4: Contents of main components

Main components	F1	F2
Tree density (x)	-0.465	0.305
Root density (y)	-0.056	0.339
Tree height (z)	-0.217	0.486
Trunk diameter (α)	0.131	0.234
Tree canopy diameter (β)	0.306	0.031
Tree stump diameter (ε)	0.265	0.068

The correlation between attributes and components of mangrove structure was shown in Figure 4. Based on the distribution of structural properties in Figure 5, these attributes could be divided into three regions. Distinct: region 1 including the canopy diameter attributes, tree stump diameter and trunk diameter closest to the X axis (primary component 1) and has a large value indicating that these attributes have an important influence on the first part zone 2, the density of separate tree attributes, and far away from the X axis indicate that this attribute did not significantly affect to the primary one. Zone 3 including attributes such as tree height and root density. Near the Y axis and great value, these two attributes have a great influence on the second major component. In addition, neighboring attributes were correlated (including: root density, stump diameter, trunk diameter and canopy diameter) and contiguous correlated (including: tree density and trunk diameter, stump diameter, canopy diameter) [13].

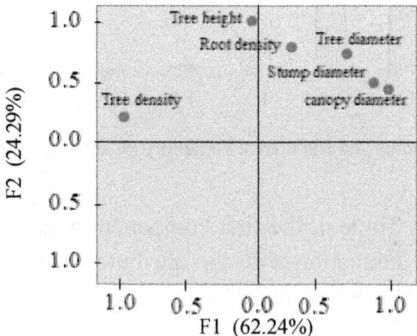

Figure 5: Distribution of mangrove structure attributes

3.3.3. *Correlation between the mangrove forests structure and wave reduction coefficients*

The main factor analysis was 2 equations (1) and (2) corresponding to F1 and F2 components. The correlation between these components with the wave reduction was shown in Figure 6. Thus, there was a correlation between the two major components and the reduced percentages of wave energy. However, for the first major component, the correlation was tighter (r = -0.86 và y = -0.0453x + 2.3705).

From the results above, there was a negative correlation (r = -0.86) between the primary component 1 and the decreasing percentage of tidal energy. However, the primary component has a negative correlation with the number of trees (Table 4). Therefore, it could be concluded that the tree density was correlation with wave reduction coefficients, in particular, the larger tree density, the lower the wave reduction coefficients were bigger. In addition, attributes such as tree height and canopy diameter have a very poor correlation with the wave reduction coefficients. The reason was that the wave amplitude in the study was relatively low, so limited by the two attributes.

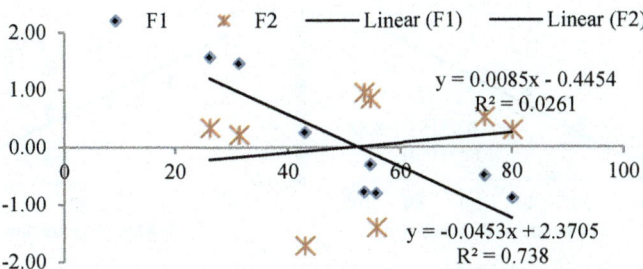

Figure 6: Correlation between mangrove forests structure and wave reduction coefficients

3.4. *Relationship between thickness of mangrove forest and wave energy*

The ability of the mangrove forest thickness to reduce the tidal power was showed by the correlation between mangrove thickness and the percentage reduction of wave energy. At the same time, the reduction rate also correlation with the thickness of the mangroves (Figure 7). The decreasing percentage of wave power was positively correlation with the thickness of the mangroves (r = 0.96) (Figure 7a). The correlation was very tight so it could be concluded that the greater the thickness of the mangrove forests, the higher the percentage reduction of wave energy. However, the reduction rate was negatively correlated with the mangrove thickness (r = -0.95) (Figure 7b). The value of "r" also showed a strongly correlation between wave reduction and mangrove thickness. Therefore, the greater the thickness of the mangrove forests, the lower the wave reduction rate and vice versa. The thickness of the mangrove forests would be determined where was capable of reduced the wave energy affecting to the coastal area, the reduction of wave energy must be greater than or equal to 80% over or equal to 0.2. Applying the regression correlation between mangrove thickness and the wave reduction coefficients, the resulted from correlation analysis showed that x = 76 (ie: mangrove thickness was 76 m). Thus, in order to protected the coastal area of Bac Lieu province from the effects of waves, mangroves should have a thickness greater or equal to approximately 76 m or more. The resulted of this study were consistent with previous research by Vuong Van Quynh (2010) [6] that the thickness of mangroves in Bac Lieu coastal area should be greater than 73 m to reduced the effects of wave energy to the coastal areas. However, the limitation of the result in normal wave conditions. In case of highly and strongly wave conditions, this length of 76 m may not be enough to achieve the 80 % of wave reduction.

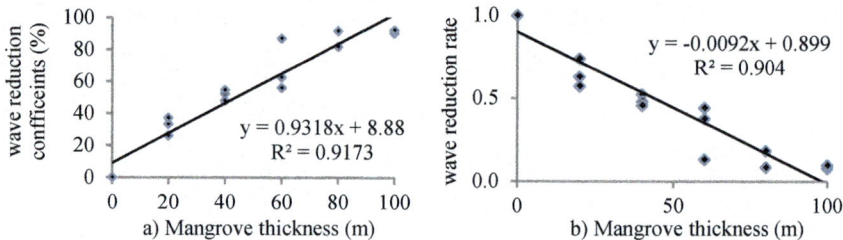

Figure 7: Correlation between mangrove forests thickness and wave reduction coefficient (a) and wave reduction rate (b)

4. Conclusions

Generally, the vertical and horizontal structure of mangrove forests were also affected by the wave energy and vice versa the wave energy would be affected by the mangrove forests structure. The wave energy reduction coefficiency passed through the mangrove forests 100 m thickness was of 91 % (corresponding to wave reduction rate about of 0.09). In addition, the wave reduction coefficient was positively correlated with the thickness of mangroves (r = 0.96). When the distance of wave energy transmission through mangroves was greater or equal to approximately 76 m, the wave energy reduction reached

80% or above, which assures the wave energy would not influence to the coastal zone. Therefore, in Bac Lieu coastal zone with dense mangrove trees, a thin band of mangroves could provide an adequate defense to protect the coastal area in this tidal range under normal wave conditions.

Acknowledgements

The present study was partially funded by JSPS KAKENHI Grant No. 25303016.

References

1. V. P. D Tri, N. H. Trung and V. Q. Thanh, 2013a. Vulnerability to Flood in Vietnamese Mekong Delta: mapping and uncertainty assessment. J. Environ. Sci. Eng. 2, 229 – 237.
2. T.R. Hashimoto, 2001. Environment issues and recent infrastructure development in the Mekong Delta: Review, Analysis and Recommendation with particular Reference to Large-Scale Water control Project and the Development of coastal area. Australian Mekong Resource Centre, University of Sydney, June 2001.
3. I. White, 2002. Water management in the Mekong Delta: Change, Conflicts and Opportunities, The Australian National University, Canberra, Australia. Centre for Resource and Environmental Studies National Institute for the Environment, Institute of Advance Studies, The Australian National University Canberra ACT 0200 Australia.
4. N. H. Trung and V. P. D. Tri, 2014. Possible Impacts of Seawater Intrusion and Strategies for Water Management in Coastal Areas in the Vietnamese Mekong Delta in the Context of Climate Change in Coastal Disasters and Climate Change in Vietnam (editors: Nguyen Danh Thao, Hiroshi Takagi and Miguel Esteban). Elsevier Inc. ISBN: 978-0-12-800007-6.
5. A. Dasgupta et al. 2007. Regulation of rRNA synthesis by TATA-binding protein-associated factor Mot1. Mol Cell Biol 27(8): 2886 - 96.
6. V. V. Quynh, 2007. Science and Technology Report "study identifies the necessary area for local". Ha Tay, 2007.
7. Y. Mazda, M. Magi, M. Kogo and P. N. Hong, 1997. Mangroves as a coastal protection from waves in the Tong King Delta, Vietnam. Mangrove Salt Marches 1, 127–135.

8. Y. Mazda, M. Magi, Y. Ikeda, T. Kurokawa and T. Asano, 2006. Wave reduction in a mangrove forest dominated by Sonneratia sp. Wetl. Ecol. Manag. 14, 365–378.

9. U. Thampanya, J. E. Vermaat, S. Sinsakul and N. Panapitukkul, 2006. Coastal erosion and mangrove progradation of Southern Thailand. Estuar. Coast. Shelf Sci. 68, 75–85.

10. T. Rasmeemasmuang and J. Sasaki, 2015. Wave Reduction in Mangrove Forests: General Information and Case Study in Thailand. In Handbook of Coastal Disaster Mitigation for Engineers and Planners (M. Esteban, H. Takagi and T. Shibayama eds.), Elsevier, Chapter 24, 511-535.

11. Ministry of Natural Resources and Environment, 2010. The National Strategy on Climate Change in Vietnam.

12. P. N. Hong et al, 2007. The role of mangrove ecosystems and coral reefs in the disaster reduction and improvement of life in coastal areas. Publisher of Agriculture, Hanoi - 2007.

13. V. Cañeque, C. Pérez, S. Velasco, M. T. Diaz, S. Lauzurica, I. Álvarez and J. De la Fuente. 2004. Carcass and meat quality of light lambs using principal component analysis. Meat Science, 67(4), 595-605.

Effect of Regulation Engineering on Tidal Currents in A Generalized Bifurcation Estuary[*]

DOU Xi-ping[†1], PAN Yun[2], GAO Xiang-yu[1], ZHANG Xin-zhou[1], DING Lei[1]

[1]*Nanjing Hydraulic Research Institute, State Key Lab of Hydrology-water Resources and Hydraulic Engineering*
Nanjing 210024, China
[†]*E-mail: xpdou@nhri.cn*
www.nhri.cn

[2]*National Engineering Research Center for Marine Aquaculture,*
Zhejiang Ocean University
Zhoushan 316022, China
E-mail: pyzjou@qq.com

Based on the Yangtze Estuary and Oujiang Estuary in China, a bifurcation estuary is generalization and its physical model and numerical model are established to study the effect of building a pair of guiding jetty and four pair of spur dikes in one branch on tidal currents of bifurcation estuary. The calculating results show that there is a critical length of guiding jetty and the velocities at ebb strength is up to the maximum value. Through the regression analysis of the predominance of ebb current, Euler unit discharge and Stokes unit discharge, the reasonable relative length of spur dike (the ratio between a spur length and 1/2 channel width) in the bifurcation estuary is 0.4~0.6.

Keywords: bifurcation estuary; regulation project; physical model; numerical model

1. Design and establishment of bifurcation estuary generalized model

A bifurcation estuary is a major form of sea estuary. With the development of economy, the demand for waterway depth is higher and higher. Regulation engineering is a main measure to improve the depth, including construction regulating structures such as spur dike, guiding jetty and so on. The two bifurcations of an estuary are a whole; after one branch is regulated with engineering, a new balance will be achieved by the hydro-sediment movement. In this paper, focusing on the construction of guiding jetty and spur dike on one

---

[*] This work is supported by National Key Research and Development Program of China (2016YFC0401500,2016YFC0401502,2016YFC0401503,2016YFC0401505,2016YFC0401506); National Natural Science Foundation of China(51479122); Central Water Resources Fee Project (1261530210283);

750

branch, a physical model and 2D tidal current numerical model of a generalized bifurcation estuary are used to study the current movement of bifurcation estuary.

1.1. *Model design*

The bifurcation estuary generalized physical model is shown in Figure 1. The estuary is divided into upstream reach, bifurcation, branch reach and offshore, with total length of 19.3km, the reaches widths of 1.2km and estuary entrance width of 7km. The upstream is connected to a reservoir through wet return and flow goes to reservoir by pump. The branch reach is divided into the south branch (S branch) and north branch (N branch) respectively. The cross section is rectangular. The elevation from the upstream reach to bifurcation is linearly from -14.7m to -15m; the elevation of shoal is +0.4m with its tail part linearly from +0.4m to -7m; the elevation of branch reach is -7m. The out sea is fan-shaped, whose elevation is -15m to -20 m with linear interpolation. Tidal currents are made by tidal maker (tail gate) at the corner of the physical model.

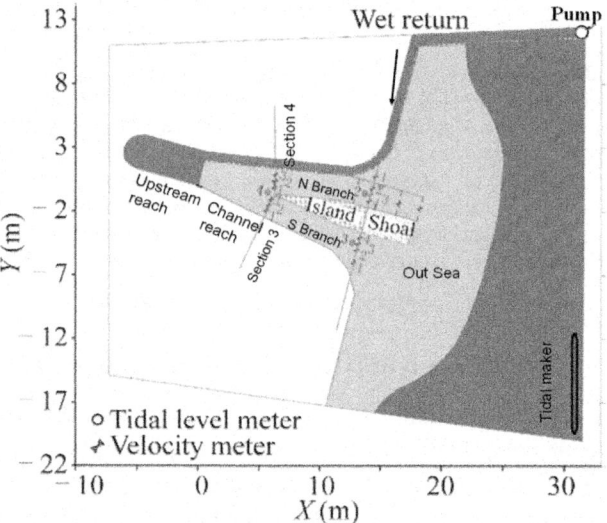

Fig. 1. Generalized model layout and measurement position

A pair of guiding jetties and four pair of spur dikes are built respectively on N branch, wherein the jetty is parallel to another, with a distance of 1.2km; the crest elevation of jetties and dikes are 0m (middle tidal level), with a width of 30m; the distance between adjacent dike is 1050m; and the dike is perpendicular to the jetty.

1.2. *Model establishment and verification*

Based on SMS and MIKE21, a non-structural grid numerical model is established, adopting three-sided and four sided grids, with a total of 68022 grid nodes and 96613 grid units. The minimum grid is 30m in the bifurcation channel; and the maximum grid is 1000m on the outer sea (Figure 2). The upstream boundary is controlled by discharge process from -6222.6~6596.5m³/s. The out sea boundary is controlled by tidal level process with a high tidal level of 3.54m, a low tidal level of -3.02m, the difference of 6.56m and tidal rise and fall durations of 6 hours.

The numerical model is verified by the experimental results of physical model. The locations of measuring points of tidal level and currents are shown in Figure 1. Figure 3 is the contrast of tidal level at 2# and 4#. 4# is located at bifurcation with good numerical simulation; 2 # is located at N branch and the high tidal level can better simulate, the low tidal level is slightly more than the physical values; the velocity of the numerical model is greater than that in physical model (Figure 4). This is because the numerical model is controlled by the tidal level.

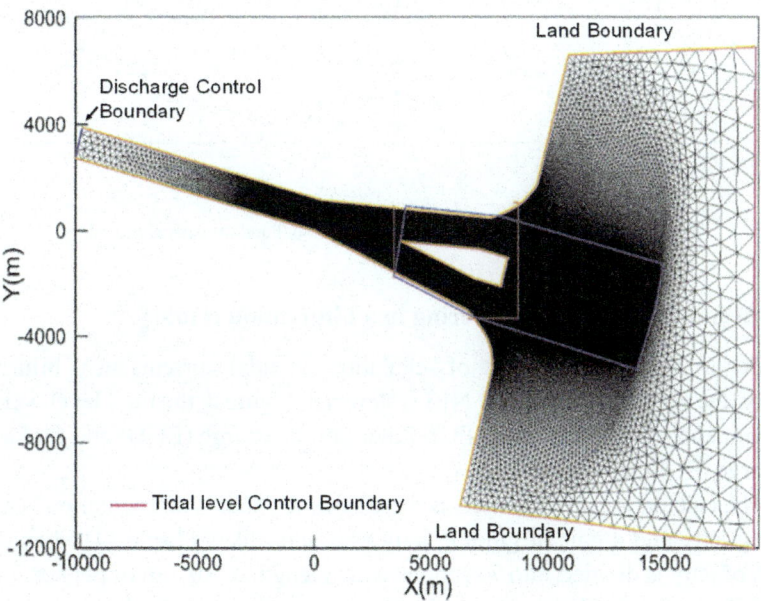

Fig. 2. Computational grid

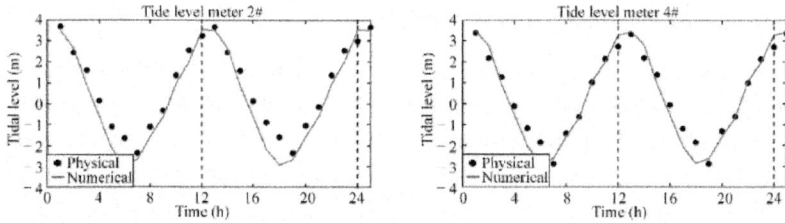

Fig. 3. Contrast of tidal level of physical model and numerical model

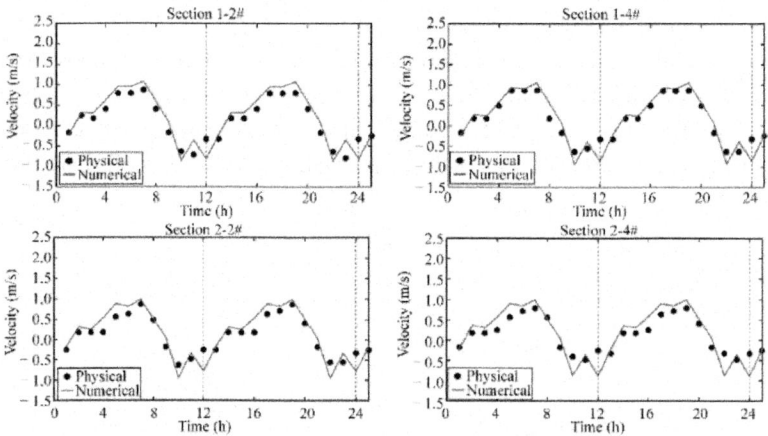

Fig. 4. Contrast of velocity of physical model and numerical model

2. Effect of regulation engineering in a bifurcation estuary

In order to analyze the effect of regulation on tidal currents in a bifurcation estuary, the sampling points P1~N13 is shown in Figure 5, the tidal level, velocity, predominance of ebb current (PEC), Euler unit discharge (EUD) and Stokes unit discharge (SUD) are studied.

The calculation are arranged as three schemes, that is no regulation, a pair of guiding jetty and a pair of guiding jetty plus spur dikes (Table 1), wherein, the length of jetty is divided into 7 sections with a length of 1050m of per section and the total length of 6714m; spur dikes are four pairs that the length of each is divided into 8 units with a length of 60m of per unit and the total length of 480m.

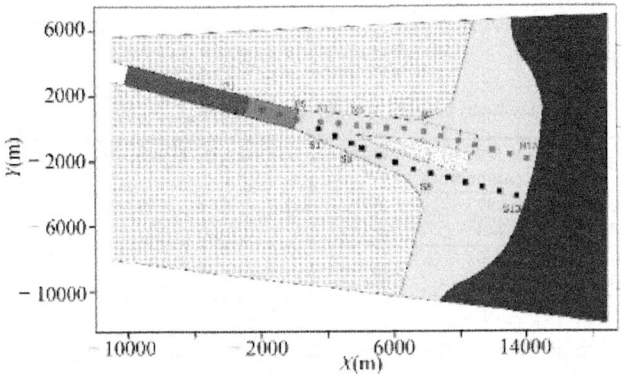

Fig.5. Sampling points in the model

Table 1. Calculation schemes

Regulation engineering	Spur dike	Guiding jetty	Regulation engineering	Spur dike	Guiding jetty
A pair of guiding jetty	0 unit	0 section	A pair of guiding jetty + four pairs of spur dikes	1 unit	4 sections
	0 unit	1 section		2 units	4 sections
	0 unit	2 sections		3 units	4 sections
	0 unit	3 sections		4 units	4 sections
	0 unit	4 sections		5 units	4 sections
	0 unit	5 sections		6 units	4 sections
	0 unit	6 sections		7 units	4 sections
	0 unit	7 sections		8 units	4 sections

2.1. *Variation of tidal currents under the effect of a pair of guiding jetty*

From the water surface profile changes longitudinally at the moment of flood (ebb) strength in the bifurcation estuary, the length of guiding jetty has a great effect on the tidal level. Namely, the longer the guiding jetty, the higher the tidal level in the lower channel; while the tidal level in the upper channel declines much more. The guiding jetty advances the ebb strength moment of N branch for 2 hours than without regulation engineering.

For N and S branch, with the increase of jetty length, compared with the velocities under the condition of no regulation engineering, the velocities of flood strength and ebb strength are much faster; when the length of guiding jetty reaches 4 sections, the velocities at ebb strength is up to the maximum value; if the guiding jetty becomes much longer, the maximum velocities of ebb strength begin to decrease (Table 2).

Table 2. The maximum velocities of flood (ebb) strength with jetty lengths (cm/s)

Branch	Moment	Without engineering	Different guiding jetty lengths (section)						
			1	2	3	4	5	6	7
N	Flood	0.037	0.7	8.0	17.7	58.9	86.3	86.3	86.6
	Ebb	0.743	0.9	21.4	48.0	87.3	86.2	72.7	62.4
S	Flood	0.460	10.3	37.9	52.0	66.7	67.6	65.4	63.6
	Ebb	0.856	9.4	20.4	24.4	26.2	23.7	19.3	13.9

With the increase of jetty length, PEC in the upstream of N branch increases constantly while PEC in the channel downstream the N branch increases firstly and then decreases. For S branch, when the jetty length reaches 4 sections, PEC in the upstream of bifurcation doesn't change too much; if the jetty becomes much longer, PEC increases significantly. PEC of S branch in the downstream of bifurcation decreases with the length increase of jetty (Figure 6).

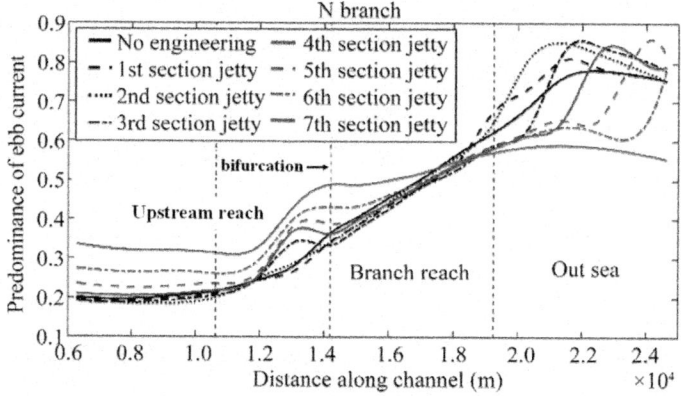

Fig. 6 Predominance of ebb (PEC) current variation

Stokes unit discharge is the net drift of current. The net drift in the upstream reach is in the direction of ebb tide while that in the outer sea is in the direction of flood tide. As the jetty length increases, SUD in the upstream reach of N branch gradually decreases while SUD in the outer sea doesn't change too much. The zero position of SUD is in the N branch so that sediment deposition easily occur here (Figure 7).

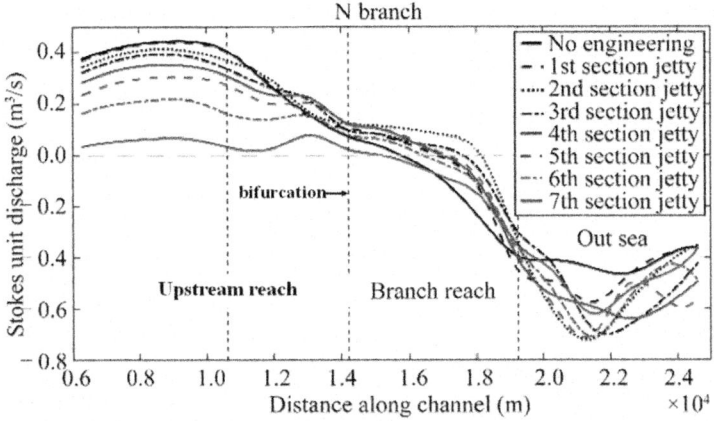

Fig. 7. Stokes unit discharge variation in N branch

2.2. Variations of tidal currents under the different length of spur dikes

Under the condition that the jetties reaches 4 sections, set four pairs of spur dikes inside them. With the increase of dike length, the tidal levels in the upstream of bifurcation significantly decrease at the moment of flood strength; at the moment of ebb strength, all the tidal levels in S branch have no obvious changes; the tidal levels in outer sea of N branch slightly increase.

At the moment of flood strength, the velocity in N branch increases with spur length and gradually decreases after the 5th unit of spur. When the spur length reaches 8 units, the velocity even closes to 0. Thus, it can be seen that the spur has a critical length (Figure 8). With the increase of spur length, the velocities in N branch at the moment of ebb strength are from increase to decrease from the upstream reach to the downstream of bifurcation and then turn to increase again. The velocities in S branch at the moment of flood strength are from decrease to increase in the bifurcation; however, they keep longitudinal increasing at the moment of ebb strength.

With the increase of spur length, PEC increases in the upstream of bifurcation reach; PEC decreases in the N branch downstream of bifurcation; however, when the spur length is 7 units and 8 units, PEC increases substantially (Figure 9); For S branch, taking the middle part of branch reach as the break point, PEC in the downstream gradually decreases with the increase of spur length.

With the increase of spur length, SUD of N branch increases significantly; the zero position of SUD in the channel reach moves seaward. SUD of S branch decreases significantly; the zero position of SUD moves toward the upstream of bifurcation (Figure 10).

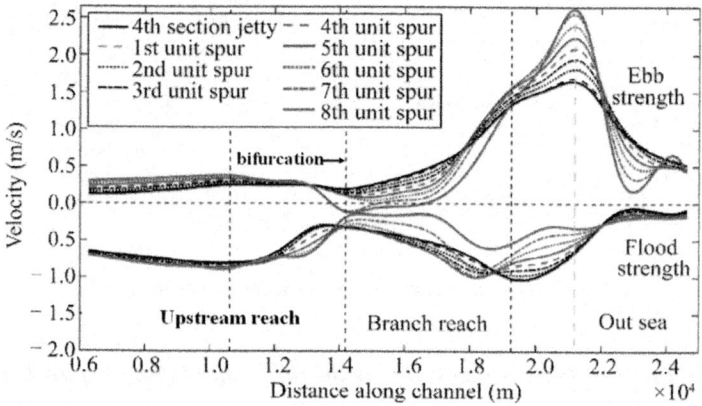

Fig. 8. Velocities variation

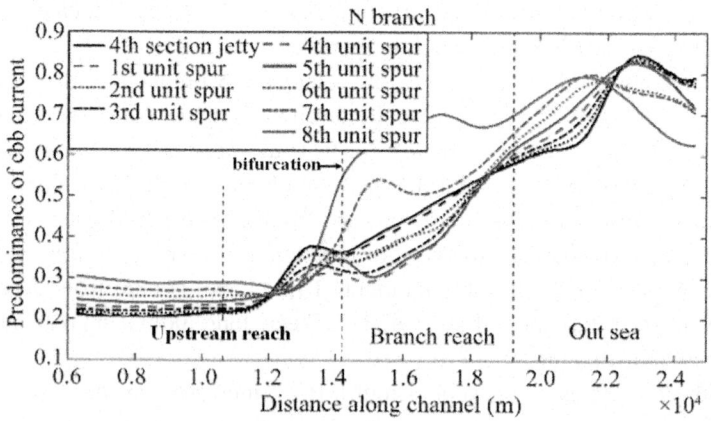

Fig. 9. Predominance of ebb current variation

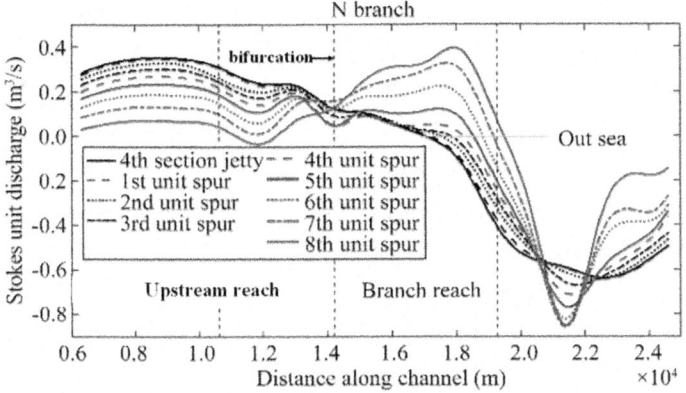

Fig. 10. Stokes unit discharge variation

2.3. *Study on the optimum length of spur dike*

The relative spur length is defined as the ratio between a spur length and 1/2 channel width (600m); namely, the relative spur lengths for 1 unit (60m) ~ 8 units (480m) respectively are 0.1~0.8. The 0 relative spur length is the calculation result of a pair of jetty without spur.

Respectively add the predominance of ebb current, Euler and Stokes unit discharge in the N branch together and take the value as such comprehensive resistance indicator (Table 3). After many attempts, respectively got two numerical expressions to fit predominance of ebb current (ϕ), Euler unit discharge (Q_e) and Stokes unit discharge (Q_s) and releative spur dike length.

Table 3. Statistics under the condition of different spur dike lengths

Relative length	ϕ	$Q_e(m^3)$	$Q_s(m^3)$
0	0.71	-0.12	0.013
0.1	0.71	-0.14	0.017
0.2	0.70	-0.21	0.022
0.3	0.69	-0.29	0.021
0.4	0.68	-0.32	0.024
0.5	0.69	-0.25	0.046
0.6	0.70	-0.09	0.089
0.7	0.75	0.16	0.127
0.8	0.81	0.46	0.146

Figure 11 is the fitting formulas with the data from Table 3. In this figure, the blue horizontal line is the value of guiding jetty without spur dike. We can get that the crossing points of ϕ, Q_e with blue line. The releative spur lengths are 0.61 and 0.59 respectively. When the releative spur length is less than 0.4, the statistics of bifurcation estuary show linear changes; when the relative length is more than

758

0.4, the statistics show curve changes; namely, the relative length 0.4 is a break point. If the statistics' break points and crossing points are taken as the reasonable spur length, then the reasonable relative length of spurs in the bifurcation estuary is 0.4~0.6.

$$\phi_{11} = -0.0917x_1 + 0.7149, \quad \phi_{12} = 0.4523x_1^3 + 0.0831x_1^2 - 0.2687x_1 + 0.7434 \qquad (1)$$

$$Q_{e11} = -0.5494x_1 - 0.1079, \quad Q_{e12} = -2.2097x_1^3 + 8.0211x_1^2 - 5.1825x_1 + 0.608 \qquad (2)$$

$$Q_{s11} = 0.0273x_1 + 0.014, \quad Q_{s12} = -3.4094x_1^3 + 6.0513x_1^2 - 3.1401x_1 + 0.5301 \qquad (3)$$

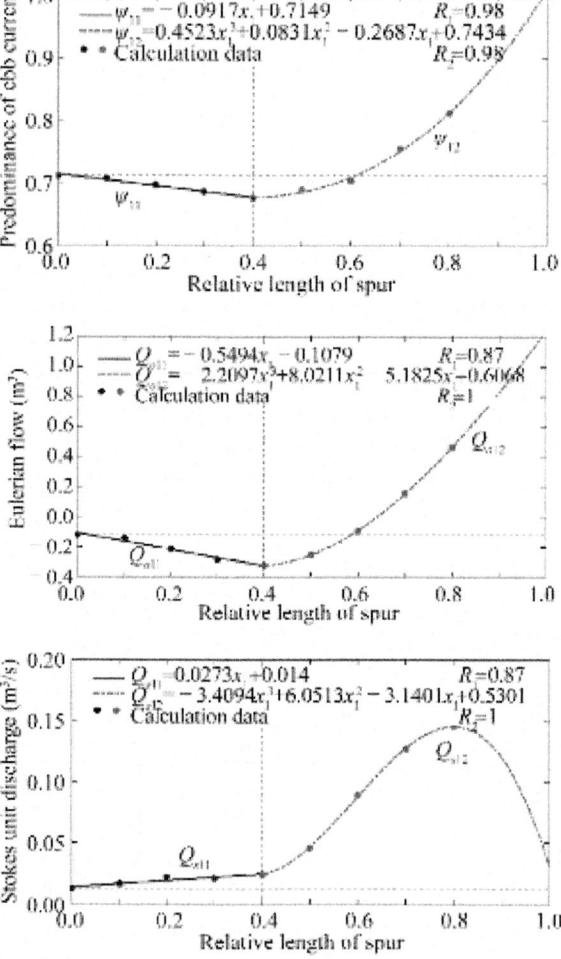

Fig.11. Calculation data fitting of N branch for characteristic parameters

3. Conclusion

Based on the fixed bed tidal current physical model of generalized bifurcation estuary, a 2D tidal current numerical model is established and verified by the experimental data of the physical model. Though the study of effect of building a pair of guiding jetty and four pair of spur dikes in one branch on tidal currents of bifurcation estuary, the following conclusions are gotten.

(1)Compared with the condition of no regulation engineering, a pair of guiding jetty can increase the flow velocity inside the jetty up to nearly 2 times. Guiding jetty's length of 4 sections is the critical length for the changes of velocities along the engineering branch in the bifurcation estuary; if the guiding jetty becomes much longer, the current velocities at flood and ebb strength in the out sea begin to decrease.

(2)The spur dikes increases the velocity inside the jetty; compared with the condition without regulation engineering, the velocities at flood and ebb strength are about three times than before.

(3)With the increase of spur dike length, the velocities at flood and ebb strength inside the regulation engineering branch gradually decreases after the 5th unit of spur dike length; when the spur dike length reaches 8 units, the velocity even closes to 0. Through the regression analysis of the predominance of ebb current, Euler unit discharge and Stokes unit discharge, the reasonable relative length of spur dike in the bifurcation estuary is 0.4~0.6.

(4)With the increase of spur dike's length, the zero position of Stokes unit discharge in the regulation engineering branch moves towards the out sea; while the zero position of Stokes unit discharge of another branch moves towards the upstream.

Acknowledgements

This work is supported by: National Key Research and Development Program of China(2016YFC0401500,2016YFC0401502,2016YFC0401503,2016YFC0401505, 2016YFC0401506); National Natural Science Foundation of China (51479122); Central Water Resources Fee Project (1261530210283)

References

Yong, H. K., George V. 2005. Effect of channel bifurcation on residual estuarine circulation: Winyah Bay, South Carolina. Estuarine, Coastal and Shelf Science, (65): 671-686.

Pan, Y., Dou X. P., Gao X. Y., et al. 2015. The effect of regulation projects on the water and sediment dynamic in bifurcation estuaries. APAC, Elsevier Ltd, 2015, pp: 786-793.

Chao, C. F., Water-sediment movement characteristics of bifurcation wstuary and 3D water flow numerical model application research. Doctoral dissertation of Hohai University, Nanjing: Hohai University, 2005. (in Chinese)

Bing, Z., Dong, F. L., Hong, W. Z., et al. 2006. 2D Numerical model about effect of guiding jetty on water-sediment movement. Journal of Hydraulic Engineering, 37(7): 880-887.

DOU X. P., LI T. L., DOU G. R., 1999. Numerical model of total sediment transport in the Yangtze Estuary. China Ocean Engineering, (3): 277-286.

Experimental and Numerical Study of Wave Loads on Stationary Pile-Supported Structures for Offshore Wind Turbine

Xuan Ni* Leiping Xue and Jinzhe Qu

Department of Engineering Mechanics, Shanghai Jiao Tong University, Shanghai 200240, China
* E-mail: nixuan1991@sjtu.edu.cn

With the rapid growth of offshore wind industry, the predition of the wave loads on different pile-supported structures is of great importance. To investigate the wave loads, a series of physical experiments and numerical simulations were conducted. As a new characteristic geometric scale, an equivalent diameter is proposed to analyze the dimensionless total wave loads. With the equivalent diameter, unified empirical formulas of wave loads versus Keulegan-Carpenter (KC) number is obtained for different complex pile-supported structures.

Keywords: pile-supported structures; equivalent diameter; Morison equation; wave loads

1. Introduction

In the recent decades, the offshore wind power industry has been growing rapidly. As estimated, the installed capacity of the offshore wind turbine (OWT) in China will reach 10GW by 2020. Compared with the onshore wind turbine, the design of the OWT requires more attention to the foundation and support structures, as it will face a more severe environment and more extreme weather. As a consequence, the prediction of wave loads on different offshore support structures, which plays a key role in the safety and stability of the OWT support structures, is of great importance.

Nowadays, we have already had some typical types of pile-supported structures, such as monopile, suction pile, pile cap and so on [1]. [2] discussed the effects of the interference parameters on the overall drag forces for pile groups. Through some small-scale and large-scale experiments, [3] and [4] reviewed that the interference parameters such as the relative spacing between piles and the number of neighbouring piles could noticeably affect the wave loads on each single pile.

For complex pile-supported structures, the total wave loads could be obtained by calculating the wave forces on each single cylinder pile with

Morison equation([5]). But the effect of different interference parameters was difficult to estimate, especially the pile group effect ([6]). The previous work about the variations of wave force with Keulegan-Carpenter ([7]) number (KC) only focused on single cylinders. Therefore, there still remains an efficient way to obtain the total wave loads on the whole complex pile-supported structures.

To analyze the variation of the hydrodynamic loads on complex pile-supported structures, both physical experiments and numerical simulations were conducted. To establish a method for calculating the wave loads on complex pile-supported structures, a new parameter called equivalent diameter was introduced in this paper. Finally, unified empirical formulas of the total wave loads along the wave direction have been obtained.

2. Material and methods

2.1. *Experimental set-up*

The experiments were conducted in the wave flume of the school of NAOCE at Shanghai Jiao Tong University. The wave flume was 0.8 m wide, 1.2 m deep and 65 m long. The wave maker was located on the left of the wave flume, and the structure model was fixed at a distance of 27.5 m from the wave maker. A wave absorber packed with stones was placed at the other end of wave flume. The layout of the apparatus is illustrated in Figure 1.

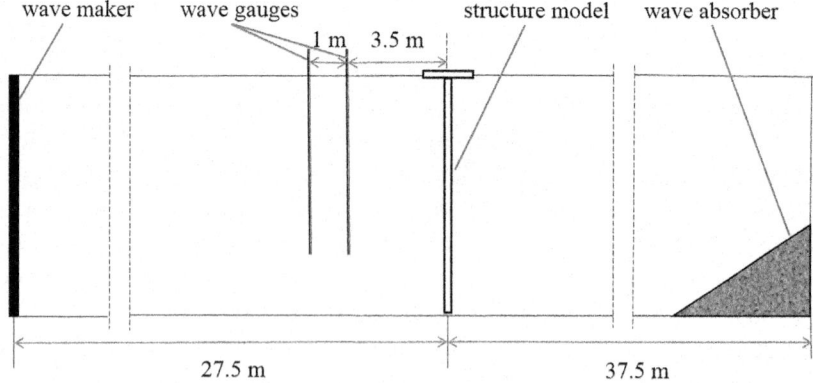

Fig. 1.   The layout of the wave flume

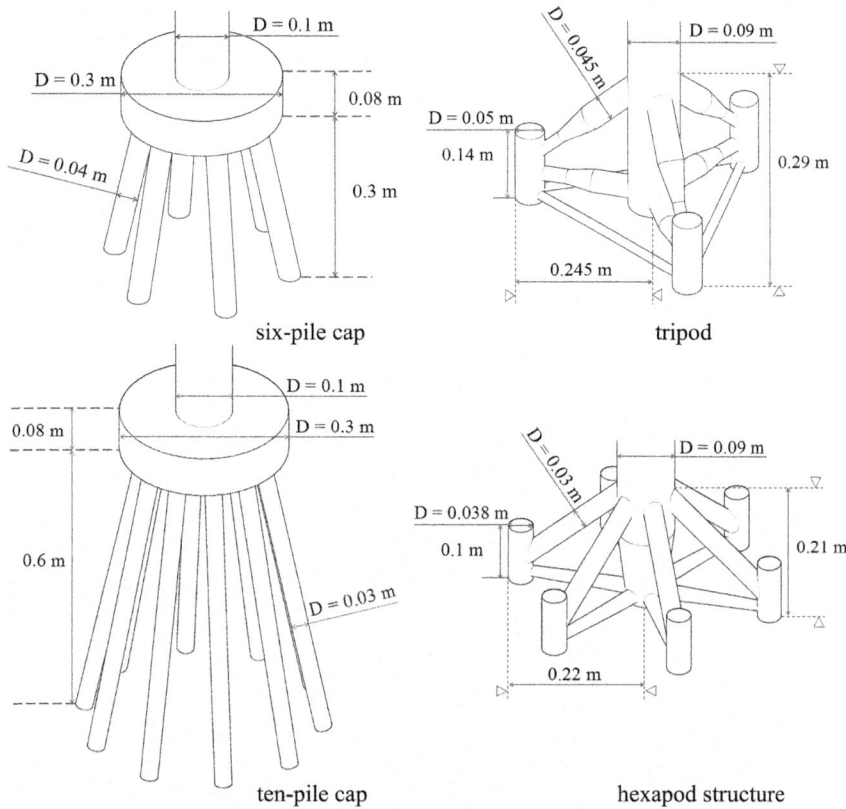

Fig. 2.   Sketches of different pile-supported structures

Two capacitance-type wave gauges, which were mounted in front of the structure model, were used to measure the wave surface elevation. The resolution of the wave gauge was 0.1 mm. Four force transducers were equipped on the top of the structure model and fixed on the stainless steel panels. The resolution of each force transducer was 0.1 N, and the relative error was 1%. The structure model was suspended in the wave flume, where the gap between the structure model base and the bottom of the wave flume was less than 5 mm. All the wave gauges and force transducers were connected to the same amplifier which ensured that the wave characteristics and wave force on the structure model were measured synchronously.

The pile-supported structures in the present study included six pile cap, ten pile cap, tripod and hexapod structures. The sketches of these structures are shown in Figure 2. The maximum vertical-sectional area of the pile is about 13% of the total sectional area of the wave flume. Therefore, the width of the flume is large enough for the ten-pile cap model.

2.2. *Numerical simulation*

The computation domain decomposed solver consists of an outer flow domain, which is governed by a three-dimensional, fully nonlinear potential flow solver, and an inner domain described by a fully nonlinear Navier-Stokes/VOF solver.

2.2.1. *Governing equations*

In the outer domain, the evolution of the free surface, , is governed by the dynamic and kinematic free surface conditions which is expressed by

$$\frac{\partial \eta}{\partial t} = -\nabla_H \, \eta \cdot \nabla_H \phi + \widetilde{\omega}(1 + \nabla_H \eta \cdot \nabla_H \eta) \tag{1}$$

$$\frac{\partial \phi}{\partial t} = -g\eta - 1/2(\nabla_H \eta \cdot \nabla_H \eta - \widetilde{\omega}^2(1 + \nabla_H \eta \cdot \nabla_H \eta)) \tag{2}$$

where $\nabla_H = (\frac{\partial}{\partial x}, \frac{\partial}{\partial y})$, and $\widetilde{\omega} = \frac{\partial \phi}{\partial z}|z = \eta$.

For the inner domain, the governing equation of the flow are the Reynolds-averaged Navier-Stocks equations:

$$\frac{\partial \rho \mathbf{u}}{\partial t} - \nabla \cdot (\rho \mathbf{u} \cdot \mathbf{u}) - \nabla \cdot ((\mu + \mu_t)\mathbf{S}) = -\nabla p + \rho g + \sigma_T \kappa_\gamma \frac{\nabla \alpha}{|\nabla \alpha|} \tag{3}$$

$$\nabla \cdot \mathbf{u} = 0 \tag{4}$$

Where $\mathbf{u}$ is velocity vector field in Cartesian coordinate system, p is the pressure field, μ is the dynamic viscosity of water, g is the acceleration of gravity, ν_t is the turbulent eddy viscosity, $\mathbf{S}$ is the strain rate tensor defined by $\mathbf{S} = 1/2[\nabla \mathbf{u} + (\nabla \mathbf{u})^T]$, k is turbulent kinetic energy, ω_T is the surface tension, κ_γ is the surface curvature, and α is the volume fraction.

Turbulence in the fluid flow model is estimated by the Splart-Allmaras (SA) one-equation model. The transport variable $\widetilde{\nu}$ in SA model is defined as the turbulence viscosity except the viscous affected area, transport equation of $\widetilde{\nu}$ can be written by

$$\frac{\partial \widetilde{\nu}}{\partial t} + \mathbf{u} \cdot \bigtriangledown \widetilde{\nu} = G_{\widetilde{\nu}} + \frac{1}{\sigma_{\widetilde{\nu}}}(\bigtriangledown \cdot (\nu + (1 + C_{b2}))\widetilde{\nu} \bigtriangledown \widetilde{\nu}) - C_{b2}\widetilde{\nu} \bigtriangledown \widetilde{\nu}) - \Upsilon_{\widetilde{\nu}} \quad (5)$$

Where $G_{\widetilde{\nu}}$ is the production of turbulence viscosity, $\upsilon_{\widetilde{\nu}}$ is the dissipation of turbulence viscosity that occurs in the near-wall region due to the wall blocking and viscous damping, $\sigma_{\widetilde{\nu}}$ and C_{b2} are constants, and ν is the molecular kinematical viscosity of water.

2.2.2. Efficient domain decomposition strategy

The relaxation zones were used for the one-way coupling the outer and inner domain. In the coupling zone ,the velocity field and the water volume fraction are at each time step updated according to

$$\Psi = \chi \Psi_{target} + (1 - \chi)\Psi_{com} \quad (6)$$

Where Ψ_{target} is the target solution in time and space given by the potential flow solver, while Ψ_{com} is the numerically computed quantity obtained by solving the governing equations(3)(4)(5), and χ is the weighting factor. This settlement ensures the solution from the potential flow solver is strongly imposed at the outer domain boundaries and avoiding the artificial re-reflections from the boundaries of the inner domain.

Generally the nodes of the inner and outer flow domains are not collocated so quantities from the outer flow domain have to be interpolated onto the grid of the inner domain. This could be done efficiently in the computational $(\mathbf{x}; \sigma)$ domain of the potential flow solver, where precomputed finite difference stencils were utilized for efficient evaluation of the following finite Taylor series expansion

$$\psi_{target}(\mathbf{x} + \delta \mathbf{e}_i) = \sum_{n=0}^{2a} \frac{\delta^n}{n!} \frac{\partial^n \psi}{\partial x_i}(x) \quad (7)$$

Details about the domain decomposition strategy can be found in[8].

2.3. Wave conditions and data analysis

Hydrodynamic loads were studied in the experiments in consideration of both regular and irregular waves. At the same time, only regular wave conditions were conducted in numerical simulations. In the experiments, each case was repeated three times to ensure the reliability of the results. The wave length was 2.67 to 10 times the length of the structures. The products of the water depth and the wave numbers ranged from 0.75 to 2.76, which means all the waves were in the finite water depths. The wave conditions in different water levels were illustrated in Table. 1, Table. 2

In both experiments and numerical simulations, the regular wave characteristics were obtained by the phase averaged method. The wave spectrums in irregular waves were virtually identical to the JONSWAP spectrum, and the sampling time was 100 times longer than the significant wave period. In the experiments, the wave condition data and load signals were all taken to a low-pass filter with a cut-off frequency of 5Hz.

3. Results and discussions

3.1. *Equivalent diameter*

To analyze the wave loads on the complex pile-supported structures, including six-pile cap, ten-pile cap, tripod and hexapod structure, a new characteristic geometric length named equivalent diameter is introduced in this study. Physically, this treatment implies that the volume of the complex pile-structures is converted to a monopile with a weight of $cosh(kz)$. The equivalent diameter is defined by

$$D_{eq} = \sqrt{\frac{4 \int_0^d A(z)cosh(kz)dz}{\pi \int_0^d cosh(kz)dz}} \tag{8}$$

which is inspired by the velocity distribution of the water particle along the water depth

$$u_x = \frac{\pi H cosh(kz)}{T sinh(kd)} \tag{9}$$

where D_{eq} is the equivalent diameter, z is the coordinate value in the vertical direction, A(z) is the cross-section area of the structure at the water depth z, and k is the wave number.

3.2. *Wave loads in regular waves*

For the complex pile-supported structures, the dimensionless coefficient of total wave force along the wave direction in regular wave conditions is

$F/(0.5\gamma D_{eq}H^2K)$. D_{eq} the equivalent diameter, H is the wave height, γ is the volumetric weight of water, and K is a function of water depth and wave number which can be expressed by

$$K = \frac{2kd + sinh(2kd)}{8sinh(2kd)} \tag{10}$$

In this study, the KC number for complex pile-supported structures ranges from 1.54 to 6.58 under regular waves, which is defined by the equivalent diameter. The dimensionless coefficient of total wave force along the wave direction versus KC number is shown in Figure 3

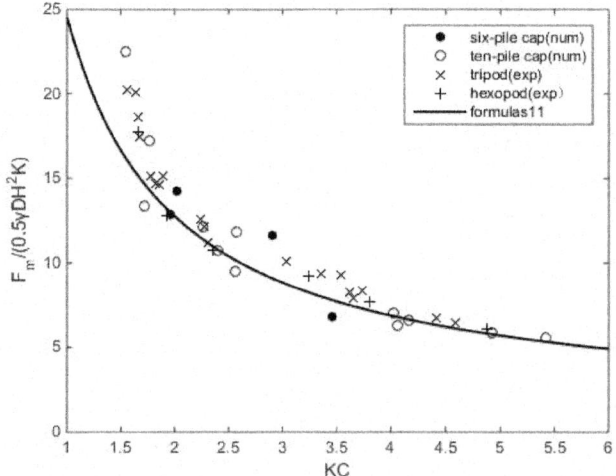

Fig. 3. Variations of dimensionless total wave force coefficient in regular waves against KC for different pile-supported structures

Using the data fitting method, the empirical formula of total wave force along the wave direction in regular waves can be presented by

$$\frac{F}{0.5\gamma D_{eq}H^2K} = 1.2 + 22.5KC^{-1} \tag{11}$$

in which the constant value on the right of the formula is determined to be 1.2, as the drag component is the dominating determinant when KC is sufficiently large.

3.3. *Wave loads in irregular wave conditions*

Under irregular wave conditions, the dimensionless coefficient is defined to be$F_{rms}/(0.5\gamma D_{eq}H_s^2 K)$, in which F_{rms} is the root-mean-square value of total wave forces, and H_s is the significant wave height. Like the regular conditions, the empirical formula in irregular waves can be expressed by

$$\frac{F_{rms}}{0.5\gamma D_{eq}H_s^2 K} = 1.2 + 22.5KC^{-1} \tag{12}$$

In irregular wave conditions, the KC number for complex pile-supported structures ranges from 0.41 to 1.98. The dimensionless coefficient of total wave force along the wave direction versus KC number is shown in Figure 4

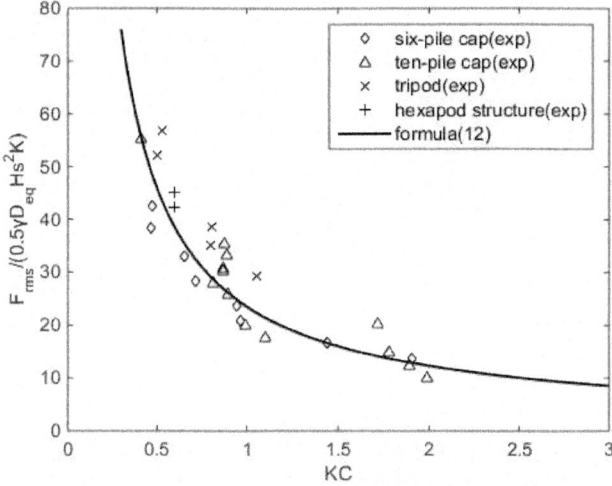

Fig. 4.   Variations of dimensionless total wave force coefficient in irregular waves with KC for different pile-supported structures

For irregular waves, zeroth moments of total wave force is another important parameter which could present the frequency characteristics and the magnitude of wave energy. The zeroth moments of total wave force along the wave direction is defined as

$$m_0 = \int_0^{+\infty} S_F(\omega)d\omega \tag{13}$$

where $S_F(\omega)$ is the spectrum density of wave force at the frequency z. The variation of $m_0/(0.5\gamma D_{eq}H_s^2K)^2$ against KC number are illustrated in Figure 5

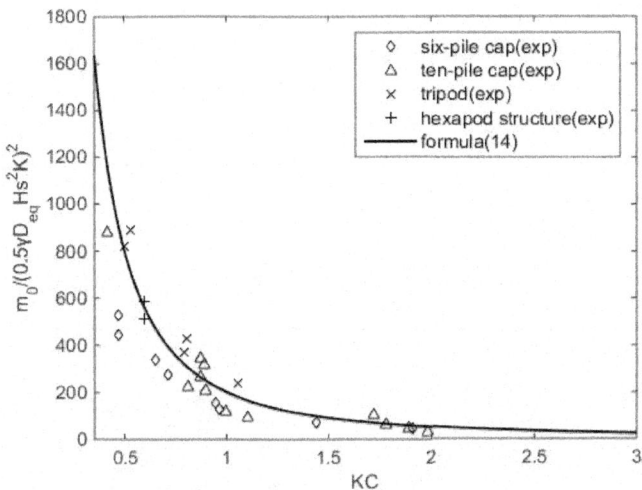

Fig. 5.   Variations of dimensionless zeroth moments of total wave force spectrum against KC for different pile-supported structures

Based on the results of $m_0/(0.5\gamma D_{eq}H_s^2K)^2$, the empirical formula can be presented by

$$\frac{m_0}{(0.5\gamma DH_s^2K)^2} = 2 + 200KC^{-2} \tag{14}$$

where the constant on the right of the formula is determined to be 2.0, as the dimensionless coefficient is close to 2.0 when KC number becomes sufficiently large. For irregular wave conditions, KC is determined to be U_mT_s/D_{eq}, in which T_s is the significant wave period, and U_m presents the maximum horizontal velocity of the water particle at the surface of still water.

4. Conclusion

In this study, the total wave loads on four different pile-supported structures are investigated in consideration of both regular and irregular wave conditions. Equivalent diameter is introduced to represent the character-istic geometric scale of different complex centrosymmetric pile-supported structures. Therefore, the dimensionless wave loads and the KC number can be obtained. The total wave loads along the wave direction on different pile-supported structures can be estimated by the unified empirical formulas for KC from 0.41 to 6.58.

In the future work, the validity of the empirical formulas can be explored with other non-centrosymmetric offshore pile-supported structures. Also, some other parameters, such as the density of the pile group, should be taken into account. Furthermore, the range of KC is limited in this study and should be expanded for future work.

Table 1. Regular wave conditions of different pile-supported structures.

six-pile cap(num)	d/m	T/s	H/m (case 1, case 2, ...)
	0.3	1.8	0.054
	0.3	1.5	0.0727, 0.0414
	0.3	1.2	0.0503
ten-pile cap(num)	0.5	1.8	0.1418
	0.5	1.5	0.1331, 0.0988, 0.0555
	0.5	1.2	0.0716, 0.0493
	0.4	1.8	0.0967
	0.4	1.5	0.0945, 0.0544
	0.4	1.2	0.0404
	0.3	1.5	0.083, 0.0517
	0.3	1.2	0.0413
tripod(exp)	0.425	1.8	0.1008, 0.0518
	0.425	1.5	0.0950, 0.0927, 0.0497
	0.425	1.2	0.0493
	0.415	1.8	0.0528
	0.415	1.5	0.0468
	0.415	1.2	0.1018, 0.0503
	0.325	1.8	0.0502
	0.325	1.5	0.0959, 0.0478
	0.315	1.8	0.1015, 0.0496
	0.315	1.5	0.0941, 0.0469
	0.315	1.2	0.0918, 0.0506, 0.0474
hexapod(exp)	0.315	1.8	0.1017, 0.0491
	0.315	1.5	0.0922, 0.0467
	0.315	1.2	0.0925, 0.0475

Note: d: the water depth; T: the significant wave period; H: the significant wave height.

Table 2. Irregular wave conditions of different pile-supported structures.

six-pile cap(exp)	d/m	T_s/s	H_s/m (case 1, case 2, ...)
	0.4	0.98	0.0528
	0.3	1.05	0.0193
	0.3	0.97	0.0536, 0.0406, 0.0267
	0.3	0.93	0.0187
ten-pile cap(exp)	0.7	1.01	0.0466, 0.0239
	0.6	1.17	0.0259
	0.6	1.01	0.0515, 0.0262
	0.5	1.15	0.0246
	0.5	1.01	0.0525, 0.0257
	0.4	1.00	0.0541, 0.0259
	0.3	1.13	0.0277
	0.3	1.01	0.0529
	0.3	0.97	0.0271
tripod(exp)	0.425	1.12	0.0248
	0.425	0.98	0.0347, 0.0263
	0.315	1.10	0.0159
	0.315	0.94	0.0181
hexapod(exp)	0.315	0.96	0.0191
	0.315	1.06	0.0182

Note: d: the water depth; T_s: the significant wave period; H_s: the significant wave height.

References

1. M. S. Ryu, K. S. Kang, J. S. Lee et al., A suggestion for the foundation type of offshore wind turbine in the test bed on the basis of economic and constructibility analysis, in *The Twenty-second International Offshore and Polar Engineering Conference*, 2012.

2. M. M. Zdravkovich, *Flow around Circular Cylinders: Volume 2: Applications* (Oxford University Press, 2003).

3. L. Bonakdar, H. Oumeraci et al., Interaction of waves and pile group-supported offshore structures: A large scale model study, in *The Twenty-second International Offshore and Polar Engineering Conference*, 2012.

4. L. Bonakdar and H. Oumeraci, Small and large scale experimental investigations of wave loads on a slender pile within closely spaced neighbouring piles, in *ASME 2014 33rd International Conference on Ocean, Offshore and Arctic Engineering*, 2014.

5. J. Morison, J. Johnson, S. Schaaf et al., The force exerted by surface waves on piles, *Journal of Petroleum Technology* **2**, 149 (1950).

6. L. Bonakdar and H. Oumeraci, Pile group effect on the wave loading of a slender pile: A small-scale model study, *Ocean Engineering* **108**, 449 (2015).

7. G. H. Keulegan and L. H. Carpenter, *Forces on cylinders and plates in an oscillating fluid* (US Department of Commerce, National Bureau of Standards, 1956).

8. B. T. Paulsen, H. Bredmose and H. B. Bingham, An efficient domain decomposition strategy for wave loads on surface piercing circular cylinders, *Coastal Engineering* **86**, 57 (2014).

Numerical Study of A Solitary Wave Impacting on the Submerged Barrier

Jiadong WANG, Guanghua HE*, Rui YOU

School of Naval Architecture and Ocean Engineering
Harbin Institute of Technology, Weihai
Weihai, Shandong 264209, China
*E-mail: ghhe@hitwh.edu.cn

Pengfei LIU

Australian Maritime College, University of Tasmania
Launceston, Tasmania 7250, Australia

The strongly nonlinear interaction between a solitary wave and the submerged barrier is investigated numerically in the present paper. Based on the Cartesian grid method of constrained interpolation profile (CIP), a two-dimensional numerical wave flume specializing to Navies-Stokes equations was established to simulate the propagation of a solitary wave over the submerged structure. The free-surface motion and hydrodynamic forces acting on the submerged breakwater were presented in this work. The comparisons between the numerical results and experimental data for the free surface elevations show a good agreement in both upstream and downstream of barrier. Complicated hydrodynamic behaviors, including wave breaking and overtopping, were captured by the numerical model which demonstrates that the CIP-based model has the ability to provide the reliable predictions for the wave transmission over the submerged structure.

Keywords: Solitary wave; Submerged barrier; CIP method; Wave forces.

1. Introduction

Offshore structures are widely employed to reflect the extreme waves and protect the coastal facilities. In recent years, the type of submerged barrier is extensively studied and has a widespread application due to its environmental advantage. So far, the experiments have been the dominant method to solve such nonlinear wave-body interaction problems, while measurements are usually restricted to their high cost. The analytical solutions are only available for a few simple cases, and most of them are based on certain assumptions and simplifications. Therefore, a numerical model which can reflect the real hydrodynamic phenomena and provide the accurate prediction on wave loads is essential to handle these complicated problems as well as has an important significance for engineering applications.

Solitary waves play an important role in simulating the behaviors of tsunami and other extreme waves. Most of studies on the solitary wave propagating over a barrier mainly focused on the transmission of wave and vortex evolution, such as works performed by Lynett et al[1]. Chang et al[2] conducted a series of experiments to study the behaviors of vortex shedding around a vertical plate, similar studies include the model experiments by Lin et al[3]. Kato et al[4] measured the characteristics of waves acting on a breakwater without the consideration of free-surface motion. Hsiao and Lin[5] studied the interaction between the tsunami-type wave and an offshore breakwater by an experimental approach. Such type of studies pay the major attention to the distribution of flow field and most of them have not included the wave forces acting on the structures.

Vertical and rigid breakwater with small thickness is considered as a desired structure to protect the coastlines due to the economical efficiency and environmental value. Earlier studies on waves propagating over this type of structure concentrated on the linear and irregular waves, while few of them considered the solitary wave, such as studies conducted by Losada et al[6]. Liu and Al-Banaa[7] studied the interaction of solitary wave with a vertical barrier without the hydrodynamic behavior of wave breaking. Most of similar works put main emphasis on the wave loads and few studies involved the extremely nonlinear hydrodynamic phenomena. It should be noted that Wu et al[8] conducted a set of experiments to study the interaction between a solitary wave and the vertical barrier, which measured distribution of flow field and wave forces acting on the barrier. Stimulated by Wu et al[8], the present paper studied the interaction of solitary wave with a barrier using a strongly nonlinear CFD numerical method.

With the rapid development of computer science and numerical computation, the CFD method based on the viscous theory is considered as an alternative to solve these complicated problems. Lin and Liu[9] performed a series of numerical simulations on the interaction of waves with vertical barrier using a COBRAS model. Shao[10] employed the SPH method to study the wave impacting on a vertical plate, simulating the propagation of solitary wave. However, most of numerical models are not able to meet the requirement of accuracy and computational efficiency in engineering applications.

Here, we established a strongly nonlinear numerical model to simulate the interaction of a solitary wave with a submerged barrier. In the present numerical model, a Cartesian grid method with constrained interpolation profile (CIP) is employed to calculate the Navier-Stokes equations. For the verification of developed model, the free-surface motion is numerically computed, and sufficient comparisons of numerical results with available experimental data are performed. Moreover, the horizontal and vertical forces impacting on the submerged barrier due to the solitary wave are computed. Finally, the distribution characteristics of flow field are presented and discussed.

2. Numerical Methods

The present computations are performed by the CIP-based Cartesian grid method [11, 12]. The CIP method is adopted as the base scheme for the flow solver on a nonuniform, staggered Cartesian grid. A turbulence model and surface tension are not included.

2.1. *Flow solver*

The unsteady, viscous and incompressible flow is considered in present model, which is governed by the following continuity and Navier-Stokes equations.

$$\frac{\partial u_i}{\partial x_i} = 0 \tag{1}$$

$$\frac{\partial u_i}{\partial t} + u_j \frac{\partial u_i}{\partial x_j} = -\frac{1}{\rho}\frac{\partial p}{\partial x_i} + \frac{1}{\rho}\frac{\partial}{\partial x_j}(2\mu S_{ij}) + f_i \tag{2}$$

where $S_{ij} = (\partial u_i/\partial x_j + \partial u_j/\partial x_i)/2$, ρ and μ are density and viscosity, respectively; u_i is the velocity component; p and f_i are pressure and body force including the gravity. The Navier-Stokes equations are differentiated with respect to the spatial coordinates as follows.

$$\frac{\partial(\partial_\zeta u_i)}{\partial t} + u_j \frac{\partial(\partial_\zeta u_i)}{\partial x_j} = -(\partial_\zeta u_j)\frac{\partial u_i}{\partial x_j} - \frac{\partial}{\partial\zeta}\left(\frac{1}{\rho}\frac{\partial p}{\partial x_i}\right) + \frac{\partial}{\partial\zeta}\left(\frac{1}{\rho}\frac{\partial}{\partial x_j}(2\mu S_{ij}) + f_i\right) \tag{3}$$

For the numerical solution of the Navier-Stokes equations, a fractional step method is employed to divide the governing equations into three calculation steps. The advection calculation is performed in advection step, which only the left hand side of Eqs. (2) and (3) are considered. The CIP method[11] is used to solve advection equations, so the following equations are obtained.

$$u_i^*(\mathbf{x}) = X_i^n(\mathbf{x} - \mathbf{u}^n \Delta t) \tag{4}$$

$$(\partial_\zeta u_i)^*(\mathbf{x}) = \frac{\partial X_i^*}{\partial \zeta}(\mathbf{x} - \mathbf{u}^n \Delta t) \tag{5}$$

where X_i is the cubic polynomial to approximate the spatial of u_i in an upwind cell for CIP scheme. The superscript 'n' denotes the present time level and '*' represents the time level after the advection step.

All terms in right hand side of Eqs. (2) and (3) except for those related to pressure are taken into the calculation in the nonadvection step (i) using a central difference scheme. The velocity-pressure coupling is treated in the nonadvection step (ii) and the Poisson equation for pressure can be obtained based on the assumption of $\partial u_i^{n+1}/\partial x_i = 0$.

$$\frac{\partial}{\partial x_i}\left(\frac{1}{\rho}\frac{\partial p^{n+1}}{\partial x_i}\right) = \frac{1}{\Delta t}\frac{\partial u_i^{**}}{\partial x_i} \tag{6}$$

Numerical solution of Eq. (6) gives the pressure distribution over the whole computation due to the assumption that Eq. (6) is valid for liquid, gas and solid phases.

2.2. Treatment of interfaces

The wave-structure interaction is treated as a multiphase problem by the present model and there are two interfaces that need to be precisely calculated, including the free surface and solid boundary. A density function φ_m ($m = 1, 2, 3$ denotes liquid, gas, solid phase, respectively) is used to recognize different phases, and it satisfies the following advection equation.

$$\frac{\partial \varphi_m}{\partial t} + u_i\frac{\partial \varphi_m}{\partial x_i} = 0 \tag{7}$$

A direct method can be used to calculate φ_3 due to the assumption of rigid body. Then we solve Eq. (7) to obtain φ_1, and the density function for gas phase is given by the relation $\varphi_2 = 1 - \varphi_1 - \varphi_3$. After the density function for all phases is determined, the physical properties can be calculated.

For the free surface, an interface capturing approach, namely THINC (Tangent of Hyperbola for INterface Capturing) scheme[13] is implemented. In the one-dimensional case, the profile of density function inside an upwind computation cell is approximated by a hyperbolic tangent function as follow.

$$F_i(x) = \frac{\alpha}{2}\left\{1 + \gamma\tanh[\beta(\frac{x - x_{i-1/2}}{\Delta x_i} - \delta)]\right\} \tag{8}$$

where $\alpha, \beta, \gamma, \delta$ are parameters to be specified. Multidimensional computations can be performed by a dimensional splitting method.

A rigid body assumption is reasonable and the fixed structure is considered in present study. To calculate the density function for the solid phase in each computation cell, a virtual particle method[12] is used to determine the body boundary. The solid body is represented by distributing particles on the surface via the virtual particle method, which can obtain the body boundary positions accurately without the problem of interface smearing.

2.3. *Computation of hydrodynamic forces*

After the pressures, velocities and density functions for all computation cells are obtained, the hydrodynamic forces acting on the solid body can be calculated by integrating the pressure and skin friction along the body surface as follow.

$$F_i = F_i^{(p)} + F_i^{(v)} = \oiint_A (-p\delta_{ik})n_k dA + \oiint_A 2\mu S_{ik} n_k dA \qquad (9)$$

where $F_i^{(p)}$ and $F_i^{(v)}$ are the forces due to pressure and friction, respectively. A denotes the surface of solid body, and n_i represents the i-th component of the outward unit normal vector.

We should note that the wave forces in present model only involve the pressure-related force, while the friction-related force is neglected due to its small magnitude.

3. Results and Discussion

Based on the experiments by Wu et al[8], we simulated the interaction of a solitary wave with a submerged barrier using developed numerical model. For the direct comparison with experimental data, the setting of numerical simulation is the same to the experiments. The schematic view of numerical model is shown in Fig. 1, a solitary wave with the amplitude of 0.07m in water with depth of 0.14m. The vertical barrier with a height of 0.1m and a width of 0.02m is mounted in bottom of wave flume and the structure center along x-axis is $X = 8$m. The length and height of whole computational field are 17.5m and 1m, respectively.

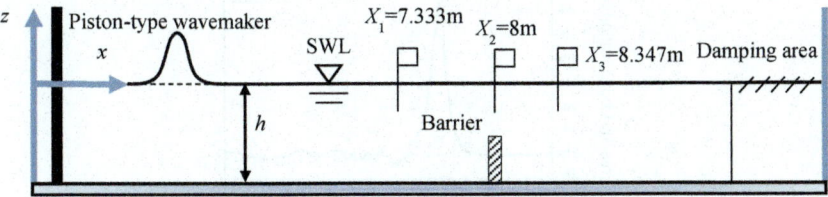

Fig. 1. Schematic view of numerical model

3.1. *Free-surface elevation*

A piston-type wavemaker is employed to generate solitary waves and the motion of paddle is controlled by the Rayleigh's solution[14]. Three wave gauges with $X_1 = 7.333$m, $X_2 = 8.0$m, $X_3 = 8.347$m, are set in the present model as shown in Fig. 1. According to the experimental setup, the time is defined as $t = 0$ s when the crest of solitary wave arrives at the first wave gauge (i.e., $X_1=7.333$m).

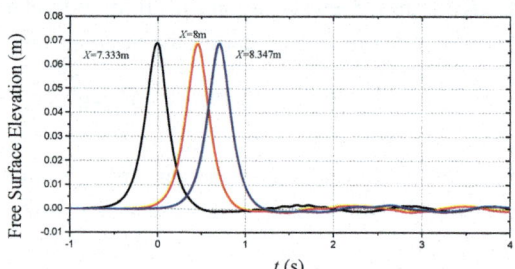

Fig. 2. Time series of free-surface elevation without barrier

Figure 2 presents the time series of the free-surface elevation without barrier at three wave gauges. As Fig. 2 shown, the numerical wave forms are consistent with the characteristic of solitary wave.

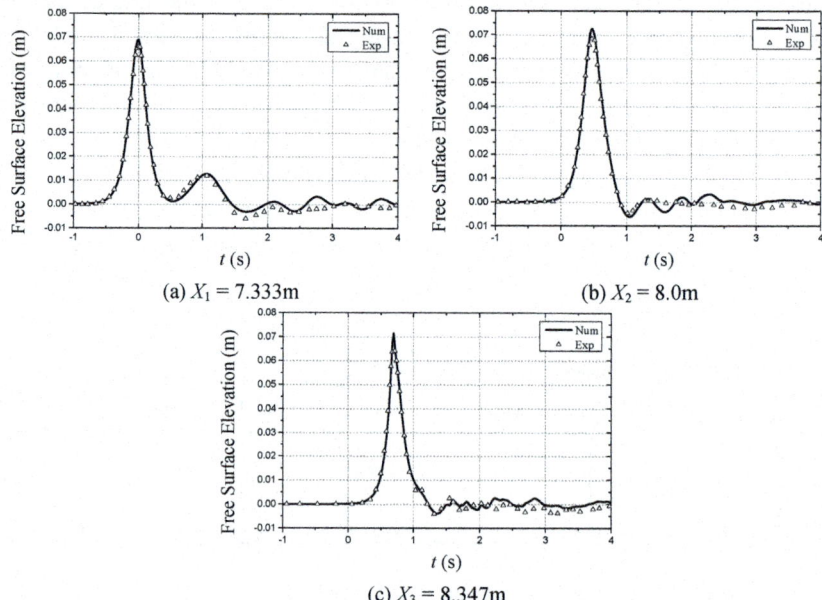

(a) $X_1 = 7.333$m

(b) $X_2 = 8.0$m

(c) $X_3 = 8.347$m

Fig. 3. Comparison of free-surface elevation between numerical results and experimental data

The comparison of free-surface elevations between the numerical results and experimental data is shown in Fig. 3. It can be observed from Fig. 3 (a) that a part of solitary wave is reflected by the barrier when the crest of wave approaches the structure. As Fig. 3 (c) presented, the free-surface motion of downstream shows a more evident nonlinearity due to the wave breaking. Overall, the numerical results have a good agreement with the experimental data, especially for the main body of the solitary wave. Compared with measurements, the maximum deviation of present results for the wave crest is $0.04a$ (a is wave amplitude) at three wave gauges. Several factors are responsible for such discrepancy, including the different resolutions of numerical model and measured facilities.

3.2. *Hydrodynamic forces*

Another interest of the present study is wave forces acting on the barrier. The time series of wave forces are shown in Fig. 4, including the horizontal force and vertical force. In Fig. 4 (a), we compared the present results with computations conducted by the COBRAS model[8] based on the RANS equations. Due to larger stressed area, the horizontal force is much bigger than vertical force.

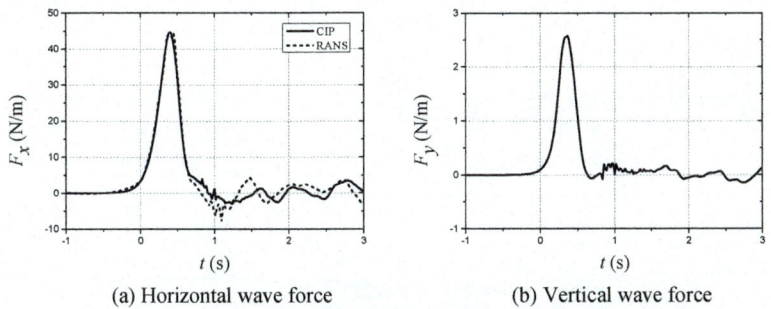

(a) Horizontal wave force   (b) Vertical wave force

Fig. 4. Time series of wave forces acting on the barrier

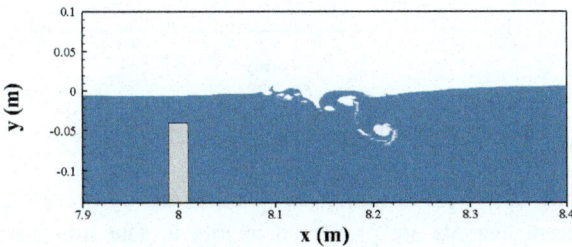

Fig. 5. Snapshot of simulated result with air-water mixing

A good agreement between the present results and other computations for the peak of horizontal wave force can be observed in the Fig. 4 (a), while the discrepancy exists in the time series of tailed stage. Strongly nonlinear phenomena, such as wave breaking, occurred when the crest of solitary wave propagated over the barrier, which lead to the complicated characteristics of hydrodynamic forces as Fig. 5 shown.

3.3. Flow fields

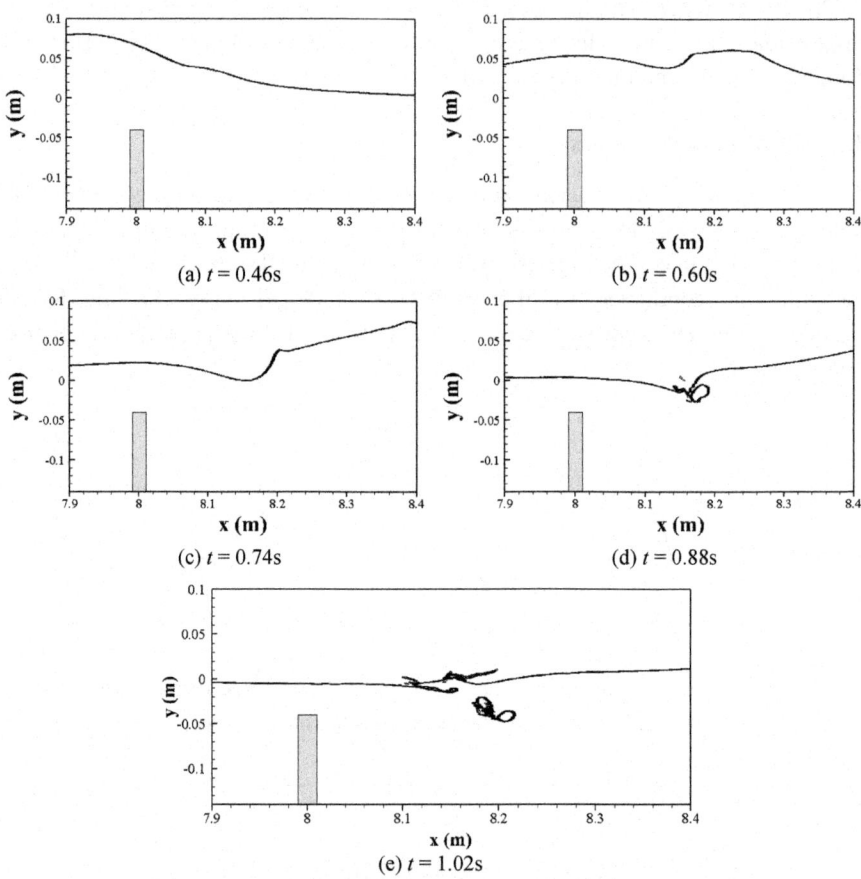

Fig. 6. Numerical results for flow fields at different instants

The evolution of flow field is numerically simulated by present model and the results at different instants are presented in Fig. 6. The free-surface motion in the flow field roughly experienced five stages during the interaction between solitary wave and barrier. Firstly, a bulge is formed near lee side of barrier as crest of solitary wave approaches the structure. Then, the bulge develops into a new crest and two peaks of wave can be observed at same time. The weather side of the new crest become abrupt as the solitary wave propagates over the barrier, and then slapping occurs on the free surface and some air is drawn into water. Finally, the phenomenon of air-water mixing can be observed along the free surface.

The same behaviors of free surface are recorded in laboratory and simulated by Wu et al [8] using the PIV technique and COBRAS model, respectively. It should be noted that the COBRAS model treats the air as numerical voids [8], while the real air is considered by the CIP-based model as presented in the section of numerical methods.

4. Conclusions

The strongly nonlinear interaction between a solitary wave and a submerged barrier is numerically studied using a developed model. The free-surface elevation is computed and numerical results fit the available experimental data fairly well. Based on the good agreement with other computations, the present model shows a capacity of providing an accurate and reliable prediction of wave forces on the submerged barrier. The free-surface motion is also simulated in this study and the extremely nonlinear phenomena, such as wave breaking and splitting, are captured and computed satisfactorily by the present numerical model. This CIP-based model is promising to solve such complicated wave-body interaction problems with a significant reliability and accuracy after a further study.

Acknowledgments

This work was gratefully supported by National Natural Science Foundation of China (51579058), Shandong Provincial Natural Science Foundation (ZR2014EEQ016), Open Research Fund Program of State Key Laboratory of Coastal & Offshore Engineering, Dalian University of Technology (LP1513), Open Research Fund Program of Key Laboratory of Water & Sediment Science and Water Hazard prevention (Changsha University of Science & Technology), Hunan Province (2015SS02).

References

1. P. J. Lynett, P. L. F. Liu, I. J. Losada, et al, Solitary wave interaction with porous breakwaters, *Journal of Waterway, Port, Coastal, and Ocean Engineering* **126**, 314 (2000).
2. K. A. Chang, T. J. Hsu and P. L. F. Liu, Vortex generation and evolution in water waves propagating over a submerged rectangular obstacle: Part I. Solitary waves, *Coastal Engineering* **44**, 13 (2001).
3. C. Lin, T. C. Ho, S. C. Chang, et al, Vortex shedding induced by a solitary wave propagating over a submerged vertical plate, *International Journal of Heat and Fluid Flow* **26**, 894 (2005).
4. F. Kato, S. Inagaki and M. Fukuhama, Wave force on coastal dike due to tsunami, *Coastal Engineering Conference.* ASCE American Society of Civil Engineers **30**, 5150 (2006).
5. S. C. Hsiao and T. C. Lin, Tsunami-like solitary waves impinging and overtopping an impermeable seawall: Experiment and RANS modeling, *Coastal Engineering* **57**, 1 (2010).
6. I. J. Losada, M. A. Losada and R. Losada, Wave spectrum scattering by vertical thin barriers, *Applied Ocean Research* **16**, 123 (1994).
7. P. L. F. Liu and K. Al-Banaa, Solitary wave runup and force on a vertical barrier, *Journal of Fluid Mechanics* **505**, 225 (2004).
8. Y. T. Wu, S. C. Hsiao, Z. C. Huang, et al, Propagation of solitary waves over a bottom-mounted barrier, *Coastal Engineering* **62**, 31(2012).
9. P. Lin and P. L. F. Liu, A numerical study of breaking waves in the surf zone, *Journal of Fluid Mechanics* **359**, 239 (1998).
10. S. Shao, SPH simulation of solitary wave interaction with a curtain-type breakwater, *Journal of Hydraulic Research* **43**, 366 (2005).
11. C. Hu and M. Kashiwagi, A CIP-based method for numerical simulations of violent free-surface flows, *Journal of Marine Science and Technology* **9**, 143 (2004).
12. C. Hu and M. Kashiwagi, Two-dimensional numerical simulation and experiment on strongly nonlinear wave–body interactions, *Journal of Marine Science and Technology* **14**, 200 (2009).
13. F. Xiao, Y. Honma and T. Kono, A simple algebraic interface capturing scheme using hyperbolic tangent function, *International Journal for Numerical Methods in Fluids* **48**, 1023 (2005).
14. G. He and M. Kashiwagi, Numerical analysis of the hydroelastic behavior of a vertical plate due to solitary waves, *Journal of Marine Science and Technology* **17**, 154 (2012).

Numerical Study on Layout Optimization of Tidal Stream Turbines in Zhoushan Demonstration Project

Ya Wang, Yanyan Zhai, Jisheng Zhang,[†] Linlong Tong and Shuang Song

*College of Harbor, Coastal and Offshore Engineering, Hohai University,
Nanjing, China
[†]E-mail: jszhang@hhu.edu.cn
www.hhu.edu.cn*

Tiantian Zhang

*China Three Gorges Corporation
Beijing, China
E-mail: zhang_tiantian@ctg.com.cn*

A tidal stream turbine is a device for harnessing energy from tidal stream, and the turbines layout has great influence on the economic benefits of tidal stream farm. In previous studies, the turbines layout is mainly determined by artificial optimization method, which is difficult to obtain the utmost utilization of tidal stream energy. Open source code OpenTidalFarm provides an automatic method to optimize the turbines layout and better utilize the tidal stream energy. In this study, the source code is applied to reproduce the tidal hydrodynamics and optimize the turbines in the Zhoushan Demonstration Project layout. Numerical results are in good agreement with the field measurements in terms of amplitude and velocity. With fixed total number of turbines, automatically optimized turbines layout can contribute to a higher electricity power compared to aligned layout.

Keywords: tidal stream energy, layout optimization, OpenTidalFarm.

1. Introduction

With gradual depletion of the traditional fossil fuels and increasing deterioration of the global environment, clean energy like ocean energy has been paid more and more attention by all over the world. Tidal stream energy, as one important part of ocean energy, is gradually being more emphasized and utilized. An advantage of tidal stream energy is that the energy extracted is predictable and reliable. Tidal stream turbines are devices for harnessing energy from tidal

* This work is supported by the NSFC grants (51479053), the Fundamental Research Funds for the Central University of China (2014B05114), the 111 Project (B12032), and the marine renewable energy research project of State Oceanic Administration (GHME2013GC03, GHME2015GC01).

stream. Similar to wind turbines, but powered by the flow of water instead of the flow of air, tidal stream turbines transform tides or deep ocean currents into electricity. In order to extract an economically amount of power, hundreds of tidal turbines must typically be deployed in an array. This naturally leads to the question of how these turbines should be configured to extract the maximum possible energy[1]. Thus, finding an optimal location for the turbines layout is significant, as it can have an important effect on the total energy, economic performance of a given site[2].

In generally, there are two common models to the design of tidal turbine layouts. First common models are simple flow models which are often defined by analytical expressions and compute fast. The models always describe a flow velocity reduction because of the energy extracted by turbines. For instance, Bryden and Couch[3] and Garrett and Cummins[4] optimized simplified models to export an estimate for the maximum energy extraction of a tidal basin. One-dimensional models were adopted by Vennell[5,6] to illustrate the significant impact of tuning each turbine individually on the channel geometry, turbine positions, and the tidal forcing. However, the simplified models can't describe the direct effects of the turbines on the velocity and the complex nonlinear flow interactions between turbines.

The second approach is to use more complex flow models to accurately predict the tidal flow, the flow interactions between turbines, and the energy extraction. These models are computationally infeasible to use expensive and an extensive search of the parameter space is prohibitive. This layout optimization is often performed manually, guided by intuition and experience[7]. In realistic domains, such optimization task becomes difficult due to the effects of complex bottom bathymetry, flow patterns and large numbers of turbines. To alleviate the problem, Funke et al. put forward the layout problem as a PDE-constrained (shallow water equations combined with adjoint method) optimization problem and solve the problem efficiently by using gradient-based optimization algorithms[2]. The approach is implemented in an open-source software framework called OpenTidalFarm.

OpenTidalFarm[8] (www.opentidalfarm.org) is a code which solves the shallow-water equations using the finite-element method. It is built on a code generation framework which facilitates efficient computation, including access to the gradient by adjointly approach, and is specifically designed for the optimization of turbine micro-siting. This work uses the OpenTidalFarm numerical modelling package to build the numerical model based upon the Zhoushan Demonstration Project.

2. Formulations

2.1. *Optimization problem*

The design optimal array of turbines is formulated as a PDE-constrained optimization problem which can be written as follows:

$$\max_{z,m} J(z,m)$$
$$\text{subject to } F(z,m) = 0,$$
$$b_l \leq m \leq b_u,$$
$$g(m) \leq 0. \tag{1}$$

where $J(z, m)$ is a function which evaluates the financial profit of the farm over its entire lifetime and is the goal quantity to be maximized; m indicates the design parameters (position and/or tuning parameters); $F(z, m)$ is a PDE operator parameterized by m and solution z; b_l and b_u are lower and upper bound constraint for the design parameters, which are used to enforce that the turbines remain in a prescribed area; and $g(m)$ enforce a minimum distance between any two turbines.

According to the fact that the equation $F(z, m)=0$, every certain m can map to a unique solution z, Hence, the relationship between z and m can be described by an implicit function $z \equiv z(m)$. So, the problem (1) can be simplified as following function:

$$\max_{z,m} J(z(m),m)$$
$$\text{subject to } b_l \leq m \leq b_u, \tag{2}$$
$$g(m) \leq 0.$$

2.2. *The PDE constraint*

The energy extraction of farm depends on the flow velocity u. In this work, the physical laws are modelled by the time-dependent, nonlinear shallow water equations:

$$\frac{\partial u}{\partial t} + u \cdot \nabla u - \nu \nabla^2 u + g\nabla\eta + \frac{c_b + c_t(m)}{H} \parallel u \parallel u = 0,$$
$$\frac{\partial \eta}{\partial t} + \nabla \cdot (Hu) = 0. \tag{3}$$

Where u denotes the depth-averaged velocity, η is the free-surface displacement, $H=\eta+h$ is the total water depth where h is the water depth at rest, whilst c_b is the drag coefficient for the natural bottom friction and $c_t(m)$ represents the additional friction induced by the turbine parameterization which depends on the turbine coordinates, g is the gravitational constant, ν denotes the viscosity coefficient. For stability reasons, we increase the viscosity at the inflow and outflow boundary conditions.

2.3. *Turbine model*

In order to be able to apply gradient-based optimization, the friction function $c_t(m)$ must be continuously differentiable with respect to the turbine positions. Therefore, the fiction function can be a bump function.

The individual turbines are represented through a locally-increased bottom friction coefficient[7]. The friction function for a single turbine centred at a point (x_i, y_i) with support radius r and friction coefficient Ki is thus given as:

$$C_i(m)(x, y) = K_i \psi_{x_i, y_i, r}(x, y) \tag{4}$$

The sum of the bottom friction of all turbines is denoted as $c_t(m)$ in equation (3) for all N turbines:

$$c_t(m) = \sum_{i=1}^{N} C_i(m) \tag{5}$$

where a single value for r is used based on the assumption that the deployed turbines are of equal size.

2.4. *The functional of interest*

The functional of interest J describes the value to be maximized and must be differentiable or gradient-based optimization. A natural choice for the functional of interest is the time-averaged energy extracted due to the increased friction by the turbines[9,10]. In the non-stationary case J is expressed as:

$$J(u, m) = \frac{1}{T} \int_0^T \int \rho c_t(m) \| u \|^3 \, dx dt \tag{6}$$

where ρ is the fluid density. For the steady-state problem considered here the energy extraction can be expressed as:

$$J(u, m) = \frac{1}{2} \int_\Omega \rho c_t(m) \| u \|^3 \tag{7}$$

where Ω is the domain of interest. For simplicity a cut-in speed and energy rating of the turbines is not considered.

Note that the J represents kinetic energy extraction rather than electrical energy generation, since it does not incorporate losses due to the turbine support structures and the conversion to electricity. This study only considers the extraction as the majorized function. More advanced functional choices could have an impact on the array of turbines in farm, such as installation and service costs. Even the potential environmental impacts could limit the turbines' positions. Those facts are not considered in this study.

3. Validations

3.1. Model setup

The model based on one demonstration project, which is recently launched in Zhoushan area, China, for exploring the tidal stream energy. The waterway between Putuoshan Island and Hulu Island has been initially selected as the project site for deploying tidal stream turbines, and one of main concerns is about the array optimization of turbines which is obviously related to energy generation.

The shallow water model of the Zhoushan area covers a domain shown in figure 5 and the mesh is in UTM coordinates. The boundaries of the model are applied in two conditions. The ocean sides are forcing by amplitude. The boundaries to the land of the domain, as well as for the islands are modelled as no-slip. All the boundaries are obtained from TPXO8-atlas global model. No wetting and drying scheme was applied here, but a minimum depth of 10 m constraint is applied to the bathymetry used in order to avoid negative water depths[11]. The whole tidal-simulation model run in OpenTidalFarm.

3.2. Data preparation

The validation shown here is based on the measurement data both amplitude and velocity provided by Second Institute of Oceanography of SOA in China. The survey area is located in the waterway between Putuoshan Island and Hulu Island (hereinafter referred to as Hulu Island Sea Area). There are five tide velocity observation station (B1~B5) and an amplitude observation station (S1) in this area. We can get velocity data from 16/8/2013 to 25/8/2013 (contain a neap tide, middle tide and spring tide) in B1~B5 station. In addition, the whole year amplitude data can be applied from S1 station. The locations of all tide stations are shown in Figure 1:

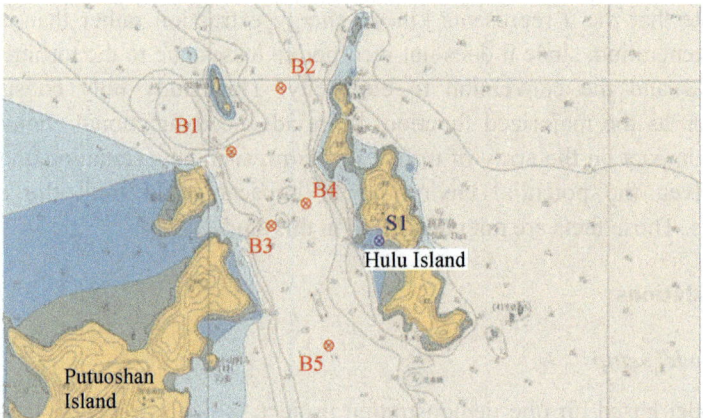

Fig. 1. Location sketch map of tide velocity stations (B1~B5) and tide amplitude station (S1).

3.3. Validations of amplitude and velocity

We will be simulating the tides in the Zhoushan area for 288 hours, starting at 00:00 am on the 13.8.2013. A point to note is that time difference between measurement data and the simulated data is 8 hours. As we can see in Figure 2, the simulated results of tide amplitude in S1 are in good agreement with the measurement data.

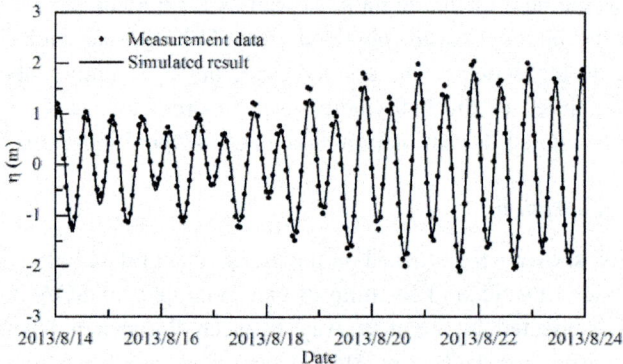

Fig. 2. Comparison of tide amplitude in S1 site.

We also compare the velocity data in observation sites, and Figure 3 shows the comparison between simulated velocity and measurement at Sites B3 and B4, indicating this model has a good ability in reproducing the local hydrodynamics.

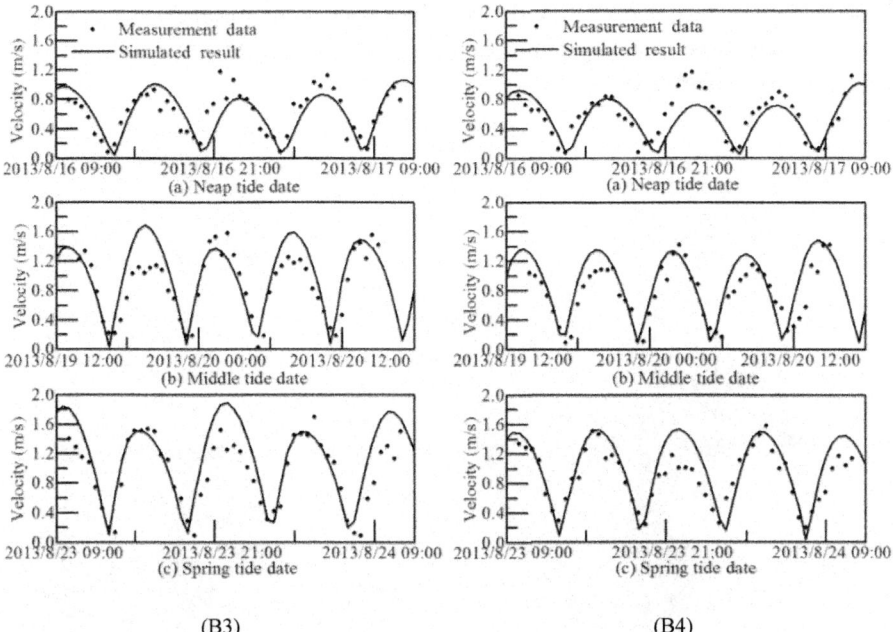

(B3) (B4)

Fig. 3. Comparison of current velocity at sites B3 and B4

4. Results

4.1. *The selection of farm site*

The farm site is a very important factor of energy extraction and a suitable site will bring greater economic profits. The hydrological factors and geography in Hulu Island Sea Area are very complicated, so many conditions should be considered when selecting farm site. The bathymetry and velocity are great factors in this study.

Firstly, the energy extracted from tide is associated with current velocity. Hence, the sea area with high velocity is always the excellent location to place turbines. It is known that tidal stream turbines have cut-in speed (1m/s) and cut-out speed (3m/s). The cut-in speed is a minimum current velocity to start turbines, and the cut-out speed is a velocity to sure the security of turbines. If the current velocity is over cut-out speed, the turbines should be closed for the protective factors. Secondly, the bathymetry suitable for turbines is between 20m and 50m. When the bathymetry is below 20m, the turbines are easy to expose to air in rectilinear tidal current. This situation can reduce extraction energy and corrode turbines. Meanwhile, when the bathymetry is over 50m, the more complex installation of turbines will cause more installation costs and maintenance costs. The tidal-simulation result is given in Figure 4, and the red area is high-velocity area.

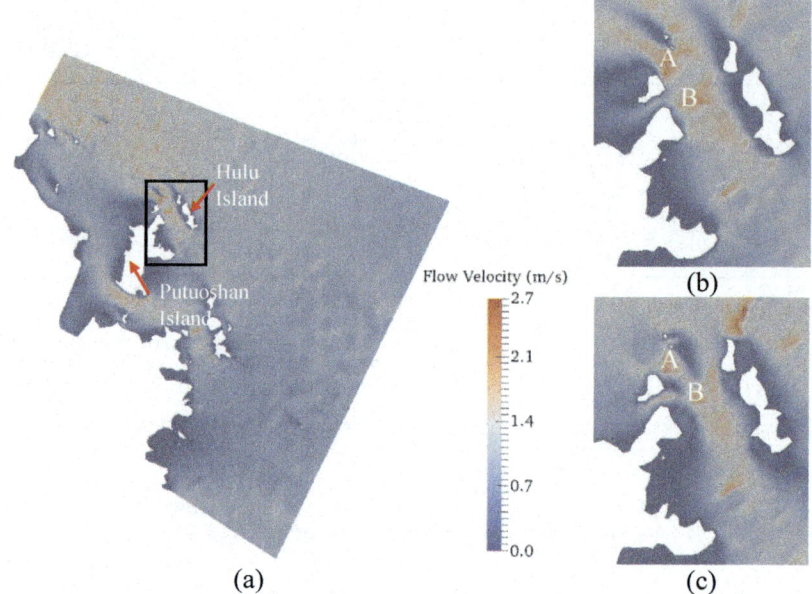

(a)   (c)

Fig. 4. Flow velocity in zhoushan area: (a) Tidal simulation of the study area (Inside rectangle box is Hulu Island Sea Area, and the red area is high velocity area); (b) and (c) The rectilinear tidal current in Hulu Island Sea Area. Site A and site B possess high flow velocity.

The waterway between Putuoshan Island and Hulu Island is the study area, and the result shows the site A and site B have high velocity in the most time of a year. According to the bathymetry, we choose the site B (hereinafter referred to as Farm B) to place turbines.

4.2. *The optimization of turbines array*

The discretized domain consists of a regular mesh in the Farm B with 5 m element size, and an unstructured mesh elsewhere with element sizes ranging from 5–500 m. The mesh was generated using GMSH. The turbines (4×4) are placed in Farm B. After 55 iterations, the final array of tidal current turbines is shown in Figure 5 (b).

After 55 iterations, the optimized farm layout has arranged the turbines to 'barrages' perpendicular to the flow. Furthermore, these barrages are aligned perpendicular to the flow field and extract the majority of the power.

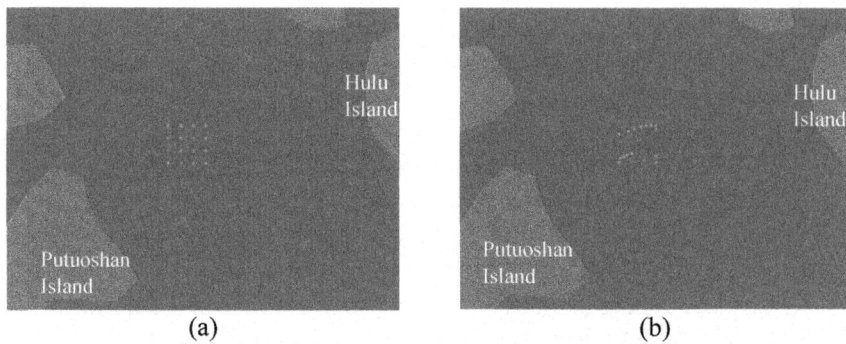

(a) (b)

Fig. 5. Layout optimization of tidal stream turbines: (a) The farm layout optimization is initialized with a regular layout of 16 turbines. The energy extraction by the farm (without taking losses into account) is 72383.18 MW; (b) The final optimized layout brings out about 80784.89 MW. That is the optimization increased the energy production by 8401.71 MW compared to the initial layout.

5. Conclusions

In this paper, a general framework for the design of tidal turbine arrays has been presented. The optimal configuration of tidal turbine farms is formulated as a constrained optimization problem with the only goal that the energy extraction is maximum. The model is constrained by partial differential equations describing the flow. Combined with the adjoint technique, the use of gradient-based optimization algorithms enables the use of physically-realistic tide models, even for a large number of turbines.

We choose the Zhoushan Demonstration Project which possess rich tidal stream energy as the study domain. Thus, putting the study domain mesh, real boundary conditions and initial layout configuration into the flow model. After running 288 hours, the tidal simulation in this area get a good agreement with the measurement data. We select a suitable site as the turbine farm according to the working conditions of tidal current turbines. The turbines layout optimization conducted with 16 turbines (A good population of turbines for its size). From the initial configuration in figures 5 (a) to the optimized design in figures 5 (b) is an improvement of energy extracted over the 8 hour which is obviously very significant. The automatic optimization of turbines layout is directly related to energy extraction of a tidal stream farm.

References

1. P. E. Farrell, D. A. Ham, S. W. Funke, and M. E. Rognes, Automated derivation of the adjoint of high-level transient finite element programs, SIAM Journal on Scientific Computing, **35(4)**, 369-393 (2013).

2. S. W. Funke, P. E. Farrell, and M. D. Piggott, Tidal turbine array optimisation using the adjoint approach, Renewable Energy, **63(1)**, 658-673 (2014).

3. I. G. Bryden, and S. J. Couch, How much energy can be extracted from moving water with a free surface: A question of importance in the field of tidal current energy, Renewable Energy, **32(11)**, 1961-1966 (2007).

4. C. Garrett, and P. Cummins, Limits to tidal current power, Renewable Energy, **33(11)**, 2485-2490 (2008).

5. R. Vennell, Tuning turbines in a tidal channel, Journal of Fluid Mechanics, **663(11)**, 253-267 (2010).

6. R. Vennell, Tuning tidal turbines in-concert to maximise farm efficiency, Journal of Fluid Mechanics, **671(3)**, 587-604 (2011).

7. T. Divett, R. Vennell, and C. Stevens, Optimization of multiple turbine arrays in a channel with tidally reversing flow by numerical modelling with adaptive mesh, Philosophical Transactions Mathematical Physical Engineering Sciences, **35 (4)**, 369-393 (2013).

8. G. L. Barnett, S. W. Funke, and M. D. Piggott, Hybrid global-local optimisation algorithms for the layout design of tidal turbine arrays, Mathematics, (2014).

9. R. Vennell, The energetics of large tidal turbine arrays, Renewable Energy, **48**, 210-219 (2012).

10. G. Sutherland, M. Foreman, and C. Garrett, Tidal current energy assessment for Johnstone Strait, Vancouver Island. Proceedings of the Institution of Mechanical Engineers, Part A: Journal of Power and Energy, **221(2)**, 147-157 (2007).

11. S. W. Funke, S. C. Kramer, and M. D. Piggott, Design optimisation and resource assessment for tidal-stream renewable energy farms using a new continuous turbine approach, Renewable Energy, **99**, 1046-1061 (2016).

Protection Performance against Storm Surge due to Vertical Telescopic Breakwater

Koki KAWAI

Civil Engineering, Chuo University,
1-13-27, Kasuga, Bunkyo-ku, Tokyo, 112-8551, Japan
E-mail: a13.ep5h@g.chuo-u.ac.jp
www.chuo-u.ac.jp

Katsumi SEKI

Research and Development Initiative, Chuo University,
1-13-27, Kasuga, Bunkyo-ku, Tokyo, 112-8551, Japan
E-mail: seki-k.15e@g.chuo-u.ac.jp

Makoto KOBAYASHI

Civil Engineering, Obayashi Corporation,
2-15-2, Konan, Minato-ku, Tokyo, 141-8604, Japan
E-mail: kobayashi.makoto@obayashi.co.jp

Taiichi OOKAWA

Offshore Steel Structure, Nippon Steel & Sumikin Engineering,
1-5-1, Osaki, Shinagawa-ku, Tokyo, 141-8604, Japan
E-mail: ookawa.taiichi.ra5@eng.nssmc.com

Hiroshi INOUE

Civil Engineering, Toa Corporation,
3-7-1, Nishishinjuku, Shinjuku-ku, Tokyo, 163-1031, Japan
E-mail: h_inoue@toa-const.co.jp

Kazuyoshi KIHARA

Engineering, MM Bridge Corporation,
9-19, Nihonbashitomizawa-cho, Chuo-ku, Tokyo, 103-0006, Japan
E-mail: kihara.kazuyoshi@mm-bridge.co.jp

Taro ARIKAWA

Faculty of Science and Engineering, Chuo University,
1-13-27, Kasuga, Bunkyo-ku, Tokyo, 112-8551, Japan
E-mail: taro.arikawa.38d @g.chuo-u.ac.jp

794

The purpose of this study is to analyze the effect of vertical telescopic breakwater on water level reduction in the event of a storm surge. This is investigated by developing an evaluation method using CADMAS-SURF/3D and STOC-ML. The flow velocity between the breakwaters corresponds to the theoretical value evaluated based on the water level difference. The results indicate that the aperture can be reduced to 5.5% - 6.0% when analyzing the breakwater with 7.2% aperture using STOC-ML. As an extension of this study, it is possible to obtain approximate water level reduction using dimensionless quantities including the conditions of wave and breakwater.

Keywords: storm surge, vertical telescopic breakwater, STOC-ML, CADMAS-SURF/3D

1. Introduction

Historically, there have been many large storm surges and tsunamis in Japan. In order to protect the region behind the breakwater from storm surges, advanced techniques have been used in the maintenance of coastal conservation facility. However, it is difficult to make the harbor safe, because the harbor entrance is used as a shipping route.

An advanced 'vertical telescopic breakwater(VTB)' is under development to protect the harbor facilities behind the breakwater against storm surges and tsunamis[1]. The breakwater is usually buried in the bottom of the sea, so as not to be an obstacle for vessel navigation. On the other hand, the breakwater is floated up quickly when storm surges and tsunamis strike. The breakwater consists of rows of steel pipe piles, each of which consists of upper and lower steel pipes. The upper steel pipe is normally encased in the lower steel pipe, which is driven into the seabed; thus, a small gap exists between the upper pipes.

A few experimental researches have been conducted on the hydraulic characteristics of this type of breakwater. This research investigates the effect of breakwater parameters like aperture on the water level reduction and the characteristics of the wave by using Super Roller Flume for Computer Aided Design of Structure (CADMAS-SURF/3D, hereinafter called CS3D) and Storm surge and Tsunami simulator in Oceans and Coastal areas (STOC)[2][3].

2. Verification of the Validity of the Flow Rate Calculation Method in the Gap of VTB using STOC-ML

VTB consists of rows of steel pipe piles, each of which consists of upper and lower steel pipes as shown in Fig. 1[4]. In order to evaluate the protection performance of VTB, it is necessary to compute the flow rate from the gap accurately. In this section, we show the validity of the calculation method in STOC-ML by comparing it with flow calculation using CS3D.

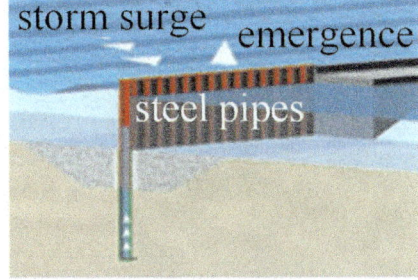

| (a) during calm | (b) during storm surge |

Fig. 1. VTB

2.1. *Study with CS3D*

2.1.1. *Analysis model and condition*

Numerical wave flume CADMAS-SURF/3D (Arikawa et al., 2005) is a numerical model developed for application to the business of the wave design of sea area facilities-resistant[2]. VTB is installed at the center of a 50 m tank as shown in Fig. 2. The calculation condition is shown in Table 1. To prevent numerical vibration, calculations were made by gradually increasing the flow velocity from 0 as shown in Fig. 3 for the inflow boundary ($x = -25$ m) and the outflow boundary ($x = 25$ m). As shown in Fig. 4, to reproduce the VTB including anti-rotation materials, a cell with zero porosity (OBST cell) is integrated with a cell including a gap (POROUS cell).

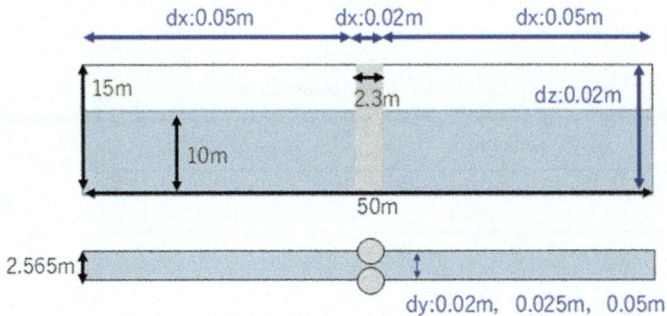

Fig. 2. Calculating area

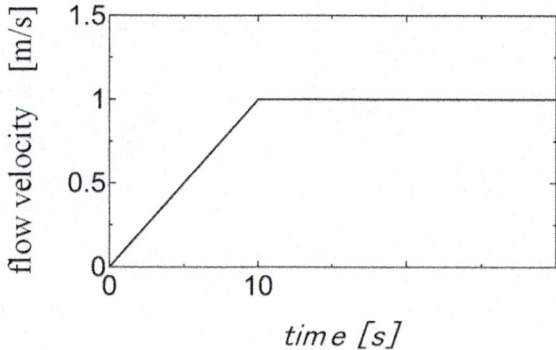

Fig. 3. Flow velocity condition

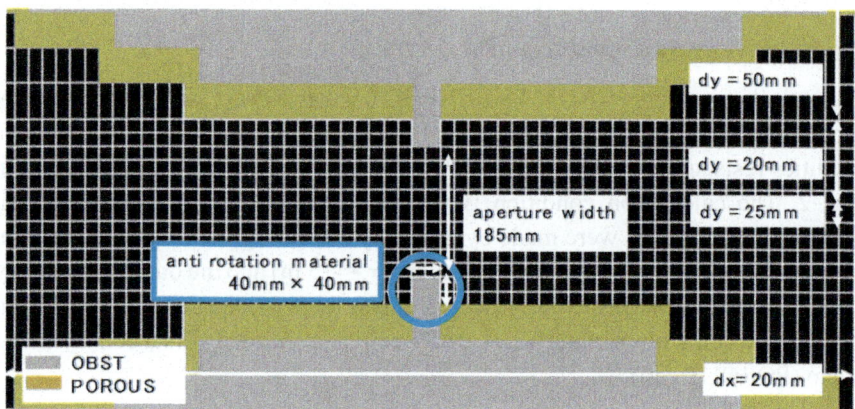

Fig. 4. Flow velocity condition

Table 1. Calculating condition

	range[m]	grid width[m]	The number of cells
X	-25. 0 ~ - 0. 8, + 0. 8 ~ + 25. 0	0. 05	1048
	- 0. 8 ~ + 0. 8	0. 02	
Y	0 ~ +1. 15, +1. 415 ~ +2. 565	0. 05	
	+1. 27 ~ + 1. 295	0. 025	59
	+1. 15 ~ + 1. 27	0. 02	
	+1. 295 ~ + 1. 415		
Z	0 ~ + 15. 0	0. 02	750

2.1.2. *Modeling the gap of VTB*

The breakwater is modeled using porosity. A cell without porosity (OBST) and a cell with porosity are combined to reproduce the breakwater with anti-rotation material.

2.1.3. *Result*

The snapshot of spatial waveform and flow velocity distribution is shown in Fig. 5 and Fig. 6. The water flow is changed by the anti-rotation material as shown in Fig. 6.

2.2. *Study with STOC-ML*

2.2.1. *Analysis model and condition*

A quasi-3-dimensional model, multi-layered static dynamics model (STOC-ML) is used for storm surge estimation.

   The area of 10 km length, 100 m width, and 10 m depth is used as shown in Fig. 7. The incident wave is introduced from the left side of the area. The grid size is 1 m along both the X and Y axes. Both the side boundary and the shore side boundary are wall boundaries. The calculation conditions are shown in Table 2.

2.2.2. *Modeling the gap of VTB*

The breakwater consists of rows of steel pipe piles and each pile consists of upper and lower steel pipes. The upper steel pipe is normally encased in the lower steel pipe, which is driven into the seabed; thus, a small gap exists between the upper pipes. The gap is modeled with a transmission coefficient[5]. The steel tube row is considered as a shieling object in a cell, and the surface transmittances are set to have the same value as the predetermined aperture.

$$\gamma_v = 1 - (1 - \gamma_x)(1 - \gamma_y) \tag{1}$$

where γ_x and γ_y are surface transmittances in the x- and y-direction, and γ_v is the porosity. It is found that, when the above-mentioned surface transmittance and porosity are set at the breakwater, numerical vibration occurred around the breakwater. As a countermeasure, cells with half the surface transmittance are set in the front and back of the breakwater as the buffer region. As a result, it is found that the occurrence of numerical vibration is suppressed by the buffer region; this is shown in Fig. 8.

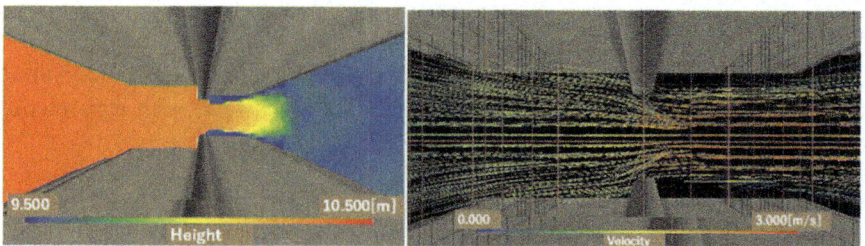

Fig. 5. Snapshot of spatial waveform   Fig. 6. Snapshot of flow velocity distribution

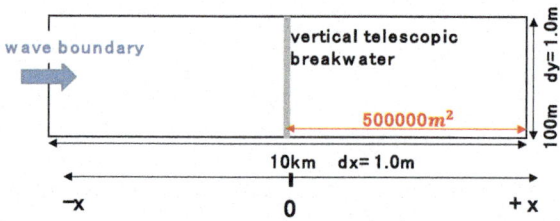

Fig. 7. Calculating area

Table 2. Calculating condition

wave height [m]	1
period [hour]	1
aperture	0. 01 , 0. 02 , 0. 03 0. 04 , 0. 05 , 0. 07 , 0. 10 no VTB
harbor area [m^2]	500000
total	8 cases

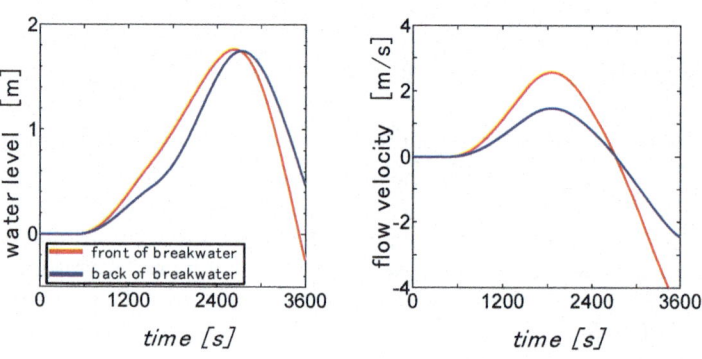

Fig. 8. Calculation data at the front and back of breakwater(T=1hour, α=0.01)

2.2.3. *Validity of calculation by STOC-ML*

We consider the validity of the calculation method of STOC-ML by comparing the flow velocity and flow rate due to water level difference with CS3D. The target is 0.07 aperture. The reference point of water level difference is at points $x=-1.0$ m and $x=+1.0$ m from the center of the waterway. In STOC-ML, the reference point of flow velocity and flow rate is the cross section passing through 0.07 aperture, and in CS3D, the one is behind the anti-rotation material. The comparison of the flow velocity calculated from the water level difference in each method and calculated value U of the flow velocity is shown in Fig. 9. The center of the waterway and the average value of the opening in CS3D are shown in Fig. 9. When the flow velocity is small, the calculation results are both smaller than the flow velocity obtained from the Bernoulli equation. However, this is because the place for comparing the water level difference should be closer to the breakwater as the flow velocity becomes smaller. It gradually becomes equal to the result of the Bernoulli equation. The aperture (apparent aperture ratio) that matches the flow rate of CS3D by an approximate straight line is shown in Fig. 10. The flow rate for STOC-ML is larger than the result of CS3D at all times. This is owing to overestimation of passing flow velocity by STOC-ML. The approximate straight line in Fig. 10 is obtained by the least squares method based on the result of STOC-ML. Here, α is the aperture in each calculation. Where It was found that the aperture input value of STOC-ML could be reduced to 0.055~0.06 in the case of the 0.072 aperture.

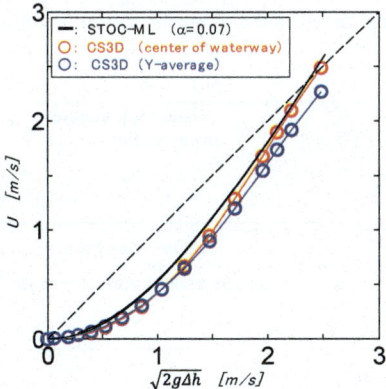

Fig. 9. Comparison of $\sqrt{2g\Delta h}$ and U

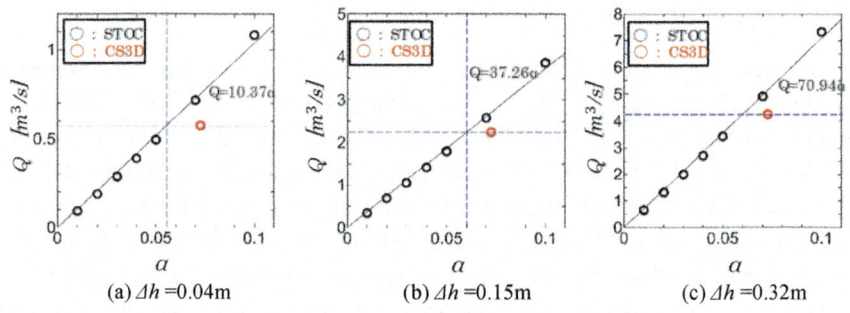

(a) Δh =0.04m   (b) Δh =0.15m   (c) Δh =0.32m

Fig. 10. Apparent aperture ratio

3. Development of Water Level Evaluation Method

In this chapter, we evaluate the protection performance against storm surge with STOC-ML.

3.1. *Numerical condition*

The regions shown in Fig. 11 and Fig. 12 are used in addition to Fig. 7. The area with 20 ha harbor is shown in Fig. 11. The area with VTB shortened to 25 m and impermeable wall at both ends is shown in Fig. 12. Hereafter, the one having a length of 100 m in the Y direction is referred to as a normal version, and one having a length of 25 m is referred to as a short version. A point 1 m from the end of the waterway, which is expected to see highest rise in water level (hereafter referred to as the bay inner part) owing to the influence of reflected waves, is taken as the measurement point. Details of the calculation case are shown in Table 3 and Table 4.

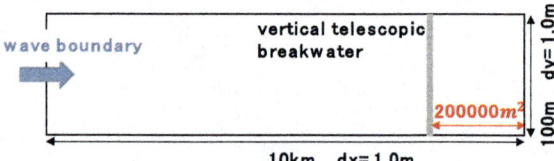

Fig. 11. Calculating area of normal version

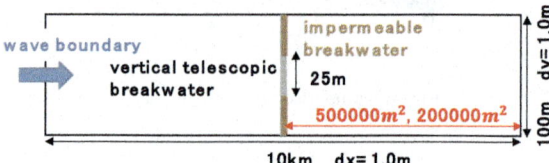

Fig. 12. Calculating area of short version

Table 3. Calculating condition of normal

wave height [m]	1
period [hour]	1,2,4,6
aperture	0. 01 , 0. 02 , 0. 03 0. 04 , 0. 05 , 0. 07 , 0. 10 no VTB
harbor area [m^2]	200000, 500000
total	60 cases

Table 4. Calculating condition of short version

wave height [m]	1
period [hour]	2
aperture	0. 01, 0. 03, 0. 05, 0.07, 0. 10 no VTB
harbor area [m^2]	200000, 500000
total	12 cases

3.2. *Results*

The water level data for a period of 1 hour in the 50 ha harbor measured at the bay inner part is shown in Fig. 13. as a main case.

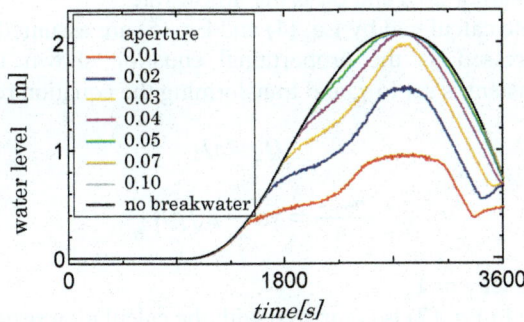

Fig. 13. Water level data at the bay back (T=1 hour, 500000 m^2)

3.3. *Approximate evaluation of the water level in the port*

The water level measured at the bay inner part is organized by dimensionless quantities. We attempt to estimate the approximate evaluation of the water level in the port by organizing the maximum water level data.

From the result of Chapter 2, it is found that the flow velocity passing through the breakwaters almost agrees with the flow velocity obtained by applying the Bernoulli equation. It is expressed by Eq. (2) in terms of the water level difference Δh. Assuming that the representative value of the water level difference is twice the wave height H as shown in Eq. (3), the inflow amount Q_1 from the gap is expressed by Eq. (4).

$$u = \sqrt{2g\Delta h} \tag{2}$$

$$\Delta h = 2H \tag{3}$$

$$Q_1 = \alpha A_1 2\sqrt{gH}\, T \tag{4}$$

Here, α is the aperture, H denotes the wave height, T is the period, A_1 is the inflow area (the product of breakwater width and depth), and Q_1 represents the inflow rate.

On the other hand, assuming that the water level rises uniformly in the harbor, the inflow into the port can be calculated from the increase in the water level as shown in Eq. (5). The inflow rate Q_2 is calculated using Eq. (5)

$$Q_2 = A_2 \eta_{max} \tag{5}$$

where A_2 is the harbor area and Q_2 is the inflow rate.

The flow rate calculated by Eq. (4) and Eq. (5) are assumed to be nearly the same, and expressed as the proportional equation shown in Eq. (6). The proportional constant is set to γ, and transforming the equation results in Eq. (7).

$$Q_1 \propto Q_2 \tag{6}$$

$$\frac{\eta_{max}}{H} = \gamma \alpha \frac{A_1}{A_2} 2\sqrt{gH}\frac{T}{H} \tag{7}$$

In this study, we set $\gamma = 1$.

The relation in Eq. (7) is compared with the calculation result of STOC-ML. Fig. 14 compares the calculated values for transmittance with the values calculated using Eq. (7). The horizontal axis represents the value of Eq. (7) and the vertical axis represents the ratio η_{max1}/η_{max2} calculated from STOC-ML, where the maximum water level in the case where VTB is installed is η_{max1}, and the maximum water level in the case where it is not is η_{max2}. There is a one-to-one plot correspondence. Therefore, it is possible to evaluate the water level reduction effect by using Eq. (7).

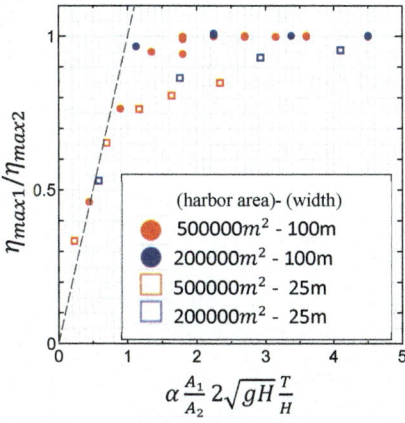

Fig. 14. Effect of water level reduction

4. Conclusions

(1) The velocity flow between the breakwaters corresponds to the theoretical value evaluated based on water level difference.

(2) The aperture can be reduced to 5.5% - 6.0% when analyzing the breakwater with 7.2% aperture using STOC-ML.

(3) It is possible to obtain approximate water level reduction using dimensionless quantities including the conditions of wave, breakwater, and harbor.

References

1. T. Yamane, T. Arikawa, M. Ito, T. Masuyama, Y. Kamei and M. Miyasaka, Development of buoyancy-driven vertical piling breakwater (in Japanese), Proceedings of civil engineering in the ocean, Vol. 21 (JSCE, Japan, 2005), pp. 115-120.
2. T. Arikawa, F. Yamada, M. Akiyama (2005): Study of Applicability of Tsunami Wave Force in a Three-dimensional Numerical Wave Flume, Collected Papers on Marine Engineering, Vol. 52, pp. 46-50.
3. T. Tomita, K. Honda and Y. Chida, Numerical simulation on tsunami inundation and debris damage STOC model (in Japanese), Technical note of the port and airport research institute, Vol.55, No.2 (PARI, Japan, 2016), pp.3-33.

804

4. OBAYASHI CORPORATION, *Service & Technology*, https://www.obayashi.co.jp/service_and_technology/ related/tech_d001.
5. T. Arikawa, H. Nomura, T. Tomita, M. Kobayashi, T. Toraishi, K. Arai and K. Kihara, Protection performance against tsunamis due to buoyancy-driven vertical piling breakwater (in Japanese), Proceedings of coastal engineering, JSCE, Vol.54 (JSCE, Japan, 2007), pp. 936-940.

Influence of Landward Slope to Evaluation of Tsunami Pressure on Seawall

K. Ando[†]

Steel Structures & Sabo Department, Engineering Business, Kobe Steel, Ltd.
2-2-4 Wakinohama-Kaigandori, Chuo-ku, Kobe, Hyogo, Japan
[†]*E-mail: ando.kei@kobelco.com*
www.kobelco.co.jp/english/

K. Suzuki and N. Tsuruta

Coastal and Ocean Engineering Research Department,
Port & Airport Research Institute
3-1-1 Nagase, Yokosuka, Kanagawa, Japan

A tsunami pressure acting on an onshore structure is generally evaluated by a formula with using an inundation depth η and a water-depth-coefficient α, which is based on the Froude number Fr in many cases. The definition of the water-depth-coefficient α is variously proposed by a semi-empirical way, and it is noteworthy that they are commonly based on only incident waves. Considering the real phenomenon and the practical procedure of design of the onshore structures, it is clear that synthetic waves including reflected waves should be considered. However, since the existing studies do not focus on the influence of the reflected waves, their applicability to the real phenomenon is not guaranteed yet. This paper performs hydraulic experiments with targeting tsunami running up the land to examine the influence of the reflected waves for more appropriate definition of the water-depth-coefficient α. Our experimental results in the cases with subcritical flow show unexpected increments of the inundation depth due to the refection of the forward tip of the run-up tsunami on the land, which implies a difficulty in obtaining the pure incident waves for the water-depth-coefficient α and would bring about overestimation of tsunami forces. Moreover, this study newly proposes a simple and effective method to evaluate the tsunami force with focusing on standing waves in front of the wall.

Keywords: tsunami; wave pressure; vertical wall; design of coastal structure; inundation depth

1. Introduction

Tsunami wave forces on an onshore structure are separately treated according to their states related to the run-up time. Figure 1 shows a typical process of a tsunami wave pressure on a vertical wall on the land. The continuous pressures are studied eagerly through hydraulic experiments, and various schemes were proposed to evaluate them. In particular, an evaluation formula with using the inundation depth η by Asakura et al.[1] has been widely applied to the following studies as a standard way.

pressure

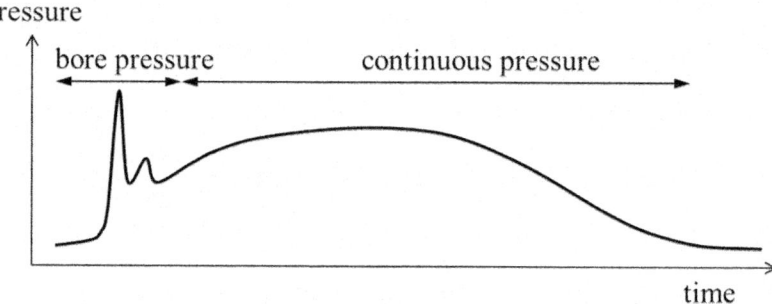

Fig. 1. A sample of time scenario about the wave pressure of tsunami on onshore structures

In [1], the wave pressures p and the inundation depths η are separately measured in individual tests. The former sets a target wall, while, the latter does not set it. That is, their conditions differ from each other. Then, in the experiments to measure the inundation depth η and the velocity v, they set a 1/5 slope behind the removed target wall in order to examine the influence of returning flows, which occur when run-up tsunamis return back to the sea. And they found that the inundation depth η has two peaks in its time series, which correspond to an incident wave (run-up flow) period and a reflected wave (returning flow) period respectively. Then, the maximum of the inundation depth η_{max} at first peak corresponding to the incident waves period is applied to their proposed formula for evaluation of the maximum tsunami force corresponding to the continuous pressure (Figure 2). Further, the velocity v and the Froude number Fr of incident waves were additionally introduced in various ways as improved formulas. Figure 3 shows some existing formulas and the difference of the definition of the water-depth-coefficient α in their formulas. The problem is that, in actual designs of the coastal structures, the inundation depth η and the velocity v are obtained by a macroscopic tsunami simulation without consideration of their accurate features depending on obstacles or the terrain such as a slope behind the removed target wall. This is one of the main causes of the inaccurate estimation of the tsunami wave pressure.

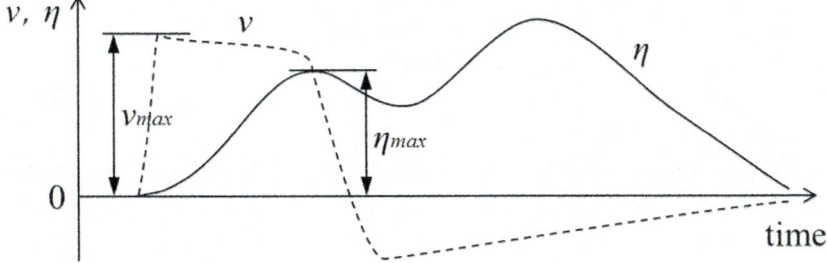

Fig. 2. Time scenario of inundation depth and velocity of run-up tsunami on land (Asakura et al.[1])

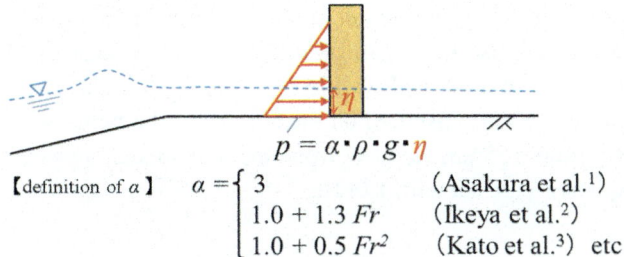

$$p = \alpha \cdot \rho \cdot g \cdot \eta$$

【definition of α】 $\alpha = \begin{cases} 3 & \text{(Asakura et al.[1])} \\ 1.0 + 1.3\, Fr & \text{(Ikeya et al.[2])} \\ 1.0 + 0.5\, Fr^2 & \text{(Kato et al.[3]) etc} \end{cases}$

Fig. 3. Proposed water-depth-coefficient α

In this study, hydraulic experiments are performed in order to examine the applicability of the existing formula with the water-depth-coefficient α to actual designs, and moreover, to improve the procedure to estimate the tsunami wave pressures. In the experiments, a slope with various gradients including a flat is set behind the removed target wall to examine the influence. Regarding the hydraulic conditions, both supercritical flows and subcritical flows are targeted to take variability of the tsunami wave pressures into consideration. From the measured water level and wave pressure of the tsunami wave, a new method to define the water-depth-coefficient α is proposed with substituting for the existing ways, which utilize the inundation depth and velocity of the incident wave.

2. Method of hydraulic experiments

Figure 4 shows a graphical presentation of a flume for our experiments. Tsunami attacks are reproduced through dam breaks by using tanks corresponding to an air valve type and a vacuum pump type. Here, several slopes are set for each test one by one to reproduce the various tsunami waveforms. By locating a slope as an uphill slope, a return flow at the developing stage and a stagnation state at the fully developed stage of a run-up tsunami is reproduced. The tank with an air valve begins flowing out the water into the flume by releasing the valve. While, the tank with a vacuum pump starts it by releasing the solenoid valves installed on the top. Table 1 shows a list of the experimental conditions.

We conducted two kinds of experiments: the one to measure the inundation depth η and the velocity v without the target wall and the one to measure the wave pressure on the target wall. In the experiment to measure the inundation depth η and the velocity v without the target wall, the flows of the run-up tsunamis are measured. The current gauges are set at the bottom of the flat floor. The details of the measurement points are shown in Figure 4. The sampling frequency of both of the current meter and the water gauges are set with 1.0kH. In the wave pressure tests, the tsunami wave pressures are measured by targeting on a forward surface of a vertical wall in front of the slopes as shown in Figure 4. The pressure gauges are installed on the surface of the target wall with 5mm, 55mm, 105mm and 155mm heights from the flat model, respectively. Their sampling frequencies are set with 1.0kHz.

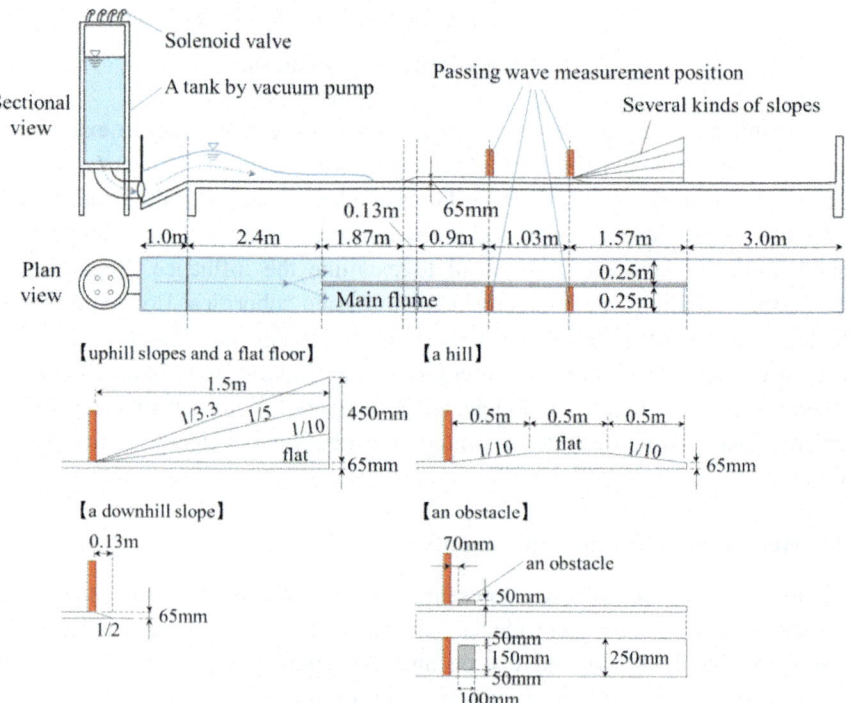

Fig. 4. Details of the experimental flume

Table 1. Experimental conditions

Floor type behind the target wall	three uphill slopes, a flat floor, a downhill slope, a hill, an obstacle	
Water volume	30 mm, 40 mm, 50 mm, 60 mm, 80 mm, 100 mm, 120 mm, 140 mm (Corresponding to water level filling the flume)	
Tank type	vacuum pump type tank	Air valve type tank
Inlet valve	full open, half open	-
Solenoid valves	Quarter release, three quarters release	-

3. Inundation depth without the target wall

Figure 5 shows schematic images of the developing processes of the inundation depth. In the case with a flat floor under a supercritical flow, the inundation depth gradually changes, that is slowing up and down. It is similar to the case with a flat floor under a subcritical flow. While, an uphill slope gives other processes including a returning flow (reflected wave). In particular, a supercritical flow includes a rapid returning flow. Consequently, its returning flow passes through the measurement points, and as a result, a rapid increment of the inundation depth is recorded. On the other hand, a subcritical flow also includes a returning flow, however, its velocity is relative slow and the inundation depth is recorded with a gradual and continuous increment. Asakura et al.[1] reported a similar tendency in a case with an uphill slope under a supercritical flow. Our study additionally discloses a tendency of that under a subcritical flow, which was so-far ignored in the actual design of the onshore structures.

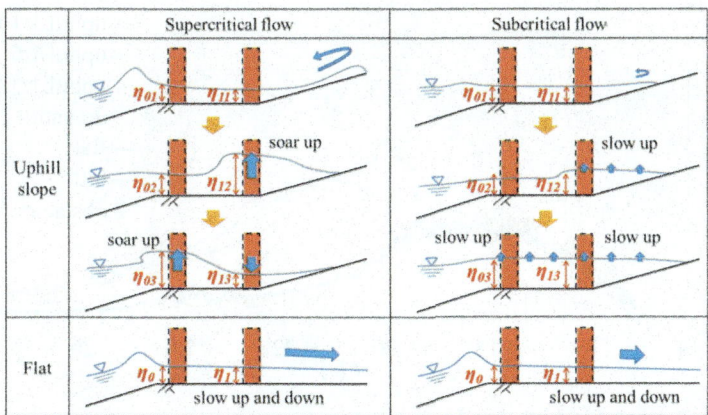

Fig. 5. Schematic images of the developing processes of the inundation depth η

810

Figure 6 and 7 show the time series of the inundation depth η at the measurement points (land side and sea side, respectively) in our experiment. The hydraulic condition corresponds to a supercritical flow by the vacuum pump type tank with releasing a solenoid valve and with a full-open inlet-valve. The water volume of the tsunami wave (inlet-flow) corresponds to 50mm water level filling the flume. The tsunami waves is generated after 10 seconds of the beginning of the measurement. The reflected waves (returning flows) are shown in the case with uphill slopes, a hill and an obstacle. Such cases clearly show sudden change of the inundation depth with implying the ease to distinguish the reflected waves from the recorded synthetic waves.

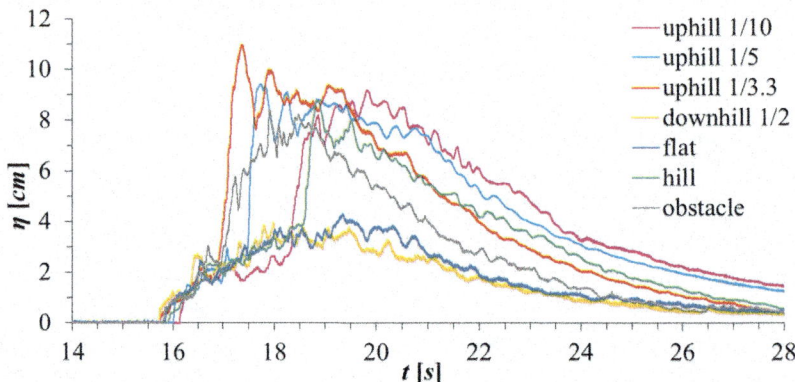

Fig. 6. Inundation depth η on the land side measurement point under supercritical wave

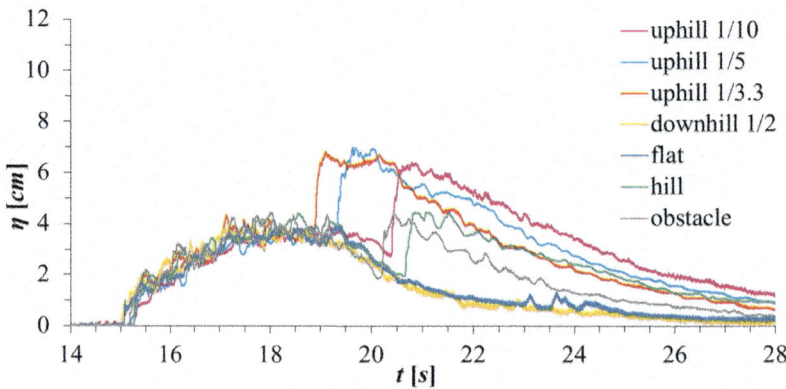

Fig. 7. Inundation depth η on the sea side measurement point under supercritical wave

Figure 8 and 9 show the inundation depth η at the measurement points with the subcritical flows by the vacuum pump type tank with releasing 3 solenoid valves and with the half-open inlet-valve of the tank, whose volume corresponds to 80 mm water level filling the water flume. In similar to the previous figures (supercritical flows), the reflected waves are shown in the cases with uphill slopes, a hill or an obstacle. However, the water level rises more slowly for long time than that of the supercritical flows. From the videos, the coming tsunamis have slow velocities and result in its stagnation. According to the figures, it is difficult to extract the pure reflected waves due to the fuzzy and continuous changes of the inundation depth.

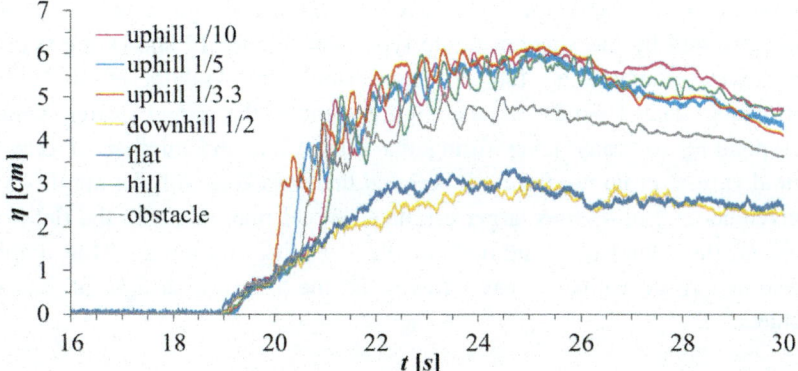

Fig. 8. Inundation depth η on the land side measurement point under subcritical wave

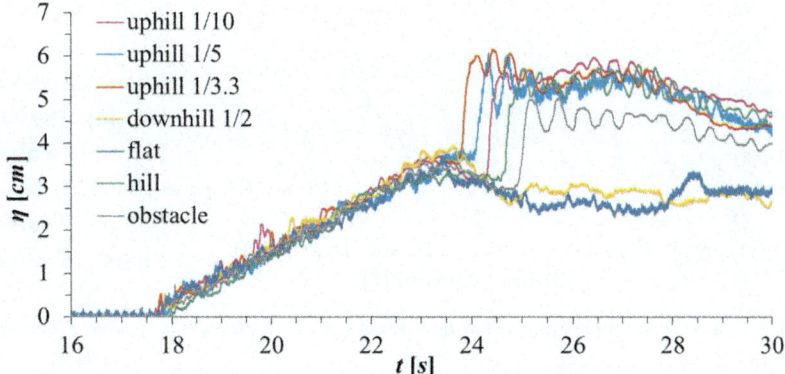

Fig. 9. Inundation depth η on the sea side measurement point under subcritical wave

Focusing on the number of the peaks of the inundation depth, it is found that Figure 6, 8 and 9 have a peak in their time series and differ from that of Asakura et al.[1], which has two peaks. As mentioned above, it is difficult to extract the pure incident waves in the figures, and thus, they have a problem to get the appropriate water-depth-coefficient α by the existing formulas based on Asakura et al.[1]. On the other hand, Figure 7 has a similar transition of the inundation depth to the Asakura et al.[1], and implies its applicability to the existing formulas.

Herewith, the measured (synthetic) waves in our experimental results are applied to the existing formula by Asakura et al.[1] in order to examine the influence of unexpected interminglement of the reflected waves. The target data is measured by the land-side water-level and current gauges under subcritical flows generated by the vacuum pump type tank. Figure 10 shows the tsunami wave forces estimated by the existing formula by Asakura et al.[1] in our experiments. The legends in the figure denote the target water volume (corresponding to water level filling the flume) and the number of released solenoid valves. From the figure, it is found that the cases with the uphill slopes are given about 2 to 4 times larger estimated forces than that with the flat floor, which is the standard condition in the existing formulas. This implies overestimation of tsunami wave forces in the existing design of onshore structures.

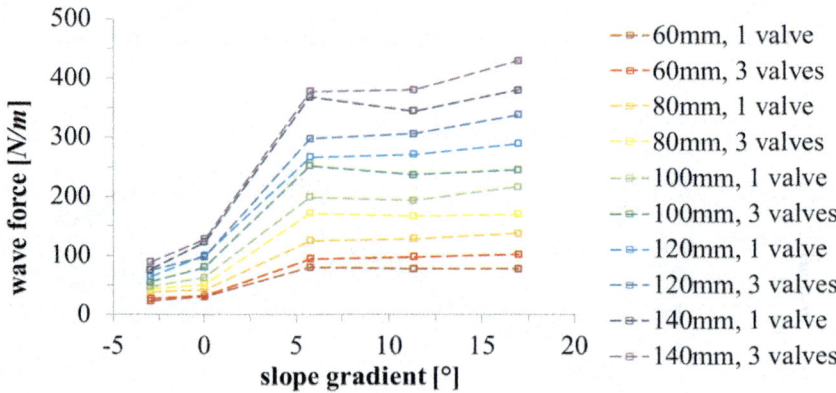

Fig. 10. Tsunami wave force estimated by an existing formula (Asakura et al.[1]) in our experiments

4. Wave pressure and inundation depth with the target wall

Figure 11 shows the time series of the inundation depth in front of the target wall in a wave pressure test under a subcritical flow by the vacuum pump type tank with releasing a solenoid valve and with a half-open inlet-valve. The water volume corresponds to 120mm water level filling the flume. The figure appends the wave pressure acting on the target wall, which are measured at 5 mm and 55mm in height from the floor. The figure shows a similar transition between the inundation depth and the wave pressure. The acting wave pressure would be estimated accurately as a hydrostatic pressure.

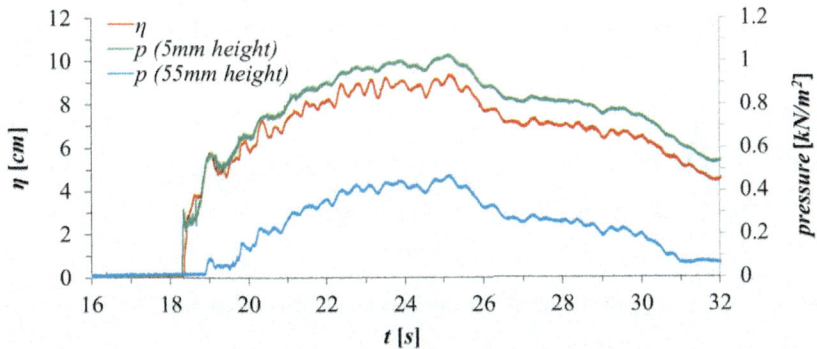

Fig. 11. Result of wave pressure test of subcritical flow

In similar to Figure 11, Figure 12 shows the inundation depth and the acting wave pressure in a wave pressure test under a supercritical flow by the vacuum pump type tank with releasing one solenoid valve and with a full-open inlet-valve. The water volume corresponds to 40mm water level filling the flume. The time series of the inundation depth and the wave pressure show different trends at the time with the first peak of them, which occurs immediately after the tsunami wave reaches to the target wall. The pressure shows a significant perturbation of pressure at the time. On the other hand, they show a similar trend at the following period with the so-called continuous pressure, which occurs after the peak of the pressure, so the continuous pressure would be estimated in accordance with the hydrostatic pressure. This gives an impressive fact that the actual inundation depth nearby the target structure would be a key to obtain the accurate evaluation of the wave pressure rather than the Froude number, which is widely applied with supposing the flat floor without the target structure in the existing studies. Here, as a next step, the quantitative relation between the maximum wave pressure p_{max} and the actual maximum inundation depth η_{max} in front of the target structure is examined. Figure 13 shows the dimensionless maximum wave pressure $(=p_{max}/\rho g \eta_{max})$ acting on the target wall at the continuous pressure period. This figure shows that the continuous wave pressure is about 1.0 to 1.2 times larger than the hydrostatic pressure of the maximum inundation depth in front of the wall under both subcritical flows and supercritical flows. It can be recognized that the tsunami wave force, including the reflected waves, can be comprehensively estimated by this way.

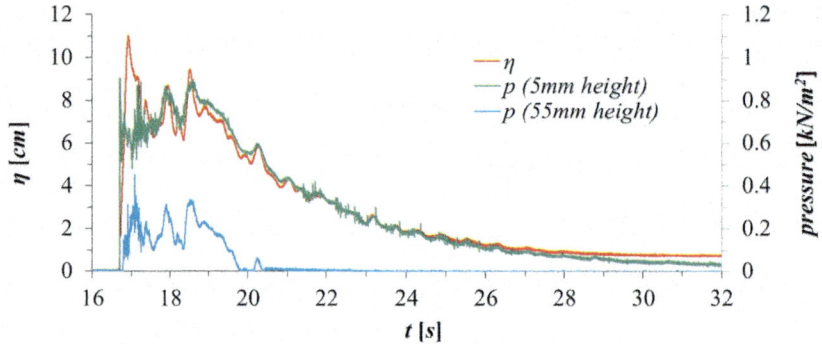

Fig. 12. Result of wave pressure test of supercritical flow

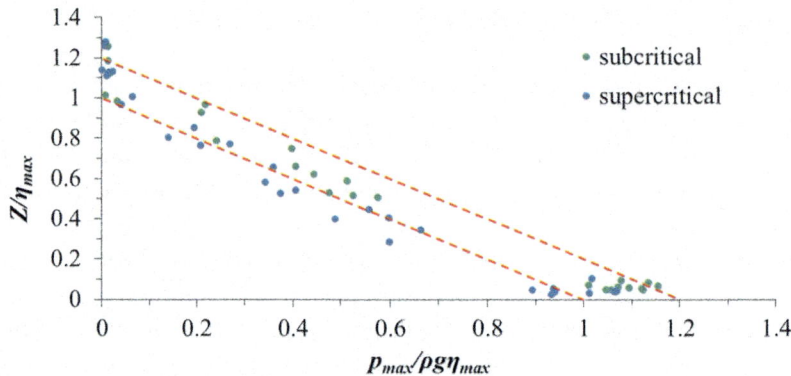

Fig. 13. Dimensionless maximum wave pressure of continuous pressure acting on the wall

5. Concluding remarks

In this study, we focused on the continuous tsunami wave pressures acting on a wall on the land. Considering the actual design of onshore structures with a tsunami simulation, the influence of a slope and an obstacle behind the target wall is examined through hydraulic experiments for more applicability of the estimation formula. From the experiments, it is newly proposed that the target walls also should be set in the pre-simulation of the tsunami, and further, the actual inundation depth of standing waves obtained by the pre-simulation should be applied to the formula to estimate the wave pressure, rather than the traditional Froude number on the flat floor without the structures. The main findings are as follows.

(1) It is difficult to extract the pure incident-wave component from the measured inundation depth if an uphill slope or an obstacle is close to the wall. Moreover, even if the uphill slope has enough distance from the target structure, supercritical flows would give overestimated inundation depth due to its large reflected wave. Consequently, the traditional estimation formulas have a possibility to bring about overestimation of the tsunami wave force in the actual design.

(2) In our experiments, the continuous wave pressures of the tsunamis get good agreement with 1.0~1.2 times hydrostatic pressure of each maximum inundation depth of the standing wave in front of the target wall for both subcritical flows and supercritical flows.

The inundation depth is obtained through a macroscopic tsunami simulation in the actual design of the structures. In the traditional models, the Froude number in their estimation formulas should be obtained by an additional tsunami simulation with omitting the target structures. On the other hand, our proposed procedure does not need such an additional procedure, and more befits the actual design of the structures.

Our experiments were performed with a relative small scale. For future work, the detail of the scaling effect should be considered.

Acknowledgments

We would like to thank Shingo Kawaguchi, Researcher of Port and Airport Research Institute, and Masatoshi Ikeuchi, student of Nagaoka University of Technology, for their kind supports in our experiments.

References

1. R. Asakura, K. Iwase, T. Ikeya, M. Takao, T. Kaneto, N. Fujii and M. Omori, Experimental Study on Wave Force by Tsunami Overflowing Seawall, *Proc. Conf. On Coastal Eng.*, JSCE, **47**, pp. 911-915 (2000).
2. T. Ikeya Y. Akiyama and N. Iwamae, On the Hydraulic Mechanism of Sustained Tsunami Wave Pressure Acting on Land Structures, *Journal of*

Japan Society of Civil Engineers, Ser. B2 (Coastal Engineering), **69**, 2, pp. I_816-I_820 (2013).

3. F. Kato, Y. Suwa, K. Fujita, H. Kishida, T. Igarashi, J. Okamura and Y. Hayashi, A method to Estimate Tsunami Setup in Front of Buildings, *Journal of Japan Society of Civil Engineers, Ser. B2 (Coastal Engineering)*, **68**, 2, pp. I_331-I_335 (2012).

Uplift and Overburden Pressure Acting on
Breakwater Caisson Under Tsunami Overflow

K. Suzuki[†] and K. Tsukasa

Maritime Structures Research Group, Port and Airport Research Institute,
Yokosuka, Kanagawa 239-0826, Japan
[†]E-mail: suzuki_k@pari.go.jp
www.pari.go.jp

Many caissons of breakwaters were slid or overturned due to tsunami overflow pressure caused by 2011 Tohoku earthquake. To prevent this sliding failure, the pressure estimation method under tsunami overflow was introduced in the new design guideline of breakwater against tsunami in 2015. In this guideline, the uplift and the overburden pressure are not considered, instead only buoyancy force acting on the caisson is considered. However, under tsunami overflow, the pressure difference between the bottom and the top of breakwater caisson, especially the caisson having a large parapet, can be extremely larger than the buoyancy force. In order to examine this excess uplift force, a series of hydraulic experiments were conducted. The experiment was conducted in an experimental flume in which the large pump was installed to produce tsunami overflow. Pressure gauges, water level gauges and velocity meter were installed at the top and the bottom of the caisson model. Through the experiments, it was clarified that the large uplift pressure and the small overburden pressure cause the upward force larger than the buoyancy force. This upward force reduces the stability of caisson, especially the caisson having the large parapet.

Keywords: Uplift pressure; Overburden pressure; Tsunami Overflow; Breakwater.

1. Indroduction

Many caissons of breakwaters were slid or overturned due to tsunami overflow pressure caused by 2011 Tohoku earthquake. To prevent this sliding failure, the pressure estimation method under tsunami overflow was introduced in the new design guideline of breakwater against tsunami in 2015 [1]. Figure 1 shows the pressure distribution under tsunami overflow written in the guideline. The front and the rear pressure are expressed as modified hydrostatic pressures. The modification factor α_f (α_r) is 1.05 (0.9) at the front (the rear). In the vertical direction, on the other hand, the guideline only accounts for the buoyancy force but not the forces due to the uplift and the overburden pressures acting on the caisson.

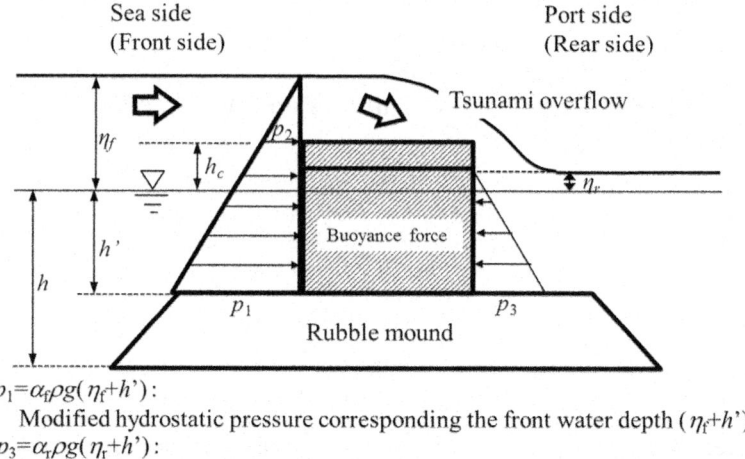

$p_1=\alpha_f\rho g(\eta_f+h')$:
    Modified hydrostatic pressure corresponding the front water depth (η_f+h')
$p_3=\alpha_r\rho g(\eta_r+h')$:
    Modified hydrostatic pressure corresponding the rear water depth (η_f+h')

Fig. 1. Pressure distribution and buoyancy defined by the design guideline of breakwater against tsunami (2015)

In the hydro static condition, buoyancy force is equivalent to the weight of the fluid that would otherwise occupy the volume of the object, i.e. the displaced fluid. The buoyancy force is equivalent to the pressure difference between the bottom and the top of the immersed object as shown in Figure 2(a), 2(c). However, under tsunami overflow, the pressure difference between the bottom and the top of breakwater caisson, especially the caisson having a large parapet, can be extremely larger than the buoyancy force as shown in Figure 2(d).

Sato et al. (2016) [2] estimated the overburden pressure p_o as a trapezoidal pressure distribution as shown in Figure 3. However, the shape of the water surface elevation at the top of caisson isn't a trapezoidal shape while the flow changes from subcritical flow to supercritical flow. Moreover, if the parapet is high and thin and overflow depth is small, mass of displaced fluid can be extremely less than the uplift pressure.

In order to clarify pressure distribution at the top and the bottom of the caisson which has a large parapet as shown in Figure 2(c) and 2(d), a series of experiments were conducted in this paper.

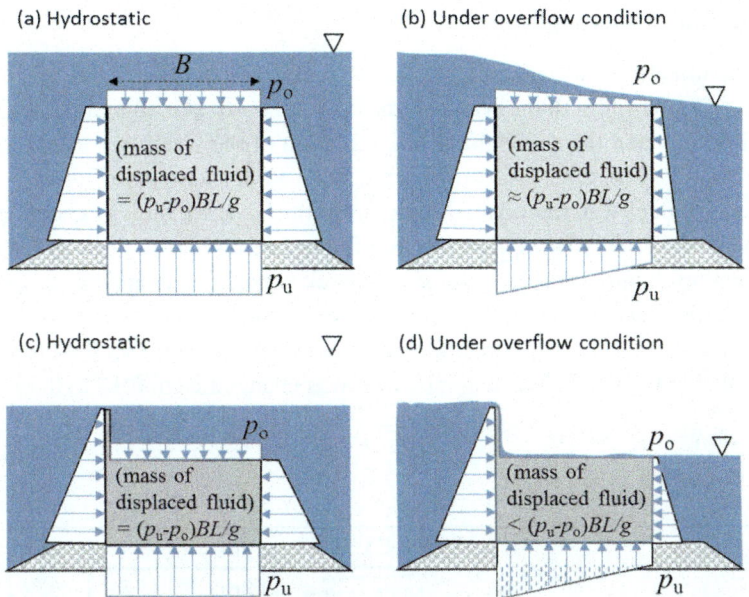

Fig. 2. Schematics of buoyancy and pressure distribution under static and overflow condition
(B is the width of the caisson, L is the length of the caisson)

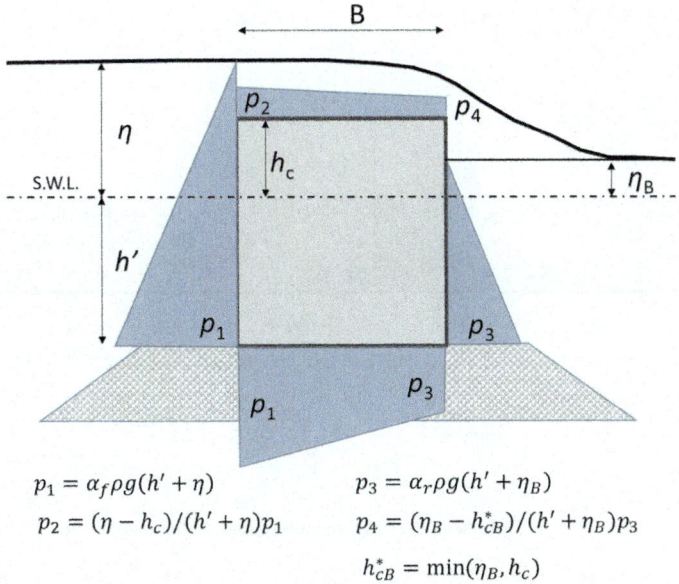

$$p_1 = \alpha_f \rho g(h' + \eta)$$

$$p_3 = \alpha_r \rho g(h' + \eta_B)$$

$$p_2 = (\eta - h_c)/(h' + \eta)p_1$$

$$p_4 = (\eta_B - h_{cB}^*)/(h' + \eta_B)p_3$$

$$h_{cB}^* = \min(\eta_B, h_c)$$

Fig. 3. Schematic figure of pressure distribution under overflow condition used by Sato et al.(2016)

2. Hydraulic Experiment

The experiments were performed using a flume with the length of 105m. As shown in Figure 4, the flume was separated by the vertical wall and the width of the main flume and the sub-flume, for circulation of the water, were set to 0.8m and 2.2m, respectively. As shown in Figure 4, tsunami overflowing on a breakwater model was reproduced under uniform flow, which was generated by a pump installed behind a weir in the main flume, and circulated through the main and secondary flumes. Pressure gauges, water elevation gauges and propeller-type current profilers were installed around breakwater model.

Initial water depth h_f (inundation depth η_f) was changed from 0.42 to 0.52m (from 0.044 to 0.144m). The height of weir was changed from 0.14 to 0.35m

(a) Plan view

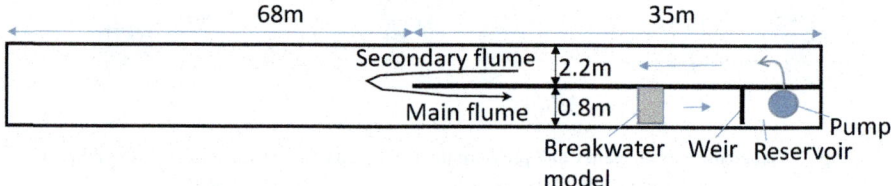

(b) Sectional view

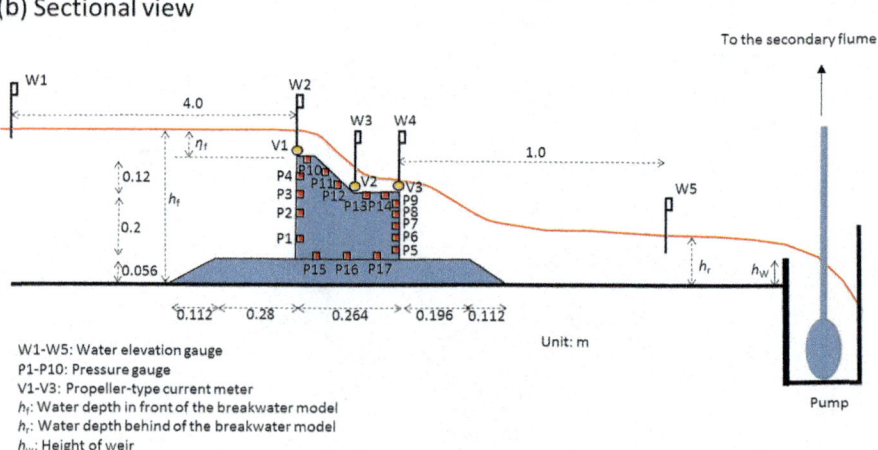

W1-W5: Water elevation gauge
P1-P10: Pressure gauge
V1-V3: Propeller-type current meter
h_f: Water depth in front of the breakwater model
h_r: Water depth behind of the breakwater model
h_w: Height of weir

Unit: m

Fig. 4. Details of the experimental flume and the model

3. Experimental Result

Figure 5 shows the time series of pressure, velocity and water surface elevation during overflow on the caisson. The pressure and the water surface elevation plotted in this figure are the changes from the initial condition. As water level decreased gradually after the pump started, the pressure behind, on, down the caisson decreased and the velocity increased on the caisson.

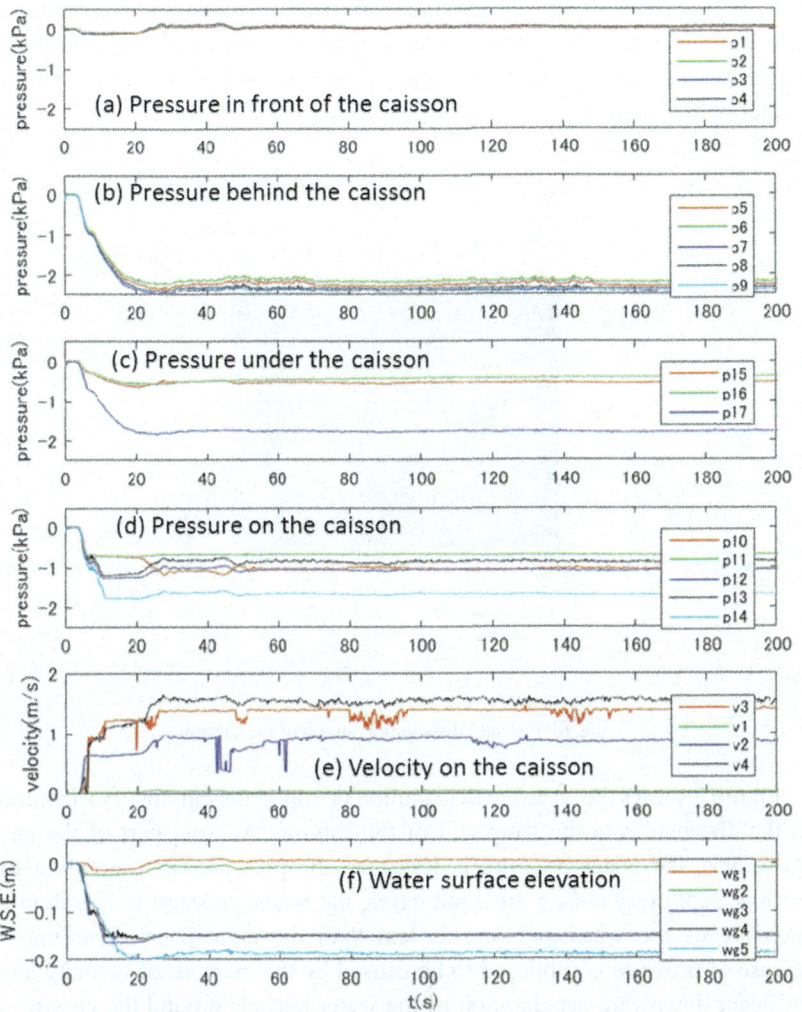

Fig. 5.   Time series of pressure, velocity and water surface elevation (The pressure and the water surface elevation plotted in this figure are the changes from the initial condition.)

822

Figure 6 shows the pressure distribution around the caisson when the current and pressure became stable. The pressure distribution in front and behind the caisson is almost static and triangle shape. In contrast, the pressure on the caisson is different from the pressure distribution estimated from the water surface elevation. The pressure at the front part (rear part) of the top of caisson measured by the pressure gauges is smaller (larger) than the pressure estimated from the water surface elevation. Decrease of pressure at the front part is supposed to be caused by the increase of velocity, the significant downward acceleration of the water particle around the curving flow and the occurrence of eddy. On the other hand, increase of pressure at the rear part is supposed to be caused by the collision of water on the caisson.

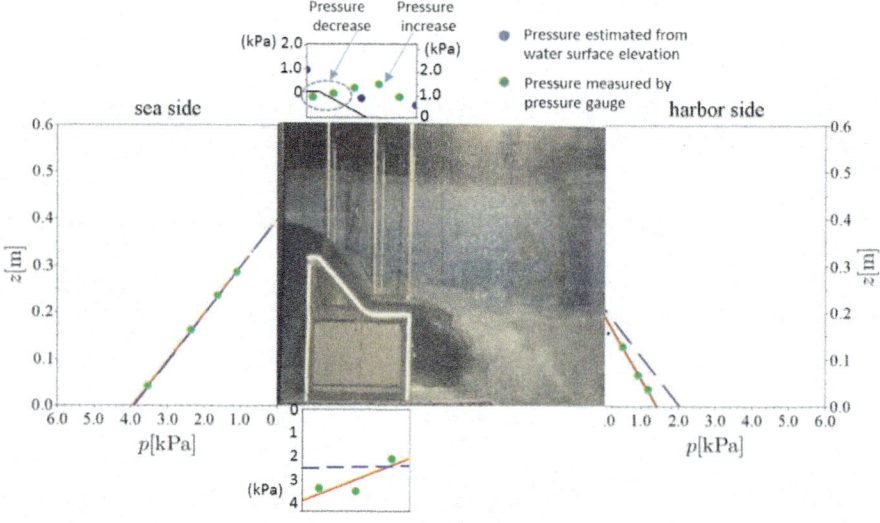

Fig. 6. Pressure distribution on top of the caisson

Figure 7 shows the pressure distribution on top of the caisson. Horizontal axis x is the distance from the front end of the caisson. At front part of the caisson $(0 < x < 0.6\text{m})$, the water pressure is less than the pressure corresponding to the overflow depth $(\rho g(\eta_f - h_c))$. In some cases, the water pressure is less than zero, which means the water pressure is less than the atmospheric pressure. This decrease of pressure is supposed to be caused by the increase of velocity and the significant downward acceleration of the water particle around the curving flow and the occurrence of eddy as mentioned before.

The large uplift pressure and the small overburden pressure cause the upward force larger than the buoyancy force. This upward force is supposed to reduce the stability of caisson.

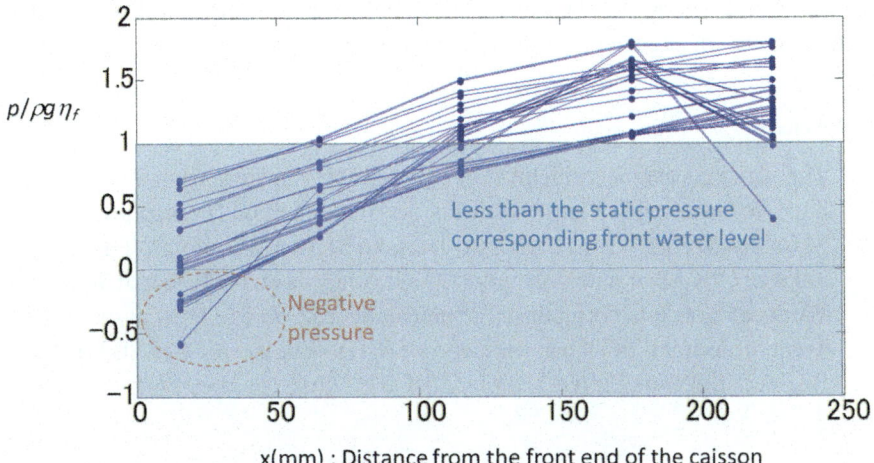

$p/\rho g \eta_f$

x(mm) : Distance from the front end of the caisson

Fig. 7. Pressure distribution on top of the caisson

4. Concluding Remarks

In order to examine this buoyancy force, a series of hydraulic experiments were conducted. The experiment was conducted in an experimental flume in which the large pump was installed to produce tsunami overflow. Pressure gauges and water level gauges, velocity meter were installed at the top and the bottom of the caisson model. Tsunami height in the experiment varies from 4 to 14 cm.

As experimental results, followings were clarified.

(1) The uplift pressure distribution is triangular shape whose front (rear) end pressure is almost as same as the static water pressure corresponding to the front (rear) water level.

(2) The front end overburden pressure is smaller than the static water pressure corresponding to the water level. The significant downward acceleration of the water prticle around the curving flow and the occurrence of eddy appear to be the main reasons of the observed pressure drop at the front end of the caisson.

(3) The large uplift pressure and the small overburden pressure cause the upward force larger than the buoyancy force. This upward force reduces the stability of caisson, especially the caisson having the large parapet.

824

Acknowledgments

We would like to express our sincere gratitude to Kenichiro Shimosako, Senior Director for Research of Port and Airport Research Institute and Masafumi Miyata, Chief of Port and Harbour Department, National Institute for Land and Infrastructure Management, for their support and insightful guidance.

References

1. The Japan design guideline of breakwater against tsunami: website; http://www.mlit.go.jp/kowan/kowan_tk5_000018.html. (in Japanese).
2. Masahiro SATO, Atsuo OMURA, Daisuke SHIBATA , Kotaro UEHARA , Takashi OIKAWA and Nobuyuki AOKI: Analysis on the Effect of Widening Works of Breakwaters against Tsunami-induced Seepage Flow in the Rubble Mound, Journal of Japan Society of Civil Engineers, Ser. B3 (Ocean Engineering), Vol.72, No.2, pp.I_533-I_538, 2016 (in Japanese).

Comparative Analysis of Non-Overtopping and Overtopping Design of Open-Deck Piers along A Storm-Tracked Coast

E.C. Cruz[†]   J.C.E.L. Santos*   M.E.L. del Rosario*

[†]*Institute of Civil Engineering, University of the Philippines,
Diliman, Quezon City 1101, Philippines*
[†]*E-mail: eccruz@upd.edu.ph*
www.upd.edu.ph

*AMH Philippines, Inc.,
Bahay ng Alumni Bldg, U.P. Diliman Campus, Quezon City 1101, Philippines
www.amhphil.com

For an open-deck pier of a coastal resort development, a comparison of 2 preliminary engineering designs involving the standard non-overtopping deck and the less conventional overtopping design is carried out. The common basis for horizontal siting and its results based on optimal prevailing waves are presented. The two design outputs are compared in terms of geometry and sizes of the pier's structural elements. The option of overtopping pier is also discussed relative to its storm tide overtopping risk and the expected pier maintenance plan. For this study coast, a non-overtopping pier design yields leaner pier geometry and smaller structural sizes. It is recommended that a non-overtopping pier design be always considered for a storm-tracked coastline.

Keywords: Overtopping, open pier, engineering, storm surge, return period

1. Introduction

With an archipelagic coastal morphology and more than 36,000 kilometers of coastlines, the Philippines crucially depends on ports to move passengers and cargo. It is then not a surprise that the country has more than 2,450 seaports all over the archipelago. A docking facility is the most important infrastructure of a port. A pier is a docking facility that is built typically perpendicular to the coastline so that one or two vessels can dock on either or both sides. Open-deck piers are designed with a deck slab supported by structural piles and framing so that waves can pass underneath the deck. The clear passing of waves below the deck and between the piles is important in order not to hamper the littoral and hydrodynamic processes (waves, tides, currents, sediments) within the foreshore area in which the pier is typically built. The open-deck pier type also does not adversely affect the natural circulation of foreshore waters.

In most cases of port master planning, engineering design criteria require the pier to be designed as non-overtopping in order to reduce the dynamic loadings on the pier structures. This criterion generally leads to longer piles and higher deck elevation which effectively reduces the pier's exposure to the natural coastal hazards. However, in other cases, decks of open-piers are vertically-limited to allow users to have an unobstructed view of the sea horizon from inland, such as in a beach pier.

A recent study of the causes of damage to the major ports of the country identified the inadequacy of freeboard as a major cause of structural damage and durability problems of Philippine ports[1]. Since a pier is the most common docking structure in the country's seaports, it is important to have a rational approach to the planning, siting, and preliminary engineering of the pier.

2. Pier's Optimal Alignment

At the planning stage, the most suitable location of a beach pier is typically determined based on the most frequent loading during the pier's operation, namely, during prevailing, non-storm conditions. Archipelagic coastlines are generally bounded by other islands, and/or by morphological features, such as bays and headlands, that provide some sheltering benefits against waves approaching from specific directions. In such case, the determination of the most suitable location and layout of the beach pier is based not only on wave conditions in deep water but also on the directions of prevailing wave approach.

For piers, waves are the primary natural hazards in terms of magnitude and frequency. Wind-generated waves are perpetual loads that all coastal structures must withstand. The location of a beach pier is optimally determined if it is in the least wave-agitated zone of a local coast. The most important function of the pier is to load or unload passengers and/or cargo. Hence, a landing area and mooring zones must be available to safely secure the vessel and allow the transfer of passengers. The location and layout of the pier are thus determined by: (1) required draft of the vessel, (2) wave height and directions, and (3) tidal range. This section discusses the methodology of determining the optimal location of the beach pier.

2.1. Open-deck pier concept

The beach pier will serve as the transfer facility from the airport to the island resorts around the airport. It will also provide on-deck commercial space at the pier end including restaurants for visitors and guests at the onshore accommodations, allowing vehicle access to these areas from land. The pier will also house docking, mooring and refueling facilities for boats, and a staging area for marine sports equipment. Hence, a floating pontoon system is included in the layout of the deck. The pier design concept is shown in Figure 1.

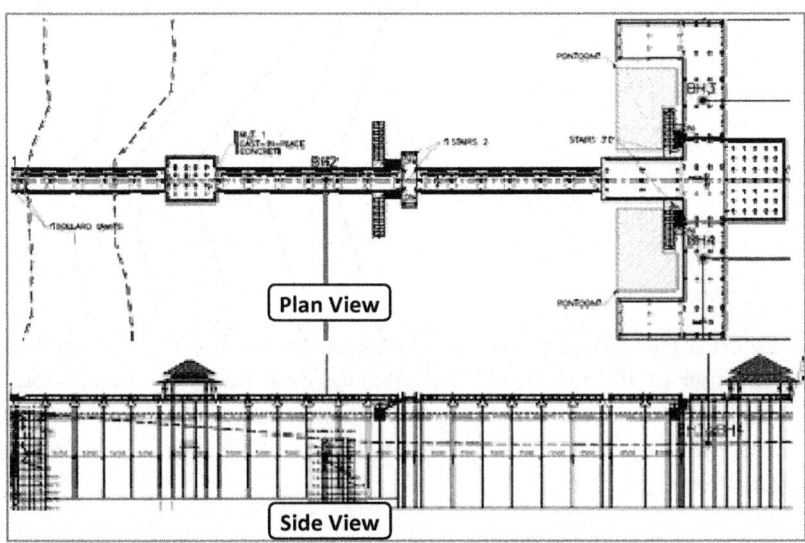

Figure 1. Plan form (top) and profile (bottom) design concepts of the open pier

2.2. *Prevailing waves*

Figure 2 shows the annual wind rose diagram at the nearest wind station of the national meteorological agency PAGASA[10] which summarizes the directional and frequency occurrence distribution of non-storm wind speeds in a typical year. These show that winds approach predominantly from the northeast and southwest and, with lower frequency but highest speeds, from the west. For the two-season climatology of the country, Figure 2 also shows the wind roses for the representative northeast monsoon month of January and for southwest month of August. These wind speeds are processed together with the effective fetches for each direction to yield a hindcast of deep water wave conditions[2] as summarized in Table 1, which indicate non-storm offshore waves are 3 to 4.4 m high, with periods of 6.9 to 8.1 s.

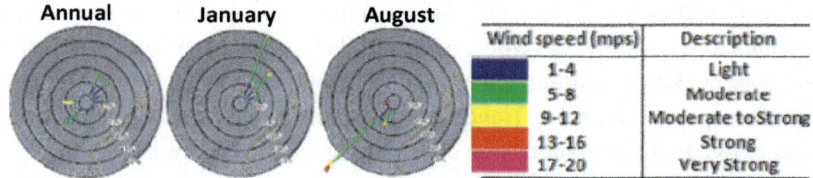

Figure 2. Wind rose diagrams for a typical year and representative months of 2 seasons

Table 1. Prevailing deepwater wave conditions

Dir.	Wind speed m/s	Annual frequency %	Fetch (km)	Deepwater waves	
				Hs (m)	Ts (s)
N	13- 16.9	0.1	231	4.38	8.07
NNW	9 - 12.9	0.2	272	3.15	7.10
NW	9 - 12.9	0.1	273	3.15	7.11
W	9 - 12.9	0.7	230	3.00	6.88

2.3. Numerical analysis of local waves

A nearshore wave model is used to determine the local waves at the shallow water location of the pier based on offshore deepwater wave conditions. A nonlinear, dispersive, phase-resolving wave model valid from deep water to the foreshore area is necessary. In this study, the following nearshore wave transformation model is used:

$$\frac{\partial \eta}{\partial t} + \nabla \cdot \left[\mathbf{u}(h + \eta)\right] = 0 \tag{1}$$

$$\frac{\partial \mathbf{u}}{\partial t} + (\mathbf{u} \cdot \nabla)\mathbf{u} + g\nabla \eta + \frac{h^2}{6}\left(\nabla \cdot \frac{\partial \mathbf{u}}{\partial t}\right) - (\tfrac{1}{2} + \gamma)h\nabla\left(h\nabla \cdot \frac{\partial \mathbf{u}}{\partial t}\right)$$
$$- \gamma g h \nabla\left[\nabla \cdot (h\nabla \eta)\right] + \mathbf{F}_b + \mathbf{F}_s + \mathbf{F}_w = 0 \tag{2}$$

where $\eta(x,y,t)$ is the water surface displacement, $\mathbf{u} = (u,v)$ the depth-averaged fluid particle horizontal velocity vector, (x, y) the horizontal coordinates, t time, $\nabla = (\partial/\partial x, \partial/\partial y)$ the horizontal gradient operator, γ the frequency dispersivity extension factor, and g the gravity acceleration. Eqs. (1) and (2) are based on the more general model equations for nonlinear wave transformation on a porous seabed [3]. In the momentum equation (2), $\mathbf{F}_b$ represents the wave breaking energy dissipation, $\mathbf{F}_s$ the energy damping by structures due to momentum exchange particularly near structures' corners, and $\mathbf{F}_w$ the bottom friction resistance due to surface roughness. The basis and formulation of these additional terms [4] and applications to wave penetration analyses in coastal harbors [5] are discussed previously.

All offshore wave conditions have been translated to nearshore wave heights through simulative analyses, using the wave model above, for two stationary tide levels, namely, MLLW and MHHW. It is necessary to consider both low and high tide levels to account for wave breaking, whose location migrates with the tide levels and thus influence the maximum wave height. Figure 3 shows the simulated wave heights at MLLW and MHHW levels for the first two cases in Table 1, which came out to be the critical cases. The synthesis of all offshore wave approach directions (note: not all cases shown in Table 1) indicated an optimal siting, i.e. the least agitated foreshore zone up to where the beach pier extends to -4 m depth, as laid out in Figure 1. At low tide, it is seen that local waves break further offshore, leading to smaller waves around the pier.

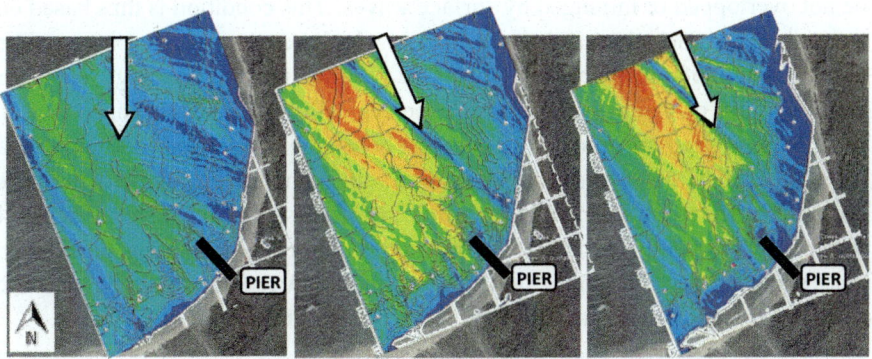

Figure 3. Local wave heights due to offshore waves from north at high tide (left), from NNW at high tide (middle) and NNW at low tide (right)

Table 2. Results of optimal siting of pier

Function	Beach pier, passenger unloading
Pier type	Open (piled)
Pier structure type	RC girder-beam, slab pier deck
Main Pier length	181.1 m
Total pier length	215.45 m
Offshore depth at MSL	~ -4.0 m
Pile length	-17.175 m (to embedment)

3. Vertical Siting of Pier Deck

Once the location, layout and orientation of the beach pier are determined, the preliminary engineering determines the proportions of the main structures. These include the pile sizes, layouts, lengths, as well as the plan areas of the superstructure. Due to the dependence of structural member size on the pile length, these preliminary proportions all depend on the vertical siting of the pier deck. The vertical siting of an open beach pier is usually based on a non-overtopping design condition of the pier deck. Pier overtopping is typically based on the extreme conditions of hazards to which the infrastructure is expected to be exposed. Hence, preliminary engineering should account for the natural hazards that could cause historical or potential overtopping at the selected location. This section discusses a methodology of determining the vertical siting of a beach pier which is tracked by typhoons.

3.1. *Storm tides*

Preliminary engineering of the pier structure starts with the vertical siting of the pier deck. The design concept requires that the deck or its supporting structures be not overtopped or impinged by surface waves. This condition is thus based on the highest displacement of the sea surface, which occurs during typhoons. The determination of the pier deck underside follows the methodology used for vertical siting of cargo piers [6], which are typically located along open coasts, such as those used for oil depots. Since the beach pier is an open type supported by marine piles, the vertical siting is based on typhoon conditions, where the combined effects of storm tides and stormy waves are considered. Figure 4 illustrates the quantities that determine the minimum vertical siting of an open pier, namely, the astronomic tide, the typhoon-induced storm surge, and the storm wave's crest elevation from the storm tide level.

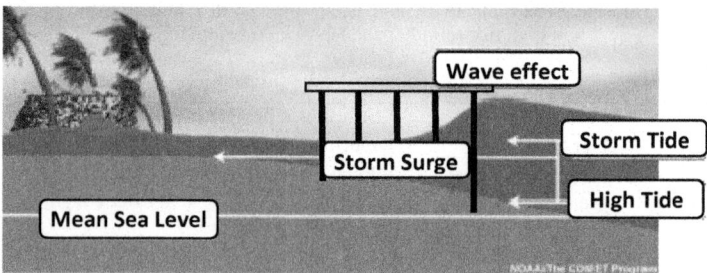

Figure 4. Components of vertical siting

3.2. *Typhoon susceptibility*

A study of historical typhoons that traversed a 150-km radius around the project coast yielded several potentially critical typhoons in the last 30 years, whose tracks and sustained wind speeds are plotted in Figure 5. Table 3 summarizes the lifetime meteorological characteristics of 4 critical typhoons in terms of wind speed and closest distance to the site. Typhoon Mike 1990 is considered the

largest (R$_{max}$ = 185 km) and strongest (V$_{max}$ = 39 km), while relatively weaker
Nelson tracked closest to the site.

Table 3. Characteristics of critical historical typhoons

Track	Typhoon/ Local name	Yr/mo/day	V$_{max}$ (mps)	R$_{max}$ (km)	P$_c$ (hPa)
1	Sarah/ Trining	1979/10/08	33.4	37	985
2	Mike/ Ruping	1990/11/13	38.6	185	960
3	Zack/ Pepang	1995/10/29	30.8	74.1	980
4	Nelson/ Bising	1982/3/27	25.7	0	990

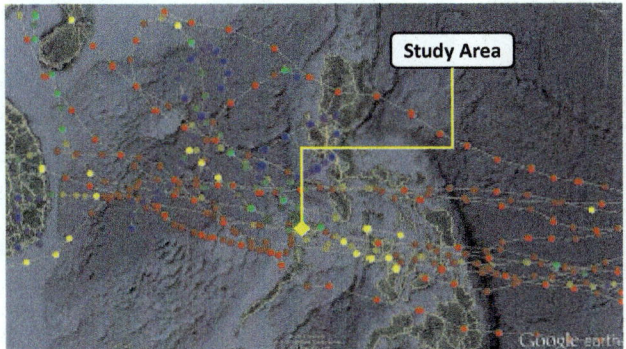

Figure 5. Tracks of historical typhoons

Table 4 summarizes the parameters of the 3 most critical typhoons above (Sarah
was found not critical). The parametric prediction method [2] is applied to
calculate the cyclone-induced wave conditions in offshore deep water, as
summarized in Table 3, and a local computational domain starting at the critical
point of approach was taken to simulate the propagation and transformation of
storm waves in the study coast's nearshore region.

Table 4. Storm wave conditions in deep water

Typhoon	Wave Height (m)	Wave Period (s)	Wave direction
Mike	5.1	12.34	WNW
Zack	4.0	10.93	WNW
Nelson	3.15	9.7	NW+15°

3.3. *Vertical siting of pier deck*

Figure 6 shows the simulated wave heights and sea surface snapshots due to
typhoons Mike and Nelson. Wave energy concentration with 2-m waves can be
seen around the pier. By computing the storm tide, i.e. astronomic tide plus

storm surge, and superimposing the wave effect, i.e. wave crest elevation and wave set-up, from the stormy wave simulation results, the required non-overtopping Pier Deck Elevation (PDE) is determined for each of the 3 historical typhoons, as summarized in Table 5. It is seen that the historical storm tide is highest under Mike (1.94 m), while the wave effect is greatest under Zack (1.34 m). The resulting minimum required PDE is governed by Mike at MTL+3.13m.

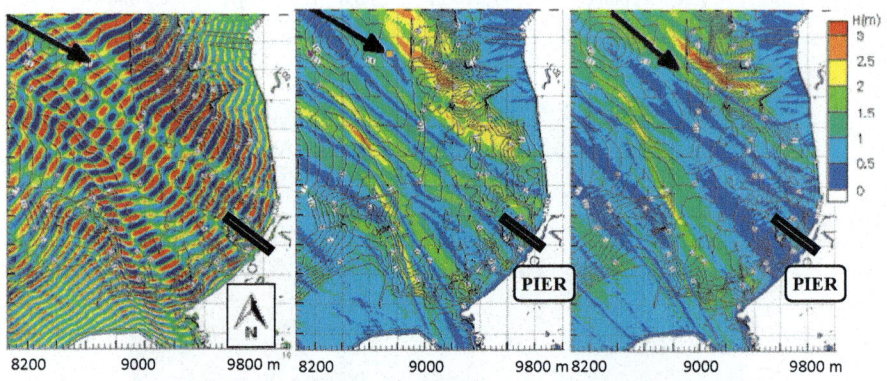

Figure 6. Nearshore wave field snapshot (left) and wave heights (middle) due to Typhoon Mike 1990; storm wave heights due to Typhoon Nelson 1982 (right)

Table 5. Synthesis of required pier deck elevation (units: m)

Typhoon	Astronomic Tide	Storm Surge	Wave crest elev.	Wave setup	Required PDE (m)
Mike	0.54	1.40	1.20	-0.02	**3.13**
Zack	0.54	0.48	1.36	-0.02	**2.37**
Nelson	0.54	0.59	0.81	-0.01	**1.93**

4. Comparison of Non-Overtopping and Overtopping Design of Pier Deck

Engineering of the pier structure depends considerably on the acceptable level of risk for storm tide overtopping. By adopting the historically highest storm tide level as the PDE, the risk of overtopping of the pier deck is minimized. Using a lower than the historical non-overtopping PDE would lower the reliability of pier design. An overtopping pier design may be a viable option in some cases such as beach piers due to the aesthetical constraint posed by an elevated pier against the view of sea horizon from shore.

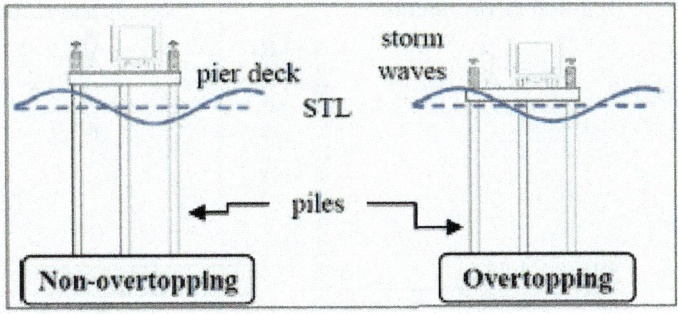

Figure 7. Illustration of non-overtopping and overtopping piers

4.1. *Evaluating storm tide overtopping*

An overtopping pier design condition (see Figure 7) may satisfy the vertical siting constraint but will have the following disadvantages: (1) reduced return period; (2) requires bigger structural elements due to the additional loadings such as wave in-deck forces (Tirendelli et al., 2003) and uplift forces (Gaeta et al., 2012); (3) imposes a pier repair and maintenance program; and (4) significantly increases its trapping action of littoral materials due to more closely spaced and bigger piles.

An overtopping PDE is evaluated by computing the exceedance probability of a lower vertical siting. Since the storm tides and wave effects are driven primarily by the wind, a frequency analysis of wind speeds of all historical cyclones that tracked within 150-km radius around the project beach was undertaken. The online database for maximum wind speeds of Japan Meteorological Agency [7] was used, drawing about 37 annual maximum speeds of 55.6-167.7 kph over 42 years.

Regression plots and coefficients of determination were used to determine the best probability density functions. The closest 2 plots of cumulative density function or CDF, indicating the non-exceedance probabilities, are shown in Figure 9. It is seen the Log-Normal PDF yields the best fit in the low-speed range, although the Weibull PDF yields the best overall fit with a coefficient of determination of 0.97, from which the return periods of wind speeds of the 3 critical typhoons, summarized in Table 7, were computed. Thus a non-overtopping PDE of (MTL+3.13m) will be exceeded every 12.8 years on the average, while an overtopping PDE of about (MTL+1.73m) will be overtopped about every 2.2 years.

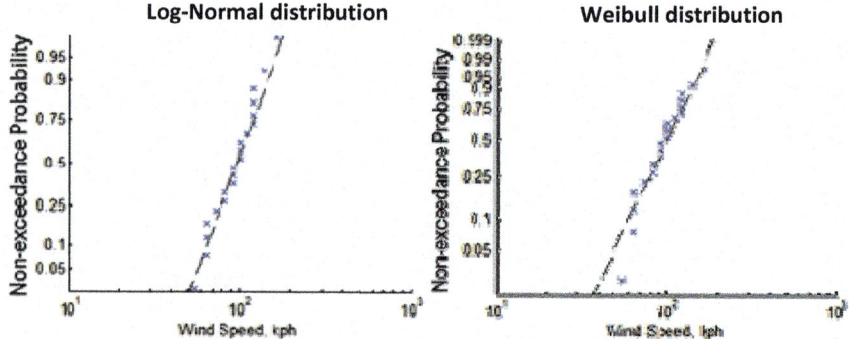

Figure 8. Plots of storm tide non-exceedance probability functions

Table 6. Computed return periods of vertical siting

Typhoon	Vertical Siting (m from MTL)	Return Period (years)
Mike/ Ruping	3.13	12.8
Zack/ Pepang	2.37	4.4
Nelson/ Bising	1.93	2.2

4.2. *Overtopping pier*

Structural computations were carried out to determine the required sizes of structural elements of the beach pier based on two options for the PDE: (1) Pier deck at non-overtopping elevation (MTL+3.13m); and (2) Pier deck at an acceptable elevation of MTL+2.0m. Table 7 summarizes the results. In general, Option 1 yields bigger sizes of the piles (0.35 m versus 0.45 m) but longer spacing of the piles (2.2m versus 1.95 m) against Option 2. Pile girders are deeper and deck slabs are thicker in overtopping PDE than overtopping ones, due mainly to wave in-deck and uplift forces as additional cycling loadings on the structural elements. Finally, the pile density comes out higher (0.16 versus 0.06 piles/m^2) in overtopping piers, which entails a higher construction cost not only for the marine foundation but for the superstructure as well.

Table 7. Computed pier structure proportions

Design output	Non-overtopping PDE	Overtopping PDE
Deck elevation	MTL + 3.13 m	MTL + 2.00 m
Pile length*	0.75 m to 4.5 m	4.5 m to 7.3 m
Pile spacing	2.2 m (along pile cap)	1.95 m (along pile cap)
Pile diameter	0.3 m and 0.35 m	0.4 m and 0.45 m
Piles density	~0.08 pile/m^2	~0.16 pile/m^2
Girder depth	0.95 m	0.95 m and 1.2 m
Slab thickness	0.35 m	0.35 m and 0.50 m

*length from seabed level to underside of pile cap

5. Conclusions

An open-deck pier requires that the structure be located in the wave-optimal site of the coast. Optimal horizontal siting leads not only to the least agitated wave conditions during normal operations in the port, but to the least construction cost of the pier structure as well.

Regardless of how the location and layout of the beach pier are determined, its preliminary engineering should be based on sustainable loading condition, i.e. non-overtopping pier deck. This vertical siting accounts for storm tides and storm wave heights during historical typhoons in order to avoid impulsive and cyclic wave loadings that lead to a significant increase in cost of the pier.

If overtopping conditions cannot be adopted due to aesthetic or environmental constraints, an evaluation of overtopping risk through the return period of the storm tides should be undertaken, consistent with the acceptable overtopping risk level for the pier design.

For the study area discussed here, an overtopping design leads to bigger structural members and a reduced reliability due to lower wave tranquility, shorter return period, and the necessary repair after historical typhoons.

References

1. Castaneda, D.C.S., Batin, K.G.F., Perdiguerra, M.R.G., Diola, N.B., and E.C. Cruz, "Analysis of the deterioration of various reinforced concrete sea ports in the Philippines using actual field inspection data", *Proc., 5th Engineering Research and Development for Technology (ERDT) Conf.*, 1-5, 2010.
2. *Coastal Engineering Manual* . United States Army Corps of Engineers, 2005.
3. Cruz, E.C., M. Isobe, and A. Watanabe, "Boussinesq equations for wave transformation over porous beds", *Coastal Eng.*, Vol.30, Nos.1-2, 125-156, 1997.
4. Cruz, E.C. and T. Aono, Simulation of nonlinear wave field induced by partially reflective harbours. *Proc., Joint 13th Australasian Coastal and Ocean Engineering Conference and 6th Australasian Port and Harbour Conference*, Vol.2, 947-952, 1997.
5. Cruz, E.C.,Wave climate studies in coastal harbors. *Proc., 33rd Phil. Inst. Civil Engineers National Convention*, Mandaue City, Cebu, 27-29 Nov 2007, WRE 1-10, 2007.
6. Cruz, E.C. and R.A.C. Luna, A methodology for rational vertical siting of marine infrastructures - application to the preliminary engineering of a

power plant along a typhoon-tracked seacoast. *Proc., National Midyear Convention and Technical Seminar*, Phil. Inst. Civil Engrs., Baguio City, 2014 June 6-7, 1-7, 2014

7. Digital Typhoon: Typhoon Images and Information (2001-2014), Kitamoto Asanobu/National Institute of Informatics (NII) http://agora.ex.nii.ac.jp/~kitamoto/.

8. Tirindelli, M., G. Cuomo, W. Allsop, A. Lamberti, Wave-in-deck forces on jetties and related structures. *Proc., 13th International Offshore and Polar Engineering* Conference, Honolulu, Hawaii, USA, May 25–30, 2003.

9. Gaeta, M.G., L. Martinelli, A. Lamberti, Uplift forces on wave exposed jetties: scale comparison and effect of venting. *Proc., Coastal Engineering Conf.,* ASCE, 2012.

10. Philippine Atmospheric, Geophysical and Astronomical Services Administration (PAGASA, 2010) Wind rose analysis 1970-2010 Pagasa Island, Palawan.

Dynamic Analysis of Different Configurations of Offshore Floating Wind Turbine

Akhila Daranikota and D. Karmakar[†]

Department of Applied Mechanics and Hydraulics,
National Institute of Technology Karnataka Surathkal, Mangalore – 575025, India
[†]E-mail: dkarmakar@nitk.edu.in
www.nitk.ac.in

The dynamic response of different types of spar and semi-submersible type floating offshore wind turbines are investigated subjected to wave and wind loads in operational conditions. The study is performed for the NREL 5MW wind turbine supported on spar and semi-submersible platforms. The study includes the analysis on the wave interaction with supporting structure and determination of the time-domain dynamic response of offshore floating wind turbine. The hydrodynamic coefficients added mass, damping, excitation forces and Response Amplitude Operators (RAO's) of the different motions of the floating platform are analyzed for different wave heading angles. The coupled dynamic analysis is performed for both regular and irregular waves and the coupled response of platform motions, tower base forces and moments for different wave and wind load conditions is analysed. The present study focuses on the overall performance of the different types of spar and semi-submersible type offshore floating wind turbines.

Keywords: Offshore Wind Turbine; aero-servo-hydro-elastic simulation; coupled dynamics; RAO.

1. Introduction

Offshore wind energy has a broad perspective of application as a green pollution free-decay application. In the offshore region the wind resources are abundant and due to the limitation of technology and cost the fixed structures are restricted to shallow waters only. Wind power is one of the fastest growing energy technologies and it appears that it will become a major generator of electricity worldwide. Floating wind farms located far out in the deep sea will draw on more wind than land-based wind farms in the near future [1]. Among many other marine renewable resources wind driven surface waves contain a great amount of energy. The wave energy level is expressed in terms of power per unit length and the typical values of good offshore locations ranges between 20 Kw/m to 70

[†] Corresponding author: Tel: +91-824-2473319, Fax: +91-824-2474039

Kw/m in moderate to high latitudes [3]. Currently many floating wind turbine concepts are evolved. In the offshore region, many floaters concepts have been proposed for the availability of abundant wind resource. Among different concepts Spar, Semi-submersible and TLP type floater concepts plays an important role in the generation of offshore wind energy.

Numerous floating concepts are available for offshore wind turbines. The research studies on different floating wind turbine concepts have been described in Wang et al. [5]. The investigations have been done on the floating structures comparing different types of floater concepts. The spar, semi-submersible and tension-leg platform type is compared with VAWTs which were originally designed for HAWTs in Borg et al. [6]. The study suggest that the spar was not able to sustain the aerodynamic loads in yaw and TLP mooring could not restrain the platform surge and sway. The mooring system should be designed particularly to restrain the surge, sway and yaw motions. The long-term analysis is performed for spar and semi-submersible comcepts in Bagbanci et al. [2]. North Atlantic wave data is considered for the long-term probability distribution in the case of surge, heave and pitch motions with side-to-side, fore-aft, and yaw tower base bending moments was analyzed. The surge, heave and pitch motion amplitudes are observed higher in case of semi-submersible floater.

Zhenya et al. [7] studied the dynamic response analysis for Hywind spar, WindFloat and NREL TLP foundation. In the case of spar platform, due to the deep draft its heave and lateral drift motion are smaller and pitch motions is higher due to small water plane area. The main challenge of semi-submersible platform is that it is more sensitive when wave height is higher, but the cost is less compared to spar and TLP platfporm. TLP can provide stiffness to decrease heave motion but it is strictly restricted to water depth. If the depth increases, the weight of the tendons increases and the design construction in complex and cost of installation is higher. Goupee et al. [8] has performed the model testing on spar-buoy, semi-submersible and TLP-type which is supported by NREL 5MW wind turbine. In wave loading the spar exhibits small surge response while TLP shows smallest pitch response and the semi-submersible is in between in spar and semi-sumbersible. In wind loading, for a TLP-type structure the wind loads increases pitch response. For spar and semi-sumbersible the operating wind turbine significantly dampens the response of the structures at pitch natural frequency. In case of semi-submersible it dampens at surge natural frequency.

The fully coupled dynamic response of a TLP floater is carried out by Ramachandran et al. [9]. The aerodynamic loads were studied with unsteady BEM theory and hydrodynamic loads are carried out using Morison's equation. The coupled dynamics is performed using an advanced aero-elastic code, Flex 5. In the present study, the dynamic analysis has been carried out for the OC3Hywind spar, DeepCwind semi-submersible and MIT/NREL TLP type floaters. These floaters are supporting the NREL 5MW Wind turbine.

2. Description of Three Floating Wind Turbines

At present, number of offshore floating wind turbine concepts are developing rapidly. The main offshore floating wind turbines are spar, semi-submersible, TLP and barge systems. In this paper spar, semi-submersible and tlp type wind turbines concepts are described. The three floating wind turbines that are considered in the present study are Hywind spar, DeepCwind semi-submersible and TLP type which is developed by National Renewable Energy Laboratory. The spar floater used in the study is Hywind spar [2] which is designed by Norway Hydro Oil & Energy Company. It consists of long and slender cylinder which is made of steel and concrete. The cylinder is filled with ballast of gravel and water to keep the centre of buoyancy above the centre of gravity to make the wind turbine float in the sea and stays upright. It has good stability because of small cross-section and deep draft. The catenary mooring lines are provided to prevent drifting of the structure. The fair leads are located at the depth of 70.0 m below MSL and the radius of fairleads are 5.2 m from the platform centerline. The angle of mooring lines is 120 degrees. The draft of the floating foundation should be high for stability and to reduce the heave motion. For the mooring system to be effective adequate keel to seabed vertical clearance is necessary.

Table 1. Specifications of the Three Floater Concepts

	Spar	Semi-submersible	Tension-Leg Platform
Draft	120 m	20 m	47.89 m
Basic Size	9.4 m	10 m	18 m
Water Depth	300 m	300 m	300 m
Platform Mass	7,466,330 kg	3,852,180 kg	8,600,410 kg
Displacement	8027.77 m³	13917 m³	12179.6 m³
Number of mooring lines	3	3	8
Length of mooring line	895.38 m	835.3 m	151.730 m
Diameter of mooring line	0.09 m	0.0766 m	0.127 m

The semi-submersible floater used in the study is OC4 DeepCwind Semi-submersible which is developed by the National Renewable Energy Laboratory and the draft of the platform is 20 m. It is considered to be buoyancy-stabilized because rotational displacements induce large buoyant-restoring forces from the volumes of water that are displaced. The platform is made up of three offset columns with larger diameter lower bases, one center support column for the turbine, and a series of horizontal and diagonal cross bracing. The 1.6 m diameter cross bracing consists of two sets of three pontoons connecting the outer columns to the center column, and three diagonal braces connecting the top of the outer column to the bottom of the center column. The mooring system used is catenary mooring lines. 3 mooring lines are used and the angle between the mooring lines is 120 degrees.

Table 2. 5MW NREL Wind Turbine Specifications

Rating	5 MW
Rotor Orientation, Configuration	Upwind,3 blades
Control	Variable speed, Collective pitch
Drivetarian	High speed, Multiple stage gearbox
Rotor, Hub Diameter	126 m,3 m
Hub Height	90 m
Cut-in Rated, Cut-out Wind Speed	3 m/s, 11.4 m/s, 25 m/s
Cut-in, Rated Rotor Speed	6.9 rpm, 12.1 rpm
Rated tip speed	80 m/s
Overhang, Shaft Tilt, Precone	5 m, 5^0, 2.5^0
Rotor Mass	110,000 kg
Tower Mass	347,460 kg
Nacelle Mass	240,000 kg
Coordinate location of Overall CM	-0.2 m, 0.0 m, 6.4 m

The Tension-leg platform used in the study is developed by National Renewable Energy Laboratory(NREL), with four spokes. The draft of the platform is 200 m. To keep the structure stable it used tensioned mooring cables as its buoyancy is greater than gravity. Sice the motions are restrained by the tendons, the most favourable response is expected in vertical motions including heave and pitch motions. These pre tensioned mooring lines contribute to a drastic reduction in the heave movement and increasing the system's horizontal stability. The main advantage lies on its low cost. 8 mooring lines are used with the diameter of 8m. TLP is positive buoyancy class platform so the tension leg is very sensitive to the topside weight and the load eccentricity ratio. The wind turbine used in the analysis is the NREL-5MW offshore baseline wind turbine model. The same wind turbine is use for all the platforms. It is because it is estimated that a 5MW wind turbine is the minimum power that a deep floating wind turbine must have in order to be cost effective to build. NREL (National Renewable Energy Laboratory) has studied conceptual versions of this size wind turbine and modelled it using the computer code FAST (Fatigue, Aerodynamics, Structures and Turbulence). Table 1 summarizes the main dimensions of the three structures.

The wind turbine used in the study is the NREL 5-MW Offshore Baseline Wind turbine. The properties of the wind turbine are described in Table 2.

3. Numerical Models and Methodology

In order to perform the fully coupled wave-wind-induced analyses, the analysis tool FAST developed by NREL is used. Initially the wet surface of the platform panel model is created. The geometric modelling of the spar, TLP and semi-submersible platforms below the design water lines were obtained as in Figure 1(a,b,c). The model surface is subdivided into patches, and all the patches represent the wetted surface. In order to provide a better continuation of the body surface discretized by these patches, a set of small elements called panels are defined with these patches. The panel size is modified depending on the accuracy requirements.

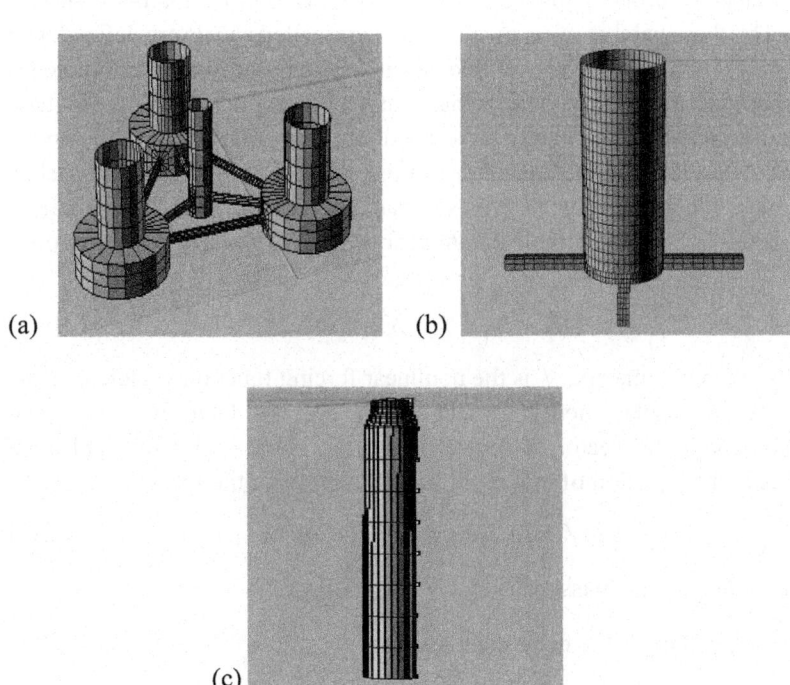

(a)   (b)

(c)

Fig. 1: Modelling of (a) DeepCwind semi-submersible, (b) Tension-Leg-platform (TLP) and (c) Spar type offshore wind turbine platform.

After an appropriate convergence study with regard to the size of the panels of the wet surface of the platform, hydrodynamic analysis in WAMIT is performed for the calculation of hydrodynamic coefficients in frequency domain. These coefficients are added mass, radiation damping and excitation wave loads. Hydrodynamic loads are calculated using a hybrid potential Morison equation approach, wherein the inertial and radiation forces are predicted using potential theory, and the viscous force is approximated by the Morison equation. The hydrodynamic analysis of six degrees of freedom is analyzed using WAMIT. The added mass, damping coefficient, hydrostatic forces, Response Amplitude Operators (RAO's) for different degrees of freedom are analyzed and compared.

3.1. Equation of Motion

The model surface is subdivided into patches, and all the patches represent the wetted surface. In order to provide a better continuation of the body surface discretized by these patches, a set of small elements called panels are defined with these patches. The panel size is modified depending on the accuracy requirements. The coupled dynamic analysis is carried out for platform RAO's, tower base forces and moments. The study is carried out for different environmental conditions. The wind turbine considered in the study is 5MW offshore baseline wind turbine model developed by National Renewable Energy Laboratory (NREL) using FAST. The complete nonlinear aero elastic equations of motion is given by

$$M(q,u,t)\ddot{q} + f(q,\dot{q},u,u_d,t) = 0, \tag{1}$$

where M is the mass matrix, f is the nonlinear forcing function vector, q is the vector of DOF displacements, $\dot{q}$ and the $\ddot{q}$ are the DOF velocities and accelerations, u is the vector of control points, u_d is the vector of wind input disturbances. The equation of motion of second order is of the form

$$M\Delta\ddot{q} + C\Delta\dot{q} + K\Delta q = F\Delta u + F_d\Delta u_d, \tag{2}$$

where $M = M\big|_{op}$ is the mass matrix,

$C = (\partial f/\partial q)\big|_{op}$ is the damping matrix,

$F = -(\partial f/\partial f_d)\big|_{op}$ is the wind input disturbance matrix,

$K = \left[(\partial M/\partial q)\ddot{q} + (\partial f/\partial q)\right]_{op}$ is the stiffness matrix,

$F = -(\partial f/\partial q)\big]_{op}$ is the control input matrix.

3.2. Response Amplitude Operator

The RAOs are calculated in the dynamic analysis module for the combined wind turbine and floating platform system. Response amplitude operators are used to determine the likely behavior of a structure operating in the offshore area. These are calculated for all motions and for different wave headings. RAO's calculated for improving the stability of the structure when operating in deep waters as well as in adverse sea conditions. The equation of motion that govern the linear dynamic motions of the system are in the form

$$\left[-\omega^2 \left\{M + A(\omega)\right\} + i\omega B(\omega) + C\right]\xi(\omega) = X(\omega), \tag{3}$$

where M is the total mass of the body, $A(\omega)$ is the added mass of the floating body, $B(\omega)$ is the damping matrix, C is the stiffness matrix and $X(\omega)$ represents the excitation forces. The hydrodynamic parameters are calculated using WAMIT module for the floating platform and RAOs for the coupled dynamic analysis are calculated using FAST module for the wind turbine. The mass matrix can be found easily, while the added mass matrix, damping matrix, and exciting forces are evaluated. The symbol $\xi(\omega)$ represents the system's dimensional response in each mode of motion. For the translational mode of motion, the RAO is given by

$$RAO_i(\omega) = \left|\frac{\xi_i(\omega)}{A_{wave}}\right| \qquad i=1,\,2,\,3, \tag{4}$$

and for the rotational modes of motion, the RAO is given by

$$RAO_i(\omega) = \left|\frac{\xi(\omega)}{A_{wave}/R}\right| \qquad i=4,\,5,\,6 \tag{5}$$

where i denotes the mode of motion, A_{wave} represents the wave amplitude. Although the RAOs are independent of the sea state, the damping and stiffness properties of the wind turbine depend on wind speed, which causes the RAOs of the combined system to depend on wind speed.

4. Result and Discussion

The hydrodynamic analysis is performed for the spar and semi-submersible platforms. The added mass, damping coefficients and Response Amplitude Operators (RAO's) for different wave heading angles are plotted against time

844

period. The RAOs are calculated for only wave loads for 0 degree and 45 degree wave heading angle. The RAOs show considerable excitation in the surge, heave and pitch modes for 0 degree wave heading angle. So only those modes are presented for 0 degree wave heading angle. The RAOs are computed considering mass, stiffness and damping matrices that account for influence of wind condition on the system. The RAOs are plotted against wave period and presented in Figure 2. The pattern of surge, heave and pitch motions for 0 degree and 45 degree wave heading angles are similar.

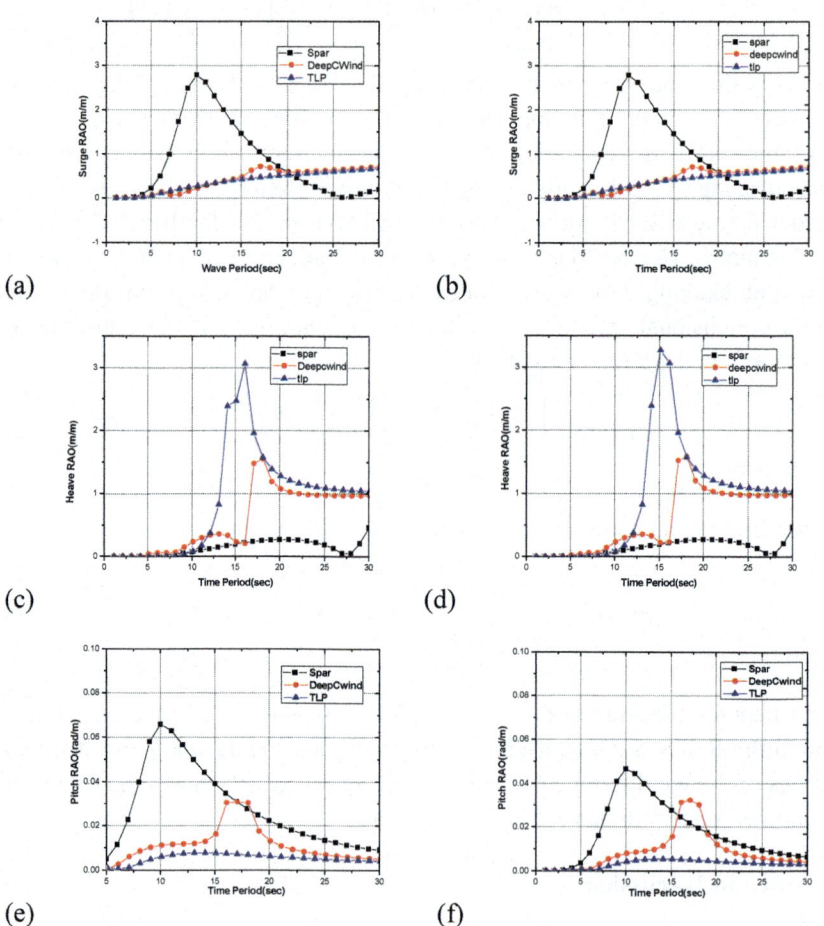

(a) (b)

(c) (d)

(e) (f)

Fig. 2: RAO for (a) Surge (c) Heave (e) Pitch for 0 degree wave heading angles and RAO for (b) Surge (d) Heave (f) Pitch for 45 degree Wave Heading angle.

In the case of surge motion, the response of spar is higher when compared to DeepCwind and TLP floaters as in Figure 2. The surge motion response is in the range of 0 to 3 for spar and 0 to 1 for semi-submersible and TLP floaters. The surge motion of spar is increasing with the wave periods and again decreasing in both the wave heading angles. The spar platform shows the highest surge because it is sensitive to the waves. The heave motion of TLP is more compare to spar and semi-submersible in wave-only loading. The heave motion is less for spar floater. The heave response also increase with the wave period and again decreases. In the case of TLP the heave motion is ranging from 0 to 3.5 and for spar 0 to 0.5, and for semi-submersible it is 0 to 1.5 sec.

The pitch response is higher for semi-submersible compared to spar and TLP-type foundations. The pitch response is very less for TLP type because of vertical stiffness. For spar type the pitch response is increasing with wave period and again decreasing. The pitch response is ranging from 0 to 0.5 for semi-submersible and 0 to 0.1 for both spar and TLP-type floaters.

In Figure 3, the RAOs for different motions for regular waves with inclusion of different quasi static and tension leg mooring lines are analyzed. The coupled response of combined wind and wave loading is presented. The RAO responses for the 6 degree of motions are plotted against wave period. The total time of 3600 sec is taken into simulation. The last 600 seconds of total time is taken into account for analysis such that irregularities will be avoided. The simulation is done for each wave period upto 30 seconds. The aerodynamic analysis is taken using AeroDyn, and the mooring system is also considered for the analysis.

The significant wave height considered is 2 m and wind speed is 8m/s which is below the rated wind speed. The wave height of 2 m which is the minimum wave height for the linear theory to be valid. The RAOs are plotted against wave period. The RAOs are calculated for 0 degree wave heading angle. The coupled loading is done in FAST an aero-servo-hydro-elastic code. The coupled responses for the surge, heave and pitch motions are larger and important for dynamic response than roll, sway and yaw motions. The surge, sway pitch motions of spar are largest when compared to semi-submersible and TLP. It means that the structure is very sensitive to the wind and wave loading. The relative amplitude of the spar platforms is high when compared with the other two, the surge and pitch of the spar platforms are the highest and it is not good for the ration of turbine. The heave response of spar is good because of its deep draft and small water plane area.

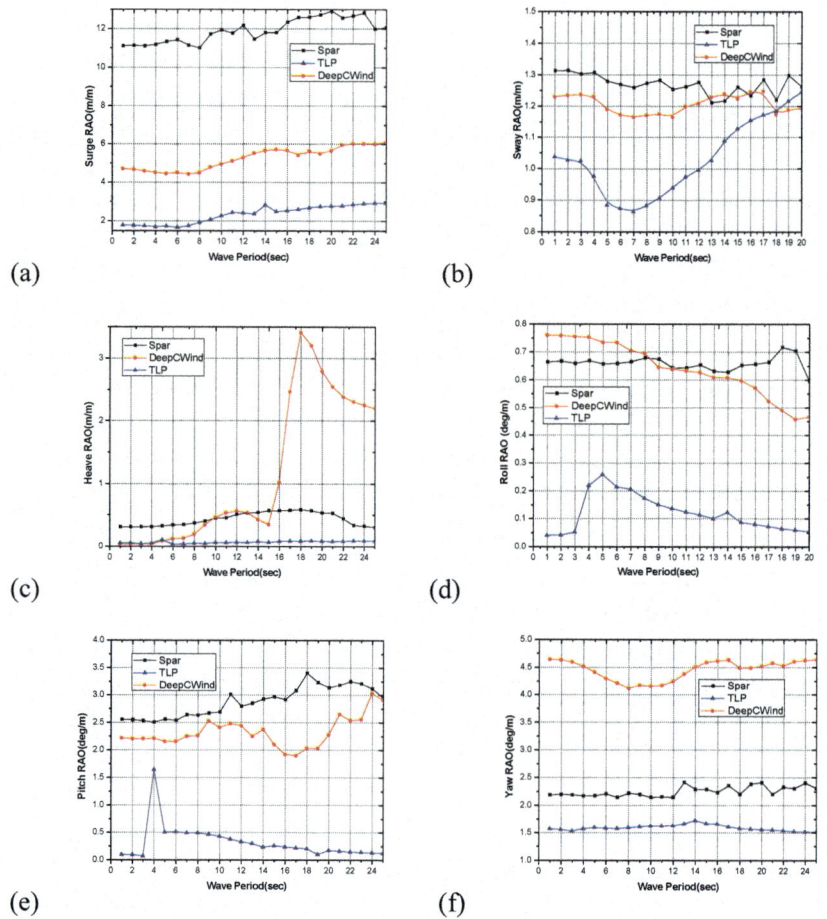

Fig. 3: RAO for (a) Surge (b) Sway (c) Heave (d) Roll (e) Pitch (f) Yaw motions for 0 degree Wave Heading angle

The pitch motion is important for floating wind turbines and it depends on the angle at which wind attacks and in the power generation. For most of the cases the surge motion values are the largest. It is because the structure is sensitive to the wind and wave loading. The relative amplitude is also high for the spar. The responses of surge, heave an pitch are higher when compared to roll, sway and yaw under specific environmental conditions but these three also important for the dynamic analysis. The pitch and heave motions of TLP are smaller than spar and semi-submersible case because of vertical stiffness.

5. Conclusion

The hydrodynamic performance and the coupled dynamic response of the spar, semi-submersible and TLP type floaters are analyzed. The hydrodynamic analysis is performed for the 0 degree and 45 degree wave heading angles and plotted against wave period. The RAOs show considerable excitation in the surge, heave and pitch modes for 0 degree wave heading angle. The surge, heave and pitch motions are the main consideration for the analysis of wind turbine. In surge, sway and roll motions the spar is showing higher excitation when compared to semi-submersible and TLP floaters. In the case of heave motion, TLP floater shows higher excitation due to the absence of mooring lines. The coupled analysis is done for wave height of 2 m and wind speed of 8 m/s for 0 degree wave heading angle. The surge, heave and pitch motions are larger when compared to sway, roll and yaw motions. The surge, heave and pitch motions of the spar platforms are highest due to deep draft and small water plane area. The heave motion of the TLP is less compared to spar and semi-submersible due to vertical stiffness. In surge, sway and pitch motions the spar response is higher. In all the cases TLP has shown the least response. This dynamic analysis is used to analyze the performance of the wind turbine for various environmental conditions. It is used to do the improvements in the design and improve the performance of the wind turbine.

Acknowledgement

The research work is supported by the Science and Engineering Research Board (SERB), Department of Science & Technology (DST), Government of India under the Young Scientist Research Grant No. YSS/2014/000812.

References

1. Bagbanci, H., Karmakar, D. and Guedes Soares, C. (2011). Review of Offshore Wind turbine Concepts, Maritime Technology and Engineering, C. Guedes Soares et al., (Eds), London, UK: Taylor and Francis Group, Vol-2, pp.553-562.

2. Bagbanci, H., Karmakar, D. and Guedes Soares, C. (2015). Comparison of spar and semi-submersible floater concepts of wind turbines using long-term analysis, Journal of Offshore Mechanics and Arctic Engineering (ASME), 137(4), pp. 061601-01-10.

3. Guedes Soares, C., Bhattachrjee, J. and Karmakar, D. (2014), Overview and prospects for offshore wave and wind energy, Brodogradnja, Vol. 65(2), pp. 91-113.

4. Liu, Y., Li, S., Yi, Q. and Chen, D. (2016). Developments in semi-submersible floating foundations supporting wind turbines: A comprehensive review, Renewable and Sustainable Energy Reviews, 60, 433-449.

5. Wang CM, Utsunomiya T, Wee SC, Choo YS, 2010, Research on floating wind turbines: a literature survey,   The IES Journal Part A: Civil & Structural Engineering, Vol. 3, No. 4, November 2010, 267-277

6. Borg,M., Collu, M. (2014), A comparison on the dynamics of a floating vertical axis wind turbines on three different floating supporting structures, EERA Deepwind'2014, 11th Deep Sea Offshore Wind R&D Conference, Energy Precedia 53, 268-279.

7. Liu, Z., Zhang, M., Wang, B., Du, J. (2014). Dynamic Response Analysis for Floating Offshore Wind Turbine Structures, Proceedings of the Eleventh Pacific / Asia Offshore Mechanics Symposium, Shanghai, China, October 12-16.

8. Goupee, A.J., Koo, B.J., Kimball, R.W., Lambrakos, K.F., Dagher, H.J. (2014). Experimental Comparison of Three Floating Wind Turbine Concepts, Journal of Offshore Mechanics and Arctic Engineering (ASME), Vol. 136, pp. 020906-1-9.

9. Ramachandran G.K.V., Bredmose, H., Sorensen J.N., Jensen J.J. (2014). Fully Coupled Three-Dimensional Dynamic Response of a Tension-Leg Platform Floating Wind Turbine in Waves and Wind, Journal of Offshore Mechanics and Arctic Engineering (ASME), Vol. 136, pp. 020901-1-12.

Discussion of Laboratory Simulation Methods
for Wind Effects on Coastal Engineering Structures

Hong YANG[†], Yu-dan WANG and Qi-hua ZUO

Nanjing Hydraulic Research Institute,
Nanjing 210029, China
[†]E-mail: hyang@nhri.cn
www.nhri.cn

Wind is one of the important dynamic factors acting on coastal engineering structures. This paper discusses the appropriateness of the methods that adopt constant wind and wind speed using gravity similarity law, which is consistent with the hydrodynamic similarity, to study the interaction of wind and coastal engineering structures for most current laboratory simulations. The results show that it is feasible to simulate random wind in laboratory so that wind effects can be more truly reflected. For the stability analysis of fixed structures assembled with slender rods, the gravity similarity law ($Fr = 1$) can be adopted and DAVENPORT or von Karman spectrum can be used as wind spectrum. For a floating structure, the wind effects include two parts. One part includes inertial forces FI and resistances FD that are generated when wind forces directly act on the structures, which can be simulated by gravity similarity law, and here API or NDP spectrum can be adopted to simulate random wind; the other is the drift velocity of flow, whose magnitude is proportional to the fetch length and wave age, caused by the wind–water surface interaction. It is pointed out that wind effects on coastal engineering structures, especially the dynamic response of floating structures, may be underestimated when constant wind and gravity similarity law are used in laboratory simulation, therefore the similarity method and its corresponding results should be properly modified.

Key words: wind spectrum; gravity similarity; laboratory simulation; coastal engineering structure

1. Introduction

The randomness of wind turbulence has long been recognized and is used earlier and more intensively in the design of civil engineering structures on land than those in coastal engineering. Many researchers have made quite a lot of numerical simulations based on some assumptions. Laboratory simulations of wind effects on civil engineering structures have been conducted for many years. In many countries, wind tunnels have been established for wind experiments with different scales, and a number of projects have also been carried out. Those countries adopt the concept of random wind to standardize the design of civil engineering structures [1]. The dynamic coefficient of random wind can be obtained from full-scale or large-scale wind tunnel tests.

However, the reliability concept is still not a strict rule in most coastal engineering designs. The effects of randomness of wind turbulence on the stability of coastal engineering structures, overtopping, and the dynamic responses of floating structures are rarely involved. Besides, owing to the limitation of experimental devices, uniform wind is mainly adopted to consider the wind effects on structures in laboratory tests. And gravity similarity is used in reduced scale tests since wind speed should be kept consistent with the hydrodynamic similarity. These may influence the reliability of experimental results of wind effect on coastal engineering structures.

In this paper, laboratory simulations of random wind affecting coastal engineering structures are discussed, involving different wind eigenvalues and wind spectra, similarities of wind effect on structures and wind–water interactions and the possible influence on the experimental results, and the choice of wind spectrum for different coastal engineering structures, and thus providing future references for coastal engineering experiments.

2. Random wind spectrum

The random wind speed can be expressed as:

$$U(t) = U_0 + \tilde{U}(t) + U'(t),$$ (1)

where U_0 is the mean wind speed, $\tilde{U}$ is the periodic fluctuating wind speed, and U' is the turbulent fluctuating wind speed. Unlike water waves, waves can be considered as up and down oscillation relative to the still water level, and the average water level is 0. However, U_0 is not equal to 0 usually. According to Eq. (1), the Fourier series expansion of the random wind speed and its spectrum $S(f)$ can be obtained as:

$$U(t) = \sum_{i=0}^{N} U_i \cos(2\pi f_i t + \varphi_i) df ;$$ (2)

$$S(f) = \frac{1}{2} U_i^2(f),$$ (3)

where f is the wind frequency, and φ is the random phase difference. Under the same conditions, the wind process satisfying the specific spectrum $S(f)$ can be generated in laboratory (see Figure 1): [2]

$$U(t) = \sum_{i=0}^{N} \sqrt{2S(f)\mathrm{d}f} \, \cos(2\pi f_i t + \varphi_i) \qquad (4)$$

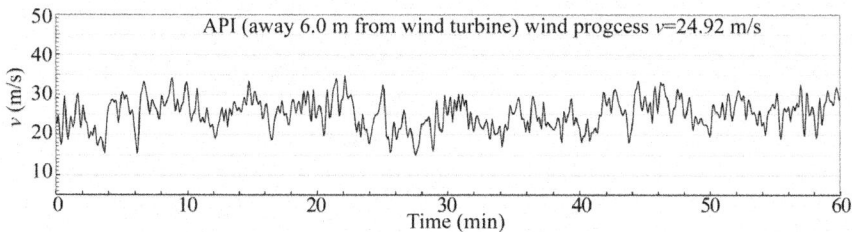

Fig. 1. Typical wind process generated in laboratory.

Up to now, various wind spectra have been applied to engineering, and they can be classified into three types[3]. The first kind is that the frequency $f=0$ with the spectral density being zero (e.g. Davenport spectrum). With its intuitiveness, similarity to that in other research fields, and some easily-determined parameters, although obtained from on-land information, this kind of wind spectrum is applied in other projects, even in coastal engineering. The second kind is that the frequency $f \neq 0$, while the spectral density is nonzero. The maximum spectral density does not occur at the lowest frequency. The spectral density gradually increases from low frequencies until reaches the peak, and then gradually decreases. The number of this kind of spectrum now is the largest, and it is also the most widely used in engineering[4-5]. For the third kind of spectrum, the maximum density occurs at frequency $f=0$ or at the lowest frequency. This kind of spectrum is mainly used in deep-sea projects. Figure 2 shows the comparisons of dimensionless spectrum densities for the first and the second kinds of spectra. Figure 3 shows the comparisons of a few typical spectra under the specific conditions of the third kind of spectrum ($U=30.9$ m/s) (including Davenport spectrum), and the spectral differences are obviously larger than those of the second kind, especially at low frequency.

In fact, it is easier to understand that the spectral density reaches the maximum at $f = 0$. Different from dynamic factors such as wave factors, wind turbulence does not vibrate around zero baseline.

Both in field observations and the practical design, the mean wind speed in Eq. (1) have different definitions. Different characteristic mean wind speeds are used in the standards of different countries[6], such as adopting the sample delay of 3 s, 3 min, 5 min, 10 min, and 1 h. Therefore, it is difficult to compare the experiments adopting the constant wind speed because its physical concept is unclear. Only using the random wind with a characteristic value or a spectral

852

form can make the physical concept clearly. Although it is usually thought that the mean wind speed of a longer sample delay is smaller than that of a shorter sample delay, it is not always supported in practical measurement. In recent years, many data have been obtained by numerical simulations. For example, Holmes and Cochran [7] carried out 5100 groups of experiments to resolve the statistical stability.

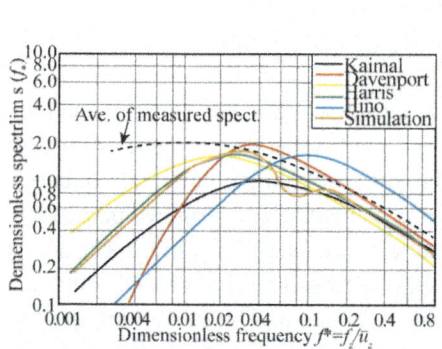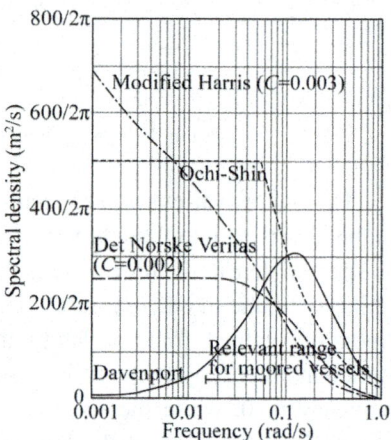

Fig. 2. Comparison of dimensionless spectrum densities of the first and the second kinds of spectra.

Fig. 3. Comparison of the third kind of wind spectrum (U=30.9 m/s).

For most experiments, the experimental groups are not enough from business view point of view. It should be noted that at least 100 groups are needed for the extreme III-type spectrum, maybe smaller than 100 for the I-type spectrum, but the safety coefficient should be adjusted when the number is too small. It also should be noted that the uncertainty of aerodynamic coefficients is still unknown, and laboratory wind tunnel tests are mostly suitable for strong frontal depression typhoon. Therefore, it is not known whether it can be extended to other types of typhoon. In addition, the extreme value will increase with the increase of sample sequence, i.e. the extreme value for 1 h may be larger than that for 10 min. So the random wind with enough long period of time is required statistically in wind experiments.

3. Wind–water interaction and simulation

The effects of wind on water include: (1) generation of wind waves; (2) wave deformation. Because of wind, the wavefront of the asymmetric wave becomes steeper, which proves that the wavefront acceleration of wind wave is larger than that obtained by the basic theory, resulting in the peak value of wind-generated waves acting on structures larger than the theoretical value. (3) Wind-induced injection will increase the run-up, overtopping, and the wind loads acting on structures in the swash zone. (4) Wind-induced surface flow, whose direction is the same as the wind direction, or in the northern hemisphere, at the right hand of the wind direction. When wind is onshore or parallel to the coastline, the wind has an onshore component. (5) Wind-induced turbulence. The wind shear stress increases the water eddies and turbulence at the water surface, the effective viscosity of water, and the effective mixing coefficient of water with heat, salt, or pollutants. Since there is little quantitative data can be used, it is useful to study proper laboratory equipments of wind and water interactions. (6) Wind shear stress. In the storm surge produced by strong winds, a large volume of water body accumulates on the shore and estuaries. In the present study, only the effects of wind-generated wave, wind increased, and wind-driven surface current are considered.

The effect of wind on fluid shear stress can be expressed as:

$$\tau = \rho_a C_{da} U^2, \tag{5}$$

where C_{da} is the drag coefficient which is related to the surface roughness, vertical gradient of air temperature, wind speed, etc. The relationship between surface waves and wind can be obtained by the dimensional analysis:

$$\frac{gT}{U} = f_1\left(\frac{gF}{U^2}, \frac{gt_r}{U}\right) \quad \frac{gH}{U^2} = f_2\left(\frac{gF}{U^2}, \frac{gt_r}{U}\right) \tag{6}$$

where T is the wave period, H is the wave height, F is the fetch length, and t_r is the time delay. If the fetch length is not considered, and L indicates the dimension of length, based on Eq. (6), one can obtain:

$$\lambda_T = \lambda_U; \tag{7}$$

$$\lambda_U = \lambda_L^{1/2}; \tag{8}$$

Eqs. (7) and (8) show that the wind similarity is consistent with the gravity similarity. In laboratory, the effect of wind fetch on the local wind-induced wave at the surface is not usually considered. However, at present there are many ports with the fetch length longer than 1000 m, and the effects of the local wind-induced waves in these small wind fetchs should be taken into account. Because the wind climate covers a large region, in general, the delay time t_r may be different for different regions. In Germany, for example, the time delay of storms on the coast is much longer than that on land, and it is the longest in the Elbe Estuary. It is 5 h in Brunsbuttel, 4 h in Bremenhr, 3 h in Bremen, and 2 h in Frankfurt.

Considering that its effects should have fully acting time, it can be assumed as a limited, fully- grown wave in the wind region.

The wind-induced drift velocity at the water surface can be qualitatively determined by assuming that the shear stress at water surface is the same as the wind-induced shear stress. The relationship between the significant wave height and wind speed can be expressed as follows:

$$\frac{gH_{1/3}}{U^2} = 0.0055 \left(\frac{gF}{U^2}\right)^{0.35} \tanh\left[30\frac{\left(\frac{gd}{U^2}\right)^{0.8}}{\left(\frac{gF}{U^2}\right)^{0.35}}\right] \tag{9}$$

For the design of an open port, the altitude of wind backwater in the limit wind fetch can be calculated as[8]:

$$e = \frac{KU^2F}{2gd}\cos\beta, \tag{10}$$

where e is the wind backwater height (m); K is the comprehensive frictional coefficient, and $K=3.6\times10^{-6}$; U is the 10 min-averaged wind speed at the height of 10 m above the water surface (m/s); F is the fetch length (m); d is the mean water depth; and β is the angle between the wind direction and the normal direction perpendicular to the coastline.

The drift velocity caused by wind can be obtained by the air–water interaction equations. The hydrodynamic equations are as follows:

$$\frac{\partial u}{\partial t} + u\frac{\partial u}{\partial x} + v\frac{\partial u}{\partial y} + \omega\frac{\partial u}{\partial z} = fv - \frac{1}{\rho}\frac{\partial p}{\partial x} + A_v\frac{\partial^2 u}{\partial z^2} + A_h\nabla^2 u; \tag{11}$$

$$\frac{\partial v}{\partial t} + u\frac{\partial v}{\partial x} + v\frac{\partial v}{\partial y} + \omega\frac{\partial v}{\partial z} = -fu - \frac{1}{\rho}\frac{\partial p}{\partial y} + A_v\frac{\partial^2 v}{\partial z^2} + A_h\nabla^2 v; \tag{12}$$

$$\frac{\partial u}{\partial x} + \frac{\partial v}{\partial y} + \frac{\partial \omega}{\partial z} = 0, \tag{13}$$

where u and v are the flow velocity components in x- and y-direction, respectively; η is the water elevation; h is the water depth; A_v and A_h are the vertical and horizontal vortical viscosity coefficients, respectively. The pressure term in the right hand takes the wind-water surface interaction into account, and the last two terms indicate the effects of vortex dynamics. According to the similarity theory, the similarities can be written as:

$$\lambda_{v_h}\frac{\lambda_U}{\lambda_{L^2}} = \text{const} \tag{14}$$

Obviously the similarity scale is largely different from the gravity similarity scale. Eqs. (11)-(13) can also be written as:

$$\frac{\partial u}{\partial t} + u\frac{\partial u}{\partial x} + v\frac{\partial u}{\partial y} = -g\frac{\partial \eta}{\partial x} + fv - \frac{1}{\rho}\frac{\partial p}{\partial x} + F_x + k_h D^2 u; \qquad (15)$$

$$\frac{\partial v}{\partial t} + u\frac{\partial v}{\partial x} + v\frac{\partial v}{\partial y} = -g\frac{\partial \eta}{\partial y} - fu + F_y + k_h D^2 v; \qquad (16)$$

$$\frac{\partial \eta}{\partial t} + \frac{\partial \left[(\eta+h)u\right]}{\partial x} + \frac{\partial \left[(\eta+h)v\right]}{\partial y} = 0, \qquad (17)$$

where F_x and F_y are the wind-induced resistance components in x- and y-directions, respectively.

$$F_x = \frac{1}{\rho_w(\eta+h)}\left(\rho_a C_D^a u_a \sqrt{u_a^2 + v_a^2} - \rho_w C_D^L u_L \sqrt{u_L^2 + v_L^2}\right); \qquad (18)$$

$$F_y = \frac{1}{\rho_w(\eta+h)}\left(\rho_a C_D^a v_a \sqrt{u_a^2 + v_a^2} - \rho_w C_D^L v_L \sqrt{u_L^2 + v_L^2}\right). \qquad (19)$$

Here C_D^a is the frictional coefficient at the water surface. Based on Eqs. (18)-(19), the similarity of the wind force F acting on the water surface can be written as:

$$\lambda_{C_D^a} \frac{\lambda_{\rho_a} \lambda_{u_a^2}}{\lambda_{\rho_w} \lambda_L} = \text{const} \qquad (20)$$

Assuming that the frictional coefficients C_D^a in both model and prototype are the same and the densities of water and air are unchanged, it can be thought from Eq. (3-11) that the wind similarity is inconsistent with the gravity similarity. However, it can be seen from Eq. (14) that the conveying to underwater (see Figure 4) is inconsistent. Assuming that the frictional coefficients at water surface and in air are the same, and the wind–air shear stress at water surface is coordinated as:

$$\frac{1}{2}C_{da}\rho_a U^2 = \frac{1}{2}C_{dw}\rho_w U_w^2. \qquad (21)$$

Thus, the drift velocity at the water surface can be obtained

$$U_w = \sqrt{\frac{\rho_a}{\rho_w}}U = K_w U \approx 0.035U. \qquad (22)$$

At the height of 10 m and wind speed U=4–16 m/s, the free surface drift velocity usually reaches 0.3 m/s. The flow velocity caused by storm is a complex function related to the wind strength, meteorological characteristics, topography, shoreline composition, and profile of water density. It is suggested in ISO TC98 that in deep water, surface flow along the open coast line can be roughly regarded as 1HR (3% of sustained wind speed). The nearer to the shore, the water is shallower, and its value is bigger.

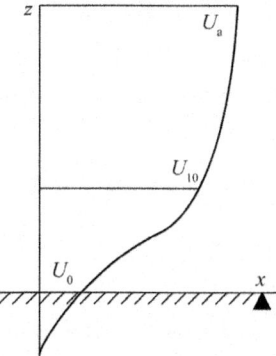

Fig. 4. Velocity distribution of the water–air boundary layer.

To sum up, owing to the limit conditions such as the laboratory site, the laboratory simulation of wind effect cannot take into account the effects of the local wind-induced waves, damming water, and drift velocity at water surface, which are related to the fetch length F, acting time t and wind speed U.

4. Wind effect on fixed coastal structures and its simulation

The dynamic response of fixed coastal engineering structures can be written as:

$$\left(M_i\right)\ddot{X}_i\left(t\right)+C_i\dot{X}_i\left(\tau\right)+K_iX_i\left(t\right)=F(t), \quad i=1,2,3 \qquad (23)$$

where X is a structure in an exercise, and can be divided into the displacement (x, y, z); M is mass; C is the damping function; K is the restoring force coefficient; t is time. The force F consisting of wave, current and wind can be written as:

$$F_j = FWV_j\left(t\right)+FCD_j(t)+FWD_j\left(t\right)+FWS_j(t) \qquad (24)$$

where FWV is wave force; FCD is flow force; FWD is the wind force directly acting on the structure; and FWS indicates the wind-induced additional forces, which can be omitted for fixed-type structures. There have been many researches on wave and current forces, taking into account the gravity similarity in laboratory simulations. The force *FWD* directly acting on the structure includes two parts: one is the inertial force consisting of fluctuating pressure, and the other is the resistance.

$$FWD = F_{\text{inertial}} + F_{\text{drag}}. \tag{25}$$

Assuming that the mean value of $u(t)$ is 0, the inertial force F_{inertial} depends on the variation of fluctuating velocity, i.e. it is proportional to $\dfrac{\partial u}{\partial t}$ as follows:

$$F_{\text{inertial}} = \rho_a A C_m \frac{\partial u}{\partial t}, \tag{26}$$

where ρ_a is the density of air, A is the frontal area, and C_m is the inertial coefficient. The inertial force will not be considered in constant wind experiments, while only considered in the random wind experiments.
The wind resistance can be written as:

$$F_{\text{drag}} = \frac{1}{2}\rho_a C_{\text{DW}} A U^2 \text{ or } F_{\text{drag}} = \frac{1}{2}\rho_a C_{\text{DW}} A\, U\, |U| \tag{27}$$

where C_{DW} is the resistance coefficient of the structure, relating to Reynolds number. The condition satisfying Reynolds number similarity is

$$Re = \lambda_L \frac{\lambda_\upsilon}{\lambda_\upsilon} = \text{const} \tag{28}$$

or

$$\lambda_L = \lambda_U. \tag{29}$$

This is inconsistent with the gravity similarity. It should be considered in experiments that the dissimilarities of flow and resistance coefficient may be induced due to the dissimilarities of structural roughness and spacing. The difference between constant wind and random wind can be obtained from Eqs. (1) and (27). In fact, the drag coefficient between wind and structure is also random.

Figure 2 shows that the first-type and the second-type spectra have more energy in high frequency part and the fixed structures with slender rods usually have high frequencies. Therefore, these two kinds of spectra should be used in experiment, such as the DAVENPOT and von Karman spectra.

5.   Wind effects on floating structures

$$\sum_{i=1}^{6}\left[M_{ii}+m_{ij}\left(\infty\right)\right]\ddot{X}_{j}\left(t\right)+\sum_{i=1}^{6}\int_{-\infty}^{t}L_{ij}\left(t-\tau\right)\dot{X}_{j}\left(\tau\right)\mathrm{d}\tau$$

$$+\sum_{i=1}^{6}\left(C_{ij}+G_{ij}\right)X_{j}\left(t\right)=F(t),\quad\left(i,j=1-6\right),\tag{30}$$

where X is a structure in an exercise, can be divided into the displacement (x, y, z) and rotate (written as pitch, roll and yaw for float structures). M is the quality or inertia moment of the structure; m is added mass; L is the damping function, C is the restoring force coefficient; and G is the additional elastic coefficient. The latter four terms are related to the displacement and consist of flow and wind forces.

The force F including wave, current and wind can be written as:

$$F_{j}=FWV_{j}\left(t\right)+FSD_{j}\left(t\right)+FCD_{j}(t)+FWD_{j}\left(t\right)+FWS_{j}(t),\tag{31}$$

where FWV is the first-order wave force; FSD is the second-order wave force; FCD is the flow force; FWD is the wind force directly acting on the structure; FWS refers the wind-induced additional forces, such as local wind-induced waves and damming water.

6.   Correction of experimental results

(1) Under the constant wind condition in laboratory, it can be assumed that the wind speed fits the Gaussian distribution, the Normal distribution or other distributions, and a characteristic value of the average wind speed is adopted in experiments. The Normal distribution can use the mean-square value, that is,

$$U_{L}=\gamma\bar{U}_{m},\tag{32}$$

where U_{L} is the actual experimental wind speed, γ is the variation coefficient of wind speed; and $\bar{U}_{m}$ is the model wind speed obtained by the gravity similarity.

(2) When the gravity similarity is used in model experiment of dynamic response or action force of slender or small spacing panel structures, it should appropriately enlarge the roughness of the structure surface or increase the spacing to modify the incoordination between the Reynolds number similarity and the gravity similarity, guaranteeing that the turbulent characteristics of wind are invariable. When the high frequency of wind is close to the natural

frequency of the structure, the elastic similarity of the structure will bring errors into the experimental results.

(3) Owing to the limited laboratory size, the influences of local wind-induced waves and damming water are unable to be considered, so experiments should be combined with numerical simulation. Based on the maximum fetch length of the practical project, the impact of these two factors are estimated by use of numerical model, superimposed with the energy of incident wave to obtain the synthetic incident wave and water level for experiments using the gravity similarity. The increase of backwater level should be smaller than $0.02U$.

(4) The wind process should be long enough in random wind experiments. When random wind is generated by adopting random phase method, numerical simulation can be used in advance to obtain the unfavorable phase spectrum against the structure, and then laboratory simulation is carried out.

(5) When conducting experiments of random wind acting on a floating structure, in addition to the interaction between the wind frequency and the resonance period of the structure, the effects of local wind-induced waves, damming water, and drift flow at the surface should also be taken into account.

(6) After the separation of wind and flow experimental results, different probability analysis of wind effects can be done under assumption of wind distributions, and superimposed with the flow results.

7. Conclusions

In this paper, the following conclutions can be drawn.

(1) In the existing experiments on coastal engineering structural safety, the effect of wind is considered to be constant. The neglection of the effects of random wind is likely to underestimate the wind force and its effects on the dynamic response of the structure.

(2) Local wind-induced waves, damming water and drift flow are usually not considered owing to limited lab conditions, so the reliability of the experimental results should be estimated for these factors, or corrective measures should be used in the experiments.

(3) At present, when the gravity similarity law is used to simulate the wind effect in laboratory, the experimental models should be modified to reduce the effect of wind–structure interaction on Reynolds number.

(4) For a fixed slender structure, the wind spectrum with a high-frequency peak should be used in experiment. The DAVENPORT spectrum or von Carmen spectrum can be used in the absence of field data.

(5) For a floating structure, the wind spectrum with a low-frequency peak should be used in experiment to study the effects of the energy when wind frequency and the structural resonance frequencies are similar. In the absence of measured data, the API or DNP spectrum can be used in the experiment.

(6) It should have enough time or adopt an appropriate phase spectrum for a particular wind spectrum in experiment to ensure that the effect of random wind on structures is fully considered.

References

1. M.Kasperski, Specification of the design wind load—A critical review of code concepts, *J. Wind Eng. Ind. Aerodyn.*, **97**(2009), 335–357.

2. Q.L. Du, *The Experimental Research on Mooring Floating System Subjected to Random Winds and Waves*. Ph.D. Thesis, Nanjing Hydraulic Research Institute, Nanjing, 2016. (in Chinese)

3. Q.H. Zuo, Q.L. Du, Y.H. Zhao, Z.B. Duan, Y.D. Wang, Review of studies on random wind spectrum and its application in coastal engineering, *The Ocean Engineering*, **34**(2016), 111–122. (in Chinese)

4. M.K. Ochi, Y.S. Shin, Wind turbulent spectra for design consideration of offshore structure, OTC5736, 1988, Houston, USA.

5. G.J. Feikema, J.E.W. Wichers, The effect of wind spectra on the low-frequency motions of a moored ranker in survival condition, OTC6605, Houston, USA, 1991.

6. D.K. Kwon, A. Kareem, Comparative study of major international wind codes and standards for winds effects on tall building, *Engineering Structures*, **51**(2013), 23–35.

7. J.D. Holmes, L.S. Cochran, Probability distributions of extreme pressure coefficients. *Journal of Wind Engineering & Industrial Aerodynamics*, **91**(2003), 893-901.

8. Ministry of Transport of the People's Republic of China, Code of Hydrology for Sea Harbour, JTS145-2015, China Communication Press, Beijing, 2015. (in Chinese)

Improvement of Water Quality by Granulated Coal Ash in the Pasig River and Its Tributaries

S. Azuma

D.N.C. Ltd., 2-12-15 Akasaka Seven Bld. 7F, Daimyo Chuo-ku, Fukuoka City, 810-0041, Japan

M. Bonga

Pasig River Rehabilitation Commission, 1608 Quezon Avenue, Quezon City, NCR Philippines

M. A. N. Tanchuling and G. Cruz

University of the Philippines Diliman, Quezon City,1101 NCR Philippines

N. Touch[§], T. Hibino, N. Pagayao

Hiroshima University, 1-4-1, Kagamiyama, Higashihiroshima, 739-8527, Japan
[§] E-mail: narong-cambodia@hiroshima-u.ac.jp

Granulated coal ash (GCA), a by-product from coal fired power plants, has been proven to be effective in improving sediment and water qualities in the littoral regions severely deteriorated by wastewater discharge. Based on the achievements of GCA application in Japan, it is found out that GCA plays an important role in restoring the ecosystem of marine environment through various mechanisms, notably neutralizing acidified sediment due to the hydrolysis of CaO, adsorbing nutrient salts and hydrogen sulfide, and decreasing the oxygen consumption in bottom water. This paper describes the potential of using GCA in improving water quality in the Pasig River through various methods based on recent study results. The pilot study at the Estero de san Miguel is currently under preparation with the Pasig River Rehabilitation Commission.

Keywords: Granulated Coal Ash; Remediation; Coastal Sediment; Polluted Water; Estero de San Miguel.

1. Introduction

The Pasig river is considered as one of the most culturally relevant bodies of water in the Philippines. Stretching at approximately 47 kilometers in length, it is known to connect Laguna de Bay to Manila Bay passing through eight cities including

Manila, Mandaluyong, Makati, Pasig, Pasay, Taguig, Caloocan and San Juan. The river is comprised of forty-seven (47) tributaries with the Marikina River and the San Juan River being the major rivers. Historically, the Pasig River was once a potential source of drinking water for locals and also known to be a flourishing habitat for abundant aquatic resources. However, deterioration of the Pasig River along with other major rivers in the metro has been observed in the last few decades due to the rapid urbanization of Manila. In 1990, the Pasig River was declared biologically dead by Danish International Development Agency (DANIDA). Based on the reports, the major culprits for the river's deterioration were cited to be industrial pollution, domestic waste dumping and urban migration.

Estuaries provide interface between community and water. Due to the close proximity of residential homes around the river, poor water quality of estuary is markedly noticeable to nearby settlers in the form of high turbidity, accumulation of garbage and pungent odor. In other words, the water quality of estuary has more bearing on the quality of life and living standards of the urban settlers in the metro.

Meanwhile, the Estero de San Miguel is one of the tributaries of the Pasig River in Manila area that directly outflows to the Pasig River. It has a total length of 2,044 meters and a width of 14.70 meters. Water level is about 0.61 to 1.5 meters at low tide and nearly 1.5 to 2.4 meters at high tide. The Estero de San Miguel is surrounded by mixed industrial/commercial warehouses, universities, government institutions (including Malacañan Palace) and residential areas. The septage water of these establishments and residents generally flow into the estero untreated. Solid wastes, mostly domestic in nature, are indiscreetly dumped in the estero making it instant garbage dumpsite. Informal settlers are also occupying the 3-meter easement of the estero which directly contribute to the pollution of the water. The location of the Estero de San Miguel is shown in Fig.1.

Granulated coal ash (GCA) has been introduced in Japan for field applications at estuarine areas. It has been reported for the effectivity of GCA in addressing environmental problems such as neutralization of acidified sediments, fixation of hydrogen sulfide which causes bad smell and consuming dissolved oxygen, and fixation of nutrients salt by adsorption onto GCA surfaces [1]. The main component of GCA is fly ash, a by-product of coal fired power plants; therefore, it is expected that the environmental load caused by the accumulation of fly ash will be reduced if GCA is used. The assessment on the efficiency and effectivity of GCA in Philippines has not yet been investigated. Thus, this study aim to investigate the effects of GCA in improving the water quality of the Estero de San Miguel through the laboratory experiments.

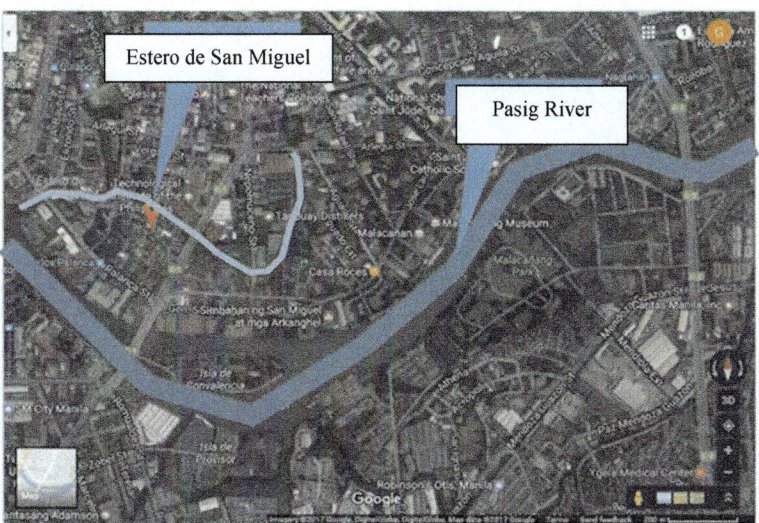

Fig. 1.   Location of the Estero de San Miguel.

2. Properties of Granulated Coal Ash

GCA was manufactured by granulating fly ash powder obtained from coal-fired power plants. The grain size distribution of GCA was similar to that of gravel. Figure 2 presents the physical appearances of GCA and fly ash. Since the main material of GCA is the fly ash achieved from coal fired power plants, therefore it is expected that the environmental loads resulting from industrial wastes will be reduced if GCA is used. Moreover, it is considered that it may be possible to transform GCA into an effective material for treating organic matter-enriched sediment. If it is possible to effectively remediate the organic matter that negatively impacts the environment, the enormous costs for the sediment disposal will be reduced considerably.

GCA consists of particles that ranged from coarse sand to gravel. The mean diameter of GCA is approximately 20 mm. Moreover, GCA is a porous material which has a small specific gravity. The particle density of GCA ranged from 0.8 to 1.1 t/m^3 in dry condition, and 1.0 to 1.4 t/m^3 in wet condition. It has been reported that GCA can fix phosphate and hydrogen sulfide [2]. And, it has already been reported that the elution of heavy metals from GCA is just a little and below the allowable heavy metal standard in Japan [3].

(a)   (b)

Fig. 2.   (a) Physical appearance of GCA; (b) Physical appearance of fly ash.

Table 1. Chemical component of granulated coal ash.

Chemical Component	Percentage (%)
Silicon dioxide (SiO_2)	44
Aluminum oxide (Al_2O_3)	13
Calcium oxide (CaO)	21
Carbon (C)	9
Others	13

Fly ash is obtained by burning the coal comprised of plant origin; thus, the components of fly ash are mainly derived from the oxidation of plant elements. Table 1 presents the chemical components of GCA. It can be seen from Table 1 that GCA principally consisted of silica (SiO_2), calcium oxide (CaO) and aluminum oxide (Al_2O_3). These components are known to be significantly involved in the purification of sediment in reduced conditions. As GCA contains calcium oxide, it also influences on the coagulation of suspended particles. In addition, since some components, e.g. silica and calcium oxide, dissolve when GCA is used in water, thus GCA can increase the solution pH and promote the growth of diatoms after the GCA application [4].

3.   Field Applications of GCA in Japan

Figure 3 shows the various application areas of using GCA for improving the estuarine environment in Hiroshima Prefecture (Japan). The use of GCA was effective in covering the sea bottom sediment to reduce the diffusion of substances from the sediment. The achievements of the using GCA in the restoration of muddy tidal flat environments has gain a reputation for high reliability from fishermen and government officials (the application of GCA were reported in 4 newspaper companies in the Chugoku area).

One of the effects of using GCA in the muddy tidal flat is the strengthening sediment strength, which allows people to walk on the GCA applied sediment (sprinkling and mixing of GCA to sediment). Another is that the soil improvement

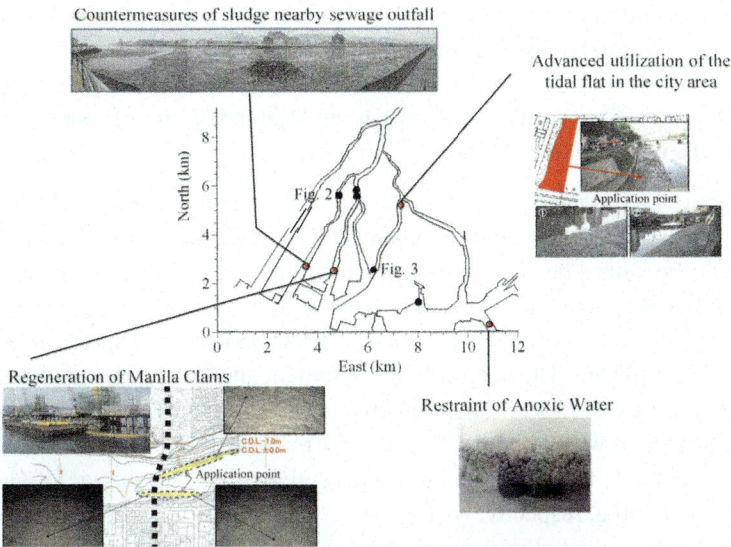

Fig. 3. Application points of the developed methods in the Ota River and Hiroshima Bay. (Different methods were applied according to sediment conditions).

Fig. 4. Growth of benthos and plants along the GCA application.

by GCA can promote the inhabitation of benthos in the tidal flat after inserting the GCA columns into the sediment (other area were applied by mixing GCA with the sediment). This is because the infiltration flow re-generated, and oxygen was supplied into ground which promotes the inhabitation of benthos (Fig. 4). The infiltration of water was caused by an increase in permeability through the inserting GCA columns. This can supply oxygen into sediment, resulting in the restoration of benthic environment, for example the presence of clams in the treated bottom sediment by GCA as shown in Fig. 4.

4. Laboratory Experiments on Improving Water Quality of Estero de San Miguel by Using GCA

The laboratory experiments were conducted at the Environmental Laboratory in the University of the Philippines Diliman. The sediment and polluted water were sampled at the Estero de San Miguel as shown in Fig. 1. The experiments were using 1.8 L wide-mouth jars having screw top lids. The sediment (300 g) and water (1 L) from the estero were placed in the jars and subsequent treatment was performed in triplicate. The heights of the sediment and overlying water (polluted water added) were 5 and 15 cm, respectively. Four treatments were conducted. The first treatment was without GCA (control). The sediments in the other three treatments (Case 1, Case 2 and Case 3) were given different dosage of GCA at 150 , 225 and 300 g, respectively (Fig. 5).

The jars were kept closed during the experiments, which were performed at room temperature. The water samples were withdrawn on 1st, 7th, 14th, 21st, 30th day of the experiments. Overlying water (or the polluted water added) samples were collected at Day 1, 7, 14, 21, 30 after mixing GCA with the sediment. These water samples were subjected for phosphates, Ammonia, Chemical Oxygen Demand and Color determination.

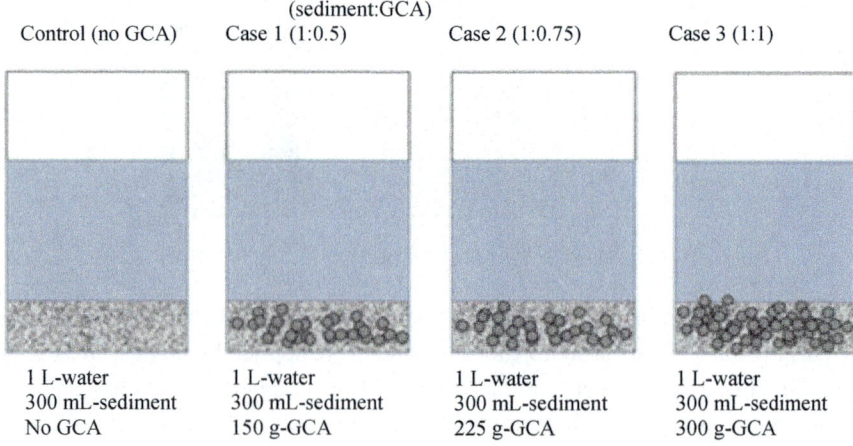

	(sediment:GCA)		
Control (no GCA)	Case 1 (1:0.5)	Case 2 (1:0.75)	Case 3 (1:1)

1 L-water	1 L-water	1 L-water	1 L-water
300 mL-sediment	300 mL-sediment	300 mL-sediment	300 mL-sediment
No GCA	150 g-GCA	225 g-GCA	300 g-GCA

Fig. 5. Experimental set-up for understanding the GCA effects on improving water quality.

5. Results and Discussion

Generally, the concentrations of NH$_3$ in the GCA containers were lower than the control. The GCA containers also showed an abrupt decrease at day 14 resulting to a very low concentrations in NH$_3$ until day 30 (Fig. 6a). This phenomena can be correlated with the continuous ammonia volatilization loss and nitrification where in ammonia could be oxidized to nitrate over time (temporal decreases of NH$_3$ in the control case). Comparing with the decreasing gradient of NH$_3$ of the control case, the cases with GCA had different tend of the decrease in NH$_3$, indicating the fixation of NH$_3$ by GCA.

The concentration of PO$_4^{3-}$ in the overlying water of the GCA containers was significantly lower compared to the control (Fig. 6b). The decrease in PO$_4^{3-}$ concentration in the overlying water is mainly due to the suppression of PO$_4^{3-}$ releasing flux from the sediment pore water to the overlying water. In relation to that, the significant decrease in the PO$_4^{3-}$ concentration is brought by the formation of calcium phosphate. PO$_4^{3-}$ is considered to be fixed by calcium ions that dissolved from the GCA components.

Figure 6c also shows clearer color of the cases with GCA compared to the control case. It is also important to understand that the true color of water is highly affected by the turbidity and the amount of dissolved organic matter present in the water. This means that GCA can decrease the amount of suspended solid through

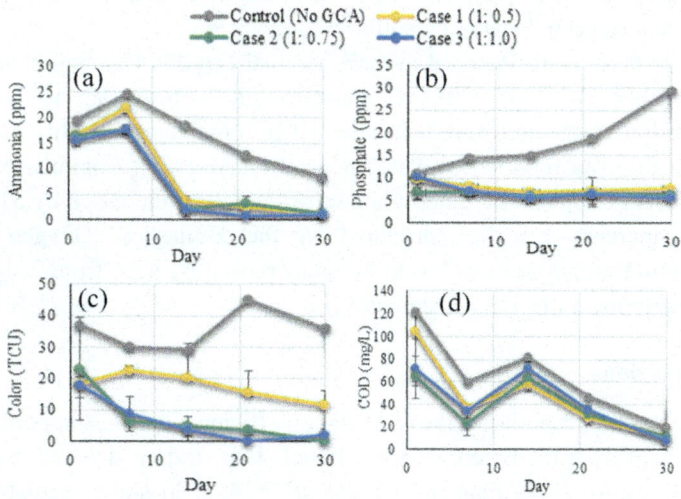

Fig. 6. Decreases in (a) ammonia, (b) phosphate, (c) color, (d) chemical oxygen demand in the overlying water influenced by additional of GCA at various dosage over time (day).

Table 2 Characteristics of water obtained from the Estero de San Miguel vs PRRC WQM data.

Parameter	Unit	Standard	2014	2015	2016	Jan. 2017	Mar. 2017	Remark
pH		6.5-8.5	7	7.13	6.93	6.85	6.62	Passed
ORP	mV	No standard				187.00	285.33	
Turbidity	NTU					98.27	82.03	
DO	mg/L	5	1.98	0.46	1.21	0.85	0.93	Failed
TDS	g/L	No standard				0.21	0.26	
EC	μS	No standard				340.00	387.00	
Odor		No standard				53.00	62.00	
Color	TCU	No standard				23.00	35.00	
COD	mg/L	No standard				119.00	128.00	
TSS	mg/L	<30 mg/L increase	15.40	44.20	15.80	32.00	37.00	Failed
Phosphate	mg/L	5			6.56	3.66	4.10	Failed
Ammonia	mg/L	No standard	10.92	1.99	7.03	10.63	11.05	Failed
Nitrate	mg/L	7			9.24	0.00	0.00	Passed
Nitrite	mg/L	No standard				0.06	0.00	
Sulfate	mg/L	No standard				28.62	24.35	

the enhancement of flocculation by calcium ions, and can adsorb dissolved organic matter in polluted water.

The adsorption of substances by GCA can be also confirmed through the lower COD concentrations of the GCA containers compared to the control. It has been reported that the application of GCA to sediment is effective in decreasing the sediment oxygen demand by sulfur oxidation [5].

While there is no standard in the Chemical Oxygen Demand concentration in the DENR DAO 34 Class C standards, it is important to monitor its concentration levels and how it is affected by other water quality parameters. Based on the data reported by Pasig River Rehabilitation Commission [6], the COD concentrations at the Estero de San Miguel have increased by 51% (Table 2). The increase can be correlated to the Biological Oxygen Demand concentration in the estero wherein it has increased by 62% from 2014 to 2015 thus failed to meet the 7 mg/L standard.

6. Conclusions

Overall, this study has shown the remediation efficiency of GCA in polluted water and sediment from the Estero de San Miguel. This study is deemed important in designing a pilot study plan on the use of GCA in in situ remediation in the Philippines. From the experiment results, it is expected that the water quality in the Estero de San Miguel can be improved through the GCA application. The following conclusions address the specific objectives of the study:

1. The water and sediment quality of the Estero de San Miguel were determined through laboratory analysis and was compared to the historical data from the water quality monitoring of Pasig River Rehabilitation Commission during dry season from the year 2014-2016. From the data, most of the water quality parameters did not pass the DENR DAO 34 Class C standards.
2. GCA has shown remediation efficiency in the improvement of physicochemical parameters and nutrient contamination in overlying water. Notably, GCA treated cases recorded significant improvement in phosphate (46% removal) and ammonia (94% removal) concentrations compared to the case without GCA.

Acknowledgments

This research is indeed a product of combined efforts of various individuals, government agencies and institutions. The authors would like to thanks the University of the Philippines for their collaboration on the research, the Chugoku Electric Company who provides laboratory logistics such as vials and other, Hiroshima University, The D.N.C. Ltd., Pasig River Rehabilitation, Barangay Local Government Unit for their participation and contribution that made this research possible.

References

1. T. Yamamoto, K. Harada, K. H. Kim, S. Asaoka and I. Yoshioka, Suppression of Phosphate Release from Coastal Sediments Using Granulated Coal Ash, *Estuar. Coast. Shelf Sci.* **116**, pp. 41-49 (2013).
2. S. Asaoka and T. Yamamoto, Characteristics of phosphate adsorption on to granulated coal ash in seawater, *Mar. Pollut. Bull.* **60**, pp. 1188-1192 (2010).
3. JSCE, *Concr. Libr.* **132**, 103 (2009).
4. H. Fukuma, T. Hibino, T. Yamamoto and T. Saito, Restoration of water environment by covering granulated coal ash in brakish water Nakaumi (Japan), *JSCE Annu. J. Coast. Eng.* **56**, pp. 1026-1030 (2009).
5. S. Asaoka and T. Yamamoto, H. Yamamoto, H. Okamura, K. Hino, K. Nakamoto, T. Saito, Estimation of Hydrogen Sulfide Removal Efficiency with Granulated Coal Ash Applied to Eutrophic Marine Sediment Using a Simplified Simulation Model, *Mar. Pollut. Bull.* **94**, pp. 55-61 (2015).
6. Department of Environment and Natural Resources, Water Quality in the Philippines (2008-2015), Vol. 4, pp. 1-64 (2015).

Physicochemical Habitability Conditions for the Genus *Halophila* in Nakagusuku Bay, Japan

T. Inoue

Marine Information and Tsunami Department, Port and Airport Research Institute,
Yokosuka, Kanagawa 239-0826, Japan
E-mail: inoue-t@ipc.pari.go.jp
www.pari.go.jp/unit/kaikj/en/member/inoue/

M. Uchimura

Research Institute on Subtropical Ecosystems, IDEA Consultants, Inc.
Nago, Okinawa 905-1631, Japan
E-mail: ucm21215@ideacon.co.jp

Field observations were conducted to investigate the effects of physicochemical conditions on the distribution of the genus *Halophila* in subtropical Nakagusuku Bay (Okinawa, Japan). Four species belonging to the genus *Halophila* (*H. ovalis*, *H. major*, *H. nipponica*, and *H. decipiens*) were found growing in the study area. *H. nipponica* was most frequently shown in Nakagusuku Bay and *H. decipiens* was encountered less frequently, while *H. major* was mainly present at the same sites as *H. ovalis*. Statistical analysis using physicochemical data obtained at each observational site revealed that the distribution patterns of *H. ovalis* and *H. major* were independent of nutrient conditions, but were strongly affected by surface wave motion. The occurrence of *H. ovalis* was confirmed up to a depth of 8.3 m, and *H. decipiens* occurred below a depth of 9 m with *H. nipponica* where the organic content was high and wave motion was smallest among the observational sites. *H. nipponica* showed a preference for relatively high sedimentation and large amounts of organic materials and its distribution was independent of flow conditions. *H. major* grew together not only with *H. ovalis* but also with *H. nipponica*.

Keywords: Spatial distribution of *Halophila* genus; Physicochemical conditions; Field observation; Statistical analysis.

1. Introduction

Seagrasses form submersed meadow communities that are among the most productive on earth [1]. Therefore, they are recognized as important components of tropical and subtropical coastal ecosystems, and have been regularly investigated both qualitatively and quantitatively. In this respect, Costanza et al. [2] indicated an economic value of US$19004 ha^{-1} yr^{-1} for 17 seagrass beds, surpassed only by estuaries and swamps/floodplains and higher than those of coral reefs, shelf, tidal marsh/mangroves, lakes/rivers, and forests.

Halophila, *Halodule*, and *Cymodocea* are the most common genera in tropical and subtropical waters, and have been the subject of many studies [3]. Ralph [4] studied the photosynthetic response of *H. ovalis* and concluded that its optimum photosynthetic range was from 25°C to 30°C. Beer et al. [5] investigated adaptation strategies of some tropical seagrasses with respect to their abilities to grow in the upper intertidal region, and demonstrated the pH compensation point of three seagrass species in the tropical intertidal region. As discussed above, many studies regarding tropical and subtropical seagrass dynamics have been conducted and quantified the effects of biochemical conditions, such as temperature, salinity, desiccation, light intensity, pH, and nutrients. On the other hand, the effects of physical conditions, such as mean flow velocity, wave amplitude, and wave period, have also been documented [6], but the majority of such studies were limited to temperate species (e.g., *Zostera marina*), whereas there have been few investigations of tropical and subtropical species.

As Hydrocharitaceae are well-known as a complex taxonomic challenge mainly due to their high morphological plasticity [7], the Japanese taxonomy of *Halophila* had been confusing. For this concern, Uchimura et al. [3] reassessed the species diversity in Japan and recommended the following four species: *H. decipiens*, *H. major*, *H. ovalis*, and *H. nipponica*. Here, we present field observations from a subtropical bay in Okinawa, Japan, to investigate the effects of physicochemical conditions on the distribution of the genus *Halophila*, following the classification recommended by Uchimura et al. [3].

2. Materials and Methods

2.1. *In situ measurement and analytical methods*

Field observations were conducted intermittently avoiding a rough weather at 9 observational sites (see Fig.1 and Table 1) in Nakagusuku Bay, Okinawa, Japan, from June 2005 to August 2008. The environmental parameters observed in this study were water temperature, salinity, depth, and flow velocity at 15 cm above the sea floor. Water temperature and salinity were measured using self-registering measurement instruments (COMPACT-CT, Alec Electronics Co., Ltd.) at 1-min or 10-min intervals. Depth was also measured using self-registering pressure measurement instruments (COMPACT-TD, Alec Electronics Co., Ltd.) at the same time intervals as water temperature and salinity. Horizontal flow velocity components were measured using self-registering electromagnetic velocimeters

(COMPACT-EM, Alec Electronics Co., Ltd.) at 0.5-s intervals with a 1-hour burst time. Wave amplitude and wave period were calculated using zero-up crossing method, and each significant value was defined as an average of the top one-third.

Table 1 Descriptions of observational sites and the distribution of *Halophila* species in the study area. (The symbols of "++", "+", and "-" mean "abundant", "scarce", and "not found", respectively.)

station	latitude	longitude	mean depth	mean velocity	significant wave-velocity amplitude	significant wave period	*H. ovalis*	*H. major*	*H. nipponica*	*H. decipiens*
1	26°17'19.4"N	127°50'17.0"E	10.3 m	1.4	2.3	10.2	-	-	++	-
2	26°17'21.3"N	127°50'16.5"E	8.3 m	2.1	3.4	10.1	+	++	+	-
3	26°17'22.6"N	127°50'17.0"E	7.8 m	1.8	3.3	9.6	-	-	-	-
4	26°17'26.6"N	127°50'17.0"E	6.0 m	2.1	4.4	9.1	+	+	-	-
5	26°17'31.0"N	127°50'17.0"E	5.3 m	2.6	4.6	9.0	+	++	-	-
6	26°18'50.8"N	127°51'50.3"E	9.7 m	2.6	1.7	13.7	-	-	+	+
7	26°17'30.7"N	127°49'05.2"E	4.0 m	2.8	4.8	9.2	+	+	+	-
8	26°17'32.3"N	127°51'31.9"E	11.8 m	2.1	2.4	12.1	-	-	++	-
9	26°16'40.6"N	127°50'18.9"E	12.7 m	1.7	2.5	12.3	-	+	-	-

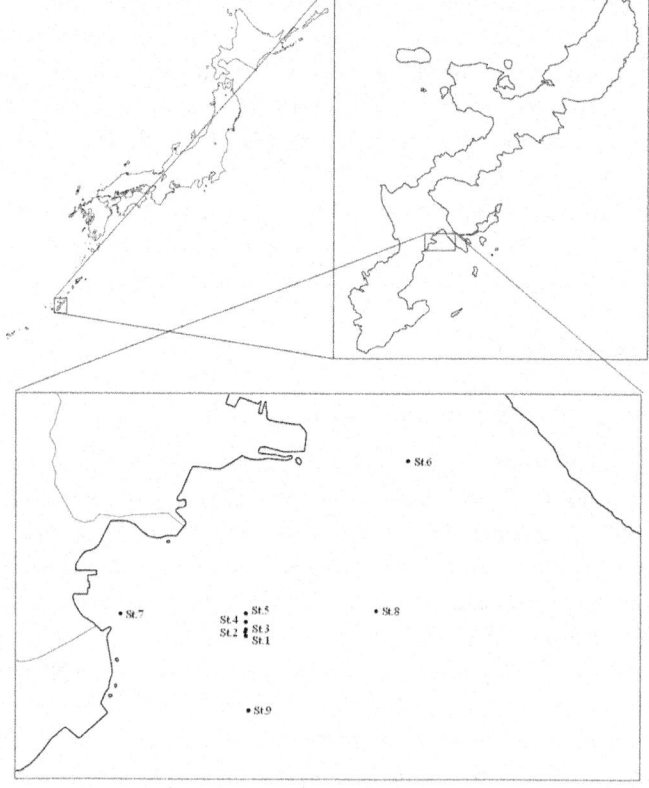

Fig. 1 Location of sampling sites.

Seagrass samples were obtained by SCUBA diving. Some shoots were desiccated in silica gel for later DNA extraction, and the remaining portions were transported to the laboratory for morphological examinations. DNA was extracted using a commercial kit (DNeasy plant mini kit, Qiagen) and protocol. The ribosomal ITS region including the 5.8S gene was selected for PCR analysis. Amplification and sequencing primers, amplification of the ITS regions of the rDNA, and automated sequencing were performed as described previously [8].

Water quality in the overlying water and the pore water and sediment quality were also measured sporadically. Sea water in the bottom layer was sampled by SCUBA diving using acid-washed polypropylene bottles taking care not to suspend the sediment. A portion of the sampled sea water was immediately filtered using a disposable filter with a pore size of 0.45 μm (Minisalt SM16555K, Saltrius K.K.). Filtrated and non-filtrated samples were bottled separately in acid-washed polypropylene bottles and stored in a cooler. Sediment was also sampled by SCUBA diving using acid-washed polypropylene bottles, and stored in a cooler. After field observation, samples were transported to the laboratory and a portion of the sediment was centrifuged in a box filled with N_2 gas to separate pore water. Obtained pore water was filtered using a disposable filter with a pore size of 0.45 μm (Minisalt SM16555K, Saltrius K.K.). All water and sediment samples were stored at $-25°C$ in a freezer until chemical analysis as described below.

Ammonium (NH_4), nitrite (NO_2), nitrate (NO_3), and soluble reactive phosphorus (SRP) of filtrated water samples were analyzed by the indophenol, colourimetric, cadmium reduction-colourimetric, and ascorbic acid methods [9] using a spectrophotometer (TRAACS800, Bran–Luebbe). Total nitrogen (TN) and phosphorus (TP) concentrations of filtrated and non-filtrated water samples were analyzed by the cadmium reduction–colourimetric and ascorbic acid methods, respectively, after persulphate digestion. Particulate total nitrogen (PTN) and phosphorus (PTP) concentrations were calculated by subtracting dissolved total nitrogen (DTN) and phosphorus (DTP) concentrations from TN and TP concentrations, respectively. Dissolved organic nitrogen (DON) and phosphorus (DOP) concentrations were also calculated by subtracting the sum of inorganic nitrogen fractions (i.e., NH_4, NO_2, and NO_3) and SRP concentrations from DTN and DTP concentrations, respectively.

Acid volatile sulphide (AVS) content in the sediment was measured by the titration method using sodium thiosulphate solution, after sulphide was distilled from the sediment under acid conditions by addition of 20% sulphuric acid solution. Sediment chemical oxygen demand (COD) was analyzed by the alkaline potassium permanganate method. Kjeldahl TN content in the sediment was analyzed by the macro-Kjeldahl method [9], and TP content in the sediment was analyzed by the ascorbic acid method after nitric acid–perchloric acid digestion. Loss on ignition (LOI) of the sediment was calculated by the weight difference of the sample before and after heating at 600°C for 2 h. Total manganese and total iron contents in the sediment were measured with atomic absorption photometry (Z-2000, Hitachi Ltd.), after decomposition using hydrochloric acid and nitric acid solution. Total organic carbon (TOC) was measured using a CHN analyzer (JM10, J-Science Lab Co., Ltd). Grain size composition was also measured by sieving method.

2.2. *Data analysis*

The obtained flow velocity data were statistically analyzed using the statistical software (Ekuseru-Toukei 2006 for Windows, Social Survey Research Information Co., Ltd.). First, Bartlett's test of homogeneity of variance was used to test the hydrodynamic similarity between the observational sites using hydrodynamic data, i.e., mean flow velocity, wave velocity amplitude, and wave period, calculated from flow velocity data that were measured continuously using self-registering electromagnetic velocimeters. If the assumption of equality of variance was rejected, the Kruskal–Wallis test was performed. Moreover, principal component analysis was carried out using nutrient concentrations in the bottom layer and the pore water and sediment quality.

It is rather difficult to quantitatively consider growth rates and/or abundances of *Halophila* species based on our observational results. Therefore, in this study, we analyzed based on whether each individual exist there or not and its relationship with the environmental conditions.

3. Results

3.1. *Distribution of Halophila*

Uchimura et al. [10] reported that the genus *Halophila* includes four distinct species in Japan: *H. ovalis*, *H. major*, *H. nipponica*, and *H. decipiens*. All of these species were found growing in Nakagusuku Bay. *H. ovalis* was found at Stns. 2, 4, 5, and 7. *H. major* was present at the same sites as *H. ovalis*, but also at Stn. 9. *H. major* especially dominated Stns. 2 and 5. *H. nipponica* was found frequently in Nakagusuku Bay and was confirmed at multiple sites in our observation, but not at Stns. 3, 4, 5, or 9. *H. decipiens* was encountered less frequently except for Stn. 6. Overall, the distribution of *Halophila* species in our observational period was rather steady and is briefly summarized in Table 1.

3.2. *Observational results and statistical analysis*

Water temperature fluctuated between 17.1°C and 30.1°C, and there were no obvious differences among the observational sites. Salinity showed a steady value from 33.0 psu to 33.6 psu at all sites. The sediment sampled from Stns. 6 and 9 contained larger portion of fine particles and Stn. 3 was rather coarse (Fig.2).

As the assumption of equality of variance was rejected for all hydrodynamic data in Bartlett's test, the Kruskal–Wallis test was performed and summarized as below. The results of mean velocity analyses showed that Stn. 1 was the slowest site, at which the mean flow velocity was 1.4 cm s^{-1}, followed by Stn. 9 (1.7 cm s^{-1}) and Stn. 3 (1.8 cm s^{-1}). Stns. 2, 4, and 8 were intermediate sites (2.1 cm s^{-1}), and there were no differences among them. Mean flow velocity at Stns. 5 and 6 were higher (2.6 cm s^{-1}) with no difference between them, and that at Stn. 7 was highest (2.8 cm s^{-1}) (see also Fig.3(a)). Concerning significant wave amplitude, that at Stn. 6 (1.7 cm s^{-1}) was smallest, and the values at Stns. 1, 8, and 9 (about 2.4 cm s^{-1}) were also rather small. Those at Stns. 2 and 3 were larger (about 3.4 cm s^{-1}) and were indistinguishable. Those at Stns. 4, 5, and 7 were largest (about 4.6 cm s^{-1}) (see also Fig.3(b)). Significant wave periods at Stns. 4, 5, and 7 were the shortest (about 9.1 s) and were indistinguishable, followed by Stn. 3 (9.6 s). The values at Stns. 1 and 2 were intermediate (about 10.2 s) and there was no significant difference between them. Those at Stns. 8 and 9 were longer (about12.2 s) with no difference between them, and that at Stn. 6 was the longest (13.7 s) (see also Fig.3(c)).

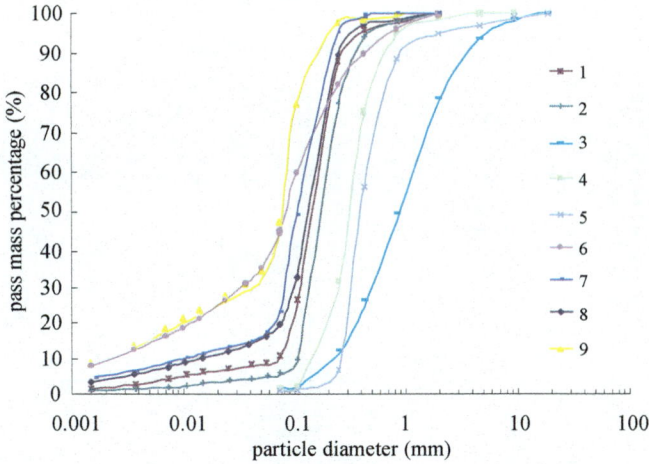

Fig. 2 Pass mass percentage of the sampled sediment.

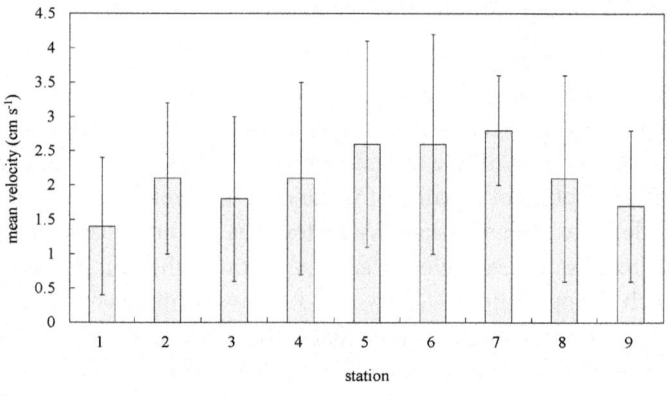

(a) mean velocity

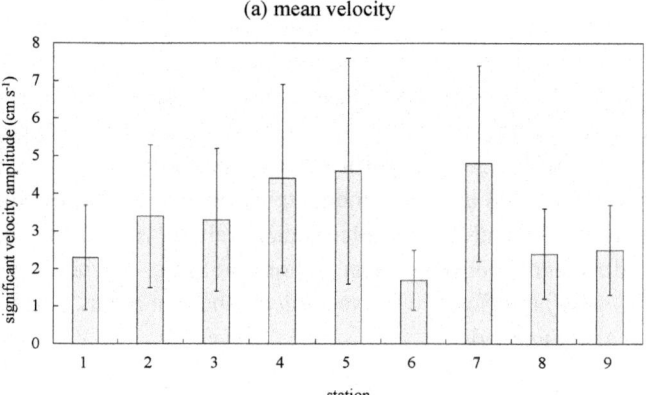

(b) significant velocity amplitude

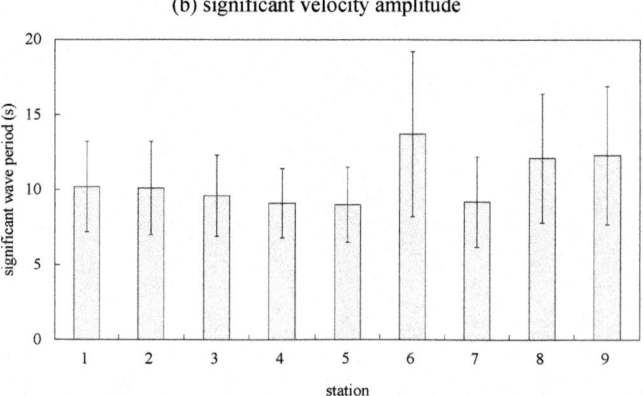

(c) significant wave period

Fig. 3 Hydrodynamic properties.

Principal component analysis regarding nutrient concentrations in the overlying water and the pore water, and sediment quality showed that the first principal component was the most significant with a contribution of 98.65% (see Fig. 4). Fig. 5 shows characteristic vectors and factor loadings of the first principal component. According to Fig. 5 (a), the characteristic vector of NH_4 concentration in the pore water showed the largest value, followed by those of NO_3 and SRP concentrations in the pore water. Moreover, Fig. 5 (b) shows that factor loadings of nutrient concentrations in the pore water had large positive values and those of organic matter, iron, and manganese content in the sediment had large negative values. Principal component scores are shown in Fig. 6. Principal component scores of Stns. 2, 6, 7, and 8 showed negative values, and those of Stns. 3, 4, and 5 showed large positive values.

4. Discussion

4.1. *Statistical analysis*

From the results of principal component analysis, factor loadings of nutrient concentrations in the pore water showed large positive values and those of organic matter, iron, and manganese content in the sediment showed large negative values for the first principal component. In addition, principal components scores of Stns. 2, 6, 7, and 8 showed negative values, and those of Stns. 3, 4, and 5 had large positive values. Therefore, in this paper, Stns. 2, 6, 7, and 8 were considered as sites with relatively high sedimentation tendencies containing large amounts of organic materials. On the other hand, Stns. 3, 4, and 5 were considered to have relatively high degradation tendencies with less sedimentation and richer nutrient conditions. Moreover, from the results of the Kruskal–Wallis test, Stns. 2, 3, 4, 5, and 7 were regarded as sites strongly affected by surface wave motion. In contrast, the other sites were less affected by surface waves. Mean flow velocities at Stns. 1 and 9 were slower, but those at Stns. 6 and 8 were intermediate and higher, respectively.

4.2. *Distribution of Halophila in Nakagusuku Bay*

The distribution patterns of *H. ovalis* and *H. major* were independent of the principal component score, i.e., nutrient conditions were not responsible for their distributions in the Awase area. On the other hand, they occurred at sites strongly affected by surface wave motion. [11] tested the effects of siltation on Southeast Asian seagrass community structure and biomass, and revealed that *H. ovalis* is the next most sensitive to siltation compared to *Enhalus acoroides*. Our results

878

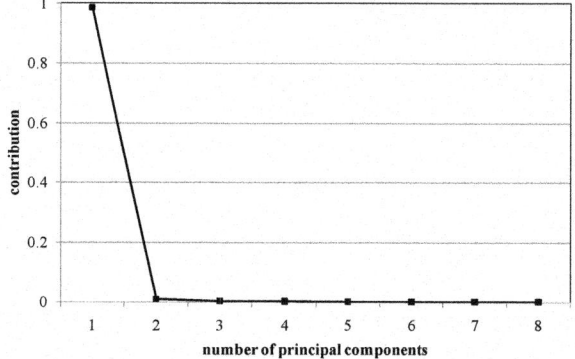

Fig. 4 Scree plot of contribution as a function of principal components.

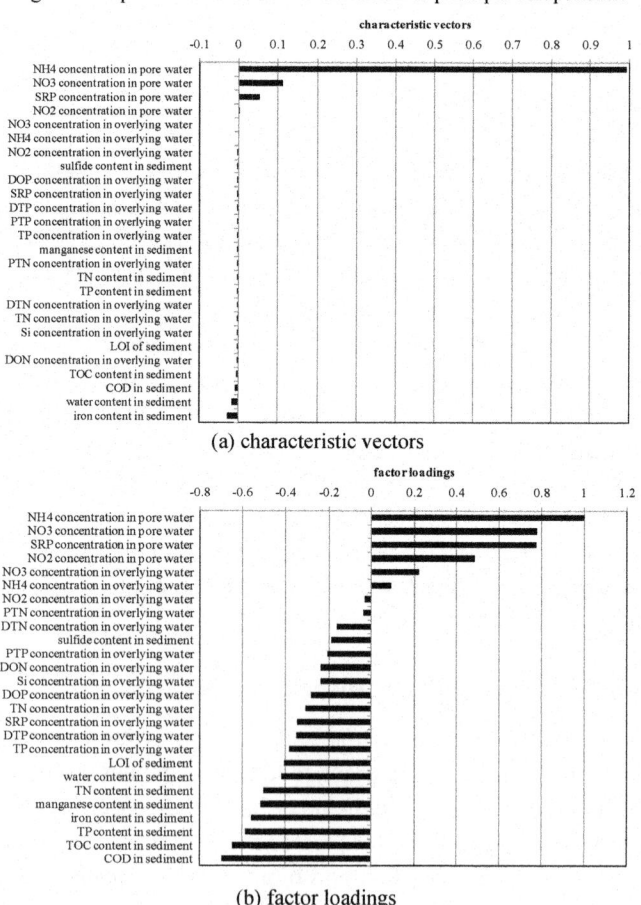

(a) characteristic vectors

(b) factor loadings

Fig. 5 Results of the first principal component.

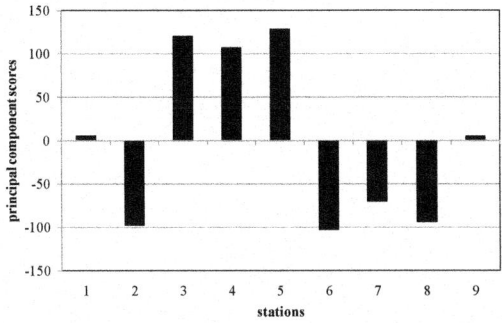

Fig. 6 Principal component scores.

agreed with their findings, but may also mean that the plants required high levels of light intensity for their growth [e.g. 12].

According to Kuo and Kirkman [13], *H. ovalis* and *H. decipiens* may grow together. However, Hulings [14] reported that *H. ovalis* occurs in a depth range of 0.5 to 2 m and there was no occurrence below a depth of 2 m, and some groups have shown that *H. decipiens* occurs in a depth range of 5 to 100 m [15]. At our observational site, the occurrence of *H. ovalis* was confirmed up to a depth of 8.3 m (Stn. 2), and *H. decipiens* occurred below a depth of 9 m (Stn. 6) with *H. nipponica* where the organic content was high and wave motion was smallest among our observational sites. These results were consistent with those reported previously [16].

H. nipponica also occurs over a wide area, but is not found at sites with high principal component scores. This suggests that *H. nipponica* has a preference for relatively high sedimentation and large amounts of organic material and its growth is independent of flow conditions. Uchimura et al. [8] reported that *H. nipponica* forms isolated small patches or is associated with *H. major* (the species described as *H. japonica* and *H. euphlebia* in their paper have been re-described as *H. nipponica* and *H. major*, respectively). According to our observations, *H. major* was associated only with *H. ovalis* at sites with high principal component scores (Stns. 4 and 5), and grew together not only with *H. ovalis* but also with *H. nipponica* at sites with low principal component scores (Stns. 2 and 7).

H. nipponica leaves found in Awase area usually had a round shape, but some had a narrow shape only at Stn. 2. Short [17] reported that seagrass morphology is closely linked to available nutrient resources. However, we could not determine the cause of the leaf deformation, and this remains to be examined in a future study.

5. Conclusions

Field observations regarding water temperature, salinity, flow velocity, water quality, sediment quality, and pore water quality were conducted to investigate the effects of physicochemical conditions on *Halophila* genus distribution in a subtropical coastal area in Japan. Nine observation sites were statistically distinguishable from each other, and their physicochemical properties were compared with *Halophila* genus distribution.

We have to note that the existence or abundance of seagrass species in coastal area must be strongly affected by catastrophic events such as the rough weather and a typhoon, which cannot be considered from our observation. Although preferable physicochemical conditions for four species of *Halophila* were briefly summarized in this paper, further studies based on physiological aspects are required to elucidate the details of these relationships.

Acknowledgments

We thank Ms. Hiromi Sekimoto for her technical assistance.

References

1. B. W. Touchette and J. M. Burkholder, Overview of the physiological ecology of carbon metabolism in seagrasses, *J. Exp. Mar. Biol. Ecol.* **250**, 169 (2000).
2. R. Costanza, R. d'Arge, R. d. Groot, S. Farber, M. Grasso, B. Hannon, K. Limburg, S. Naeem, R. V. O'Neill, J. Paruelo, R. G. Raskin, P. Sutton and M. v. d. Belt, The value of the world's ecosystem services and natural capital, *Nature* **387**, 253 (1997).
3. M. Uchimura, E. J. Faye, S. Shimada, S. Arai, T. Inoue and Y. Nakamura, A re-evaluation of the taxonomic status of *Halophila euphlebia* Makino (Hydrocharitaceae) based on morphological features and ITS sequence data, *Bot. Mar.* **49**, 111 (2006).
4. P. J. Ralph, Photosynthetic response of laboratory-cultured *Halophila ovalis* to thermal stress, *Mar. Ecol. Prog. Ser.* **171**, 123 (1998).
5. S. Beer, M. Mtolera, T. Lyimo and M. Björk, The photosynthetic performance of the tropical seagrass *Halophila ovalis* in the upper intertidal, *Aquat. Bot.* **84**, 367 (2006).
6. M. S. Fonseca and M. A. R. Koehl, Flow in seagrass canopies: The influence of patch width, *Estuar., Coast. Shelf Sci.* **67**, 1 (2006).

7. V. X. Nguyen, M. Detcharoen, P. Tuntiprapas, U. Soe-Htun, J. B. Sidik, M. Z. Harah, A. Prathep and J. Papenbrock, Genetic species identification and population structure of *Halophila* (Hydrocharitaceae) from the Western Pacific to the Eastern Indian Ocean, *BMC Evol. Biol.* **14**, 92 (2014).

8. M. Uchimura, E. J. Faye, S. Shimada, G. Ogura, T. Inoue and Y. Nakamura, A taxonomic study of the seagrass genus *Halophila* (Hydrocharitaceae) from Japan: Description of a new species *Halophila japonica* sp. nov. and characterization of *H. ovalis* using morphological and molecular data, *Bull. Nat. Sci. Mus. Tokyo, Ser. B* **32**, 129 (2006).

9. APHA-AWWA-WEF: Standard Methods for the Examination of Water and Wastewater, 20th edition. American Public Health Association/American Water Works Association/Water Environment Federation, New York, USA (1998).

10. M. Uchimura, E. J. Faye, S. Shimada, T. Inoue and Y. Nakamura, A reassessment of *Halophila* species (Hydrocharitaceae) diversity with special reference to Japanese representatives, *Bot. Mar.* **51**, 258 (2008).

11. J. Terrados, C. M. Duarte, M. D. Fortes, J. Borum, N. S. R. Agawin, S. Bach, U. Thampanya, L. Kamp-Nielsen, W. J. Kenworthy, O. Geertz-Hansen and J. Vermaat, Changes in community structure and biomass of seagrass communities along gradients of siltation in SE Asia, *Estuar., Coast. Shelf Sci.* **46**, 757 (1998).

12. S. Beer and M. Björk, Measuring rates of photosynthesis of two tropical seagrasses by pulse amplitude modulated (PAM) fluorometry. *Aquat. Bot.* **66**, 69 (2000).

13. J. Kuo and H. Kirkman, *Halophila decipiens* Ostenfeld in estuaries of southwestern Australia, *Aquat. Bot.* **51**, 335 (1995).

14. N. C. Hulings, The ecology, biometry and biomass of the seagrass *Halophila stipulacea* along the Jordanian coast of the Gulf of Aqaba, *Bot. Mar.* **22**, 425 (1979).

15. F. E. C. de Oliveira, J. Pirani and A. M. Giulietti, The Brazilian seagrasses, *Aquat. Bot.* **16**, 251 (1983).

16. E. Rasmussen, The wasting disease of eelgrass (*Zostera marina*) and its effects on environmental factors and fauna, in *Seagrass Ecosystems: A Scientific Perspective*, eds. C. P. McRoy and C. Helfferich (M. Dekker, New York, 1977) pp. 1-52.

17. F. T. Short, Effects of sediment nutrients on seagrasses: Literature review and mesocosm experiment, *Aquat. Bot.* **27**, 41 (1987).

Impact on Dissolved Oxygen of Intermittent Deep Water Intrusion from the Outer Bay after the 2011 Tohoku Earthquake Tsunami in Ofunato Bay, Japan

Toshiyuki Takao[†‡], Keiji Kozawa[§], Kazuo Murakami[⊥], Ikuo Abe[#]
Toshinori Ogasawara[¶], Hidenori Shibaki[†] and Yoshiyuki Nakamura[‡]

[†]*ECOH Corporation, Taito-ku, Tokyo 1100014, Japan*
E-mail: takao@ecoh.co.jp
[‡]*Yokohama National University, Yokohama, Kanagawa 2408501, Japan*
[§]*Tohoku Regional Bureau Ministry of Land, Transport and Tourism, Kamaishi, Iwate 0260011, Japan*
[⊥]*Japan Sediments Management Association, Chuo-ku, Tokyo 104-0042, Japan*
[#]*Tokoha University, Fuji-shi, Shizuoka 417-0801, Japan*
[¶]*Iwate University, Morioka, Iwate 020-8551, Japan*

Tsunami protection breakwaters constructed at the mouth of Ofunato Bay were collapsed by the 2011 Tohoku Earthquake Tsunami. Before the 2011 Tsunami, anoxic/hypoxic water masses developed every summer in the bay. To understand anoxic/hypoxic environment after the Tsunami event and to inform appropriate reconstruction of the breakwaters, field measurements of water temperature, salinity, dissolved oxygen (DO) and currents were conducted in 2012 and 2013. DO concentrations in the Bay in 2012 and 2013 were not anoxic/hypoxic as they were before the 2011 Tsunami. We clarified that the primary factor supplying DO into the lower layer of the inner bay was intermittent intrusion of low temperature, high density, and high DO water masses from the outer bay.

Keywords: Tsunami protection breakwater; Anoxic/hypoxic water; Water exchange; The 2011 Tohoku Earthquake Tsunami.

1. Introduction

Ofunato Bay located on the Sanriku Coast of Japan's Tohoku region is one of the several partially-enclosed bays along a ria coast facing the Pacific Ocean (Fig. 1). Due to repeated damage caused by tsunami events in the past, tsunami protection breakwaters (mound height DL-16.0m, open width 200m) were constructed at the mouth of Ofunato Bay in 1967. However, almost all of those breakwaters, including their mounds, were collapsed by the unexpected massive tsunami generated by the 2011 Tohoku Earthquake in March 2011. Before the

2011 Tsunami, the water quality deterioration inside Ofunato Bay was serious due to limited seawater exchange which was impeded by the breakwaters and increased pollutant loadings from the drainage area. Therefore, anoxic water masses in which the amount of dissoloved oxygen (DO) was less than 1 mg/l were developed in the lower layer of the inner bay during the stratified season from summer to autumn every year[1,2].

Okazaki[3] reported that the thermocline zone in the outer ocean in the Sanriku Coast becomes shallower than the water depth at the mouth of the bay due to the fact that the internal gravity waves propagate from north to south along the Sanriku Coast in the stratified season, and it causes that the offshore cold seawater intrude into the low layer of the inner bay as density currents. Okada and Nakayama[4] also reported that the seawater exchange due to density intrusion at the mouth of the Ofunato Bay in the stratified season is important for the water quality in the inner region of the bay, supported by field experiments and simulations.

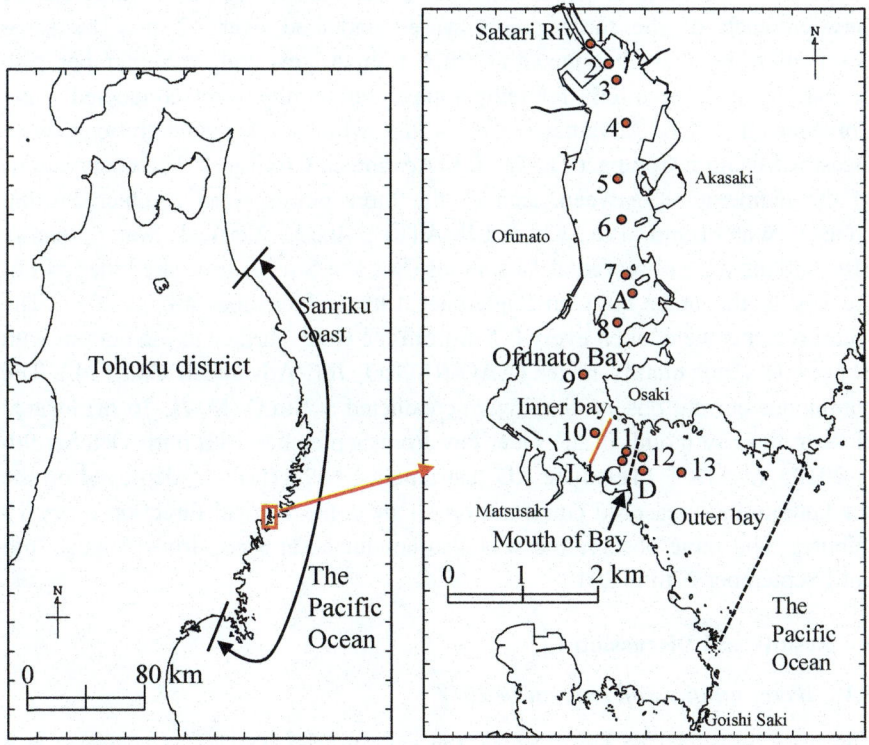

Fig. 1. Map of Ofunato Bay (left) and field measurement locations (right).

The objective of this study is to evaluate the water quality in the inner region of Ofunato Bay, and thereby reveal the primary factors of DO concentrations shifting before and after the collapse of the breakwaters at the mouth of the bay, based on field observations of DO concentrations conducted in 2012 and 2013.

2. Field Measurements

Field measurements on water quality and currents[5] were conducted at several stations inside and outside Ofunato Bay in 2012 and 2013 (Fig. 1).

For water quality, both continuous observations and vertical distribution observations for temperature, salinity and DO concentrations were conducted. Continuous water quality observations were conducted at two stations, Stn.A (MWL -32 m) located in the inner bay and Stn.D (MWL -36 m) located in the outer bay and three points of those stations, 2 m below sea surface, 18 m below MWL and 1 m above the seabed respectively, in every 10 minutes from September 18th until October 23th in 2012. Observations of temperature between each of the three points were conducted every 2 m. The same observation were also conducted at Stn.A from August 26th to September 27th in 2013. The observations for temperature and salinity were conducted using Compact-CT (JFE Advantech Co., Ltd.) while DO concentrations were measured using a Compact-DOW (JFE Advantech Co., Ltd.). The observations of the temperature between each of the three points were conducted using HOBO WaterTemp Pro v2 (CLIMATEC, Inc.). Vertical water quality distributions were observed at 13 stations (Stn.1 to Stn.13, from the inner part of the bay to the outer bay) on September 19th and October 4th in 2012. The measurements were taken every 0.1 m from the water surface to the bottom with a multiple water quality meter (AAQ-RINKO, JFE Advantech Co., Ltd.). The continuous current observations were conducted at Stn.C (MWL-36 m) located close to the bay mouth in the inner bay from September 18th until October 9th in 2012. An ADCP (Teledyne RD Instruments, 600 kHz) was deployed on the sea bottom and measured currents every 1 m in the vertical direction every 10 minutes. The same observation was also conducted at Stn.C from August 26th until September 27th in 2013.

3. Results and Discussion

3.1. *Water quality and current in 2012*

The time series of the temperature, the salinity, the density of seawater at a given temperature and salinity (σ_t) and the DO concentrations of the upper layer (represented by the measurement located at 2 m under sea surface), the middle layer (represented by the measurement at 18 m below MWL) and the lower layer (represented by the measurement at 1 m above seabed) at Stn.A and Stn.D observed in 2012 are shown in Fig. 2.

The observed data at Stn.A indicated the following features. The temperatures were stratified during the observation period and the one in the upper layer was the highest of those three layers. The difference of the temperature between the middle and the lower layer was larger compared to the one between the upper and the middle layer around the time of the cold water intrusion. On the other hand, the difference of the salinity between the upper and the middle layer was larger than the one between the middle and the lower layer, while the absolute differences among the three layers were smaller than the one in the temperature. The difference of the σ_t between the middle and the lower layer was larger than the one between the upper and the middle layer due to the influence of the structure of the temperature distribution. The temperature of the lower layer occasionally decreased significantly, which caused the σ_t in the lower layer to increase. The DO concentrations of the upper and the middle layer were high at all times and the ones of the lower layer were also kept relatively high most of the time: only 23 hrs out of the whole 36 period days did the DO become less than 4 mg/l, which indicates the hypoxic condition. The minimum concentration at the lower layer was 3.4 mg/l at 16:10 on October 8th, 2012. It is worth to mention that the DO concentration in the lower layer increased abruptly when the temperature in this layer decreased rapidly, which happened occasionally during the measurement. After these occasional decreases of the temperature in the lower layer, the temperature increased gradually towards the original temperature. The same can be applicable to the σ_t: after these occasional increases of the σ_t in the lower layer, the σ_t decreased gradually to the original level of the σ_t.

The observed data at Stn.D indicated the following features. The temperature in the middle and lower layer at Stn.D occasionally decreased in a short duration when the temperature in the middle and lower layer at Stn.A decreased. As for the salinity, there was not so much change, while it shows low value in the upper layer. It was observed that the σ_t in the middle and lower layer became high for a short moment due to the influence of the temperature. The DO concentrations in the middle and the lower layer showed high value during the measurement.

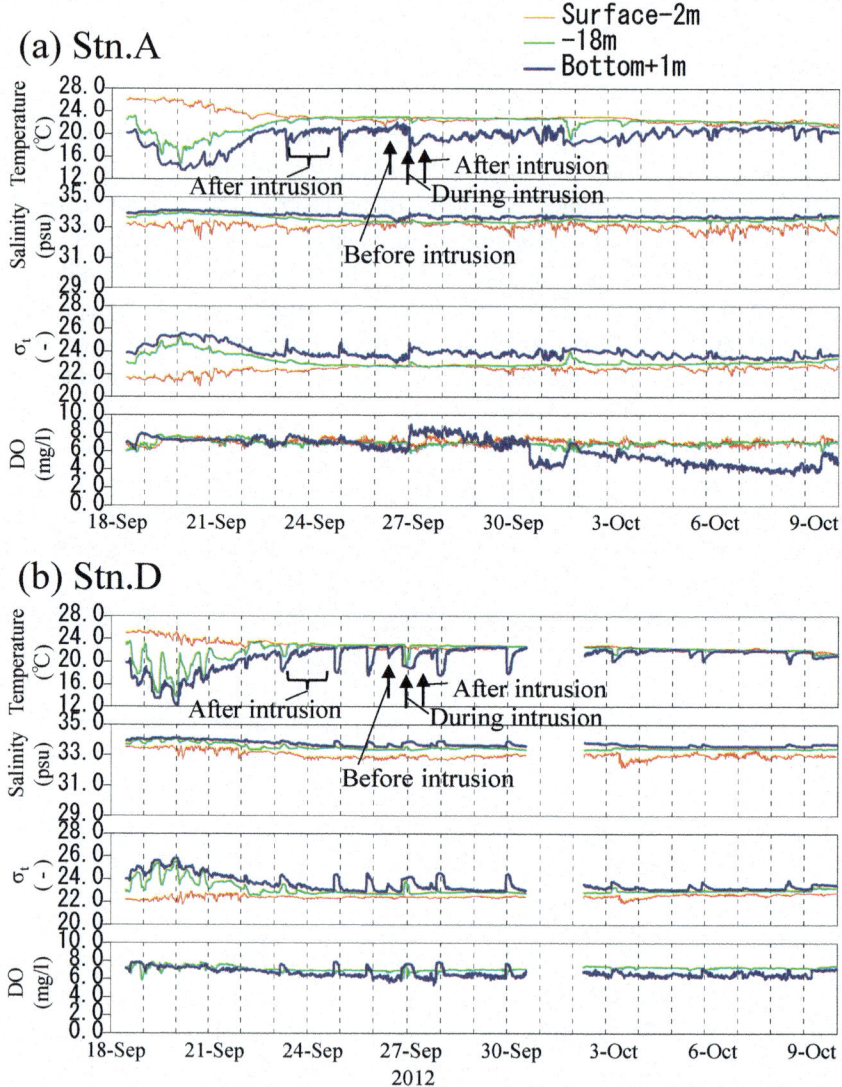

Fig. 2. Time series of temperature, salinity, σ_t and dissolved oxygen concentration at 2 m below sea surface, 18 m depth, 1 m above the seabed from September 18th, 2012 until October 9th, 2012 at station A (a) and station D (b).

Fig. 3 (a) shows a time series of velocity profile at the mouth of the bay (Stn.C). Note that the velocity component is normal to the cross-section at the mouth of the bay (101 degrees clockwise from North). Fig. 3 (b) shows time series of velocity at 31 m from MSL at Stn.C and temperatures at 31 m from MSL at Stn.A and Stn.D. The time series of the velocity profile indicated that there were occasional inflows at the lower layer (31 m from MSL), for instance, on September 18th-19th, September 23rd, September 26th -27th, September 30th -October 1st, October 3rd and October 9th. Other than these periods, there were inflows at the middle layer and outflows in the upper and lower layers.

(a) Stn.C

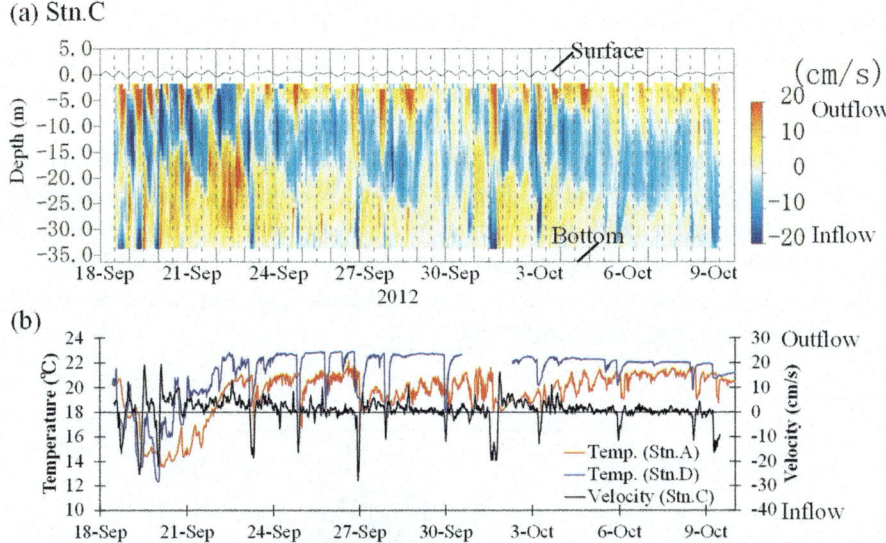

(b)

Fig. 3. Time series of velocity profile and temperature at Stn.A, C and D from September 18th, 2012 to October 9th, 2012. (a) velocity components normal to the cross-section at the mouth of the bay (101 degrees clockwise from North) at station C. (b) temperature at station A and D, velocity (each data were obtained at -31 m from MSL).

The timings of the inflow in the lower layer correspond to the timings of temperature drop in the lower layer at Stn.D. Right after the temperature drop at Stn.D, the temperature drop can be seen at Stn.A. Fig. 4 shows the relationship between the velocities at 31 m below MSL at Stn.C and the differences of the bottom temperature between Stn.D and Stn.A. The figure indicates that when the difference is negative (when temperatures of the lower layer at the outer bay are less than those at the inner bay), the velocity at the lower layer at Stn.C is also negative (i.e. inflow). In addition, the DO concentrations of the lower layer at Stn.A increased when there is a temperature drop at the same location (Fig. 2 (a)). In order to confirm that such changes are caused by inflow from the outer bay, TS diagram of the lower layer at Stn.A and Stn.D before and after intermittent cold water intrusion on September 26th and 27th were shown (Fig. 5 (a)). Before the cold water intrusion, the temperatures at Stn.D were higher than the ones at Stn.A, and the densities were lower than the ones at Stn.A. During the cold water intrusion, the temperatures at Stn.D were lower than ones at Stn.A, and the densities were higher than ones at Stn.A. During the cold water intrusion, the temperatures and salinities at Stn.A did not show a lot of changes compared to the ones before the intrusion. After the cold water intrusion, the temperatures at Stn.A decreased and the temperatures and salinities at Stn.A became closer to the ones at Stn.D during the intrusion. On top, the densities at Stn.A became higher than the ones at Stn.A before the cold water intrusion. From those results, it can be considered that the low temperature (high density) water masses entered to the lower layer of the inner bay through the mouth of the bay. Fig. 5(b) shows The TS diagram of the lower layer at Stn.A and Stn.D after the intermittent cold water intrusion on September 23rd and 24th. The figure shows that the temperatures of the lower layer at Stn.A increased gradually to the values at the outer bay, and the densities at Stn.A also decreased gradually to the values of the outer bay.

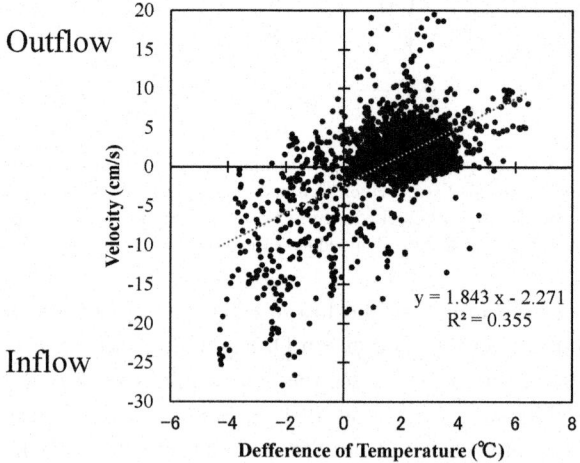

Fig. 4. Velocity at 31 m below MSL and difference of bottom temperature between Stn.D and the Stn.A.

From those results, the reason that the DO concentration in the lower layer at Stn.A remained high during the measurement can be explained as follows. First of all, the low temperature and high density water masses occasionally appeared at the outer bay. They eventually intruded the middle and the lower layers of the inner bay through the bay's mouth under the form of density currents. This made the temperatures of the middle and the lower layers of the inner bay decrease (while the densities of the same layers of the inner bay increased). Since the DO concentrations at the outer bay were always high, the ones in the lower layer of the inner bay were increased by the cold water intrusion. After that, those low temperature (high density) water masses gradually flow out to the outer bay and the temperatures returned gradually to their original level. It encouraged the next intermittent cold water intrusion. It can be considered that this mechanism would be responsible for the high DO concentration during the measurement in 2012.

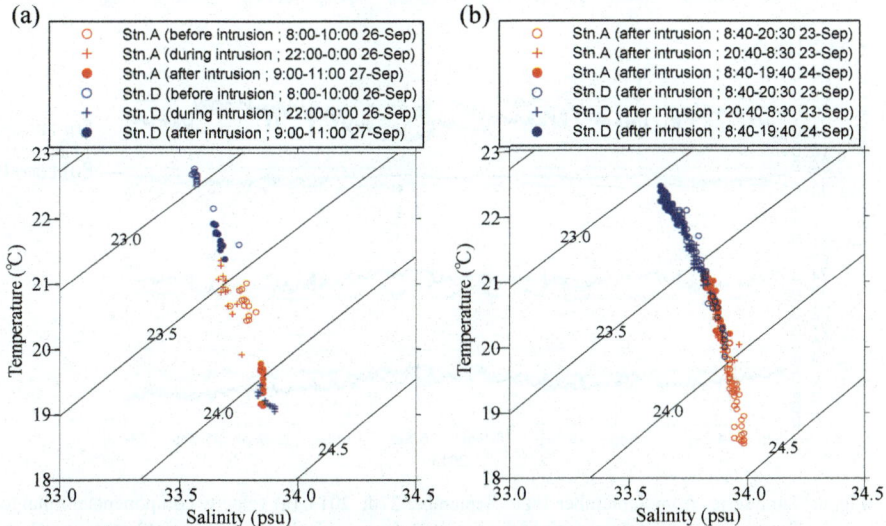

Fig. 5. TS diagram at Stn.A and Stn.D in September 2012. (a) before, during and after the cold water intrusion on September 26th and 27th. (b) after cold water intrusion on September 23th and 24th.

3.2. *Effects of intermittent cold water intrusion in 2013*

Fig. 6 (a) shows a time series of velocity profile at the mouth of the bay (Stn.C) in September 2013. Equal to the measurement settings in 2012, the velocity component is normal to the cross-section at the mouth of the bay. Similar to the characteristics of the currents in 2012, inflows (i.e. cold water intrusion) at the lower layer from the outer bay were observed (on September 4th-7th, September

9th-10th, September 13th, September 15th-17th, September 20th-22nd and September 24th). After those inflows at the lower layer, cold water intrusions at the middle layer from the outer bay were also observed. These were returned to the outer bay from the lower layer.

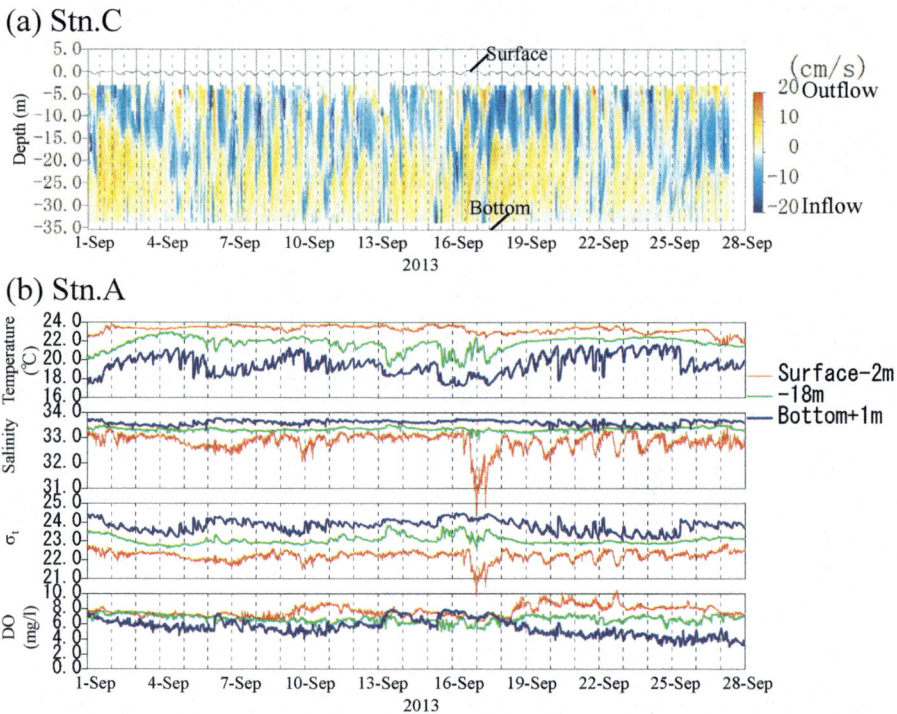

(a) Stn.C

(b) Stn.A

Fig. 6. Time series from September 1st to September 27th, 2013. (a) velocity components normal to the cross-section at the mouth of the bay (101 degrees clockwise from North) at station C. (b) temperature, salinity, σ_t and dissolved oxygen concentration at 2 m below sea surface, 18 m from MSL and 1 m above seabed.

Fig. 6 (b) shows a time series of temperature, salinity, σ_t and DO concentrations of the upper, the middle and the lower layer at Stn.A in 2013. Same as the observed characteristics in 2013, the temperatures at the lower layer were decreased, and the σ_t and the DO concentrations at the lower layer were increased during the period in which there were inflows in the lower layer at Stn. A. Afterwards the temperatures of the lower layer at Stn.A were gradually increased and σ_t were gradually decreased. Due to those processes, high DO concentration at the lower layer of the inner bay was kept high in 2013.

3.3. *Water quality before and after the collapse of the breakwaters due to the 2011 Tsunami*

The vertical cross sectional DO concentration from the river mouth to the outer bay obtained before and after the collapse of the breakwaters due to the 2011 Tsunami are shown in Fig.7 (a) and (b), respectively. The DO concentration at 1999 before the collapse of the breakwaters shows that a huge anoxic/hypoxic water mass exists below the middle layer in the inner bay. However, the DO concentration at 2012 after the collapse of the breakwaters shows that the anoxic/hypoxic water mass cannot be seen in the inner bay.

Due to the damage from the 2011 Tsunami, the cross-sectional area at the mouth of Ofunato Bay was increased from 3,200 m² to 17,800 m², and the depth at the mouth of the bay was increased from 16 m to 31 m. The primary factor of the high DO concentration of the lower layer at the inner bay is the increase of the cross-sectional area and the depth at the mouth of the bay. Because of this geomorphological change, low temperature (high density) water masses can inflow easier to the inner bay, and also outflow easier to the outer bay.

(a) Before the collapse of the breakwaters

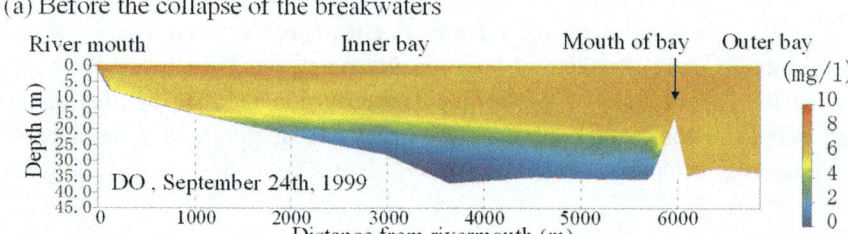

(b) After the collapse of the breakwaters

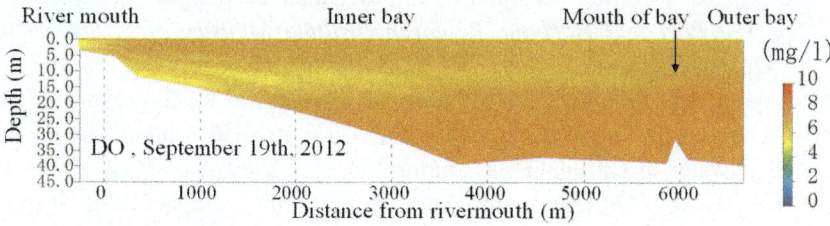

Fig. 7. DO distribution along river mouth to outer bay. (a) September 24th, 1999 shown in Okada and Nakayama (2007)[4]. (b) September 19th, 2012.

4. Conclusion

Based on the field measurement data obtained in 2012 and 2013, it is confirmed that the DO concentrations at the inner bay have had been kept high. The primary factor of such high DO concentrations were considered as follows. First, low temperature (high density) water masses were occasionally appeared in outer bay in summer. These intruded the middle and lower layers in the inner bay through the bay mouth as density currents. As a result, DO were supplied from the outer to the inner bay. Since those low temperature water masses gradually flowed out to the outer bay, the temperature would gradually return to its original level. Because of this, low temperature water mass can easily intrude intermittently, and high DO concentration is eventually kept in the lower layer of the inner bay.

Acknowledgments

This research is part of a work on water quality conservation against reconstruction of tsunami protection breakwaters at the mouth of the Ofunato Bay. We would like to express our sincere gratitude to late Prof. Shigeki Sakai for his contribution and Dr. Tomonari Okada for his valuable advices and comments. We also appreciate members of the meeting from Iwate Fisheries Technology Center, Prefectural land development section of Iwate Prefectural Office and Ofunato City Office for the data provision and their helpful advices, and people in the local fishing industry for their helpful suggestions on the environments at Ofunato Bay.

References

1. Murakami, K., 1998. Study on Hydraulics and Water Quality Environments in Enclosed Coastal Seas and Its Improvement Techniques, *Technical Note of The Port and Harbour Research Institute Ministry of Transport* 900, 142p. (in Japanese)
2. Toyota, M., Hibino T., Nishimori D., Hosokawa Y., Tsuruya H., 1999. Characteristics of Water Quality and Current Field around the Tsunami Breakwater in Ohfunato Bay during Summer, *Technical Note of The Port and Harbour Research Institute Ministry of Transport* 938, 17p. (in Japanese)
3. Okazaki, M., 1990. Internal tidal waves and internal long period waves in the Sanriku coastal seas, eastern coast of northern Japan, *La mer* 28, 5-29.

4. Okada, T. & Nakayama, K., 2007. Density Intrusion and Variation in Dissolved Oxygen Concentrations in a Bay with a Sill at Its Mouth, *ASCE, J. Environ. Eng.* 133(4), 447–453.

5. Murakami, A., Sakai, S., Murakami, K., Nakamura, Y., Okada, T., Takao, T., Shibaki, H., 2013. Change in Water and Sediment Quality in Ohfunato Bay induced by the 2011 Great East Japan Tsunami, *J. JSCE B2 (Coastal Eng.)* 69 (2), 496-500. (in Japanese)

Numerical Modeling of Thermal Discharge
of Lamu Power Plant, Kenya[*]

LI Ti-lai,[†] CHEN Li-ming, GAO Xiang-yu, ZHANG Xin-zhou

Nanjing Hydraulic Research Institute, State Key Lab of Hydrology-water Resources
and Hydraulic Engineering
Nanjing 210024, China
[†]*E-mail: tlli@nhri.cn*
www.nhri.cn

Kenya is located in the eastern Africa. Currently, among 35 million people in Kenya, there are only about 5 million people can use electricity. The problem in power shortage is serious. Kenya Lamu Power Co., Ltd. plans to construct Kenya Lamu3×350MW (supercritical) Independent Power Station Project in LAMU Region. The power plant uses seawater as cooling medium. In order to demonstrate the feasibility of the power plant and the effect of drainage thermal discharge on surrounding water area, a 2D and 3D numerical models of tidal currents and thermal discharge are established and the thermal discharge of two drainage schemes are simulated. The research results show that the influencing area of temperature rise in open channel drainage scheme are larger than that in pipe drainage and the temperature rise in water intake in pipe drainage scheme is larger than that in open channel drainage.

Keywords: Kenya Lamu Power Plant; thermal discharge; tidal current; numerical model

1. Overview of Natural Conditions

The Power Station is located in Wasase on the west coast of Manda Bay, about 800m away from the coast. There is Wange Creek and Dodori Creek flowing into the bay and there are some shoals exposed intertidal flat in low tide (Figure 1). The 7m isobaths runs through from the site to open seas. There is a large quantity of mangrove forests growing on both coasts.

It can be known from the measuring data in Lamu Meteorological Station (2° 16′ S 40° 54′ E) about 16 km in the southwest of the power plant during 1993 to 2013 that the highest air temperature is 35.5 ℃ in March and the lowest air temperature is 22.2 ℃ in February. The statistical analysis is conducted

---

[*] This work is supported by National Key Research and Development Program of China(2016YFC0401500,2016YFC0401502,2016YFC0401503,2016YFC0401505,2016YFC0401506); National Natural Science Foundation of China(51479122); Central Water Resources Fee Project (1261530210283);

according to wind data day by day in Lamu Meteorological Station during January 1, 2001 to July 31, 2013 with south wind mainly in April to May, northeast wind during December to March and the prevailing wind direction for many years as the south wind; the larger value of the average wind velocity appears in June to September while the minimum wind velocity appears in November. The maximum average wind velocity is about 4.6 m/s and the maximum wind velocity is 15.6 m/s (June 26, 2006, SSW).

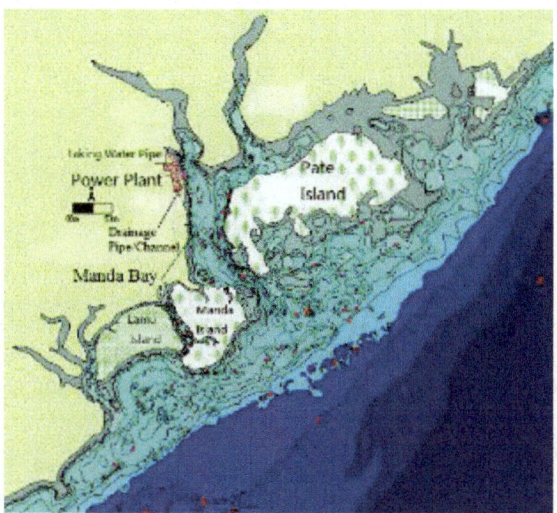

Fig. 1. Power plant location

According to the drainage area of Wange Creek (about 1,460km²) and Dodori Creek (about9, 348km²), the mean annual precipitation (about 1,000 mm), the runoff coefficient 0.05, the annual runoffs of Wange Creek and Dodori Creek are respectively 73 million m³ and 467.40 million m³.

The tides in the region is semi-diurnal tides. It is analyzed according to measurement from April to May, 2015 (Figure 2 shows the positions of measuring points) that the highest high tide level is 4.22 m and the lowest low tide level is 0.81 m. The largest depth average velocity is 0.87 m/s at A9 and about 0.60 m/s at other points. The maximum velocity of neap tides is within 0.50 m/s and the maximum depth average velocity is less than 0.43 m/s. The tides in Manda Bay are obvious reversing currents and the tides have the nature of standing waves. The positions of measuring water temperature is the same as velocity (Figure 2). The temperature of sea water is between 29.0°C and 30.0°C. The water temperature in depth is distributed uniformly and the difference is no more than 0.5 °C.

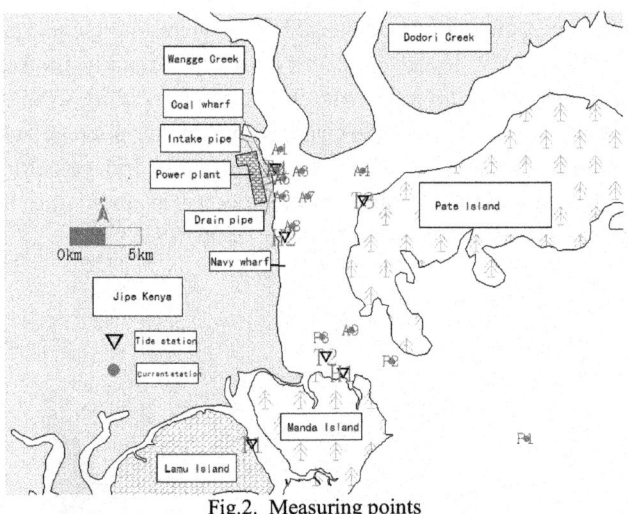

Fig.2. Measuring points

2. Numerical Model Establishment and Verification

In order to know distribution of thermal rise in horizontal and vertical directions, 2D and 3D model for currents and thermal discharge are used.

2.1. *2D tidal model*

In the estuarine and coastal regions, the following governing equations can be adopted.

$$\frac{\partial z}{\partial t} + \frac{1}{c_\xi c_\eta}\left[\frac{\partial(c_\eta H u)}{\partial \xi} + \frac{\partial(c_\xi H v)}{\partial \eta}\right] = q \tag{1}$$

$$\frac{\partial(Hu)}{\partial t} + \frac{1}{c_\xi c_\eta}\left[\frac{\partial(c_\eta H u^2)}{\partial \xi} + \frac{\partial(c_\xi H u v)}{\partial \eta} + H u v \frac{\partial c_\xi}{\partial \eta} - H v^2 \frac{\partial c_\eta}{\partial \xi}\right] = -\frac{gH}{c_\xi}\frac{\partial z}{\partial \xi} +$$
$$\frac{1}{c_\xi c_\eta}\left[\frac{\partial}{\partial \xi}\left(c_\eta H \sigma_{\xi\xi}\right) + \frac{\partial}{\partial \eta}\left(c_\xi H \sigma_{\eta\xi}\right) + H \sigma_{\xi\eta}\frac{\partial c_\xi}{\partial \eta} - H \sigma_{\eta\eta}\frac{\partial c_\eta}{\partial \xi}\right] + fHv + \frac{\tau_{s\xi}-\tau_{b\xi}}{\rho} -$$
$$\frac{1}{\rho}\left[\frac{\partial s_{\xi\xi}}{c_\xi\,\partial \xi} + \frac{\partial s_{\xi\eta}}{c_\eta\,\partial \eta}\right] + qu_*$$

$$\tag{2}$$

$$\frac{\partial(Hv)}{\partial t} + \frac{1}{c_\xi c_\eta}\left[\frac{\partial(c_\eta H u v)}{\partial \xi} + \frac{\partial(c_\xi H v^2)}{\partial \eta} + H u v \frac{\partial c_\eta}{\partial \xi} - H u^2 \frac{\partial c_\xi}{\partial \eta}\right] = -\frac{gH}{c_\eta}\frac{\partial z}{\partial \eta} +$$
$$\frac{1}{c_\xi c_\eta}\left[\frac{\partial}{\partial \xi}\left(c_\eta H \sigma_{\xi\eta}\right) + \frac{\partial}{\partial \eta}\left(c_\xi H \sigma_{\eta\eta}\right) + H \sigma_{\xi\eta}\frac{\partial c_\eta}{\partial \xi} - H \sigma_{\xi\xi}\frac{\partial c_\xi}{\partial \eta}\right] - fHu + \frac{\tau_{s\eta}-\tau_{b\eta}}{\rho} -$$
$$\frac{1}{\rho}\left[\frac{\partial s_{\eta\xi}}{c_\xi\,\partial \xi} + \frac{\partial s_{\eta\eta}}{c_\eta\,\partial \eta}\right] + qv_*$$

$$\tag{3}$$

Where, u and v are respectively flow velocity components in the directions of ξ and η; Z is the tidal level; H is the water depth; C is the Chezy coefficient; f is the Coriolis force coefficient; q is the source/sink strength on the unit area and u* and $v*$ are the velocity components surrounding source sink node. $(\tau_{s\xi}, \tau_{s\eta})$ is the stress of surface wind and $(\tau_{b\xi}, \tau_{b\eta})$ is the resistance of the bottom; $S_{\xi\xi}$, $S_{\xi\eta}$, $S_{\eta\xi}$ and $S_{\eta\eta}$ are components of wave radiation stress; $\sigma_{\xi\xi}$, $\sigma_{\eta\eta}$, $\sigma_{\xi\eta}$ and $\sigma_{\eta\xi}$ are turbulent shear stresses. SIMPLE method (Van Doormaal 1984) is adopted to solve the discrete equations.

For initial time t=0, the initial velocity is 0 and the initial tidal level is a given value. Tidal level processes are given on the open boundaries of seas while flow discharge is given on the boundaries of rivers. The freezing method is adopted to deal with moving boundaries for shallows. When the depth is less than 0.01 m, the roughness is a positive infinity; when the depth is more than 0.01 m, it is included into normal computation.

The computation scope of the model includes two rivers emptying into the sea and partial regions of sea; the model is 54.5 km in the length from the west to the east and 84 km in the width from the north to the east. The computational grids are unequal grids and the number of grids is 430×473(Figure 3). The grids in the region of the project is made dense with the minimum size of grids of 2m and the maximum is 500m in whole region. The topography of the model is the newest measurement in 2014 while the open seas is adopted in the newest sea chart.

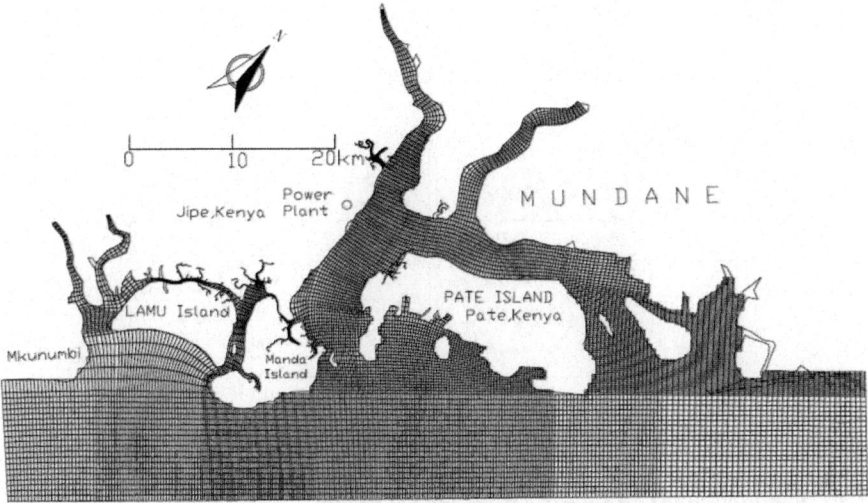

Fig.3. Computational grid

When the depth is no less than 1.5m, $n = n_0 + n'/H$. n_0 is the foundation roughness which is 0.012 in open seas through verification computation; n'/H is the correction term of roughness, n' is 0.008; when the depth is less than 1.5 m, $n = n_0 + 0.005e^{(1.5-h)}$ is adopted for computation. $v_t = ku*H$, where $u*$ is a friction velocity, H is the water depth and k is the constant which is 1.0 in the paper.

The measured data of spring and neap tides during April to May, 2015 are adopted to calibrate the model. There are four tidal level stations including T1, T2, T3 and T4 and nine flow velocity stations including A1 to A9. It can be seen from the verification results that the computation values of tidal levels (Figure 4), current velocities and directions (Figure 5) is basically corresponding with the measured values, except the flow direction has slight deviation at the moment of turning of tidal currents. The error of high and low tides for spring tides shall be -3.3% to 4.7% while the error of the maximum flow velocity is -10% to 19% and the model has well demonstrated characteristics of flow fields in engineering areas.

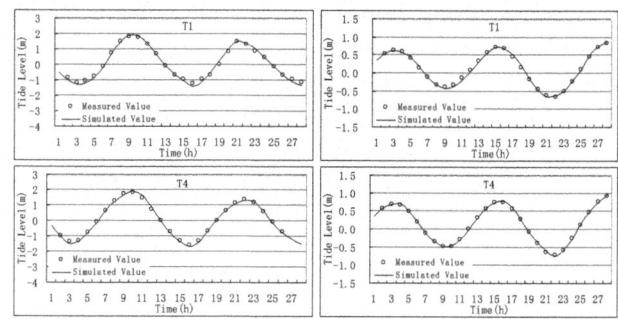

Fig. 4. Verification of tidal levels(the left is spring tide and the right is neap tide)

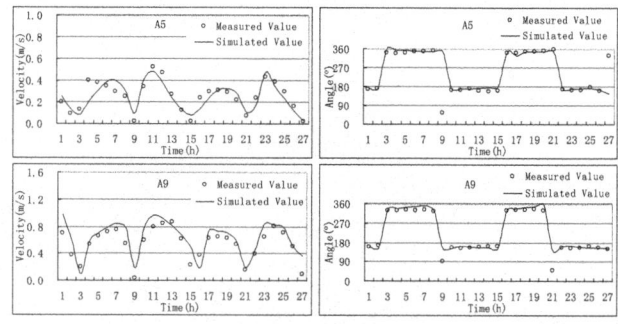

Fig.5. Comparison of velocities and directions in spring tides

2.2. Establishment and verification for 3D model

The following governing equations can be adopted for 3D fluid movement.

$$\frac{\partial h}{\partial t} + \frac{\partial hu}{\partial x'} + \frac{\partial hv}{\partial y'} + \frac{\partial hw}{\partial \sigma} = hS \tag{4}$$

$$\frac{\partial hu}{\partial t} + \frac{\partial hu^2}{\partial x'} + \frac{\partial hvu}{\partial y'} + \frac{\partial hwu}{\partial \sigma} = fvh - gh\frac{\partial \eta}{\partial x'} - \frac{h}{\rho_0}\frac{\partial p_a}{\partial x'} - \frac{hg}{\rho_0}\int_z^\eta \frac{\partial \rho}{\partial x}dz -$$
$$\frac{1}{\rho_0}\left(\frac{\partial s_{xx}}{\partial x} + \frac{\partial s_{xy}}{\partial y}\right) + hF_u + \frac{\partial}{\partial \sigma}\left(\frac{v_v}{h}\frac{\partial u}{\partial \sigma}\right) + hu_s S \tag{5}$$

$$\frac{\partial hv}{\partial t} + \frac{\partial hvu}{\partial x'} + \frac{\partial hv^2}{\partial y'} + \frac{\partial hwv}{\partial \sigma} = -fvh - gh\frac{\partial \eta}{\partial y'} - \frac{h}{\rho_0}\frac{\partial p_a}{\partial x'} - \frac{hg}{\rho_0}\int_z^\eta \frac{\partial \rho}{\partial x}dz -$$
$$\frac{1}{\rho_0}\left(\frac{\partial s_{yx}}{\partial x} + \frac{\partial s_{yy}}{\partial y}\right) + hF_v + \frac{\partial}{\partial \sigma}\left(\frac{v_v}{h}\frac{\partial v}{\partial \sigma}\right) + hu_s S \tag{6}$$

where σ is the vertical Sigma conversion, $\sigma = (z - z_b)/h$; d is the water ; η is the tidal level; $h=\eta+d$ is the total water depth; u, v, and ω are components of flow velocity at the directions x, y and z; ρ is the density of water while ρ_0 is the reference; P_a is the local barometric pressure; F_u and F_v are the accelerated velocity caused by the earth rotation; v_v is the vertical turbulent viscosity coefficient; S is a source sink term and us and v_s are current flow velocities of source sink terms.

Triangular non-structure body-fitted grids are adopted to subdivide the computational domain. There are 10,690 nodes and 20,770 grids. The size of grids is between 40m and 300m. Sigma coordinate system is adopted vertically and is divided into five layers. The bed resistance coefficient is 0.005 and the eddy viscosity coefficient is gained from Smagorinsky Formula. The computed time step is 30s.

Since the temperature rise of thermal discharge in neap tide has a larger influence for water intake in power plant, the 3D model is verified by the measured tidal level, current velocity and direction of neap tide. The computation values are basically corresponding with the measured tidal levels, current velocities at surface-layer, middle-layer and bottom-layer from 12:00 on April 28 to 14:00 on April 29, 2015 (Figure 6).

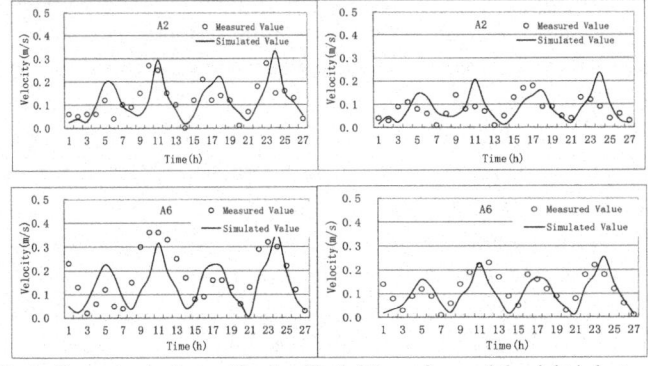

Fig.6. Current velocity verification(the left is surface and the right is bottom)

3. Temporal and Spatial Distribution of Thermal Discharge Prediction

3.1. *Foundation of numerical model of thermal discharge*

The following governing equations can be adopted for 3D thermal discharge diffusion.

$$\frac{\partial \Delta TH}{\partial t} + \frac{1}{c_\xi c_\eta}\frac{\partial(uH\Delta Tc_\eta)}{\partial \xi} + \frac{1}{c_\xi c_\eta}\frac{\partial(vH\Delta Tc_\xi)}{\partial \eta} = \frac{D_\xi H}{g_\xi^2}\frac{\partial^2 \Delta T}{\partial \xi^2} + \frac{D_\eta H}{g_\eta^2}\frac{\partial^2 \Delta T}{\partial \eta^2} - \frac{k\Delta T}{\rho c_P} + q\Delta T_i \quad (7)$$

where ΔT is temperature rise (℃); ΔTi is temperature of the source item (℃); D_ξ and D_η are coefficients of thermal diffusion; k is the coefficient of comprehensive heat dissipation; C_p is specific heat; ρ is density of water (kg/m³); C_η and C_ξ are Lami numbers; q is discharge per unit area of source item (m³/s·m²).

$$\frac{\partial h\Delta T}{\partial t} + \frac{\partial hu\Delta T}{\partial x} + \frac{\partial hv\Delta T}{\partial y} + \frac{\partial hw\Delta T}{\partial \sigma} = hF_T + \frac{\partial}{\partial \sigma}\left(\frac{Dv}{h}\frac{\partial \Delta T}{\partial \sigma}\right) + h\widehat{H} + h\Delta T_s S \quad (8)$$

$$F_T = \left(\frac{\partial}{\partial x}\left(D_h\frac{\partial}{\partial x}\right) + \frac{\partial}{\partial y}\left(D_h\frac{\partial}{\partial y}\right)\right)\Delta T \quad (9)$$

where F_T is the horizontal diffusion item; D_h and D_v are the coefficient of turbulent diffusion in horizontal and vertical direction respectively; S is the discharge of source item and ΔT_s is the temperature of it.

3.2. *Calculation of thermal diffusion*

The water intake is located in 1km downstream from the Wange mouth, the distance to the coast is about 685m and the elevation of it is between -6.1m and -3.1m. There are two schemes of the outfall: One is pipe drainage, which is about 1265m downstream from the intake, 525m from the coast and the height of drainage is between -2.1m and -0.1m. The another is open channel drainage, which lies in the same position as that of the pipe (Figure 7).

The maximum discharge of recycled water of the intake is 35 m³/s and the temperature rise of drainage is 9℃ in any season. According to the tide level from 1995 to 2015 of LAMU Port close to the power plant, a typical type of tides is designed that a spring tide with the guarantee rate of tidal range of 10%, a middle tide with the guarantee rate 50%, a neap tide with the guarantee rate 90% and a lowest tide with the guarantee rate 97%. The tidal ranges of typical spring tide, middle tide, neap tide and the lowest tide are respectively 3.113m, 2.206m, 1.117m and 3.545m. The temperature of sea water is taken as 28.7 ℃. The wind speed at 1.5m above water surface is taken as 3.53 m/s.

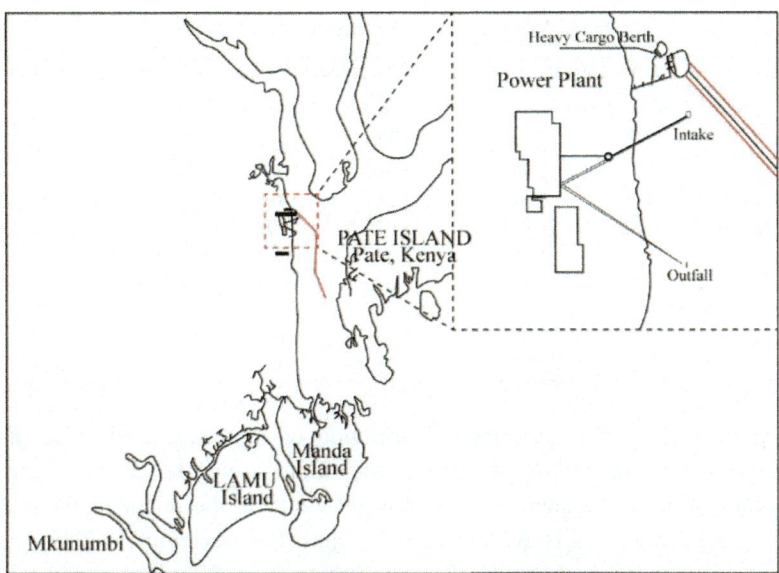

Fig.7 Schemes of drainage and intake of LAMU Power Plant

Comparing two drainage schemes, the influencing area of temperature rise of each centigrade in the open channel drainage scheme is greater than that in the pipe drainage scheme, except the situation in a typical neap tide below 1℃ (Table 1). Table 2 shows the maximum area influenced by temperature rise in pipe drainage and open channel drainage scheme under the action of different tide types. In a whole, the influencing area of temperature rise in pipe drainage scheme is smaller than that in open channel drainage.

Table 1.Influencing area of temperature rise

Plans	Tide type	Area of Temperature Rise (km²)						
		≥0.1℃	≥0.3℃	>0.5℃	≥1℃	≥2℃	≥3℃	≥4℃
Pipe Drainage	Spring	37.63	17.93	12.37	6.21	0.96	0.35	0.12
	Middle	34.65	15.93	10.43	5.43	2.22	0.75	0.26
	Neap	26.94	12.75	8.47	4.49	2.03	0.86	0.37
	Lowest	41.92	16.85	10.52	3.37	0.36	0.10	0.04
Open Channel Drainage	Spring	39.75	20.12	14.77	8.80	4.92	2.75	1.54
	Middle	33.14	16.73	11.80	6.88	3.58	1.79	1.09
	Neap	21.57	11.40	8.36	5.48	2.92	1.51	0.90
	Lowest	58.56	27.09	17.59	10.21	5.11	2.61	1.55

Table 2. Maximum area influenced by temperature rise

		Upper Stream Alongside (km)		Down Stream Alongside (km)		Offshore Direction (km)	
		Pipe	Open Channel	Pipe	Open Channel	Pipe	Open Channel
Spring	1℃	4.81	5.60	6.20	5.93	0.88	1.11
Tide	3℃	0.80	3.50	1.26	2.98	0.24	0.86
Middle	1℃	3.62	3.84	5.52	4.69	0.80	0.98
Tide	3℃	2.08	2.47	2.23	1.56	0.25	0.69
Neap	1℃	3.35	3.44	4.03	4.41	0.86	0.88
Tide	3℃	2.13	2.02	2.37	1.77	0.33	0.67
Lowest	1℃	4.02	5.71	3.07	6.11	1.04	1.18
Tide	3℃	0.43	2.92	0.35	3.00	0.18	0.86

Figure 8 show the horizontal distribution of the temperature rise in two schemes under the action of spring tide. Comparing the average and the maximum values of temperature rise in water intake of the schemes, the average value in pipe drainage is within 1.33 ℃, and the maximum is 2.26 ℃; the average value in open channel scheme is within 0.76 ℃, and the maximum is 1.85 ℃. Whether during spring, middle or neap tide, the average and the maximum value of water intake in pipe is larger than that in open channel. Table 3 shows vertical temperature rise value in each level of water intake in two schemes. The difference of temperature rise from the surface to the bottom is small, basically within 0.1℃, and the vertical stratification is not obvious.

4. Conclusion

Based on investigation and analyses of measured data, and 2D and 3D numerical models of tide currents and thermal drainage, following conclusions are made:

- Through verification of on-site observed data, it can be shown that the established models can reflect the hydropower in engineering area.
- The influencing area of temperature rise in open channel drainage scheme is larger than that in pipe drainage scheme. For pipe, the influence area more than 1℃ is 6.21km², 3℃ is 0.86km², 4℃ is 0.37km²; For open channel, the influence area more than 1℃ is 10.21 km², 3℃ is 2.75 km², 4℃ is 1.55 km².
- The average and maximum value of temperature rise in water intake in pipe drainage scheme are larger than those in open channel drainage scheme. The average value of the temperature rise of the pipe drainage is within 1.33 ℃, the maximum is 2.26℃. The average value of the temperature rise of the open channel drainage is within 0.76 ℃, and the maximum is 1.85 ℃.

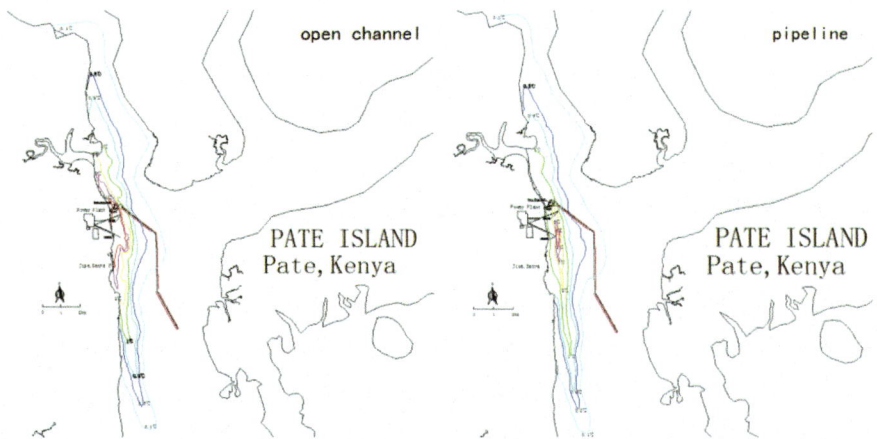

Fig.8. Distribution of temperature rise

Table 3. Vertical temperature rise in each level of water intake

Water Depth	Maximum /℃(pipe)				Maximum /℃(open channel)			
	Spring Tide	Middle Tide	Neap Tide	Lowest Tide	Spring Tide	Middle Tide	Neap Tide	Lowest Tide
0.1H	1.54	2.22	2.31	0.75	1.64	1.89	1.67	2.39
0.3H	1.53	2.20	2.29	0.74	1.62	1.88	1.65	2.38
0.5H	1.52	2.18	2.27	0.73	1.61	1.86	1.64	2.35
0.7H	1.49	2.14	2.23	0.72	1.58	1.83	1.61	2.31
0.9H	1.49	2.14	2.23	0.72	1.58	1.82	1.61	2.31

Acknowledgements

This work is supported by: National Key Research and Development Program of China (2016YFC0401500, 2016YFC0401502, 2016YFC0401503, 2016YFC0401505, 2016YFC0401506); National Natural Science Foundation of China (51479122); Central Water Resources Fee Project (1261530210283)

References

Liu, L. F., 2004. Experiment study on attenuation law of residual chlorine in cooling water discharged from thermal power plants[J]. Journal of Hydraulic, 35(5): 94-98.

Van Doormaal J P. RaithbyG D. Enhancement of SIMPLE method for predicting incompressible fluid flows [J]. Numer Heat Transfer, 1984 (7): 147-163.

Thermal Analysis (TG-DTA) for Identify Organic Matter Conditions in Sediment

H. Takata[†], N. Touch, S. Yamaji, T. Hibino

Hiroshima University, 1-4-1, Kagamiyama, Higashihiroshima, 739-0046, Japan
[†]E-mail: m175472@hiroshima-u.ac.jp

Sediment quality strongly depends on organic matter conditions in the sediment, thus identifying organic matter conditions in sediment is very necessary. The thermal analysis (TG/DTA) is widely used to understand the ignition characteristic of sediment. It has been reported that different types of organic matter have different exoergic temperatures. Unfortunately, no any TG/DTA analysis standard has been proposed for sediment analysis. In other words, TG/DTA analysis of sediment is different from a study to another study. This study is aiming at proposal of TG/DTA analysis for identifying organic matter conditions in sediment. Different sediments collected from different littoral regions in Japan were used in the TG/DTA analysis. The sediments were first oven dried, and then were passed through 75-μm sieve. The sediments with and without passing through 75-μm sieve were used in the analysis. In the analysis, different weights (10 to 30 mg) of samples were ignited from 25 to 850°C with heating rates of 1, 5, 10, and 20°C /min. It was observed that a more accurate analysis could be obtained when the sediment less than 75-μm was used, specifically for the sediment collected from the littoral region exposed to wastewater discharge. The exothermic peak easily appeared with increasing weight of the sample, which easily identified organic matter conditions in the sediment. Moreover, mass loss on ignition of the sediment burned at ≤5°C /min was higher than that at ≥10°C /min. This ensured that organic matter was incompletely burned at≥10°C /min. It can be concluded from our results that 20-30 mg of sediment passed through 75-μm sieve should be burned at 5°C /min to obtain a more accurate analysis of organic matter conditions in sediment.

Keywords: TG-DTA; Sediment; Pre-Treatment; Sediment Weight; Heating Rate.

1. Introduction

Organic matter present in marine sediments plays an important role in the ecosystem of marine environment. However, organic matter also impacts marine environment because of its decomposition. Hence, identifying organic matter conditions (e.g., amount, state) in sediments is required for evaluating both sediment and water qualities. However, many difficulties appear during the identification of organic matter conditions in littoral sediments.

The Walkley-Black method [1] has been used to evaluate amount of organic matter in soils and sediments [2, 3]. Loss on ignition (LOI) method has also been used for a few decades because of its simplicity and low cost [4-7]. Automated CHNS-analyzer method has been used because of its high precision and time saving [8, 9]. However, these methods are unable to directly evaluate changes in the state of organic matter without mineralization. To understand the organic matter state in terrestrial soils and sediments, thermal gravimetric and differential thermal analysis (TG-DTA) has been widely used [10, 11]. Unfortunately, no any TG-DTA analysis standard has been proposed for sediment analysis. In other words, TG-DTA analysis of sediment is different from a study to another study.

This study is aiming at proposal of TG-DTA analysis for identifying organic matter conditions in sediment. Namely, sediments collected from different littoral regions were used to examine the effects of drying methods, sediment sieving, sediment weight, and heating rate on changes in the results of mass loss on ignition and DTA characteristics.

2. Materials and Methods

2.1. *Sediment used in experiments*

Sediments collected from a coastal region exposed to wastewater discharge (organic matter derived from sewage), from a coastal region that no direct inflow of wastewater (organic matter derived from primary production), and freshwater (lake) sediment were used in the experiments. The sediments were collected by using a HR type core sampler (Rigo, Cat. No. 5172) and an Ekman-Berge bottom sampler (Rigo, Cat. No. 5114-B). The properties of collected sediments are shown in Table 1.

The collected sediments were passed through a 2-mm sieve to remove coarse debris and other large terrestrial deposits (only FUS). The sediments were stored in a room with a temperature of $25°C \pm 1°C$, and homogenized before being use in the TG-DTA analysis.

Table 1. Properties of collected sediments

Sediment	pH	Redox potential mV vs Ag/AgCl
From coastal region exposed to wastewater discharge (FUS)	7.40	-319
From coastal region no direct inflow of by wastewater discharge (KUS)	7.11	-340
Freshwater sediment (KTS)	7.16	-201

2.2. *Sediment pre-treatment*

It is considered that ignition characteristics (thermal conductivity) of each substance are different from one to another according to different structures. In TG-DTA analysis, sediment must be dried before being used in the analysis. As such, the method of drying sediment should be considered to obtain a precise analysis. In this study, the sieved sediment was dried by air drying, freeze-drying method (Yamato, DC401), 50°C-drying method (Yamato, DSV402), and 100°C-drying method using an electric furnace (As One, DOV-600A), for understanding the effects of drying methods on TG-DTA analysis.

In TG-DTA analysis, only up to 50 mg of sediment can be analysed. Thus pre-treatment of dried sediment should be conducted to remove small inorganic particles and terrestrial deposits to obtain a better understanding of organic matter conditions in sediment. Therefore, the dried sediment was passed through 75-μm sieve. TG-DTA analyses of the sediment without sieving and the sediment passed through 75-μm sieve were conducted to examine the effects of sieving sediment on the accuracy of TG-DTA analysis.

2.3. *TG-DTA analysis*

TG-DTA analysis was conducted using a thermal analyser (Shimadzu, DTG-60H). In the TG-DTA analysis, the ignition characteristic of substance is considered to depend on sample weight and heating rate. For instance, sediment having large amounts of humic substances may not be completely burned at a high heating rate. As sample weight is limited (up to 50 mg), so that the adequate weight of sediment should be considered for achieving more accurate analysis. To reveal the effects of sample weight and heating rate on the ignition characteristic of sediment, the sediment weight was varied from 5 to 30 mg and was heated at 1 to 20°C/min.

3. Results and Discussion

3.1. *Sediment sieving for a better TG-DTA analysis of sediment*

Figure 1 shows the mass loss on ignition (IL) of FUS without sieving compared with FUS passed through 75-μm sieve. FUS was the sediment collected from the coastal region exposed to wastewater discharge, thus terrestrial deposits present in the sediment. Here, air-dried FUS was used in the analysis. In the case without sieving the sediment, variances of mass loss on ignition were found, especially the mass loss at over than 300°C (Fig. 1a). The maximum difference of mass loss was approximately 78 mg/g at the ignition temperature of 600°C.

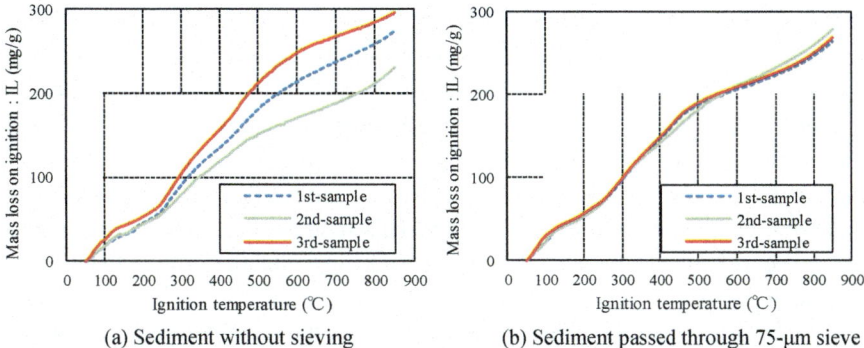

Fig. 1. Mass loss on ignition of the air-dried FUS without sieving compared with that passed through 75-μm sieve.

Conversely, the mass loss at 600°C was of the same order among each sample in the case of using the sediment passed through 75-μm sieve (Fig. 1b).

The difference in the mass loss over than 300°C may be partly due to the presence of small terrestrial deposits and inorganic matter. As sample weight is too small, so that mass loss decreases when a small of inorganic matter present in the sample. On the other hand, mass loss increases when a small amount of woody and plant deposits present in the sample. Such deposits are known as humic substances which burn at the temperature higher than over 370°C according to Cuypers et al. [10].

In summary, the analysis results suggest that the TG-DTA analysis of sediment become more accurate when the sediment passed through 75-μm sieve, particularly in the case of small terrestrial deposits and inorganic matter present in the sediment.

3.2. *Effects of drying method on the ignition characteristic of sediment*

The thermal conductivity is dependent on the property (i.e. soil structure) of sediment. Sediment burns easily when it has loose structure because heat diffuses into the sediment easily. It is thought that soil structure is different according to different drying methods of sediment. Commonly, soil structure is hardened (compressed) by oven drying method, while it becomes loose in the case of using freeze-drying method. Figure 2 presents the changes in ignition characteristic of FUS due to different drying methods. Here, FUS passed through 75-μm sieve was used in the analysis.

As expected, the ignition characteristic of sediments was different according to drying method. Compared with the ignition characteristic of 100°C-dried sediment, the mass loss at 100°C was larger in the freeze-dried and 50°C-dried sediments, and was largest in the air-dried sediment. The mass loss at 100°C is considered to be the mass loss of volatile substances such as water and sulfur compounds. Compared with the freeze-dried sediment, the mass loss at 100°C of the air-dried sediment may be partly due to mass loss of water presented in the sediment. Interestingly, the mass loss at 100°C of the freeze-dried sediment and the air-dried sediment was of the same order, indicating that the mass loss of volatile substances was of the same order. This suggests that 50°C-drying method can be used instead of freeze-drying method for evaluating volatile substances burned at 100°C.

The difference of ignition characteristic in the temperature range of 100-500°C was not found in spite of different drying method. However, the ignition characteristic over 500°C of the sediment dried by freeze-drying method was different from those of other drying methods. The increasing gradient of mass loss over 500°C of the freeze-dried sediment was larger than that of the sediments dried by other methods. This suggests that humic substances present in the sediment dried with freeze-dying method burn easily.

In summary, the ignition characteristic in the temperature range of 100-500° of 50°C-dried sediment was similar to that of the freeze-dried sediment. However, humic substances present in the freeze-dried sediment burned easily compared with those in the 50°C-dried sediment, suggesting that attention should be paid to the sediment having high content of humic substances. Hibino et al. [6] examined the decreases in carbon content in sediments at different ignition temperatures. The authors reported that around 98% of carbon content in sediments decreased when the sediments were burned at 500°C. As our results showed that 50°C-dried sediment and freeze-dried sediment had the same ignition characteristic in the temperature range of 100-500°C. it can be said that 50°C-drying method can be used instead of freeze-drying method for understanding organic matter conditions in sediment.

3.3. *Sample weight for a better TG-DTA analysis of sediment*

The heat diffusion into sediment is also dependent on sediment weight and heating rate. It is thought that sediment is not completely burned in the case that large weights of sample are used. Figure 3 depicts the results of TG-DTA analysis of 50℃-dried FUS, KUS, and KTS. No difference of mass loss below 400℃ was found in spite of different sample weights. However, the mass loss over 400℃ of 5 and 10 mg-samples was higher than that of 20 and 30 mg-samples, suggesting

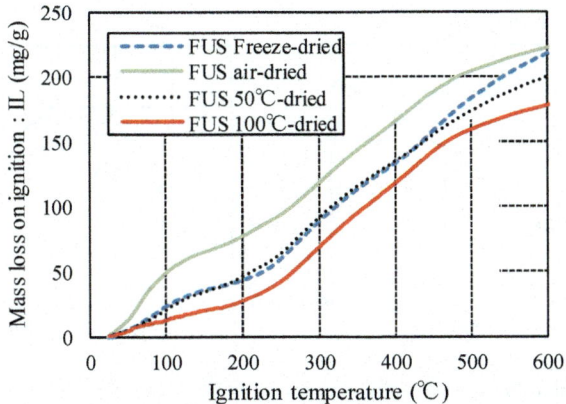

Fig. 2. Changes in ignition characteristic of FUS due to different drying methods.

that the sediment containing much of humic substances should be burned at a weight of less than 10 mg.

On the other hand, the DTA results showed that exothermic and endoergic peaks clearly appeared with increasing the weight of sample. Exothermic and endothermic peaks play important roles in identifying burned substances. Therefore, sample weight should be varied to obtain exothermic and endothermic peaks.

In summary, 10 mg of sediment should be burned to obtain a more accurate mass loss on ignition. However, the mass of sediment should be increased to obtain a clearer exothermic and endothermic peaks for identifying burned substances in sediment. As discussed above, the organic matter in sediment was mostly burned at 500°C and a little difference of mass loss was found at 500°C in spite of different sediment weights (5-30 mg). Therefore, 20 to 30 mg of sediment should be used in the TG-DTA analysis for examining organic matter conditions in sediment.

3.4. *Heating speed for a better TG-DTA analysis of sediment*

50°C-dried FUS was used in the TG-DTA analysis to examine changes in the ignition characteristic of sediment according to the variation of heating rate, as shown in Fig. 4. It is thought that each substance present in sediment is completely when a small heating rate is used. As can be seen in Fig. 4, no difference of mass loss below 500°C was found in spite of different heating rates. However, a difference was found over 500°C. The mass loss of samples burned at 1 and 5°C/min was higher than that burned at 10 and 20°C/min. This means that the sediment contained much of humic substances should be burned at a heating rate

910

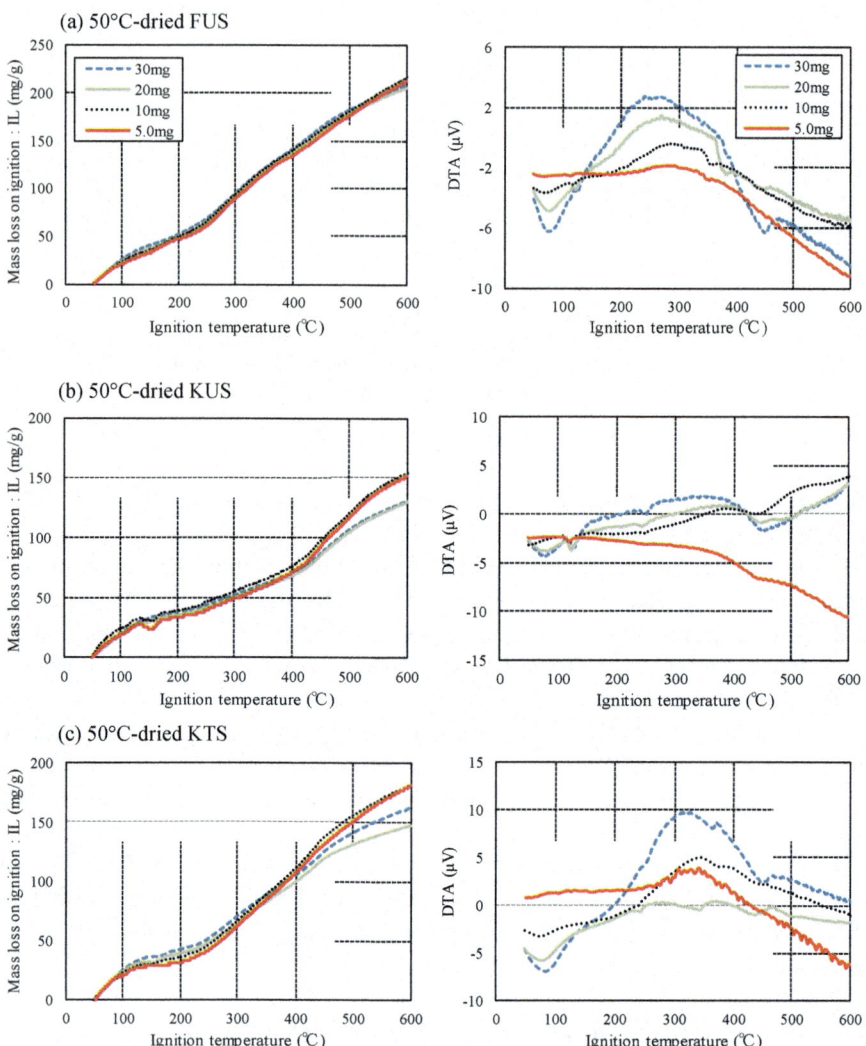

Fig. 3. Results of TG-DTA analysis of 50°C-dried FUS, KUS, and KTS.

of less than 5°C/min.

On the other hand, the DTA results showed that exothermic and endothermic peaks seemed to clearly appear with increasing heating rate. However, no large difference of exothermic and endothermic peaks was found when the heating rate was increased from 5 to 20°C/min. It can be said from these results that the heating rate of 5°C/min should be used in the TG-DTA analysis for evaluating organic matter conditions in sediment.

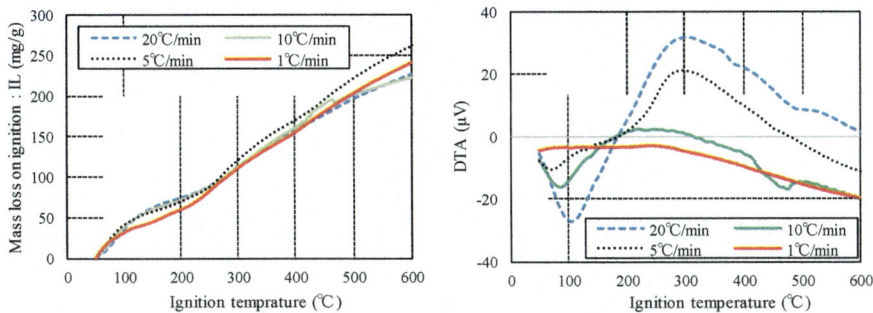

Fig. 4. Results of TG-DTA analysis of 50°C-dried FUS.

4. Conclusions

In this study, sediments collected from different littoral regions were used in TG-DTA analysis for proposing sediment pre-treatment and analysis method to identify organic matter conditions in the sediments. The analysis accuracy of the sediment passed through 75-μm sieve was found to be better than that of the sediment without sieving. The ignition characteristic of sediments was different according to drying methods of the sediments, specifically for the sediments with and without much content of humic substances. It was found out that 50°C-drying method can be used instead of freeze-drying method for understanding organic sediment. The ignition characteristic of sediments was also found to be dependent on sediment weight and heating rate that used in the analysis, and attention should be paid for the sediment having much content of humic substances which burned over 370°C.It can be concluded from our results that 20-30 mg of 50°C-dried sediment that passed through 75-μm sieve should be burned at 5°C /min to obtain a more accurate analysis of organic matter conditions in sediment.

Acknowledgments

The authors gratefully acknowledge partial funding from Japan Society for the Promotion of Science: Grant-in-Aid for Science Research (Grant No. 16K14311 and 15H05221). The constructive comments of anonymous reviewers are also appreciated.

References

1. A. J. Walkley and I. A. Black, An estimation of the Degtjareff method for determining soil organic matter, and a proposed modification of the chromic acid titration method, *Soil Sci.* **37**, pp. 29-38 (1934).

2. F. E. Broadbent, The soil organic matter fraction, *Adv. Agron.* **5**, pp. 153-183 (1953).

3. D. L. Heanes, Determination of total organic C in soils by an improved chromic acid digestion and spectrophotometric procedure, *Commun. Soil Sci. Plant Anal.* **15**(10), pp. 1191-1213 (1984).

4. C. A. Cambardella, A. M. Gajda, J. W. Doran, B. J. Wienhold and T. A. Kettler, Estimation of particulate and total organic matter by weight loss-on-ignition. In R. Lal et al. (ed.) Assessment methods for soil carbon, *Lewis Publ.*, New York, pp. 349-359 (2001).

5. M. E. Konen, P. M. Jacobs, C. L. Burras, B. J. Talaga and J. A. Mason, Equation for predicting soil organic carbon using loss-on-ignition for north central U.S. soils. *Soil Sci. Soc. Am. J.* **66**, pp. 1878-1881 (2002).

6. H. Tadashi, T. Kohei, T. Narong and N. Shinya, Classification of the organic matter existing in littoral sediments, *J. Japan Soc. Civ. Eng, Ser. B2 (Coast. Eng.).* **70**(2), pp. 1101-1105 (2014).

7. T. Narong, K. Nobutaka, N. Shinya and H. Tadashi, Dry combustion based method for evaluating variation of organic matter state in sediment, *J. Japan Soc. Civ. Eng, Ser. B2 (Coast. Eng.),* **71**(2), pp. 1417-1422 (2015).

8. G. Telek and N. Marshall, Using a CHN analyzer to reduce carbonate interference in particulate organic carbon analysis, *Mar. Biol.* **24**, pp. 219-221 (1974).

9. J. H. Sharp, Improved analysis for particulate organic carbon and nitrogen from seawater. *Limnol. Oceanogr.* **19**, pp. 984-989 (1974).

10. C. Cuypers, T. Grotenhuis, K.G.J Nierop, E.M. Franco, A.D. Jager and W. Rulkens, Amorphous and condensed organic matter domains- the effect of persufate oxidation on the composition of soil/sediment organic matter, *Chemosphere.* **48**, pp. 919-931 (2002).

11. G. Galbacs, T. Kantor, L. Moens and R. Dams, Mass spectrometric studies of thermal decomposition prod-ucts of reference materials for use in solid sampling atomic spectrometry, *Spectrochimica Acta.* **53**, Part B, pp. 1335-1346 (1998).

Development of Electrochemical Method for Improving Water Quality

S. Yamaji, N. Touch, H. Takata and T. Hibino

Hiroshima University, 1-4-1, Kagamiyama, Higashihiroshima, 739-0046, Japan
†E-mail: m172401@hiroshima-u.ac.jp

In the coastal region exposed to wastewater discharge, large amounts of sediment accumulates on the seafloor. The decomposition of large amounts of organic matter releases an excess amount of reduced substances (e.g., Fe^{2+}, H_2S) in sediment. The diffusion of reduced substances into water body consumes oxidants, leading to the aggravation of water quality (the redox potential was around -200 mV vs Ag/AgCl) which influences on marine ecosystem. In the literature, the improvement of water quality by lowering the diffusion of substances from sediment under the uses of steel slag and granulated coal ash has been studied. This study is aiming at developing an electrochemical method from improving water quality. Laboratory experiments were conducted to examine the change of water quality after applying the developed method. In the experiment, the carbon electrodes were installed near sediment surface (anode electrode) and water surface (cathode electrode). The current (recovering electron from bottom water) was made by connecting the electrodes with an external resistance of 100 Ω and a potentiostat (for fixing current). To understand the improvement of water quality, the electrode potential of bottom water was measured continuously during the experiment. It was found that the electrode potential of bottom water increased according to the electron recovery, indicating the oxidation of reduced substances existing in bottom water at the anode electrode. Interestingly, a more improvement of water quality could be obtained with increasing the current of electron recovery. It can be concluded that an electrochemical method can be a mean for improving water quality, and a larger current of electron recovery leads to a more improvement of water quality.

Keywords: Electron Recovery; Bottom Water; Redox Potential; External Power Supply; Nutrient Salts

1. Introduction

In a coastal region that is greatly influenced by the inflow of wastewater, high rates of organic matter decomposition within sediments releases many reduced substances, such as hydrogen sulfide (H_2S) and nutrient salts. These reduced substances decrease not only the sediment quality but also the bottom water quality due to diffusion [1], [2]. Therefore, methods for improving and preserving water quality are required.

Sediment microbial fuel cell (SMFC), an electrochemistry technology, has also received much attention as an alternative technology for improving both water and sediment environments. SMFC is possible to transfer the electrons from oxidizing reduce substances to react with oxygen in the surface water that is continuously supplied from the air. This can lower the deterioration of bottom water quality. Some studies demonstrate decreases in organic matter content and a positive increase in redox potential (ORP) when SMFC was applied into sediments [3], [4]. Few experimental studies have addressed the improvement of bottom water quality when SMFC was applied into sediments [5], [6]. However, to the best of our knowledge, no publication can be found in the literature that discuss the direct application of SMFC into deteriorated bottom water.

The first objective of this present study is to understand the characteristics of current generation when deteriorated bottom water is used to fuel an SMFC. The another objective is to assesses the improvement of water quality through the SMFC application. Laboratory experiments were conducted and changes in redox potential and water quality along with electricity generation were examined.

2. Materials and Methods

2.1. *Experimental device and materials*

Figure 1 shows the experimental device used in the experiments. The device comprised a container with 655 mm in height, 525 mm in length, and 375 mm in width (New select, C-70) (Fig. 1a). Sediment was filled up to 50 mm from the bottom of the container. The sediment used in the experiments was collected from the riverbank of a tidal river (Eba, Hiroshima, Japan). Approximately 30 cm of surface sediment from the deposited sediment on the riverbank was collected using a scope. The sediment was placed in a clean plastic buckets and transported to the laboratory, and was homogenized by mixing before use in the experiments. No pre-treatment of the sediment was conducted before being used in the experiments.

The brush type-anode and cathode electrodes were then installed in the container at 150 mm from the bottom of the container and near water surface, respectively (Fig. 1b). The carbon cloth (News Company, PL200-E) with a surface area of 0.6 m^2 was served as each electrode material. The carbon cloth was heated to 500 °C for 1 h to improve its performance before being used in the experiments [7]. The heated carbon cloth was broken into fibers to create a brush-type electrode. The heated carbon cloth with a surface area of 2500 mm^2

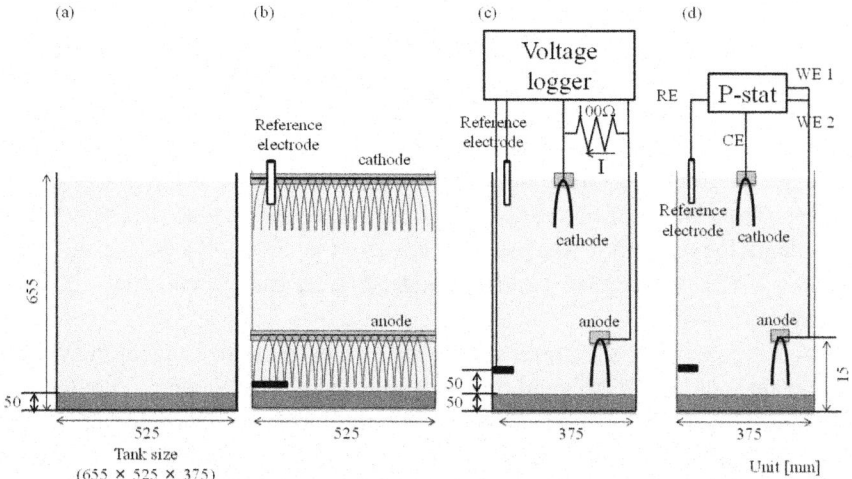

Fig. 1. Experimental device used in the experiments, (a) Cross-sectional view in the length direction without electrodes, (b) Cross-sectional view in the length direction with electrodes, (c) Cross-sectional view in the width direction with an external resistance, (d) Cross-sectional view in the width direction with a potentiostat.

Table 1. Initial characteristics of collected sediment and tap water used in the experiments.

Parameter	Abbreviation		Sediment	Tap water
pH	pH		7.24	7.5
Redox potential	ORP	(mV vs. Ag/AgCl)	−425	640
Electrical conductivity	EC	(mS/m)	4480	9.53
Dissolved oxygen	DO	(mg/L)	N/A	8.72
Ammonium ion	NH_4^+	(mg/L)	27.92	N/A
Phosphate ion	PO_4^{3-}	(mg/L)	7.08	N/A
Manganese ion	Mn	(mg/L)	2.02	N/A
Total iron ion	Fe	(mg/L)	0.08	N/A
Ferric ion	Fe^{2+}	(mg/L)	0.08	N/A
Hydrogen sulfide ion	H_2S	(mg/L)	26.5	N/A
Sulfate ion	SO_4^{2-}	(mg/L)	1450	N/A

(50 mm ×50 mm) was installed at 50 mm from the sediment surface for measuring the redox potential (ORP) of overlying water (Fig.1c). After the installation of the electrodes, top water was filled by tap water. The initial characteristics of sediment and tap water are listed in Table 1.

2.2. *Operation and analysis*

The electrode and an Ag/AgCl saturated KCl reference electrode (Toyo Co., TRE-7) were connected to the positive end and negative end of a voltage logger (Graphtec, GL840-M), respectively (Fig. 1d). The system was put in open circuit conditions (OCV: without current flow) for 4 days, and an external resistance of 100 Ω was then loaded in the circuit to create electrical current for 50 days. The voltage of two ends of the external resistance was also recorded by the voltage logger, and it was used to calculate the circuit current based Ohm's law, $I = U/Rex$, where U (V) is the voltage, I (A) is the current, and Rex (Ω) is the external resistance.

After generating the current by using the external resistance, the system was again put in OCV conditions for 10 days. And then, a potentiostat (Hokutodenko, HSV-110) was fixed the circuit current at 1 mA for 18 days, 2 mA for 11 days, and 3 mA for 5 days. During the experiments, the overlying water (50 mm from the sediment surface) was collected and its quality was analyzed. The Ca, Na, and K ion concentrations in the extracted water were measured using Horiba-laqua twin electrodes. Then, the concentration of nutrient salts (i.e., NH_4^+ and PO_4^{3-}) in the extracted water was measured using a digital water quality analyzer (Kyoritsu Chemical-Check Lab. Corp., Λ9000). It should be noted here that no pre-treatment of samples was conducted.

3. Results and Discussions

3.1. *Characteristics of current generation*

As noted earlier, the first objective of this study was to understand the current generation when SMFC is applied into deteriorated bottom water. Figure 2 shows the temporal changes in current density during loading the external resistance of 100 Ω. Before creating the circuit current, the cathode potential and anode potential were around 200 mV and –250 mV, respectively. This potential difference generated a maximum current density of around 6 mA/m², which was 20-fold smaller that of the SMFC fueled by the used sediment.

This means that the current produced by deteriorated water is lower than that produced by sediment when they are used as fuel in an SMFC. In the case of using sediment, a larger potential difference between the anode and the cathode, and a higher rate of electron transfer may lead to the increase in current.

The current then decreased temporally, and finally was stable at 1 mA/m². The current depletion may be partly due to the lack of electron donors near the anode. This is in good agreement with previous studies, which reported a drop in

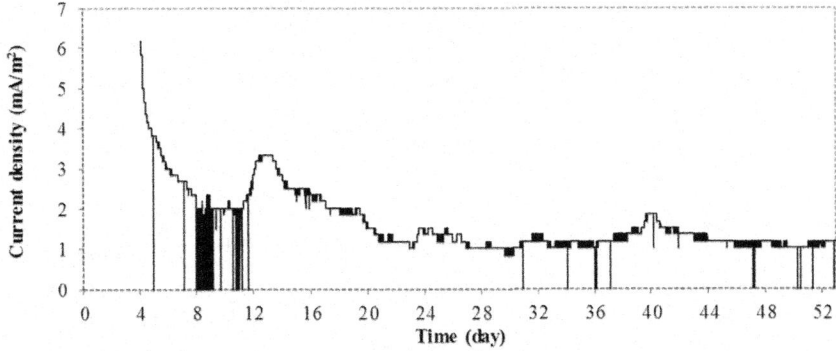

Fig. 2. Temporal changes in current density during loading the external resistance of 100 Ω.

current that they attributed to the shortage of substrates near the anode that are used in the microbial metabolism [8], [9].

3.2. *Changes in the electrode potential in overlying water along with electricity generation*

After reduced substances diffuse from sediment into overlying water, they react with dissolved oxygen (DO) in the overlying water, resulting in the depletion of DO concentration in the overlying water. In these reactions, reduced substances are converted into oxidants, as shown in Eqs. (1) - (3). These oxidants are reused in reduction oxidation reactions after DO is completely consumed, this in turns change the ORP of overlying water. As each reduction oxidation reaction has its own ORP, it can be understood the diffusion and oxidation of reduced substances based on changes in ORP.

$$Mn^{2+} + O_2 + 2e^- \Leftrightarrow MnO_2 \qquad E^0 = 1.224 \text{ V, Eh (pH= 7) = 0.398 V} \quad (1)$$
$$2Fe^{2+} + 3O_2 + 6H^+ + 10e^- \Leftrightarrow 2Fe(OH)_3 \quad E^0 = 1.057 \text{ V, Eh (pH= 7) = } -0.182 \text{ V} \quad (2)$$
$$H_2S + 2O_2 \Leftrightarrow SO_4^{2-} + 2H^+ \qquad E^0 = 0.303 \text{ V, Eh (pH= 7) = } -0.215 \text{ V} \quad (3)$$

Figure 3 depicts the temporal changes in the electrode potential in overlying water. The electrode potential largely decreased from −80 mV to −340 mV during the first 1.5 day, and reached a minimum of −360 mV before stating electricity generation. As discussed above, the decreased in ORP reflects to the diffusion of reduced substances from the sediment. In other words, the water quality of overlying water was deteriorated by discussion of reduced substances from the sediment.

After stating the electricity generation by using an external resistance (at day 4), the electrode potential increased from −360 mV to −300 mV. This

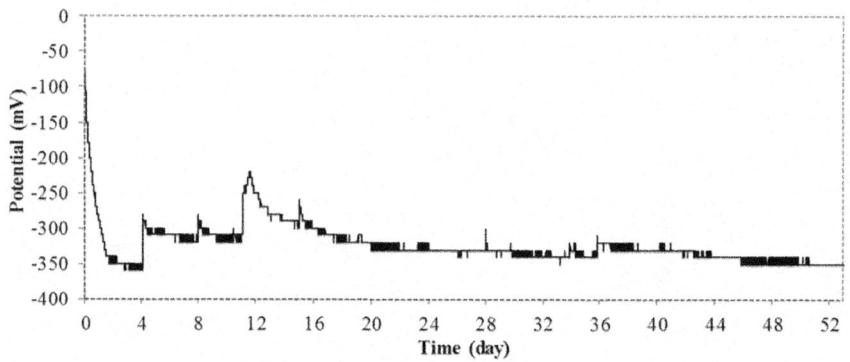

Fig. 3. Temporal changes in the electrode potential in overlying water.

suggested that the electrons present in reduced substances are oxidized at the anode that installed in overlying water. This ensures that SMFC can be used to improve the water quality deteriorated by the diffusion of reduced substances from sediment. Unfortunately, the electrode potential decreased temporally, and reached –350 mV at day 53. It can be understood from this results that the diffusion rate of reduced substances from the sediment is higher than the oxidation of reduced substances at the anode. In other words, the current produced by SMFC alone is still low that is not enough for a complete restoration of water quality (increase ORP to that of tap water, approximately 300 mV). Therefore, an external power supply is required for increasing the SMFC current to restore the water quality completely.

3.3. Improving the SMFC ability by addition of external power supply

Figure 4 presents the temporal changes in the electrode potential in overlying water after increasing the SMFC current by a potentiostat. When the SMFC current increased to 1 mA (current density: 1.67 mA/m^2), the electrode potential started to increase from –350 mV to a maximum of –110 mV at day 72. The electrode potential was not immediately increased after the increasing current. This is because electron donors are continuously provided from the sediment. In other words, not only the reduced substances present in overlying water are oxidized at the anode. However, the electrode potential then decreased to –350 mV at day 81. This indicates that the current of 1 mA was also not enough for a complete restoration of water quality.

When the SMFC current was increased to 2 mA (current density: 3.34 mA/m^2) and 3 mA (current density: 5.01 mA/m^2), the electrode potential was

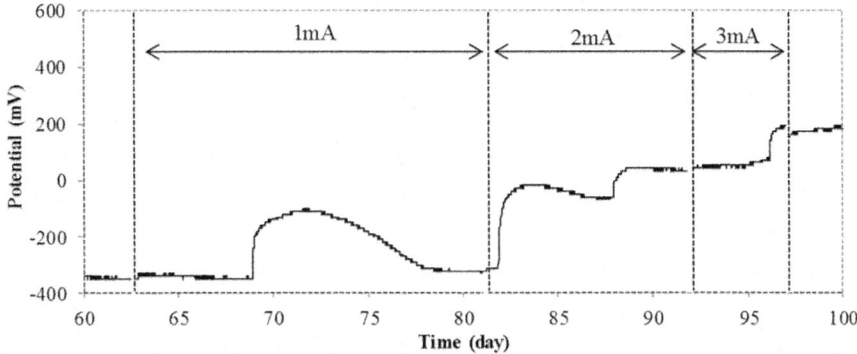

Fig. 4. Temporal changes in the electrode potential in overlying water after increasing the SMFC current by a potentiostat.

stabilized at 30 mV and 170 mV, respectively. These indicate that larger amounts of reduced substances in overlying water are oxidized at the anode, leading to the increases in the electrode potential. It can be said from these results that an increase in the current of SMFC by addition of external power supply provides a more effective improvement of water quality.

3.4. Removal of nutrient salts by electricity generation

Figure 5 shows the temporal changes in the water quality along with electricity generation. After day 30, the concentration of ammonium ion (NH_4^+) in the case with electricity generation became lower compared with that of the case without electricity generation. The NH_4^+ concentration was 2.28 mg/L on average between days 36 and 53 in the case without electricity generation, while it was 0.39 mg/L on average in the case with electricity generation. This suggested that the electricity generation by SMFC could remove NH_4^+ from overlying water. The reasons behind the removal of NH_4^+ are, one is considered to be due to the oxidation of NH_4^+ at the anode, and another is the enhancement of denitrification by SMFC. It is thought that electricity generation increases the redox potential of overlying water; this in turns improves the activity of organisms in overlying water. Similarly, large differences in phosphate (PO_4^{3-}) were also found after day 53 (after increasing the SMFC current by addition of external power supply). As PO_4^{3-} could not be removed through the anode oxidation, it is thought that PO_4^{3-} was removed through the formation of phosphate complex with other metal ions.

920

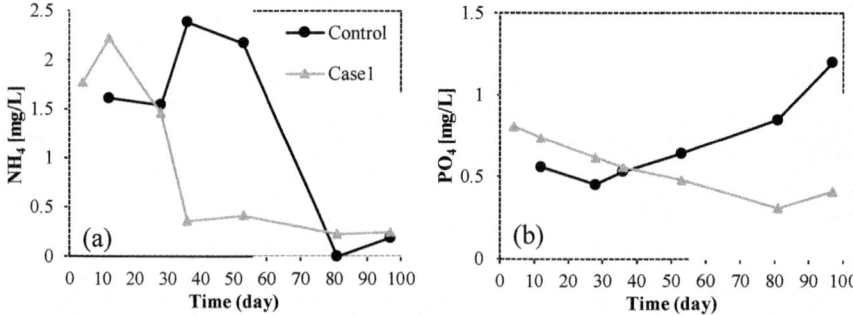

Fig. 5. Temporal changes in the concentration of nutrient salts along with electricity generation (a) NH_4^+ and (b) PO_4^{3-}.

4. Conclusions

In this study, a series of laboratory experiments was conducted to examine the electricity generation when bottom water deteriorated by reduced substances was used to fuel an SMFC, and the potential of SMFC in improving bottom water quality. When bottom water deteriorated was used, the current density produced by SMFC was approximately 6 mA/m^2, 20-fold smaller than that of the SMFC fueled by the used sediment. From the variations of the electrode potential in overlying water, the current produced by SMFC alone could not recovery deteriorated water quality completely. It was found that an increase in the SMFC current by addition of external power supply could provide a more improvement of water quality. A more increase in the electrode potential in overlying water was obtained with a more higher SMFC current. Furthermore, the SMFC application into bottom water could also remove NH_4^+ and PO_4^{3-} from the bottom water.

Acknowledgments

The authors gratefully acknowledge partial funding from Japan Society for the Promotion of Science: Grant-in-Aid for Science Research (Grant No. 16K18158 and 16H04418). The constructive comments of anonymous reviewers are also appreciated.

References

1. R. C. Aller, Diagenetic processes near the sediment-water interface of Long Island sound: I. Decomposition and nutrient element chemistry (S, N, P), *Adv. Geophys.* **22**, pp. 237-350 (1980).

2. J. J. Wright, K. M. Konwar and S. J. Hallam, Microbial ecology of expanding oxygen minimum zones, *Nat. Rev. Microbiol.* **10**, pp. 381-394 (2012).

3. N. Touch, T. Hibino. Y. Nagatsu and K. Tachiuchi, Characteristics of electricity generation and biodegradation in tidal river sludge-used microbial fuel cells, *Bioresour. Technol.* **158**, pp. 225-230 (2014).

4. N. Touch, T. Hibino, S. Nakashita and K. Nakamoto, Variation in properties of the sediment following electrokinetic treatments, *Environ. Technol.* **38**(3), pp.277-284 (2017).

5. N. Touch, N. Kinjo, T. Hibino and H. Takata, Effects of electron recover from sediment on sediment and water environments, In the Proc. Of SICEAS2017, Seoul, pp. 46-57, 2017.

6. N. Touch, T. Hibino, Y. Morimoto and N. Kinjo: Relaxing the formation of hypoxic bottom water with sediment microbial fuel cells, *Environ. Technol.*, DOI: 10.1080/09593330.2017.1285965 (2017).

7. Y. Nagatsu, K. Tachiuchi, N. Touch and T. Hibino, Factors for improving the performance of sediment microbial fuel cell, *J. Jpn. Soc. Civ. Eng. Ser. B2 (Coast. Eng.)* **70**(2), pp. 1066-1070 (2014). in Japanese

8. N. J. Sacco, E. L. M. Figuerola, G. Pataccini, M. C. Bonetto, L. Erijman and E. Cortón, Performance of planar and cylindrical carbon electrodes at sedimentary microbial fuel cells, *Bioresour. Technol.***126**, pp. 328-335 (2012).

9. D. R. Bond, D. E. Holmes, L. M. Tender and D. R. Lovley, Electrode-reducing microorganisms that harvest energy from marine sediments, *Science* **295**, pp. 483-485 (2002).

Using Real Cases to Elaborate Chinese Graded Feasibility Assessment on the Sea Area Usage[*]

L. Teng[†], Y. H. Wang and M. X. Qian

Nanjing Hydraulic Research Institute, Hujuguan 34,
Nanjing, 210024, China
†E-mail: tengling@vip.sina.com
www.nhri.cn

Natural resource management and socio-economic planning in coastal zone to implement sustainable development have become a quite challenging topic. The State Oceanic Administration of China has published the management regulation of the feasibility assessment on the sea area usage since 2008. It clarifies that the feasibility assessment must be carried out for the exclusive usage of sea area in Chinese inland and territorial waters. After that, the technical guide of this assessment has been issued. By taken the usage purpose, scale and characteristics of the sea area into consideration, the feasibility assessment can be categorized into three grades. If a project has different usage purposes on sea areas, the assessment grade should be consistent with the highest one. In this paper, three real cases i.e., Hong Kong-Zhuhai-Macau Bridge (HZMB), Supporting Facility Project for International Cruise Port at Phoenix Island and Ningde Doumi central fishing port with the first, second and third assessment grade, respectively, are introduced. On the basis of detailed understanding and investigation of the marine resources and the present development situation of each case, this paper analyzes the necessity of the project construction and need of sea area usage, and evaluates the rationality of the site selection, area and period, and the project consistency with local marine functional zoning and relevant planning. After that, the negative influences of the projects are presented, and in terms of the relationship between the project owner and stakeholders, the solution and suggestion of the project sea area usage are proposed. By comparison of these cases, we can conclude that there exist obvious differences on the study scope, evaluating emphasis and data requirement for different assessment grades. It provides a clear view on the feasibility assessment on the sea area usage in China. The system of the graded feasibility assessment which is judged by how a specified project uses sea, the occupied area of the project in the sea (sea area usage) and the characteristics of the area (for example, sensitive or insensitive), balances the rights and interests of the project owner and other stakeholders, and promotes the reasonable usage and sustainable development of the ocean.

---

[*] This work is supported by the project of Water and Sediment Change of the Yangtze Estuary and Safety of Key Projects by grant 2017YFC0405406.

Keywords: Sea Area Usage; Graded Feasibility Assessment; Stakeholder.

1. Introduction

Due to the population boom and land shortage, people have turned their eyes to the coastal area. Ocean has rich resources and vast space, which is of vital importance to the future development of mankind. However, natural resource management and physical planning in urbanized coastal areas are challenged through "complex processes, uncertain understandings, multiple threats, multiple jurisdictions and scales, and multiple stakeholders and perspectives" [1]. Therefore, we can present the following needs to meet abovementioned challenges [2]: (1) the need to provide an integrated policy and decision-making process, including all involved stakeholders, to promote compatibility and a balance of uses; (2) the need to apply preventive and precautionary approaches in the project planning and implementation, including prior assessment and systematic observation of the impacts of each individual project; (3) the need to promote the development and application of methods to relieve or eliminate pollution, erosion, loss of resources and habitat destruction; (4) the need to provide access for concerned individuals, groups and organizations to relevant information and opportunities for consultation and participation in planning and decision-making.

Managing sustainability in coastal zones is usually referred to as Integrated Coastal Zone Management (ICZM). To implement ICZM in coastal areas, two methods are widely used, i.e. marine environmental impact assessment and feasibility assessment on the sea area usage. Even though their ultimate goals are identical to realize the overall coordination of sustainable economic and social development with natural resources and environment, there exist apparent differences between these two methods in terms of their legislation basis, focus points and methodology. The essential contents of the feasibility assessment on the sea area usage are to analyze the rationality of marine resource exploitation, to coordinate among stakeholders, and to discuss the usage way and plane arrangement [3].

In terms of China, on October 29, 1998, the State Oceanic Administration of China clearly pointed out that the feasibility assessment on the sea area usage must be carried out to examine and approve the sea area using application. And, Law of the People's Republic of China on the Administration of the Use of Sea Areas put into effect in 2002, which provides the legal basis for the feasibility assessment on the sea area usage. After that, a series of laws, regulations and methods about this assessment system have been proposed. The State Oceanic Administration (SOA) of China formally published the management regulation of the feasibility assessment on the sea area usage in 2008. It clarifies that the feasibility assessment must be carried out for the exclusive usage of sea area in Chinese inland and territorial waters. Subsequently, the technical guidance of this assessment has been issued [4].

2. Methodology

2.1. *Principles and contents of the assessment*

The feasibility assessment of sea area usage should follow the scientific, objective and justice principles, which emphasize both development and protection to achieve harmony of economic efficiency, social benefits and environmental health. It insists the rational exploitation and sustainable utilization of sea area, insists the overall balance to promote regional coordination, insists the people-oriented policy to ensue economic and social harmonious development at coastal region, and insists the priority of national interest to safeguard national defense and marine interests.

The contents of the assessment include: (1) the necessity of sea usage of the project; (2) environmental impact analysis of the project on marine resources; (3) coordinated analysis on sea area development and utilization; (4) compliance analysis between sea area usage of the project and marine functional zoning, or other related plans; (5) rationality analysis on sea area usage of the project; (6) countermeasure analysis on sea area usage of the project.

2.2. *Procedures of the assessment*

The assessment work is divided into four stages: preparation; field investigation, analysis and evaluation, and report compilation.

(1) Preparation phase: to study the technical documents and basic information related to the project, to collect historical and present data, to carry out preliminary analysis on the project to determine the grade, scope and contents of the assessment, filter and decide the emphases of the assessment, and formulate the work program of the assessment.

(2) Field investigation phase: according to the sea usage application of the project, investigate the project site to understand the characteristics of topography and geology, the location of the coastline and the current situation of development and utilization; visit the relevant government departments, organizations and individuals to acquire the authentic right certificates and actual situation of existed projects. According to the

collected relevant information, carry out the necessary investigation on the present status.

(3) Analysis and evaluation phase: based on the obtained data and information, analyze the necessity of sea area usage of the project, the influence of the project on marine resources and environment, the coordination with marine development and utilization, and the compliance between sea area usage of the project and marine functional zoning, or other related plans; and put forward the countermeasures of the project and assessment conclusions.

(4) Report compilation phase: on the basis of the analyzed contents and conclusions, compile the report of the sea area usage assessment.

2.3. Grade criterion for the assessment

For detailed grade criterion of a specific project for the feasibility assessment of sea area usage, please see Table 1 issued by the State Oceanic Administration of China [4].

2.4. Scope and key points of the assessment

The scope of the assessment should be determined in terms of sea area usage of the project, characteristics of the project site, the present development and utilization of the surrounding waters, which must cover all regions where the project will influence.

In general, the scope of the assessment takes the outline of the sea area usage of the project as beginning points, and extends outwards 15 km and 8 km for Grade first and second assessments, respectively. Specially, the scope of the assessment of sea-crossing bridge and submarine pipeline expands to each side of 5 km and 3 km for Grade first and second, respectively.

The scope of the assessment should be illustrated in a plan view, indicating its geographical location, scope, area and so on.

The key points of assessment should be decided on the basis of the usage type, way, and scale, combined be the present situation of marine resources and environment, stakeholders, etc. the following aspects should be paid attention:

(1) If the project has large scale reclamation, the analyses on hydrodynamic and scouring changes, the rationality of the project plane layout, and the rationality of the usage area should be taken as the key points of the assessment.

(2) If the site selection is not unique, the rationality of the site selection should be listed as a key point.

(3) If the project includes large area seabed excavation or underwater blasting, the impacts on resources and environment should be taken as key points.

(4) If the project involves marine oil and gas exploration, petrochemical and other dangerous goods, the risk analysis should be listed as a key point.

(5) If the project belongs coastal sand mining, the coastal erosion and influence on topography should be taken as key points.

Table 1 Grade criterion for the feasibility assessment of sea area usage

Level 1 usage way	Level 2 usage purpose		Sea area usage scale (area or length)	Characteristics of the area	Grade
Reclamation with filled materials	Metallurgy, petrochemical industry; paper industry; thermal power; nuclear power; waste disposal		All scales	All types	First
	Other construction; agriculture		Area ≥10 hm²	All types	First
			Area: 5~10 hm²	Sensitive type	First
				Other type	Second
			Area≤5 hm²	All types	Second
Structures	Sea-crossing bridge and undersea tunnel	Non-permeable structures	Total length ≥500 m; area≥10 ha	All types	First
			Total length: 250~500 m; area: 5~10 hm²	Sensitive	First
				Other type	Second
			Total length ≤250 m; Area≤5 hm²	All types	Second
		Sea-crossing bridge	Length ≥2000 m	All types	First
			Length: 800~2000 m	Sensitive type	First
				Others	Second
			Length ≤800m	All types	Second
			Single span sea-crossing bridge	All types	Third
		Open-digging undersea tunnel	All scales	All types	First
		Excavated undersea tunnel	All scales	All types	Second
		Subsea storage	All scales	All types	First
		Sea aquarium	All scales	All types	Second
	Permeable structures	Artificial fish reef	Area ≥50 hm²	All types	First
			Area ‹ 50 hm²	All types	Second
		Other permeable structures	Total length ≥2000 m or total area ≥30 hm²	All types	First
			Total length: 400~2000 m ; total area: 10~30 hm²	Sensitive type	First
				Other type	Second
			Total length ≤400 m or total area ≤10 hm²	All types	Third

Table 1 (*Continued*)

Level 1 usage way	Level 2 usage purpose		Sea area usage scale (area or length)	Characteristics of the area	Grade	Level 1 usage way
Usage with enclosed structure	Harbor basin			Area ≥100 hm²	All types	Second
				Area ‹ 100 hm²	All types	Third
	Water storage			Area ≥100 hm²	All types	First
				Area: 20~100 hm²	Sensitive type	First
					Other type	Second
				Area ≤20 hm²	All types	Third
	Salt field			Area ≥100 hm²	All types	First
				Area: 20~100 hm²	Sensitive type	First
					Other type	Second
				Area ≤20 hm²	All types	Third
	Enclosed mariculture			Area ≥100 hm²	All types	First
				Area: 10~100 hm²	All types	Second
				Area: ≤10 hm²	All types	Third
Open sea area usage	Open mariculture			Area: ≥700 hm²	All types	Second
				Area: ‹ 700 hm²	All types	Third
	Bathing field, fairground			Area: ≥30 hm² or occupied coast line ≥ 500 m	All types	Second
				Area ‹ 30 hm² or occupied coast line <500 m	All types	Third
	Navigation channel			Length ≥10 km	All types	First
				Length	All types	Second
	Anchorage	Hazardous material		All scales	All types	Second
		Other		All scales	All types	Third
	Warm water discharge	Warm water discharge for thermal and nuclear power		All scales	All types	First
		Other warm water discharge		All scales	All types	Second

......

Remark: (1) When calculating the length of parallel installed submarine cables, pipelines and so on, use the longest one.

(2) The original discharge scale should be taken into consideration when judging newly warm water discharge and standardized sewage discharge.

(3) Sensitive sea regions include marine nature reserves, marine special protection areas, important estuaries and bays.

3. Case Study

3.1. *Hong Kong-Zhuhai-Macau bridge*

3.1.1. *Introduction of the project*

The Hong Kong-Zhuhai-Macau Bridge (HZMB), being situated at the waters of Lingdingyang of Pearl River Estuary, is a large sea crossing linking the Hong

Kong Special Administrative Region (HKSAR), Zhuhai City of Guangdong Province and Macao Special Administrative Region (see Figure 1). It consists of a Main Bridge in Mainland waters together with the boundary crossing facilities and link roads within the three places. The functions of the Bridge are to meet the demand of passenger and freight land transport among Hong Kong, the Mainland (particularly the region of Pearl River West) and Macao, to establish a new land transport link between the east and west banks of the Pearl River, and to enhance the economic and sustainable development of the three places.

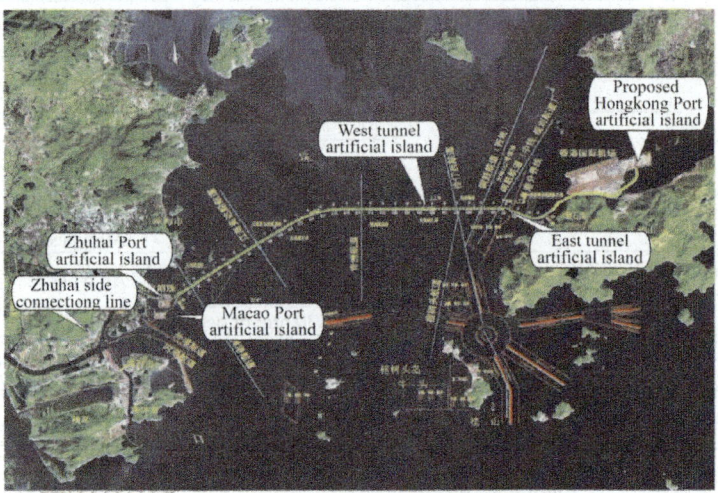

Fig. 1. Layout of the HZMB project

3.1.2. *Grade judgment of the assessment*

The sea area used for the sea-crossing bridge section of the HZMB is 155.69 hm^2, for the submarine tunnel crossing Tungu Channel and Western Lingdingyang Channel is 40.36 hm^2. The reclamation area for two tunnel artificial islands is 36.28 hm^2. Besides, the sea area used for the port and tunnel artificial islands at Zhuhai side is 124.8 hm^2, among which the reclamation area for the port is 111.3 hm^2 and the sea area used for the tunnel artificial island is 13.9 hm^2, for the port artificial island at Macao side is 96.37 hm^2, and for the submarine tunnel at the Gongbei is 5.69 hm^2. The sea area used for the Zhuhai connecting bridge is 4.49 hm^2, and for the Macao connecting bridge is 35.83 hm^2. According to Table 1, if the reclamation area is larger than 10 hm^2, or the length of sea crossing bridge is longer than 2000 m, the assessment grade will belongs to the first grade. Obviously, the assessment for this project will be categorized Grade first.

3.1.3. Key points of the assessment

The Indo-Pacific humpback dolphin (Sousa chinensis), or commonly known as Chinese White Dolphin (CWD), is found throughout the western Pacific and Indian oceans, from southern China and northern Australia in the east to South Africa in the west. There are about 1,300 to 1,500 dolphins found in the eastern part of the Pearl River Estuary and Hong Kong waters. The highest dolphin density in Hong Kong is found in the west Lantau waters.

The locations and alignments of the HZMB Hong Kong Projects are carefully chosen to avoid the CWD major active areas. During the construction, a series of mitigation measures are implemented to minimize potential impacts to the Chinese White Dolphin and the major ones include: (1) Non-dredged Reclamation Method: to adopt non-dredging methods for reclamation and seawall construction to reduce the amount of dredging and marine disturbance; (2) No underwater percussive piling; (3) No underwater percussive piling; (4) No formation of underwater sockets into rock (bored-piling) during peak dolphin calving season in May and June; (5) Acoustic decoupling measures: noisy equipment mounted on construction vessels is acoustically-decoupled to minimize underwater noise; (6) Marine vessels follow pre-defined regular routes: construction vessel will follow regular travel routes and avoid CWD active areas; (7) Marine vessel speed control: <10 knots in work site, marine park and proposed marine park; (8) Skipper Training: captains of construction vessels working in the areas have undergone training on local dolphins and porpoises; (9) Dolphin Exclusion Zone (DEZ): a dolphin exclusion zone of 250 meters radius is implemented during the silt curtain installation and the bored pile casing installation works. Works suspended if dolphin is found within the DEZ. (10) Dolphin Watching Plan (DWP): a silt curtain enclosed areas are regularly checked and works will be suspended if dolphin is found within the enclosed area.

3.2. Supporting facility project for international cruise port at the Phoenix Island

3.2.1. Introduction of the project

Sanya city, the southernmost island of Hainan Island, is the only tropical city with attractive coastline in China. It is located between Hong Kong and Singapore, which is a transfer and supply station for the international cruise ships in Southeast Asia. It has an excellent geographical position at the cruise travelling route. With the recovery of the world economy, the eastward shift of the cruise industry and the resumption of the growth of cruise passengers, the existing berths and port facilities of the Phoenix Island at Sanya cannot meet an overall development goal as an international cruise ships' home port, which directly restricts the cruise port development of the Phoenix Island.

However, after Phase II of The Phoenix Island International Cruise Port completion, it still needs lands to build the management office, dressing room, bathroom, sewage treatment station and other supporting services, which are necessary for marine tourism and entertainment projects. The reclamation land of 1.1246 hm^2 at the south side of the White Row Reef project will be used to build these facilities (see Figure 2)

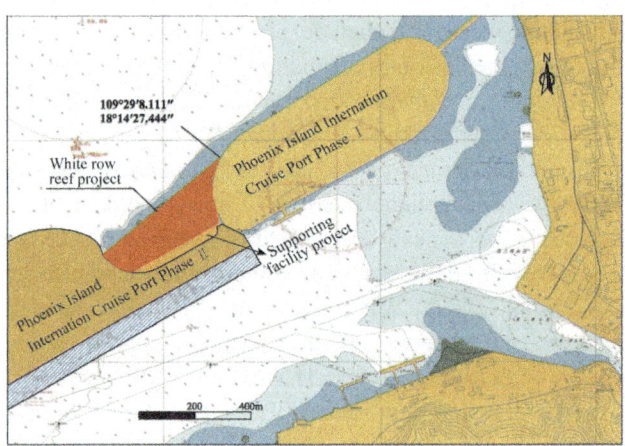

Fig. 2. Position of the supporting facility project for international cruise port at the Phoenix Island.

3.2.2. *Grade judgment of the assessment*

The sea usage way of the project is other type construction with land reclamation. According to Table 1, if the reclamation area≤ 5 hm^2, a project in all type areas will conduct Grade second assessment, therefore, the assessment of this project will be classified into Grade second.

3.2.3. *Key points of the assessment*

(1) Rationalities of the site selection and plain layout

The water depth of the project site is small, and the underwater bathymetry is relatively gentle, which is beneficial to the construction of the project. The foundation layers of the site are stable and suitable for the construction. The hydrodynamic impact after the construction is limited within 400 m of the site. Thus, it has little influence on Sanya coral reef protection region at the south side and Sanya tourism and recreation area at the north side. The plain layout of the project follows the principle of intensive and economical usage of sea, minimizes the impacts on the hydrodynamic and erosion/siltation environment, and is compatible with other sea using activities in adjacent waters.

(2) Impacts on resources and environment

Due to a minor effect of the project on the flow and sediment conditions, the impact of the project on water quality, ecosystem and natural resources in the surrounding waters is acceptable. Table 2 provides the average loss of marine biotic resources after the project construction.

Table 2 Average loss of marine biotic resources after the project construction

	Fish eggs	Fish larvae	Zooplankton	Phytoplankton
Density of biotic resources	4.45 indiv /m^3	1.12 indiv /m^3	51.2 mg/m^3	452.3×103 cell/m^3
Permanent losses	9.95×105 indiv	2.50×105 indiv	11.44kg	1.01×1011 cell/m^3
Continuous losses	1.99×106 indiv	5.01×105 indiv	22.89kg	2.02×1011 cell/m^3

3.3. Ningde Doumi central fishing port

3.3.1. Introduction of the project

Due to historical reasons, fishing port construction in Fujian Province is relatively under-developed. In 2008, the rate of nearby sheltered against storm for fishing ships in the province was 54.93%, of which Ningde City was 45%. There still exists a quite gap compared with the requirements of fishery production. Especially in recent years, strong typhoons and other natural disasters have made the problems, such as small sheltered waters, low construction standards, incomplete supporting facilities and unreasonable layout, more prominent.

With the rapid development of local fishery economy, the production scale of Doumi is increasing year by year, and the throughput of aquatic products and aquaculture bait are also greatly increased. The existed old pier cannot meet the increasing capacity demands. Thus, the construction of the central fishing port is definitely required.

3.3.2. *Grade judgment of the assessment*

If the assessment grades according to different usage ways and scales are different, it will adopt higher grade rather than lower one. Thus, the project will conduct Grade second assessment (see Table 3). However, the Fujian Provincial development and reform commission issued a supplementary notice on further accelerating the construction of fishing port projects, in which claimed if the project does not involve reclamation and ecological sensitive areas, it can be classified into Grade three assessment.

To sum up, the project can be compiled a report form to complete the assessment work.

Table 3   Judgment of assessment grade for the project

Sea usage unit	Level 2 usage purpose	Sea area usage scale	Characteristics of the area	Grade
Wharf (0.0204 hm²)	Non-permeable structures	Total length ≤250 m; Area≤5 hm²	All types	Second
Harbor basin (1.4734 hm²)	Harbor basin	Area ‹ 100 hm²	All types	Third
Breakwaters (1.2108 hm²)	Non-permeable structures	Total length ≤250 m; Area≤5 hm²	All types	Second

3.3.3. *Key points of the assessment*

Seaweed farming has a strong seasonality which from lunar August to January next year. The field investigation in May, 2014 found that the seaweed farming area was large with a close distance to the project. Therefore, the project owner promised to avoid seaweed farming period, and give reasonable compensation for the losses of farmers caused by improper construction. At the same time, this

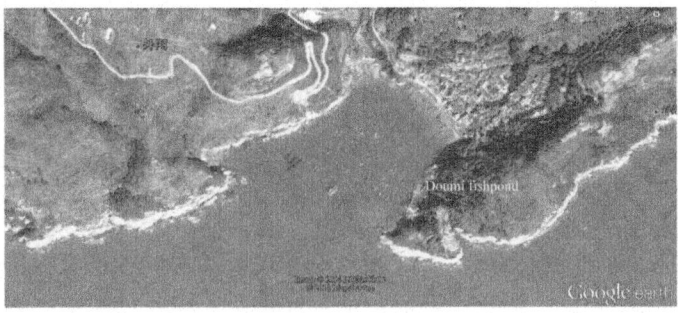

Fig. 3. Satellite image of the project site.

project will occupy the existed old dock and the fishpond of a villager (see Figure 3). The dock owner, Doumi village committee, promised to sign an agreement to prove Doumi fishing port construction, and supervised the compensation agreement between the project owner and the fish pond owner.

In summary, the relationships between the project owner and related stakeholders were basically clear, and the coordination has been acceptable.

4. Conclusions

The management on Chinese sea usage has achieved remarkable results to sort the order of sea area development, to ensure coastal economy, and to improve the marine environmental quality.

Our paper uses three real cases to elaborate different study scopes, evaluating emphases and data requirements for different assessment grades. It exhibits a clear view on the feasibility assessment on the sea area usage in China. The system of the graded feasibility assessment balances the rights and interests of the project owner and other stakeholders, and promotes the reasonable usage and sustainable development of the ocean.

With the increasing demands for sea area, the usage purposes and ways are varied, and new environmental and ecological problems have consequently occurred. The guidance needs to be amended to meet these challenges.

References

1. B. Coffey, and O' K. Toole, Towards an improved understanding of knowledge dynamics in integrated coastal zone management: a knowledge systems framework, *Conserv. Soc.* **10**(4), 318–329 (2012).

2. J. P. Ducrotoy, Integrated coastal management: Lessons learned to address new challenges, *Marine Pollution Bulletin*, **102**, 241–242 (2016).

3. H. Yang, Relationship between sea area use demonstration and marine environmental impact assessment, *Marine Environmental Science*, **26**(2), 197–200. (in Chinese)

4. SOA (State Oceanic Administration), *Technical Guidance of the Feasibility Assessment on the Sea Area Usage* (2010). (in Chinese)

CPSIA information can be obtained
at www.ICGtesting.com
Printed in the USA
LVOW05*1724250118
563992LV00002B/17/P

9 789813 233805